AF819951

All in One

Recommended by Teachers Trusted by Toppers

MATHEMATICS

Class 11

EVERYTHING YOU NEED TO SCORE

- Simplified Theory
- Topic & Chapter Exercises
- All Types of Exam Pattern Questions
- CBQs; MCQs, A-R, Case Based
- Sample Question Papers

CBSE Exams 2026

All in One®

Recommended by **Teachers** Trusted by **Toppers**

MATHEMATICS

Class 11

EVERYTHING YOU NEED TO SCORE

- Simplified Theory
- Topic & Chapter Exercises
- All Types of Exam Pattern Questions
- CBQs; MCQs, A-R, Case Based
- Sample Question Papers

CBSE Exams 2026

Author
Er. Prem Kumar

Co-authors
Jitender Gupta
Brijesh Dwevedi

arihant
SUCCESS AND NOTHING LESS

arihant
SUCCESS AND NOTHING LESS

ARIHANT PRAKASHAN
(School Division Series)

卐 ADMINISTRATIVE & PRODUCTION OFFICES

Regd. Office
'Ramchhaya' 4577/15, Agarwal Road, Darya Ganj, New Delhi -110002
Tele: 011- 47630600, 43518550; Fax: 011- 23280316

卐 *Head Office*
Kalindi, TP Nagar, Meerut (UP) - 250002, Tel: 0121-7156203, 7156204

卐 SALES & SUPPORT OFFICES
Agra, Ahmedabad, Bengaluru, Bareilly, Chennai, Delhi, Guwahati, Hyderabad, Jaipur, Jhansi, Kolkata, Lucknow, Nagpur & Pune.

PO No : TXT-XX-XXXXXXX-X-XX

Published By Arihant Publications (India) Ltd.

For further information about the books published by Arihant, log on to www.arihantbooks.com or e-mail at info@arihantbooks.com

Follow us on

卐 PRICE ₹500.00

PRODUCTION TEAM

Publishing Managers	Cover Designer	Inner Designer
Mahendra Singh Rawat & Keshav Mohan	Salman Shaikh	Ankit Tyagi
Project Head	**Page Layouting**	**Proof Reader**
Ashwani	Shivam Prajapati	Mukul Kaushik, Manish Rathore

Index *cum* Study Planner

One Book- Endless Features

to Help you Succeed

Chapter Intro

A quick information about the coverage of the chapter to explain students about what they are going to study.

01

CHAPTER

Sets

There are things which seem incredible to most men who have not studied Mathematics. —*Archimedes*

YOUR STUDY GOALS

1.1 Set
1.2 Types of Sets
1.3 Subset
1.4 Universal Set
1.5 Venn Diagram
1.6 Operations on Sets

YOUR PRACTICE GOALS

- Objective Type Questions (25 Qs; 14 MCQs, 10 A-R, 1 Case Based)
- Subjective Type Questions (25 Qs; 11 VSA, 10 SA, 4 LA)

1.1 Set

A well-defined collection of objects is called set. Sets are usually denoted by capital letters A, B, X, Y etc. The elements of set are represented by small letters a, b, x, y, z etc.

Suppose, a is an element of set A, then we write $a \in A$ (a belongs to A) and if b is not an element of set A, then we write $b \notin A$ (b does not belong to A). The word 'belongs to' denoted by the Greek symbol $\in$ (epsilon).

e.g. if $A = \{1, 2, 3, 4\}$, then $2 \in A$ and $5 \notin A$.

Some examples of sets are

- $N \to$ The set of natural numbers.
- $Z \to$ The set of all integers.
- $Q \to$ The set of all rational numbers.

Simplified Theory

It contains the necessary study material well supported by Definitions, Facts, Figures, Flow Charts, etc.

TEACHER TIPS

- In roster form, the order of the elements does not matter.
- The number of elements belonging to a finite set is called **cardinal number** of that set and denoted by $n(A)$.
  e.g. Cardinal number for set $\{1, 2, 3\}$ is 3.
- Do not write any repeated element in roster form.
- All the infinite sets cannot be written in roster form.

Teacher Tips

Important Points related to given concept to bring to help students focus on important points.

Quick Test 1

1. Which of the following are sets?
   (i) The collection of all planets in the solar system.
   (ii) A team of eleven best cricket batsmen of the world.
   (iii) The collection of all boys in your class.
   (iv) The collection of all natural numbers less than 100.
   (v) The collection of all good athletes of India.
2. Describe the following sets in roster form.
   (i) The set of all letters in the word 'ALGEBRA'.
   (ii) The set of all natural numbers less than 7.
   (iii) $\{x : x \text{ is a positive integer and } x^2 = x\}$
   (iv) If $Y = \{x : x \text{ is a positive factor of the number } 2^{p-1}(2^p - 1)$, where $2^p - 1$ is a prime number$\}$. *Competency Based Que*
3. Describe the following sets in set-builder form.
   (i) The set of all letters in the word PROBABILITY.
   (ii) The set of all even natural numbers.
   (iii) $\{5, 25, 125, 625\}$ (iv) $\{10, 11, 12, 13, 14, 15\}$
   (v) $\{1, 4, 9, 16, \ldots, 100\}$ (vi) $\left\{1, \frac{1}{2}, \frac{1}{3}, \frac{1}{4}, \frac{1}{5}, \ldots\right\}$
4. Which of the following sets are empty sets?
   (i) $A = \{x : 4 < x < 5, x \in N\}$
   (ii) $D = \{x : x^2 = 25 \text{ and } x \text{ is an odd integer}\}$
   (iii) $\{x : x \in N \text{ and } x^2 = 9\}$
   (iv) $\{x : x^2 - 3 = 0, x \text{ is rational}\}$
5. State which of the following sets are finite and which are infinite? *Competency Based Que*
   (i) $A = \{x : x \in Z \text{ and } x^2 - 5x + 6 = 0\}$
   (ii) $B = \{x : x \in Z \text{ and } x^2 \text{ is even}\}$
   (iii) $C = \{x : x \in Z \text{ and } x^2 = 36\}$
   (iv) $D = \{x : x \in Z \text{ and } x > -10\}$
   (v) $A = \{x : x \in Z \text{ and } x^2 - 2x - 3 = 0\}$
6. Find the cardinal number of the set of letters of the word 'MATHEMATICS'.
7. Are the following sets are equal?
   $A = \{x : x \text{ is a letter in the word 'WOLF'}\}$
   $B = \{x : x \text{ is a letter in the word 'FOLLOW'}\}$

Answers

1. (i), (iii), (iv)
2. (i) {A, L, G, E, B, R} (ii) {1, 2, 3, 4, 5, 6}
   (iii) {1} (iv) $Y = \{1, 2, 2^2, 2^3, \ldots, 2^{p-1}, 2^p - 1\}$
3. (i) $\{x : x \text{ is a letter of the word PROBABILITY}\}$
   (ii) $\{x : x = 2n, n \in N\}$ (iii) $\{x : x = 5^n, n \in N \text{ and } n < 5\}$
   (iv) $\{x : x \in N, 9 < x < 16\}$ (v) $\{x : x = n^2, 1 \le n \le 10, n \in N\}$
   (vi) $\left\{x : x = \frac{1}{n}, n \in N\right\}$
4. (i) and (iv)
5. (i) Finite, (ii) Infinite (iii) Finite (iv) Infinite (v) Finite
6. 8 7. Yes

For Solutions, See QR Code on Pg. 18

Topical Exercises

To make students understand the chapter in a smooth part-by-part manner, Topical Exercises have been given in each chapter. These Exercises help students to assess the understanding of related concepts at regular interval.

One Book- Endless Features
to Help you Succeed

Illustrative Figures

To make the concepts completely visible, illustrative diagrams have been given at various place.

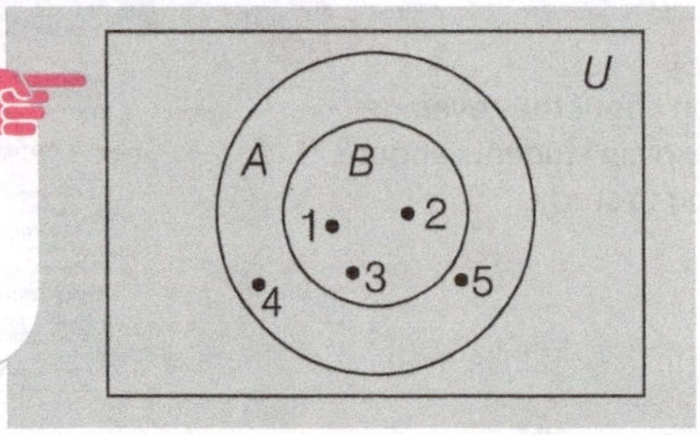

Example 6. **What universal sets would you propose for each of the following?** *NCERT*

(i) The set of right triangles.

(ii) The set of equilateral triangles.

Sol. (i) The universal set for the set of right triangles is set of triangles.

(ii) The universal set for the set of equilateral triangle is set of triangles.

NCERT Questions

To make students completely understand NCERT books, solutions of selected NCERT questions have been given in each chapter.

13. **Let F_1 be the set of parallelograms, F_2 be the set of rectangles, F_3 be the set of rhombus, F_4 be the set of squares and F_5 be the set of trapeziums in a plane. Then, F_1 may be equal to** **NCERT Exemplar**

(a) $F_2 \cap F_3$
(b) $F_3 \cap F_4$
(c) $F_2 \cup F_5$
(d) $F_2 \cup F_3 \cup F_4 \cup F_1$

14. **If R be the set of points inside a rectangle of sides a and b $(a, b > 1)$ with two sides along the positive direction of X-axis and Y-axis, then** **NCERT Exemplar**

(a) $R = \{(x, y): 0 \le x \le a, 0 \le y \le b\}$
(b) $R = \{(x, y): 0 \le x < a, 0 \le y \le b\}$
(c) $R = \{(x, y): 0 \le x \le a, 0 < y < b\}$
(d) $R = \{(x, y): 0 < x < a, 0 < y < b\}$

NCERT Exemplar Problems

NCERT Exemplar Problems are considered to be the best problem books for School/Board Examinations. To make students practice of exemplary questions given in this book, each chapter has selected problems from NCERT Exemplar

One Book- Endless Features

to Help you Succeed

Exam Pattern Questions

To make students completely prepared for examination, this Exercise contains Exam Pattern Questions. This exercise covers all types of questions, like; MCQs, Case Based Questions, Assertion-Reason, VSA, SA & LA Type Questions. All the questions have hints and answers.

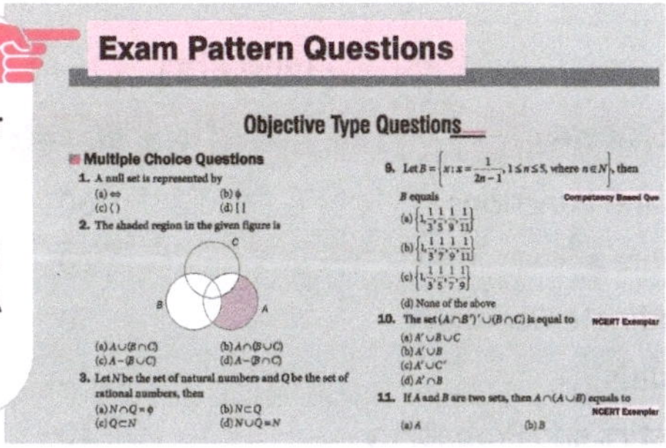

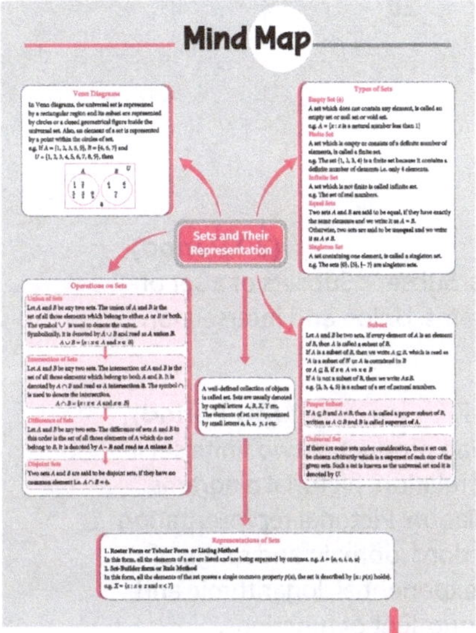

Internal Assessment of Practical Work

To make this book complete in all aspects, Internal Assessment based on latest syllabus is also given.

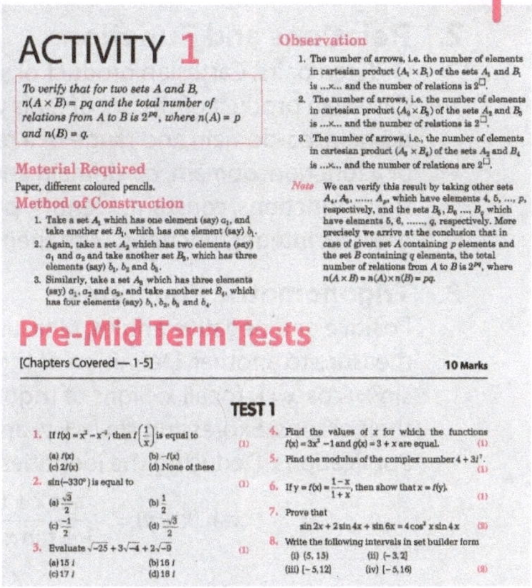

Mind Map

A quick revision visual tool for the complete chapter.

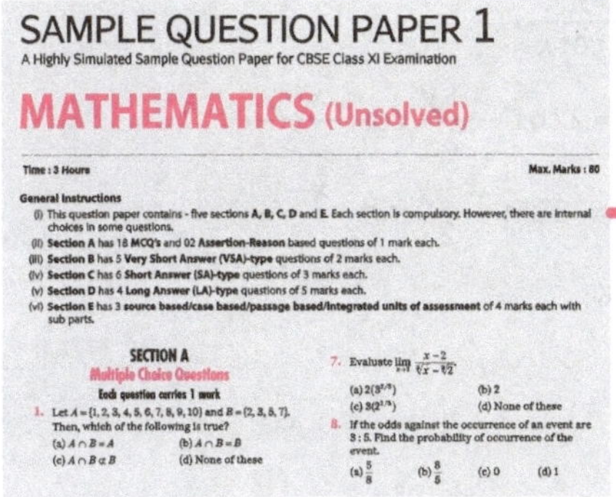

Sample Papers

To make students practice of examination in real sense, there are 3 Sample Question Papers at the end. All these Sample Questions Papers are exactly prepared on the latest syllabus & pattern of examination.

Latest Course Structure

One Paper **Time: 3 Hrs** **Total Marks : 80**

Unit	Unit Name	No. of Periods	Marks
I	Sets and Functions	60	23
II	Algebra	50	25
III	Coordinate Geometry	50	12
IV	Calculus	40	08
V	Statistics and Probability	40	12
	Total	**240**	**80**
	Internal Assesseent		**20**

**No Chapter/Unit-wise Weightage. Care to be taken to cover all the chapters.*

UNIT I SETS AND FUNCTIONS

1. Sets **(20 Period)**

Sets and their representations, Empty set, Finite and Infinite sets, Equal sets, Subsets. Subsets of a set of real numbers especially intervals (with notations). Universal set. Venn diagrams. Union and Intersection of sets. Difference of sets. Complement of a set. Properties of Complement.

2. Relations and Functions **(20 Period)**

Ordered pairs, Cartesian product of sets. Number of elements in the Cartesian product of two finite sets. Cartesian product of the set of reals with itself (upto $R \times R \times R$). Definition of relation, pictorial diagrams, domain, co-domain and range of a relation. Function as a special type of relation. Pictorial representation of a function, domain, co-domain and range of a function. Real valued functions, domain and range of these functions, constant, identity, polynomial, rational, modulus, signum, exponential, logarithmic and greatest integer functions, with their graphs. Sum, difference, product and quotient of functions.

3. Trigonometric Functions **(20 Period)**

Positive and negative angles. Measuring angles in radians and in degrees and conversion from one measure to another. Definition of trigonometric functions with the help of unit circle. Truth of the identity $\sin^2 x+\cos^2 x=1$, for all x. Signs of trigonometric functions. Domain and range of trignometric functions and their graphs. Expressing $\sin(x \pm y)$ and $\cos(x \pm y)$ in terms of $\sin x$, $\sin y$, $\cos x$ & $\cos y$ and their simple applications. Deducing the identities like the following:

$$\tan(x \pm y) = \frac{\tan x \pm \tan y}{1 \mp \tan x \tan y}, \cot(x \pm y) = \frac{\cot x \cot y \mp 1}{\cot y \pm \cot x},$$

$$\sin x + \sin y = 2 \sin \frac{x+y}{2} \cos \frac{x+y}{2}, \cos x + \cos y = 2 \cos \frac{x+y}{2} \cos \frac{x-y}{2},$$

$$\sin x - \sin y = 2 \cos \frac{x+y}{2} \sin \frac{x-y}{2}, \cos x - \cos y = -2 \sin \frac{x+y}{2} \sin \frac{x-y}{2}$$

Identities related to $\sin 2x$, $\cos 2x$, $\tan 2x$, $\sin 3x$, $\cos 3x$ and $\tan 3x$.

UNIT II ALGEBRA

2. Complex Numbers and Quadratic Equations (10 Period)

Need for complex numbers, especially $\sqrt{-1}$, to be motivated by inability to solve some of the quadratic equations. Algebraic properties of complex numbers. Argand Plane.

3. Linear Inequalities (10 Period)

Linear inequalities. Algebraic solutions of linear inequalities in one variable and their representation on the number line.

4. Permutation and Combinations (10 Period)

Fundamental principle of counting. Factorial *n*. (*n*!). Permutation and combinations, derivation of formulae for ${}^nP_r$ and ${}^nC_r$ and their connections, simple applications.

5. Binomial Theorem (10 Period)

Historical, perspective, statement and proof of the binomial theorem for positive integral indices. Pascal's triangle, simple applications.

6. Sequence and Series (10 Period)

Sequence and Series. Arithmetic Mean (AM), Geometric Progression (GP), general term of a GP, sum of first *n* terms of a GP, infinite GP and its sum. Geometric mean (GM), relation between AM and GM.

UNIT III COORDINATE GEOMETRY

1. Straight Lines (15 Period)

Brief recall of two dimensional geometry from earlier classes. Slope of a line and angle between two lines. Various forms of equations of a line: parallel to axis, point-slope form, slope-intercept form, two-point form, intercept form. Distance of a point from a line.

2. Conic Sections (25 Period)

Sections of a cone: circles, ellipse, parabola, hyperbola, a point, a straight line and a pair of intersecting lines as a degenerated case of a conic section. Standard equations and simple properties of parabola, ellipse and hyperbola. Standard equation of a circle.

3. Introduction To Three-dimensional Geometry (10 Period)

Coordinate axes and coordinate planes in three dimensions. Coordinates of a point. Distance between two points.

UNIT IV CALCULUS

1. Limits And Derivatives (40 Period)

Derivative introduced as rate of change both as that of distance function and geometrically. Intuitive idea of limit. Limits of polynomials and rational functions, trigonometric, exponential and logarithmic functions. Definition of derivative relate it to scope of tangent of the curve, Derivative of sum, difference, product and quotient of functions. Derivatives of polynomial and trigonometric functions.

UNIT V STATISTICS AND PROBABILITY

1. Statistics (20 Period)

Measures of dispersion: Range, Mean deviation, variance and standard deviation of ungrouped/grouped data.

2. Probability (20 Period)

Events; occurrence of events, 'not', 'and' and 'or' events, exhaustive events, mutually exclusive events, Axiomatic (set theoretic) probability, connections with other theories of earlier classes. Probability of an event, probability of 'not', 'and' and 'or' events.

Topic/Chapters to be Assessed Formatively

The following topics are included in the syllabus but will be assessed only formatively to reinforce understanding without adding to summative assessments. This reduces academic stress while ensuring meaningful learning. Schools can integrate these with existing chapters as they align well. Relevant NCERT textual material is enclosed for reference.

S.No.	Content
	Unit-I: Sets and Functions
1.	**Sets**
	Practical problems on Union and Intersection of two sets.
2.	**Relations and Functions**
	Composition of Functions
3.	**Trigonometric Functions**
	General solution of trigonometric equations of the type $\sin \gamma = \sin \alpha$, $\cos \gamma = \cos \alpha$ and $\tan \gamma = \tan \alpha$.

	Unit-II: Algebra
1.	**Principle of Mathematical Induction**
	Process of the proof by induction, motivating the application of the method by looking at natural numbers as the least inductive subset of real numbers. The principle of mathematical induction and simple applications.
2.	**Complex Numbers and Quadratic Equations**
	Polar representation of complex numbers. Statement of Fundamental Theorem of Algebra, solution of quadratic equations (with real coefficients) in the complex number system.
3.	**Linear Inequalities**
	Graphical solution of linear inequalities in two variables. Graphical method of finding a solution of system of linear inequalities in two variables.
4.	**Binomial Theorem**
	General and middle term in binomial expansion.
5.	**Sequence and Series**
	Formulae for the following special sums $\sum_{k=1}^{n} k, \sum_{k=1}^{n} k^2, \sum_{k=1}^{n} k^3,$
	Unit-III: Coordinate Geometry
1.	**Straight Lines**
	Normal form. General equation of a line.
2.	**Introduction to Three-dimensional Geometry**
	Section formula.
	Unit-IV: Calculus
1.	**Limits and Derivatives**
	Derivatives of composite functions (Chain rule).
	Unit-V Statistics and Probability
1.	**Probability**
	Random experiments; outcomes, sample space (set representation).

Question Paper Design

SN	Typology of Questions	Marks	Percentage
1.	**Remembering: and Understanding:** Remembering: Exhibit memory of previously learned material by recalling facts, terms, basic concepts, and answers. Demonstrate understanding of facts and ideas by organizing, comparing, translating, interpreting, giving descriptions, and stating main ideas	43	54%
2.	**Applying:** Solve problems to new situations by applying acquired knowledge, facts, techniques and rules in a different way.	19	24%
3.	**Analysing, Evaluating and Creating:** Examine and break information into parts by identifying motives or causes. Make inferences and find evidence to support generalizations. Present and defend opinions by making judgments about information, validity of ideas, or quality of work based on a set of criteria. Compile information together in a different way by combining elements in a new pattern or proposing alternative solutions.	18	22%
	Total	**80**	**100%**

INTERNAL ASSESSMENT	20 Marks
Pen Paper Test and Multiple Assessment (5+5)	10 Marks
Portfolio	05 Marks
Lab Practical (Lab activities to be done from the prescribed books)	05 Marks

01

CHAPTER

There are things which seem incredible to most men who have not studied Mathematics. —*Archimedes*

▼ **YOUR STUDY GOALS**

▼ **YOUR PRACTICE GOALS**

- Objective Type Questions (25 Qs; 14 MCQs, 10 A-R, 1 Case Based)
- Subjective Type Questions (25 Qs; 11 VSA, 10 SA, 4 LA)

Sets

1.1 Set

A well-defined collection of objects is called set. Sets are usually denoted by capital letters A, B, X, Y etc. The elements of set are represented by small letters a, b, x, y, z etc.

Suppose, a is an element of set A, then we write $a \in A$ (a belongs to A) and if b is not an element of set A, then we write $b \notin A$ (b does not belong to A). The word 'belongs to' denoted by the Greek symbol $\in$ (epsilon).

e.g. if $A = \{1, 2, 3, 4\}$, then $2 \in A$ and $5 \notin A$.

Some examples of sets are

- $N \rightarrow$ The set of natural numbers.
- $Z \rightarrow$ The set of all integers.
- $Q \rightarrow$ The set of all rational numbers.
- $R \rightarrow$ The set of real numbers (rational and irrational numbers).
- $Z^+ \rightarrow$ The set of positive integers.
- $Q^+ \rightarrow$ The set of positive rational numbers.
- $R^+ \rightarrow$ The set of positive real numbers (positive rational and irrational numbers).

Representations of Sets

1. Roster Form or Tabular Form or Listing Method

In this form, all the elements of a set are listed, the elements are being separated by commas and are enclosed within curly braces { }.

e.g. The set of all natural numbers less than 10 is represented in roster form as $\{1, 2, 3, 4, 5, 6, 7, 8, 9\}$.

2. Set-Builder Form or Rule Method

In this form, all the elements of set possess a single common property $p(x)$, which is not possessed by any other element outside the set.

In such a case, the set is described by $\{x : p(x) \text{ holds}\}$.

e.g. Set of all natural numbers less than 10,
$A = \{x : x \in N \text{ and } x < 10\}$.

1.2 Types of Sets

Empty Set

A set which does not contain any element, is called an empty set or null set or void set. It is denoted by ϕ or { }.

e.g. $A = \{x : x \text{ is a natural number less than } 1\}$.

Finite Set

A set which is empty or consists of a definite number of elements, is called a finite set.

e.g. The set $\{1, 2, 3, 4\}$ is a finite set because it contains a definite number of elements i.e. only 4 elements.

Infinite Set

A set which is not finite is called infinite set.

e.g. The set of real numbers.

Equal Sets

Two sets A and B are said to be equal, if they have exactly the same elements and we write it as $A = B$.

Otherwise, two sets are said to be **unequal** and we write $A \neq B$.

e.g. Let $A = \{a, b, c, d\}$ and $B = \{c, d, b, a\}$. Then, $A = B$ because each element of A is in B and *vice-versa*.

Singleton Set

A set containing one element, is called a singleton set.

e.g. The sets $\{0\}, \{5\}, \{-7\}$ are singleton sets.

TEACHER TIPS

- In roster form, the order of the elements does not matter.
- The number of elements belonging to a finite set is called **cardinal number** of that set and denoted by $n(A)$.
  e.g. Cardinal number for set $\{1, 2, 3\}$ is 3.
- Do not write any repeated element in roster form.
- All the infinite sets cannot be written in roster form.

Example 1. **Which of the following are sets? Justify your answer.** *NCERT*

(i) The collection of all the months of a year beginning with the letter J.
(ii) The collection of ten most talented writers of India.
(iii) A collection of novels written by the writer Munshi Premchand.
(iv) A collection of most dangerous animals of the world.

Sol. (i) We are sure that members of this collection are January, June and July.
So, this collection is well-defined.
Hence, it is a set.

(ii) A writer may be most talented for one person and may not be for other. Therefore, we cannot definitely decide which writer will be there in the collection.
So, this collection is not well-defined.
Hence, it is not a set.

(iii) Here, we can definitely decide whether a given novel is written by Munshi Premchand or not.
So, this collection is well-defined.
Hence, it is a set.

(iv) The term most dangerous is vague term. An animal may be most dangerous for one person and may not be for the other.
So, it is not well-defined.
Hence, it is not a set.

Example 2. **If $A = \{0, 1, 2, 3, 4, 5, 6, 7, 8, 9, 10\}$, then insert appropriate symbol $\in$ or $\notin$ in each of the following blank spaces.**

(i) $4 \ldots A$ (ii) $-4 \ldots A$ (iii) $12 \ldots A$
(iv) $9 \ldots A$ (v) $0 \ldots A$ (vi) $-2 \ldots A$

Sol. Given, $A = \{0, 1, 2, 3, 4, 5, 6, 7, 8, 9, 10\}$

(i) Since, 4 is an element of A, therefore $4 \in A$.
(ii) Since, -4 is not an element of A, therefore $-4 \notin A$.
(iii) Since, 12 is not an element of A, therefore $12 \notin A$.
(iv) Since, 9 is an element of A, therefore $9 \in A$.
(v) Since, 0 is an element of A, therefore $0 \in A$.
(vi) Since, -2 is not an element of A, therefore $-2 \notin A$.

Example 3. **Write the following sets in roster form.**

(i) Set of all vowels in english alphabet which precedes 's'.

(ii) Set of all natural numbers x such that $4x+9<50$.

(iii) Set of all letters in MATHEMATICS.

(iv) $\{x : x = \frac{n}{1+n^2}$ and $1 \le n \le 3$, where $n \in N\}$.

Competency Based Que

(v) $\{a_n : n \in N, a_{n+2} = a_{n+1} + a_n$ and $a_1 = a_2 = 1\}$.

Sol. (i) The vowels which precedes 's' are *a, e, i* and *o*.
So, the required set is $\{a, e, i, o\}$.

(ii) We have, $4x+9<50$

$\Rightarrow 4x+9-9<50-9$ [subtract 9 from both sides]

$\Rightarrow 4x<41 \Rightarrow x<\frac{41}{4}$

$\therefore x<10.25$

Since, x is a natural number, so x can take values 1, 2, 3, 4, 5, 6, 7, 8, 9, 10.

$\therefore$ Required set $=\{1, 2, 3, 4, 5, 6, 7, 8, 9, 10\}$

(iii) The letters in the word MATHEMATICS (without repeating any letter) are M, A, T, H, E, I, C, S.
So, the required set is {M, A, T, H, E, I, C, S}.

(iv) Given, $\{x : x = \frac{n}{1+n^2}$ and $1 \le n \le 3$, where $n \in N\}$.

Here, $x = \frac{n}{1+n^2}, \ 1 \le n \le 3, \ n \in N$

$\Rightarrow x = \frac{n}{1+n^2}, \ n = 1, 2, 3$

$\Rightarrow x = \frac{1}{1+1^2}, \ \frac{2}{1+2^2}, \ \frac{3}{1+3^2} = \frac{1}{2}, \frac{2}{5}, \frac{3}{10}$

$\therefore$ Required set is $\left\{\frac{1}{2}, \frac{2}{5}, \frac{3}{10}\right\}$.

(v) We have, $a_1 = 1, a_2 = 1$

and $a_{n+2} = a_{n+1} + a_n, \ \forall n \in N$

On putting $n = 1, 2, 3, \ldots$ in $a_{n+2} = a_{n+1} + a_n$, we get

$a_3 = a_2 + a_1 = 1+1 = 2$

$a_4 = a_3 + a_2 = 2+1 = 3$

$a_5 = a_4 + a_3 = 3+2 = 5$

$a_6 = a_5 + a_4 = 5+3 = 8$ and so on.

$\therefore B = \{2, 3, 5, 8, \ldots\}$

Example 4. **Write the following in set-builder form.**

(i) $A = \{14, 21, 28, 35, 42, \ldots 98\}$

(ii) $X = \left\{1, \frac{1}{4}, \frac{1}{9}, \frac{1}{16}, \frac{1}{25}, \ldots\right\}$

Sol. (i) Let x represent the elements of given set.
Given numbers are natural numbers greater than 13, less than 99 and multiples of 7.

Thus, $A = \{x : x$ is a natural number greater than 13, less than 99 and a multiple of 7}, which is the required set-builder form of given set.

This can also be written as

$A = \{x : x$ is a natural number, a multiple of 7 and $13 < x < 99\}$

or $A = \{x : x = 7n, \ n \in N$ and $2 \le n \le 14\}$

(ii) We observe that the elements of set X are the reciprocals of the squares of all natural numbers.
So, the set X in the set-builder form is

$$X = \left\{x : x = \frac{1}{n^2}, \ n \in N\right\}$$

Example 5. **Which of the following sets are empty sets?**

(i) Set of all even natural numbers divisible by 5.

(ii) Set of all even prime numbers. NCERT

(iii) $\{x : x^2 - 2 = 0$ and x is rational$\}$

(iv) $\{x : x$ is a point common to any two parallel lines$\}$ NCERT

Sol. (i) There are infinite even natural numbers which are divisible by 5. e.g. 10, 20, 30, 40, 50, etc.
Therefore, it is not an empty set.

(ii) We know that 2 is only even prime.
Therefore, it is not an empty set.

(iii) Given, $\{x : x^2 - 2 = 0$ and x is rational$\}$

Here, $x^2 - 2 = 0 \Rightarrow x^2 = 2 \Rightarrow x = \pm\sqrt{2} \notin Q$

Therefore, it is an empty set.

(iv) Given,
$\{x : x$ is a point common to any two parallel lines$\}$
We know that no point is common to any two parallel lines. Therefore, it is an empty set.

Example 6. **Classify the given sets are finite or infinite.**

(i) The set $\{x \in Z : -1 < x < 3\}$

(ii) A set of rational numbers.

(iii) The set of positive integers greater than 101.

(iv) Set of prime numbers less than 99.

(v) The set of animals living on Earth. NCERT

Sol. (i) There are three numbers 0, 1, 2 between -1 and 3.
$\therefore \{0, 1, 2\}$ is a finite set.

(ii) There are infinite rational numbers.
$\therefore$ It is an infinite set.

(iii) There are infinite positive integers greater than 101.
$\therefore$ It is an infinite set.

(iv) There are 25 prime numbers which are less than 99.
$\therefore$ It is a finite set.

(v) There are finite numbers of animals living on Earth.
$\therefore$ It is a finite set.

Example 7. **Which of the following pairs of sets are equal? Justify your answer.**

(i) $A = \{x : x$ is a letter of the word 'LOYAL'$\}$,
$B = \{x : x$ is a letter of the word 'ALLOY'$\}$

(ii) $A = \{x : x \in Z$ and $x^2 \le 8\}$,
$B = \{x : x \in R$ and $x^2 - 4x + 3 = 0\}$

Sol. (i) Given, $A = \{x : x$ is a letter of the word 'LOYAL'$\}$
$= \{L, O, Y, A\}$
and $B = \{x : x$ is a letter of the word 'ALLOY'$\}$
$= \{A, L, O, Y\}$

Here, we see that both sets have exactly the same elements.
$\therefore \quad A = B$

(ii) Given, $A = \{x : x \in Z$ and $x^2 \le 8\} = \{-2, -1, 0, 1, 2\}$
[$\because$ the squares of integers $0, \pm 1, \pm 2$ are less than 8]
and $B = \{x : x \in R$ and $x^2 - 4x + 3 = 0\} = \{1, 3\}$
[$\because x^2 - 4x + 3 = 0 \Rightarrow (x-1)(x-3) = 0 \Rightarrow x = 1, 3$]

Here, we see that set A has 5 distinct elements and set B has 2 distinct elements.
So, they do not have same elements.
$\therefore \quad A \ne B$

Quick Test 1

1. Which of the following are sets?
(i) The collection of all planets in the solar system.
(ii) A team of eleven best cricket batsmen of the world.
(iii) The collection of all boys in your class.
(iv) The collection of all natural numbers less than 100.
(v) The collection of all good athletes of India.

2. Describe the following sets in roster form.
(i) The set of all letters in the word 'ALGEBRA'.
(ii) The set of all natural numbers less than 7.
(iii) $\{x : x$ is a positive integer and $x^2 = x\}$
(iv) If $Y = \{x : x$ is a positive factor of the number $2^{p-1}(2^p - 1)$, where $2^p - 1$ is a prime number$\}$. *Competency Based Que*

3. Describe the following sets in set-builder form.
(i) The set of all letters in the word PROBABILITY.
(ii) The set of all even natural numbers.
(iii) $\{5, 25, 125, 625\}$ (iv) $\{10, 11, 12, 13, 14, 15\}$
(v) $\{1, 4, 9, 16, \ldots, 100\}$ (vi) $\left\{1, \frac{1}{2}, \frac{1}{3}, \frac{1}{4}, \frac{1}{5}, \ldots\right\}$

4. Which of the following sets are empty sets?
(i) $A = \{x : 4 < x < 5, x \in N\}$
(ii) $D = \{x : x^2 = 25$ and x is an odd integer$\}$
(iii) $\{x : x \in N$ and $x^2 = 9\}$
(iv) $\{x : x^2 - 3 = 0, x$ is rational$\}$

5. State which of the following sets are finite and which are infinite? *Competency Based Que*
(i) $A = \{x : x \in Z$ and $x^2 - 5x + 6 = 0\}$
(ii) $B = \{x : x \in Z$ and x^2 is even$\}$
(iii) $C = \{x : x \in Z$ and $x^2 = 36\}$
(iv) $D = \{x : x \in Z$ and $x > -10\}$
(v) $A = \{x : x \in Z$ and $x^2 - 2x - 3 = 0\}$

6. Find the cardinal number of the set of letters of the word 'MATHEMATICS'.

7. Are the following sets are equal?
$A = \{x : x$ is a letter in the word 'WOLF'$\}$
$B = \{x : x$ is a letter in the word 'FOLLOW'$\}$

Answers

1. (i), (iii), (iv)
2. (i) $\{A, L, G, E, B, R\}$ (ii) $\{1, 2, 3, 4, 5, 6\}$
(iii) $\{1\}$ (iv) $Y = \{1, 2, 2^2, 2^3, \ldots, 2^{p-1}, 2^p - 1\}$
3. (i) $\{x : x$ is a letter of the word PROBABILITY$\}$
(ii) $\{x : x = 2n, n \in N\}$ (iii) $\{x : x = 5^n, n \in N$ and $n < 5\}$
(iv) $\{x : x \in N, 9 < x < 16\}$ (v) $\{x : x = n^2, 1 \le n \le 10, n \in N\}$
(vi) $\left\{x : x = \frac{1}{n}, n \in N\right\}$
4. (i) and (iv)
5. (i) Finite, (ii) Infinite (iii) Finite (iv) Infinite (v) Finite
6. 8 7. Yes

For Solutions, See QR Code on Pg. 18

1.3 Subset

Let A and B be two sets. If every element of A is an element of B, then A is called a subset of B.

If A is a subset of B, then we write $A \subseteq B$, which is read as 'A is a subset of B' or A is contained in B

or $\quad A \subseteq B$, if $x \in A \Rightarrow x \in B$

If A is not a subset of B, then we write $A \not\subseteq B$.

e.g. $\{2, 3, 4, 5\}$ is a subset of a set of natural numbers.

Proper Subset

If $A \subset B$ and $A \neq B$, then A is called a proper subset of B, written as $A \subset B$ and B is called **superset of A.**

e.g. Let $A = \{x : x \text{ is an even natural number}\}$

and $B = \{x : x \text{ is a natural number}\}$

Then, $A = \{2, 4, 6, 8, ...\}$ and $B = \{1, 2, 3, 4, 5, ...\}$

$\therefore \quad A \subset B$

Subsets of the Set of Real Numbers

Some important subsets of set of real numbers R are

- the set of natural numbers $N = \{1, 2, 3 \ldots\}$
- the set of whole numbers $W = \{0, 1, 2, \ldots\}$
- the set of integers $Z = \{\ldots -2, -1, 0, 1, 2, \ldots\}$
- the set of rational numbers Q $= \left\{x : x = \dfrac{p}{q}, p, q \in Z \text{ and } q \neq 0\right\}$
- the set of irrational numbers $T = \{x : x \in R \text{ and } x \notin Q\}$

Intervals as Subsets of R

Let a, b belong to R and $a < b$. Then, the set of real numbers $\{x : a < x < b\}$ is called an **open interval** and is denoted by (a, b).

The interval which contains the end points (a and b) also, is called **closed interval** and it is denoted by $[a, b]$.

Hence, $[a, b] = \{x : a \leq x \leq b\}$

Semi-closed or Semi-open Interval

Some intervals are closed at one end and open at the other, such intervals are called semi-closed or semi-open interval.

$[a, b) = \{x : a \leq x < b\}$ is a semi-closed interval from a to b which includes a but excludes b.

$(a, b] = \{x : a < x \leq b\}$ is a semi-open interval from a to b which excludes a but includes b.

On real number line, we can draw the interval, which is shown by the dark portion on the number line.

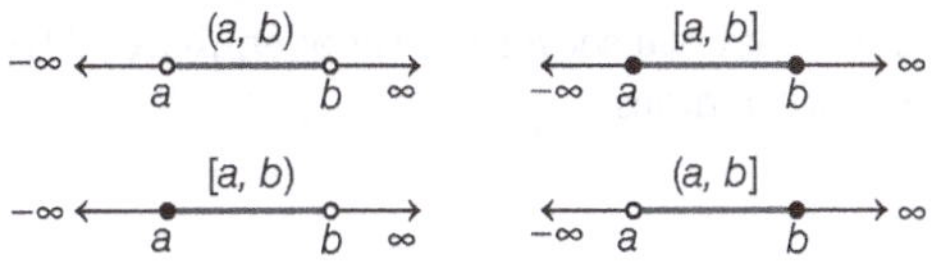

- The set $[0, \infty)$ defines the set of non-negative real numbers.
- The set $(-\infty, 0)$ defines the set of negative real numbers.
- $(-\infty, \infty)$ is the set of real numbers.

Length of an Interval

The number $(b - a)$ is called the length of any of the intervals $(a, b), [a, b], [a, b)$ or $(a, b]$.

1.4 Universal Set

If there are some sets under consideration, then a set can be chosen arbitrarily which is a **superset** of each one of the given sets. Such a set is known as the universal set and it is denoted by U.

e.g. Let $A = \{2, 4, 6\}, B = \{1, 3, 5\}$ and $C = \{0, 7\}$.

Then, $U = \{0, 1, 2, 3, 4, 5, 6, 7\}$ is an universal set.

TEACHER TIPS

- Every set is a subset of itself.
- The empty set ϕ is a subset of every set.
- The total number of subsets and proper subsets of a finite set containing n elements is 2^n and $2^n - 1$, respectively.
- If $A \subseteq B$ and $B \subseteq C$, then $A \subseteq C$.
- $A = B$ if and only if $A \subseteq B$ and $B \subseteq A$.
- Some obvious results among subsets are $N \subset Z \subset Q, Q \subset R, T \subset R, N \not\subset T$

Example 1. **If $S = \{1, 3, 5, 7, 9, 11, 13\}$, then which of the following set is a subset of S?**

(i) $A = \{3, 5, 7\}$ (ii) $B = \{2, 5, 7\}$

Sol. We have, $S = \{1, 3, 5, 7, 9, 11, 13\}$

(i) Given, $A = \{3, 5, 7\}$

Here, we see that each element of set A lies in set S. Hence, $A \subseteq S$.

(ii) Given, $B = \{2, 5, 7\}$

Since, element 2 is not belonging to the set S.

$\Rightarrow \quad 2 \notin S$

$\therefore \quad B \not\subseteq S$

Example 2. **Insert the correct symbol $\subset$ or $\not\subset$ between each of the following pair of sets.**

(i) $\{x : x \text{ is a student of class XI of your school}\}$ ...... $\{x : x \text{ is a student of your school}\}$ NCERT

(ii) $\{x : x \text{ is a triangle in the plane}\}$ ...... $\{x : x \text{ is a rectangle in the plane}\}$

(iii) {x: x is an equilateral triangle in the plane} ......
{x: x is a triangle in the plane} *NCERT*

(iv) {x: x is an even natural number} ......
{x: x is an integer}

(v) {1, 4, 8}......{1, 2, 4, 6, 8}

Sol. (i) {x : x is a student of class XI of your school}
$\subset${x: x is a student of your school}

(ii) {x : x is a triangle in the plane}
$\not\subset${x : x is a rectangle in the plane}
[$\because$ all elements of first set are not there in the second set]

(iii) {x : x is an equilateral triangle in the plane}
$\subset${x : x is a triangle in the plane}
[$\because$ elements in the second set are triangles, so there will also be equilateral triangle, thus all elements of first set are there in the second set]

(iv) {x : x is an even natural number}$\subset$ {x : x is an integer}
[$\because$ integers have both even and odd natural numbers, so all elements of first set will be there in the second set]

(v) {1, 4, 8}$\subset${1, 2, 4, 6, 8}

Example 3. Write the following as intervals and also represent on the number line.

(i) $\{x : x \in R, -5 < x \le 6\}$
(ii) $\{x : x \in R, -11 < x < -9\}$
(iii) $\{x : x \in R, 2 \le x < 8\}$

Sol. (i) $\{x : x \in R, -5 < x \le 6\}$ is the set that does not contain -5 but contains 6. So, it can be written as a semi-closed interval whose first end point is open and last end point is closed i.e. $(-5, 6]$.
On the number line, $(-5, 6]$ can be graphed as shown in figure given below.

The dark portion on the number line represents $(-5, 6]$.

(ii) $\{x : x \in R, -11 < x < -9\}$ is the set that neither contains -11 nor -9, so it can be represented as open interval i.e. $(-11, -9)$.
On the number line, $(-11, -9)$ can be graphed as shown in figure given below.

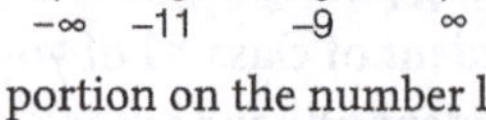

The dark portion on the number line represents $(-11, -9)$.

(iii) $\{x : x \in R, 2 \le x < 8\}$ is the set that contains 2 but does not contain 8. So, it can be represented as a semi-open interval whose first end point is closed and the other end point is open i.e. [2, 8).
On the number line, [2, 8) can be graphed as shown in figure given below.

The dark portion on the number line represents [2, 8).

Example 4. Write the following intervals in set-builder form.
(i) $(-3, 0)$ (ii) $[6, 12]$ (iii) $(6, 12]$ (iv) $[-23, 5)$
Also, find the length of each interval.

Sol. We know that closed interval contains end points while open interval does not contain end points.
(i) $(-3, 0) = \{x : x \in R, -3 < x < 0\}$, Length $= 0-(-3) = 3$
(ii) $[6, 12] = \{x : x \in R, 6 \le x \le 12\}$, Length $= 12-6 = 6$
(iii) $(6, 12] = \{x : x \in R, 6 < x \le 12\}$, Length $= 12-6 = 6$
(iv) $[-23, 5) = \{x : x \in R, -23 \le x < 5\}$,
Length $= 5-(-23) = 28$

Example 5. Two finite sets have m and n elements. The number of subsets of the first set is 112 more than that of the second set. Find the values of m and n. *NCERT Exemplar*

Sol. Let the two sets be A and B such that $n(A) = m$ and $n(B) = n$.
Then, the number of subsets of set $A = 2^m$ and number of subsets of set $B = 2^n$.
According to the given condition,
$$2^m = 112 + 2^n$$
$$\Rightarrow \quad 2^m - 2^n = 2^7 - 2^4$$
On comparing both sides, we get
$$2^m = 2^7 \text{ and } 2^n = 2^4$$
$$\Rightarrow \quad m = 7 \text{ and } n = 4$$

Example 6. What universal sets would you propose for each of the following? *NCERT*
(i) The set of right triangles.
(ii) The set of equilateral triangles.

Sol. (i) The universal set for the set of right triangles is set of triangles.
(ii) The universal set for the set of equilateral triangle is set of triangles.

Quick Test 2

1. Write down the subsets of the following sets.
(i) $\{1, 2, 3\}$ (ii) $\{\phi\}$ *NCERT*

2. Given that $N = \{1, 2, 3, \ldots 50\}$, then write the subset A of N, whose elements are represented by $x + 3$, where $x \in N$. *Competency Based Que*

3. Consider the following sets ϕ, $A = \{2, 5\}$, $B = \{1, 2, 3, 4\}$ and $C = \{1, 2, 3, 4, 5\}$.
Insert the correct symbol $\subset$ or $\not\subset$ between each pair of sets.
(i) $\phi \ldots B$ (ii) $A \ldots B$
(iii) $A \ldots C$ (iv) $B \ldots C$

4. Write the following intervals in set-builder form.
(i) $(-6, 0)$ (ii) $[3, 21)$
(iii) $[2, 21]$ (iv) $(-20, 5]$

5. Write the following as intervals and also represent on real line.
(i) $\{x : x \in R, -3 < x \le 7\}$ (ii) $\{x : x \in R, -11 < x < -7\}$
(iii) $\{x : x \in R, 0 \le x < 11\}$ (iv) $\{x : x \in R, 2 \le x \le 9\}$

6. Determine whether the given statements are true or false.
(i) If $A = \{3, 6, 7\}$ and $B = \{2, 3, 7, 8, 10\}$, then $A \subset B$.
(ii) If $A = \{x : x^2 + 4x - 21 = 0, x \in N\}$ and $B = \{-7, 3\}$, then $A \subseteq B$. *Competency Based Que*
(iii) If $A = \{1, 7, 9\}$, $B = \{2, 4, 6\}$ and $C = \{0, 2, 4\}$, then $U = \{0, 1, 2, 3, 4, 5, 6, 7, 8, 9\}$ is the universal set for all three sets.
(iv) $\{6\} \in \{4, 5, \{6\}, 7\}$

7. Given, the sets $A = \{1, 3, 5\}$, $B = \{2, 4, 6\}$ and $C = \{0, 2, 4, 6, 8\}$.
Which of the following may be considered as universal set(s) for all three sets A, B and C? *NCERT*
(i) ϕ (ii) $\{0, 1, 2, 3, 4, 5, 6, 7, 8, 9, 10\}$

8. Write the following subset of R as interval. Also, find the length of interval and represent on number line.
$$\{x : x \in R, -12 \le x \le -10\}$$

9. Prove that $A \subseteq \phi$ implies $A = \phi$.

Answers

1. (i) ϕ, $\{1\}$, $\{2\}$, $\{3\}$, $\{1, 2\}$, $\{2, 3\}$, $\{1, 3\}$, $\{1, 2, 3\}$.
(ii) The subsets are ϕ and $\{\phi\}$.

2. $\{4, 5, 6, \ldots 50\}$

3. (i) $\subset$ (ii) $\not\subset$ (iii) $\subset$ (iv) $\subset$

4. (i) $\{x : x \in R, -6 < x < 0\}$ (ii) $\{x : x \in R, 3 \le x < 21\}$
(iii) $\{x : x \in R, 2 \le x \le 21\}$ (iv) $\{x : x \in R, -20 < x \le 5\}$

5. (i) $(-3, 7]$
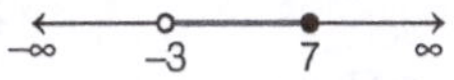

(ii) $(-11, -7)$
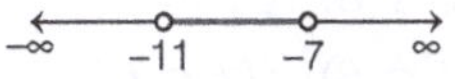

(iii) $[0, 11)$

(iv) $[2, 9]$

6. (i) False (ii) True (iii) True (iv) True

7. (i) ϕ cannot be considered as universal set.
(ii) $\{0, 1, 2, 3, 4, 5, 6, 7, 8, 9, 10\}$ is the universal set.

8.

$[-12, -10]$, 2

For Solutions, See QR Code on Pg. 18

1.5 Venn Diagrams

In Venn diagrams, the universal set is represented by a **rectangular region** and its subset are represented by **circles** or a **closed geometrical figure** inside the universal set. Also, an element of a set is represented by a point within the circles of set.

e.g. (i) If $A = \{1, 2, 3, 5, 9\}$, $B = \{4, 6, 7\}$ and $U = \{1, 2, 3, 4, 5, 6, 7, 8, 9\}$, then

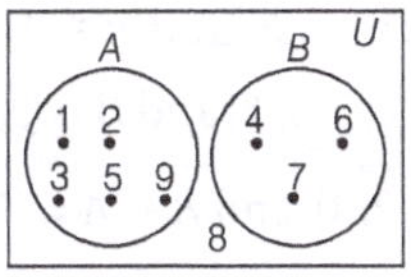

Here, there is no element common in set A and B.

(ii) If $A = \{1, 2, 3, 4, 5\}$ and $B = \{1, 2, 3\}$, then

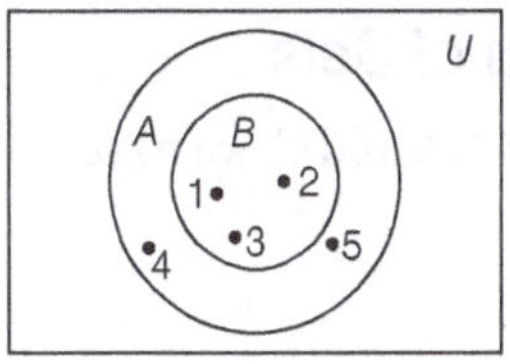

Here, all the elements of B are common in set A.

$\therefore \quad B \subset A$

(iii) If $A=\{1,2\}$, $B=\{1,2,3\}$ and $C=\{1,2,3,4,5\}$, then

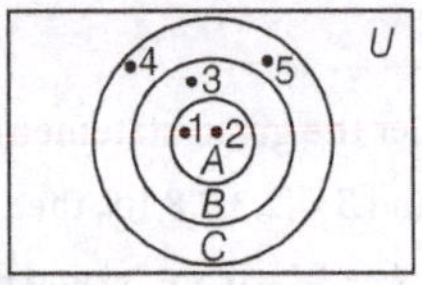

Here, $A \subset B \subset C$

(iv) If $A=\{3,4,6,7\}$ and $B=\{4,6,10,12\}$, then

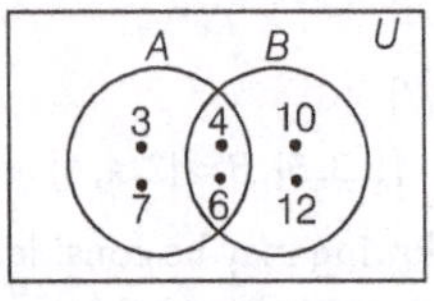

Here, elements 4 and 6 are common in both sets.

1.6 Operations on Sets

Union of Sets

Let A and B be any two sets. The union of A and B is the set of all those elements which belong to either A or B or both. The symbol '$\cup$' is used to denote the union.

Symbolically, it is denoted by $A \cup B$ and read as **A union B.**

$\therefore \quad A \cup B = \{x : x \in A \text{ or } x \in B\}$

e.g. Let $A=\{2,3\}$ and $B=\{3,4,5\}$

Then, $\quad A \cup B = \{2,3,4,5\}$

The union of sets A and B is represented by the following Venn diagram,

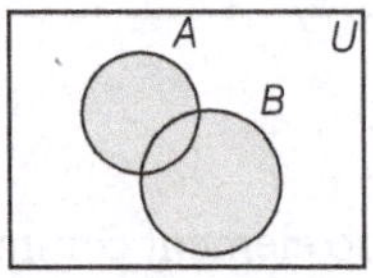

The shaded portion represents $A \cup B$.

Laws of Union of Sets

For any three sets A, B and C, we have

- $A \cup \phi = A$ [Identity law]
- $U \cup A = U$ [Universal law]
- $A \cup A = A$ [Idempotent law]
- $A \cup B = B \cup A$ [Commutative law]
- $(A \cup B) \cup C = A \cup (B \cup C)$ [Associative law]

Intersection of Sets

Let A and B be any two sets. The intersection of A and B is the set of all those elements which belong to both A and B. It is denoted by $A \cap B$ and read as **A intersection B.** The symbol '$\cap$' is used to denote the intersection.

$\therefore \quad A \cap B = \{x : x \in A \text{ and } x \in B\}$

e.g. Let $A=\{2,3,4,5\}$ and $B=\{1,3,6,4\}$

Then, $A \cap B = \{3,4\}$

The intersection of sets A and B is represented by the following Venn diagram,

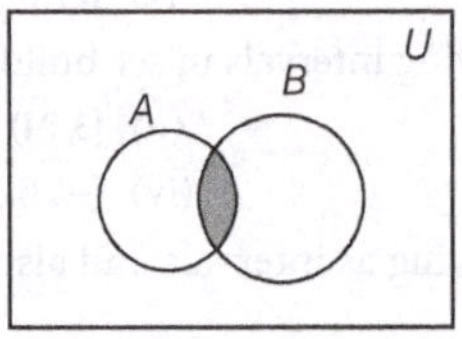

The shaded portion represents $A \cap B$.

Laws of Intersection of Sets

For any three sets A, B and C, we have

- $A \cap \phi = \phi$ [Identity law]
- $U \cap A = A$ [Universal law]
- $A \cap A = A$ [Idempotent law]
- $A \cap B = B \cap A$ [Commutative law]
- $(A \cap B) \cap C = A \cap (B \cap C)$ [Associative law]
- $A \cap (B \cup C) = (A \cap B) \cup (A \cap C)$ [intersection distributes over union]
- $A \cup (B \cap C) = (A \cup B) \cap (A \cup C)$ [union distributes over intersection]

Difference of Sets

Let A and B be any two sets. The difference of sets A and B in this order is the set of all those elements of A which do not belong to B.

It is denoted by $A - B$ and read as **A minus B.**

The symbol '–' is used to denote the difference of sets.

$\therefore \quad A - B = \{x : x \in A \text{ and } x \notin B\}$

Similarly, $B - A = \{x : x \in B \text{ and } x \notin A\}$

e.g. Let $\quad A=\{1,2,3,4,5\}$ and $B=\{3,5,7,9\}$

Then, $\quad A - B = \{1,2,4\}$ and $B - A = \{7,9\}$

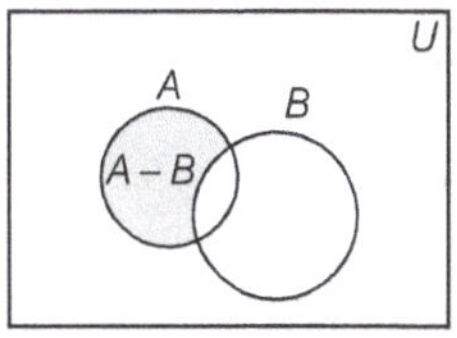

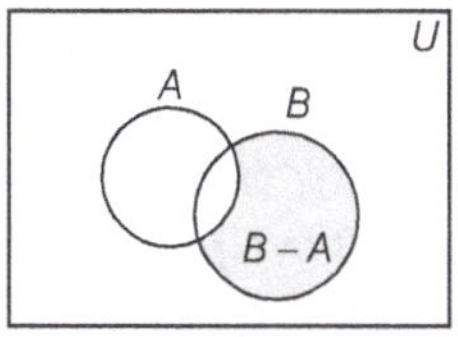

The shaded portion represents the difference of two sets A and B.

Law of Difference of Sets

(i) For any two sets A and B, we have

- $A - B = A \cap B'$
- $B - A = B \cap A'$
- $A - B \subseteq A$
- $B - A \subseteq B$
- $A - B = A \Leftrightarrow A \cap B = \phi$
- $(A - B) \cup B = A \cup B$
- $(A - B) \cap B = \phi$
- $(A - B) \cup (B - A) = (A \cup B) - (A \cap B)$

(ii) If A, B and C are any three sets, then

- $A - (B \cap C) = (A - B) \cup (A - C)$
- $A - (B \cup C) = (A - B) \cap (A - C)$
- $A \cap (B - C) = (A \cap B) - (A \cap C)$
- $A \cap (B \Delta C) = (A \cap B) \Delta (A \cap C)$

Where, Δ represents symmetric difference and read as 'delta'. $A \Delta B$ is read as 'A delta B' or 'set A symmetric difference set B'.

Disjoint Sets

Two sets A and B are said to be disjoint sets, if they have **no common element** i.e. $A \cap B = \phi$.

e.g. Let $A = \{2, 4, 6\}$ and $B = \{1, 3, 5\}$

Then, $A \cap B = \phi$.

Hence, A and B are disjoint sets.

The disjoint of two sets A and B can be represented by the Venn diagram,

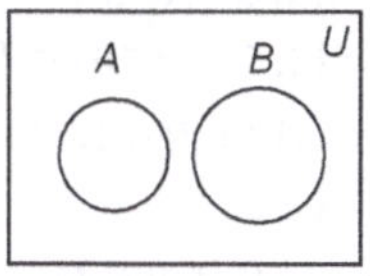

Complement of a Set

Let U be the universal set and A be any subset of U, then complement of A with respect to U is the set of all those elements of U which are not in A. It is denoted by $\overline{A}$ or A' and read as, '**A complement**'.

Thus, $\overline{A} = \{x : x \in U \text{ and } x \notin A\}$ or $A' = U - A$

e.g. Let $U = \{1, 2, 3, 4, 5, 6\}$ and $A = \{2, 4\}$

Then, $A' = U - A = \{1, 2, 3, 4, 5, 6\} - \{2, 4\} = \{1, 3, 5, 6\}$

The complement of set A is represented by the following Venn diagram,

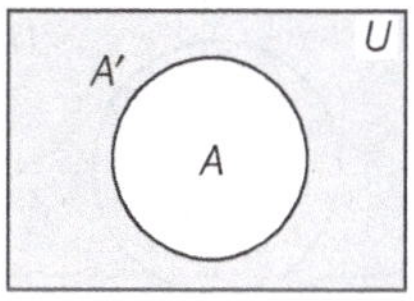

Here, the shaded portion represents the complement of set A.

Some Properties of Complement of Sets

- $(A')' = A = U - A'$ [law of double complementation]
- (a) $A \cup A' = U$ (b) $A \cap A' = \phi$ [complement laws]
- (a) $\phi' = U$ (b) $U' = \phi$ [laws of empty set and universal set]
- $(A \cup B)' = U - (A \cup B)$

De-Morgan's Law:

If A and B are any two sets, then

(i) $(A \cup B)' = A' \cap B'$ (ii) $(A \cap B)' = A' \cup B'$

Example 1. Draw the Venn diagrams to illustrate the following relationship among sets E, M and U, where E is the set of students studying English in a school, M is the set of students studying Mathematics in the same school and U is the set of all students in that school.

(i) All the students who study Mathematics also study English but some students who study English do not study Mathematics.

(ii) Not all students study Mathematics but every student studying English studies Mathematics.

NCERT Exemplar, Competency Based Que

Sol. Given, E = Set of students studying English

M = Set of students studying Mathematics

and U = Set of all students

(i) Since, all of the students who study Mathematics, also study English but some students who study English do not study Mathematics.

$\therefore$ $M \subset E \subset U$

Through Venn diagram, we represent it as

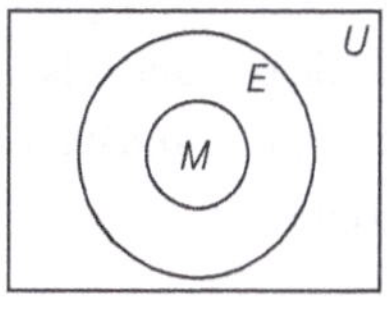

(ii) Since, every student studying English studies Mathematics.

$\therefore \quad E \subset M \subset U$

Through Venn diagram, we represent it as

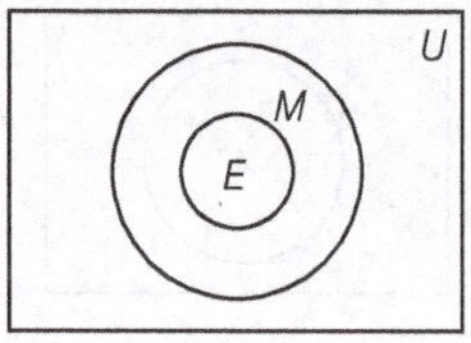

Example 2. Find the union of each of the following pairs of sets.

(i) $A = \{a, e, i, o, u\}$, $B = \{a, c, d\}$

(ii) $A = \{1, 3, 5\}$, $B = \{2, 4, 6\}$

(iii) $A = \{x : x$ is a natural number and $1 < x \le 5\}$ and $B = \{x : x$ is a natural number and $5 < x \le 10\}$

Sol. (i) Given, $A = \{a, e, i, o, u\}$ and $B = \{a, c, d\}$

$\therefore \quad A \cup B = \{a, e, i, o, u\} \cup \{a, c, d\}$
$= \{a, c, d, e, i, o, u\}$

(ii) Given, $A = \{1, 3, 5\}$ and $B = \{2, 4, 6\}$

$\therefore \quad A \cup B = \{1, 3, 5\} \cup \{2, 4, 6\} = \{1, 2, 3, 4, 5, 6\}$

(iii) Given, $A = \{x : x$ is a natural number and $1 < x \le 5\}$
$= \{2, 3, 4, 5\}$

and $B = \{x : x$ is a natural number and $5 < x \le 10\}$
$= \{6, 7, 8, 9, 10\}$

$\therefore \quad A \cup B = \{2, 3, 4, 5\} \cup \{6, 7, 8, 9, 10\}$
$= \{2, 3, 4, 5, 6, 7, 8, 9, 10\}$
$= \{x : x$ is a natural number and $1 < x \le 10\}$

Example 3. Find the intersection of each of the following pairs of sets.

(i) $A = \{1, 3, 5, 7, 9\}$, $B = \{2, 3, 6, 8, 9\}$

(ii) $A = \{e, f, g\}$, $B = \phi$

(iii) $A = \{x : x = 3n, n \in Z\}$, $B = \{x : x = 4n, n \in Z\}$

Competency Based Que

Sol. (i) Given, $A = \{1, 3, 5, 7, 9\}$ and $B = \{2, 3, 6, 8, 9\}$

$\therefore \quad A \cap B = \{3, 9\}$

[$\because$ 3 and 9 are only elements which are common]

(ii) Given, $A = \{e, f, g\}$ and $B = \phi$

$\therefore \quad A \cap B = \phi$ [$\because$ there is no common element]

(iii) Let $x \in A \cap B$

$\Rightarrow x \in A$ and $x \in B$

$\Rightarrow x$ is a multiple of 3 and x is a multiple of 4

$\Rightarrow x$ is a multiple of 3 and 4 both

$\Rightarrow x$ is a multiple of 12

$\Rightarrow x = 12n, n \in Z$

$\therefore A \cap B = \{x : x = 12n, n \in Z\}$

Example 4. If $A = \{4, 5, 7, 8, 10\}$, $B = \{4, 5, 9\}$ and $C = \{1, 4, 6, 9\}$, then verify that

(i) $(A \cap B) \cap C = A \cap (B \cap C)$

(ii) $A \cup (B \cap C) = (A \cup B) \cap (A \cup C)$

(iii) $A \cap (B \cup C) = (A \cap B) \cup (A \cap C)$

Sol. Given, $A = \{4, 5, 7, 8, 10\}$,

$B = \{4, 5, 9\}$ and $C = \{1, 4, 6, 9\}$

(i) Now, $A \cap B = \{4, 5, 7, 8, 10\} \cap \{4, 5, 9\} = \{4, 5\}$

$\therefore \quad$ LHS $= (A \cap B) \cap C$
$= \{4, 5\} \cap \{1, 4, 6, 9\} = \{4\}$ ...(i)

Now, $B \cap C = \{4, 5, 9\} \cap \{1, 4, 6, 9\} = \{4, 9\}$

$\therefore \quad$ RHS $= A \cap (B \cap C)$
$= \{4, 5, 7, 8, 10\} \cap \{4, 9\} = \{4\}$ ...(ii)

From Eqs. (i) and (ii), we get

LHS = RHS = $\{4\}$

Hence, $(A \cap B) \cap C = A \cap (B \cap C)$

(ii) Here, $B \cap C = \{4, 9\}$

$\therefore \quad$ LHS $= A \cup (B \cap C) = \{4, 5, 7, 8, 10\} \cup \{4, 9\}$

$\Rightarrow \quad$ LHS $= \{4, 5, 7, 8, 9, 10\}$ ...(iii)

Now, $A \cup B = \{4, 5, 7, 8, 10\} \cup \{4, 5, 9\}$
$= \{4, 5, 7, 8, 9, 10\}$

and $A \cup C = \{4, 5, 7, 8, 10\} \cup \{1, 4, 6, 9\}$
$= \{1, 4, 5, 6, 7, 8, 9, 10\}$

$\therefore \quad$ RHS $= (A \cup B) \cap (A \cup C)$
$= \{4, 5, 7, 8, 9, 10\} \cap \{1, 4, 5, 6, 7, 8, 9, 10\}$
$= \{4, 5, 7, 8, 9, 10\}$ ...(iv)

From Eqs. (iii) and (iv), we get

LHS = RHS = $\{4, 5, 7, 8, 9, 10\}$

Hence, $A \cup (B \cap C) = (A \cup B) \cap (A \cup C)$

(iii) Now, $B \cup C = \{4, 5, 9\} \cup \{1, 4, 6, 9\} = \{1, 4, 5, 6, 9\}$

$\therefore \quad$ LHS $= A \cap (B \cup C)$
$= \{4, 5, 7, 8, 10\} \cap \{1, 4, 5, 6, 9\} = \{4, 5\}$ ...(v)

Now, $A \cap B = \{4, 5, 7, 8, 10\} \cap \{4, 5, 9\} = \{4, 5\}$

and $A \cap C = \{4, 5, 7, 8, 10\} \cap \{1, 4, 6, 9\} = \{4\}$

$\therefore \quad$ RHS $= (A \cap B) \cup (A \cap C) = \{4, 5\} \cup \{4\}$
$= \{4, 5\}$ ...(vi)

From Eqs. (v) and (vi), we get

LHS = RHS = $\{4, 5\}$

Hence, $A \cap (B \cup C) = (A \cap B) \cup (A \cap C)$

Example 5.

(i) If $X = \{a, b, c, d, e, f\}$ and $Y = \{f, b, d, g, h, k\}$, then find $X - Y$ and $Y - X$.

(ii) If $A = \{1, 2, 3, 4, 5\}$ and $B = \{2, 4, 6\}$, then find $A - B$ and $B - A$.

Also, represent each of these by Venn diagram.

Sol. (i) Given, $X = \{a, b, c, d, e, f\}$ and $Y = \{f, b, d, g, h, k\}$

$\therefore\ X - Y = \{a, b, c, d, e, f\} - \{f, b, d, g, h, k\} = \{a, c, e\}$

[only those elements of X which do not belong to Y]

and $Y - X = \{f, b, d, g, h, k\} - \{a, b, c, d, e, f\}$

$= \{g, h, k\}$

[only those elements of Y which do not belong to X]

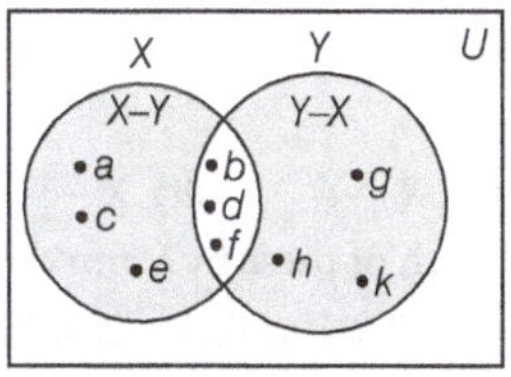

(ii) Given, $A = \{1, 2, 3, 4, 5\}$ and $B = \{2, 4, 6\}$

$\therefore\ A - B = \{1, 2, 3, 4, 5\} - \{2, 4, 6\}$

$A - B = \{1, 3, 5\}$

[only those elements of A which do not belong to B]

and $B - A = \{2, 4, 6\} - \{1, 2, 3, 4, 5\} = \{6\}$

[only those elements of B which do not belong to A]

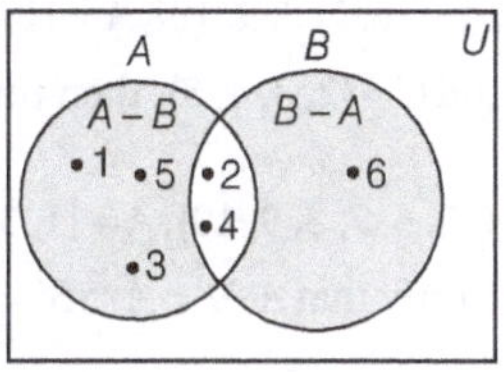

Example 6. **Which of the following pairs of sets are disjoint?**

(i) $A = \{1, 2, 3, 4, 5, 6\}$ and $B = \{x : x$ is a natural number and $4 \le x \le 6\}$

(ii) $A = \{x : x$ is the boy of your school$\}$ and $B = \{x : x$ is the girl of your school$\}$

Sol. (i) Given, $A = \{1, 2, 3, 4, 5, 6\}$ and $B = \{4, 5, 6\}$

$\therefore\ A \cap B = \{1, 2, 3, 4, 5, 6\} \cap \{4, 5, 6\}$

$= \{4, 5, 6\} \ne \phi$

Hence, this pair of set is not disjoint.

(ii) Here, $A = \{b_1, b_2, \dots b_n\}$ and $B = \{g_1, g_2, \dots g_m\}$, where $b_1, b_2, \dots, b_n$ are the boys and $g_1, g_2, \dots, g_m$ are the girls of school.

Clearly, $A \cap B = \phi$

Hence, this pair of set is disjoint set.

Example 7. **Let $U = \{1, 2, 3, 4, 5, 6, 7, 8, 9\}$, $A = \{1, 2, 3, 4\}$, $B = \{2, 4, 6, 8\}$ and $C = \{3, 4, 5, 6\}$. Find**

(i) A' (ii) B' (iii) $(A \cap C)'$
(iv) $(A \cup B)'$ (v) $(A')'$ (vi) $(B - C)'$

Sol. Given, $U = \{1, 2, 3, 4, 5, 6, 7, 8, 9\}$, $A = \{1, 2, 3, 4\}$,

$B = \{2, 4, 6, 8\}$ and $C = \{3, 4, 5, 6\}$

(i) $A' = U - A = \{1, 2, 3, 4, 5, 6, 7, 8, 9\} - \{1, 2, 3, 4\}$
$= \{5, 6, 7, 8, 9\}$

(ii) $B' = U - B = \{1, 2, 3, 4, 5, 6, 7, 8, 9\} - \{2, 4, 6, 8\}$
$= \{1, 3, 5, 7, 9\}$

(iii) $A \cap C = \{3, 4\}$

$\therefore\ (A \cap C)' = U - (A \cap C)$
$= \{1, 2, 3, 4, 5, 6, 7, 8, 9\} - \{3, 4\}$
$= \{1, 2, 5, 6, 7, 8, 9\}$

(iv) $A \cup B = \{1, 2, 3, 4, 6, 8\}$

$(A \cup B)' = U - (A \cup B) = \{1, 2, 3, 4, 5, 6, 7, 8, 9\}$
$- \{1, 2, 3, 4, 6, 8\} = \{5, 7, 9\}$

(v) $(A')' = U - A' = \{1, 2, 3, 4, 5, 6, 7, 8, 9\} - \{5, 6, 7, 8, 9\}$ [using part (i)]
$= \{1, 2, 3, 4\}$

Alternate Method

We know that $(A')' = A$

$\therefore\ (A')' = \{1, 2, 3, 4\}$

(vi) Now, $B - C = \{2, 8\}$

$(B - C)' = U - (B - C)$
$= \{1, 2, 3, 4, 5, 6, 7, 8, 9\} - \{2, 8\}$
$= \{1, 3, 4, 5, 6, 7, 9\}$

Example 8. **Prove that**

(i) $A - (B \cap C) = (A - B) \cup (A - C)$

(ii) $A \cap (B - C) = (A \cap B) - (A \cap C)$

Sol. (i) Let $x \in A - (B \cap C)$

$\Rightarrow\ x \in A$ and $x \notin (B \cap C)$

$\Rightarrow\ x \in A$ and $(x \notin B$ or $x \notin C)$

$\Rightarrow\ (x \in A$ and $x \notin B)$ or $(x \in A$ and $x \notin C)$

$\Rightarrow\ x \in (A - B)$ or $x \in (A - C)$

$\Rightarrow\ x \in (A - B) \cup (A - C)$

$\therefore\ A - (B \cap C) \subseteq (A - B) \cup (A - C)$ ...(i)

Again, let $y \in (A - B) \cup (A - C)$

$\Rightarrow\ y \in (A - B)$ or $y \in (A - C)$

$\Rightarrow\ (y \in A$ and $y \notin B)$ or $(y \in A$ and $y \notin C)$

$\Rightarrow\ y \in A$ and $(y \notin B$ or $y \notin C)$

$\Rightarrow\ y \in A$ and $y \notin (B \cap C) \Rightarrow y \in A - (B \cap C)$

$\therefore\ (A - B) \cup (A - C) \subseteq A - (B \cap C)$ ...(ii)

From Eqs. (i) and (ii), we get

$A - (B \cap C) = (A - B) \cup (A - C)$ **Hence proved.**

(ii) Let x be any arbitrary element of $A \cap (B - C)$.

Then, $x \in A \cap (B - C)$

$\Rightarrow\ x \in A$ and $x \in (B - C)$

$\Rightarrow\ x \in A$ and $(x \in B$ and $x \notin C)$

$\Rightarrow\ (x \in A$ and $x \in B)$ and $(x \in A$ and $x \notin C)$

$\Rightarrow\ x \in (A \cap B)$ and $x \notin (A \cap C)$

$\Rightarrow\ x \in (A \cap B) - (A \cap C)$

$\Rightarrow\ A \cap (B - C) \subseteq (A \cap B) - (A \cap C)$ ...(i)

Now, let y be any arbitrary element of $(A \cap B) - (A \cap C)$.

Then, $y \in (A \cap B) - (A \cap C)$

$\Rightarrow \quad y \in (A \cap B)$ and $y \notin (A \cap C)$

$\Rightarrow \quad (y \in A$ and $y \in B)$ and $(y \in A$ and $y \notin C)$

$\Rightarrow \quad y \in A$ and $(y \in B$ and $y \notin C)$

$\Rightarrow \quad y \in A$ and $y \in (B - C)$

$\Rightarrow \quad y \in A \cap (B - C)$

$\Rightarrow \quad (A \cap B) - (A \cap C) \subseteq A \cap (B - C)$ ...(ii)

From Eqs. (i) and (ii), we get

$$A \cap (B - C) = (A \cap B) - (A \cap C)$$

Quick Test 3

1. Find the union of each of the following pair of sets.
(i) $A = \{a, e, i, o, u\}, B = \{a, b, c\}$
(ii) $A = \{1, 3, 5\}, B = \{1, 2, 3\}$
(iii) $A = \{2, 4, 6, 8\}, B = \{6, 8, 10, 12\}$

2. Let $A = \{x : x \in N\}, B = \{x : x = 2n, n \in N\}$, $C = \{x : x = 2n - 1, n \in N\}$ and $D = \{x : x$ is a prime natural number$\}$. Find
(i) $A \cap B$ (ii) $A \cap C$ (iii) $B \cap C$ (iv) $B \cap D$

3. Let $A = \{1, 2, 4, 5\}, B = \{2, 3, 5, 6\}, C = \{4, 5, 6, 7\}$. Verify that
(i) $A \cup (B \cap C) = (A \cup B) \cap (A \cup C)$
(ii) $A \cap (B \cup C) = (A \cap B) \cup (A \cap C)$

4. Let $A = \{3, 6, 12, 15, 18, 21\}, B = \{4, 8, 12, 16, 20\}$, $C = \{2, 4, 6, 8, 10, 12, 14, 16\}$ and $D = \{5, 10, 15, 20\}$. Find
(i) $A - B$ (ii) $A - C$ (iii) $B - C$ (iv) $B - D$

5. If $X = \{a, b, c, d\}$ and $Y = \{f, b, d, g\}$, then find *NCERT*
(i) $X \cup Y$ (ii) $X \cap Y$ (iii) $X - Y$ (iv) $Y - X$
Also, represent each of these by Venn diagram.

6. If $U = \{2, 3, 4, 5, 6, 7, 8, 9, 10, 11\}, A = \{2, 4, 7\}, B = \{3, 5, 7, 9, 11\}$ and $C = \{7, 8, 9, 10, 11\}$, then compute
(i) $(A \cap U) \cap (B \cup C)$ (ii) $C - B$ (iii) $B - C$

7. Which of the following pairs of sets are disjoint? *NCERT*
(i) $A = \{1, 2, 3, 4\}$ and $B = \{x : x$ is a natural number and $5 \le x \le 6\}$
(ii) $A = \{a, e, i, o, u\}$ and $B = \{c, d, e, f\}$
(iii) $A = \{x : x$ is an even integer$\}$ and $B = \{x : x$ is an odd integer$\}$

8. Let $U = \{1, 2, 3, 4, 5, 6, 7, 8, 9, 10\}, A = \{1, 2, 3\}$, $B = \{2, 4, 6, 7\}$ and $C = \{2, 3, 4, 8\}$, then find $(B \cup C)'$, $(A \cup B)'$. *NCERT Exemplar*

9. If $U = \{1, 2, 3, 4, 5, 6, 7, 8, 9, 10\}, A = \{1, 3, 4\}$ and $B = \{5, 6\}$, then verify that $A - B = A \cap B' = B' - A'$.

Answers

1. (i) $\{a, b, c, e, i, o, u\}$ (ii) $\{1, 2, 3, 5\}$ (iii) $\{2, 4, 6, 8, 10, 12\}$

2. (i) $\{x : x = 2n, n \in N\} = B$ (ii) $\{x : x = 2n - 1, n \in N\} = C$
(iii) $\{\ \} = \phi$ (iv) $D = \{2\}$

4. (i) $\{3, 6, 15, 18, 21\}$ (ii) $\{3, 15, 18, 21\}$
(iii) $\{20\}$ (iv) $\{4, 8, 12, 16\}$

5. (i) $\{a, b, c, d, f, g\}$

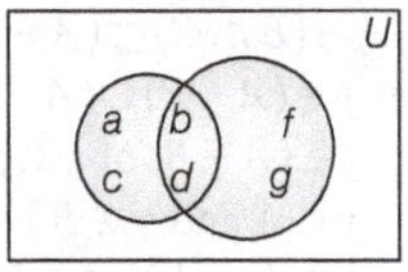

(ii) $\{b, d\}$

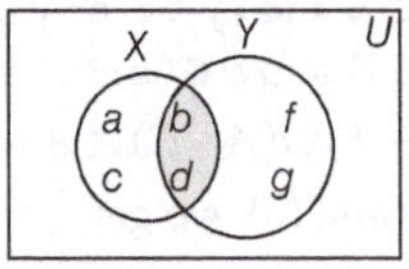

(iii) $\{a, c\}$

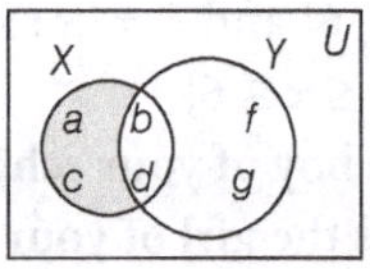

(iv) $\{f, g\}$

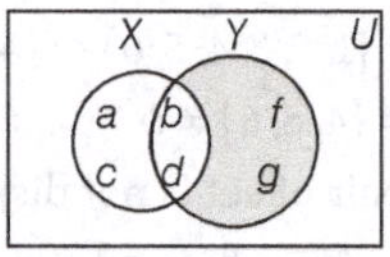

6. (i) $\{7\}$ (ii) $\{8, 10\}$ (iii) $\{3, 5\}$

7. (i) Disjoint sets (ii) Not disjoint sets (iii) Disjoint sets

8. $(B \cup C)' = \{1, 5, 9, 10\}$ and $(A \cup B)' = \{5, 8, 9, 10\}$

For Solutions, See QR Code on Pg. 18

Exam Pattern Questions

Objective Type Questions

Multiple Choice Questions

1. A null set is represented by

(a) $\Leftrightarrow$ (b) ϕ
(c) ( ) (d) [ ]

2. The shaded region in the given figure is

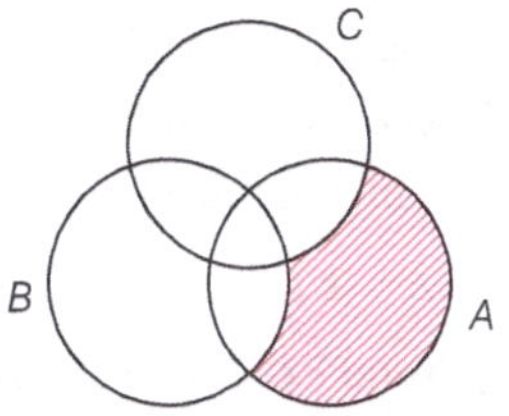

(a) $A \cup (B \cap C)$ (b) $A \cap (B \cup C)$
(c) $A - (B \cup C)$ (d) $A - (B \cap C)$

3. Let N be the set of natural numbers and Q be the set of rational numbers, then

(a) $N \cap Q = \phi$ (b) $N \subset Q$
(c) $Q \subset N$ (d) $N \cup Q = N$

4. If $A = \{a, b, c, d, e\}$, then the number of proper subsets of A is

(a) 31 (b) 120
(c) 64 (d) 10

5. If $U = [-4, 4]$ and $A = (-2, 4]$, then A' is **Competency Based Que**

(a) $[-4, -2)$ (b) $(-4, -2)$
(c) $[-2, 2]$ (d) $[-4, -2]$

6. The set $A = \{x \in N : x^2 - 9 = 0\}$ is **Competency Based Que**

(a) $\{3\}$ (b) $\{0\}$
(c) $\{-3\}$ (d) $\{-3, 3\}$

7. Which of the following shows De-Morgan's Laws?

(a) $(A')' = A$
(b) $A \cup A' = U$
(c) $A \cap A' = \phi$
(d) $(A \cup B)' = A' \cap B'$

8. If $n(P) = 24$, $n(Q) = 20$ and $n(P \cup Q) = 24$ then, $n(P \cap Q)$ is

(a) 4 (b) 6
(c) 20 (d) 24

9. Let $B = \left\{x : x = \frac{1}{2n-1}, 1 \le n \le 5, \text{ where } n \in N\right\}$, then B equals **Competency Based Que**

(a) $\left\{1, \frac{1}{3}, \frac{1}{5}, \frac{1}{9}, \frac{1}{11}\right\}$
(b) $\left\{1, \frac{1}{3}, \frac{1}{7}, \frac{1}{9}, \frac{1}{11}\right\}$
(c) $\left\{1, \frac{1}{3}, \frac{1}{5}, \frac{1}{7}, \frac{1}{9}\right\}$
(d) None of the above

10. The set $(A \cap B')' \cup (B \cap C)$ is equal to **NCERT Exemplar**

(a) $A' \cup B \cup C$
(b) $A' \cup B$
(c) $A' \cup C'$
(d) $A' \cap B$

11. If A and B are two sets, then $A \cap (A \cup B)$ equals to **NCERT Exemplar**

(a) A (b) B
(c) ϕ (d) $A \cap B$

12. If X and Y are two sets and X' denotes the complement of X, then $X \cap (X \cup Y)'$ is equal to

(a) X (b) Y
(c) ϕ (d) $X \cap Y$

13. Let F_1 be the set of parallelograms, F_2 be the set of rectangles, F_3 be the set of rhombus, F_4 be the set of squares and F_5 be the set of trapeziums in a plane. Then, F_1 may be equal to **NCERT Exemplar**

(a) $F_2 \cap F_3$
(b) $F_3 \cap F_4$
(c) $F_2 \cup F_5$
(d) $F_2 \cup F_3 \cup F_4 \cup F_1$

14. If R be the set of points inside a rectangle of sides a and b $(a, b > 1)$ with two sides along the positive direction of X-axis and Y-axis, then **NCERT Exemplar**

(a) $R = \{(x, y) : 0 \le x \le a, 0 \le y \le b\}$
(b) $R = \{(x, y) : 0 \le x < a, 0 \le y \le b\}$
(c) $R = \{(x, y) : 0 \le x \le a, 0 < y < b\}$
(d) $R = \{(x, y) : 0 < x < a, 0 < y < b\}$

Assertion-Reason Based Questions

Directions (Q. Nos. 15-24) *In the questions given below are two statements labelled as Assertion (A) and Reason (R). In the context of the two statements, which one of the following is correct?*

(a) Both A and R are correct; R is the correct explanation of A.
(b) Both A and R are correct; R is not the correct explanation of A.
(c) A is correct; R is incorrect.
(d) R is correct; A is incorrect.

15. **Assertion** (A) If $A = \{1, 3, 5, 7, 9\}$ and $B = \{2, 3, 5, 7, 11\}$, then A and B are intersecting sets.

Reason (R) Sets A and B are intersecting if $A \cap B \neq \phi$.

16. **Assertion** (A) If $A = \{x : x \in N, x$ is a factor of $6\}$ and $B = \{x \in N : x$ is a factor of $8\}$, then $A - B$ is $\{4, 6\}$.

Reason (R) For any sets A and B, their difference $(A - B)$ is defined as $(A - B) = \{x : x \in A$ and $x \notin B\}$.

17. **Assertion** (A) If $A = \{x : x \in N, x$ is a factor of $12\}$ and $B = \{x : x \in N, x$ is a factor of $18\}$, then $A \cap B$ is $\{1, 3, 4, 6\}$.

Reason (R) If two sets are A and B, then $A \cap B = \{x : x \in A$ and $x \in B\}$. **Competency Based Que**

18. **Assertion** (A) If $A = \{x : x \in N, x$ is the prime number less than $7\}$ and $B = \{x : x$ is the prime factor of $6\}$, then $B \subset A$.

Reason (R) If there are two sets A and B such that $A \subseteq B$ and $B \subseteq A \Rightarrow A = B$.

19. **Assertion** (A) For any two sets A and B, if $A - B$, then the given Venn diagram represents the relationship between A and B.

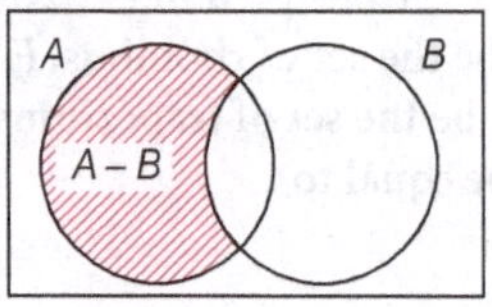

Reason (R) $A - B$ is the collection of those elements of A which are not in B.

20. **Assertion** (A) Set of all composite numbers less than 50 is finite set.

Reason (R) Set of all prime numbers greater than 100 is infinite set. **Competency Based Que**

21. **Assertion** (A) 'Set of difficult topic in Physics' is a well-defined set.

Reason (R) A collection of object which is well defined is called a set.

22. **Assertion** (A) Set of real numbers is a universal set.

Reason (R) Set of real numbers is a superset for the set of natural numbers, set of whole numbers, set of integers, the set of rational/irrational numbers.

23. **Assertion** (A) Sets $A = \{1, 2, 3\}$ and $B = \{2, 1, 3\}$ are equal sets.

Reason (R) If $A \subseteq B$ and $B \subseteq A$, then $A = B$.

24. **Assertion** (A) The intersection of sets $A = \{2, 4, 6\}$ and $B = \{3, 5, 7\}$ is empty.

Reason (R) The union of sets A and B is denoted by $A \cup B$.

Case Study Based Questions

25. Aarti explained operations on sets to her younger sister 'Pooja', then wrote three sets as $A = \{2, 3, 6, 7\}$, $B = \{4, 5, 8\}$ and $C = \{x : x$ is a prime number less than $9\}$. She asked her sister 'Pooja' that the following question will judge how much you have understood. She asked her younger sister to solve and write down answer.

(i) $A \cup B$
(ii) $A \cap B$
(iii) $(A \cup B) \cap C$
(iv) $(A \cap C) - B$

Subjective Questions

Very Short Answer Type Questions

1. What is the difference between a collection and a set? Give reason to support your answer.

2. Find the number of subsets of the set $A = \{1, 4, 5\}$.

3. List all the proper subsets of the set $A = \{a, b\}$.

4. Write the following set in roster form.
$\{x : x$ is a prime number and a divisor of $6\}$

Competency Based Que

5. Write $\{2, 4, 8, 16\}$ in set-builder form.

6. Write the set $\{x \in R : -3 < x \leq 5\}$ as an interval.

7. If set X = set of rational numbers and set Y = set of irrational numbers, then find $X \cup Y$ and $X \cap Y$.

8. Find number of subsets if $A = \{x : x^2 + 3 < 2, x \in R\}$.

Competency Based Que

9. State true or false
$7747 \in \{x : x$ is a multiple of $37\}$. **NCERT Exemplar**

10. If $U = \{1, 2, 3 \dots 10\}$ and $A = \{2, 4, 6\}$, then find A'.

11. If $A = \{3, 5\}$ and $B = \{4, 6\}$, then find $A \cap B$.

Short Answer Type Questions

12. Which of the following sets are empty set, equal set, finite set and infinite set. **Competency Based Que**

$A = \{x : x$ is a solution of $x^2 - 5x + 6 = 0\}$

$B = \{x : x^2 - 16x + 55 = 0$ and $x^2 = 25\}$

$C = \left\{x : \frac{-1}{2} \leq x \leq \frac{1}{2}\right\}, D = \{x : 0 \leq 4x^2 \leq 1\}$

$E = \{x : x$ is a natural number and $x \geq 100\}$

$F = \{x : x$ is a whole number and $2 < x < 3\}$

13. If $U = \{a, b, c, d, e, f\}, A = \{a, b, c\}, B = \{c, d, e, f\}, C = \{c, d, e\}$ and $D = \{d, e, f\}$, then tabulate the following sets.

(i) $A \cap D$
(ii) $A \cap C$
(iii) $U \cap D$
(iv) $A \cup \phi$
(v) $(U \cap \phi)'$
(vi) $(U \cup A)'$

14. If $A = \{1, 2, 3, 4, 5\}, B = \{1, 3, 5, 8\}$ and $C = \{2, 5, 7, 8\}$, then verify that $A - (B \cup C) = (A - B) \cap (A - C)$.

15. Let A and B be two sets. Then, prove that $A = B \Leftrightarrow A \subseteq B$ and $B \subseteq A$.

16. Express the shaded region of the following Venn diagram in terms of union and difference of sets A, B and C.

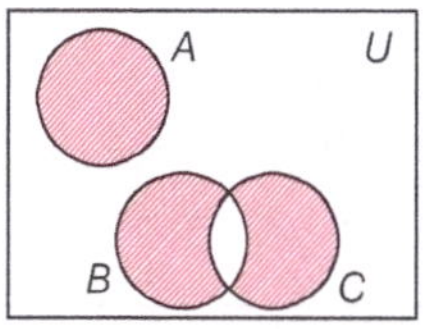

17. If $A = \{4^n - 3n - 1, n \in N\}$ and $B = \{9(n-1) : n \in N\}$, then show that $A \subset B$. **Competency Based Que**

18. Two finite sets have m and n elements. The total number of subsets of the first set is 56 more than the total number of subsets of the second set. Find the values of m and n. **Competency Based Que**

19. If $X = \{1, 2, 3\}$ and n represents any member of X, write the following sets containing all numbers represented by

(i) $4n$
(ii) $n + 6$
(iii) $\frac{n}{2}$
(iv) $n - 1$

20. Given set $A = \{1, 2, 3, 4, \dots, 100\}$, write the subset

(i) X of A, whose elements are multiple of 7.
(ii) Y of A, whose elements are represented by $x + 3$, where $x \in A$.

21. Use properties of set to prove that for all the sets A and B, $A - (A \cap B) = A - B$.

Long Answer Type Questions

22. Let $T = \left\{x : \frac{x+5}{x-7} - 5 = \frac{4x - 40}{13 - x}\right\}$. **Competency Based Que**

Is T an empty set? Justify your answer.

23. Let $U = \{1, 2, 3, 4, 5, 6, 8\}, A = \{2, 3, 4\}, B = \{3, 4, 5\}$.

Show that $(A \cup B)' = A' \cap B'$ and $(A \cap B)' = A' \cup B'$.

24. Let P be the set of prime numbers and let $S = \{t : 2t - 1$ is a prime$\}$. Prove that $S \subset P$. **NCERT Exemplar**

Hints & Solutions

Objective Type Questions

1. (b) Null set is represented by ϕ.
2. (c) $A-(B \cup C)$
3. (b) Since, all natural numbers come in Q.

   Hence, $N \subset Q$ [all elements of N are in set Q]
4. (a) Use formula $2^n - 1$,

   $$A = \{a, b, c, d, e\}, n = 5$$
   $$2^5 - 1 = 32 - 1 = 31$$
5. (d) Given, $U = [-4, 4]$,

   and $A = (-2, 4]$,

   $A' = U - A$

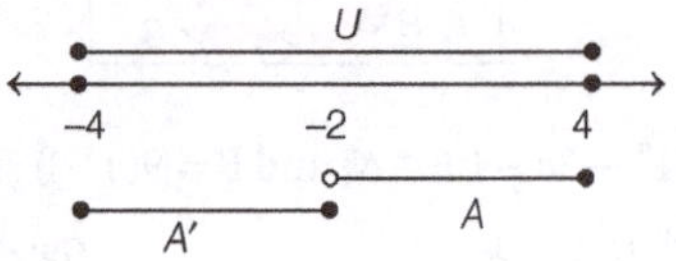

   $$A' = [-4, -2]$$
6. (a) $A = \{x \in N : x^2 - 9 = 0\}$

   $\Rightarrow \quad x^2 - 9 = 0$

   $\Rightarrow \quad x^2 = 9$

   $\Rightarrow \quad x = \pm\sqrt{9}$

   $\Rightarrow \quad x = \pm 3 \Rightarrow \{3\} \qquad [\because x \in N]$
7. (d) $(A \cup B)' = A' \cap B'$
8. (c) Given, $n(P) = 24, n(Q) = 20$

   and $\quad n(P \cup Q) = 24$

   Then, $\quad n(P \cup Q) = n(P) + n(Q) - n(P \cap Q)$

   $\Rightarrow \quad 24 = 24 + 20 - n(P \cap Q)$

   $\Rightarrow \quad 24 = 44 - n(P \cap Q)$

   $\Rightarrow \quad n(P \cap Q) = 44 - 24$

   $\therefore \quad n(P \cap Q) = 20$
9. (c) Given, $B = \{x : x \leq \frac{1}{2n-1}, 1 \leq n \leq 5 \text{ where } n \in N\}$

   $$n = 1; x = \frac{1}{2-1} = \frac{1}{1} = 1$$
   $$n = 2; x = \frac{1}{4-1} = \frac{1}{3}$$
   $$n = 3; x = \frac{1}{6-1} = \frac{1}{5}$$
   $$n = 4; x = \frac{1}{8-1} = \frac{1}{7}$$
   $$n = 5; x = \frac{1}{10-1} = \frac{1}{9} = \left\{1, \frac{1}{3}, \frac{1}{5}, \frac{1}{7}, \frac{1}{9}\right\}$$
10. (b) Here, $(A \cap B')' \cup (B \cap C)$

    $= (A' \cup B) \cup (B \cap C) \quad$ [using De-Morgan's law]

    $= A' \cup B$
11. (a) Since, $A \cup B$ is a set of those elements which belongs to either A or B and A is a set of elements which belongs to only A.

    $\therefore A \cap (A \cup B)$ is set of common elements which belongs to both sets A and $A \cup B$ i.e. equal to A.
12. (c) Here, $X \cap (X \cup Y)' = X \cap (X' \cap Y')$

    [by De-Morgan's law, $(A \cup B)' = A' \cap B'$]

    $= (X \cap X') \cap (X \cap Y')$

    [using distributive law]

    $= \phi \cap (X \cap Y') = \phi \qquad [\because \phi \cap A = \phi]$
13. (d) Every rectangle, rhombus, square in a plane is a parallelogram but every trapezium is not a parallelogram.

    Hence, $\quad F_1 = F_2 \cup F_3 \cup F_4 \cup F_1$
14. (d) Since, R is the set of points inside a rectangle.

    $\therefore$ (0, 0), $(a, 0)$, $(0, b)$ and (a, b) points are excluded.

    $\therefore R = \{(x, y) : 0 < x < a \text{ and } 0 < y < b\}$
15. (a) Clearly, Assertion and Reason are correct and Reason gives correct explanation of Assertion.
16. (d) Clearly, Assertion is incorrect because

    set $A = \{1, 2, 3, 6\}$ and set $B = \{1, 2, 4, 8\}$

    $A - B = \{3, 6\} \neq \{4, 6\}$ and Reason is correct.
17. (d) Assertion is incorrect because

    $$A = \{1,2,3,4,6,12\} \text{ and } B = \{1,2,3,6,9,18\}$$
    $$A \cap B = \{1,2,3,6\} \neq \{1,3,4,6\}$$

    Clearly, Reason is correct.
18. (b) Clearly, Assertion and Reason are correct and Reason is not the correct explanation of Assertion.
19. (a) Clearly, Assertion and Reason both are correct and Reason is the correct explanation of Assertion.
20. (b) Clearly, Assertion and Reason both are correct and Reason is not the correct explanation of Assertion.
21. (d) Clearly, Assertion is incorrect because the term "difficult" is not well defined and Reason is correct.
22. (a) Clearly, Assertion and Reason both are correct and Reason is the correct explanation of Assertion.
23. (a) Clearly, Assertion and Reason both are correct and Reason is the correct explanation of Assertion.
24. (b) Clearly, Assertion and Reason both are correct and Reason is not the correct explanation of Assertion.

25. We have, $A = \{2,3,6,7\}$, $B = \{4,5,8\}$ and $C = \{2,3,5,7\}$

(i) $A \cup B = \{2,3,6,7\} \cup \{4,5,8\} = \{2,3,4,5,6,7,8\}$

(ii) $A \cap B = \{2,3,6,7\} \cap \{4,5,8\} = \phi$

(iii) $(A \cup B) \cap C$

$(A \cup B) \cap C = \{2,3,4,5,6,7,8\} \cap \{2,3,5,7\} = \{2,3,5,7\}$

(iv) $(A \cap C) - B$

$A \cap C = \{2,3,6,7\} \cap \{2,3,5,7\} = \{2,3,7\}$

$(A \cap C) - B = \{2,3,7\} - \{4,5,8\} = \{2,3,7\}$

Subjective Questions

1. Every set is a collection but a collection is not necessarily a set. Only well-defined collection are sets. e.g. Collection of most talented writers of India is a collection but it is not a set.

2. Given, set $A = \{1,4,5\}$

Number of elements = 3

Then, number of subsets of set $A = 2^n = 2^3 = 8$ subsets

3. The proper subsets of the set $A = \{a, b\}$ are ϕ, $\{a\}$, $\{b\}$.

4. Given, $\{x : x$ is a prime number and a divisor of $6\}$

2 and 3 are the only prime numbers which are divisor of 6, so in roster form $\{2,3\}$.

5. The set builder form is $\{x : x = 2^n, n \in N, 1 \le n \le 4\}$

6. The interval of the given set is $(-3, 5]$.

7. Given, X = set of rational numbers

and Y = set of irrational numbers

$X \cup Y$ = set of rational numbers $\cup$ set of irrational numbers

= set of real numbers (rational + irrational) and

$X \cap Y = \phi$

8. Given, $A = \{x : x^2 + 3 < 2, x \in R\}$

Set A is a null set, so there will be only one subset ϕ itself.

So, number of subset is 1.

9. 7747 is not the multiple of 37.

So, given statement is false.

10. Given, $U = \{1, 2, 3......10\}$ and $A = \{2,4,6\}$

Now, $A' = U - A$

$= \{1,2,3,.....10\} - \{2,4,6\} = \{1,3,5,7,8,9,10\}$

11. Given, $A = \{3,5\}$ and $B = \{4,6\}$

$A \cap B = \phi$

12. We have, $A = \{x : x$ is a solution of $x^2 - 5x + 6 = 0\}$

Now, $x^2 - 5x + 6 = 0$

$\Rightarrow x^2 - 3x - 2x + 6 = 0$

$\Rightarrow x(x-3) - 2(x-3) = 0$

$\Rightarrow (x-3)(x-2) = 0 \Rightarrow x = 2, 3$

$A = \{3, 2\}$

So, it is a finite set.

$B = \{x : x^2 - 16x + 55 = 0$ and $x^2 = 25\}$

$\Rightarrow x^2 - 16x + 55 = 0$ and $x^2 = 25$

$\Rightarrow x^2 - 11x - 5x + 55 = 0$ and $x = \pm\sqrt{25}$

$\Rightarrow x(x-11) - 5(x-11) = 0$ and $x = \pm 5$

$\Rightarrow (x-11)(x-5) = 0$ and $x = \pm 5$

$\therefore x = 5$

So, it is also a finite and singleton set.

$C = \left\{x : \frac{-1}{2} \le x \le \frac{1}{2}\right\}$ and $D = \{x : 0 \le 4x^2 \le 1\}$

$= \left\{x : 0 \le x^2 \le \frac{1}{4}\right\} = \left\{x : -\frac{1}{2} \le x \le \frac{1}{2}\right\}$

$\Rightarrow C = D$

They are equal sets.

$E = \{x : x \in N$ and $x \ge 100\}$

$\therefore$ It is a infinite set.

$F = \{x : x$ is a whole number and $2 < x < 3\}$

There is no whole number between 2 and 3.

$\therefore$ It is an empty set.

13. Given, $U = \{a, b, c, d, e, f\}$, $A = \{a, b, c\}$

$B = \{c, d, e, f\}$, $C = \{c, d, e\}$ and $D = \{d, e, f\}$

(i) $A \cap D = \{a, b, c\} \cap \{d, e, f\} = \phi$

(ii) $A \cap C = \{a, b, c\} \cap \{c, d, e\} = \{c\}$

(iii) $U \cap D = \{a, b, c, d, e, f\} \cap \{d, e, f\} = \{d, e, f\}$

(iv) $A \cup \phi = \{a, b, c\} \cup \phi = \{a, b, c\}$

(v) $(U \cap \phi)' = U - (U \cap \phi)$

Now, $U \cap \phi = \{a, b, c, d, e, f\} \cap \phi = \phi$

$\therefore U - (U \cap \phi) = \{a, b, c, d, e, f\} - \phi$

$= \{a, b, c, d, e, f\} = U$

(vi) $(U \cup A)' = U - (U \cup A)$

Now, $(U \cup A) = \{a, b, c, d, e, f\} \cup \{a, b, c\}$

$= \{a, b, c, d, e, f\}$

$\therefore U - (U \cup A) = \{a, b, c, d, e, f\} - \{a, b, c, d, e, f\}$

$= \phi$

14. Given, $A = \{1, 2, 3, 4, 5\}$, $B = \{1, 3, 5, 8\}$ and $C = \{2, 5, 7, 8\}$

Now, $B \cup C = \{1, 2, 3, 5, 7, 8\}$

$\therefore A - (B \cup C) = \{4\}$

and $(A - B) \cap (A - C) = \{2, 4\} \cap \{1, 3, 4\} = \{4\}$

15. To prove that two sets A and B are equal if and only if $A \subseteq B$ and $B \subseteq A$, we need to show two implications

(i) If $A = B$, then $A \subseteq B$ and $B \subseteq A$

(ii) If $A \subseteq B$ and $B \subseteq A$ then $A = B$

(i) **If $A = B$**

By definition of set equality, A and B have same elements. Then, $A \subseteq B$ (every element of A is in B) and $B \subseteq A$ (every element of B is in A).

Thus, if $A = B$, then $A \subseteq B$ and $B \subseteq A$.

(ii) **If $A \subseteq B$ and $B \subseteq A$**

By definition of subset, $A \subseteq B$ means every element of A is also an element of B and $B \subseteq A$ means every element of B is in A.

So, $A = B$ (by definition of equality)

Thus, $A = B$ if and only if $A \subseteq B$ and $B \subseteq A$.

16. $A \cup \{(B - C) \cup (C - B)\}$

17. We have, $A = \{0, 9, 54 \ldots\}$, $n \in N$

and $B = \{0, 9, 18, 27, 36 \ldots\}$

$\because$ All the elements of A belong to B.

$\therefore \quad A \subset B$

18. Let A and B be such sets i.e. $n(A) = m$ and $n(B) = n$

So, total number of subsets of $A = 2^m$

and total number of subsets of $B = 2^n$

Given, $\quad 2^m - 2^n = 56$

$\Rightarrow \quad 2^n (2^{m-n} - 1) = 2^3 \cdot 7$

$\Rightarrow \quad 2^n (2^{m-n} - 1) = 2^3 (8 - 1) = 2^3 (2^3 - 1)$

$\Rightarrow \quad n = 3$ and $m - n = 3$

$\Rightarrow \quad n = 3$ and $m = 6$

19. (i) $\{4, 8, 12\}$ (ii) $\{7, 8, 9\}$ (iii) $\left\{\frac{1}{2}, 1, \frac{3}{2}\right\}$ (iv) $\{0, 1, 2\}$

20. (i) $\{7, 14, 21, 28, 35, 42, 49, 56, 63, 70, 77, 84, 91, 98\}$

(ii) $\{4, 5, 6, \ldots, 100\}$

21. Given, LHS $= A - (A \cap B) = A \cap (A \cap B)'$

$[\because A - B = A \cap B']$

$= A \cap (A' \cup B')$ [by De-Morgan's law]

$= (A \cap A') \cup (A \cap B')$ [by distributive law]

$= \phi \cup (A \cap B') = A \cap B' = A - B$

$A - (A \cap B) = A - B$

LHS = RHS **Hence proved.**

22. Given, $T = \left\{x : \frac{x+5}{x-7} - 5 = \frac{4x - 40}{13 - x}\right\}$

Since, $\quad \frac{x+5}{x-7} - 5 = \frac{4x-40}{13-x}$

$\Rightarrow \quad \frac{x + 5 - 5x + 35}{x - 7} = \frac{4x - 40}{13 - x}$

$\Rightarrow \quad \frac{-4x + 40}{x - 7} = \frac{4x - 40}{13 - x}$

$\Rightarrow (4x - 40)(x - 7) + (4x - 40)(13 - x) = 0$

$\Rightarrow \quad 24(x - 10) = 0 \Rightarrow x = 10$

$\Rightarrow \quad T = \{10\}$

Hence, T is not an empty set.

23. We have, $U = \{1, 2, 3, 4, 5, 6, 8\}$, $A = \{2, 3, 4\}$ and $B = \{3, 4, 5\}$

(i) $A \cup B = \{2, 3, 4\} \cup \{3, 4, 5\} = \{2, 3, 4, 5\}$

$\therefore \quad (A \cup B)' = U - (A \cup B)$

$= \{1, 2, 3, 4, 5, 6, 8\} - \{2, 3, 4, 5\}$

$= \{1, 6, 8\}$ ...(i)

$A' = U - A = \{1, 2, 3, 4, 5, 6, 8\} - \{2, 3, 4\}$

$= \{1, 5, 6, 8\}$

$B' = U - B = \{1, 2, 3, 4, 5, 6, 8\} - \{3, 4, 5\}$

$= \{1, 2, 6, 8\}$

$A' \cap B' = \{1, 5, 6, 8\} \cap \{1, 2, 6, 8\} = \{1, 6, 8\}$ ...(ii)

From Eqs. (i) and (ii), we get $(A \cup B)' = A' \cap B'$

(ii) $A \cap B = \{2, 3, 4\} \cap \{3, 4, 5\} = \{3, 4\}$

$\therefore \quad (A \cap B)' = U - (A \cap B)$

$= \{1, 2, 3, 4, 5, 6, 8\} - \{3, 4\}$

$= \{1, 2, 5, 6, 8\}$ ...(iii)

Now, $\quad A' \cup B' = \{1, 5, 6, 8\} \cup \{1, 2, 6, 8\}$

$= \{1, 2, 5, 6, 8\}$ ...(iv)

From Eqs. (iii) and (iv), we get $(A \cap B)' = A' \cup B'$

24. We will prove the statement by contradiction method.

Let $\quad x \notin P$

$\Rightarrow x$ is a composite number.

Let us now assume that $x \in S$

$\Rightarrow \quad 2^x - 1 = m$ (where m is a prime number)

$\Rightarrow \quad 2^x = m + 1$

which is not true for all composite numbers, say for $x = 4$ because $2^4 = 16$ which cannot be equal to the sum of any prime numbers m and 1.

Thus, we arrive at a contradiction

$\Rightarrow \quad x \notin S.$

Thus, when $x \notin P$, we arrive at $x \notin S$.

So, $\quad S \subset P$

QR Code for Quick Test 1

QR Code for Quick Test 2

QR Code for Quick Test 3

Mind Map

Sets and Their Representation

Venn Diagrams

In Venn diagrams, the universal set is represented by a rectangular region and its subset are represented by circles or a closed geometrical figure inside the universal set. Also, an element of a set is represented by a point within the circles of set.

e.g. If $A = \{1, 2, 3, 5, 9\}$, $B = \{4, 6, 7\}$ and $U = \{1, 2, 3, 4, 5, 6, 7, 8, 9\}$, then

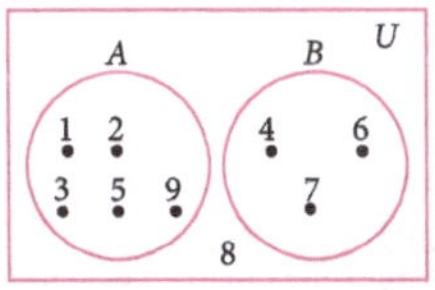

Types of Sets

Empty Set (ϕ)

A set which does not contain any element, is called an empty set or null set or void set.

e.g. $A = \{x : x \text{ is a natural number less than } 1\}$

Finite Set

A set which is empty or consists of a definite number of elements, is called a finite set.

e.g. The set $\{1, 2, 3, 4\}$ is a finite set because it contains a definite number of elements i.e. only 4 elements.

Infinite Set

A set which is not finite is called infinite set.

e.g. The set of real numbers.

Equal Sets

Two sets A and B are said to be equal, if they have exactly the same elements and we write it as $A = B$.

Otherwise, two sets are said to be **unequal** and we write it as $A \neq B$.

Singleton Set

A set containing one element, is called a singleton set.

e.g. The sets $\{0\}$, $\{5\}$, $\{-7\}$ are singleton sets.

Operations on Sets

Union of Sets

Let A and B be any two sets. The union of A and B is the set of all those elements which belong to either A or B or both. The symbol '$\cup$' is used to denote the union.

Symbolically, it is denoted by $A \cup B$ and read as ***A* union *B***.

$$A \cup B = \{x : x \in A \text{ and } x \in B\}$$

Intersection of Sets

Let A and B be any two sets. The intersection of A and B is the set of all those elements which belong to both A and B. It is denoted by $A \cap B$ and read as ***A* intersection *B***. The symbol $\cap$ is used to denote the intersection.

$$A \cap B = \{x : x \in A \text{ and } x \in B\}$$

Difference of Sets

Let A and B be any two sets. The difference of sets A and B in this order is the set of all those elements of A which do not belong to B. It is denoted by **$A - B$ and read as *A* minus *B***.

Disjoint Sets

Two sets A and B are said to be disjoint sets, if they have **no common element i.e. $A \cap B = \phi$.**

A well-defined collection of objects is called set. Sets are usually denoted by capital letters A, B, X, Y etc. The elements of set are represented by small letters a, b, x, y, z etc.

Subset

Let A and B be two sets. If every element of A is an element of B, then A is called a subset of B.

If A is a subset of B, then we write $A \subseteq B$, which is read as 'A is a subset of B' or A is contained in B or $A \subseteq B$, if $x \in A \Rightarrow x \in B$

If A is not a subset of B, then we write $A \not\subset B$.

e.g. $\{2, 3, 4, 5\}$ is a subset of a set of natural numbers.

Proper Subset

If $A \subseteq B$ and $A \neq B$, then A is called a proper subset of B, written as $A \subset B$ and B is called **superset of *A***.

Universal Set

If there are some sets under consideration, then a set can be chosen arbitrarily which is a **superset** of each one of the given sets. Such a set is known as the universal set and it is denoted by U.

Representations of Sets

1. Roster Form or Tabular Form or Listing Method

In this form, all the elements of a set are listed and are being separated by commas. e.g. $A = \{a, e, i, o, u\}$

2. Set-Builder form or Rule Method

In this form, all the elements of the set posses a single common property $p(x)$, the set is described by $\{x : p(x) \text{ holds}\}$.

e.g. $X = \{x : x \in z \text{ and } x < 7\}$

Topics to be Assessed Formatively

Applications of Set Theory

In this topic, we will discuss some word problems related to our daily life, which are based on union and intersection of two sets. Before solving these types of problem, we should know the following formulae.

If A and B are two finite sets, then two cases arise

***Case* I** If A and B are **disjoint sets**, i.e. there is no common element in A and B i.e. $A \cap B = \phi$. Then,

$$n(A \cup B) = n(A) + n(B)$$

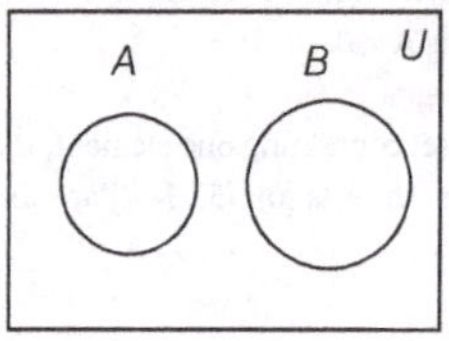

***Case* II** If A and B are **not disjoint sets**, then there are common elements in A and B. Then,

$$n(A \cup B) = n(A) + n(B) - n(A \cap B)$$

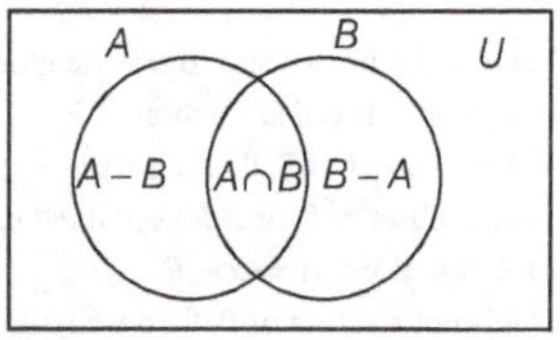

Note Sets $A - B$, $A \cap B$ and $B - A$ are disjoint and their union is $A \cup B$.

Important Results

(i) $n(A \cup B) = n(A - B) + n(B - A) + n(A \cap B)$

(ii) $n(A) = n(A - B) + n(A \cap B)$

(iii) $n(B) = n(B - A) + n(A \cap B)$

(iv) $n(A \Delta B) = n[(A - B) \cup (B - A)] = n(A - B) + n(B - A)$ [since, $(A - B)$ and $(B - A)$ are disjoint sets] $= n(A) + n(B) - 2n(A \cap B)$

(v) $n(A' \cup B') = n[(A \cap B)'] = n(U) - n(A \cap B)$

(vi) $n(A' \cap B') = n[(A \cup B)'] = n(U) - n(A \cup B)$

(vii) $n(A - B) = n(A \cap B)' = n(A) - n(A \cap B)$

(viii) If A, B and C are finite sets, then

(a) $n(A \cup B \cup C)$
$= n(A) + n(B) + n(C) - n(A \cap B) - n(B \cap C) - n(A \cap C) + n(A \cap B \cap C)$

(b) $n(A \text{ only}) = n(A) - n(A \cap B) - n(A \cap C) + n(A \cap B \cap C)$

(c) $n(\overline{A} \cap \overline{B} \cap \overline{C}) = n(U) - n(A \cup B \cup C)$

TEACHER TIPS

Let A, B and C be any three sets, then the meaning of different operations on these sets are given below.

(i) $A' \to$ Set of those objects which does not belong to A (similar meaning for B' and C').

(ii) $A \cup B \to$ Set of those objects which lies either in A or in B (similar meaning for $B \cup C$ and $A \cup C$).

(iii) $A \cap B \to$ Set of those objects which belongs to both the sets A and B (similar meaning for $B \cap C$ and $C \cap A$).

(iv) $A \cap B \cap C \to$ Set of those objects which belongs to A, B and C.

(v) $A \cup B \cup C \to$ Set of those objects which belongs to atleast one of the sets A, B and C.

Example 1 If X and Y are two sets such that X has 40 elements, $X \cup Y$ has 60 elements and $X \cap Y$ has 10 elements, then how many elements does Y have?

Sol. Given, $n(X) = 40$, $n(X \cup Y) = 60$

and $n(X \cap Y) = 10$

Clearly, $n(X \cup Y) = n(X) + n(Y) - n(X \cap Y)$

$60 = 40 + n(Y) - 10$

$\Rightarrow \quad 60 = 30 + n(Y)$

$\Rightarrow \quad n(Y) = 60 - 30$

$\therefore \quad n(Y) = 30$

Hence, Y have 30 elements.

Example 2 If X and Y are two sets such that $n(X) = 17$, $n(Y) = 23$ and $n(X \cup Y) = 38$, then find $n(X \cap Y)$.

Sol. Given, $n(X) = 17$, $n(Y) = 23$

and $n(X \cup Y) = 38$

Clearly, $n(X \cup Y) = n(X) + n(Y) - n(X \cap Y)$

$\Rightarrow \quad n(X \cap Y) = n(X) + n(Y) - n(X \cup Y)$

$\Rightarrow \quad n(X \cap Y) = 17 + 23 - 38 = 40 - 38 = 2$

Example 3 If $n(A) = 4$, $n(B) = 5$, $n(U) = 7$ and $n(A \cap B) = 2$, then find the value of $n(A \cup B)'$.

Sol. Given, $n(A) = 4$, $n(B) = 5$, $n(U) = 7$

and $n(A \cap B) = 2$

$\because \quad n(A \cup B) = n(A) + n(B) - n(A \cap B)$

$\therefore \quad n(A \cup B) = 4 + 5 - 2 = 7$

Now, $n(A \cup B)' = n(U) - n(A \cup B) = 7 - 7 = 0$

Example 4 In a class of 60 students, 25 students play Cricket, 20 students play Tennis and 10 students play both the games. Then, find the number of students who play neither games.

Sol. Let C and T respectively denotes the set of students who play Cricket and Tennis and set U denotes the set of all students in a class.

Then, $n(C) = 25, n(T) = 20, n(C \cap T) = 10$ and $n(U) = 60$

We know that $n(C \cup T) = n(C) + n(T) - n(C \cap T)$

$\Rightarrow \quad n(C \cup T) = 25 + 20 - 10 = 45 - 10 = 35$

Now, the number of students who play neither game

$n(C' \cap T') = n(C \cup T)'$ [by De Morgan's law]

$= n(U) - n(C \cup T) = 60 - 35 = 25$

Hence, 25 students play neither games.

Example 5 In a town of 840 persons, 450 persons read Hindi, 300 read English and 200 read both newspapers. Then, find the number of persons who read neither of the newspapers.

Sol. Let H and E respectively denote the set of persons who read Hindi and English newspapers and let U set of all perseon in a town.

Then, we have $n(U) = 840,$

$n(H) = 450,$

$n(E) = 300$

and $n(H \cap E) = 200$

Clearly, $n(H \cup E) = n(H) + n(E) - n(H \cap E)$

$\therefore \quad n(H \cup E) = 450 + 300 - 200$

$= 750 - 200 = 550$

Now, the number of persons who read neither of the newspaper is given by

$n(H' \cap E') = n(H \cup E)'$ [by De Morgan's law]

$= n(U) - n(H \cup E)$

$= 840 - 550 = 290$

Hence, 290 persons read neither of the newspapers.

Example 6 Each student in a class of 40 students study atleast one of the subjects English, Mathematics and Economics. 16 students study English, 22 Economics and 26 Mathematics, 5 study English and Economics, 14 Mathematics and Economics and 2 English, Economics and Mathematics. Find the number of students who study English and Mathematics.

Sol. Let A, B and C denote the set of students who study English, Economics and Mathematics, respectively.

Then, we have

Total number of students, $n(A \cup B \cup C) = 40$

Number of students who study English, $n(A) = 16$

Number of students who study Economics, $n(B) = 22$

Number of students who study Mathematics, $n(C) = 26$

Number of students who study English and Economics, $n(A \cap B) = 5$

Number of students who study Mathematics and Economics, $n(B \cap C) = 14$

and number of students who study all subjects,

$n(A \cap B \cap C) = 2$

Clearly, $n(A \cup B \cup C) = n(A) + n(B) + n(C) - n(A \cap B) - n(B \cap C) - n(A \cap C) + n(A \cap B \cap C)$

$40 = 16 + 22 + 26 - 5 - 14 - n(C \cap A) + 2$

$\Rightarrow \quad 40 = 66 - 19 - n(C \cap A)$

$\Rightarrow \quad n(C \cap A) = 47 - 40$

$= 7$

Hence, number of students who study English and Mathematics are 7.

Example 7 A market research group conducted a survey of 2000 consumers and reported that 1720 consumers liked "Patanjali Ghee" and 1450 consumers liked "Amul Ghee". What is the least number that must have liked both the product?

Sol. Let U be the set of consumers who were participated in the survey.

Then, $n(U) = 2000$

Let P be the set of consumers who liked "Patanjali Ghee".

Then, $n(P) = 1720$

Let A be the set of consumers who liked "Amul Ghee".

Then, $n(A) = 1450$

$\because \quad n(P \cup A) = n(P) + n(A) - n(P \cap A)$

$\therefore \quad n(P \cup A) = 1720 + 1450 - n(P \cap A) = 3170 - n(P \cap A)$

Now, as $n(P \cup A) \leq n(U)$

$\therefore \quad 3170 - n(P \cap A) \leq 2000$

$\Rightarrow \quad n(P \cap A) \geq 1170$

Hence, the least number of consumers who liked both the product is 1170.

Example 8 In a survey of 500 customer in an electronic shop, 400 were purchases LG refrigerator and 200 purchases SAMSUNG refrigerator, 50 purchase both refrigerators. Is this data correct?

Sol. Let U be the set of all customers who were participated in the survey. Then, $n(U) = 500$

Let denotes the set of customers who purchased LG refrigerator.

Then, $n(L) = 400$

Let S denotes the set of customers who purchased SAMSUNG refrigerator.

Then, $n(S) = 200$

and $n(L \cap S) = 50$

$\because \quad n(L \cup S) = n(L) + n(S) - n(L \cap S)$

$\therefore \quad n(L \cup S) = 400 + 200 - 50$

$\Rightarrow \quad n(L \cup S) = 550$

But $\quad L \cup S \subseteq U$

$\Rightarrow \quad n(L \cup S) \leq n(U)$

$\Rightarrow \quad 550 \leq 500$, which is not true.

This is a contradiction.

So, the given data is incorrect.

Example 9 If A and B are two sets containing 3 and 6 elements respectively, what can be the minimum number of elements in $A \cup B$? Also, find the maximum number of elements in $A \cup B$.

Sol. Given, $n(A) = 3$ and $n(B) = 6$

We know that

$$n(A \cup B) = n(A) + n(B) - n(A \cap B)$$

Clearly, $n(A \cup B)$ will be maximum when $n(A \cap B)$ is minimum and it will be minimum when $n(A \cap B)$ is maximum. Therefore, two cases arise

***Case* I When $n(A \cap B)$ is minimum**

The minimum value of $n(A \cap B) = 0$

$\therefore \quad n(A \cup B) = n(A) + n(B) = 3 + 6 = 9$

Then, maximum value of $n(A \cup B)$ is 9.

***Case* II When $n(A \cap B)$ is maximum**

$n(A \cap B)$ will be maximum, if $A \subseteq B$

In this case, $n(A \cap B) = 3$

$\therefore \quad n(A \cup B) = 3 + 6 - 3 = 6$

Thus, the minimum number of elements in $A \cup B$ is 6.

Example 10 A survey shows that 63% of the Indian like cheese whereas 76% like apples. If x% of the Indian like both cheese and apples, then find the values of x.

Sol. Let the population of India be 100 crore.

Let C denotes the set of Indians who like cheese.

Then, $\quad n(C) = 63$ crore

Again, let A denotes the set of Indians who like apple.

Then, $\quad n(A) = 76$ crore

and $n(C \cap A) = x$ crore

Now, $n(C \cup A) = n(C) + n(A) - n(C \cap A)$

$\Rightarrow \quad n(C \cup A) = 63 + 76 - n(C \cap A)$

$\Rightarrow \quad n(C \cup A) + n(C \cap A) = 139$

$\Rightarrow \quad n(C \cap A) = 139 - n(C \cup A) \quad$ ...(i)

But $n(C \cup A) \leq 100$

$\therefore \quad -n(C \cup A) \geq -100$

$\Rightarrow \quad 139 - n(C \cup A) \geq 139 - 100$

$\Rightarrow \quad 139 - n(C \cup A) \geq 39$

$\Rightarrow \quad n(C \cap A) \geq 39 \quad$ [using Eq. (i)] ...(ii)

Again, $C \cap A \subseteq C$ and $C \cap A \subseteq A$

$\Rightarrow n(C \cap A) \leq n(C)$ and $n(C \cap A) \leq n(A)$

$\Rightarrow n(C \cap A) \leq 63$ and $n(C \cap A) \leq 76$

$\Rightarrow \quad n(C \cap A) \leq 63 \quad$ ...(iii)

From Eqs. (ii) and (iii), we get

$39 \leq n(C \cap A) \leq 63$

$\Rightarrow \quad 39 \leq x \leq 63$

Exercise

Multiple Choice Questions

1. If X and Y are two sets such that $X \cup Y$ has 50 elements, X has 28 elements and Y has 32 elements, then number of elements $X \cap Y$ have

(a) 10 (b) 46 (c) 110 (d) 54

2. In a class of 35 students, 24 like to play cricket and 16 like to play football. Also, each student likes to play atleast one of the two games. The number of students who like to play both cricket and football, is

(a) 27 (b) 43 (c) 5 (d) 75

3. In a committee, 50 people speak French, 20 speak Spanish and 10 speak both Spanish and French. The number of people who speak atleast one of these two languages, is

(a) 40 (b) 60 (c) 20 (d) 80

4. Each set X_r contains 5 elements and each set Y_r contains 2 elements and $\bigcup_{r=1}^{20} X_r = S = \bigcup_{r=1}^{n} Y_r$. If each element of S belongs to exactly 10 of the X_r's and to exactly 4 of the Y_r's, then n is equal to

(a) 10 (b) 20 (c) 100 (d) 50

5. Let A and B be two sets such that $n(A) = 0.16$, $n(B) = 0.14$ and $n(A \cup B) = 0.25$. Then, $n(A \cap B)$ is equal to

(a) 0.3 (b) 0.5
(c) 0.05 (d) None of these

Very Short Answer Type Questions

6. If A and B are two sets, such that $n(A) = 28$, $n(B) = 32$ and $n(A \cap B) = 10$, then find the value of $n(A \cup B)$.

7. If X and Y are two sets such that $n(X) = 28$, $n(Y) = 32$ and $n(X \cup Y) = 45$, then find $n(X \cap Y)$.

Short Answer Type I Questions

8. In a group of 400 people, 250 can speak Hindi and 200 can speak English. How many can speak both Hindi and English?

9. Let $n(U) = 700$, $n(A) = 200$, $n(B) = 300$ and $n(A \cap B) = 100$. Find $n(A' \cap B')$.

10. Let F_1 be the set of parallelograms, F_2 be the set of rectangles, F_3 be the set of rhombus and F_4 be the set of squares. Then, show that F_1 is equal to union of all sets.

11. In a town with a population of 5000, 3200 people are egg-eaters, 2500 meat-eaters and 1500 eat both egg and meat. How many are pure vegetarians?

12. In a school, there are 20 teachers who teach Maths or Physics. Out of these, 12 teach Maths and 4 teach Physics and Maths. How many teach Physics?

13. In a committee, 50 people speak French, 20 speak Spanish and 10 speak both Spanish and French. How many people speak atleast one of these two languages?

14. The members of a group of 400 people speak either Hindi or English or both. If 270 speak Hindi only and 50 speak both Hindi and English, then how many of them speak English only?

Short Answer Type II Questions

15. In a group of 70 people, 37 like coffee, 52 like tea and each person like atleast one of the two drinks. How many people like both coffee and tea?

16. In a survey of 400 students in a school, 100 were listed as taking apple juice, 150 as taking orange juice and 75 were listed as taking both apple as well as orange juice. Find how many sutdents were taking neither apple juice nor orange juice.

17. Out of 100 students, 15 passed in English, 12 passed in Mathematics, 8 in Science, 6 in English and Mathematics, 7 in Mathematics and Science, 4 in English and Science, 4 in all the three.

Find how many students passed in

(i) English and Mathematics but not in Science?

(ii) Mathematics and Science but not in English?

Long Answer Type Questions

18. In a group of 100 people, 65 like to play Cricket, 40 like to play Tennis and 55 like to play Volleyball. All of them like to play atleast one of the three games. If 25 like to play both Cricket and Tennis, 24 like to play both Tennis and Volleyball and 22 like to play both Cricket and Volleyball, then

(i) how many like to play all the three games?

(ii) how many like to play Cricket only?

(iii) how many like to play Tennis only?

Represent above information in a Venn diagram.

19. A college awarded 38 medals in Football, 15 in Basketball and 20 in Cricket. If these medals went to a total of 58 men and only three men got medals in all the three sports, then how many received medals in exactly two of the three sports?

Answers

1. (a) **2.** (c) **3.** (b) **4.** (b) **5.** (c)
6. 50 **7.** 15 **8.** 50 **9.** 300 **11.** 800
12. 12 **13.** 60 **14.** 80 **15.** 19 **16.** 225
17. (i) 2 (ii) 3 **18.** (i) 11 (ii) 29 (iii) 2
19. 9

02

CHAPTER

Mathematics knows no races or geographic boundaries; for mathematics, the cultural world is one country.

—David Hilbert

▼ YOUR STUDY GOALS

2.1 Ordered Pairs
2.2 Relations
2.3 Functions
2.4 Real Functions
2.5 Some Standard Real Function and Their Graphs
2.6 Algebra of Real Functions

▼ YOUR PRACTICE GOALS

- Objective Type Questions (32 Qs ; 20 MCQs, 10 A-R, 2 Case Based)
- Subjective Type Questions (38 Qs; 10 VSA, 16 SA, 12 LA)

Relations and Functions

2.1 Ordered Pairs

The ordered pair of two elements a and b is denoted by (a, b) : a is the first element and b is the second element.

Equality of Two Ordered Pairs

Two ordered pairs are equal, if their corresponding elements are equal.

i.e. $(a, b) = (c, d) \Rightarrow a = c$ and $b = d$

Cartesian Product of Two Sets

For two non-empty sets A and B, the cartesian product $A \times B$ is the set of all ordered pairs of elements from sets A and B.

In symbolic form, it can be written as

$$A \times B = \{(a, b) : a \in A, b \in B\}$$

Number of Elements in Cartesian Product of Two Sets

- If there are p elements in set A and q elements in set B, then there will be pq elements in $A \times B$.
  i.e. if $n(A) = p$ and $n(B) = q$, then
  $$n(A \times B) = pq$$

- If A and B are non-empty sets and either A or B is an infinite set, then $A \times B$ will also be an infinite set.
- If A or B is the null set or an empty set, then $A \times B$ will also be an empty set.
  i.e. $A \times B = \phi$
- If there are p elements in set A and q elements in set B, then the number of subsets of $A \times B$ is 2^{pq}.

Ordered Triplet

An ordered triplet is a list of 3 elements written in a particular order and enclosed within the circular bracket. e.g. (1, 2, 3) is an ordered triplet. Ordered triplet is also called 3-tuple.

Cartesian Product of Three Sets

If A, B and C are three sets, then

$$A \times B \times C = \{(a, b, c) : a \in A, b \in B, c \in C\}$$

In general, if $A_1, A_2, ..., A_n$ are n sets, then $(a_1, a_2, \ldots, a_n)$ is called a ***n*-tuple**, where $a_i \in A_i, i = 1, 2, \ldots, n$ and the set of all such n-tuples is called the cartesian product of $A_1, A_2, ..., A_n$ and denoted by $A_1 \times A_2 \times ... \times A_n$.

TEACHER TIPS

- If A, B and C are three sets, then
  $A \times B \times C = \{(a, b, c) : a \in A, b \in B, c \in C\}$
  For any three non-empty sets A, B and C, we get
  (i) $A \times (B \cup C) = (A \times B) \cup (A \times C)$
  (ii) $A \times (B \cap C) = (A \times B) \cap (A \times C)$
- $A \times A \times A = \{(a, b, c) : a, b, c \in A\}$.
  Here, (a, b, c) is called an ordered triplet.

Example 1. **If $(x + 2, y - 5) = (4, 10)$. Find the values of x and y.**

Sol. Since, the ordered pairs are equal, the corresponding elements are equal.

$\therefore$ $x + 2 = 4$ and $y - 5 = 10$

$\Rightarrow$ $x = 2$ and $y = 15$

Example 2. **If $A \times B = \{(a, 1), (b, 3), (a, 3), (b, 1), (a, 2), (b, 2)\}$. Then, find A and B.**

Sol. Here, first element of each ordered pair of $A \times B$ gives the elements of set A and corresponding second element gives the elements of set B.

$\therefore$ $A = \{a, b\}$ and $B = \{1, 3, 2\}$

Example 3. **A and B are two sets given in such a way that $A \times B$ contains 6 elements. If three elements of $A \times B$ are (1, 3), (2, 5) and (3, 3), then find A, B and remaining elements of $A \times B$.** *Competency Based Que*

Sol. It is given that (1, 3), (2, 5) and (3, 3) are in $A \times B$.
It follows that 1, 2 and 3 are elements of A and 3, 5 are elements of B.

$\therefore$ $A = \{1, 2, 3\}$ and $B = \{3, 5\}$

$\therefore$ $A \times B = \{1, 2, 3\} \times \{3, 5\}$

$\Rightarrow$ $= \{(1, 3), (1, 5), (2, 3), (2, 5), (3, 3), (3, 5)\}$

Hence, the remaining elements of $(A \times B)$ are (1, 5), (2, 3), (3, 5).

Example 4. **If the set A has 3 elements and set B has 4 elements, then find the number of elements in $A \times B$.**

Sol. Given, $n(A) = 3$ and $n(B) = 4$.

$\therefore$ The number of elements in $A \times B$ is

$$n(A \times B) = n(A) \times n(B) = 3 \times 4 = 12$$

Example 5. **If $A = \{1, 2\}$, then find $A \times A \times A$.**

Sol. We have, $A \times A = \{1, 2\} \times \{1, 2\}$
$= \{(1, 1), (1, 2), (2, 1), (2, 2)\}$

Now, $(A \times A) \times A = A \times A \times A$
$= \{(1, 1), (1, 2), (2, 1), (2, 2)\} \times \{1, 2\}$
$= \{(1, 1, 1), (1, 1, 2), (1, 2, 1), (1, 2, 2), (2, 1, 1), (2, 1, 2), (2, 2, 1), (2, 2, 2)\}$

Example 6. **If $A = \{1, 4\}$, $B = \{2, 3, 6\}$ and $C = \{2, 3, 7\}$, then verify that**

(i) $A \times B \neq B \times A$

(ii) $A \times (B \cup C) = (A \times B) \cup (A \times C)$

Sol. (i) $A \times B = \{1, 4\} \times \{2, 3, 6\}$
$= \{(1, 2), (1, 3), (1, 6), (4, 2), (4, 3), (4, 6)\}$ ...(i)

$B \times A = \{2, 3, 6\} \times \{1, 4\}$
$= \{(2, 1) (2, 4), (3, 1), (3, 4), (6, 1), (6, 4)\}$ ...(ii)

From Eqs. (i) and (ii), we get
$A \times B \neq B \times A$

(ii) To find $A \times (B \cup C)$
$B \cup C = \{2, 3, 6\} \cup \{2, 3, 7\}$
$\Rightarrow$ $B \cup C = \{2, 3, 6, 7\}$
$\therefore$ $A \times (B \cup C) = \{1, 4\} \times \{2, 3, 6, 7\}$
$= \{(1, 2), (1, 3), (1, 6), (1, 7), (4, 2), (4, 3), (4, 6), (4, 7)\}$ ...(i)

To find $(A \times B) \cup (A \times C)$,
$A \times B = \{(1, 2), (1, 3), (1, 6), (4, 2), (4, 3), (4, 6)\}$
and $A \times C = \{(1, 2), (1, 3), (1, 7), (4, 2), (4, 3), (4, 7)\}$
$\therefore$ $(A \times B) \cup (A \times C) = \{(1, 2), (1, 3), (1, 6), (1, 7), (4, 2), (4, 3), (4, 6), (4, 7)\}$ ...(ii)

From Eqs. (i) and (ii), we get
$A \times (B \cup C) = (A \times B) \cup (A \times C)$

Quick Test 1

1. Find x and y, if $(x+2, 4) = (5, 2x+y)$.
2. If $A = \{a, b, c\}$ and $B = \{r, s\}$, then find
   (i) $A \times B$ (ii) $B \times A$
3. If $A \times B = \{(a, x), (a, y), (b, x), (b, y)\}$, then find A and B.
4. The cartesian product $A \times A$ has 9 elements among which are found $(-1, 0)$ and $(0, 1)$. Find the set A and the remaining element of $A \times A$. *Competency Based Que*
5. If $P = \{1, 4\}$, then find $P \times P \times P$.
6. Let A and B be two sets such that $n(A) = 5$ and $n(B) = 2$. If a, b, c, d and e are distinct and $(a, 2), (b, 3), (c, 2), (d, 3), (e, 2)$ are elements of $A \times B$, then find A and B.
7. If $A = \{x : x^2 - 5x + 6 = 0\}$, $B = \{2, 4\}$ and $C = \{4, 5\}$, then find the value of $A \times (B \cap C)$. *Competency Based Que*
8. If $A = \{1, 2, 3\}$, $B = \{3, 4\}$ and $C = \{4, 5, 6\}$, then find the value of $(A \times B) \cap (B \times C)$.
9. If $A = \{1, 2, 3\}$, $B = \{1, 2, 3, 4\}$, $C = \{5, 6\}$ and $D = \{5, 6, 7, 8\}$, then verify that $(A \times C) \subset (B \times D)$. *Competency Based Que*
10. If $A = \{1, 2, 3\}$, $B = \{3, 4\}$ and $C = \{4, 5, 6\}$, then find
    (i) $A \times (B \cap C)$ (ii) $(A \times B) \cap (A \times C)$

Answers

1. $x = 3$, $y = -2$
2. (i) $\{(a, r), (a, s), (b, r), (b, s), (c, r), (c, s)\}$
   (ii) $\{(r, a), (r, b), (r, c), (s, a), (s, b), (s, c)\}$
3. $A = \{a, b\}$, $B = \{x, y\}$
4. $A = \{-1, 0, 1\}$ and remaining elements of $A \times A$ are $(-1, -1), (-1, 1), (0, -1), (0, 0), (1, -1), (1, 0), (1, 1)$
5. $P \times P \times P = \{(1, 1, 1), (1, 1, 4), (1, 4, 1), (1, 4, 4), (4, 1, 1), (4, 1, 4), (4, 4, 1), (4, 4, 4)\}$
6. $A = \{a, b, c ,d\ e\}$, $B = \{2, 3\}$
7. $\{(2, 4), (3, 4)\}$
8. $\{(3, 4)\}$
10. (i) $\{(1, 4), (2, 4), (3, 4)\}$
    (ii) $\{(1, 4), (2, 4), (3, 4)\}$

For Solutions, See QR Code on Pg. 48

2.2 Relations

Let A and B be two non-empty sets. Then, a relation R from set A to set B is a subset of cartesian product $A \times B$

i.e. $$R \subseteq A \times B.$$

If $(a, b) \in R$, then we write it as aRb and we say that **a is related to b by the relation R.**

If $(a, b) \notin R$, then we write it as $a \not R b$ and we say that **a is not related to b by the relation R.**

Now, we discuss the relation between two sets.

e.g. Let find the relation between a set of country to set of currency, where, R is a relation from set A = {India, America, England, Canada, France} to set B = {Pound, Rupee, Euro, Dollar} and defined by aRb.

Relations among Domain, Codomain and Range

Let us consider the relation R from A to B.

- **Domain** The set of first elements of all the ordered pairs present in R is called the domain of relation.
- **Range** The set of second elements of all the ordered pairs present in R is called range of the relation.
- **Codomain** The set B is called the codomain of relation R.

i.e. Domain $(R) = \{a : (a, b) \in R\}$

and Range $(R) = \{b : (a, b) \in R\}$

$\therefore$ **Range $\subseteq$ Codomain**

Method of Representation

Let R be a relation between a set of country to set of currency. Then, the relation R can be represented algebraically by roster form or by set-builder form and visually it can be represented by an arrow diagram which are given below.

(i) Roster form

In roster form, the relation between A to B is

R = {(India, Rupee), (America, Dollar), (England, Pound), (France, Euro), (Canada, Dollar)}

(ii) Set-Builder form

$$R = \{(x, y) : x \in A,\ y \in B;\ y \text{ is currency of } x\}$$

(iii) Arrow diagram

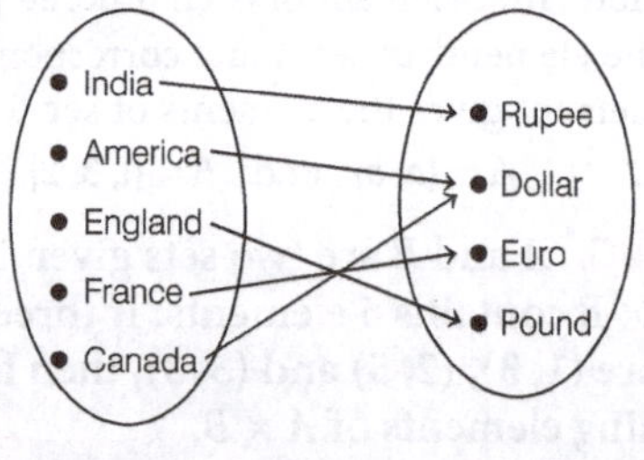

Number of Relations

Let A and B be any two non-empty finite sets containing m and n elements, respectively. Then, number of ordered pairs in $A \times B$ is mn.

$\therefore$ Total number of subsets of $A \times B$ is 2^{mn}.

Hence, the total number of relations from set A to set B is 2^{mn} because each relation from set A to set B is a subset of $A \times B$.

Among these 2^{mn} relation, the void relation ϕ and the universal relation $A \times B$ are **trivial relations** from set A to set B.

TEACHER TIPS

- If $R = \phi$, then R is called an empty relation.
- If $R = A \times B$, then R is called the universal relation.
- If R_1 and R_2 are two relations from A to B, then $R_1 \cup R_2$, $R_1 \cap R_2$ and $R_1 - R_2$ are also relations from A to B.

Example 1. If $A = \{1, 2, 3, 4\}$ and $B = \{5, 6, 7, 8\}$, then which of the following are relations from A to B? Give answer.

(i) $R_1 = \{(1, 5), (2, 7), (3, 8)\}$

(ii) $R_2 = \{(6, 2), (3, 7), (4, 7)\}$

Sol. We have, $A = \{1, 2, 3, 4\}$ and $B = \{5, 6, 7, 8\}$

$A \times B = \{(1, 5), (1, 6), (1, 7), (1, 8), (2, 5), (2, 6), (2, 7), (2, 8), (3, 5), (3, 6), (3, 7), (3, 8), (4, 5), (4, 6), (4, 7), (4, 8)\}$

(i) Since, $R_1 \subseteq A \times B$, therefore R_1 is a relation from A to B.

(ii) Since, $(6, 2) \in R_2$ but $(6, 2) \notin A \times B$, therefore $R_2 \not\subseteq A \times B$. Thus, R_2 is not a relation from A to B.

Example 2. Write the relation between sets P and Q given by an arrow diagram in

(i) roster form (ii) set-builder form

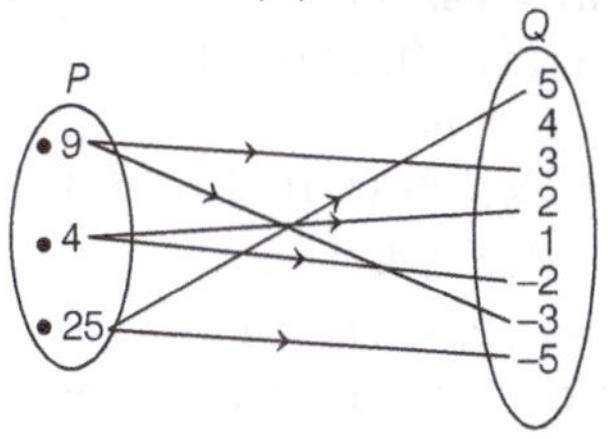

Sol. From arrow diagram, we have

$P = \{9, 4, 25\}$ and $Q = \{5, 4, 3, 2, 1, -2, -3, -5\}$

Here, the relation R is 'x is the square of y', where $x \in P$ and $y \in Q$.

In roster form, it can be written as

$R = \{(9, 3)(9, -3)(4, 2)(4, -2)(25, 5)(25, -5)\}$

In set-builder form, R can be written as

$R = \{(x, y) : x \in P, y \in Q \text{ and } x \text{ is the square of } y\}$

Example 3. A relation R is defined, from a set $A = \{2, 3, 4, 5\}$ to set $B = \{3, 6, 7, 10\}$, as $(x, y) \in R \Leftrightarrow x$, is relatively prime to y. Express R as a set of ordered pairs and depict this relations by arrow diagram.

Sol. Given,

$R = \{(x, y): x \in A, y \in B, x \text{ is relatively prime to } y\}$

Here, x is relatively prime to y means $\gcd(x, y) = 1$.

$\therefore R = \{(2, 3), (2, 7), (3, 7), (3, 10), (4, 3), (4, 7), (5, 3), (5, 6), (5, 7)\}$

Visual representation is shown below.

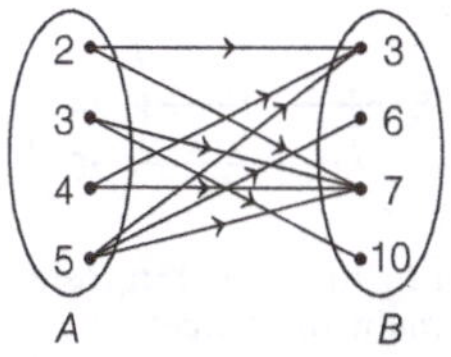

Example 4. If $n(A) = 3$ and $B = \{2, 3, 4, 6, 7, 8\}$, then find the number of relations from A to B.

Sol. Given, $n(A) = 3$ and $B = \{2, 3, 4, 6, 7, 8\} \Rightarrow n(B) = 6$

$\therefore$ Number of relations from A to B $= 2^{n(A) \times n(B)}$

$= 2^{3 \times 6} = 2^{18}$

Example 5. If $A = \{5, 7\}$, $B = \{9, 13\}$ and $R = \{(a, b): a \in A, b \in B, a - b \text{ is odd}\}$, then prove that relation R is a empty relation from sets A to B.

Sol. We have, $A = \{5, 7\}$ and $B = \{9, 13\}$

When $a = 5$, $b = 9$; $a - b = 5 - 9 = -4$, which is not odd.

When $a = 5$, $b = 13$; $a - b = 5 - 13 = -8$, which is not odd.

When $a = 7$, $b = 9$; $a - b = 7 - 9 = -2$, which is not odd.

When $a = 7$, $b = 13$; $a - b = 7 - 13 = -6$, which is not odd.

Since, $(a - b)$ is not odd for any ordered pair (a, b), where $a \in A$, $b \in B$.

$\therefore \quad R = \phi$

Hence, R is a empty relation from sets A to B.

Hence proved.

Example 6. Determine the domain and range of the relation $R=\{(a,b): a \in N, a<5, b=4\}$

Sol. Given, $R=\{(a,b): a \in N, a<5, b=4\}$

$\Rightarrow$ $R=\{(1,4),(2,4),(3,4),(4,4)\}$

Domain $(R)=\{a:(a,b)\in R\}=\{1,2,3,4\}$

Range $(R)=\{b:(a,b)\in R\}=\{4\}$

Example 7. Let $A=\{1,2,3,4,5\}$. Define a relation R from A to A by $R=\{(x,y): y=x-1, x, y \in A\}$

(i) Depict this relation using an arrow diagram.

(ii) Write down the domain, codomain and range of R. Competency Based Que

Sol. (i) Given, $R=\{(x,y): y=x-1, x, y\in A\}$

$\therefore$ $R=\{(2,1),(3,2),(4,3),(5,4)\}$

The corresponding arrow diagram is shown in following figure.

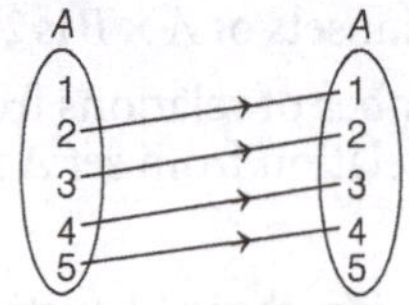

(ii) It is clear from the arrow diagram, domain $=\{2,3,4,5\}$, codomain $=\{1,2,3,4,5\}$ and range $=\{1,2,3,4\}$.

Quick Test 2

1. The given figure shows a relationship between the sets P and Q. Write this relation

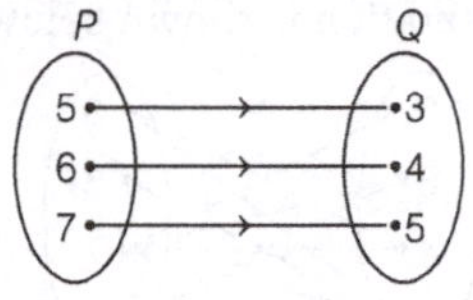

(i) in set- builder form (ii) in roster form
(iii) What is its domain and range?

2. Let $A=\{1,2,3,4\}$ and $B=\{1,4,9,16,25\}$ and R be a relation defined from A to B, as $R=\{(x,y): x\in A, y\in B \text{ and } y=x^2\}$.

(i) Depict this relation using arrow diagram.
(ii) Find domain of R.
(iii) Find range of R.
(iv) Write codomain of R.

3. Two finite sets A and B have m and n elements, respectively. If the total number of relations from A to B is 64, then find the possible values of m and n.

4. The relation on the set $A=\{x:|x|<3, x\in Z\}$ is defined by $R=\{(x,y): y=|x|, x\neq -1\}$. Then, find the number of elements in the power set of R. Competency Based Que

5. Let $A=\{1,2,3,5\}$ and $B=\{4,6,9\}$. If a relation R from A to B is defined by $R=\{(x,y)$: the difference between x and y is odd, $x\in A, y\in B\}$, then write R in roster form. NCERT

6. If $A=\{1,2,3,4,6\}$ and R be the relation defined on A by $R=\{(a,b): a\in A, b\in A \text{ and } a \text{ divides } b\}$, then find domain and range of R. Competency Based Que

7. Determine the domain and range of the relation R, where

(i) $R=\{(-3,1),(-1,1),(1,0),(3,0)\}$
(ii) $R=\{(x-2,x^2): x$ is a prime number less than 15$\}$

8. Let $A=\{2,4,6,9\}$ and $B=\{4,6,18,27,54\}$, then find the set of ordered pair (a,b) such that $a\in A, b\in B$, a is factor of b, and $a<b$. Competency Based Que

9. If $A=\{x,y,z\}$ and $B=\{1,2\}$, then find the number of relation from A to B. NCERT

Answers

1. (i) $\{(x,y): x\in P, y\in Q, x=y+2\}$
(ii) $\{(5,3),(6,4),(7,5)\}$
(iii) Domain $(R)=\{5,6,7\}$
Range $(R)=\{3,4,5\}$

2. (i) The relation $R=\{(1,1),(2,4),(3,9),(4,16)\}$

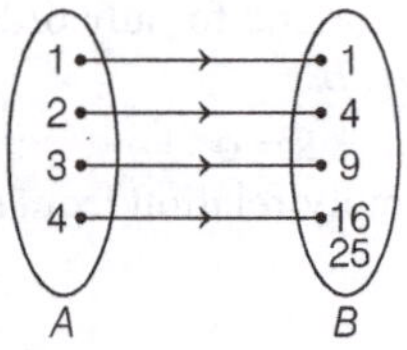

(ii) Domain $(R)=\{1,2,3,4\}$
(iii) Range $(R)=\{1,4,9,16\}$
(iv) Codomain $(R)=\{1,4,9,16,25\}$

3. 2 and 3 **4.** 16

5. $R=\{(1,4),(1,6),(2,9),(3,4),(3,6),(5,4),(5,6)\}$

6. Domain $(R)=\{1,2,3,4,6\}=A$
Range $(R)=\{1,2,3,4,6\}=A$

7. (i) Domain $(R)=\{-3,-1,1,3\}$ and Range $(R)=\{1,0\}$
(ii) Domain $(R)=\{0,1,3,5,9,11\}$
Range $(R)=\{4,9,25,49,121,169\}$

8. $R=\{(2,4),(2,6),(2,18),(2,54),(6,18),(6,54),(9,18),(9,27),(9,54)\}$

9. 2^6 or 64

For Solutions, See QR Code on Pg. 48

2.3 Functions

A special type of relation is called **function**.

Function as a Relation

A relation f from a non-empty set A to a non-empty set B is said to be a function, if every element of set A has one and only one image in set B.

If f is a function from set A to set B, then we write

$$f: A \rightarrow B \text{ or } A \xrightarrow{f} B$$

and it is read as f is a function from A to B or f maps A to B.

If $(a, b) \in f$, then we can also write it as $f(a) = b$.

Here, b is called the **image** of a under f and a is called the **pre-image** of b under f.

e.g. (i) Let $A = \{2, 3, 4\}$, $B = \{3, 4, 5\}$ and f_1, f_2 and f_3 be three subsets of $A \times B$ defined as

$$f_1 = \{(2, 3), (3, 4), (4, 5)\},$$
$$f_2 = \{(2, 3), (2, 4), (3, 4), (4, 5)\}$$
and $$f_3 = \{(2, 4), (3, 5)\}$$

Then, f_1 is a function from A to B, since every element of A has one and only one image in set B. f_2 is not a function from A to B, since $2 \in A$ has two images 3 and 4 in set B. Also, f_3 is not a function from A to B, since $4 \in A$ has no image in set B.

(ii) In the form of arrow diagram,

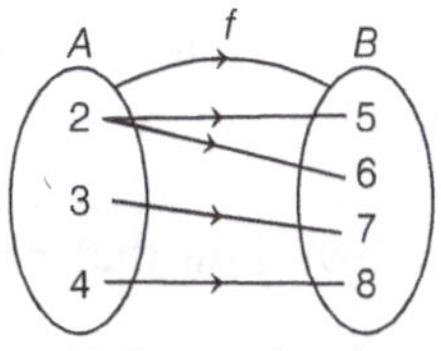

(a) f is not a function.

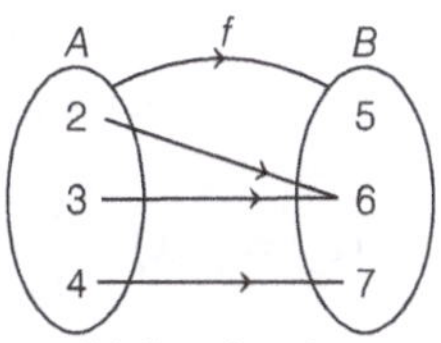

(b) f is a function.

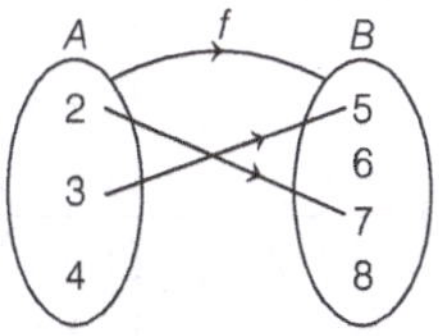

(c) f is not a function.

Domain, Codomain and Range of a Function

If f is a function from A to B, i.e. each element of A corresponds to one and only one element of B, whereas every element in B need not be the image of some x in A.

Then, the set A is called the **domain** of function f and the set B is called the **codomain** of f.

The subset of B containing the images of elements of A is called the **range** of the function.

Thus, if a function f is expressed as the set of ordered pairs, then the domain of f is the **set of first elements** of all ordered pairs of f and the range of f is the **set of second elements** of all ordered pairs of f i.e.

$$\textbf{Domain of } f = \{a : (a, b) \in f\}$$
$$\textbf{Range of } f = \{b : (a, b) \in f\}$$

See the arrow diagram given below for more understanding.

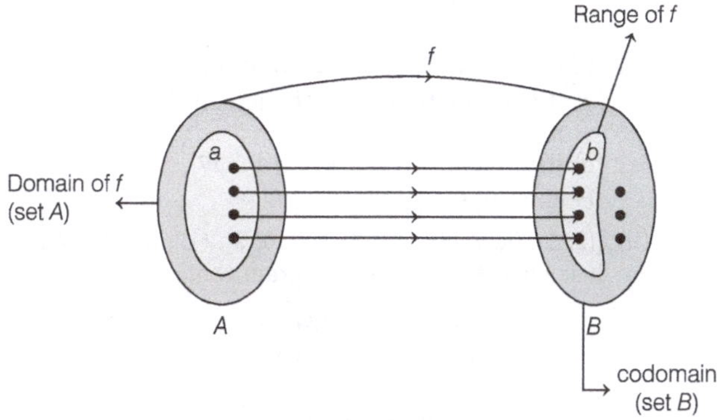

2.4 Real Functions

A function which has either R (set of real numbers) or one of its subsets as its range is called a **real valued function** and if its domain is also either R or a subset of R i.e. range and domain both are either R or a subset of R, then function is called real function.

For real functions, generally domain and codomain both are infinite subsets of R. Therefore, a real function is generally described by some general formula.

e.g. Let f be a function define by $f(x) = 2x + 1$ from N to N. Then,

x	1	2	3	4	5
$f(x) = y$	$f(1) = 3$	$f(2) = 5$	$f(3) = 7$	$f(4) = 9$	$f(5) = 11$

Thus, range $= \{3, 5, 7, 9, 11, \ldots\} \subseteq R$, codomain $= N$ and domain $= N \subseteq R$.

Value of a Real Function

If any function x is defined from $f : A \rightarrow B$, then the value of function at any point a of A will be $f(a)$.

Example 1. **Which of the following relations are functions? Give reasons. If it is a function, then determine its domain and range.** NCERT

(i) $\{(2, 1), (5, 1), (8, 1), (11, 1), (14, 1), (17, 1)\}$

(ii) $\{(2, 1), (4, 2), (6, 3), (8, 4), (10, 5), (12, 6), (14, 7)\}$

(iii) $\{(1, 3), (1, 5), (2, 5)\}$

Sol. Here,

(i) $\{(2, 1), (5, 1), (8, 1), (11, 1), (14, 1), (17, 1)\}$

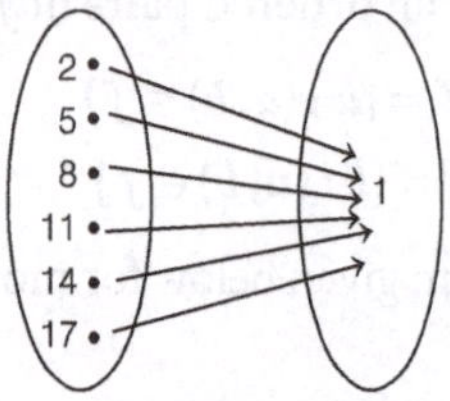

It is clear from arrow diagram that

(*a*) it is a function, as each element have a unique image.

(*b*) its domain is $\{2, 5, 8, 11, 14, 17\}$.

(*c*) its range is $\{1\}$.

(ii) $\{(2, 1), (4, 2), (6, 3), (8, 4), (10, 5), (12, 6), (14, 7)\}$.
It is clear from arrow diagram that

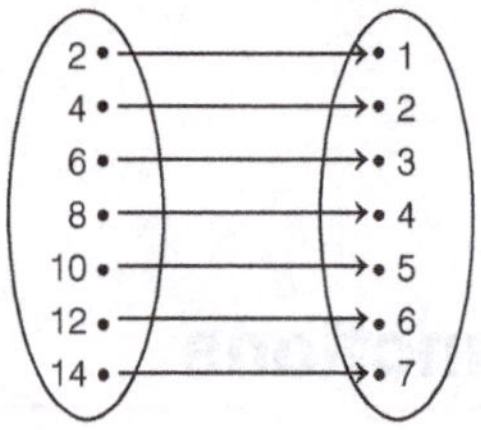

(a) it is a function, as each element have a unique image.

(b) its domain is $\{2, 4, 6, 8, 10, 12, 14\}$.

(*c*) its range is $\{1, 2, 3, 4, 5, 6, 7\}$.

(iii) $\{(1, 3), (1, 5), (2, 5)\}$
It is clear from the arrow diagram that

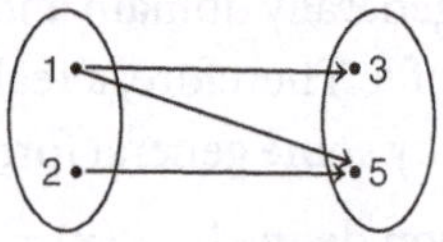

It is not a function because 1 has two images 3 and 5.

Example 2. **Let $A = \{-2, -1, 0, 1, 2\}$ and $f : A \to Z$ be given by $f(x) = x^2 - 2x - 3$. Find**

(i) the range of f.

(ii) pre-images of 6, – 3 and 5. Competency Based Que

Sol. (i) We have, $f(x) = x^2 - 2x - 3$

$\therefore \quad f(-2) = (-2)^2 - 2(-2) - 3 = 5,$

$f(-1) = (-1)^2 - 2(-1) - 3 = 0,$

$f(0) = -3,\ f(1) = 1^2 - 2 \times 1 - 3 = -4,$

and $f(2) = 2^2 - 2 \times 2 - 3 = -3.$

So, range $(f) = \{f(-2), f(-1), f(0), f(1), f(2)\}$
$= \{0, 5, -3, -4\}$

(ii) Let x be a pre-image of 6, then

$f(x) = 6 \Rightarrow x^2 - 2x - 3 = 6$

$\Rightarrow x^2 - 2x - 9 = 0 \Rightarrow x = 1 \pm \sqrt{10}$

Since, $x = 1 \pm \sqrt{10} \notin A$.

So, there is no pre-image of 6.

Let x be a pre-image of -3.

Then, $f(x) = -3$

$\Rightarrow \quad x^2 - 2x - 3 = -3$

$\Rightarrow \quad x^2 - 2x = 0 \Rightarrow x = 0, 2$

Clearly, $0, 2 \in A$.

So, 0 and 2 are pre-images of -3.

Let x be a pre-image of 5.

Then, $f(x) = 5$

$\Rightarrow \quad x^2 - 2x - 3 = 5$

$\Rightarrow \quad x^2 - 2x - 8 = 0$

$\Rightarrow \quad (x-4)(x+2) = 0 \Rightarrow x = 4, -2$

Since, $-2 \in A$ but $4 \notin A$.

So, -2 is a pre-image of 5.

Example 3. **If $f(x) = x^2 + 2x + 3$, then find $f(1)$, $f(2)$ and $f(3)$.**

Sol. We have, $f(x) = x^2 + 2x + 3$ ...(i)

$f(1) = (1)^2 + 2(1) + 3$ [put $x = 1$ in Eq. (i)]
$= 1 + 2 + 3 = 6$

$f(2) = (2)^2 + 2(2) + 3$ [put $x = 2$ in Eq. (i)]
$= 4 + 4 + 3 = 11$

$f(3) = (3)^2 + 2(3) + 3$ [put $x = 3$ in Eq. (i)]
$= 9 + 6 + 3 = 18$

Example 4. **If $y = f(x) = \dfrac{1-x}{1+x}$, then show that $x = f(y)$.**

Sol. We have, $y = f(x) = \dfrac{1-x}{1+x}$ ...(i)

Then, $f(y) = \dfrac{1-y}{1+y} = \dfrac{1 - \left(\dfrac{1-x}{1+x}\right)}{1 + \left(\dfrac{1-x}{1+x}\right)}$ [from Eq. (i)]

$= \dfrac{\dfrac{1+x-(1-x)}{1+x}}{\dfrac{1+x+1-x}{1+x}} = \dfrac{2x}{2} \Rightarrow f(y) = x$

$\therefore \quad x = f(y)$, when $y = f(x) = \dfrac{1-x}{1+x}$

Example 5. **Find the domain of real function**

$$f(x)=\frac{x^2+3x+5}{x^2-5x+4}$$

Sol. Here, denominator $= x^2-5x+4$

On putting it equal to zero, we get

$$x^2-5x+4=0$$
$$\Rightarrow \quad x^2-4x-x+4=0$$
$$\Rightarrow \quad x(x-4)-1(x-4)=0$$
$$\Rightarrow \quad (x-4)(x-1)=0$$
$$\Rightarrow \quad x=4 \text{ and } x=1$$

Clearly, at $x=1$ and $x=4$ denominator becomes zero and thus given function will not be define for these values.

So, the required domain of given function $= R-\{1, 4\}$.

Example 6. **Find the domain and range of the function** $f(x)=\dfrac{1}{\sqrt{x-5}}$.

Sol. We have, $f(x)=\dfrac{1}{\sqrt{x-5}}$

Domain of f

Clearly, $f(x)$ takes real values, if

$$x-5>0 \Rightarrow x>5 \Rightarrow x\in(5, \infty)$$

Domain of $f=(5, \infty)$

Range of f

For $x>5$, we have $x-5>0$

$$\Rightarrow \quad \sqrt{x-5}>0$$
$$\Rightarrow \quad \frac{1}{\sqrt{x-5}}>0 \Rightarrow f(x)>0$$

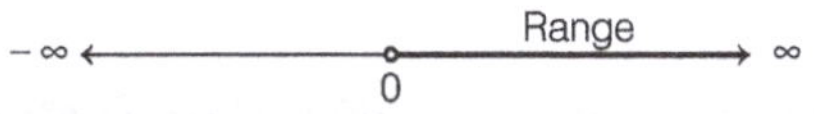

Thus, $f(x)$ takes all real values greater than zero.

Hence, range of $f=(0, \infty)$.

Example 7. **Find the domain and the range of the function** $f(x)=\sqrt{16-x^2}$.

Sol. Given, $f(x)=\sqrt{16-x^2}$

Domain of f We observe that $f(x)$ is defined for all x satisfying $16-x^2\geq 0$

$$\Rightarrow \quad x^2-16\leq 0$$
$$\Rightarrow \quad (x-4)(x+4)\leq 0 \quad [\because a^2-b^2=(a-b)(a+b)]$$
$$\Rightarrow \quad -4\leq x\leq 4 \Rightarrow x\in[-4, 4]$$

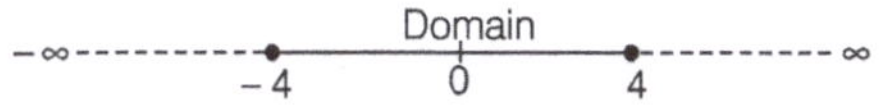

Range of f Let $y=f(x)$.

Then, $\quad y=\sqrt{16-x^2} \Rightarrow y^2=16-x^2$

$$\Rightarrow \quad x^2=16-y^2 \Rightarrow x=\pm\sqrt{16-y^2}$$

Clearly, x will take real values, if

$$16-y^2\geq 0$$
$$\Rightarrow \quad y^2-16\leq 0$$
$$\Rightarrow \quad (y-4)(y+4)\leq 0$$
$$\Rightarrow \quad -4\leq y\leq 4 \Rightarrow y\in[-4, 4]$$

Also, $\quad y=\sqrt{16-x^2}\geq 0$ for all $x\in[-4, 4]$

$\therefore \quad y\in[0, 4]$ for all $x\in[-4, 4]$

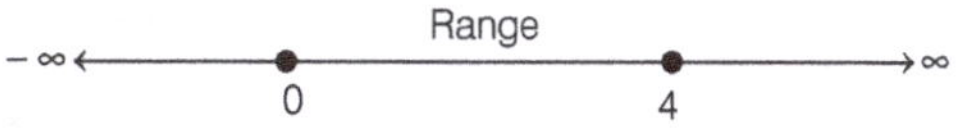

Hence, range of $f=[0, 4]$.

Example 8. **Find the domain and range of the real valued function $f(x)$ is given by** $f(x)=\dfrac{4-x}{x-4}$.

Sol. Given, $f(x)=\dfrac{4-x}{x-4}$

Domain of f Here, we see that $f(x)$ is defined for all x except 4.

$\therefore \quad$ Domain $(f)=R-\{4\}$

$\because \quad f(x)=\dfrac{4-x}{x-4}=\dfrac{-(x-4)}{x-4}=-1$

$\therefore \quad$ Range $(f)=\{-1\}$

Example 9. **Let $f: R\rightarrow R$ be such that $f(x)=2^x$. Determine**

(i) range of f.

(ii) whether $f(x+y)=f(x)\cdot f(y)$ holds.

(iii) $\{x: f(x)=1\}$

Sol. (i) Given, $f(x)=2^x$, which is well defined for all $x\in R$.

$\therefore \quad$ Domain $(f)=R$

Now, $\quad$ let $y=f(x)=2^x$

On taking log with base 2 both sides, we get

$$\log_2 y=\log_2 2^x$$
$$\Rightarrow \quad \log_2 y=x\log_2 2=x \quad [\because \log_a a=1]$$
$$\Rightarrow \quad x=\log_2 y,$$

which is defined for all $y>0$.

Hence, the range of f is set of all positive real numbers.

(ii) We have, $\quad f(x+y)=2^{x+y}=2^x\cdot 2^y=f(x)\cdot f(y)$

$\therefore \quad f(x+y)=f(x)\cdot f(y)$ holds for all $x, y\in R$.

(iii) $\because \quad f(x)=1$

$$\Rightarrow \quad 2^x=1 \Rightarrow x=0$$

$\therefore \quad \{x: f(x)=1\}=\{0\}$

Example 10. Find the domain and range of the function $f(x) = \dfrac{1}{2-\sin 3x}$.

Competency Based Que

Sol $\because$ $-1 \le \sin 3x \le 1, \forall\, x \in R$

$\Rightarrow$ $-1 \le -\sin 3x \le 1, \forall x \in R$

$\Rightarrow$ $1 \le 2-\sin 3x \le 3, \forall x \in R$

$\Rightarrow$ $2-\sin 3x \ne 0$, for any $x \in R$

$\Rightarrow$ $f(x) = \dfrac{1}{2-\sin 3x}$ is defined, $\forall\, x \in R.$

$\therefore$ Domain $(f) = R$

From the above discussion, $1 \le 2-\sin 3x \le 3, \forall\, x \in R$

$\Rightarrow$ $\dfrac{1}{3} \le \dfrac{1}{2-\sin 3x} \le 1, \forall x \in R$

$\Rightarrow$ $\dfrac{1}{3} \le f(x) \le 1, \forall\, x \in R$

$\Rightarrow$ $f(x) \in \left[\dfrac{1}{3}, 1\right]$

$\therefore$ Range $(f) = \left[\dfrac{1}{3}, 1\right]$

Quick Test 3

1. Which of the relations are functions?

(i) $f_1 = \{(2,1),(3,1),(4,1),(3,2),(4,2),(4,3)\}$

(ii) $f_2 = \{(1,4),(2,3),(3,2),(4,1)\}$

2. If $f(x) = 3x^4 - 5x^2 + 9$, then find $f(x-1)$.

3. Find the domain of $\sqrt{9^2 - x^2}$

4. Let f be a relation on the set N of natural numbers defined by $f = \{(n, 3n) : n \in N\}$. Is f a function from N to N? If so, then find the range of f. Competency Based Que

5. Find the domain of the following functions.

(i) $f(x) = \sqrt{x^2+4}$ (ii) $f(x) = x^2$

(iii) $f(x) = \dfrac{x}{x^2+5x+6}$ (iv) $f(x) = \dfrac{1}{\sqrt{x-2}}$

(v) $f(x) = \dfrac{1}{\sqrt{x+|x|}}$

6. Find the range of the following functions.

(i) $f(x) = \dfrac{x-2}{3-x}$ (ii) $f(x) = \dfrac{1}{4-x^2}$

(iii) $f(x) = x^2+2$ (iv) $f(x) = |2x-3|-3$

(v) $f(x) = \dfrac{x}{1+x^2}$

7. Find the domain and range of the real valued function f defined by $f(x) = \sqrt{x-1}$. Competency Based Que

8. If $f(x) = x + \dfrac{1}{x}$, then prove that $[f(x)]^3 = f(x^3) + 3f\left(\dfrac{1}{x}\right)$.

9. If $f(x) = \dfrac{1}{2x+1}, x \ne -\dfrac{1}{2}$, then show that $f(f(x)) = \dfrac{2x+1}{2x+3}$, provided that $x \ne -\dfrac{3}{2}$.

Competency Based Que

10. Find the domain and range of $f(x) = \dfrac{1}{2-\sin 8x}$.

Answers

1. (i) Not a function. (ii) It is a function.

2. $3x^4 - 12x^3 + 13x^2 - 2x + 7$ **3.** $x \in [-9, 9]$

4. Yes, f is a function from N to N.

$\therefore$ Range $(f) = \{3n : n \in N\}$

5. (i) Domain $(f) = R$ (ii) $D(f) = R$

(iii) $D(f) = R - \{-2, -3\}$ (iv) $D(f) = (2, \infty)$

(v) $D(f) = (0, \infty)$

6. (i) $R(f) = R - \{-1\}$ (ii) $R(f) = (-\infty, 0) \cup [1/4, \infty)$

(iii) $R(f) = [2, \infty)$ (iv) $R(f) = [-3, \infty)$

(v) $R(f) = \left[-\dfrac{1}{2}, \dfrac{1}{2}\right]$

7. $D(f) = [1, \infty)$ and $R(f) = [0, \infty)$

10. $D(f) = R$ and $R(f) = \left[\dfrac{1}{3}, 1\right]$

For Solutions, See QR Code on Pg. 48

2.5 Some Standard Real Functions and Their Graphs

Identity Function

The real function $f : R \to R$ defined by $f(x) = x, \forall\, x \in R$ is called an identity function. Its domain is R and range is also R.

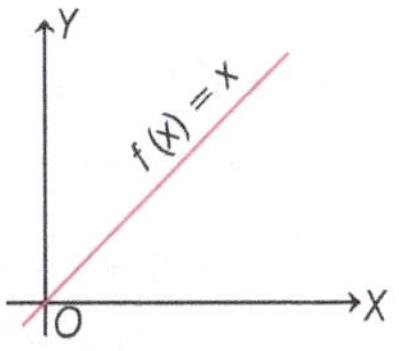

It may be observed that

- Graph of identity function is a straight line.
- It passes through the origin.

Constant Function

A function $f : R \to R$ is said to be a constant function, if there exists a **real number k** such that

$$f(x) = k, \forall\, x \in R$$

Here, domain of $f = R$ and its range $= \{k\}$

e.g. $f(x) = 5, \forall\, x \in R$ (symbol $\forall$ means for all)

Here, $f(x)$ is a constant function whose

domain $= \{x : x \in R\}$ and range $= \{5\}$

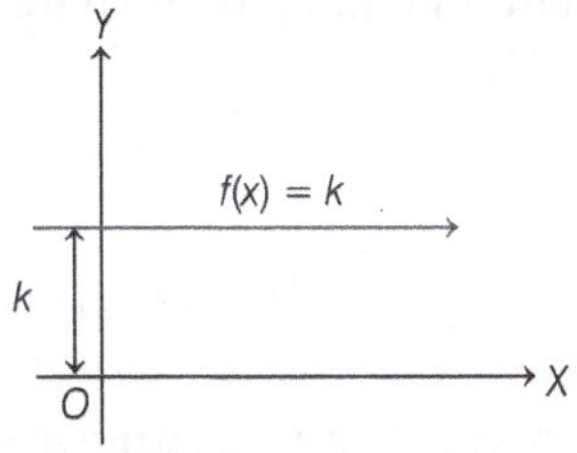

It may be observed that

- The graph of a constant function is a line parallel to X-axis.
- If $k > 0$, then the line will be above X-axis.
- If $k = 0$, then graph coincides with the X-axis.
- If $k < 0$, then the line will be below X-axis.

Polynomial Function

A function $f : R \to R$ defined by

$f(x) = a_0 + a_1 x + a_2 x^2 + \ldots + a_n x^n$, where n is a non-negative integer and $a_0, a_1, a_2, \ldots, a_n \in R$ is called a polynomial function. If $a_n \neq 0$, then n is called the **degree of the polynomial.**

The domain of a polynomial function is R and range depends on the polynomial representing the function.

e.g. (i) $f(x) = x^3 - x^2 + 9$ is polynomial function in x.

(ii) $h(x) = x^5 + \sqrt{3}x$ is polynomial function in x.

(iii) $g(x) = x^{\frac{3}{2}} + 3x$ is not a polynomial function.

Since, power $\frac{3}{2}$ is not an integer.

Rational Function

A function of the form $f(x) = \dfrac{P(x)}{Q(x)}$, where $P(x)$ and $Q(x)$ are polynomial functions of x defined in a domain and $Q(x) \neq 0$, is called a rational function.

The domain of a rational function is $R - \{x : Q(x) = 0\}$ and range depends on the expression representing the function.

Thus, $f(x) = \left(\dfrac{x^2 + 4}{x^3 - 6x + 4}\right)$ is called a rational function, where $x^3 - 6x + 4 \neq 0$.

Its domain is $R - \{x : x^3 - 6x + 4 = 0\}$.

Modulus Function or Absolute Value Function

The function $f : R \to R$ defined by

$$f(x) = |x| = \begin{cases} x, & \text{if } x \geq 0 \\ -x, & \text{if } x < 0 \end{cases}$$

is called the modulus function. It is also called absolute value function. Its domain is R and its range is $[0, \infty)$.

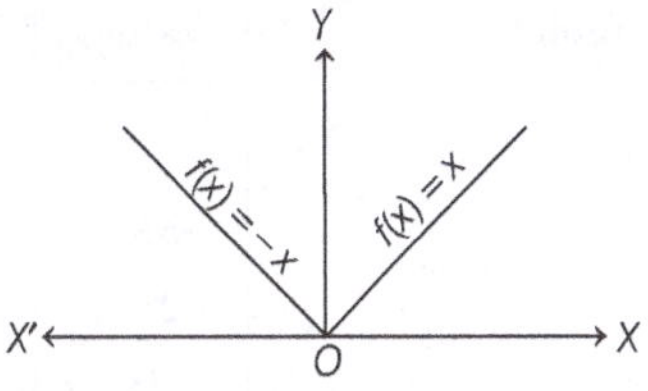

It may be observed that

- The graph is symmetrical with respect to Y-axis.
- Graph lies above the X-axis.
- It passes through the origin.
- In the first quadrant, it coincides with the graph of the identity function.

Signum Function

The function $f: R \to R$ defined by

$$f(x) = \begin{cases} \dfrac{|x|}{x}, & \text{when } x \neq 0 \\ 0, & \text{when } x = 0 \end{cases}$$

or $$f(x) = \begin{cases} 1, & \text{when } x > 0 \\ 0, & \text{when } x = 0 \\ -1, & \text{when } x < 0 \end{cases}$$

is called the signum function.

Its domain is R and range is $\{-1, 0, 1\}$.

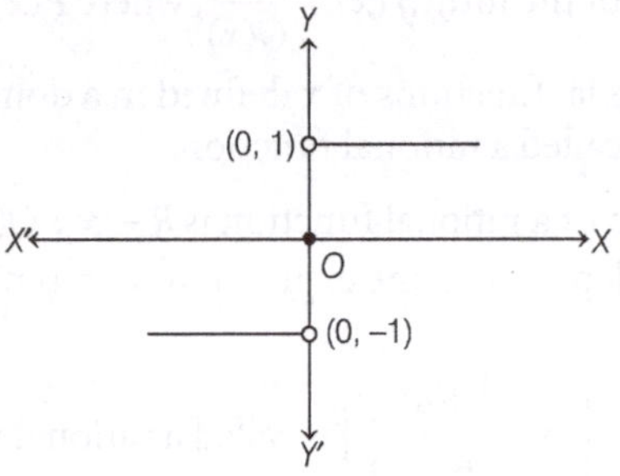

Greatest Integer Function (Step Function)

The function $f: R \to R$ defined by $f(x) = [x]$ is called the greatest integer function, where $[x]$ = **integral** part of x or greatest integer less than or equal to x.

Its domain is R and its range is also I (I = Set of integer).

e.g. $[7.4] = [7 + 0.4] = 7$, $[0.6] = [0 + 0.6] = 0$

and $[-1.5] = [-2 + 0.5] = -2$

Graph	Values of x	$f(x) = [x]$
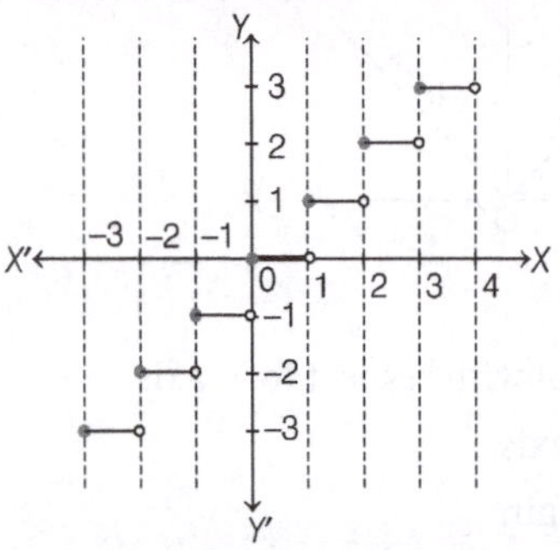	⋮	⋮
	$-3 \le x < -2$	-3
	$-2 \le x < -1$	-2
	$-1 \le x < 0$	-1
	$0 \le x < 1$	0
	$1 \le x < 2$	1
	$2 \le x < 3$	2
	$3 \le x < 4$	3
	⋮	⋮

2.6 Algebra of Real Functions

Let $f: D_1 \to R$ and $g: D_2 \to R$ be two real functions with domain D_1 and D_2, respectively.

Then, algebraic operations such as addition, subtraction, multiplication, division and scalar multiplication on two real functions are given below.

(i) **Addition of two real functions**

The sum function $(f + g)$ is defined by

$$(f + g)(x) = f(x) + g(x), \forall x \in D_1 \cap D_2$$

The domain of $(f + g)$ is $D_1 \cap D_2$.

(ii) **Subtraction of two real functions**

The difference function $(f - g)$ is defined by

$$(f - g)(x) = f(x) - g(x), \forall x \in D_1 \cap D_2$$

The domain of $(f - g)$ is $D_1 \cap D_2$.

(iii) **Multiplication of two real functions**

The product function (fg) is defined by

$$(fg)(x) = f(x) \cdot g(x), \forall x \in D_1 \cap D_2$$

The domain of (fg) is $D_1 \cap D_2$.

(iv) **Quotient of two real functions**

The quotient function is defined by

$$\frac{f}{g}(x) = \frac{f(x)}{g(x)}, \forall x \in D_1 \cap D_2 - \{x : g(x) = 0\}$$

The domain of $\left(\dfrac{f}{g}\right)$ is $D_1 \cap D_2 - \{x : g(x) = 0\}$.

(v) **Multiplication of a real function by a scalar**

The scalar multiple function cf is defined by

$$(cf)(x) = c \cdot f(x), \forall x \in D_1,$$

where c is a scalar (real number).

The domain of cf is D_1.

Example 1. **Let R be the set of real numbers. Define the real function $f: R \to R$ by $f(x) = x + 5$ and sketch the graph of this function.**

Sol. Here, $f(0) = 5$, $f(1) = 6$, $f(2) = 7$, ...

$f(-1) = 4$, $f(-2) = 3$, ..., and so on.

Hence, the graph of the given function is

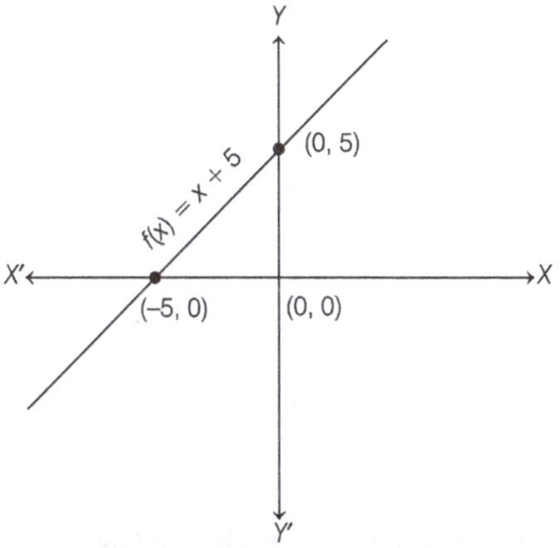

Example 2. **Draw the graph of the fucntion $f(x) = 5$. Also, write the name of the graph.**

Sol. (i) Given, $f(x) = 5, \forall\ x \in R$

Clearly, domain of f is R and range of f is $\{5\}$.

Now, we prepare a table for different values of x as shown below.

x	–3	–2	–1	0	1	2	3
$f(x)$	5	5	5	5	5	5	5

Locate the points (–3, 5), (–2, 5), (–1, 5), (0, 5), (1, 5), (2, 5) and (3, 5) on the graph paper and join them, we obtained the required graph as shown below.

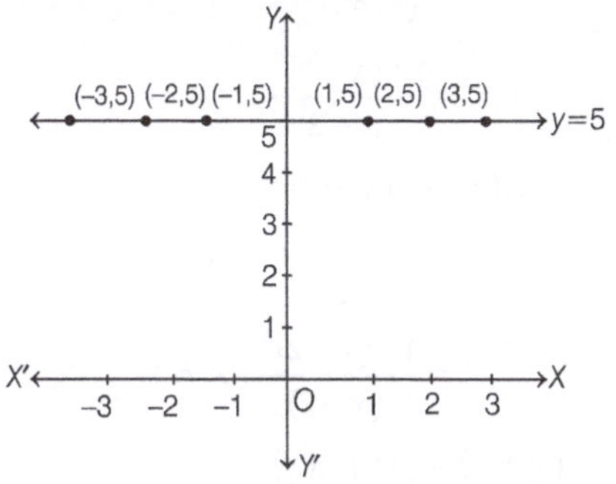

Clearly, the graph of $f(x) = 5$ represents a straight line parallel to X-axis and at a distance of 5 units above the X-axis and symmetric with respect to Y-axis.

$\therefore$ It is the graph of a constant function.

Example 3. **Let R be the set of real numbers. Define the real function $f: R \to R$ be $f(x) = x^2$. Sketch the graph of this function.** *Competency Based Que*

Sol. We have, $f(x) = x^2, \forall x \in R$

Here, domain of $f = R$ and range of $f = \{x : x \in R,\ x \geq 0\}$

For different values of x from domain f, the corresponding values of x^2 are given below.

x	–2	–1	0	1	2
$f(x) = x^2$	4	1	0	1	4

Take X and Y-axes, 1 cm = 1 unit

Plot the points (–2, 4), (–1, 1), (0, 0), (1, 1) and (2, 4) on a graph paper and join them to get parabolic curve, which is shown below.

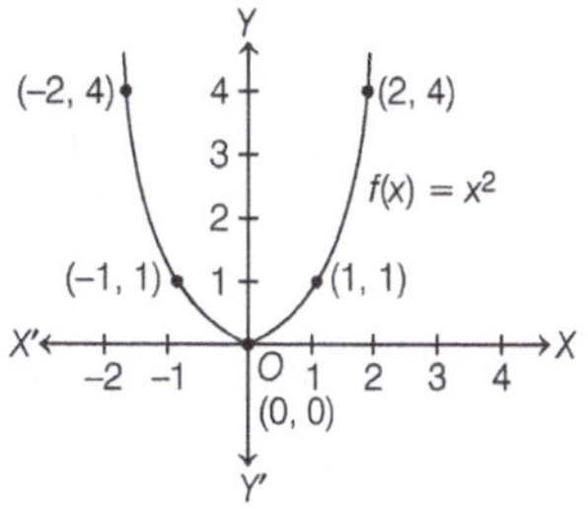

Example 4. **The function f is defined by**

$$f(x) = \begin{cases} 2 - x, & x < 0 \\ 2, & x = 0 \\ x + 2, & x > 0 \end{cases}$$

Draw the graph of $f(x)$.

Sol. Here, $f(x) = 2 - x,\ x < 0,$

$\therefore\quad f(-1) = 2 - (-1) = 3$

$f(-2) = 2 - (-2) = 4$

$f(-3) = 2 - (-3) = 5$ etc.

and $f(x) = x + 2,\ x > 0$

$f(1) = 3,\ f(2) = 4,\ f(3) = 5$ etc.

Thus, the graph of f is as shown below.

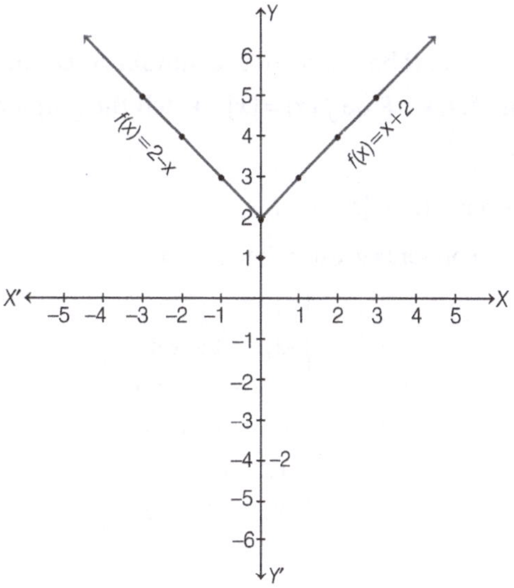

Example 5. **Draw the graph of the function $f: R \to R$ such that $f(x) = |x-2|$.** *Competency Based Que*

Sol. Clearly, $y = |x-2| = \begin{cases} x-2, & \text{if } x-2 \ge 0 \\ -(x-2), & \text{if } x-2 < 0 \end{cases}$

$= \begin{cases} x-2, & \text{if } x \ge 2 \\ 2-x, & \text{if } x < 2 \end{cases}$

We know that a linear equation in x and y represents a line. For drawing a line, we need only two points.

For $y = x-2$,

x	2	4
y	0	2

So, plot the points $P(2, 0)$, $Q(4, 2)$ and join PQ to get the graph of $y = x-2$.

For $y = 2-x$,

x	1	0
y	1	2

Plot the points $R(1, 1)$, $S(0, 2)$ and join RS to get the graph of $y = 2-x$.

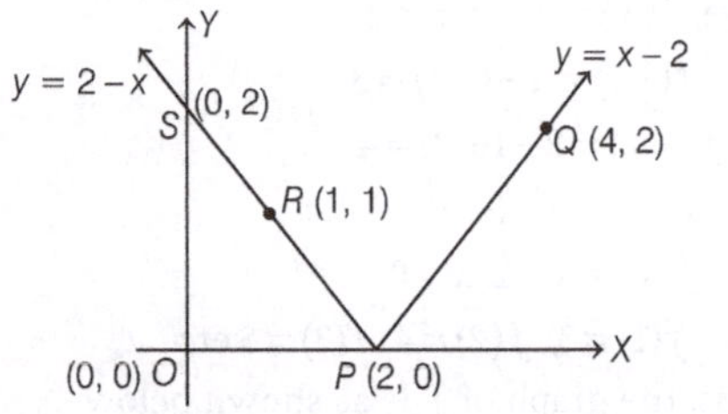

Example 6. **Let R be the set of real numbers. Define the function $f: R \to R$ be $f(x) = [x]$. Sketch the graph of this function.** *Competency Based Que*

Sol. Given, $f(x) = [x]$, $x \in R$

$[x]$ represents greatest integer $\le x$.

$$\Rightarrow \quad f(x) = \begin{cases} \vdots & \vdots \\ -2, & -2 \le x < 1 \\ -1, & -1 \le x < 0 \\ 0, & 0 \le x < 1 \\ 1, & 1 \le x < 2 \\ 2, & 2 \le x < 3 \\ \vdots & \vdots \end{cases}$$

The graph is shown below.

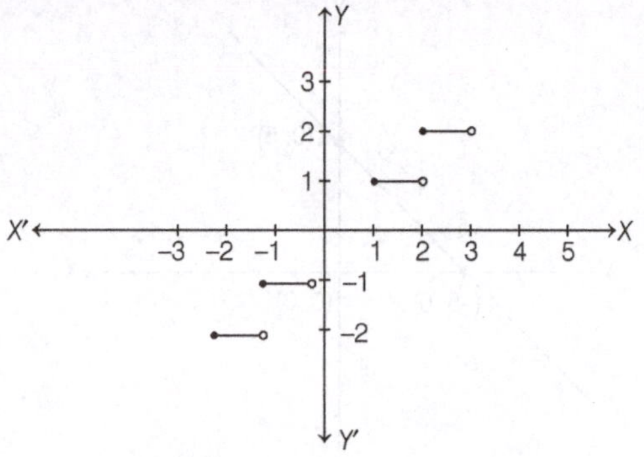

Example 7. **If two functions are defined as $f(x) = \frac{1}{(x-2)}$, $x \ne 2$ and $g(x) = (x-2)^2$, then find**

(i) $f+g$ **(ii) $f-g$**

(iii) fg **(iv) $\frac{f}{g}$**

Sol. Given functions are

$$f(x) = \frac{1}{x-2}, \; x \ne 2$$

and $$g(x) = (x-2)^2$$

(i) $(f+g)(x) = f(x) + g(x)$

$$= \frac{1}{x-2} + (x-2)^2, \; x \ne 2$$

$$= \frac{1+(x-2)^3}{x-2}, \; x \ne 2$$

(ii) $(f-g)(x) = f(x) - g(x)$

$$= \frac{1}{(x-2)} - (x-2)^2, \; x \ne 2$$

$$= \frac{1-(x-2)^3}{(x-2)}, \; x \ne 2$$

(iii) $(fg)(x) = f(x) \cdot g(x)$

$$= \frac{1}{(x-2)} \times (x-2)^2, \; x \ne 2$$

$$= (x-2), \; x \ne 2$$

(iv) $\left(\frac{f}{g}\right)(x) = \frac{f(x)}{g(x)}$

$$= \frac{\left(\frac{1}{x-2}\right)}{(x-2)^2}, \; x \ne 2$$

$$= \frac{1}{(x-2)^3}, \; x \ne 2$$

Example 8. **Find the sum and difference of the identity function and the modulus function.**

Sol. We know that

$f : R \to R$ defined by $f(x) = x$ is the identity function.

$g : R \to R$ defined by $g(x) = |x|$ is the modulus function.

$\therefore$ Sum of functions, $(f + g) : R \to R$ defined by

$$(f+g)(x) = f(x) + g(x) = x + |x|$$
$$= \begin{cases} x + x, & \text{if } x \ge 0 \\ x - x, & \text{if } x < 0 \end{cases} = \begin{cases} 2x, & \text{if } x \ge 0 \\ 0, & \text{if } x < 0 \end{cases}$$

Difference of functions, $(f - g) : R \to R$ defined by

$$(f-g)(x) = f(x) - g(x) = x - |x|$$
$$= \begin{cases} x - x, & \text{if } x \ge 0 \\ x - (-x), & \text{if } x < 0 \end{cases} = \begin{cases} 0, & \text{if } x \ge 0 \\ 2x, & \text{if } x < 0 \end{cases}$$

Example 9. **Find the quotient of the identity function by the modulus function.**

Sol. Let f and g denote, the identity function and the modulus function respectively. Then,

$f : R \to R$ is defined as $f(x) = x$

and $g : R \to R$ is defined as $g(x) = |x|$.

Clearly, f and g have the same domain.

Also, $g(x) = 0 \Rightarrow |x| = 0 \Rightarrow x = 0$.

Therefore, the quotient of f by g i.e. $\frac{f}{g}$ is a function from $R - \{0\} \to R$ and it is defined as

$$\left(\frac{f}{g}\right)(x) = \frac{f(x)}{g(x)} = \frac{x}{|x|} = \begin{cases} \frac{x}{x} = 1, & x > 0 \\ \frac{x}{-x} = -1, & x < 0 \end{cases}$$

Quick Test 4

Directions (Q. Nos. 1-4) *Draw the graph of the following functions.*

1. $f(x) = \frac{1}{x^2}$

2. $f(x) = x^3$

3. $f(x) = |x - 3|$ *Competency Based Que*

4. $f(x) = -2$

5. Let $f : R \to R$ defined by $f(x) = 1 - x^2$ for all $x \in R^+$. Find its domain and range. Also, draw its graph. *Competency Based Que*

6. Draw the graph of $f(x) = \frac{1}{2}x^3$. Also, find its domain and range.

7. Let f and g be real functions defined by $f(x) = \sqrt{x+2}$ and $g(x) = \sqrt{4 - x^2}$. Then, find each of the following functions. *Competency Based Que*

(i) $f + g$ (ii) $f - g$

(iii) $f \cdot g$ (iv) $\frac{f}{g}$

8. If $f(x) = x^3 - \frac{1}{x^3}$, then find $f(x) + f\left(\frac{1}{x}\right)$.

9. If $f(x) = y = \frac{ax - b}{cx - a}$, then prove that $f(y) = x$. *NCERT Exemplar*

10. Let f and g be two real functions given by

$f = \{(0, 1), (2, 0), (3, -4), (4, 2), (5, 1)\}$

and $g = \{(1, 0), (2, 2), (3, -1), (4, 4), (5, 3)\}$.

Then, find the domain of $f \cdot g$. *NCERT Exemplar, Competency Based Que*

Answers

1.

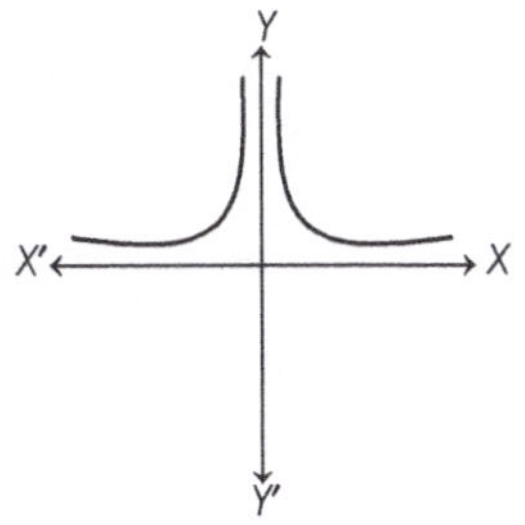

2.

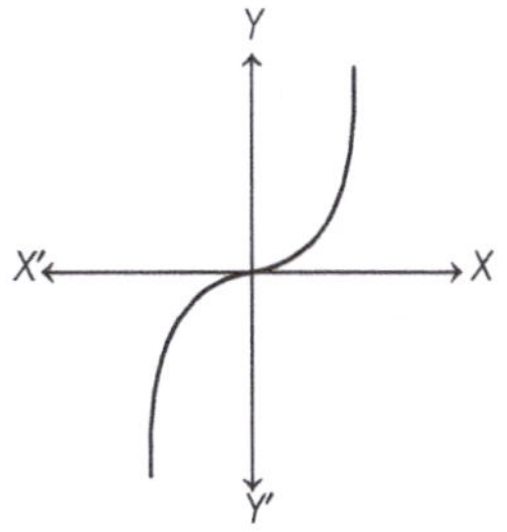

3.

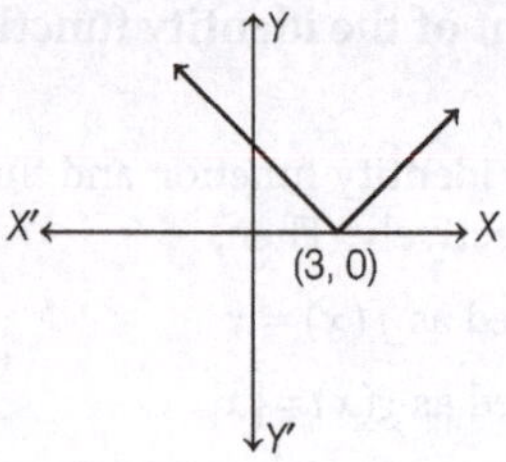

4.

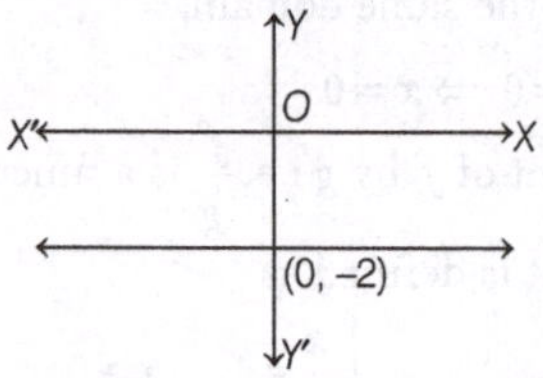

5. Domain of $f = (0, \infty)$ and Range of $f = (-\infty, 1)$

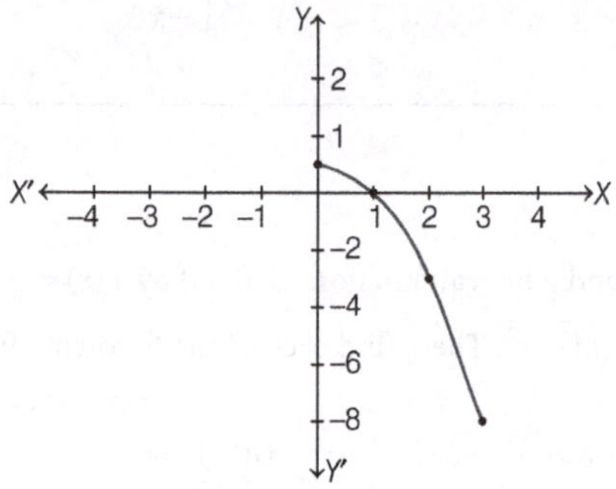

For Solutions, See QR Code on Pg. 48

6.

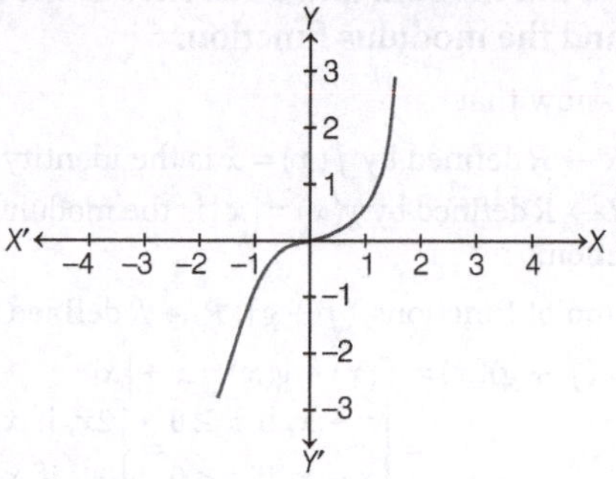

Domain of $f = (-\infty, \infty)$;
Range of $f = (-\infty, \infty)$

7. (i) $(\sqrt{x+2})(1+\sqrt{2-x})$
(ii) $\sqrt{x+2}\ (1-\sqrt{2-x})$
(iii) $(x+2)\sqrt{2-x}$
(iv) $\dfrac{1}{\sqrt{2-x}}$

8. 0

10. Domain of $(f \cdot g) = \{2, 3, 4, 5\}$

Exam Pattern Questions

Objective Type Questions

Multiple Choice Questions

1. If two sets A and B are having 99 elements in common, then the number of elements common to each of the sets $A \times B$ and $B \times A$ are

(a) 2^{99} (b) 99^2 (c) 100 (d) 18

2. If $A = \{1, 2, 5, 6\}$ and $B = \{1, 2, 3\}$, then $(A \times B) \cap (B \times A)$ is equal to

(a) $\{(1,1),(2,1),(6,1),(3,2)\}$
(b) $\{(1,1),(1,2),(2,1),(2,2)\}$
(c) $\{(1,1),(2,2)\}$
(d) $\{(1,1),(1,2),(2,5),(2,6)\}$

3. If $A = \{7, 8, 9\}$ and $B = \{9, 5\}$, then $(A \cup B) \times (A \cap B)$ is

(a) $\{(7, 9), (7, 5), (8, 9), (8, 5), (9, 9), (9, 5)\}$
(b) $\{(5, 9), (7, 9), (8, 9), (9, 9)\}$
(c) $\{(9, 5), (9, 7), (9, 8), (9, 9)\}$
(d) None of the above

4. Let $R = \{x \mid x \in N, x$ is a multiple of 3 and $x \le 100\}$
$S = \{x \mid x \in N, x$ is a multiple of 5 and $x \le 100\}$.
What is the number of elements in $(R \times S) \cap (S \times R)$?

Competency Based Que

(a) 36 (b) 33 (c) 20 (d) 6

5. If A is a finite set having n elements, then the number of relations which can be defined in A is

(a) 2^n (b) n^2
(c) 2^{n^2} (d) n^n

6. Let R be a relation in N defined by

$$R = \{(1 + x, 1 + x^2) : x \le 5, x \in N\}.$$

Which of the following is false? Competency Based Que

(a) $R = \{(2,2),(3,5),(4,10),(5,17),(6,25)\}$
(b) Domain of $R = \{2,3,4,5,6\}$
(c) Range of $R = \{2,5,10,17,26\}$
(d) None of the above

7. The range of the function $f(x) = \dfrac{1}{\sqrt{9 - x^2}}$ is

(a) $\left(\dfrac{1}{3}, \infty\right)$ (b) $\left[\dfrac{-1}{3}, \dfrac{1}{3}\right]$
(c) $\left[\dfrac{1}{3}, \infty\right)$ (d) $(-\infty, \infty)$

8. Let $f : R \to R$ be such that $f(x) = 2^x$. Then, consider the following statements. Competency Based Que

I. Range of $f = (0, \infty)$ II. $\{x : f(x) = 1\} = \{0\}$
III. $f(x + y) = f(x) \cdot f(y)$

Which of the above statement(s) is/are correct?

(a) Only I (b) I and II
(c) II and III (d) All of these

9. The domain of the function $f(x) = \sqrt{a^2 - x^2}$ $(a > 0)$ is

NCERT Exemplar

(a) $(-a, a)$ (b) $[-a, a]$ (c) $[0, a]$ (d) $(-a, 0]$

10. The domain of the real function

$f(x) = \dfrac{x}{x^2 + 3x + 2}$ is

(a) R (b) $R - \{-2, -1\}$
(c) $\{-2, -1\}$ (d) $R - \{2\}$

11. Range of $f(x) = \dfrac{1}{1 - 2\cos x}$ is

NCERT Exemplar

(a) $\left[\dfrac{1}{3}, 1\right]$ (b) $\left[-1, \dfrac{1}{3}\right]$
(c) $[-\infty, -1] \cup \left[\dfrac{1}{3}, \infty\right)$ (d) $\left[\dfrac{-1}{3}, 1\right]$

12. The range of the function $f(x) = 9^x - 3^x + 1$ is

(a) $(-\infty, \infty)$ (b) $(-\infty, 0)$ (c) $(0, \infty)$ (d) $\left[\dfrac{3}{4}, \infty\right)$

13. The domain and range of the real function f defined by $f(x) = |x - 1|$ is

(a) $R, [0, \infty)$ (b) $R, (-\infty, 0)$
(c) R, R (d) $(-\infty, 0), R$

14. Let $A = \{x \in R \mid x \ne 0, -4 \le x \le 4\}$ and $f : A \to R$ defined as $f(x) = \dfrac{|x|}{x}$ for all $x \in A$. Then, $R(f)$ is

Competency Based Que

(a) $\{-1, 1\}$ (b) $\{x \mid 0 \le x \le 4\}$
(c) $\{1\}$ (d) $\{x \mid -4 \le x \le 0\}$

15. If $[x^2] - 5[x] + 6 = 0$, where $[\cdot]$ denotes the greatest integer function, then NCERT Exemplar

(a) $x \in [3, 4]$ (b) $x \in (2, 3]$
(c) $x \in [2, 3]$ (d) $x \in [2, 4)$

16. The graph of the function defined by

$$f(x)=\begin{cases}1-x\ , & x<0\\ 1\ , & x=0 \text{ is}\\ 1+x\ , & x>0\end{cases}$$

Competency Based Que

(a)

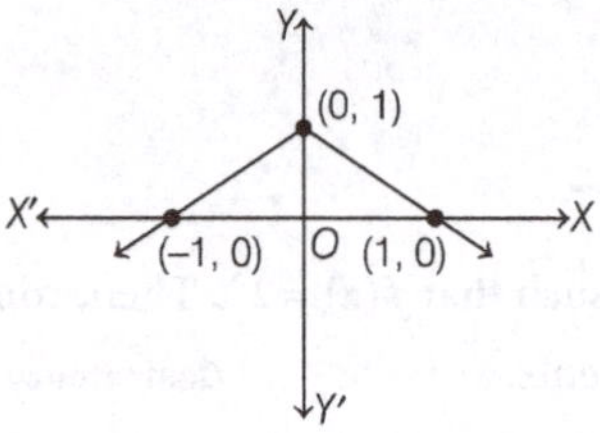

(b)

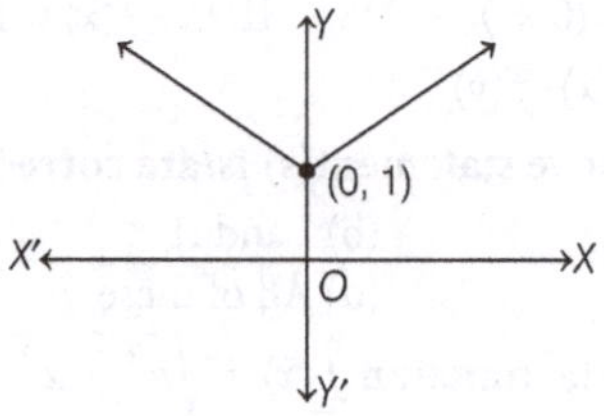

(c)

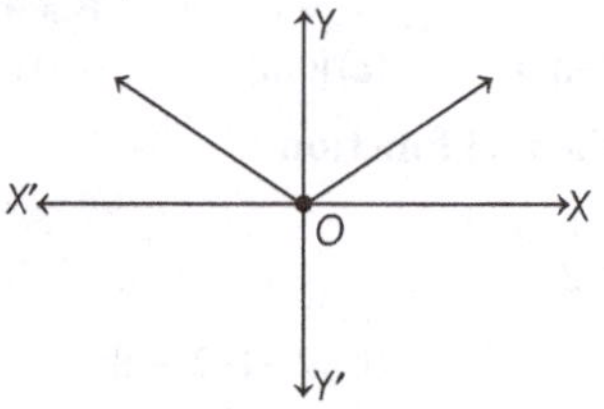

(d)

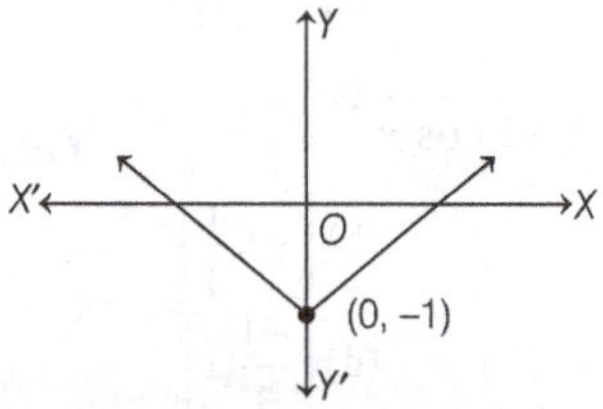

17. If $f(x)=\dfrac{2x}{1-x^2}$, then $f(\tan\theta)$ is

(a) $\cos 2\theta$ (b) $\sin 2\theta$
(c) $\tan 2\theta$ (d) $\cot 2\theta$

18. If $f(x)=\dfrac{1+x}{1-x}$, then $\dfrac{f(x)\cdot f(x^2)}{1+[f(x)]^2}$ is equal to

Competency Based Que

(a) $\dfrac{1}{4}$ (b) $\dfrac{1}{6}$ (c) $\dfrac{1}{8}$ (d) $\dfrac{1}{2}$

19. Let $f(x)=1+2x$ and $g(x)=x^2$ be two real functions. Then, $f+g$ is equal to

(a) $(1+x)$ (b) $(1-x)$
(c) $(1+x)^2$ (d) $(1-x)^2$

20. If $f(x)=\dfrac{4^x}{4^x+2}$, then the value of $f(x)+f(1-x)$ is equal to

(a) 1 (b) 2
(c) 3 (d) 4

Assertion-Reason Based Questions

Directions (Q. Nos. 21-30) *In the questions given below are two statements labelled as Assertion (A) and Reason (R). In the context of the two statements, which one of the following is correct?*

(a) Both A and R are correct; R is the correct explanation of A.
(b) Both A and R are correct; R is not the correct explanation of A.
(c) A is correct; R is incorrect.
(d) R is correct; A is incorrect.

21. **Assertion** (A) If $(a+2, b-8)=(7,3)$, then $a=5$ and $b=11$.

Reason (R) Two ordered pairs are equal, if their corresponding elements are equal.

22. **Assertion** (A) The ordered pair with entries 5 and 4 is (5, 4).

Reason (R) For ordered pair (a, b), a represents first entry and b represents second entry.

23. **Assertion** (A) If $n(A)=4$ and $B=\{3,4,5,6,8,9\}$, then the number of relations from A to B is 2^{20}.

Reason (R) If A and B are any two non-empty finite sets containing m and n elements respectively, then total number of relations from set A to B is 2^{mn}.

Competency Based Que

24. **Assertion** (A) If $A=\{1,2\}$ and $B=\{a,b\}$, then $A\times B=\{(1,a),(1,b),(2,a)\ (2,b)\}$.

Reason (R) The cartesian product of two non-empty sets X and Y is denoted by $X\times Y$ and $X\times Y=\{(x,y): x\in X, y\in Y\}$.

25. **Assertion** (A) In the given arrow diagram range of relation from set A to set B is set B.

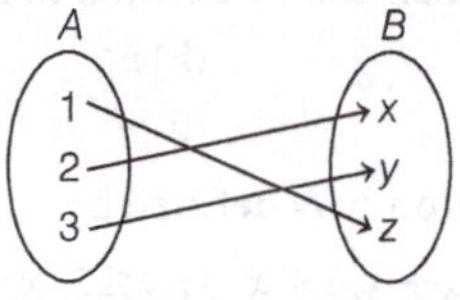

Reason (R) Second entries of ordered pair in given relation is range.

Competency Based Que

26. **Assertion** (A) If $f: R-\{4\}\rightarrow R$ be defined by $f(x)=\dfrac{x^2-16}{x-4}$ and $g: R\rightarrow R$ be defined by $g(x)=x+4$, then $f=g$.

Reason (R) Two functions f and g are said to be equal, iff

(i) Domain of f = Domain of g
(ii) Codomain of f = Codomain of g
(iii) $f(x)=g(x)$ for every x belonging to their common domain.

27. **Assertion** (A) The range of the real function $f(x) = |x - 1|$ is the set of all non-negative real numbers.

Reason (R) The function $f: R \to R$ defined by $f(x) = |x|$ for each $x \in R$ is called modulus function.

28. **Assertion** (A) Domain of function $f(x) = \dfrac{1}{x-3}, x \neq 3$ in set of real numbers is set of real numbers.

Reason (R) For the function $f(x) = \dfrac{P(x)}{Q(x)}, Q(x) \neq 0$, $P(x)$ and $Q(x)$ being polynomial in set of real numbers range is $R - \{\text{zeroes of } Q(x)\}$. **Competency Based Que**

29. **Assertion** (A) Let $A = \{1, 2, 3\}$ and $f: A \to R$ defined by $f(x) = x + 2$, then the range of f is $\{3, 4, 5\}$.

Reason (R) The elements which are present in domain is known as range.

30. **Assertion** (A) If f and g be two real functions defined by $f(x) = \dfrac{1}{x+5}$ and $g(x) = (x+5)^3$, then

$$(f+g) = \frac{(x+5)^4 + 1}{x+5}.$$

Reason (R) The sum of function $(f + g)$ is defined by $(f+g)x = f(x) + g(x), \forall\, x \in D_1 \cap D_2$.

Competency Based Que

Case Study Based Questions

31. Read the following passage and answer the question given below.

A company conducted interview under 2 different categories that is male and female. Totally there were 100 participants. Among all of them finally three from category 1 and two from category 2 were selected for final race. Rahul forms two sets M and F with these participants for his college project.

Let $M = \{m_1, m_2, m_3\}$ and $F = \{f_1, f_2\}$, where M represents the set of males selected and F the set of females who were selected for the final. Rahul decides to explore these sets for various types of relations and functions.

Competency Based Que

(i) Rahul wants to form all the relations possible from M to F. How many such relations are possible?

(a) 2^6 (b) 2^5
(c) 0 (d) 2^3

(ii) The elements of set M represent

(a) domain (b) range
(c) codomain (d) None of these

(iii) If $M \times M \times M = \{m_1, m_2, m_3 \mid m_1, m_2, m_3 \in M\}$, then m_1, m_2, m_3 is called

(a) range
(b) constant function
(c) ordered triplet
(d) rational function

(iv) If five elements of $A \times B$ are $(m_1, f_1), (m_1, f_2) (m_2, f_1), (m_3, f_1), (m_3, f_2)$, then the remaining elements of $A \times B$

(a) (m_2, f_1) (b) (m_2, f_2)
(c) (m_3, f_1) (d) (m_3, f_2)

32. Read the following passage and answer the questions given below.

A teacher asked randomly his two students to draw graph (figure) on the black board. Two students Raman and Prashant draw figure (graphs) (a) and (b) one by one on the board as shown below. **Competency Based Que**

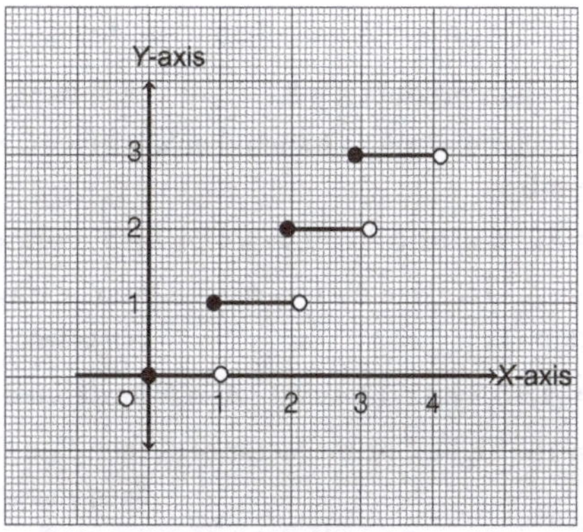

Figure (a)

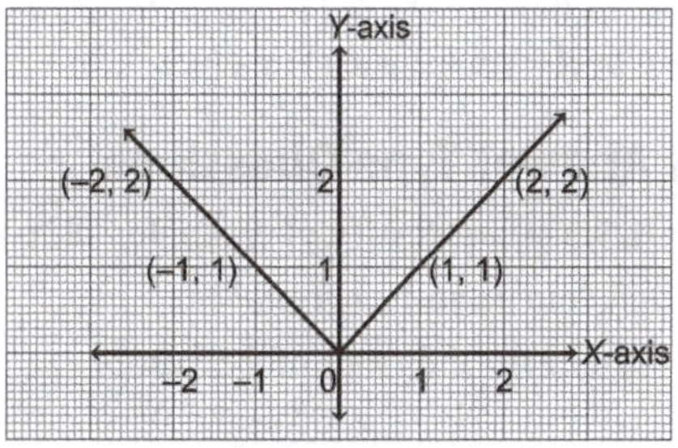

Figure (b)

(i) For the figure (a), write the function in terms of x.

(ii) (a) For the figure (b), write the function in terms of x.

or (b) Write the range of the function $f(x)$ of the figure (a).

(iii) The function $f: R \to R$ defined by $f(x) = |x|$, then write the value of $f(-3)$.

Subjective Questions

Very Short Answer Type Questions

1. If $\left(\frac{x}{3}+1, y-\frac{2}{3}\right)=\left(\frac{5}{3},\frac{1}{3}\right)$, then find the values of x and y. **NCERT**

2. If $n(A)=m$ and $n(B)=n$, then find the total number of non-empty relations that can be defined from A to B. **NCERT Exemplar**

3. If $A\times B=\{(a, 1),(a, 5),(a, 2),(b, 2),(b, 5),(b, 1)\}$, then find A, B and $B\times A$.

4. If $A=\{a, b\}$, $B=\{2, 3, 5, 6, 7\}$ and $C=\{5, 6, 7, 8, 9\}$, then find $A\times(B\cap C)$.

5. If $A=\{-1, 1\}$, then find $A\times A\times A$. **Competency Based Que**

6. If $A=\{1, 2\}$ and $B=\{3, 4, 5\}$, then write down

(i) cartesian product $A\times B$. **Competency Based Que**

(ii) the relation $R: A\rightarrow B$ such that $(a+b)$ is an even number.

7. If $A=\{0, 1, 2\}$ and $B=\{1, 2, 3, 4, 5\}$, then represent the rule $f: A\rightarrow B$ given by $f(x)=x+2$ by an arrow diagram. **Competency Based Que**

8. If $A=\{9, 10, 11, 12, 13\}$ and $f: A\rightarrow N$ be defined by $f(n)=$ the highest prime factor of n, then find the range of f. **Competency Based Que**

9. If $f(x)=[x]$, then find $(3f)(x)$.

10. Determine the domain and range of the relation R defined by $R=\{(x+2, x+4): x\in(0, 1, 2, 3)\}$.

Short Answer Type Questions

11. If $U=\{1, 2, 3, 4\}$ and $R=\{(x, y): y>x$ for all, $x, y\in U\}$, then find domain and range of R.

12. Find the domain and range of the relation
$$R=\{(x, y): x+y=8\,;\, x, y\in N\}$$

13. If $N=\{1, 2, 3\}$, then find the relation $R=\{(x, y): x\in N, y\in N$ and $2x+y=10\}$ in $N\times N$. Also, find its domain and range.

14. Which of the following are functions, if $X=\{a, b, c, d\}$ and $Y=\{1, 2, 3, 4, 5\}$?

(i) $f_1=\{(a,1),(b,1),(c,3),(d,4)\}$

(ii) $f_2=\{(a, 1),(b, 2),(c, 4),(a,2),(d,5)\}$

15. If $f(x)=\dfrac{x^2-x+1}{x^2+x+1}$, then find $f(1+h)$. **Competency Based Que**

16. If $f(x)=\dfrac{x+1}{x-1}$, then find $f(x^2)$ and $[f(x)]^2$.

17. Let A and B be two sets, such that $A\times B$ consists of 6 elements. If 3 elements of $A\times B$ are (1, 4), (2, 6) and (3, 6), then find $A\times B$ and $B\times A$. **Competency Based Que**

18. .Find the range of the following.

(i) $f(x)=\dfrac{|x-1|}{x-1}$

(ii) $f(x)=2-3x, x\in R, x>0$

(iii) $f(x)=x^2+2, x\in R$

19. If a function $f: R\rightarrow R$ defined by
$$f(x)=\begin{cases}3x-2, & x<0\\ 1, & x=0,\\ 4x+1, & x>0\end{cases}$$
then find $f(1)$, $f(-1)$, $f(0)$ and $f(2)$.

20. If $f(x)=\begin{cases}x^2, & x<0\\ x, & 0\le x<1\\ \dfrac{1}{x}, & x\ge 1\end{cases}$

Find (i) $f\left(\frac{1}{2}\right)$ (ii) $f(-2)$ (iii) $f(1)$ (iv) $f(\sqrt{3})$

21. Determine the domain and range of the relation R, where $R=\{(2x+3, x^3): x$ is a prime number less than 10$\}$.

22. If $X=\{1, 2, 3, 4, 5\}$, $Y=\{1, 2, 5, 6, 7, 9, 10, 11, 12, 13, 14\}$ and $f: X\rightarrow Y$ defined by $f(x)=2x+3$, then find the domain and range of f. **Competency Based Que**

23. Is $g=\{(2, 3), (4, 5), (6, 7)\}$ a function? If this is described by the formula $g(x)=\alpha x+2\beta$, then what values would be assigned to α and β? **Competency Based Que**

24. Let A and B be two sets, such that $n(A)=3$ and $n(B)=2$. If $(x, 1)$, $(y,2)$ and $(z,1)$ are in $A\times B$, then find A, B and $A\times B$, where x, y and z are distinct elements. **NCERT**

25. Find the domain of the function f defined by
$$f(x)=\sqrt{4-x}+\frac{1}{\sqrt{x^2-1}}.$$
NCERT Exemplar

26. Let $A=\{1, 2, 3, \ldots, 14\}$. Define a relation R from A to A by $R=\{(x, y): 3x-y=0$, where $x, y\in A\}$. Depict this relationship using an arrow diagram. Write down its domain, codomain and range. **Competency Based Que**

Long Answer Type Questions

27. Find the domain and range of the following relations.

(i) $R = \{(x,y): x, y \in N, y = x^2 + 3 \text{ and } 0 < x < 5\}$

(ii) $R = \{(x,y): x, y \in N, y = \frac{1}{1+x}$ and x is an odd natural number$\}$

(iii) $R_1 = \left\{\left(x, \frac{1}{x}\right): 0 < x < 6, x \in N\right\}$

(iv) $R_2 = \{(x, x^2): x$ is prime number less than $10\}$

Competency Based Que

28. Let $A = \{-3, -2, -1, 4\}$ and $f: A \to Z$ given by $f(x) = x^2 + x + 2$. Find **Competency Based Que**

(i) the range of f. (ii) pre-images of 6 and 4.

29. Find the domain and range of the following real functions. **NCERT, Competency Based Que**

(i) $f(x) = -|x|$ (ii) $f(x) = \sqrt{9 - x^2}$

30. Find the range of the function $f(x) = 1 - |x + 2|$.

31. Find the range of the function $f(x) = \dfrac{|x+4|}{x+4}$.

32. (i) Let $A = \{8, 11, 12, 15, 18, 23\}$ and f is a function from $A \to N$ such that $f(x)$ = highest prime factor of x, find f and its range. **Competency Based Que**

(ii) Find the domain of the function

$$f(x) = \frac{x^2 + 2x + 1}{x^2 - 8x + 12}.$$

33. (i) If $f(x) = \sqrt{x^2 + 1}$ and $g(x) = 2x^2 + 3$, then find $(f + g)(x)$. **Competency Based Que**

(ii) Find the product of the identity function and the modulus function.

34. The relation f is defined by $f(x) = \begin{cases} x^2, & 0 \le x \le 3 \\ 3x, & 3 \le x \le 10 \end{cases}$ and the relation g is defined by $g(x) = \begin{cases} x^2, & 0 \le x \le 2 \\ 3x, & 2 \le x \le 10 \end{cases}$.

Show that f is a function and g is not a function.

Competency Based Que

35. If $f(x) = \begin{cases} 1 + x, & -1 \le x < 0 \\ x^2 - 1, & 0 < x < 2 \\ 2x, & 2 \le x \end{cases}$

Then, find $f(3)$, $f(-2)$, $f(0)$, $f\left(\frac{1}{2}\right)$, $f(2 - h)$ and $f(-1 + h)$, where $h > 0$ is very small. **Competency Based Que**

36. Write the following after removing the modulus $f(x) = |2x - 1|, -1 \le x \le 1$.

37. Find domain and range of the real function $f(x)$ defined by $f(x) = \begin{cases} 1 - x, & x < 0 \\ 1, & x = 0 \\ x - 1, & x > 0 \end{cases}$ and draw its graph.

Competency Based Que

Hints & Solutions

Objective Type Questions

1. (b) $n(A \cap B) = 99$

Now, $n[(A \times B) \cap (B \times A)] = n[(A \cap B) \times (B \cap A)]$

$= n(A \cap B) \times n(B \cap A)$

$= 99 \times 99 = 99^2$

2. (b) $A \times B = \{(1, 1), (1, 2), (1, 3), (2, 1), (2, 2), (2, 3), (5, 1), (5, 2), (5, 3), (6, 1), (6, 2), (6, 3)\}$

and $B \times A = \{(1, 1), (1, 2), (1, 5), (1, 6), (2, 1), (2, 2), (2, 5), (2, 6), (3, 1), (3, 2), (3, 5), (3, 6)\}$

$\therefore (A \times B) \cap (B \times A) = \{(1, 1), (1, 2), (2, 1), (2, 2)\}$

3. (b) We have, $A = \{7, 8, 9\}$ and $B = \{9, 5\}$

$\therefore A \cup B = \{5, 7, 8, 9\}$ and $A \cap B = \{9\}$

$\therefore (A \cup B) \times (A \cap B) = \{(5, 9), (7, 9), (8, 9), (9, 9)\}$

4. (a) $R = \{3, 6, 9, 12, \ldots, 99\}$

$S = \{5, 10, 15, 20, \ldots, 100\}$

$\because (R \times S) \cap (S \times R) = (R \cap S) \times (S \cap R)$ ...(i)

Now, $(R \cap S) = \{15, 30, 45, 60, 75, 90\}$ = 6 elements

and $(S \cap R) = \{15, 30, 45, 60, 75, 90\}$ = 6 elements

From Eq. (i), we get

$(R \times S) \cap (S \times R) = 6 \times 6 = 36$ elements.

Thus, total number of elements is 36.

5. (a) If A is finite set having n elements, then the number of relations in A is 2^n.

6. (a) $R = \{(1 + x, 1 + x^2): x \le 5, x \in N\}$

Let $R = \{(2, 2), (3, 5), (4, 10), (5, 17), (6, 26)\}$

Then, domain = $\{2, 3, 4, 5, 6\}$ and range = $\{2, 5, 10, 17, 26\}$

7. (c) **Hint** Let $y = \frac{1}{\sqrt{9-x^2}} \Rightarrow y^2(9-x^2) = 1$

$\Rightarrow \quad 9y^2 - x^2y^2 = 1 \Rightarrow x^2y^2 = 9y^2 - 1$

$\Rightarrow \quad x = \frac{\sqrt{9y^2-1}}{y}$

$\Rightarrow \quad y \neq 0$ and $9y^2 - 1 \geq 0$

$\Rightarrow \quad 9y^2 \geq 1 \Rightarrow y^2 \geq \frac{1}{9}$

$\Rightarrow \quad y \leq -\frac{1}{3}$ or $y \geq \frac{1}{3}$

$[x^2 \geq a^2 \Rightarrow x \leq -a, x \geq a]$

As y or $f(x) \geq 0$, $y \geq \frac{1}{3}$

$\therefore$ Range of the function $\left[\frac{1}{3}, \infty\right)$.

8. (d) Since, the range of exponential function is $(0, \infty)$.
So, Statement I is true.

For $f(x) = 1 \Rightarrow 2^x = 1 = 2^0 \Rightarrow x = 0$

So, Statement II is true.

$f(x+y) = 2^{x+y} = 2^x \cdot 2^y = f(x) \cdot f(y)$

$\therefore f(x+y) = f(x) \cdot f(y)$ holds for all $x, y \in R$

9. (b) Let $y = \sqrt{a^2 - x^2}$ the function y is defined, if

$a^2 - x^2 \geq 0 \Rightarrow x^2 - a^2 \leq 0$ or $x^2 \leq a^2$

$\therefore -a \leq x \leq a$

So, domain of $y = [-a, a]$

10. (b) **Hint** For domain of f, $x^2 + 3x + 2 \neq 0$

$\Rightarrow \quad (x+1)(x+2) \neq 0 \Rightarrow x \neq -1, -2$

$\therefore \quad$ Domain $= R - \{-2, -1\}$

11. (b) **Hint** $-1 \leq \cos x \leq 1$

$\Rightarrow \quad 2 \geq -2 \cos x \geq -2$

$\Rightarrow \quad 1+2 \geq 1-2 \cos x \geq 1-2$

$\Rightarrow \quad 3 \geq 1-2 \cos x \geq -1$

$\Rightarrow \quad -1 \leq \frac{1}{1-2 \cos x} \leq \frac{1}{3} \Rightarrow -1 \leq f(x) \leq \frac{1}{3}$

$\therefore \quad R(f) = \left[-1, \frac{1}{3}\right]$

12. (d) **Hint** Let $f(x) = y = 3^{2x} - 3^x + 1 = (3^x)^2 - 3^x + 1$

$= \left(3^x - \frac{1}{2}\right)^2 + 1 - \frac{1}{4} = \left(3^x - \frac{1}{2}\right)^2 + \frac{3}{4}$

Now, $\left(3^x - \frac{1}{2}\right)^2 \geq 0$

$\therefore \left(3^x - \frac{1}{2}\right)^2 + \frac{3}{4} \geq \frac{3}{4}$

$\therefore \quad f(x) \geq \frac{3}{4}$

$\therefore$ Range of $f(x)$ is $\left[\frac{3}{4}, \infty\right)$.

13. (a) We have, $f(x) = |x-1|$
Here, $f(x)$ is a modulus function and since modulus of a real number is uniquely defined for all real positive numbers.

$\therefore$ The domain of $f(x)$ is R.

We see that $f(x) = |x-1|$

$f(x) = \begin{cases} x-1, & \text{if } x \geq 1 \\ -(x-1), & \text{if } x < 1 \end{cases} \Rightarrow f(x) = \begin{cases} x-1, & \text{if } x \geq 1 \\ 1-x, & \text{if } x < 1 \end{cases}$

From above, we observe that in both cases $f(x) \geq 0$.
Hence, range of $f(x)$ is $[0, \infty)$.

14. (a) **Hint** $f(x) = \frac{|x|}{x} = \begin{cases} 1, & \text{if } x \geq 0 \\ -1, & \text{if } x < 0 \end{cases}$

$\therefore \quad$ Range $(f) = \{-1, 1\}$

15. (d) Let $[x] = y$, then $[x]^2 - 5[x] + 6 = 0 \Rightarrow y^2 - 5y + 6 = 0$

$\Rightarrow \quad (y-2)(y-3) = 0 \Rightarrow y = 2$ or 3

$\Rightarrow x \in [2, 3)$ or $x \in [3, 4)$

Hence, $x \in [2, 4)$

16. (b) Here, $f(x) = 1 - x$, $x < 0$

$f(-1) = 1-(-1) = 2$, $f(-2) = 1-(-2) = 3$

$f(-3) = 1-(-3) = 4$ etc.

and $\quad f(x) = x+1, x > 0$,

$f(1) = 2$, $f(2) = 3$, $f(3) = 4$

and when $x = 0$, $f(x) = 1$

graph (b) is the correct answer.

17. (c) $f(\tan\theta) = \frac{2\tan\theta}{1-\tan^2\theta} = \tan 2\theta$

18. (d) Since, $f(x) = \frac{1+x}{1-x}$ and $f(x^2) = \frac{1+x^2}{1-x^2}$.

Then, $\frac{f(x) \cdot f(x^2)}{1+[f(x)]^2} = \frac{\left(\frac{1+x}{1-x}\right) \cdot \left(\frac{1+x^2}{1-x^2}\right)}{1+\left(\frac{1+x}{1-x}\right)^2}$

$= \frac{\frac{(1+x)(1+x^2)}{(1-x)^2(1+x)}}{\frac{1-2x+x^2+1+2x+x^2}{(1-x)^2}}$

$= \frac{(1+x)(1+x^2)(1-x)^2}{(1-x)^2(1+x) \cdot 2(1+x^2)} = \frac{1}{2}$

19. (c) We have, $f(x) = 1+2x$ and $g(x) = x^2$

$\therefore \quad f + g = f(x) + g(x) = 1 + 2x + x^2 = (1+x)^2$

20. (a) Hint We have, $f(x)=\dfrac{4^x}{4^x+2}$

$$\therefore f(1-x)+f(x)=\frac{4^{1-x}}{4^{1-x}+2}+\frac{4^x}{4^x+2}$$
$$=\frac{4}{4+2\cdot 4^x}+\frac{4^x}{4^x+2}$$
$$=\frac{2}{2+4^x}+\frac{4^x}{4^x+2}=\frac{2+4^x}{2+4^x}=1$$

21. (a) **Hint** Since, $(a+2, b-8)=(7,3)$

$\Rightarrow \quad a+2=7,\ b-8=3 \Rightarrow a=5,\ b=11$

22. (a) Both Assertion and Reason are correct; Reason is the correct explanation of Assertion.

23. (d) **Hint** $n(A)=4$ and $n(B)=6$

$n(A\times B)=4\times 6=24$

Total number of relations from set A to set B is 2^{24}.

$\therefore$ Assertion is incorrect but Reason is correct.

24. (a) Assertion and Reason both are correct and Reason is the correct explanation of Assertion.

25. (a) Set $B=\{x, y, z\}$ is the range of the Set $A=\{1, 2, 3\}$

$\therefore$Assertion is correct and Reason is the correct explanation of Assertion.

26. (a) Since, $f(x)=\dfrac{x^2-16}{x-4}=\dfrac{(x-4)(x+4)}{(x-4)}=(x+4)$

and $g(x)=x+4$ have the same domain, then $f=g$.

$\therefore$ Reason is correct for the given Assertion.

27. (a) Both Assertion and Reason are correct; Reason is the correct explanation of Assertion.

28. (a) For Assertion, $f(x)=\dfrac{1}{x-3}=\dfrac{P(x)}{Q(x)}$

Since, $\quad Q(x)=x-3\neq 0 \Rightarrow x=3$

Domain is $R-[3]$.

Assertion is correct and Reason is the correct explanation of Assertion.

29. (c) $A=\{1, 2, 3\}$

$f(1)=1+2=3,\ f(2)=2+2=4,\ f(3)=3+2=5$

$\Rightarrow$Range of $f(x)$ is $\{3, 4, 5\}$.

$\therefore$ Assertion is correct but Reason is incorrect.

30. (a) Given, $\quad f(x)=\dfrac{1}{x+5}$ and $g(x)=(x+5)^3$

$$\therefore \quad (f+g)(x)=\frac{1}{x+5}+(x+5)^3=\frac{1+(x+5)^4}{x+5}$$

Assertion is correct and Reason is the correct explanation of Assertion.

31. (i) (a) Since, $n(M)=3$ and $n(F)=2$

The number of relations is 2^6.

(ii) (a) The elements of the set M is called elements of domain.

(iii) (c) Ordered triplet.

(iv) (b) Since, there are 6 subsets of $A\times B$ and 5 are already given, then the remaining one is (m_2, f_2).

32. (i) From the given figure, we have

when $0\le x<1 \Rightarrow y=0$

when $1\le x<2 \Rightarrow y=1$

$2\le x<3 \Rightarrow y=2$

So, it is greatest integer function.

$\therefore f(x)=[x]\le x$, where x is an integer.

(ii) (a) From the given figure (b), we have

For $x=1,-1 \Rightarrow y=1$

$x=2,-2 \Rightarrow y=2$

$\therefore y=|x|$

or

(b) For the figure (a), we have

$f(x)=[x]\le x$, where x is an integer.

$\therefore$ Its range is the set of integers.

(iii) We have, $f(x)=|x| \Rightarrow f(-x)=|-3|=3$

Subjective Questions

1. **Hint** $\dfrac{x}{3}+1=\dfrac{5}{3}$ and $y-\dfrac{2}{3}=\dfrac{1}{3}$

$\Rightarrow \dfrac{x}{3}=\dfrac{2}{3}$ and $y=1$

Hence, $x=2$ and $y=1$

2. $2^{mn}-1$ (non empty relations)

3. $A=\{a, b\}, B=\{1, 5, 2\}$

$\therefore B\times A=\{(1, a), (1, b), (5, a), (5, b), (2, a), (2, b)\}$

4. **Hint** $B\cap C=\{5, 6, 7\}$, also given $A=\{a, b\}$

$\therefore A\times(B\cap C)=\{(a, 5), (a, 6), (a, 7), (b, 5), (b, 6), (b, 7)\}$

5. **Hint** First find $A\times A$. Then, find $A\times(A\times A)$.

Ans. $A\times A\times A=\{(-1,-1,-1), (-1,-1,1), (-1,1,-1), (-1,1,1), (1,-1,-1), (1,-1,1), (1,1,-1), (1,1,1)\}$

6. (i) $A\times B=\{(1,3),(1,4),(1,5),(2,3),(2,4),(2,5)\}$

(ii) $R: A\rightarrow B$ is $R=\{(1,3),(1,5),(2,4)\}$

7. **Hint** Relation $R=\{(0, 2), (1, 3), (2, 4)\}$

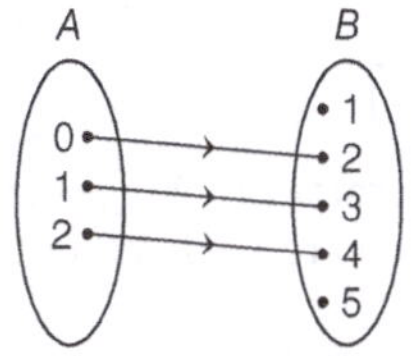

8. Hint We have,

$f(n)$ = the highest prime factor of n.

Now,

$$9 = 3 \times 3, 10 = 5 \times 2$$
$$11 = 11 \times 1, 12 = 3 \times 4$$
$$13 = 13 \times 1$$

$\therefore$ $f = \{(9, 3), (10, 5), (11, 11), (12, 3), (13, 13)\}$

Clearly, range $(f) = \{3, 5, 11, 13\}$

9. Hint We have, $f(x) = [x]$

Now, $(3f)(x) = 3f(x) = 3[x]$

10. $R = \{(2, 4), (3, 5), (4, 6), (5, 7)\}$

Domain $(R) = \{2, 3, 4, 5\}$

Range $(R) = \{4, 5, 6, 7\}$

11. $R = \{(1, 2), (2, 3), (3, 4)\}$

Domain $(R) = \{1, 2, 3, 4\}$, Range $(R) = \{2, 3, 4\}$

12. Hint $R = \{(1, 7), (2,6), (3, 5), (4, 4), (5, 3), (6, 2), (7, 1)\}$

Ans. Domain $(R) = \{1, 2, 3, 4, 5, 6, 7\}$

Range $(R) = \{7, 6, 5, 4, 3, 2, 1\}$

13. Hint $R = \{(1, 8), (2, 6), (3, 4)\}$

Ans. Domain $(R) = \{1, 2, 3\}$ and Range $(R) = \{8, 6, 4\}$

14. (i) f_1 is a function, since every element of X has one and only one image in set Y.

(ii) Here, $f(a) = 1$ and $f(a) = 2$

So, f_2 is not a function, since $a \in X$ and has two images 1 and 2 in set Y.

15. Hint Substitute $(1+h)$ in $f(x)$,

$$\therefore f(1+h) = \frac{(1+h)^2 - (1+h) + 1}{(1+h)^2 + (1+h) + 1}$$

Ans. $$\frac{h^2 + h + 1}{h^2 + 3h + 3}$$

16. Hint Substitute x^2 in $f(x)$ to find $f(x^2)$

Ans. $f(x^2) = \dfrac{x^2 + 1}{x^2 - 1}$ and $[f(x)]^2 = \left(\dfrac{x+1}{x-1}\right)^2$

17. Hint $A = \{1, 2, 3\}$ and $B = \{4, 6\}$

Ans. $A \times B = \{(1, 4), (1, 6), (2, 4), (2, 6), (3, 4), (3, 6)\}$

and $B \times A = \{(4, 1), (4, 2), (4, 3), (6, 1), (6, 2), (6, 3)\}$

18. (i) $f(x) = \dfrac{|x-1|}{x-1}$

The numerator $|x - 1|$ gives the absolute values of $x - 1$, which are $x - 1$. When $x > 1, -(x - 1) = 1 - x$ when $x < 1$

The denominator is $x - 1$.

Therefore, when $x > 1$, $f(x) = \dfrac{x-1}{x-1} = 1$

and when $x < 1$, $f(x) = \dfrac{1-x}{x-1} = -1$

At $x = 1$, $f(x)$ is undefined due to division by 0.

Range of $f(x)$ The function takes only two values -1 and 1, so the range is $\{-1, 1\}$.

(ii) $f(x) = 2 - 3x, x > 0$

Since, $x > 0$. Therefore, $f(x)$ is linear function decreasing as x increases.

As $x \to 0^+, f(x) \to 2$

As $x \to \infty, f(x) \to -\infty$

Range of $f(x) = (-\infty, 2)$

(iii) $f(x) = x^2 + 2, x \in R$

$x^2 \geq 0$ for all $x \in R$, so $x^2 + 2 \geq 2$

The minimum value of $f(x)$ occurs when $x = 0$, giving $f(0) = 2$.

Range of $f(x) = [2, \infty)$.

19. We have, $f(x) = 4x + 1$ $\quad [\because 4x + 1, x > 0]$

$$f(1) = 4(1) + 1 = 5,$$

$f(-1) = 3x - 2$ $\quad [\because 3x - 2, x < 0]$

$$= 3(-1) - 2 = -5,$$

$f(0) = 1$ $\quad [\because 1, x = 0]$

and $f(2) = 4x + 1$ $\quad [\because 4x + 1, x > 0]$

$$= 4(2) + 1 = 9$$

20. (i) $f\left(\dfrac{1}{2}\right) = x$ $\quad [\because x, 0 \leq x < 1]$

$$= \frac{1}{2}$$

(ii) $f(-2) = (-2)^2$ $\quad [\because x^2, x < 0]$

$$= 4$$

(iii) $f(1) = \dfrac{1}{1}$ $\quad \left[\because \dfrac{1}{x}, x \geq 1\right]$

$$= 1$$

(iv) $f(\sqrt{3}) = \dfrac{1}{\sqrt{3}}$ $\quad \left[\because \dfrac{1}{x}, x \geq 1\right]$

21. Hint Since, x is a prime number less than 10.

$R = \{(2x + 3, x^3)\}$

$\therefore R = \{(7, 8), (9, 27), (13, 125), (17, 343)\}$

Then, domain $= \{7, 9, 13, 17\}$ and range $= \{8, 27, 125, 343\}$

22. $X = \{1, 2, 3, 4, 5\}, Y = \{1, 2, 5, 6, 7, 9, 10, 11, 12, 13, 14\}$

and $f : X \to Y$ defined by $f(x) = 2x + 3$

$R = \{(1, 5), (2, 7), (3, 9), (4, 11), (5, 13)\}$

Domain $= \{1, 2, 3, 4, 5\}$ and range $= \{5, 7, 9, 11, 13\}$

23. We have, $g = \{(2, 3), (4, 5), (6, 7)\}$

Yes, it is a function, since every element has one and only one image.

Formula is $g(x) = \alpha\, x + 2\beta$ ...(i)

Now, put $x = 2$ and $g(x) = 3$ in Eq. (i), we get

$$3 = 2\alpha + 2\beta \quad ...(ii)$$

Put $x = 4$ and $g(x) = 5$, in Eq. (i) we get

$$5 = 4\alpha + 2\beta \quad ...(iii)$$

On solving Eqs. (ii) and (iii), we get

$$\alpha = 1, \beta = \frac{1}{2}$$

24. $n(A) = 3$ and $n(B) = 2$

If $(x, 1), (y, 2)$ and $(z, 1)$ are in $(A \times B)$

So, $A = \{x, y, z\}$ and $B = \{1, 2\}$

Hence, $A \times B = \{x, y, z\} \times \{1, 2\}$

$= \{(x, 1), (x, 2), (y, 1), (y, 2), (z, 1), (z, 2)\}$

25. We have, $f(x) = \sqrt{4 - x} + \dfrac{1}{\sqrt{x^2 - 1}}$ ...(i)

For Eq. (i) to be defined,

$4 - x \geq 0$ or $x^2 - 1 > 0$

$\Rightarrow$ $x - 4 \leq 0$ or $(x + 1)(x - 1) > 0$

$\Rightarrow$ $x \leq 4$ or $x < -1$ and $x > 1$

$\therefore$ Domain of $f = (-\infty, -1) \cup (1, 4]$.

26. Relation $R = \{(1, 3), (2, 6), (3, 9), (4, 12)\}$

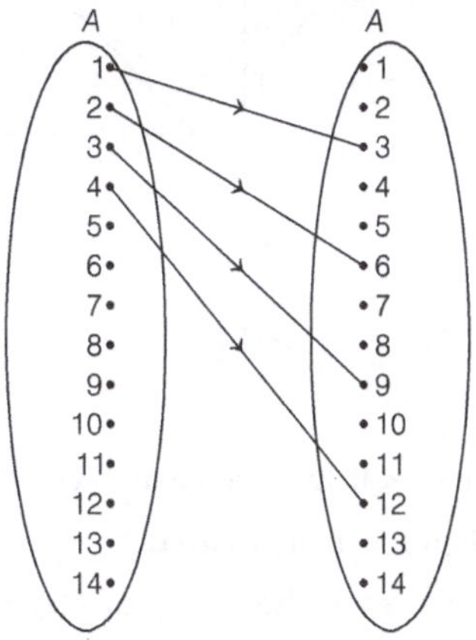

Domain = $\{1, 2, 3, 4\}$, codomain = A and range = $\{3, 6, 9, 12\}$

27. (i) $R = \{(x, y) : x, y \in N, y = x^2 + 3$ and $0 < x < 5\}$

$R = \{(1, 4), 2, 7), (3, 12), (4, 19)\}$

Domain = $\{1, 2, 3, 4\}$ and range = $\{4, 7, 12, 19\}$

(ii) $R = \{(x, y) : x, y \in N, y = \dfrac{1}{1+x}$ and x is an odd natural number$\}$

$$\therefore R = \left\{\left(1, \frac{1}{2}\right), \left(3, \frac{1}{4}\right), \left(5, \frac{1}{6}\right), \left(7, \frac{1}{8}\right), \left(9, \frac{1}{10}\right) .....\right\}$$

Therefore, Domain = $\{1, 3, 5, 7, 9, ...\}$

and range = $\left\{\dfrac{1}{2}, \dfrac{1}{4}, \dfrac{1}{6}, \dfrac{1}{8}, \dfrac{1}{10} ....\right\}$

(iii) $R_1 = \left\{\left(x, \dfrac{1}{x}\right) : 0 < x < 6, x \in N\right\}$

$$R_1 = \left\{(1, 1), \left(2, \frac{1}{2}\right), \left(3, \frac{1}{3}\right), \left(4, \frac{1}{4}\right), \left(5, \frac{1}{5}\right)\right\}$$

Domain = $\{1, 2, 3, 4, 5\}$ and range = $\left\{1, \dfrac{1}{2}, \dfrac{1}{3}, \dfrac{1}{4}, \dfrac{1}{5}\right\}$

(iv) $R_2 = \{(x, x^2) : x$ is prime number less than $10\}$

$R_2 = \{(2, 4), (3, 9), (5, 25), (7, 49)\}$

Domain = $\{2, 3, 5, 7\}$ and range = $\{4, 9, 25, 49\}$

28. **Hint** $A = \{-3, -2, -1, 4\}$ and $f : A \to Z$

$f(x) = x^2 + x + 2$

On substituting $x = -3, -2, -1$ and 4 in $f(x)$, we get

$f(-3) = 8, f(-2) = 4, f(-1) = 2$ and $f(4) = 22$, respectively.

$\therefore R = \{(-3, 8), (-2, 4), (-1, 2), (4, 22)\}$

(i) Range $(f) = \{8, 4, 2, 22\}$

(ii) There is no pre-image of 6, pre-image of 4 is -2.

29. (i) $f(x) = -|x|, x \in R$

We know that $-|x| = \begin{cases} -x, & x \geq 0 \\ x, & x < 0 \end{cases}$

Since, $f(x)$ is defined for $x \in R$, therefore the domain of f is R.

It can be observed that the range of $f(x) = -|x|$ is all real numbers except positive real numbers.

$\therefore$ The range of f is $(-\infty, 0]$.

(ii) $f(x) = \sqrt{9 - x^2}$

Since, $\sqrt{9 - x^2}$ is defined for all real numbers that are greater than or equal to -3 and less than or equal to 3, the domain of $f(x)$ is $\{x : -3 \leq x \leq 3\}$ or $[-3, 3]$.

For any value of x such that $-3 \leq x \leq 3$, the value of $f(x)$ will lie between 0 and 3.

$\therefore$ The range of $f(x)$ is $\{x : 0 \leq x \leq 3]$ or $[0, 3]$.

30. $f(x) = 1 - |x + 2|$

The term $|x + 2|$ represents the absolute value of $x + 2$.

Absolute values are always non-negative.

So, $|x + 2| \geq 0$

Therefore, $f(x) = 1 - |x + 2| \leq 1$

The maximum value of $f(x)$ occurs when $|x + 2| = 0$, which happens at $x = -2$.

Thus, $f(-2) = 1$

As $|x + 2| \to \infty$, $f(x) \to -\infty$

Range of $f(x) = (-\infty, 1]$.

31. $f(x) = \dfrac{|x+4|}{x+4}$

When $x + 4 > 0$ (i.e. $x > -4$), $f(x) = \dfrac{x+4}{x+4} = 1$

and when $x + 4 < 0$ (i.e. $x < -4$), $f(x) = \dfrac{-(x+4)}{x+4} = -1$

At $x = -4$, $f(x)$ is undefined due to division by zero.

Range of $f(x) = \{-1, 1\}$.

32. (i) $f = \{(8, 2), (11, 11), (12, 3), (15, 5), (18, 3), (23, 23)\}$

Range $= \{2, 3, 5, 11, 23\}$

(ii) $f(x) = \dfrac{x^2 + 2x + 1}{x^2 - 8x + 12}$

Now, $x^2 - 8x + 12 \neq 0 \Rightarrow x^2 - 6x - 2x + 12 \neq 0$

$\Rightarrow x(x-6) - 2(x-6) \neq 0 \Rightarrow (x-6)(x-2) \neq 0$

$\Rightarrow \quad x \neq 6, 2$

$\therefore$ Domain $= R - \{2, 6\}$

33. (i) $(f+g)(x) = \sqrt{x^2+1} + 2x^2 + 3$

(ii) The identity function is $f(x) = x$.

The modulus function is $g(x) = |x|$.

The product of the identity function and modulus function is

$$f(x) \cdot g(x) = x \cdot |x|$$

Now, break into cases based on definition of $|x|$

When $x \geq 0$, $|x| = x$, so $f(x) = x \cdot x = x^2$

When $x < 0$, $|x| = -x$, so $f(x) = x \cdot (-x) = -x^2$

$$(f \cdot g)(x) = \begin{cases} x^2, & x \geq 0 \\ -x^2, & x < 0 \end{cases}$$

34. $f(x) = \begin{cases} x^2, & 0 \leq x \leq 3 \\ 3x, & 3 \leq x \leq 10 \end{cases}$

Checking whether $f(x)$ is a function.

$f(x)$ gives a different output for each x in its domain, as the intervals [0, 3] and (3, 10] do not overlap.

Therefore, $f(x)$ is a function.

$g(x)$ is defined as $g(x) = \begin{cases} x^2, & 0 \leq x \leq 2 \\ 3x, & 2 \leq x \leq 10 \end{cases}$

At $x = 2$, $g(x) = x^2 = 4$ and $g(x) = 3x = 6$

Thus, $g(x)$ gives two different outputs for $x = 2$, so it is not a function.

35. $f(x) = \begin{cases} 1 + x, & -1 \leq x < 0 \\ x^2 - 1, & 0 < x < 2 \\ 2x, & 2 \leq x \end{cases}$

Now, $f(3) = 2(3) = 6, \quad [\because 2x, 2 \leq x]$

$f(-2) =$ Not defined,

$f\left(\dfrac{1}{2}\right) = \left(\dfrac{1}{2}\right)^2 - 1 \quad [\because x^2 - 1, 0 < x < 2]$

$= \dfrac{1}{4} - 1 = \dfrac{-3}{4},$

$f(2-h) = x^2 - 1 \quad [\because x^2 - 1, 0 < x < 2]$

$= (2-h)^2 - 1$

$= 4 + h^2 - 4h - 1 = h^2 - 4h + 3$

and $f(-1+h) = 1 + x \quad [\because 1 + x, -1 \leq x < 0]$

$= 1 + (-1+h) = 1 - 1 + h = h$

36. The modulus function $|2x-1|$ is piecewise defined as

$$|2x-1| = \begin{cases} 2x - 1, & 2x - 1 \geq 0 \text{ or } x \geq \dfrac{1}{2} \\ -(2x-1), & 2x - 1 < 0 \text{ or } x < \dfrac{1}{2} \end{cases}$$

Thus, $f(x) = \begin{cases} 1 - 2x, & -1 \leq x < \dfrac{1}{2} \\ 2x - 1, & \dfrac{1}{2} \leq x \leq 1 \end{cases}$

37. Domain $= R$

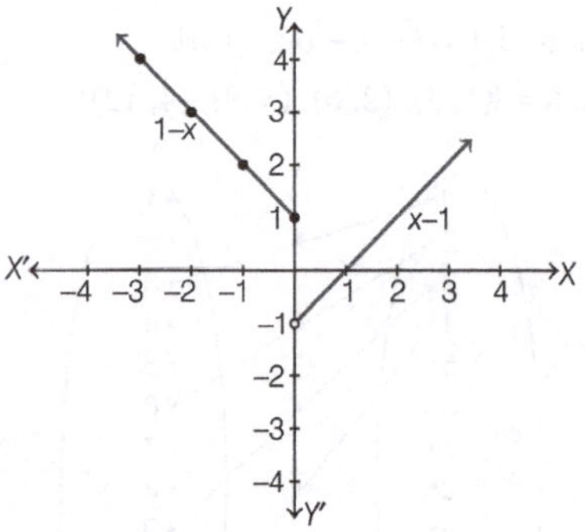

For range,

Case I When $x < 0$, then range of $f(x) = (1, \infty)$

Case II When $x = 0$, then $f(0) = 1$

Case III When $x > 0$, then range of $f(x) = (-1, \infty)$

Now, from all cases, we get Range $= (-1, \infty)$

QR Code for Quick Test 1

QR Code for Quick Test 2

QR Code for Quick Test 3

QR Code for Quick Test 4

Mind Map

Relations and Function

Domain, Range and Codomain

Domain The set of first elements of all the ordered pairs present in R is called the domain of relation.

Range The set of second elements of all the ordered pairs present in R is called range of the relation.

Codomain The set B is called the codomain of relation R.

i.e. Domain $(R) = \{a : (a, b) \in R\}$

and Range $(R) = \{b : (a, b) \in R\}$

$\therefore$ **Range $\subseteq$ Codomain**

Relations

If A and B be two non-empty sets. Then, a relation R from set A to set B is a subset of cartesian product $A \times B$

i.e. $\mathbf{R \subseteq A \times B}$

Number of Relations

Let A and B be any two non-empty finite sets containing m and n elements, respectively.

Then, the total number of relations from set A to set B is 2^{mn}.

Cartesian Product of Two Sets

For two non-empty sets A and B, the cartesian product $A \times B$ is the set of all ordered pairs of elements from sets A and B.

i.e. $A \times B = \{(a, b) : a \in A, b \in B\}$

Number of elements in cartesian product of two set

If $n(A) = p$ and $n(B) = q$, then $n(A \times B) = pq$

Method of Representation

Let R be a relation between a set of country to set of currency. Then,

(i) **Roster form**

R = {(India, Rupee), (America, Dollar), (England, Pound), (France, Euro), (Canada, Dollar)}

(ii) **Set-Builder form**

$R = \{(x, y) : x \in A, y \in B ; y \text{ is currency of } x\}$

(iii) **Arrow diagram**

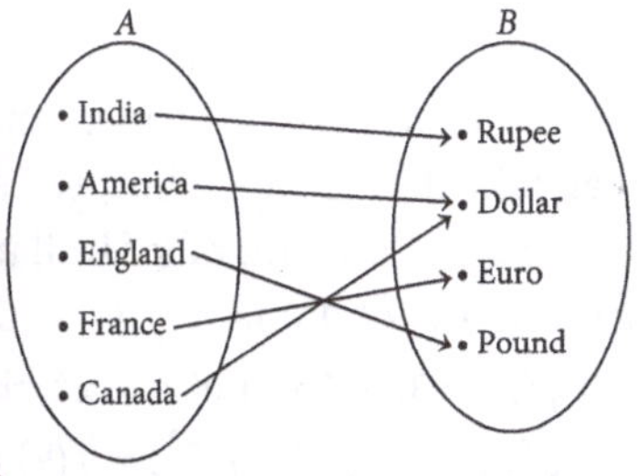

Ordered Pairs

The ordered pair of two elements a and b is denoted by (a, b): a is the first element and b is the second element.

Two ordered pairs are equal, if their corresponding elements are equal.

i.e. $(a, b) = (c, d) \Rightarrow a = c$ and $b = d$

Functions

A relation f from a non-empty set A to a non-empty set B is said to be a functions, if every element of set A has one and only one image in set B.

Cartesian Product of Three Sets

If A, B and C are three sets, then

$A \times B \times C = \{(a, b, c) : a \in A, b \in B, c \in C\}$

Real Functions

If the range and domain both are either R or a subset of R, then function is called real function.

Domain, Codomain and Range of a Function

If f is a function from A to B, i.e. each element of A corresponds to one and only one element of B, whereas every element in B need not be the image of some x in A. Then,

Domain The set A is called the domain of function f.

Codomain The set B is called the codomain of function f.

Range The subset of B containing the images of elements of A is called the range of the function.

Algebra of Real Number

Let $f : D_1 \to R$ and $g : D_2 \to R$ be two real functions with domain D_1 and D_2, respectively.

(i) **Addition of two real functions**

$(f + g)(x) = f(x) + g(x), \forall x \in D_1 \cap D_2$

(ii) **Subtraction of two real functions**

$(f - g)(x) = f(x) - g(x), \forall x \in D_1 \cap D_2$

(iii) **Multiplication of two real functions**

$(fg)(x) = f(x) \times g(x), \forall x \in D_1 \cap D_2$

(iv) **Quotient of two real functions**

$\frac{f}{g}(x) = \frac{f(x)}{g(x)} \forall x \in D_1 \cap D_2 - \{x : g(x) = 0\}$

(v) **Multiplication of a real function by a scalar**

$(cf)(x) = c \times f(x), \forall x \in D_1$

Some Standard Real Functions

Rational Function

A function of the form $f(x) = \frac{P(x)}{Q(x)}$, where $P(x)$ and $Q(x)$ are polynomial functions of x defined in a domain and $Q(x) \neq 0$, is called a rational function.

Identity Function

The real function $f : R \to R$ defined by $f(x) = x, \forall x \in R$ is called an identity function. Its domain is R and range is also R.

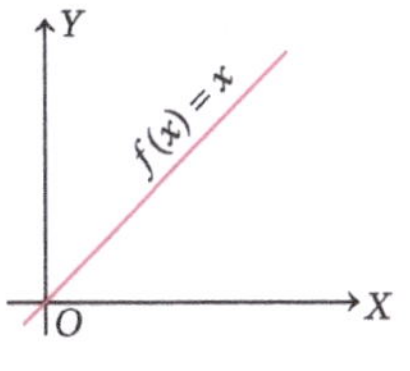

Modulus Function or Absolute Value Function

The function $f : R \to R$ defined by

$$f(x) = |x| = \begin{cases} x, & \text{if } x \geq 0 \\ -x, & \text{if } x < 0 \end{cases}$$

is called the modulus function. It is also called absolute value function.

Polynomial Function

A function $f : R \to R$ defined by

$f(x) = a_0 + a_1x + a_2x^2 + ... + a_nx^n$, where n is a non-negative integer and $a_0, a_1, a_2, ... a_n \in R$ is called a polynomial function.

If $a_n \neq 0$, then n is called the degree of the polynomial.

Greatest Integer Function (Step Function)

The function $f : R \to R$ defined by $f(x) = [x]$ is called the greatest integer function, where $[x]$ = **integral** part of x or greatest integer less than or equal to x.

Constant Function

A function $f : R \to R$ is said to be a constant function, if there exists a **real number** k such that $f(x) = k, \forall x \in R$

Here, domain of $f = R$ and its range $= \{k\}$

Topics to be Assessed Formatively

Composition of Functions

Let $f: A \rightarrow B$ and $g: B \rightarrow C$ be any two functions. Then, the composition of f and g, denoted by gof is defined as function $gof: A \rightarrow C$ given by $gof(x) = g[f(x)], \forall\, x \in A$.
Clearly, domain (gof) = domain (f).

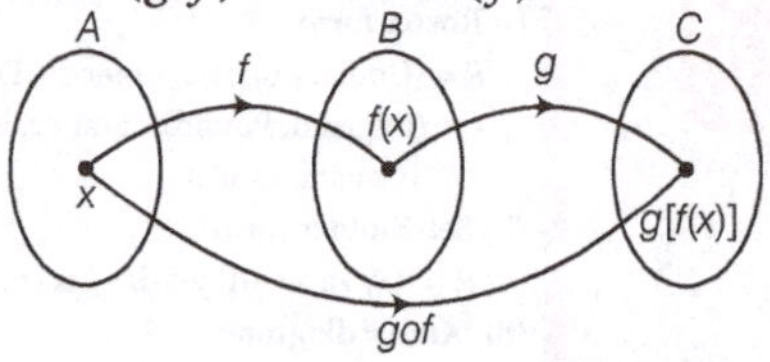

It is clear from the definition that gof is defined only, if for each $x \in A, f(x)$ is an element of domain of g, so that we can take its g-image. Thus, gof exists, iff the range of f is a subset of domain of g. Similarly, fog exists, iff range of g is a subset of domain of f.

e.g. If $f(x) = 3x + 5$ and $g(x) = x^2$, then

$$fog\,(3) = f[g(3)] = f(3^2) = f(9) = 3 \times 9 + 5 = 32$$

Note

(i) $gof\,(x) = g\{f(x)\}$, i.e. first 'f' rule is applied, then 'g' rule is applied.
(ii) $fog\,(x) = f\{g(x)\}$, i.e. first 'g' rule is applied, then 'f' rule is applied.
(iii) Generally, the composition of functions is not commutative, i.e. $fog \neq gof$.
(iv) If $f: R \rightarrow R$ and $g: R \rightarrow R$ are real functions, then both fog and gof exists.

Example 1 **Let $f: \{1, 3, 4\} \rightarrow \{1, 2, 5\}$ and $g: \{1, 2, 5\} \rightarrow \{1, 3\}$, given by $f = \{(1, 2), (3, 5), (4, 1)\}$ and $g = \{(1, 3), (2, 3), (5, 1)\}$. Write down gof.**

Sol. Given, $f: \{1, 3, 4\} \rightarrow \{1, 2, 5\}$ and $g: \{1, 2, 5\} \rightarrow \{1, 3\}$

Also, $f(1) = 2,\ f(3) = 5,\ f(4) = 1$

and $g(1) = 3,\ g(2) = 3,\ g(5) = 1$

Now, $gof\,(1) = g\{f(1)\} = g(2) = 3$

$gof\,(3) = g\{f(3)\} = g(5) = 1$

and $gof\,(4) = g\{f(4)\} = g(1) = 3$

$\therefore\quad gof = \{(1, 3), (3, 1), (4, 3)\}$

Example 2 **Find gof and fog, if $f: R \rightarrow R$ and $g: R \rightarrow R$ are given by**

(i) $f(x) = \sin x$ and $g(x) = 4x^2$.
(ii) $f(x) = x^2$ and $g(x) = 2x + 1$.

Sol. Here, range of f is a subset of domain of g and range of g is a subset of domain of f, so fog and gof both exist.

(i) $gof(x) = g\{f(x)\} = g(\sin x) = 4(\sin x)^2 = 4\sin^2 x$

and $fog(x) = f\{g(x)\} = f(4x^2) = \sin(4x^2)$

(ii) $gof(x) = g\{f(x)\} = g(x^2) = 2x^2 + 1$

and $fog(x) = f\{g(x)\} = f(2x + 1) = (2x + 1)^2$

Example 3 **Let $f: R \rightarrow R$ be the signum function defined as $f(x) = \begin{cases} 1, & x > 0 \\ 0, & x = 0 \\ -1, & x < 0 \end{cases}$ and $g: R \rightarrow R$ be the greatest integer function given by $g(x) = [x]$. Then, prove that fog and gof coincide in $[-1, 0)$.**

Sol. As both f and g are real functions from R to R, so both fog and gof will exist. Let $x \in [-1, 0)$ be any arbitrary real number.

Then, $fog\,(x) = f\,\{g(x)\} = f([x]) = f(-1) = -1$ ...(i)

$[\because -1 \le x < 0 \Rightarrow [x] = -1]$

and $gof\,(x) = g\,\{f(x)\} = g\,(-1) = [-1] = -1$ ...(ii)

From Eqs. (i) and (ii), we get $fog\,(x) = gof\,(x)$

$\because\ x \in [-1, 0)$ was arbitrary.

$\therefore\ fog\,(x) = gof\,(x), \forall\, x \in [-1, 0)$

Thus, $fog = gof$ in $[-1, 0)$.

Note Two functions, say f and g, having the same domain D, are said to be equal, if $f(x) = g(x), \forall\, x \in D$.

Theorem If $f: X \rightarrow Y, g: Y \rightarrow Z$ and $h: Z \rightarrow S$ are functions, then $ho\,(gof) = (hog)of$.

Example 4 **Consider $f: N \rightarrow N,\ g: N \rightarrow N$ and $h: N \rightarrow R$, defined as $f(x) = 3x,\ g(y) = 2y + 3$ and $h(z) = \cos z, \forall x, y, z \in N$. Show that $(hog)of = ho(gof)$.**

Sol. Given functions are

$f(x) = 3x,\ g(y) = 2y + 3$ and $h(z) = \cos z$.

Now, $((hog)of)(x) = (hog)f(x) = (hog)(3x)$

$= h(g(3x)) = h(2(3x) + 3)$

$= h(6x + 3) = \cos(6x + 3), \forall x \in N$

and $(ho(gof))(x) = h(gof(x)) = h(g(f(x))) = h(g(3x))$

$= h(2(3x) + 3) = h(6x + 3)$

$= \cos(6x + 3), \forall x \in N$

Hence, $(hog)\,of = ho(gof)$.

Some Important Properties

1. If $f: A \to B$ and $g: B \to C$ are one-one, then $gof: A \to C$ is also one-one.
2. If $f: A \to B$ and $g: B \to C$ are onto, then $gof: A \to C$ is also onto.
3. Let $f: A \to B$ and $g: B \to C$ be two functions. Then,
   (i) $gof: A \to C$ is onto $\Rightarrow g: B \to C$ is onto.
   (ii) $gof: A \to C$ is one-one $\Rightarrow f: A \to B$ is one-one.
   (iii) $gof: A \to C$ is onto and $g: B \to C$ is one-one $\Rightarrow f: A \to B$ is onto.
   (iv) $gof: A \to C$ is one-one and $f: A \to B$ is onto $\Rightarrow g: B \to C$ is one-one.

Exercise

Objective Type Questions

1. If $f: R \to R$ and $g: R \to R$ are given by $f(x) = \cos x$ and $g(x) = 3x^2$, then

(a) $gof(x) = \cos^2 x$ (b) $fog(x) = \cos x^2$
(c) $gof \neq fog$ (d) $fog = gof$

2. Let f, g and h be functions from R to R. Then,

(a) $(f+g)oh = fog + goh$ (b) $(f+g)oh = foh + goh$
(c) $(f \cdot g)oh = (foh) + (goh)$ (d) $(f \cdot g)oh = (fog) \cdot (goh)$

3. Which of the following options is correct?

(a) gof is one one $\Rightarrow g$ is one-one
(b) gof is one-one $\Rightarrow f$ is one-one
(c) gof is onto $\Rightarrow g$ is not onto
(d) gof is onto $\Rightarrow f$ is onto

4. Let $f(y) = \dfrac{ay}{y+1}, y \neq -1$. Then, for what value of a is $f(f(y)) = y$?

(a) 1 (b) -1
(c) 0 (d) 2

5. Let $f(x) = \begin{cases} -1, & x < 0 \\ 0, & x = 0 \\ 1, & x > 0 \end{cases}$ and $g(x) = 1 + x - [x]$,

where $[x]$ denotes the greatest integer less than or equal to x.
Then, for all x, $f(g(x))$ is equal to

(a) x (b) 1 (c) $g(x)$ (d) $f(x)$

Very short answer Type Questions

6. If $f(x) = x + 7$ and $g(x) = x - 7$, $x \in R$ then find $fog(6)$.

7. Let $f: R \to R$ be defined by $f(x) = 2x - 3$ and $g: R \to R$ be defined by $g(x) = \dfrac{x+3}{2}$, show that $fog = I_R$.

8. If $f: R \to R$ is given by $f(x) = (3 - x^3)^{1/3}$, then find $(fof)x$.

Short Answer Type I Questions

9. If f is the greatest integer function and g is the modulus function.
Find the value of $gof\left(\dfrac{-1}{3}\right) - fog\left(\dfrac{-1}{3}\right)$.

10. If $g\{f(x)\} = |\sin x|$ and $f\{g(x)\} = (\sin\sqrt{x})^2$, then find $f(x)$ and $g(x)$.

Short Answer Type II Questions

11. If the function $f: R \to R$ is given by $f(x) = x^2 + 2$ and $g: R \to R$ is given by $g(x) = \dfrac{x}{x-1}$; $x \neq 1$, then find fog and gof and hence find $fog(2)$ and $gof(-3)$.

12. Let $f: Z \to Z$ be a function defined by $f(n) = 3n, \forall n \in Z$ and $g: Z \to Z$ be defined by

$$g(n) = \begin{cases} \dfrac{n}{3}, & \text{if } n \text{ is a multiple of } 3 \\ & \text{if } n \text{ is not a multiple of } 3 \end{cases}.$$

Show that $gof = I_Z$ and $fog \neq I_Z$.

13. Let $f: R \to R$ be defined as $f(x) = 10x + 7$. Find the function $g: R \to R$ such that $gof = fog = I_R$.

Long answer Type Questions

14. Let $f: R \to R$ be a function, given by $f(x) = px + q$, $\forall x \in R$. Find constants p and q such that $fof = I_R$.

15. Show that the function $f: R \to R$ defined by $f(x) = \dfrac{x}{x^2+1}, \forall x \in R$ is neither one-one nor onto. Also, if $g: R \to R$ is defined as $g(x) = 2x - 1$, find $fog(x)$.

Answers

1. (c) 2. (b) 3. (b) 4. (b)
5. (b) 6. 6 8. x 9. 1
10. $f(x) = \sin^2 x$ and $g(x) = \sqrt{x}$.
11. $fog(2) = 6$ and $gof(-3) = \dfrac{11}{10}$
15. $fog(x) = \dfrac{2x-1}{4x^2 - 4x + 2}$

03

CHAPTER

Beauty is the first step, there is no permanent place in the world for ugly mathematics.
—*Godfrey Harold Hardy*

Trigonometric Functions

YOUR STUDY GOALS

3.1 Angle
3.2 Measure of an Angle
3.3 Trigonometric Functions of Any Angle
3.4 Trigonometric Identities
3.5 Sign of Trigonometric Functions
3.6 Domain and Range of Trigonometric Functions
3.7 Trigonometric Functions of Allied Angles
3.8 Trigonometric Functions
3.9 Transformation Formulae
3.10 Trigonometric Function of Multiple Angles

YOUR PRACTICE GOALS

- Objective Type Questions (27 Qs ; 15 MCQs, 10 A-R, 2 Case Based)
- Subjective Type Questions (32 Qs ; 10 VSA, 22 SA)

3.1 Angle

In Geometry, an angle can be defined as the figure formed by two rays meeting at a common end point. The common end point is called **vertex**.

When a ray OA starting from its initial position OA rotates about its end point O and takes the final position OB, we say that $\angle AOB$ has been formed.

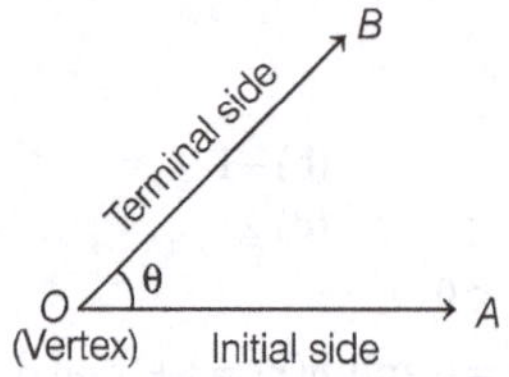

Here, initial position OA and final position OB are respectively, known as the **initial side** and the **terminal side** of $\angle AOB$ and the point O, is called its **vertex.**

Sign of Angles

Angle formed by anti-clockwise rotation is said to be **positive angle.**

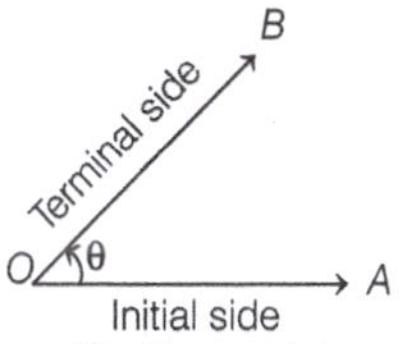

(Positive angle)

Angle formed by clockwise rotation is said to be **negative angle**.

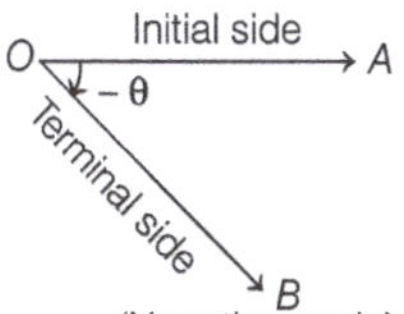

(Negative angle)

3.2 Measure of an Angle

The amount of rotation from the initial side to the terminal side, is called the measure of an angle. There are two systems for measuring an angle, which are given below

Degree Measure (Sexagesimal System)

If the rotation from initial side to terminal side is $\left(\frac{1}{360}\right)$th of a complete revolution, then the angle is said to have a measure of one degree (1°).

One degree is divided into 60 equal parts, called **minute** and 1 min is denoted by $1'$. One minute is divided into 60 equal parts, called **second** and 1 sec is denoted by $1''$.

Thus, $$1° = 60'; \quad 1' = 60''$$

Some of the angles, whose measures are 360°, 180°, 270°, 430°, – 40°, – 430°, are shown below.

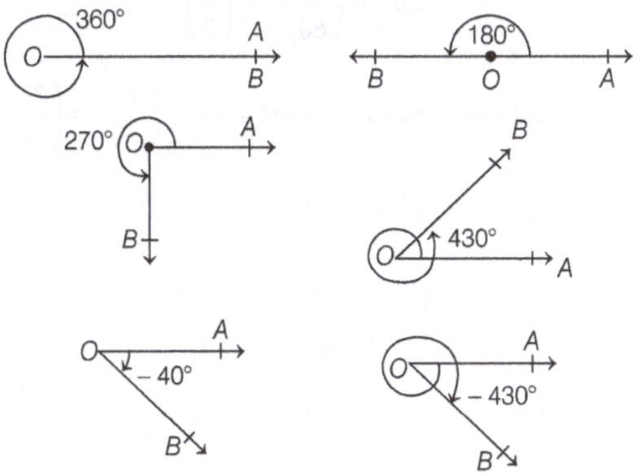

Radian Measure (Circular System)

Angle subtended at the centre by an arc of length one unit in a circle of radius 1 unit is said to have a measure of 1 radian.

In general, a circle of radius r having an arc of length r will subtend an angle of 1 radian at the centre.

1 radian is written as 1^c or 1 rad.

Also, a circle of radius r having an arc of length l will subtend an angle θ radian at the centre, where

$$\theta = \frac{l}{r} = \frac{\textbf{Length of arc}}{\textbf{Radius}} \text{ or } l = r\theta$$

The figures show the angles, whose measures are 1 radian (1^c) and –1 radian (-1^c).

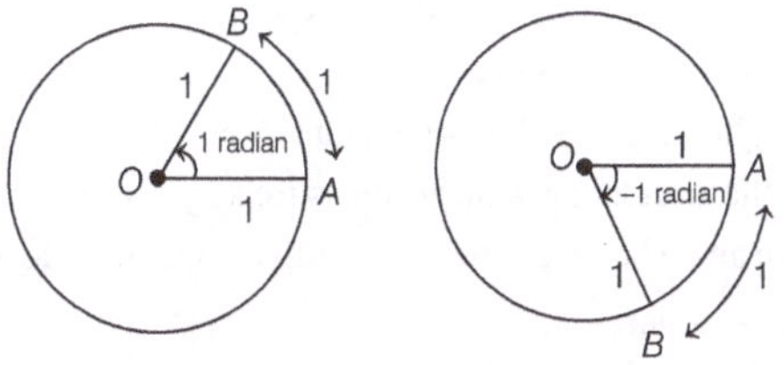

Relation between Degree and Radian

A circle subtends at its centre an angle whose radian measure is 2π and its degree measure is 360°. It follows that

$$2\pi \text{ radian} = 360° \quad \text{or} \quad \pi \text{ radian} = 180°$$

We can express a radian measure in terms of degree measure and a degree measure in terms of radian measure using these relations. Using approximate value of π as $\frac{22}{7}$, we have

$$1 \text{ rad} = \frac{180°}{\pi} = 57°16'22'' \text{ (approx.)}$$

Also, $$1° = \frac{\pi}{180} \text{ rad} = 0.01746 \text{ rad (approx.)}$$

Generally, when an angle is expressed in radians, the word **radian** is frequently omitted. Thus, 45° = π / 4, is written with the understanding that π / 4 is radian measure. Thus,

$$\text{Radian measure} = \frac{\pi}{180} \times \text{Degree measure}$$

$$\boxed{\textbf{Degree measure} = \frac{180}{\pi} \times \textbf{Radian measure}}$$

If D denotes degree measure and R denotes radian measure, then relation between degree and radian is given by $\boxed{\frac{D}{180} = \frac{R}{\pi}}$

The relation between degree measure and radian measure of some common angles are given in the following table.

Degree	30°	45°	60°	90°	180°	270°	360°
Radian	$\frac{\pi}{6}$	$\frac{\pi}{4}$	$\frac{\pi}{3}$	$\frac{\pi}{2}$	π	$\frac{3\pi}{2}$	2π

Points to Remember

(i) In the sexagesimal system, we measure angles in degree, minutes and seconds.
1 right angle $= 90°, 1° = 60'$ and $1' = 60''$

(ii) In the circular system, we measure angles in radians.

(iii) The angle between two consecutive digits in a clock is 30° i.e. $\frac{\pi}{6}$ radian.

(iv) The hour hand rotates through an angle of 30° in one hour i.e. $\left(\frac{1}{2}\right)^\circ$ in one minute.

(v) The minute hand rotates through an angle of 6° in one minute.

Example 1. **Convert 6 radian into degree measure.**

Sol. We know that π radian $= 180°$

Hence, 6 radian $= \frac{180}{\pi} \times 6$ degree $\quad \left[\because 1 \text{ rad} = \left(\frac{180}{\pi}\right)^\circ\right]$

$= \frac{1080 \times 7}{22} = 343\frac{7}{11}$ degree

$= 343° + \frac{7 \times 60}{11}$ minute $\quad [\because 1° = 60']$

$= 343° + 38' + \frac{2}{11}$ minute $\quad [\because 1' = 60'']$

$= 343° + 38' + 10.9'' = 343° \, 38' \, 10.9''$

Hence, 6 radian $= 343° \, 38' \, 11''$ (approx.)

Example 2. **Convert the following into radian.** *NCERT*

(i) 105° (ii) 40°20′
(iii) −47°30′ (iv) 48° 37′30″

Sol. We know that $180° = \pi$ radian

$\therefore \quad 1° = \frac{\pi}{180}$ radian

(i) $105° = \left(\frac{\pi}{180} \times 105\right)$ rad $= \frac{7\pi}{12}$ rad

(ii) Here, given degree measure has minutes in it. Firstly, convert minutes into degree.

$\because \quad 1' = \left(\frac{1}{60}\right)^\circ$

$\therefore \quad 20' = \left(\frac{20}{60}\right)^\circ = \left(\frac{1}{3}\right)^\circ$

Now, total degree to convert $= 40° + \frac{1}{3}^\circ$

$= \left(\frac{120+1}{3}\right)^\circ = \left(\frac{121}{3}\right)^\circ$

$\because$ Radian measure $= \frac{\pi}{180} \times$ Degree measure

$\therefore$ Required radian measure $= \frac{\pi}{180} \times \frac{121}{3} = \frac{121\pi}{540}$ radian

(iii) We have, $-47° \, 30' = -\left[47° + \left(\frac{30}{60}\right)^\circ\right] \quad \left[\because 1' = \left(\frac{1}{60}\right)^\circ\right]$

$= -\left[47° + \left(\frac{1}{2}\right)^\circ\right] = -\left(\frac{95}{2}\right)^\circ$

Now, required radian measure

$= \frac{\pi}{180} \times$ Degree measure

$= -\left(\frac{\pi}{180} \times \frac{95}{2}\right)$ rad $= \frac{-19\pi}{72}$ rad

Hence, radian measure of $-47° \, 30'$ is $-\frac{19\pi}{72}$ rad.

(iv) Here, given degree measure has seconds and minutes. So, firstly convert seconds into minutes.

$\because \quad 1'' = \left(\frac{1}{60}\right)'$

$\Rightarrow \quad 30'' = \left(\frac{30}{60}\right)' = \left(\frac{1}{2}\right)'$

$\therefore$ Total minutes to convert $= 37' + \left(\frac{1}{2}\right)' = \left(\frac{75}{2}\right)'$

$\because \quad 1' = \left(\frac{1}{60}\right)^\circ$

$\Rightarrow \quad \left(\frac{75}{2}\right)' = \left(\frac{75}{2 \times 60}\right)^\circ = \left(\frac{5}{8}\right)^\circ$

$\Rightarrow$ Total degree to convert $= 48° + \left(\frac{5}{8}\right)^\circ = \left(\frac{389}{8}\right)^\circ$

$\because$ Radian measure $= \dfrac{\pi}{180} \times$ Degree measure

$\therefore \left(\dfrac{389}{8}\right)^\circ = \dfrac{\pi}{180^\circ} \times \dfrac{389}{8} = \dfrac{389}{1440} \pi$ rad

Hence, $48^\circ 37' 30''$ is $\dfrac{389\pi}{1440}$ rad.

Example 3. Find the length of an arc of a circle of radius 3 cm, if the angle subtended at the centre is 30°. [use $\pi = 3.14$]

Sol. Given, $r = 3$ cm and $\theta = 30^\circ = 30 \times \dfrac{\pi}{180} = \dfrac{\pi}{6}$ rad

$\therefore$ By $\theta = \dfrac{l}{r}$, we get

$$\frac{\pi}{6} = \frac{l}{3} \Rightarrow l = \frac{\pi}{2} = \frac{3.14}{2} = 1.57 \text{ cm}$$

Example 4. Find in degrees and radians, the angle between the hour hand and the minute hand of a clock at half past three.

Sol. The angle traced by the hour hand in 12 h $= 360^\circ$

$\therefore$ The angle traced by the hour hand in 3 h 30 min

i.e. in $\dfrac{7}{2}$ h $= \left(\dfrac{360}{12} \times \dfrac{7}{2}\right)^\circ = 105^\circ$

The angle traced by the minute hand in 60 min $= 360^\circ$

$\therefore$ The angle traced by the minute hand in 30 min

$$= \left(\frac{360}{60} \times 30\right)^\circ = 180^\circ$$

Hence, the required angle between the hour hand and the minute hand

$$= 180^\circ - 105^\circ = 75^\circ$$

or $\left(75 \times \dfrac{\pi}{180}\right) = \dfrac{5\pi}{12}$ rad

Example 5. The minute hand of a watch is 1.5 cm long. How far does its tip move in 40 min? [use $\pi = 3.14$]

Sol. In one complete rotation, the minute hand of a watch rotates through 360° in 60 min.

$\therefore$ Angle traced out by minute hand in 1 min

$$= \left(\frac{360^\circ}{60}\right) = 6^\circ$$

$\therefore$ Angle traced by the minute hand in 40 min

$$= (40 \times 6)^\circ = 240^\circ$$

$$= 240 \times \frac{\pi}{180} = \left(\frac{4\pi}{3}\right) \text{rad}$$

We know that $\theta = \dfrac{l}{r} \Rightarrow \dfrac{4\pi}{3} = \dfrac{l}{1.5}$

$\Rightarrow l = \dfrac{4\pi}{3} \times 1.5 = 2\pi$

$= 2 \times 3.14$

$= 6.28$ cm

Hence, the tip of the minute hand moves 6.28 cm in 40 min.

Example 6. Find the degree measure of the angle subtended at the centre of a circle of radius 100 cm by an arc of length 22 cm as shown in figure. [use $\pi = 22/7$] Competency Based Que

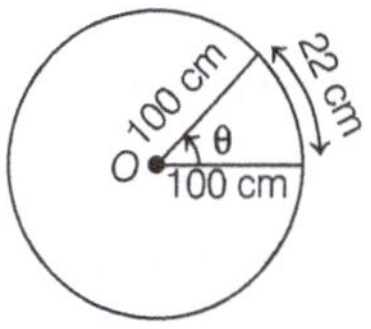

Sol. Given, radius, $r = 100$ cm and arc length, $l = 22$ cm

We know that $l = r\theta$

$\therefore \theta = \dfrac{l}{r} = \dfrac{22}{100} = 0.22 \text{ rad} = 0.22 \times \dfrac{180}{\pi}$ degree

$= \left(0.22 \times \dfrac{180 \times 7}{22}\right)^\circ = \left(\dfrac{22}{100} \times \dfrac{180 \times 7}{22}\right)^\circ$

$= \left(\dfrac{126}{10}\right)^\circ = \left(12\dfrac{6}{10}\right)^\circ$

$= 12^\circ + \dfrac{6}{10} \times 60'$ $\quad [\because 1^\circ = 60']$

$= 12^\circ + 36' = 12^\circ 36'$

Hence, the degree measure of the required angle is $12^\circ 36'$.

Example 7. In a circle of diameter 44 cm and the length of chord is 22 cm. Find the length of minor arc of the chord. Competency Based Que

Sol. Given, diameter = 44 cm

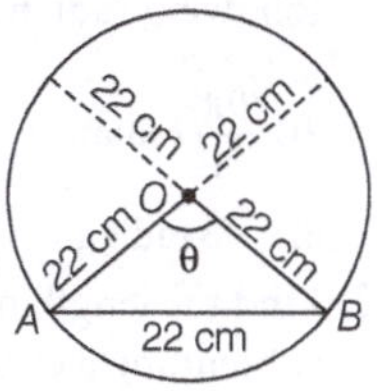

$\therefore$ Radius, $r = \dfrac{44}{2} = 22$ cm

Let chord $AB = 22$ cm [given]

Since, $OA = OB = AB = 22$ cm

$\therefore \Delta OAB$ is an equilateral.

$\Rightarrow \theta = \angle AOB = 60^\circ$

Now, we convert 60° into radian measure.

$\therefore 60^\circ = \left(60 \times \dfrac{\pi}{180}\right)$ rad

$\left[\because \text{radian measure} = \dfrac{\pi}{180} \times \text{degree measure}\right]$

$\Rightarrow 60^\circ = \dfrac{\pi}{3}$ radian

$\therefore$ Length of minor arc $AB = r\theta$ $\left[\because \theta = \dfrac{\text{arc length } (l)}{\text{radius } r)} \quad \therefore \text{arc length} = r\theta\right]$

$$= 22\times\frac{\pi}{3} = \frac{22}{3}\times\frac{22}{7} \quad \left[\because \pi = \frac{22}{7}\right]$$

$$= \frac{484}{21}\text{ cm}$$

Example 8. **A wheel makes 360 revolutions in one minute. Through how many radians does it turn in one second?** NCERT

Sol. Given, number of revolutions in one minute = 360

$\because$ Number of revolutions in 60 sec = 360

$\therefore$ Number of revolutions in one second $= \dfrac{360}{60} = 6$

$\because$ Angle moved in one revolution $= 2\pi$ rad

$\therefore$ Angle moved in 6 revolutions $= (2\pi \times 6)$ rad

$= 12\pi$ rad

Example 9. **A horse is tied to a post by a rope. If the horse moves along a circular path always keeping the rope tight and describes 88 m when it has traced out 54° at the centre. Find the length of the rope.** Competency Based Que

Sol. Let the post be at point P and PA be the length of the rope in tight position.

Suppose the horse moves along the arc AB so that $\angle APB = 54°$ and arc AB of length, $l = 88$ m.

Let r be the length of the rope i.e. $PA = r$ m

Now, $\theta = 54° = \left(54 \times \dfrac{\pi}{180}\right) = \dfrac{3\pi}{10}$ rad

$\left[\because 1° = \dfrac{\pi}{180}\text{ rad}\right]$

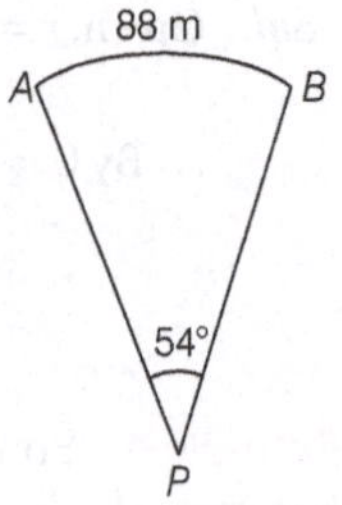

$\because \quad \theta = \dfrac{l}{r}$

$\therefore \quad \dfrac{3\pi}{10} = \dfrac{88}{r} \Rightarrow r = \dfrac{88\times 10}{3\pi}$

$\Rightarrow \quad r = \dfrac{880}{9.42} = 93.41$ m $\quad [\because \pi = 3.14]$

Quick Test 1

1. Find the angle in radian through which a pendulum swings, if its length is 60 cm and the tip describes an arc of length
(i) 12 cm (ii) 15 cm (iii) 10 cm (iv) 14 cm

2. Find the radian measure corresponding to the following degree measures.
(i) 75° (ii) 135°
(iii) −22°30′ (iv) −300°
(v) 5°37′30″ (vi) −14°20′15″

3. Find the degree measure corresponding to the following radian measures. $\left[\text{use } \pi = \dfrac{22}{7}\right]$
(i) $\dfrac{9\pi}{5}$ radian (ii) $-\dfrac{\pi}{3}$ radian
(iii) −3 radian (iv) 11 radian

4. Find the length of arc of a circle of radius 5 cm, subtending a central angle measuring 15°.

5. Find the angle in degree subtended at the centre of a circle by an arc whose length is 2.2 times the radius.

6. In a circle of diameter 40 cm, a chord is 20 cm. Find the length of the minor arc of the chord.

7. If the arc of the same length in two circles subtend angles 65° and 110° at the centre, then find the ratio of their radii. Competency Based Que

8. The angles of a triangle are in AP and the greatest angle is double the least, find all angles in degree and radians.

9. Find the angle in degree and radians, between the hour hand and the minute hand of a clock at half past four. Competency Based Que

10. A horse is tied to a post by a rope. If the horse moves along a circular path always keeping the rope tight and describe 70 m when it has traced out 80° at the centre, then find the length of the rope. Competency Based Que

Answers

1. (i) $\left(\dfrac{1}{5}\right)$ radian (ii) $\dfrac{1}{4}$ radian (iii) $\dfrac{1}{6}$ radian (iv) $\dfrac{7}{30}$ radian

2. (i) $\dfrac{5\pi}{12}$ (ii) $\dfrac{3\pi}{4}$ (iii) $-\dfrac{\pi}{8}$ (iv) $\dfrac{-5\pi}{3}$ (v) $\dfrac{\pi}{32}$ (vi) $\dfrac{-3441\pi}{43200}$

3. (i) 324° (ii) − 60° (iii) − 171°49′5″ (iv) 630°

4. 1.30 cm **5.** 126° **6.** $\dfrac{20}{3}\pi$ cm **7.** 22 : 13

8. 40°, 60°, 80° and $\dfrac{2\pi}{9}, \dfrac{\pi}{3}, \dfrac{4\pi}{9}$ radian **9.** $\left(45°, \dfrac{\pi}{4}\text{ radian}\right)$

10. 50m (approx)

For Solution, See QR Code on Pg. 84

3.3 Trigonometric Functions of Any Angle

In XY-plane, taking O as centre and unit radius draw a circle. Let $P(x, y)$ be any point on the circle with angle $AOP = \theta$ radian, then length of an arc $AP = \theta$.

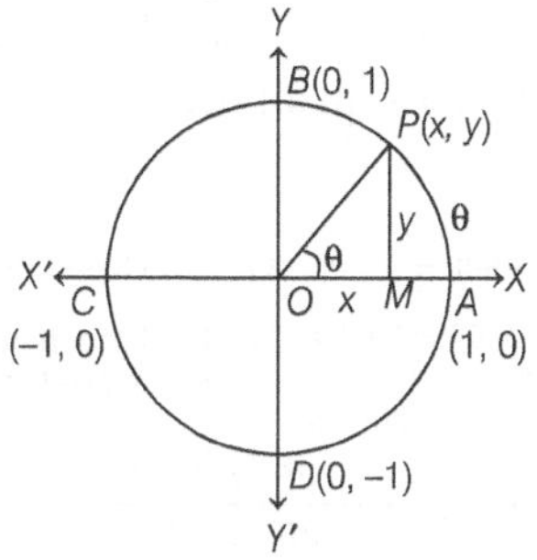

Now, the six circular functions may be defined as

(i) $\cos\theta = x$ (ii) $\sin\theta = y$

(iii) $\tan\theta = \dfrac{y}{x}, x \neq 0$ (iv) $\cot\theta = \dfrac{x}{y}, y \neq 0$

(v) $\sec\theta = \dfrac{1}{x}, x \neq 0$ (vi) $\operatorname{cosec}\theta = \dfrac{1}{y}, y \neq 0$

Since, one complete revolution subtends an angle of 2π radian at the centre of the circle. If we take complete revolution from the point $P(x, y)$, we come back to the same point $P(x, y)$.

Thus, it is clear that if θ increases or decreases by an integral multiple of 2π, the value of trigonometric ratios will not change.

Thus, $\sin(2n\pi + \theta) = \sin\theta, n \in Z$

and $\cos(2n\pi + \theta) = \cos\theta, n \in Z$

Further, if θ is an integral multiples of π i.e. $\theta = 0, \pm\pi, \pm 2\pi, \pm 3\pi, \ldots$, then $\sin\theta = 0$

and if θ is an odd multiples of $\dfrac{\pi}{2}$,

i.e. $\theta = \pm\dfrac{\pi}{2}, \pm\dfrac{3\pi}{2}, \pm\dfrac{5\pi}{2}, \ldots$, then $\cos\theta = 0$.

Thus, we can say that

$\sin\theta = 0 \Rightarrow \theta = n\pi$, where n is any integer.

and $\cos\theta = 0 \Rightarrow \theta = (2n+1)\dfrac{\pi}{2}$, where n is any integer.

Note *All angles which are integral multiples of* $\dfrac{\pi}{2}$ *are called quadrant angles.*

Trigonometric Functions of Some Standard Angles

Angle θ	$0°$	$30°\left(\frac{\pi}{6}\right)$	$45°\left(\frac{\pi}{4}\right)$	$60°\left(\frac{\pi}{3}\right)$	$90°\left(\frac{\pi}{2}\right)$	$120°\left(\frac{2\pi}{3}\right)$	$135°\left(\frac{3\pi}{4}\right)$	$150°\left(\frac{5\pi}{6}\right)$	$180° (\pi)$	$270°\left(\frac{3\pi}{2}\right)$	$360° (2\pi)$
$\sin\theta$	0	$\frac{1}{2}$	$\frac{1}{\sqrt{2}}$	$\frac{\sqrt{3}}{2}$	1	$\frac{\sqrt{3}}{2}$	$\frac{1}{\sqrt{2}}$	$\frac{1}{2}$	0	-1	0
$\cos\theta$	1	$\frac{\sqrt{3}}{2}$	$\frac{1}{\sqrt{2}}$	$\frac{1}{2}$	0	$-\frac{1}{2}$	$-\frac{1}{\sqrt{2}}$	$-\frac{\sqrt{3}}{2}$	-1	0	1
$\tan\theta$	0	$\frac{1}{\sqrt{3}}$	1	$\sqrt{3}$	∞	$-\sqrt{3}$	-1	$-\frac{1}{\sqrt{3}}$	0	∞	0
$\cot\theta$	∞	$\sqrt{3}$	1	$\frac{1}{\sqrt{3}}$	0	$-\frac{1}{\sqrt{3}}$	-1	$-\sqrt{3}$	$-\infty$	0	∞
$\sec\theta$	1	$\frac{2}{\sqrt{3}}$	$\sqrt{2}$	2	∞	-2	$-\sqrt{2}$	$-\frac{2}{\sqrt{3}}$	-1	∞	1
$\operatorname{cosec}\theta$	∞	2	$\sqrt{2}$	$\frac{2}{\sqrt{3}}$	1	$\frac{2}{\sqrt{3}}$	$\sqrt{2}$	2	∞	-1	∞

Note *Here, the symbol '∞' is used for not defined.*

Trigonometric Functions in Terms of sine and cosine Functions

- $\operatorname{cosec} x = \dfrac{1}{\sin x}$, $x \neq n\pi$, where n is any integer.
- $\sec x = \dfrac{1}{\cos x}$, $x \neq (2n+1)\dfrac{\pi}{2}$, where n is any integer.
- $\tan x = \dfrac{\sin x}{\cos x}$, $x \neq (2n+1)\dfrac{\pi}{2}$, where n is any integer.
- $\cot x = \dfrac{\cos x}{\sin x}$, $x \neq n\pi$, where n is any integer.

3.4 Trigonometric Identities

The relation involving trigonometric functions, which are true for all values of the angles involved are called trigonometric identities.

$$\cos^2\theta + \sin^2\theta = 1;\ 1 + \tan^2\theta = \sec^2\theta;\ 1 + \cot^2\theta = \operatorname{cosec}^2\theta$$

3.5 Sign of Trigonometric Functions

In anti-clockwise order, these quadrants are numbered as I, II, III and IV. All angles from 0° to 90° are taken in Ist quardant, 90° to 180° in II quardant, 180° to 270° in III quadrant and 270° to 360° in IV quadrant.

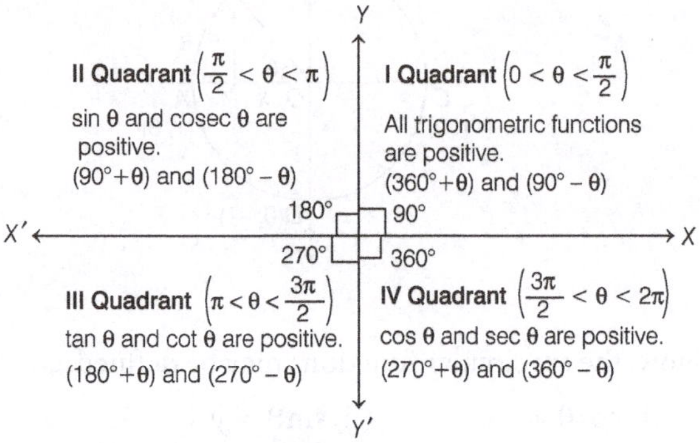

3.6 Domain and Range of Trigonometric Functions

The domain of a trigonometric function consists of all the values that you can use as input into the angles i.e. the values for which trigonometric function is defined. The range of trigonometric functions consists of all its output value i.e. the values which are obtained by trigonometric functions.

e.g. $\sin x$ is define for all real values of x, so its domain is R and obtained value lies between 1 and –1, so its range is the interval $[-1, 1]$.

The domain, range and graph of each one of the six trigonometric functions are given below.

T-function	Domain	Range	Graph
$\sin\theta$	R	$[-1, 1]$	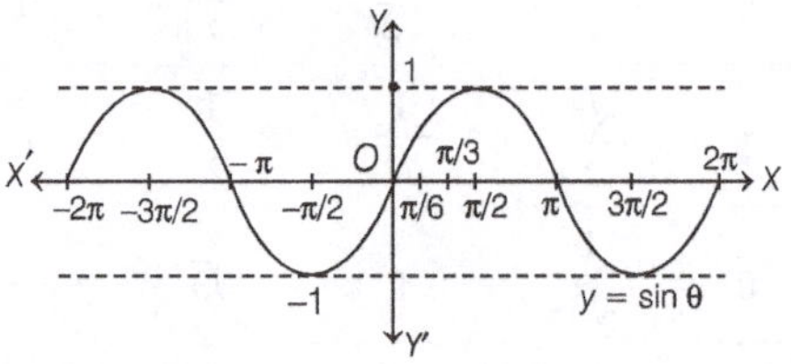
$\cos\theta$	R	$[-1, 1]$	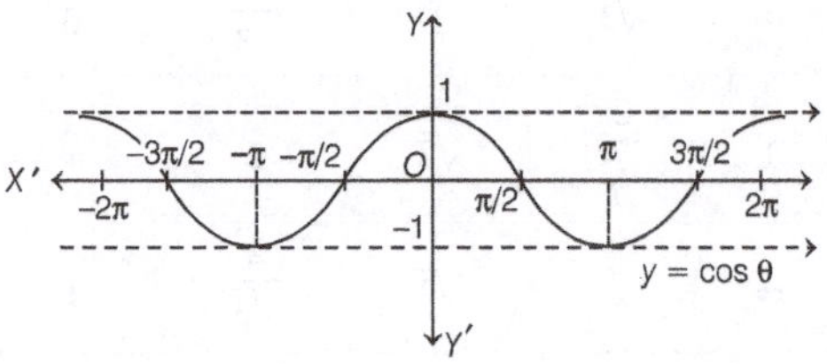

T-function	Domain	Range	Graph
$\tan\theta$	$\{x : x \in R \text{ and } x \neq (2n+1)\,\pi/2,\, n \in I\}$	R	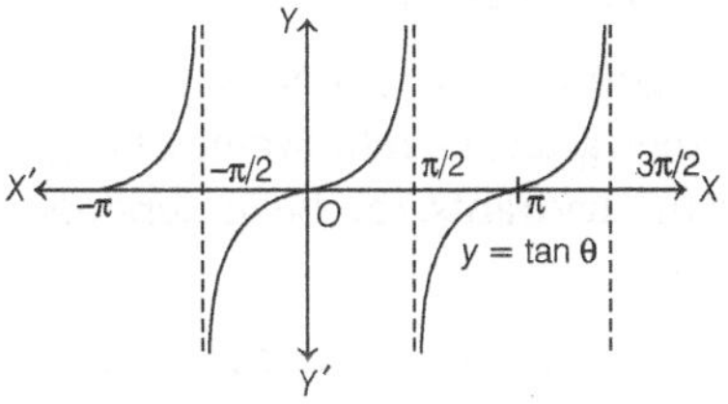
$\cot\theta$	$\{x : x \in R \text{ and } x \neq n\pi,\, n \in I\}$	R	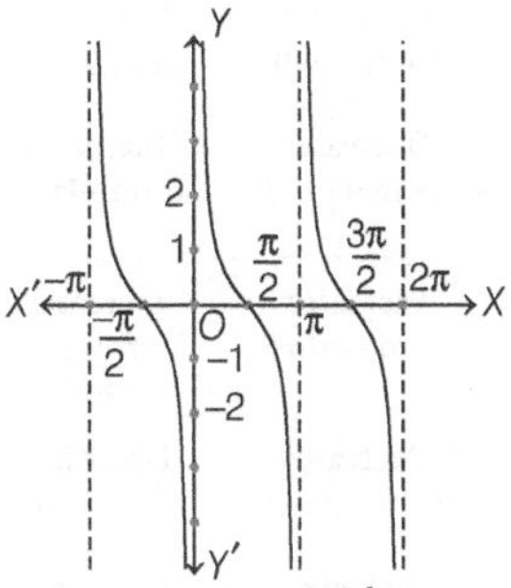
$\operatorname{cosec}\theta$	$\{x : x \in R \text{ and } x \neq n\pi,\, n \in I\}$	$R - (-1, 1)$	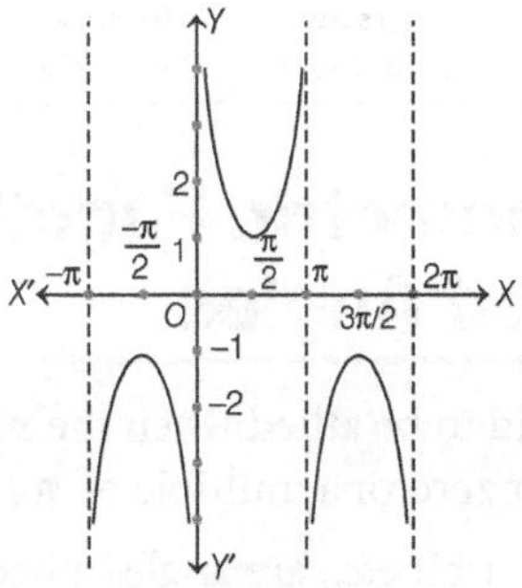
$\sec\theta$	$\{x : x \in R \text{ and } x \neq (2n+1)\,\pi/2,\, n \in I\}$	$R - (-1, 1)$	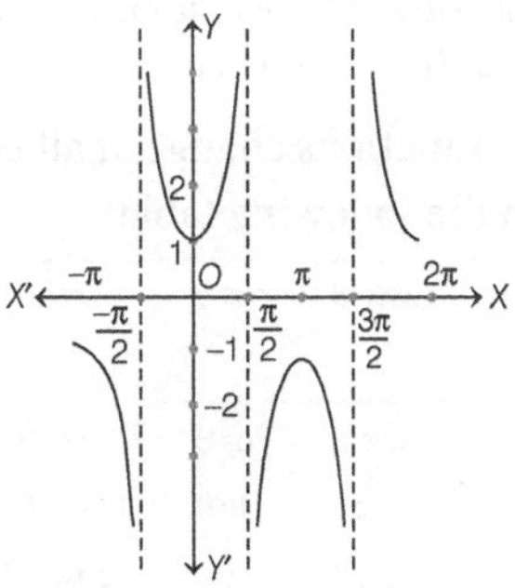

Behaviour of Trigonometric Functions in Different Quadrants

Behaviour of trigonometric functions in different quadrants i.e. the variation in the value of trigonometric ratio in different quadrants, can be understood from the table given below.

	I quadrant	II quadrant	III quadrant	IV quadrant
sin	Increases from 0 to 1	Decreases from 1 to 0	Decreases from 0 to −1	Increases from −1to 0
cos	Decreases from 1 to 0	Decreases from 0 to −1	Increases from −1to 0	Increases from 0 to 1
tan	Increases from 1 to 0	Increases from −∞ to 0	Increases from 0 to ∞	Increases from −∞ to 0
cot	Decreases from ∞ to 0	Decreases from 0 to −∞	Decreases from ∞ to 0	Decreases from 0 to −∞
sec	Increases from 0 to ∞	Increases from −∞ to −1	Decreases from −1to −∞	Decreases from ∞ to 1
cosec	Decreases from ∞ to 1	Increases from 1 to ∞	Increases from −∞ to −1	Decreases from −1to −∞

3.7 Trigonometric Functions of Allied Angles

Two angles are said to be **allied** when their sum or difference is either zero or a multiple of $\pi/2$. The angles $-\theta, \frac{\pi}{2} \pm \theta, \pi \pm \theta, 2\pi \pm \theta$ etc., are angles allied to the angle θ, if θ is measured in radians.

Note *If θ is measured in degrees, then the angles allied to θ are $-\theta, 90^\circ \pm \theta, 180^\circ \pm \theta, 270^\circ \pm \theta$, etc.*

The trigonometric functions change at allied angles which are given in the following table.

Allied angle/ Trigonometric functions	$\sin\theta$	$\text{cosec}\,\theta$	$\cos\theta$	$\sec\theta$	$\tan\theta$	$\cot\theta$
$-\theta$	$-\sin\theta$	$-\text{cosec}\,\theta$	$\cos\theta$	$\sec\theta$	$-\tan\theta$	$-\cot\theta$
$\pi/2-\theta$	$\cos\theta$	$\sec\theta$	$\sin\theta$	$\text{cosec}\,\theta$	$\cot\theta$	$\tan\theta$
$\pi/2+\theta$	$\cos\theta$	$\sec\theta$	$-\sin\theta$	$-\text{cosec}\,\theta$	$-\cot\theta$	$-\tan\theta$
$\pi-\theta$	$\sin\theta$	$\text{cosec}\,\theta$	$-\cos\theta$	$-\sec\theta$	$-\tan\theta$	$-\cot\theta$
$\pi+\theta$	$-\sin\theta$	$-\text{cosec}\,\theta$	$-\cos\theta$	$-\sec\theta$	$\tan\theta$	$\cot\theta$
$3\pi/2-\theta$	$-\cos\theta$	$-\sec\theta$	$-\sin\theta$	$-\text{cosec}\,\theta$	$\cot\theta$	$\tan\theta$
$3\pi/2+\theta$	$-\cos\theta$	$-\sec\theta$	$\sin\theta$	$\text{cosec}\,\theta$	$-\cot\theta$	$-\tan\theta$
$2\pi-\theta$	$-\sin\theta$	$-\text{cosec}\,\theta$	$\cos\theta$	$\sec\theta$	$-\tan\theta$	$-\cot\theta$
$2\pi+\theta$	$\sin\theta$	$\text{cosec}\,\theta$	$\cos\theta$	$\sec\theta$	$\tan\theta$	$\cot\theta$

Note

(i) Trigonometric functions for $(2\pi-\theta)$ and $(2n\pi-\theta)$ are same as those for $(-\theta)$, where $n \in Z$.

(ii) Trigonometric functions for $(2\pi+\theta)$ and $(2n\pi+\theta)$ are same as those for θ, where $n \in Z$.

Even and Odd Functions

- **Even Function** A function $f(x)$ is said to be an even function, if $f(-x) = f(x)$ for all x in its domain. e.g. $\cos x$ is an even function, as $\cos(-x) = \cos x$.
- **Odd Function** A function $f(x)$ is said to be an odd function, if $f(-x) = -f(x)$ for all x in its domain. e.g. $\sin x$ is an odd function, as $\sin(-x) = -\sin x$.

TEACHER TIPS

- Check whether the given angle is positive or negative. If it is negative, then make it positive using the following formulae.
  $\sin(-\theta) = -\sin\theta, \cos(-\theta) = \cos\theta,$
  $\tan(-\theta) = -\tan\theta$, etc.
- If the given angle is greater than 360° (or 2π), then split it in the integral multiple of 360° (or 2π) and then use the results mentioned in the above note.
- If the given angle is less than 360° (or 2π), then split it as $90^\circ \pm \theta, 180^\circ \pm \theta, 270^\circ \pm \theta, 360^\circ - \theta$ $\left(\text{or}\ \frac{\pi}{2} \pm \theta, \pi \pm \theta, \frac{3\pi}{2} \pm \theta, 2\pi - \theta\right)$, where θ is an acute angle.

Example 1. **Find the value of $\sin^2 \frac{\pi}{6} + \cos^2 \frac{\pi}{6} + \sin^2 \frac{\pi}{4}$.**

Sol. We have, $\sin^2 \frac{\pi}{6} + \cos^2 \frac{\pi}{6} + \sin^2 \frac{\pi}{4}$

$$= \left(\frac{1}{2}\right)^2 + \left(\frac{\sqrt{3}}{2}\right)^2 + \left(\frac{1}{\sqrt{2}}\right)^2 = \frac{1}{4} + \frac{3}{4} + \frac{1}{2}$$

$$= \frac{1+3+2}{4} = \frac{6}{4} = \frac{3}{2}$$

Example 2. **Prove that**

$$\frac{1+\tan^2 A}{1+\cot^2 A} = \left(\frac{1-\tan A}{1-\cot A}\right)^2 = \tan^2 A.$$

Sol. $$\text{LHS} = \frac{1+\tan^2 A}{1+\cot^2 A} = \frac{1+\dfrac{\sin^2 A}{\cos^2 A}}{1+\dfrac{\cos^2 A}{\sin^2 A}}$$

$$\left[\because \tan\theta = \frac{\sin\theta}{\cos\theta}, \cot\theta = \frac{\cos\theta}{\sin\theta}\right]$$

$$= \frac{\dfrac{\cos^2 A + \sin^2 A}{\cos^2 A}}{\dfrac{\sin^2 A + \cos^2 A}{\sin^2 A}} = \frac{\dfrac{1}{\cos^2 A}}{\dfrac{1}{\sin^2 A}}$$

$[\because \sin^2\theta + \cos^2\theta = 1]$

$$= \frac{1}{\cos^2 A} \times \frac{\sin^2 A}{1} = \frac{\sin^2 A}{\cos^2 A} = \tan^2 A = \text{RHS}$$

$\left[\because \tan\theta = \dfrac{\sin\theta}{\cos\theta}\right]$

$$\text{Mid term} = \left(\frac{1-\tan A}{1-\cot A}\right)^2 = \left(\frac{1-\dfrac{\sin A}{\cos A}}{1-\dfrac{\cos A}{\sin A}}\right)^2$$

$\left[\because \tan\theta = \dfrac{\sin\theta}{\cos\theta} \quad \cot\theta = \dfrac{\cos\theta}{\sin\theta}\right]$

$$= \frac{\dfrac{(\cos A - \sin A)^2}{\cos^2 A}}{\dfrac{(\sin A - \cos A)^2}{\sin^2 A}}$$

$$= \frac{(\cos A - \sin A)^2 \times \sin^2 A}{(\sin A - \cos A)^2 \times \cos^2 A}$$

$$= \frac{\sin^2 A}{\cos^2 A} = \tan^2 A = \text{RHS}$$

$\therefore$ LHS = Mid term = RHS **Hence proved.**

Example 3. Prove the following identities.

(i) $(1+\cot\theta - \text{cosec}\,\theta)(1+\tan\theta + \sec\theta) = 2$

(ii) $\dfrac{\tan\theta+\sec\theta-1}{\tan\theta-\sec\theta+1} = \dfrac{1+\sin\theta}{\cos\theta}$

Sol. (i) LHS $= (1+\cot\theta - \text{cosec}\,\theta)(1+\tan\theta+\sec\theta)$

$$= \left(1+\frac{\cos\theta}{\sin\theta} - \frac{1}{\sin\theta}\right)\left(1+\frac{\sin\theta}{\cos\theta}+\frac{1}{\cos\theta}\right)$$

$$= \left(\frac{\sin\theta+\cos\theta-1}{\sin\theta}\right)\left(\frac{\sin\theta+\cos\theta+1}{\cos\theta}\right)$$

$[\because (a+b)(a-b) = a^2 - b^2]$

$$= \frac{(\sin\theta+\cos\theta)^2 - 1}{\sin\theta\cos\theta}$$

$$= \frac{\sin^2\theta + \cos^2\theta + 2\sin\theta\cos\theta - 1}{\sin\theta\cos\theta}$$

$$= \frac{1+2\sin\theta\cos\theta - 1}{\sin\theta\cos\theta} \quad [\because \sin^2 A + \cos^2 A = 1]$$

$$= \frac{2\sin\theta\cos\theta}{\sin\theta\cos\theta}$$

$= 2 =$ RHS **Hence proved.**

(ii) LHS $= \dfrac{\tan\theta+\sec\theta-1}{\tan\theta-\sec\theta+1}$

$$= \frac{(\tan\theta+\sec\theta)-(\sec^2\theta-\tan^2\theta)}{\tan\theta-\sec\theta+1}$$

$[\because \sec^2 A - \tan^2 A = 1]$

$$= \frac{(\tan\theta+\sec\theta)-(\sec\theta-\tan\theta)(\sec\theta+\tan\theta)}{\tan\theta-\sec\theta+1}$$

$[\because a^2 - b^2 = (a-b)(a+b)]$

$$= \frac{(\tan\theta+\sec\theta)\,\{1-(\sec\theta-\tan\theta)\}}{\tan\theta-\sec\theta+1}$$

$$= \frac{(\tan\theta+\sec\theta)(\tan\theta-\sec\theta+1)}{\tan\theta-\sec\theta+1}$$

$= \tan\theta + \sec\theta$

$$= \frac{\sin\theta}{\cos\theta} + \frac{1}{\cos\theta} = \frac{\sin\theta+1}{\cos\theta} = \text{RHS}$$ **Hence proved.**

Example 4. **If $\cos\theta = -\dfrac{1}{2}$, θ lies in III quadrant, then find other five trigonometric functions.** *NCERT*

Sol. We have, $\cos\theta = -\dfrac{1}{2}$ and θ lies in III quadrant.

We know that $\sin^2\theta + \cos^2\theta = 1$

$\Rightarrow \quad \sin\theta = \pm\sqrt{1-\cos^2\theta}$

In III quadrant, $\sin\theta$ is negative, therefore

$$\sin\theta = -\sqrt{1-\cos^2\theta} = -\sqrt{1-\left(-\frac{1}{2}\right)^2}$$

$$= -\sqrt{1-\frac{1}{4}} = -\sqrt{\frac{4-1}{4}}$$

$\Rightarrow \quad \sin\theta = \dfrac{-\sqrt{3}}{2}$

Now, $\tan\theta = \dfrac{\sin\theta}{\cos\theta} = \dfrac{-\sqrt{3}/2}{-1/2} = \sqrt{3}$

$$\text{cosec}\,\theta = \frac{1}{\sin\theta} = \frac{1}{\left(\dfrac{-\sqrt{3}}{2}\right)} = \frac{-2}{\sqrt{3}} \quad \left[\because \sin\theta = \frac{-\sqrt{3}}{2}\right]$$

$$\sec\theta = \frac{1}{\cos\theta} = \frac{1}{\left(-\dfrac{1}{2}\right)} = -2 \quad \left[\because \cos\theta = -\frac{1}{2}\right]$$

and $\cot\theta = \dfrac{1}{\tan\theta} = \dfrac{1}{\sqrt{3}}$ $[\because \tan\theta = \sqrt{3}]$

Hence, other five trigonometric functions are

$\sin\theta = \dfrac{-\sqrt{3}}{2}$, cosec$\theta$ $\dfrac{-2}{\sqrt{3}}$, $\sec\theta = -2$,

$\tan\theta = \sqrt{3}$ and $\cot\theta = \dfrac{1}{\sqrt{3}}$.

Example 5. If $\sin\theta = -\frac{3}{5}$ and $\pi < \theta < \frac{3\pi}{2}$, then find the value of $\frac{\sec\theta - \cot\theta}{\tan\theta - \text{cosec}\,\theta}$.

Competency Based Que

Sol. Here, $\pi < \theta < \frac{3\pi}{2}$, it means θ lies in III quadrant.

We have, $\sin\theta = -\frac{3}{5}$, then

$$\text{cosec}\,\theta = \frac{1}{\sin\theta} = \frac{1}{-3/5} = -\frac{5}{3}$$

Now, $\sin^2\theta + \cos^2\theta = 1 \Rightarrow \left(-\frac{3}{5}\right)^2 + \cos^2\theta = 1$

$$\Rightarrow \quad \cos^2\theta = 1 - \frac{9}{25} = \frac{16}{25} \Rightarrow \cos\theta = \pm\sqrt{\frac{16}{25}}$$

$$\Rightarrow \quad \cos\theta = \pm\frac{4}{5}$$

Since, θ lies in III quadrant. So, $\cos\theta$ is negative.

$$\therefore \quad \cos\theta = -\frac{4}{5} \Rightarrow \sec\theta = \frac{1}{\cos\theta} = -\frac{5}{4}$$

$$\tan\theta = \frac{\sin\theta}{\cos\theta} = \frac{-3/5}{-4/5} = \frac{3}{5}\times\frac{5}{4} = \frac{3}{4}$$

and $\cot\theta = \frac{1}{\tan\theta} = \frac{1}{3/4} = \frac{4}{3}$

On putting these values in given expression, we get

$$\frac{\sec\theta - \cot\theta}{\tan\theta - \text{cosec}\,\theta} = \frac{(-5/4 - 4/3)}{(3/4 + 5/3)} = \frac{(-15 - 16/12)}{(9 + 20/12)} = -\frac{31}{29}$$

Example 6. Which of the following is not correct?

NCERT Exemplar

(i) $\sin\theta = -\frac{1}{5}$ (ii) $\cos\theta = 1$

(iii) $\sec\theta = \frac{1}{2}$ (iv) $\tan\theta = 20$

Sol. (i) $\sin\theta = -\frac{1}{5} \in [-1, 1]$.

Therefore, the given equation is correct.

(ii) $\cos\theta = 1 \in [-1, 1]$.

Therefore, the given equation is correct.

(iii) Since, $\sec\theta \in [R - (-1, 1)]$

But here, $\sec\theta = \frac{1}{2} \in (-1, 1)$

Therefore, $\sec\theta$ can not be equal to $\frac{1}{2}$.

$$\therefore \quad \sec\theta \neq \frac{1}{2},$$

Hence, $\sec x = \frac{1}{2}$ is not correct.

(iv) $\tan\theta = 20 \in (-\infty, \infty)$. Therefore, the given equation is correct.

Example 7. Evaluate

(i) $\sin(-1125^\circ)$ (ii) $\sin 930^\circ$

(iii) $\cos(-870^\circ)$ (iv) $\tan(-2025^\circ)$

Sol. (i) We have, $\sin(-1125^\circ) = -\sin(1125^\circ)$ $[\because \sin(-\theta) = -\sin\theta]$

$= -\sin(1080^\circ + 45^\circ) = -\sin(3\times360^\circ + 45^\circ)$

$= -\sin 45^\circ$ $[\because \sin(n\times360^\circ + \theta) = \sin\theta, n \in Z]$

$= -\frac{1}{\sqrt{2}}$

(ii) We have, $\sin 930^\circ = \sin(3\times360^\circ - 150^\circ) = -\sin 150^\circ$ $[\because \sin(n\times360^\circ - \theta) = -\sin\theta, n \in Z]$

$= -\sin(180^\circ - 30^\circ)$

$= -\sin 30^\circ$ $[\because \sin(180^\circ - \theta) = \sin\theta]$

$= -\frac{1}{2}$

(iii) We have, $\cos(-870^\circ) = \cos 870^\circ$ $[\because \cos(-\theta) = \cos\theta]$

$= \cos(720^\circ + 150^\circ) = \cos(2\times360^\circ + 150^\circ)$

$= \cos 150^\circ$ $[\because \cos(n\times360^\circ + \theta) = \cos\theta, n \in Z]$

$= \cos(180^\circ - 30^\circ) = -\cos 30^\circ$ $[\because \cos(180^\circ - \theta) = -\cos\theta]$

$= -\frac{\sqrt{3}}{2}$

(iv) We have, $\tan(-2025^\circ) = -\tan(2025^\circ)$ $[\because \tan(-\theta) = -\tan\theta]$

$= -\tan(6\times360^\circ - 135^\circ)$

$= \tan 135^\circ$ $[\because \tan(n\times360^\circ - \theta) = -\tan\theta, n \in Z]$

$= \tan(90^\circ + 45^\circ) = -\cot 45^\circ$ $[\because \tan(90^\circ + \theta) = -\cot\theta]$

$= -1$

Example 8. Find the values of

(i) $\sin\left(\frac{25\pi}{3}\right)$ (ii) $\cos\left(\frac{41\pi}{4}\right)$

Sol. (i) We have, $\sin\left(\frac{25\pi}{3}\right) = \sin\left(8\pi + \frac{\pi}{3}\right) = \sin\left(2\pi\times4 + \frac{\pi}{3}\right)$

$= \sin\frac{\pi}{3}$ $[\because \sin(2n\pi + \theta) = \sin\theta, n \in Z]$

$= \frac{\sqrt{3}}{2}$

(ii) We have, $\cos\left(\frac{41\pi}{4}\right) = \cos\left(10\pi + \frac{\pi}{4}\right)$

$= \cos\left(2\pi\times5 + \frac{\pi}{4}\right) = \cos\frac{\pi}{4}$

$[\because \cos(2n\pi + \theta) = \cos\theta, n \in Z]$

$= \frac{1}{\sqrt{2}}$

Example 9. Prove that

$$\cot^2\left(\frac{\pi}{6}\right)+\text{cosec}\frac{5\pi}{6}+3\tan^2\frac{\pi}{6}=6.$$

Sol. LHS $=\cot^2\left(\frac{\pi}{6}\right)+\text{cosec}\frac{5\pi}{6}+3\tan^2\frac{\pi}{6}$

$$=\left(\cos\frac{\pi}{6}\right)^2+\text{cosec}\left(\pi-\frac{\pi}{6}\right)+3\left(\tan\frac{\pi}{6}\right)^2$$

$$=\left(\cos\frac{\pi}{6}\right)^2+\text{cosec}\frac{\pi}{6}+3\left(\tan\frac{\pi}{6}\right)^2$$

$[\because \text{cosec}(\pi-\theta)=\text{cosec}\,\theta]$

$$=(\sqrt{3})^2+2+3\left(\frac{1}{\sqrt{3}}\right)^2$$

$$=3+2+1=6$$

$=$ RHS **Hence proved.**

Quick Test 2

1. Find the value of $\tan^2\frac{\pi}{3}+2\cos^2\frac{\pi}{4}+3\sec^2\frac{\pi}{6}+4\cos^2\frac{\pi}{2}$.
2. If $\tan x=\frac{3}{4}$ and x lies in III quadrant, then find the value of $\sec x$.
3. If $\cot x=-\frac{5}{12}$, x lies in II quadrant, then find the value of other five trigonometric functions. *NCERT*
4. If $\sec x=-2$ and $\pi<x<\frac{3\pi}{2}$, then find the values of all other five trigonometric functions.
5. If $\cos\theta=\frac{-3}{5}$ and $\pi<\theta<\frac{3\pi}{2}$, then find the value of $\frac{\text{cosec}\,\theta+\cot\theta}{\sec\theta-\tan\theta}$.
6. Evaluate the value of the following.
   (i) $\sec\frac{15\pi}{4}$ (ii) $\sin 390°$
   (iii) $\sin 690°$ (iv) $\text{cosec}(-1170°)$
   (v) $\cos 135°$ (vi) $\sec 120°$
   (vii) $\text{cosec}\,150°$ (viii) $\tan(-945°)$
7. Prove that
   (i) $\frac{\cos(90°+\theta)\sec(-\theta)\tan(180°-\theta)}{\sec(360°-\theta)\sin(180°+\theta)\cot(90°-\theta)}=-1$
   (ii) $\frac{\tan(90°-\theta)\sec(180°-\theta)\sin(-\theta)}{\sin(180°+\theta)\cot(360°-\theta)\text{cosec}(90°-\theta)}=1$
8. If $\sin\theta+\text{cosec}\,\theta=2$, then find the value of $\sin^2\theta+\text{cosec}^2\theta$. *NCERT Exemplar*

Answers

1. 8
2. $\sec x=-\frac{5}{4}$
3. $\tan x=\frac{-12}{5}$, $\sec x=\frac{-13}{5}$, $\cos x=\frac{-5}{13}$, $\sin x=\frac{12}{13}$, $\text{cosec}\,x=\frac{13}{12}$
4. $\sin x=-\frac{\sqrt{3}}{2}$, $\cos x=-\frac{1}{2}$, $\tan x=\sqrt{3}$, $\cot x=\frac{1}{\sqrt{3}}$, $\text{cosec}\,x=\frac{-2}{\sqrt{3}}$
5. 1/6
6. (i) $\sqrt{2}$ (ii) 1/2 (iii) $-1/2$ (iv) -1 (v) $-1/\sqrt{2}$ (vi) -2 (vii) 2 (viii) -1
8. 2

For Solutions, See QR Code on Pg. 84

3.8 Trigonometric Functions of Compound Angles

- $\cos(A+B)=\cos A\cos B-\sin A\sin B$
- $\cos(A-B)=\cos A\cos B+\sin A\sin B$
- $\sin(A+B)=\sin A\cos B+\cos A\sin B$
- $\sin(A-B)=\sin A\cos B-\cos A\sin B$
- $\tan(A+B)=\frac{\tan A+\tan B}{1-\tan A\tan B}$,
  where A, B and $(A+B)$ are not odd multiples of $\frac{\pi}{2}$.
- $\tan(A-B)=\frac{\tan A-\tan B}{1+\tan A\tan B}$,
  where A, B and $(A-B)$ are not odd multiples of $\frac{\pi}{2}$.
- $\cot(A+B)=\frac{\cot A\cot B-1}{\cot B+\cot A}$,
  where A, B and $(A+B)$ are not integral multiples of π.
- $\cot(A-B)=\frac{\cot A\cot B+1}{\cot B-\cot A}$,
  where A, B and $(A-B)$ are not integral multiples of π.

Some Important Results

(i) $\sin(A+B)\sin(A-B)=\sin^2 A-\sin^2 B=\cos^2 B-\cos^2 A$

(ii) $\cos(A+B)\cos(A-B)=\cos^2 A-\sin^2 B=\cos^2 B-\sin^2 A$

(iii) $\sin(A+B+C)=\sin A\cos B\cos C+\cos A\sin B\cos C + \cos A\cos B\sin C-\sin A\sin B\sin C$

(iv) $\cos(A+B+C)=\cos A\cos B\cos C-\cos A\sin B\sin C - \sin A\cos B\sin C-\sin A\sin B\cos C$

(v) $\tan(A+B+C)=\dfrac{\tan A+\tan B+\tan C-\tan A\tan B\tan C}{1-\tan A\tan B-\tan B\tan C-\tan C\tan A}$

Example 1. **If $\sin A=\frac{4}{5}$ and $\cos B=\frac{5}{13}$, where $0<A, B<\frac{\pi}{2}$, then find the values of the following.**

(i) $\sin(A+B)$ (ii) $\cos(A-B)$

Sol. Given, $\sin A=\frac{4}{5}, 0<A<\frac{\pi}{2}$

$\therefore \quad \cos A=\sqrt{1-\sin^2 A}$ [$\because A$ lies in I quadrant]

$=\sqrt{1-\left(\frac{4}{5}\right)^2}=\sqrt{1-\frac{16}{25}}$

$\Rightarrow \quad \cos A=\sqrt{\frac{9}{25}}=\frac{3}{5}$ and $\cos B=\frac{5}{13}, 0<B<\frac{\pi}{2}$

$\therefore \quad \sin B=\sqrt{1-\cos^2 B}$ [$\because B$ lies in I quadrant]

$=\sqrt{1-\left(\frac{5}{13}\right)^2}=\sqrt{1-\frac{25}{169}}$

$\Rightarrow \quad \sin B=\sqrt{\frac{144}{169}}=\frac{12}{13}$

(i) $\sin(A+B)=\sin A\cos B+\cos A\sin B$

$=\frac{4}{5}\times\frac{5}{13}+\frac{3}{5}\times\frac{12}{13}=\frac{20}{65}+\frac{36}{65}=\frac{56}{65}$

(ii) $\cos(A-B)=\cos A\cos B+\sin A\sin B$

$=\frac{3}{5}\times\frac{5}{13}+\frac{4}{5}\times\frac{12}{13}=\frac{15+48}{65}=\frac{63}{65}$

Example 2. **Show that**

$$\sin(40^\circ+\theta)\cos(10^\circ+\theta)-\cos(40^\circ+\theta)\sin(10^\circ+\theta)=\frac{1}{2}.$$

NCERT

Sol. LHS

$=\sin(40^\circ+\theta)\cos(10^\circ+\theta)-\cos(40^\circ+\theta)\cdot\sin(10^\circ+\theta)$

$=\sin\{(40^\circ+\theta)-(10^\circ+\theta)\}$

[$\because \sin(x-y)=\sin x\cdot\cos y-\cos x\cdot\sin y$]

$=\sin 30^\circ=\frac{1}{2}$ $\left[\because \sin 30^\circ=\frac{1}{2}\right]$

= RHS **Hence proved.**

Example 3. **Find the value of $\sin 75^\circ$.**

Sol. We have, $\sin 75^\circ=\sin(45^\circ+30^\circ)$

$=\sin 45^\circ\cos 30^\circ+\cos 45^\circ\sin 30^\circ$

[$\because \sin(A+B)=\sin A\cos B+\cos A\sin B$]

$=\frac{1}{\sqrt{2}}\cdot\frac{\sqrt{3}}{2}+\frac{1}{\sqrt{2}}\cdot\frac{1}{2}$ $\left[\because \sin 30^\circ=\frac{1}{2}, \cos 30^\circ=\frac{\sqrt{3}}{2}\right.$ and $\left.\sin 45^\circ=\cos 45^\circ=\frac{1}{\sqrt{2}}\right]$

$=\frac{\sqrt{3}}{2\sqrt{2}}+\frac{1}{2\sqrt{2}}=\frac{\sqrt{3}+1}{2\sqrt{2}}$

Example 4. **Prove that $\dfrac{\cos 15^\circ+\sin 15^\circ}{\cos 15^\circ-\sin 15^\circ}=\sqrt{3}$.**

Competency Based

Sol. LHS $=\dfrac{\cos 15^\circ+\sin 15^\circ}{\cos 15^\circ-\sin 15^\circ}=\dfrac{\frac{\cos 15^\circ}{\cos 15^\circ}+\frac{\sin 15^\circ}{\cos 15^\circ}}{\frac{\cos 15^\circ}{\cos 15^\circ}-\frac{\sin 15^\circ}{\cos 15^\circ}}$

[dividing numerator and denominator by $\cos 15^\circ$]

$=\dfrac{1+\tan 15^\circ}{1-\tan 15^\circ}$ $\left[\because \tan\theta=\frac{\sin\theta}{\cos\theta}\right]$

$=\dfrac{\tan 45^\circ+\tan 15^\circ}{1-\tan 45^\circ\tan 15^\circ}$ [$\because \tan 45^\circ=1$]

$=\tan(45^\circ+15^\circ)$ $\left[\because \tan(A+B)=\frac{\tan A+\tan B}{1-\tan A\tan B}\right]$

$=\tan 60^\circ=\sqrt{3}=$ RHS **Hence proved.**

Example 5. **If $A+B=\frac{\pi}{4}$, then prove that**

(i) $(1+\tan A)(1+\tan B)=2$

(ii) $(\cot A-1)(\cot B-1)=2$.

Sol. (i) We have, $A+B=\frac{\pi}{4}$

$\therefore \quad \tan(A+B)=\tan\frac{\pi}{4} \Rightarrow \frac{\tan A+\tan B}{1-\tan A\tan B}=1$

$\Rightarrow \quad \tan A+\tan B=1-\tan A\tan B$

$\Rightarrow \quad \tan A+\tan B+\tan A\tan B=1$ ...(i)

On adding 1 both sides of Eq. (i), we get

$1+\tan A+\tan B+\tan A\tan B=2$

$\Rightarrow \quad (1+\tan A)+\tan B(1+\tan A)=2$

$\therefore \quad (1+\tan A)(1+\tan B)=2$

Hence proved.

(ii) On dividing both sides of Eq. (i) by $\tan A\tan B$, we get

$$\frac{\tan A+\tan B+\tan A\tan B}{\tan A\tan B}=\frac{1}{\tan A\tan B}$$

$$\Rightarrow \quad \cot B+\cot A+1=\cot A\cot B$$
$$\Rightarrow \quad \cot A\cot B-\cot A-\cot B=1$$
$$\Rightarrow \quad \cot A\cot B-\cot A-\cot B+1=2$$
$$\Rightarrow \quad \cot A(\cot B-1)-(\cot B-1)=2$$
$$\therefore \quad (\cot B-1)(\cot A-1)=2$$

Hence proved.

Example 6. Prove that

$$\frac{\tan\left(\frac{\pi}{4}+x\right)}{\tan\left(\frac{\pi}{4}-x\right)}=\left(\frac{1+\tan x}{1-\tan x}\right)^2.$$

NCERT

Sol.
$$\text{LHS}=\frac{\tan\left(\frac{\pi}{4}+x\right)}{\tan\left(\frac{\pi}{4}-x\right)}=\frac{\left(\dfrac{\tan\frac{\pi}{4}+\tan x}{1-\tan\frac{\pi}{4}\cdot\tan x}\right)}{\left(\dfrac{\tan\frac{\pi}{4}-\tan x}{1+\tan\frac{\pi}{4}\cdot\tan x}\right)}$$

$$\left[\because \tan(A+B)=\frac{\tan A+\tan B}{1-\tan A\cdot\tan B} \text{ and } \tan(A-B)=\frac{\tan A-\tan B}{1+\tan A\cdot\tan B}\right]$$

$$=\frac{\left(\dfrac{1+\tan x}{1-\tan x}\right)}{\left(\dfrac{1-\tan x}{1+\tan x}\right)}=\left(\frac{1+\tan x}{1-\tan x}\right)^2=\text{RHS}$$

Hence proved.

Example 7. Prove that $\tan 70^\circ=\tan 20^\circ+2\tan 50^\circ$.

Sol. Clearly, $\tan 70^\circ=\tan(20^\circ+50^\circ)$

$$\Rightarrow \quad \tan 70^\circ=\frac{\tan 20^\circ+\tan 50^\circ}{1-\tan 20^\circ\tan 50^\circ}$$
$$\Rightarrow \tan 70^\circ-\tan 70^\circ\tan 20^\circ\tan 50^\circ=\tan 20^\circ+\tan 50^\circ$$
$$\Rightarrow \tan 70^\circ-\tan 70^\circ\tan(90^\circ-70^\circ)\tan 50^\circ=\tan 20^\circ+\tan 50^\circ$$
$$\Rightarrow \tan 70^\circ-\tan 70^\circ\cot 70^\circ\tan 50^\circ=\tan 20^\circ+\tan 50^\circ$$
$$\Rightarrow \quad \tan 70^\circ-\tan 50^\circ=\tan 20^\circ+\tan 50^\circ$$
$$\Rightarrow \quad \tan 70^\circ=\tan 20^\circ+\tan 50^\circ+\tan 50^\circ$$
$$\therefore \quad \tan 70^\circ=\tan 20^\circ+2\tan 50^\circ$$

Hence proved.

Example 8. Find the value of $\tan(\alpha+\beta)$, if

$$\cot\alpha=\frac{1}{2}, \alpha\in\left(\pi,\frac{3\pi}{2}\right) \text{ and } \sec\beta=\frac{-5}{3}, \beta\in\left(\frac{\pi}{2},\pi\right).$$

Sol. We have, $\tan(\alpha+\beta)=\dfrac{\tan\alpha+\tan\beta}{1-\tan\alpha\cdot\tan\beta}$

$$\left[\because \tan(x+y)=\frac{\tan x+\tan y}{1-\tan x\cdot\tan y}\right]$$

Given, $\cot\alpha=\dfrac{1}{2} \Rightarrow \tan\alpha=2$ $\quad\left[\because \tan x=\dfrac{1}{\cot x}\right]$

Also, $\sec\beta=\dfrac{-5}{3}$

Then, $\tan\beta=\pm\sqrt{\sec^2\beta-1}$

$$\Rightarrow \quad \tan\beta=\pm\sqrt{\frac{25}{9}-1}=\pm\sqrt{\frac{16}{9}} \quad \left[\because \sec\beta=\frac{-5}{3}\right]$$

$$\therefore \quad \tan\beta=\pm\frac{4}{3}$$

But $\tan\beta\neq\dfrac{4}{3}$ $\quad\left[\because \beta\in\left(\dfrac{\pi}{2},\pi\right) \text{ and } \tan\beta \text{ is negative in II quadrant}\right]$

$$\therefore \quad \tan\beta=-\frac{4}{3}$$

$$\therefore \tan(\alpha+\beta)=\frac{\tan\alpha+\tan\beta}{1-\tan\alpha\cdot\tan\beta}=\frac{2+\left(-\frac{4}{3}\right)}{1-(2)\left(\frac{-4}{3}\right)}$$

$$\left[\because \tan\alpha=2 \text{ and } \tan\beta=\frac{-4}{3}\right]$$

$$=\frac{\left(2-\frac{4}{3}\right)}{\left(1+\frac{8}{3}\right)}=\frac{\left(\frac{2}{3}\right)}{\left(\frac{11}{3}\right)}=\frac{2}{11}$$

Example 9. Prove that $\sin^2 A=\cos^2(A-B)+\cos^2 B-2\cos(A-B)\cos A\cos B$. *Competency Based Que*

Sol. RHS

$=\cos^2(A-B)+\cos^2 B-2\cos(A-B)\cos A\cos B$

$=\cos^2 B+\cos^2(A-B)-2\cos(A-B)\cos A\cos B$

$=\cos^2 B+\cos(A-B)\{\cos(A-B)-2\cos A\cos B\}$

$=\cos^2 B+\cos(A-B)\{\cos A\cos B+\sin A\sin B-2\cos A\cos B\}$

$=\cos^2 B+\cos(A-B)[\sin A\sin B-\cos A\cos B]$

$=\cos^2 B+\cos(A-B)[-(\cos A\cos B-\sin A\sin B)]$

$=\cos^2 B-\cos(A-B)\cos(A+B)$

$=\cos^2 B-(\cos^2 A-\sin^2 B)$

$=\cos^2 B+\sin^2 B-\cos^2 A$

$=1-\cos^2 A=\sin^2 A=\text{LHS}$ **Hence proved.**

Quick Test 3

1. If $\sin x = \frac{8}{17}$ and $\cos y = \frac{7}{25}$, where $0 < x, y < \frac{\pi}{2}$, then find the values of the following.

(i) $\sin(x+y)$

(ii) $\cos(x+y)$

(iii) $\sin(x-y)$

(iv) $\cos(x-y)$

2. Find the value of the following trigonometric functions.

(i) $\sin(105°)$ (ii) $\cos(-75°)$

(iii) $\tan(435°)$ (iv) $\sec(105°)$

(v) $\sin 15°$

3. Find the value of the following.

(i) $\tan\frac{7\pi}{12}$ (ii) $\tan\frac{13\pi}{12}$

4. Find the value of $\sin(45° + \theta) - \cos(45° - \theta)$.

NCERT Exemplar

5. Show that

$$\cos\left(\frac{\pi}{4}-\theta\right)\cos\left(\frac{\pi}{4}-\phi\right)-\sin\left(\frac{\pi}{4}-\theta\right)\sin\left(\frac{\pi}{4}-\phi\right) = \sin(\theta+\phi).$$

Competency Based Que

6. Prove that $\frac{\sqrt{3}\cos 23° - \sin 23°}{2} = \cos 53°$.

Competency Based Que

7. Prove that

(i) $\tan 15° + \cot 15° = 4$ (ii) $\cos 15° - \sin 15° = \frac{1}{\sqrt{2}}$

8. Prove that $\cos\left(\frac{3\pi}{4}+x\right)-\cos\left(\frac{3\pi}{4}-x\right) = -\sqrt{2}\sin x$.

9. Prove that $\sin^2\left(\frac{\pi}{8}+\frac{A}{2}\right)-\sin^2\left(\frac{\pi}{8}-\frac{A}{2}\right) = \frac{1}{\sqrt{2}}\sin A$.

10. If $\cos(\theta+\phi) = m\cos(\theta-\phi)$, then find the value of $\frac{1-m}{1+m}\cot\phi$.

Competency Based Que

Answers

1. (i) $\frac{416}{425}$ (ii) $\frac{-87}{425}$ (iii) $\frac{-304}{425}$ (iv) $\frac{297}{425}$

2. (i) $\frac{\sqrt{3}+1}{2\sqrt{2}}$ (ii) $\frac{\sqrt{3}-1}{2\sqrt{2}}$ (iii) $\frac{\sqrt{3}+1}{\sqrt{3}-1}$ (iv) $\frac{2\sqrt{2}}{1-\sqrt{3}}$ (v) $\frac{\sqrt{3}-1}{2\sqrt{2}}$

3. (i) $-(2+\sqrt{3})$ (ii) $(2-\sqrt{3})$

4. 0

For Solutions, See QR Code on Pg. 84

3.9 Transformation Formulae

Transformation of Product into Sum or Difference

(i) $2\cos A\cos B = \cos(A+B)+\cos(A-B)$

(ii) $2\sin A\sin B = \cos(A-B)-\cos(A+B)$

(iii) $2\sin A\cos B = \sin(A+B)+\sin(A-B)$

(iv) $2\cos A\sin B = \sin(A+B)-\sin(A-B)$

Transformation of Sum or Difference into Product

- $\cos A + \cos B = 2\cos\left(\frac{A+B}{2}\right)\cdot\cos\left(\frac{A-B}{2}\right)$
- $\cos A - \cos B = -2\sin\left(\frac{A+B}{2}\right)\cdot\sin\left(\frac{A-B}{2}\right)$
- $\sin A + \sin B = 2\sin\left(\frac{A+B}{2}\right)\cdot\cos\left(\frac{A-B}{2}\right)$
- $\sin A - \sin B = 2\cos\left(\frac{A+B}{2}\right)\cdot\sin\left(\frac{A-B}{2}\right)$

Example 1. Find the value of the following functions.

(i) $2\cos 45°\sin 15°$

(ii) $2\sin 15°\cos 75°$

Sol. (i) $2\cos 45°\sin 15°$

$= \sin(45°+15°)-\sin(45°-15°)$

$[\because 2\cos A\sin B = \sin(A+B)-\sin(A-B)]$

$= \sin 60° - \sin 30°$

$= \frac{\sqrt{3}}{2}-\frac{1}{2}$ $\left[\because \sin 60° = \frac{\sqrt{3}}{2} \text{ and } \sin 30° = \frac{1}{2}\right]$

$= \frac{\sqrt{3}-1}{2}$

(ii) $2\sin 15°\cos 75°$

$= \sin(15°+75°)+\sin(15°-75°)$

$[\because 2\sin A\cdot\cos B = \sin(A+B)+\sin(A-B)]$

$= \sin 90° + \sin(-60°)$

$= \sin 90° - \sin 60°$ $[\because \sin(-\theta) = -\sin\theta]$

$= 1-\frac{\sqrt{3}}{2} = \frac{2-\sqrt{3}}{2}$

Example 2. **Find the value of $2\sin 52\frac{1^\circ}{2}\sin 7\frac{1^\circ}{2}$.**

Sol. We have, $2\sin 52\frac{1^\circ}{2}\sin 7\frac{1^\circ}{2} = 2\sin\frac{105^\circ}{2}\sin\frac{15^\circ}{2}$

$$= \cos\left(\frac{105^\circ}{2}-\frac{15^\circ}{2}\right) - \cos\left(\frac{105^\circ}{2}+\frac{15^\circ}{2}\right)$$

$[\because 2\sin A\sin B = \cos(A-B) - \cos(A+B)]$

$$= \cos\left(\frac{90^\circ}{2}\right) - \cos\left(\frac{120^\circ}{2}\right) = \cos 45^\circ - \cos 60^\circ$$

$$= \frac{1}{\sqrt{2}} - \frac{1}{2} \qquad \left[\because \cos 45^\circ = \frac{1}{\sqrt{2}} \text{ and } \cos 60^\circ = \frac{1}{2}\right]$$

$$= \frac{2-\sqrt{2}}{2\sqrt{2}} = \frac{\sqrt{2}(\sqrt{2}-1)}{2\sqrt{2}}$$

$$= \frac{\sqrt{2}-1}{2} \qquad [\because \sqrt{2}\times\sqrt{2} = 2]$$

Example 3. **Prove that**

$$\cos\left(\frac{\pi}{6}-x\right)\cos\left(\frac{\pi}{6}+x\right) = \frac{1}{4}+\frac{1}{2}\cos 2x.$$

Sol. LHS $= \cos\left(\frac{\pi}{6}-x\right)\cos\left(\frac{\pi}{6}+x\right)$

$$= \frac{1}{2}\left[2\cos\left(\frac{\pi}{6}+x\right)\cos\left(\frac{\pi}{6}-x\right)\right]$$

$$= \frac{1}{2}\left[\cos\left\{\left(\frac{\pi}{6}+x\right)+\left(\frac{\pi}{6}-x\right)\right\} + \cos\left\{\left(\frac{\pi}{6}+x\right)-\left(\frac{\pi}{6}-x\right)\right\}\right]$$

$[\because 2\cos A\cos B = \cos(A+B) + \cos(A-B)]$

$$= \frac{1}{2}\left(\cos\frac{\pi}{3} + \cos 2x\right)$$

$$= \frac{1}{2}\left(\frac{1}{2} + \cos 2x\right) \qquad \left[\because \cos\frac{\pi}{3} = \frac{1}{2}\right]$$

$$= \frac{1}{4} + \frac{1}{2}\cos 2x = \text{RHS}$$

Hence proved.

Example 4. **Prove that**

$$4\cos 12^\circ\cos 48^\circ\cos 72^\circ = \cos 36^\circ.$$

Sol. LHS $= 4\cos 12^\circ\cos 48^\circ\cos 72^\circ$

$$= 2(2\cos 12^\circ\cos 48^\circ)\cos 72^\circ$$

$$= 2[\cos(12^\circ+48^\circ) + \cos(12^\circ-48^\circ)]\cos 72^\circ$$

$[\because 2\cos A\cos B = \cos(A+B) + \cos(A-B)]$

$$= 2[\cos 60^\circ + \cos(-36^\circ)]\cos 72^\circ$$

$$= 2[\cos 60^\circ + \cos 36^\circ]\cos 72^\circ \qquad [\because \cos(-\theta) = \cos\theta]$$

$$= 2\left[\frac{1}{2}\cos 72^\circ + \cos 36^\circ\cos 72^\circ\right] \qquad \left[\because \cos 60^\circ = \frac{1}{2}\right]$$

$$= \cos 72^\circ + 2\cos 36^\circ\cos 72^\circ$$

$$= \cos 72^\circ + \cos(36^\circ+72^\circ) + \cos(36^\circ-72^\circ)$$

$[\because 2\cos A\cos B = \cos(A+B) + \cos(A-B)]$

$$= \cos 72^\circ + \cos 108^\circ + \cos 36^\circ$$

$$= \cos 72^\circ + \cos(180^\circ-72^\circ) + \cos 36^\circ$$

$$= \cos 72^\circ - \cos 72^\circ + \cos 36^\circ$$

$[\because \cos(\pi-\theta) = -\cos\theta]$

$$= \cos 36^\circ = \text{RHS}$$

Hence proved.

Example 5. **Prove that**

$$\sin 10^\circ\sin 30^\circ\sin 50^\circ\sin 70^\circ = \frac{1}{16}.$$

Sol. LHS $= \sin 10^\circ\sin 30^\circ\sin 50^\circ\sin 70^\circ$

$$= \sin 30^\circ(\sin 50^\circ\sin 10^\circ)\sin 70^\circ$$

$$= \frac{1}{2}\times\frac{1}{2}(2\sin 50^\circ\sin 10^\circ)\sin 70^\circ$$

$$= \frac{1}{4}[\cos(50^\circ-10^\circ) - \cos(50^\circ+10^\circ)]\sin 70^\circ$$

$[\because 2\sin A\sin B = \cos(A-B) - \cos(A+B)]$

$$= \frac{1}{4}[\cos 40^\circ - \cos 60^\circ]\sin 70^\circ$$

$$= \frac{1}{4}[\sin 70^\circ\cos 40^\circ - \sin 70^\circ\cos 60^\circ]$$

$$= \frac{1}{4}\left[\sin 70^\circ\cos 40^\circ - \frac{1}{2}\sin 70^\circ\right]$$

$$= \frac{1}{8}[2\sin 70^\circ\cos 40^\circ - \sin 70^\circ]$$

$$= \frac{1}{8}[\sin(70^\circ+40^\circ) + \sin(70^\circ-40^\circ) - \sin 70^\circ]$$

$[\because 2\sin A\cos B = \sin(A+B) + \sin(A-B)]$

$$= \frac{1}{8}[\sin 110^\circ + \sin 30^\circ - \sin 70^\circ]$$

$$= \frac{1}{8}\left[\sin(180^\circ-70^\circ) + \frac{1}{2} - \sin 70^\circ\right]$$

$[\because \sin(\pi-\theta) = \sin\theta]$

$$= \frac{1}{8}\left[\sin 70^\circ + \frac{1}{2} - \sin 70^\circ\right] = \frac{1}{8}\times\frac{1}{2} = \frac{1}{16} = \text{RHS}$$

Hence proved.

Example 6. **Prove that**

$$\tan\theta\tan(60^\circ-\theta)\tan(60^\circ+\theta) = \tan 3\theta.$$

Sol. LHS $= \tan\theta\tan(60^\circ-\theta)\tan(60^\circ+\theta)$

$$= \frac{\sin\theta}{\cos\theta}\cdot\frac{\sin(60^\circ-\theta)}{\cos(60^\circ-\theta)}\cdot\frac{\sin(60^\circ+\theta)}{\cos(60^\circ+\theta)}$$

$$= \frac{\sin\theta[2\sin(60^\circ-\theta)\sin(60^\circ+\theta)]}{\cos\theta[2\cos(60^\circ-\theta)\cos(60^\circ+\theta)]}$$

$$= \frac{\sin\theta(\cos 2\theta - \cos 120^\circ)}{\cos\theta(\cos 120^\circ + \cos 2\theta)}$$

$\left[\begin{array}{l}\because 2\sin A\sin B = \cos(A-B) - \cos(A+B) \\ \text{and } 2\cos A\cos B = \cos(A+B) + \cos(A-B)\end{array}\right]$

$$= \frac{\sin\theta\left(\cos 2\theta + \frac{1}{2}\right)}{\cos\theta\left(\cos 2\theta - \frac{1}{2}\right)}$$

$$\left[\because \cos 120^\circ = \cos(180^\circ - 60^\circ) = -\cos 60^\circ = -\frac{1}{2}\right]$$

$$= \frac{\sin\theta\cos 2\theta + \frac{1}{2}\sin\theta}{\cos\theta\cos 2\theta - \frac{1}{2}\cos\theta} = \frac{2\sin\theta\cos 2\theta + \sin\theta}{2\cos\theta\cos 2\theta - \cos\theta}$$

$$= \frac{\sin(\theta + 2\theta) + \sin(-\theta) + \sin\theta}{\cos(\theta + 2\theta) + \cos(-\theta) - \cos\theta}$$

$$\left[\begin{array}{l}\because 2\sin A\cos B = \sin(A+B) + \sin(A-B) \\ \text{and } 2\cos A\cos B = \cos(A+B) + \cos(A-B)\end{array}\right]$$

$$= \frac{\sin 3\theta - \sin\theta + \sin\theta}{\cos 3\theta + \cos\theta - \cos\theta} \quad \left[\begin{array}{l}\because \sin(-\theta) = -\sin\theta \\ \text{and } \cos(-\theta) = \cos\theta\end{array}\right]$$

$$= \frac{\sin 3\theta}{\cos 3\theta} = \tan 3\theta$$

= RHS **Hence proved.**

Example 7. **Prove that $\cos 55^\circ + \cos 65^\circ + \cos 175^\circ = 0$.**

Sol LHS = $\cos 55^\circ + \cos 65^\circ + \cos 175^\circ$

$$= 2\cos\left(\frac{55^\circ + 65^\circ}{2}\right)\cos\left(\frac{55^\circ - 65^\circ}{2}\right) + \cos 175^\circ$$

$$\left[\because \cos A + \cos B = 2\cos\frac{A+B}{2}\cos\frac{A-B}{2}\right]$$

$$= 2\cos 60^\circ \cos(-5^\circ) + \cos 175^\circ$$

$$= 2 \times \frac{1}{2}\cos 5^\circ + \cos(180^\circ - 5^\circ) \quad [\because \cos(-\theta) = \cos\theta]$$

$$= \cos 5^\circ - \cos 5^\circ = 0 \quad [\because \cos(\pi - \theta) = -\cos\theta]$$

= RHS **Hence proved.**

Example 8. **Prove that**

$$\frac{\sin 5x - 2\sin 3x + \sin x}{\cos 5x - \cos x} = \operatorname{cosec} 2x - \cot 2x.$$

Competency Based Que

Sol LHS $= \dfrac{\sin 5x - 2\sin 3x + \sin x}{\cos 5x - \cos x} = \dfrac{\sin 5x + \sin x - 2\sin 3x}{\cos 5x - \cos x}$

$$= \frac{2\sin\left(\frac{5x+x}{2}\right)\cos\left(\frac{5x-x}{2}\right) - 2\sin 3x}{-2\sin\left(\frac{5x+x}{2}\right)\sin\left(\frac{5x-x}{2}\right)}$$

$$\left[\begin{array}{l}\because \sin A + \sin B = 2\sin\frac{A+B}{2}\cos\frac{A-B}{2} \\ \text{and } \cos A - \cos B = -2\sin\frac{A+B}{2}\sin\frac{A-B}{2}\end{array}\right]$$

$$= \frac{2\sin 3x\cos 2x - 2\sin 3x}{-2\sin 3x\sin 2x}$$

$$= \frac{2\sin 3x(\cos 2x - 1)}{-2\sin 3x\sin 2x}$$

$$= \frac{-\cos 2x}{\sin 2x} + \frac{1}{\sin 2x}$$

$$= -\cot 2x + \operatorname{cosec} 2x$$

$$= \operatorname{cosec} 2x - \cot 2x$$

= RHS **Hence proved.**

Example 9. **Prove that**

$$\cos 2A\cos\frac{A}{2} - \cos 3A\cos\frac{9A}{2} = \sin 5A\sin\frac{5A}{2}.$$

Competency Based Que

Sol. LHS $= \cos 2A\cos\dfrac{A}{2} - \cos 3A\cos\dfrac{9A}{2}$

$$= \frac{1}{2}\left[2\cos 2A\cos\frac{A}{2}\right] - \frac{1}{2}\left[2\cos\frac{9A}{2}\cos 3A\right]$$

$$= \frac{1}{2}\left[\cos\left(2A + \frac{A}{2}\right) + \cos\left(2A - \frac{A}{2}\right)\right] - \frac{1}{2}\left[\cos\left(\frac{9A}{2} + 3A\right) + \cos\left(\frac{9A}{2} - 3A\right)\right]$$

$$[\because 2\cos A\cos B = \cos(A+B) + \cos(A-B)]$$

$$= \frac{1}{2}\left[\cos\left(\frac{4A+A}{2}\right) + \cos\left(\frac{4A-A}{2}\right)\right] - \frac{1}{2}\left[\cos\left(\frac{9A+6A}{2}\right) + \cos\left(\frac{9A-6A}{2}\right)\right]$$

$$= \frac{1}{2}\left(\cos\frac{5A}{2} + \cos\frac{3A}{2}\right) - \frac{1}{2}\left(\cos\frac{15A}{2} + \cos\frac{3A}{2}\right)$$

$$= \frac{1}{2}\cos\frac{5A}{2} + \frac{1}{2}\cos\frac{3A}{2} - \frac{1}{2}\cos\frac{15A}{2} - \frac{1}{2}\cos\frac{3A}{2}$$

$$= \frac{1}{2}\cos\frac{5A}{2} - \frac{1}{2}\cos\frac{15A}{2}$$

$$= \frac{1}{2}\left(\cos\frac{5A}{2} - \cos\frac{15A}{2}\right)$$

$$= \frac{1}{2}\left[-2\sin\left(\frac{\frac{5A}{2} + \frac{15A}{2}}{2}\right)\sin\left(\frac{\frac{5A}{2} - \frac{15A}{2}}{2}\right)\right]$$

$$\left[\because \cos A - \cos B = -2\sin\frac{A+B}{2}\sin\frac{A-B}{2}\right]$$

$$= -(\sin 5A)\sin\left(\frac{-5A}{2}\right)$$

$$= \sin 5A\sin\frac{5A}{2} = \text{RHS}$$

Hence proved.

Example 10. **Prove that**

$$(\cos A-\cos B)^2+(\sin A-\sin B)^2=4\sin^2\frac{A-B}{2}.$$

Sol. $\text{LHS}=(\cos A-\cos B)^2+(\sin A-\sin B)^2$

$$=\left(-2\sin\frac{A+B}{2}\sin\frac{A-B}{2}\right)^2+\left(2\cos\frac{A+B}{2}\sin\frac{A-B}{2}\right)^2$$

$$\left[\because \cos C-\cos D=-2\sin\left(\frac{C+D}{2}\right)\sin\left(\frac{C-D}{2}\right) \text{ and } \sin C-\sin D=2\cos\left(\frac{C+D}{2}\right)\sin\left(\frac{C-D}{2}\right)\right]$$

$$=4\sin^2\frac{A+B}{2}\sin^2\frac{A-B}{2}+4\cos^2\frac{A+B}{2}\sin^2\frac{A-B}{2}$$

$$=4\sin^2\frac{A-B}{2}\left(\sin^2\frac{A+B}{2}+\cos^2\frac{A+B}{2}\right)$$

$$=4\sin^2\frac{A-B}{2}\times 1 \qquad [\because \sin^2\theta+\cos^2\theta=1]$$

$$=4\sin^2\frac{A-B}{2}=\text{RHS}$$

Hence proved.

Example 11. **If $m\sin\theta=n\sin(\theta+2\alpha)$, then prove that $\tan(\theta+\alpha)\cot\alpha=\frac{m+n}{m-n}$.**

NCERT Exemplar

Sol. Given, $m\sin\theta=n\sin(\theta+2\alpha)$

$$\Rightarrow \quad \frac{m}{n}=\frac{\sin(\theta+2\alpha)}{\sin\theta}$$

On applying componendo and dividendo rule, we get

$$\frac{m+n}{m-n}=\frac{\sin(\theta+2\alpha)+\sin\theta}{\sin(\theta+2\alpha)-\sin\theta}$$

$$=\frac{2\sin\left(\frac{\theta+2\alpha+\theta}{2}\right)\cos\left(\frac{\theta+2\alpha-\theta}{2}\right)}{2\cos\left(\frac{\theta+2\alpha+\theta}{2}\right)\sin\left(\frac{\theta+2\alpha-\theta}{2}\right)}$$

$$\left[\because \sin A+\sin B=2\sin\left(\frac{A+B}{2}\right)\cos\left(\frac{A-B}{2}\right) \text{ and } \sin A-\sin B=2\cos\left(\frac{A+B}{2}\right)\sin\left(\frac{A-B}{2}\right)\right]$$

$$=\frac{2\sin(\theta+\alpha)\cos\alpha}{2\cos(\theta+\alpha)\sin\alpha}=\tan(\theta+\alpha)\cot\alpha$$

Hence proved.

Quick Test 4

1. Find the value of the following function.

(i) $2\cos 75^\circ\cos 15^\circ$ (ii) $\sin\frac{5\pi}{12}\sin\frac{\pi}{12}$

(iii) $\sin 75^\circ\cos 15^\circ$ (iv) $2\cos 22\frac{1}{2}^\circ\cos 67\frac{1}{2}^\circ$

2. Prove that

(i) $\sin 10^\circ\sin 50^\circ\sin 70^\circ=\frac{1}{8}$

(ii) $\cos 20^\circ\cos 40^\circ\cos 80^\circ=\frac{1}{8}$

(iii) $\tan 10^\circ\tan 50^\circ\tan 70^\circ=\frac{1}{\sqrt{3}}$

3. Prove that

(i) $\sin 20^\circ\sin 40^\circ\sin 60^\circ\sin 80^\circ=\frac{3}{16}$

(ii) $\cos 20^\circ\cos 40^\circ\cos 60^\circ\cos 80^\circ=\frac{1}{16}$

4. Prove that

(i) $\frac{\sin 9A-\sin 7A}{\cos 7A-\cos 9A}=\cot 8A$

(ii) $\frac{\cos A+\cos B}{\cos B-\cos A}=\cot\left(\frac{A+B}{2}\right)\cot\left(\frac{A-B}{2}\right)$

5. If $\cos x+\cos y=\frac{1}{3}$ and $\sin x+\sin y=\frac{1}{4}$, then prove that $\tan\left(\frac{x+y}{2}\right)=\frac{3}{4}$.

6. Prove that

$\sin A\cdot\sin(60^\circ-A)\cdot\sin(60^\circ+A)=\frac{1}{4}\sin 3A$.

7. If $x\cos\theta=y\cos\left(\theta+\frac{2\pi}{3}\right)=z\cos\left(\theta+\frac{4\pi}{3}\right)$, then show that $xy+yz+zx=0$.

8. Prove that

$(\cos A+\cos B)^2+(\sin A+\sin B)^2=4\cos^2\left(\frac{A-B}{2}\right)$.

9. Prove that $\cos x+\cos y+\cos z+\cos(x+y+z)$

$=4\cos\left(\frac{x+y}{2}\right)\cos\left(\frac{y+z}{2}\right)\cos\left(\frac{z+x}{2}\right)$.

Answers

1. (i) $\frac{1}{2}$ (ii) $\frac{1}{4}$ (iii) $\frac{2+\sqrt{3}}{4}$ (iv) $\frac{1}{\sqrt{2}}$

For Solutions, See QR Code on Pg. 84

3.10 Trigonometric Function of Multiple Angles

Some important trigonometric functions of angles $2x$, $3x$,... etc, in terms of the trigonometric functions of angle x are given below.

- $\cos 2x = \cos^2 x - \sin^2 x = 2\cos^2 x - 1$

$$= 1 - 2\sin^2 x = \frac{1-\tan^2 x}{1+\tan^2 x}$$

- $\sin 2x = 2\sin x \cos x = \dfrac{2\tan x}{1+\tan^2 x}$
- $\tan 2x = \dfrac{2\tan x}{1-\tan^2 x}$
- $\cos 3x = 4\cos^3 x - 3\cos x$
- $\sin 3x = 3\sin x - 4\sin^3 x$
- $\tan 3x = \dfrac{3\tan x - \tan^3 x}{1-3\tan^2 x}$

Trigonometric Ratios of the Angle x in Terms of $\frac{x}{2}$

- $\sin x = 2\sin\dfrac{x}{2}\cos\dfrac{x}{2} = \dfrac{2\tan\frac{x}{2}}{1+\tan^2\frac{x}{2}}$
- $\cos x = \cos^2\left(\dfrac{x}{2}\right) - \sin^2\left(\dfrac{x}{2}\right) = 2\cos^2\left(\dfrac{x}{2}\right) - 1$

$$= 1 - 2\sin^2\left(\frac{x}{2}\right) = \frac{1-\tan^2\left(\frac{x}{2}\right)}{1+\tan^2\left(\frac{x}{2}\right)}$$

- $\tan x = \dfrac{2\tan\left(\frac{x}{2}\right)}{1-\tan^2\left(\frac{x}{2}\right)}$

Trigonometric Ratios of the Angle x in Terms of $\frac{x}{3}$

- $\sin x = 3\sin\left(\dfrac{x}{3}\right) - 4\sin^3\left(\dfrac{x}{3}\right)$
- $\cos x = 4\cos^3\left(\dfrac{x}{3}\right) - 3\cos\left(\dfrac{x}{3}\right)$
- $\tan x = \dfrac{3\tan\left(\frac{x}{3}\right) - \tan^3\left(\frac{x}{3}\right)}{1-3\tan^2\left(\frac{x}{3}\right)}$

Trigonometric Ratios of Some More Angles

- $\sin 18^\circ = \cos 72^\circ = \dfrac{\sqrt{5}-1}{4}$
- $\cos 18^\circ = \sin 72^\circ = \dfrac{\sqrt{10+2\sqrt{5}}}{4}$
- $\cos 36^\circ = \sin 54^\circ = \dfrac{\sqrt{5}+1}{4}$
- $\sin 36^\circ = \cos 54^\circ = \dfrac{\sqrt{10-2\sqrt{5}}}{4}$

Example 1. If $\sin x = \frac{1}{3}$, where $0 < x < \frac{\pi}{2}$, then find the value of

(i) $\sin 3x$ (ii) $\cos 3x$ (iii) $\tan 3x$

Sol. Given, $\sin x = \frac{1}{3}$

$$\therefore \quad \cos x = \sqrt{1-\sin^2 x}$$

$$= \sqrt{1-\left(\frac{1}{3}\right)^2} = \frac{2\sqrt{2}}{3}$$

and $\tan x = \dfrac{\sin x}{\cos x} = \dfrac{1/3}{2\sqrt{2}/3} = \dfrac{1}{2\sqrt{2}}$

(i) $\sin 3x = 3\sin x - 4\sin^3 x$

$$= 3\left(\frac{1}{3}\right) - 4\left(\frac{1}{3}\right)^3 = 1 - \frac{4}{27} = \frac{23}{27}$$

(ii) $\cos 3x = 4\cos^3 x - 3\cos x$

$$= 4\left(\frac{2\sqrt{2}}{3}\right)^3 - 3\left(\frac{2\sqrt{2}}{3}\right)$$

$$= \frac{64\sqrt{2}}{27} - 2\sqrt{2} = \frac{10\sqrt{2}}{27}$$

(iii) $\tan 3x = \dfrac{3\tan x - \tan^3 x}{1-3\tan^2 x} = \dfrac{3\left(\frac{1}{2\sqrt{2}}\right) - \left(\frac{1}{2\sqrt{2}}\right)^3}{1-3\left(\frac{1}{2\sqrt{2}}\right)^2}$

$$= \frac{\frac{3}{2\sqrt{2}} - \frac{1}{16\sqrt{2}}}{1-\frac{3}{8}} = \frac{24-1}{16\sqrt{2}} \times \frac{8}{5} = \frac{23}{10\sqrt{2}}$$

Example 2. **Find the value of $\sqrt{3}$ cosec 20° − sec 20°.**

Sol. We have, $\sqrt{3}$ cosec 20° − sec 20°

$$= \frac{\sqrt{3}}{\sin 20^\circ} - \frac{1}{\cos 20^\circ} = \frac{\sqrt{3}\cos 20^\circ - \sin 20^\circ}{\sin 20^\circ \cos 20^\circ}$$

$$= \frac{2\left(\frac{\sqrt{3}}{2}\cos 20^\circ - \frac{1}{2}\sin 20^\circ\right)}{\sin 20^\circ \cos 20^\circ}$$

$$= \frac{2(\sin 60^\circ \cos 20^\circ - \cos 60^\circ \sin 20^\circ)}{\sin 20^\circ \cos 20^\circ}$$

$$\left[\because \sin 60^\circ = \frac{\sqrt{3}}{2} \text{ and } \cos 60^\circ = \frac{1}{2}\right]$$

$$= \frac{2\sin(60^\circ - 20^\circ)}{\sin 20^\circ \cos 20^\circ} = \frac{2\sin 40^\circ}{\sin 20^\circ \cos 20^\circ}$$

$$= \frac{4\sin 20^\circ \cos 20^\circ}{\sin 20^\circ \cos 20^\circ} \quad [\because \sin 2\theta = 2\sin\theta\cos\theta]$$

$$= 4$$

Example 3. **Prove that**

(i) $\cos 4x = 1 - 8\sin^2 x \cos^2 x$ *NCERT*

(ii) $\frac{1-\cos 2x}{\sin 2x} = \tan x$

(iii) $\frac{1-\sin 2x}{1+\sin 2x} = \tan^2\left(\frac{\pi}{4} - x\right)$

(iv) $\frac{1+\sin 2x - \cos 2x}{1+\sin 2x + \cos 2x} = \tan x$

Sol. (i) LHS $= \cos 4x = 1 - 2\sin^2 2x$ $[\because \cos 2\theta = 1 - 2\sin^2\theta]$

$$= 1 - 2(\sin 2x)^2$$

$$= 1 - 2(2\sin x \cdot \cos x)^2 \quad [\because \sin 2\theta = 2\sin\theta\cos\theta]$$

$$= 1 - 8\sin^2 x \cdot \cos^2 x = \text{RHS}$$ **Hence proved.**

(ii) $$\text{LHS} = \frac{1-\cos 2x}{\sin 2x} = \frac{1-(1-2\sin^2 x)}{2\sin x \cos x}$$

$$= \frac{2\sin^2 x}{2\sin x \cos x} = \tan x = \text{RHS}$$ **Hence proved.**

(iii) $$\text{LHS} = \frac{1-\sin 2x}{1+\sin 2x} = \frac{\cos^2 x + \sin^2 x - 2\sin x\cos x}{\cos^2 x + \sin^2 x + 2\sin x \cos x}$$

$$= \frac{(\cos x - \sin x)^2}{(\cos x + \sin x)^2} = \left(\frac{\cos x - \sin x}{\cos x + \sin x}\right)^2$$

$$= \left(\frac{1 - \frac{\sin x}{\cos x}}{1 + \frac{\sin x}{\cos x}}\right)^2 = \left(\frac{1-\tan x}{1+\tan x}\right)^2$$

$$= \left(\frac{\tan\frac{\pi}{4} - \tan x}{1 + \tan\frac{\pi}{4}\tan x}\right)^2 = \tan^2\left(\frac{\pi}{4} - x\right) = \text{RHS}$$

Hence proved.

(iv) $$\text{LHS} = \frac{1+\sin 2x - \cos 2x}{1+\sin 2x + \cos 2x} = \frac{1-\cos 2x + \sin 2x}{1+\cos 2x + \sin 2x}$$

$$= \frac{1-(1-2\sin^2 x) + 2\sin x\cos x}{1+(2\cos^2 x - 1) + 2\sin x \cos x}$$

$$= \frac{2\sin^2 x + 2\sin x\cos x}{2\cos^2 x + 2\sin x\cos x}$$

$$= \frac{2\sin x(\sin x + \cos x)}{2\cos x(\sin x + \cos x)} = \tan x = \text{RHS}$$

Hence proved.

Example 4. **If $\cos\alpha + \cos\beta = 0 = \sin\alpha + \sin\beta$, then find the value of $\cos 2\alpha + \cos 2\beta$.** *NCERT Exemplar*

Sol. Given, $\cos\alpha + \cos\beta = 0$ and $\sin\alpha + \sin\beta = 0$

On squaring both equations, we get

$$(\cos\alpha + \cos\beta)^2 = 0 \quad \text{...(i)}$$

and $$(\sin\alpha + \sin\beta)^2 = 0 \quad \text{...(ii)}$$

On subtracting Eq. (ii) from Eq. (i), we get

$$(\cos\alpha + \cos\beta)^2 - (\sin\alpha + \sin\beta)^2 = 0$$

$$\Rightarrow (\cos^2\alpha + \cos^2\beta + 2\cos\alpha\cos\beta) - (\sin^2\alpha + \sin^2\beta + 2\sin\alpha\sin\beta) = 0$$

$$[\because (a+b)^2 = a^2 + b^2 + 2ab]$$

$$\Rightarrow \cos^2\alpha + \cos^2\beta + 2\cos\alpha\cos\beta - \sin^2\alpha - \sin^2\beta - 2\sin\alpha\sin\beta = 0$$

$$\Rightarrow (\cos^2\alpha - \sin^2\alpha) + (\cos^2\beta - \sin^2\beta) + 2[\cos\alpha\cos\beta - \sin\alpha\sin\beta] = 0$$

$$\Rightarrow \cos 2\alpha + \cos 2\beta + 2\cos(\alpha+\beta) = 0$$

$$\left[\begin{array}{l}\because \cos 2A = \cos^2 A - \sin^2 A \text{ and} \\ \cos A\cos B - \sin A \sin B = \cos(A+B)\end{array}\right]$$

$$\therefore \cos 2\alpha + \cos 2\beta = -2\cos(\alpha+\beta)$$

Example 5. **Prove that**
$\sin 2x + 2\sin 4x + \sin 6x = 4\cos^2 x \cdot \sin 4x$. *NCERT*

Sol. LHS $= \sin 2x + 2\sin 4x + \sin 6x$

$$= \sin 2x + \sin 6x + 2\sin 4x$$

$$= 2\sin\left(\frac{2x+6x}{2}\right)\cos\left(\frac{2x-6x}{2}\right) + 2\sin 4x$$

$$\left[\because \sin A + \sin B = 2\sin\left(\frac{A+B}{2}\right)\cdot\cos\left(\frac{A-B}{2}\right)\right]$$

$$= 2\sin 4x \cdot \cos(-2x) + 2\sin 4x$$

$$= 2\sin 4x \cdot \cos 2x + 2\sin 4x \quad [\because \cos(-\theta) = \cos\theta]$$

$$= 2\sin 4x(\cos 2x + 1)$$

$$= 2\sin 4x(2\cos^2 x) = 4\sin 4x \cdot \cos^2 x$$

$$[\because \cos 2\theta = 2\cos^2\theta - 1]$$

$= \text{RHS}$ **Hence proved.**

Example 6. **Prove that**
$$\cos^2 x+\cos^2\left(x+\frac{2\pi}{3}\right)+\cos^2\left(x-\frac{2\pi}{3}\right)=\frac{3}{2}.$$
Competency Based Que

Sol. LHS $=\cos^2 x+\cos^2\left(x+\frac{2\pi}{3}\right)+\cos^2\left(x-\frac{2\pi}{3}\right)$

$=\frac{1}{2}\left[2\cos^2 x+2\cos^2\left(x+\frac{2\pi}{3}\right)+2\cos^2\left(x-\frac{2\pi}{3}\right)\right]$

$=\frac{1}{2}\left[1+\cos 2x+1+\cos 2\left(x+\frac{2\pi}{3}\right)+1+\cos 2\left(x-\frac{2\pi}{3}\right)\right]$

$[\because 2\cos^2\theta=1+\cos 2\theta]$

$=\frac{1}{2}\left[3+\cos 2x+\cos\left(2x+\frac{4\pi}{3}\right)+\cos\left(2x-\frac{4\pi}{3}\right)\right]$

$=\frac{1}{2}\left[3+\cos 2x+2\cos 2x\cos\frac{4\pi}{3}\right]$

$[\because \cos(A+B)+\cos(A-B)=2\cos A\cos B]$

$=\frac{1}{2}\left[3+\cos 2x+2\cos 2x\cos\left(\pi+\frac{\pi}{3}\right)\right]$

$=\frac{1}{2}\left[3+\cos 2x+2\cos 2x\left(-\cos\frac{\pi}{3}\right)\right]$

$=\frac{1}{2}\left[3+\cos 2x+2\cos 2x\left(-\frac{1}{2}\right)\right]$

$=\frac{1}{2}[3+\cos 2x-\cos 2x]=\frac{3}{2}=$ RHS **Hence proved.**

Example 7. **Prove that**
$$\cos\frac{2\pi}{15}\cos\frac{4\pi}{15}\cos\frac{8\pi}{15}\cos\frac{16\pi}{15}=\frac{1}{16}$$
NCERT Exemplar

Sol. LHS $=\cos\frac{2\pi}{15}\cos\frac{4\pi}{15}\cos\frac{8\pi}{15}\cos\frac{16\pi}{15}$

$=\cos 24^\circ\cos 48^\circ\cos 96^\circ\cos 192^\circ$

$=\frac{1}{16\sin 24^\circ}[(2\sin 24^\circ\cos 24^\circ)(2\cos 48^\circ)(2\cos 96^\circ)(2\cos 192^\circ)]$

$=\frac{1}{16\sin 24^\circ}[2\sin 48^\circ\cos 48^\circ(2\cos 96^\circ)(2\cos 192^\circ)]$

$[\because 2\sin\theta\cos\theta=\sin 2\theta]$

$=\frac{1}{16\sin 24^\circ}[(2\sin 96^\circ\cos 96^\circ)(2\cos 192^\circ)]$

$=\frac{1}{16\sin 24^\circ}(2\sin 192^\circ\cos 192^\circ)$

$=\frac{1}{16\sin 24^\circ}\sin 384^\circ=\frac{\sin(360^\circ+24^\circ)}{16\sin 24^\circ}$

$=\frac{1}{16}=$ RHS **Hence proved.**

Example 8. **Prove that**
$$\cos 6x=32\cos^6 x-48\cos^4 x+18\cos^2 x-1.$$
NCERT

Sol. We know that $\cos 3x=4\cos^3 x-3\cos x$

On replacing x by $2x$, we get

$\cos 3(2x)=4\cos^3(2x)-3\cos 2x$

$\Rightarrow \cos 6x=4(2\cos^2 x-1)^3-3(2\cos^2 x-1)$

$[\because \cos 2\theta=2\cos^2\theta-1]$

$=4[8\cos^6 x-12\cos^4 x+6\cos^2 x-1]-6\cos^2 x+3$

$[\because (a-b)^3=a^3-3a^2b+3ab^2-b^3]$

$=32\cos^6 x-48\cos^4 x+24\cos^2 x-4-6\cos^2 x+3$

$\therefore \cos 6x=32\cos^6 x-48\cos^4 x+18\cos^2 x-1$

Hence proved.

Example 9. **Prove that**
$$\cos 5A=16\cos^5 A-20\cos^3 A+5\cos A.$$
Competency Based Que

Sol. LHS $=\cos 5A=\cos(3A+2A)$

$=\cos 3A\cdot\cos 2A-\sin 3A\cdot\sin 2A$

$[\because \cos(A+B)=\cos A\cdot\cos B-\sin A\cdot\sin B]$

$=(4\cos^3 A-3\cos A)(2\cos^2 A-1)-(3\sin A-4\sin^3 A)(2\sin A\cdot\cos A)$

$[\because \cos 3x=4\cos^3 x-3\cos x,$
$\sin 3x=3\sin x-4\sin^3 x,$
$\cos 2x=2\cos^2 x-1$ and $\sin 2x=2\sin x\cdot\cos x]$

$=(4\cos^3 A-3\cos A)(2\cos^2 A-1)-(3-4\sin^2 A)(2\sin^2 A\cdot\cos A)$

$=(4\cos^3 A-3\cos A)(2\cos^2 A-1)-[3-4(1-\cos^2 A)]\times[2(1-\cos^2 A)\cdot\cos A]$

$[\because \sin^2 x+\cos^2 x=1\Rightarrow\sin^2 x=1-\cos^2 x]$

$=(8\cos^5 A-4\cos^3 A-6\cos^3 A+3\cos A)-(3-4+4\cos^2 A)\times 2\cos A(1-\cos^2 A)$

$=(8\cos^5 A-10\cos^3 A+3\cos A)-2\cos A(1-\cos^2 A)\times(4\cos^2 A-1)$

$=(8\cos^5 A-10\cos^3 A+3\cos A)-2\cos A(4\cos^2 A-1-4\cos^4 A+\cos^2 A)$

$=(8\cos^5 A-10\cos^3 A+3\cos A)-2\cos A(5\cos^2 A-4\cos^4 A-1)$

$=8\cos^5 A-10\cos^3 A+3\cos A-10\cos^3 A+8\cos^5 A+2\cos A$

$=16\cos^5 A-20\cos^3 A+5\cos A$

$=$ RHS **Hence proved.**

Example 10. **Find the value of** $\sin 22\frac{1^\circ}{2}$.

Sol. We know that $\sin\frac{A}{2}=\pm\sqrt{\frac{1-\cos A}{2}}$

$$\therefore\ \sin\frac{45^\circ}{2}=\sqrt{\frac{1-\cos 45^\circ}{2}}\quad\left[\because \frac{45^\circ}{2}\text{ lies in I quadrant}\right]$$

$$=\sqrt{\frac{1-\frac{1}{\sqrt{2}}}{2}}\quad\left[\because \cos 45^\circ=\frac{1}{\sqrt{2}}\right]$$

$$=\sqrt{\frac{\frac{\sqrt{2}-1}{\sqrt{2}}}{2}}=\sqrt{\frac{\sqrt{2}-1}{2\sqrt{2}}}$$

$$=\sqrt{\frac{\sqrt{2}-1}{2\sqrt{2}}\times\frac{\sqrt{2}}{\sqrt{2}}}=\sqrt{\frac{2-\sqrt{2}}{4}}$$

$$\therefore\ \sin 22\frac{1^\circ}{2}=\frac{\sqrt{2-\sqrt{2}}}{2}$$

Example 11. **If** $\cos\theta=\cos\alpha\cos\beta$**, then prove that**

$$\tan\frac{\theta+\alpha}{2}\tan\frac{\theta-\alpha}{2}=\tan^2\frac{\beta}{2}.$$

Sol. Given, $\cos\theta=\cos\alpha\cos\beta$

$$\Rightarrow\quad \cos\beta=\frac{\cos\theta}{\cos\alpha}$$

$$\Rightarrow\quad \frac{1-\tan^2\frac{\beta}{2}}{1+\tan^2\frac{\beta}{2}}=\frac{\cos\theta}{\cos\alpha}$$

$$\Rightarrow\quad \frac{1-\tan^2\frac{\beta}{2}+1+\tan^2\frac{\beta}{2}}{1-\tan^2\frac{\beta}{2}-1-\tan^2\frac{\beta}{2}}=\frac{\cos\theta+\cos\alpha}{\cos\theta-\cos\alpha}$$

[applying componendo and dividendo]

$$\Rightarrow\frac{2}{-2\tan^2\frac{\beta}{2}}=\frac{2\cos\left(\frac{\theta+\alpha}{2}\right)\cos\left(\frac{\theta-\alpha}{2}\right)}{-2\sin\left(\frac{\theta+\alpha}{2}\right)\sin\left(\frac{\theta-\alpha}{2}\right)}$$

$$\Rightarrow\quad \frac{1}{\tan^2\frac{\beta}{2}}=\cot\left(\frac{\theta+\alpha}{2}\right)\cot\left(\frac{\theta-\alpha}{2}\right)$$

$$\Rightarrow\quad \tan^2\frac{\beta}{2}=\tan\left(\frac{\theta+\alpha}{2}\right)\tan\left(\frac{\theta-\alpha}{2}\right)$$

Hence proved.

Example 12. **Prove that**

(i) $\sin^2 24^\circ-\sin^2 6^\circ=\frac{\sqrt{5}-1}{8}$

(ii) $\sin\frac{\pi}{10}\sin\frac{13\pi}{10}=-\frac{1}{4}$

Sol. (i) LHS $=\sin^2 24^\circ-\sin^2 6^\circ$

$=\sin(24^\circ+6^\circ)\sin(24^\circ-6^\circ)$

$[\because \sin^2 A-\sin^2 B=\sin(A+B)\sin(A-B)]$

$=\sin 30^\circ\sin 18^\circ$

$=\frac{1}{2}\times\frac{\sqrt{5}-1}{4}=\frac{\sqrt{5}-1}{8}=$ RHS **Hence proved**

(ii) LHS $=\sin\frac{\pi}{10}\sin\frac{13\pi}{10}=\sin\frac{\pi}{10}\sin\left(\pi+\frac{3\pi}{10}\right)$

$=-\sin\frac{\pi}{10}\sin\frac{3\pi}{10}=-\sin 18^\circ\sin 54^\circ$

$=-\left(\frac{\sqrt{5}-1}{4}\right)\left(\frac{\sqrt{5}+1}{4}\right)=\frac{-(5-1)}{16}=\frac{-4}{16}$

$=-\frac{1}{4}=$ RHS **Hence proved.**

Example 13. **Find the value of**
$\cos 12^\circ+\cos 84^\circ+\cos 156^\circ+\cos 132^\circ$.

NCERT Exemplar

Sol. We have, $\cos 12^\circ+\cos 84^\circ+\cos 156^\circ+\cos 132^\circ$

$=\cos 12^\circ+\cos 156^\circ+\cos 84^\circ+\cos 132^\circ$

$=2\cos\left(\frac{12^\circ+156^\circ}{2}\right)\cdot\cos\left(\frac{12^\circ-156^\circ}{2}\right)$

$+2\cos\left(\frac{84^\circ+132^\circ}{2}\right)\cdot\cos\left(\frac{84^\circ-132^\circ}{2}\right)$

$=2\cos 84^\circ\cos 72^\circ+2\cos 108^\circ\cdot\cos 24^\circ$

$[\because \cos(-72^\circ)=\cos 72^\circ \text{ and } \cos(-24^\circ)=\cos 24^\circ]$

$=2\cos 84^\circ\cos(90^\circ-18^\circ)+2\cos(90^\circ+18^\circ)\cdot\cos 24^\circ$

$=2\cos 84^\circ\sin 18^\circ-2\sin 18^\circ\cdot\cos 24^\circ$

$=2\sin 18^\circ(\cos 84^\circ-\cos 24^\circ)$

$=2\sin 18^\circ\left[-2\sin\left(\frac{84^\circ+24^\circ}{2}\right)\cdot\sin\left(\frac{84^\circ-24^\circ}{2}\right)\right]$

$=-4\sin 18^\circ\cdot\sin 54^\circ\sin 30^\circ$

$=-4\left(\frac{\sqrt{5}-1}{4}\right)\cdot\cos 36^\circ\cdot\frac{1}{2}$ $[\because \sin(90^\circ-36^\circ)=\cos 36^\circ]$

$=-(\sqrt{5}-1)\left(\frac{\sqrt{5}+1}{4}\right)\cdot\frac{1}{2}$

$=-\left(\frac{5-1}{8}\right)=\frac{-4}{8}=\frac{-1}{2}$

Quick Test 5

1. If $\tan A = \frac{2}{5}$ and $0 < A < \frac{\pi}{2}$, then find the value of

   (i) $\sin 2A$ (ii) $\cos 2A$

   (iii) $\tan 2A$

2. Prove that $\operatorname{cosec} 10° - \sqrt{3} \sec 10° = 4$.

3. Prove that $\sin 4A = 4 \sin A \cos^3 A - 4 \cos A \sin^3 A$.

   Competency Based Que

4. Prove that $4 \cos\theta \cos\left(\frac{\pi}{3}+\theta\right)\cos\left(\frac{\pi}{3}-\theta\right) = \cos 3\theta$.

5. Prove that $4 \sin\theta \sin\left(\frac{\pi}{3}+\theta\right)\sin\left(\frac{2\pi}{3}+\theta\right) = \sin 3\theta$.

   Competency Based Que

6. Prove that $\cot 7\frac{1°}{2} = \tan 82\frac{1°}{2} = (\sqrt{3}+\sqrt{2})(\sqrt{2}+1)$.

   Competency Based Que

7. If $\tan x = -\frac{4}{3}$, x lies in quadrant II, then find the value of

   (i) $\sin\frac{x}{2}$ (ii) $\cos\frac{x}{2}$ (iii) $\tan\frac{x}{2}$

8. Find the value of the expression

$$\cos^4\frac{\pi}{8} + \cos^4\frac{3\pi}{8} + \cos^4\frac{5\pi}{8} + \cos^4\frac{7\pi}{8}.$$

   NCERT Exemplar

9. Prove that

$$\cos^3 A + \cos^3(120° + A) + \cos^3(240° + A) = \frac{3}{4}\cos 3A.$$

   Competency Based Que

Answers

1. (i) $\frac{20}{29}$ (ii) $\frac{21}{29}$ (iii) $\frac{20}{21}$

7. (i) $\frac{2}{\sqrt{5}}$ (ii) $\frac{1}{\sqrt{5}}$ (iii) 2

8. $\frac{3}{2}$

For Solutions, See QR Code on Pg. 84

Exam Pattern Questions

Objective Type Questions

Multiple Choice Questions

1. Second quadrant is represented by

(a) $0 < \theta < \frac{\pi}{2}$ (b) $\frac{\pi}{2} < \theta < \pi$
(c) $\pi < \theta < \frac{3\pi}{2}$ (d) $\frac{3\pi}{2} < \theta < 2\pi$

2. $\sin(-330^\circ)$ is equal to

(a) $\frac{\sqrt{3}}{2}$ (b) $\frac{1}{2}$ (c) $\frac{-1}{2}$ (d) $\frac{-\sqrt{3}}{2}$

3. The value of $\frac{1-\tan^2 15^\circ}{1+\tan^2 15^\circ}$ is

(a) 1 (b) $\sqrt{3}$ (c) $\frac{\sqrt{3}}{2}$ (d) 2

4. The value of $\tan\frac{\pi}{8}$ is

(a) $\sqrt{2}+1$ (b) $\sqrt{3}-1$ (c) $\sqrt{3}+1$ (d) $\sqrt{2}-1$

5. If $\tan(A+B) = m$ and $\tan(A-B) = n$, then value of $\tan 2B$ is **Competency Based Que**

(a) $\frac{m+n}{1+mn}$ (b) $\frac{m-n}{1+mn}$ (c) $\frac{m+n}{1-mn}$ (d) $\frac{m-n}{1-mn}$

6. The radian of $48^\circ\ 37'\ 30''$ is

(a) $\frac{389\pi}{1440}$ rad (b) $\frac{389}{1440}$ rad
(c) $\frac{752\ \pi}{1440}$ rad (d) $\frac{752}{1440}$ rad

7. The value of $\cos 1^\circ \cos 2^\circ \cos 3^\circ \ldots \cos 179^\circ$ is **NCERT Exemplar, Competency Based Que**

(a) $\frac{1}{\sqrt{2}}$ (b) 0 (c) 1 (d) -1

8. The value of $\sin\frac{\pi}{10}\sin\frac{13\pi}{10}$ is **NCERT Exemplar**

(a) $\frac{1}{2}$ (b) $-\frac{1}{2}$ (c) $-\frac{1}{4}$ (d) 1

9. If $\tan x = \frac{a}{b}$, then $b\cos 2x + a\sin 2x$ is equal to

(a) a (b) b (c) ab (d) $\frac{a}{b}$

10. The minimum value of $3\cos x + 4\sin x + 8$ is **Competency Based Que**

(a) 5 (b) 9 (c) 7 (d) 3

11. Which of the following is correct? **Competency Based Que**

(a) $\sin 1^\circ > \sin 1$ (b) $\sin 1^\circ < \sin 1$
(c) $\sin 1^\circ = \sin 1$ (d) $\sin 1^\circ = \frac{\pi}{180^\circ}\sin 1$

12. If $\tan\alpha = \frac{m}{m+1}$ and $\tan\beta = \frac{1}{2m+1}$, then $\alpha+\beta$ is equal to **NCERT Exemplar, Competency Based Que**

(a) $\frac{\pi}{2}$ (b) $\frac{\pi}{3}$ (c) $\frac{\pi}{6}$ (d) $\frac{\pi}{4}$

13. $\frac{(\sin 7x + \sin 5x) + (\sin 9x + \sin 3x)}{(\cos 7x + \cos 5x) + (\cos 9x + \cos 3x)}$ is equal to

(a) $\tan 3x$ (b) $\cot 3x$ (c) $\tan 6x$ (d) $\cot 6x$

14. If $\sin\theta + \cos\theta = 1$, then value of $\sin 2\theta$ is

(a) 1 (b) 1/2 (c) 0 (d) -1

15. Which of the following is not correct? **Competency Based Que**

(a) $\sin\theta = \frac{-1}{5}$ (b) $\cos\theta = 1$
(c) $\sec\theta = \frac{1}{2}$ (d) $\tan\theta = 20$

Assertion-Reason Based Questions

Directions (Q. Nos. 16-25) *In the questions given below are two statements labelled as Assertion (A) and Reason (R). In the context of the two statements, which one of the following is correct?*

(a) Both A and R are correct; R is the correct explanation of A.
(b) Both A and R are correct; R is not the correct explanation of A.
(c) A is correct; R is incorrect.
(d) R is correct; A is incorrect.

16. **Assertion** (A) Degree measure of $\frac{\pi}{6}$ is 60°.

Reason (R) Relationship between degree measure and radian measure is given by $\frac{D}{180} = \frac{C}{\pi}$, where D is degree measure and C is radian measure.

17. **Assertion** (A) $\theta = \frac{l}{r}$

Reason (R) A circle of radius r having an arc of length subtends an angle θ radian at the centre.

18. Assertion (A) $\sin x$ is negative in third and fourth quadrants. **Competency Based Que**

Reason (R) $\cot x$ decreases from 0 to $-\infty$ in first quadrant and increase from 0 to ∞ in third quadrant.

19 Assertion (A) The degree measure corresponding to (–2) radian is – 114° 19 min.

Reason (R) The degree measure of a given radian measure $= \frac{180}{\pi} \times$ Radian measure.

20. Assertion (A) If $A = \cos^2\theta + \sin^4\theta$ for all values of θ, then $\frac{3}{4} \le A \le 1$.

Reason (R) $\sin^2\theta + \cos^2\theta = 1$. **Competency Based Que**

21. Assertion (A) If $3\tan(\theta - 15°) = \tan(\theta + 15°)$ $0° < \theta < 90°$, then $\theta = \frac{\pi}{4}$.

Reason (R) $\tan 3\theta = \frac{3\tan\theta - \tan^3\theta}{1 - 3\tan^2\theta}$.

Competency Based Que

22. Assertion (A) If $\sin A = \frac{1}{2}$, then $\cos A = \frac{\sqrt{3}}{4}$.

Reason (R) $\cos^2 A = 1 - \sin^2 A$

23. Assertion (A) $\cos^2 33° + \cos^2 57° = 0$

Reason (R) $\cos^2\theta + \sin^2\theta = 1$

24. Assertion (A) $\text{cosec}^2 25° - \tan^2 65° = 2$

Reason (R) $\text{cosec}^2\theta - \cot^2\theta = 1$

25. Assertion (A) $\sin A = -\frac{1}{3}$

Reason (R) All the output values of $\sin A$ lie in $[-1, 1]$.

Case Study Based Questions

26. Divya made a figure of a compass for her school project. She pointed East direction along the positive X-axis (0°angle) and North direction is along the the Y-axis (90° angle). Initially, she pointed the pointer towards North-East direction. Pointer can deflected in a magnetic field by some angles. **Competency Based Que**

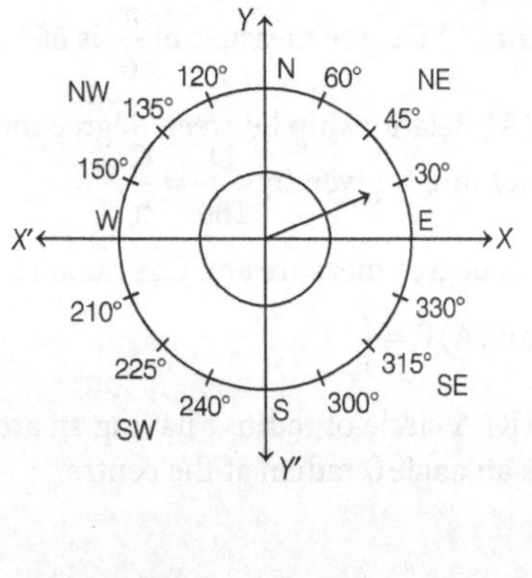

On the basis of the above information, answer the following questions.

(i) If pointer moves in anti-clockwise direction by an angle of 90°, then find the value of sine of angle made by pointer from East direction.

(ii) If pointer moves an angle of 210° from its initial position in anti-clockwise direction, then find the value of cosine of angle made by pointer from East direction.

(iii) If the sine and cosine of angle made by pointer with East direction is $\frac{-1}{\sqrt{2}}$, then find where the pointer pointed.

(iv) How much angle will pointer move in anti-clockwise direction, if tangent of angle made by pointer with X-axis is -1?

27. In the figure, it is given that $\angle C = 90°$, $AD = DB$, ED is perpendicular to AB, $AB = 20$ cm and $AC = 12$ cm.

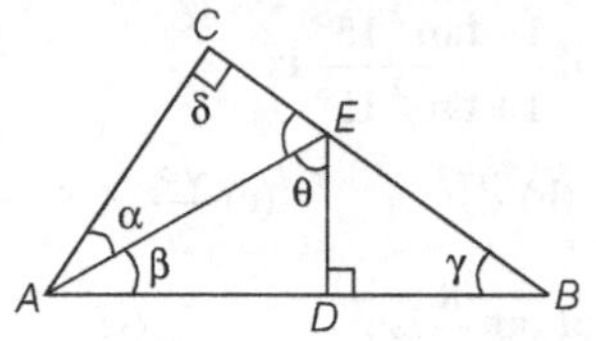

Based on the above information, answer the following questions. **Competency Based Que**

(i) The value of $\cos(\alpha + \theta)$ is

(a) $\frac{4}{5}$ (b) $\frac{3}{5}$ (c) $\frac{117}{125}$ (d) $\frac{44}{125}$

(ii) The value of $\tan 2\alpha$ is

(a) $\frac{3}{4}$ (b) $\frac{7}{24}$

(c) $\frac{336}{527}$ (d) $\frac{24}{25}$

(iii) Which of the following is true?

(a) $\alpha + \beta = \theta + \delta$

(b) $\alpha + \beta > \theta + \delta$

(c) $\alpha + \beta < \theta + \delta$

(d) None of the above

(iv) The value of $\sin\left(\frac{\alpha + \theta + \gamma + \delta}{2}\right)\sin\left(\frac{\alpha + \theta - \gamma - \delta}{2}\right)$ is

(a) $\frac{4}{5}$ (b) $\frac{3}{5}$

(c) $\frac{117}{125}$ (d) $\frac{-119}{250}$

(v) The value of $\tan\beta + \tan\theta$ is

(a) $\frac{16}{9}$ (b) $\frac{12}{25}$

(c) $\frac{25}{12}$ (d) $\frac{9}{16}$

Subjective Questions

Very Short Answer Type Questions

1. Find the degree measure of the angle subtended at the centre of a circle of radius 100 cm by an arc of length 22 cm. **NCERT**

2. Find the radian measure corresponding the degree measure $125° 30'$.

3. Find the degree measure corresponding the radius measure -2^c. **Competency Based Que**

4. Find the value of the following trigonometric ratios.

(i) $\sin\left(\frac{-11\pi}{3}\right)$ (ii) $\cot\left(\frac{-15\pi}{4}\right)$ **NCERT**

(iii) $\text{cosec}(-1200°)$

5. A wheel makes 270 revolutions in one minute. Through how many radians does it turn in one second. **Competency Based Que**

6. Evaluate $\sin 78° \cos 18° - \cos 78° \sin 18°$. **Competency Based Que**

7. Find the value of $\tan \frac{19\pi}{3}$.

8. Find the value of $\sin^2 24° - \sin^2 6°$. **Competency Based Que**

9. If $\tan A = \frac{1-\cos B}{\sin B}$, then find the value of $\tan 2A$. **Diksha**

10. Find the value of $\frac{\tan 55° + \tan 35°}{1 - \tan 55° \tan 35°}$. **Competency Based Que**

Short Answer Type Questions

11. The minute hand of a watch is 1.8 cm long. How far does its tip move in 30 min?

12. A rail road curve is to be laid out on a circle. What radius should be used, if the track is to change direction by 30° at a distance of 88 m? **Competency Based Que**

13. Find $\sin\theta$ and $\tan\theta$, if $\cos\theta = -\frac{3}{5}$ and θ lies in the third quadrant.

14. Find the value of $\tan 15°$.

15. If $\tan(A+B) = p$, $\tan(A-B) = q$, then show that $\tan 2A = \frac{p+q}{1-pq}$. **Competency Based Que**

16. If $\frac{2\sin\alpha}{1+\cos\alpha+\sin\alpha} = y$, then prove that $\frac{1-\cos\alpha+\sin\alpha}{1+\sin\alpha} = y$.

17. Prove that

(i) $2\sin^2 \frac{\pi}{6} + \text{cosec}^2 \frac{7\pi}{6} \cos^2 \frac{\pi}{3} = \frac{3}{2}$

(ii) $\cot^2 \frac{\pi}{6} + \text{cosec}\frac{5\pi}{6} + 3\tan^2 \frac{\pi}{6} = 6$

(iii) $\tan 720° - \cos 70° - \sin 150° \cos 120° = \frac{1}{4}$

18. If $\sin A = \frac{3}{5}$, $0 < A < \frac{\pi}{2}$ and $\cos B = -\frac{12}{13}$, $\pi < B < \frac{3\pi}{2}$, then find $\tan(A-B)$. **Competency Based Que**

19. If $\sin A = \frac{3}{5}$ and $\cos B = \frac{-12}{13}$, $0 < A < \frac{\pi}{2}$, $\pi < B < \frac{3\pi}{2}$, then find $\cos(A+B)$.

20. Prove that

(i) $\frac{1+\cos 4x}{\cot x - \tan x} = \frac{1}{2}\sin 4x$

(ii) $8\cos^3 \frac{\pi}{9} - 6\cos\frac{\pi}{9} = 1$

(iii) $108 \sin\frac{\pi}{18} - 144 \sin^3 \frac{\pi}{18} = 18$ **Competency Based Que**

21. Prove that $\sin 4A = 4\sin A \cos^3 A - 4\cos A \sin^3 A$.

22. If $\tan(\alpha+\theta) = n\tan(\alpha-\theta)$, then show that $(n+1)\sin 2\theta = (n-1)\sin 2\alpha$. **Competency Based Que**

23. If $2\alpha + 2\beta = 90°$, then find the maximum and minimum values of $\sin 2\alpha \sin 2\beta$.

24. Prove that

$$\cos\left(\frac{3\pi}{2}+x\right)\cos(2\pi+x)\left\{\cot\left(\frac{3\pi}{2}-x\right)+\cot(2\pi+x)\right\} = 1$$

Competency Based Que

25. Prove that $\cos^2 2x - \cos^2 6x = \sin 4x \sin 8x$.

26. Prove that $\frac{\cos(90°+\theta)\sec(-\theta)\tan(180°-\theta)}{\sec(360°-\theta)\sin(180°+\theta)\cot(90°-\theta)} = -1$

27. The perimeter of a certain sector of a circle is equal to the length of the arc of a semi-circle having the same radius. Express the angle of the sector in degrees, minutes and seconds. **Competency Based Que**

28. The angles of a triangle are in AP. The ratio of measure of degrees in the smallest angle and the measure of radians in the greatest angle is as $60 : \pi$. Find the angles in degree. **Competency Based Que**

29. Find the value of $\tan\frac{\pi}{8}$.

30. If $\cos A=\frac{2}{3}$ and $0<A<\frac{\pi}{2}$, then find the value of

(i) $\sin 2A$

(ii) $\cos 2A$

(iii) $\tan 2A$

31. Prove that
$$\frac{\sin x}{\cos 3x}+\frac{\sin 3x}{\cos 9x}+\frac{\sin 9x}{\cos 27x}=\frac{1}{2}(\tan 27x-\tan x).$$

Competency Based Que

32. If $\tan x=\frac{3}{4}$, $\pi<x<\frac{3\pi}{2}$, find the values of $\sin\frac{x}{2}$, $\cos\frac{x}{2}$, $\tan\frac{x}{2}$.

Competency Based Que

Hints & Solutions

Objective Type Questions

1. (b) Second quadrant is represented by
$$\frac{\pi}{2}<\theta<\pi$$

2. (b) $\sin(-330^\circ)=-\sin 330^\circ$
$$=-\sin(360^\circ-30^\circ)$$
$$=-(-\sin 30^\circ)$$
$$=\frac{1}{2}$$

3. (c) We know that $\cos 2\theta=\dfrac{1-\tan^2\theta}{1+\tan^2\theta}$
$$\Rightarrow \frac{1-\tan^2 15^\circ}{1+\tan^2 15}=\cos 2\times 15^\circ$$
$$=\cos 30^\circ=\frac{\sqrt{3}}{2}$$

4. (d) $\tan\frac{\pi}{8}=\sqrt{\dfrac{1-\cos\frac{\pi}{4}}{1+\cos\frac{\pi}{4}}}=\sqrt{\dfrac{1-\frac{1}{\sqrt{2}}}{1+\frac{1}{\sqrt{2}}}}$
$$=\sqrt{\frac{\sqrt{2}-1}{\sqrt{2}+1}}=\sqrt{2}-1$$

5. (b) $\tan 2B=\tan\{(A+B)-(A-B)\}$
$$=\frac{\tan(A+B)-\tan(A-B)}{1+\tan(A+B)\tan(A-B)}=\frac{m-n}{1+mn}$$

6. (a) $48^\circ\,37'\,30''=48^\circ+37\times\frac{1^\circ}{60}+30\times\frac{1^\circ}{3600}$

$[\because 1^\circ=60'$ and $1'=60'']$
$$=48^\circ+\frac{37^\circ}{60}+\frac{1^\circ}{120}$$
$$=\frac{389^\circ}{8}$$

The radian measure of angle $=\frac{389}{8}\times\frac{\pi}{180}=\frac{389\pi}{1440}$.

7. (b) **Hint** We have, $\cos 1^\circ\cos 2^\circ\cos 3^\circ \ldots \cos 179^\circ$

Since, above expression is multiple of $\cos 90^\circ$.

$\because\ \cos 90^\circ=0$

Hence, the value of given expression became zero.

8. (c) $\sin\frac{\pi}{10}\sin\frac{13}{10}\pi=\sin\frac{\pi}{10}\sin\left(\pi+\frac{3\pi}{10}\right)$
$$=-\sin\frac{\pi}{10}\sin\frac{3\pi}{10}\quad[\because \sin(\pi+\theta)=-\sin\theta]$$
$$=-\sin 18^\circ\sin 36^\circ$$
$$=-\left(\frac{\sqrt{5}-1}{4}\right)\left(\frac{\sqrt{5}+1}{4}\right)=-\frac{1}{4}$$

9. (b) Given, $\tan x=\frac{a}{b}$

Now, we have

$b\cos 2x+a\sin 2x$
$$=b\left[\frac{1-\tan^2 x}{1+\tan^2 x}\right]+a\left[\frac{2\tan x}{1+\tan^2 x}\right]$$
$$=b\left[\frac{1-\left(\frac{a}{b}\right)^2}{1+\left(\frac{a}{b}\right)^2}\right]+a\left[\frac{2\cdot\frac{a}{b}}{1+\left(\frac{a}{b}\right)^2}\right]\quad\left[\because \tan x=\frac{a}{b}\right]$$
$$=b\left[\frac{\frac{b^2-a^2}{b^2}}{\frac{b^2+a^2}{b^2}}\right]+a\left[\frac{2\cdot\frac{a}{b}}{\frac{b^2+a^2}{b^2}}\right]$$
$$=b\left(\frac{b^2-a^2}{b^2+a^2}\right)+a\left(\frac{2ab}{b^2+a^2}\right)$$
$$=\frac{b^3-a^2b+2a^2b}{b^2+a^2}=\frac{b^3+a^2b}{b^2+a^2}$$
$$=\frac{b(b^2+a^2)}{(b^2+a^2)}=b$$

10. (d) **Hint** Let $y = 3\cos x + 4\sin x + 8$

$\Rightarrow \quad y - 8 = 3\cos x + 4\sin x$

$\therefore$ Minimum value of $y - 8 = -\sqrt{9+16}$

[$\because$ minimum value is $-\sqrt{A^2+B^2}$ for the expression $A\cos\theta + B\sin\theta$]

$\therefore \quad y - 8 = -5$

$\Rightarrow \quad y = 3$

11. (b) We know that if θ is increasing, then $\sin\theta$ is also increasing.

$\therefore \quad \sin 1^\circ < \sin 1 \quad [\because 1 \text{ rad} = 57^\circ 30']$

12. (d) $\tan(\alpha+\beta) = \dfrac{\tan\alpha + \tan\beta}{1 - \tan\alpha\tan\beta}$

$$= \frac{\frac{m}{m+1} + \frac{1}{2m+1}}{1 - \left(\frac{m}{m+1}\right)\left(\frac{1}{2m+1}\right)}$$

$$= \frac{m(2m+1) + (m+1)}{(m+1)(2m+1) - m} = 1$$

$\Rightarrow \tan(\alpha+\beta) = \tan\dfrac{\pi}{4} \quad \left[\because \tan\dfrac{\pi}{4} = 1\right]$

$\therefore \quad \alpha + \beta = \dfrac{\pi}{4}$

13. (c) $\dfrac{(\sin 7x + \sin 5x) + (\sin 9x + \sin 3x)}{(\cos 7x + \cos 5x) + (\cos 9x + \cos 3x)}$

$$= \frac{2\sin\left(\frac{12x}{2}\right).\left(\cos\frac{2x}{2}\right) + 2\sin\left(\frac{12x}{2}\right)\cos\left(\frac{6x}{2}\right)}{2\cos\left(\frac{12x}{2}\right)\cos\left(\frac{2x}{2}\right) + 2\cos\left(\frac{12x}{2}\right)\cos\left(\frac{6x}{2}\right)}$$

$$= \frac{2[\sin 6x\cos x + \sin 6x\cos 3x]}{2[\cos 6x\cos x + \cos 6x\cos 3x]}$$

$$= \frac{4\sin 6x\cos 3x}{4\cos 6x\cos 3x} = \tan 6x$$

14. (c) Given, $\quad \sin\theta + \cos\theta = 1$

$\therefore \quad (\sin\theta + \cos\theta)^2 = 1^2$

$\Rightarrow \sin^2\theta + \cos^2\theta + 2\sin\theta\cos\theta = 1$

$\Rightarrow \quad 1 + 2\sin\theta\cos\theta = 1$

$\Rightarrow \quad 2\sin\theta\cos\theta = 0$

$\Rightarrow \quad \sin 2\theta = 0$

15. (c) Since, $-1 \le \sin\theta < 1, 1 \le \cos\theta \le 1$ and $\tan\theta \in (-\infty, \infty)$.

$\therefore$ *(a)*, *(b)*, *(d)* are correct, since the range of $\sec\theta$ is $R - (-1, 1)$. Therefore, $\sec\theta \ne \dfrac{1}{2}$.

16. (d) Degree measure of $\dfrac{\pi}{6} = \dfrac{\pi}{6} \times \dfrac{180^\circ}{\pi} = 30^\circ$

R is correct but A is incorrect.

17. (a) Assertion and Reason both are correct and Reason is the correct explanation of Assertion.

18. (c) A is correct but R is incorrect.

19. (d) Degree measure $= \dfrac{180}{\pi} \times$ Radian measure

$$= \frac{180}{\pi} \times (-2) = -[114^\circ\, 32'\, 44'']$$

$\therefore$ Assertion is incorrect but Reason is correct.

20. (b) $\because A = \cos^2\theta + (\sin^2\theta)^2$

We know that $-1 \le \sin\theta \le 1$

$\Rightarrow \quad 0 \le \sin^2\theta \le 1$

For any x, $0 \le x \le 1, x^n \le x$

$\therefore \quad (\sin^2\theta)^2 \le \sin^2\theta$

$\Rightarrow \quad \cos^2\theta + (\sin^2\theta)^2 \le \cos^2\theta + \sin^2\theta$

$\Rightarrow \quad A \le 1 \quad \ldots(i)$

So, $\quad A = \cos^2\theta + (\sin^2\theta)^2$

$\Rightarrow \quad A = 1 - \sin^2\theta + (\sin^2\theta)^2$

$\Rightarrow \quad A = 1 - (1/4) + (1/4) - \sin^2\theta + (\sin^2\theta)^2$

$\Rightarrow \quad A = \dfrac{3}{4} + [\sin^2\theta - (1/2)]^2$

Now, $\quad [\sin^2\theta] - (1/2)]^2 \ge 0$

$\Rightarrow \quad (3/4) + [\sin^2\theta - (1/2)]^2 \ge 3/4 \Rightarrow A \ge 3/4 \quad \ldots(ii)$

From Eqs. (i) and (ii), we get

$$\frac{3}{4} \le A \le 1.$$

Hence, both are correct but Reason is not the correct explanation of Assertion.

21. (b) $3\tan(\theta - 15^\circ) = \tan(\theta + 15^\circ)$

$\Rightarrow \quad 3 = \dfrac{\tan(\theta + 15^\circ)}{\tan(\theta - 15^\circ)}$

On applying componendo and dividendo rule, we get

$$\frac{3+1}{3-1} = \frac{\tan(\theta+15^\circ) + \tan(\theta-15^\circ)}{\tan(\theta+15^\circ) - \tan(\theta-15^\circ)}$$

$$\Rightarrow \quad 2 = \frac{\frac{\sin(\theta+15^\circ)}{\cos(\theta+15^\circ)} + \frac{\sin(\theta-15^\circ)}{\cos(\theta-15^\circ)}}{\frac{\sin(\theta+15^\circ)}{\cos(\theta+15^\circ)} - \frac{\sin(\theta-15^\circ)}{\cos(\theta-15^\circ)}}$$

$$\Rightarrow 2 = \frac{\sin(\theta+15^\circ)\cos(\theta-15^\circ) + \sin(\theta-15^\circ)\cos(\theta+15^\circ)}{\sin(\theta+15^\circ)\cos(\theta-15^\circ) - \sin(\theta-15^\circ)\cos(\theta+15^\circ)}$$

$$\Rightarrow 2 = \frac{\sin(\theta+15^\circ+\theta-15^\circ)}{\sin(\theta+15^\circ-\theta+15^\circ)}$$

$\Rightarrow 2 = \dfrac{\sin 2\theta}{\sin 30^\circ} \quad \Rightarrow \quad 2 = 2\sin 2\theta$

$\Rightarrow \quad 1 = \sin 2\theta$

$\because \quad 0 < \theta < 90^\circ$

$\therefore \quad 2\theta = 90^\circ$

$\Rightarrow \quad \theta = 45° = \frac{\pi}{4}$

Reason is also correct but not correct explanation of Assertion.

22. (d) $\sin A = \frac{1}{2}, \cos A = \sqrt{1-\sin^2 A}$

$\Rightarrow \quad \cos A = \sqrt{1-\frac{1}{4}} = \frac{\sqrt{3}}{2}$

∴ Assertion is incorrect but Reason is correct.

23. (a) $\cos^2 33°+\sin^2(90°-33°) = \cos^2 33°+\sin^2 33° = 1$

∴ Assertion and Reason both are correct and Reason is correct explanation of Assertion.

24. (d) $\because \operatorname{cosec}^2 25° - \tan^2 65°$

$= \operatorname{cosec}^2 25° - \tan^2(90°-25°)$

$= \operatorname{cosec}^2 25° - \cot^2 25° = 1$

∴ Assertion is incorrect but Reason is correct.

25. (a) $\sin A = \frac{-1}{3} \in [-1, 1]$

∴ Assertion and Reason both are correct and Reason is the correct explanation of Assertion.

26. (i) Angle made by pointer with East direction

$= 30° + 90° = 120°$

$\therefore \quad \sin 120° = \sin(180°-60°)$

$= \sin 60°$

$= \frac{\sqrt{3}}{2}$

(ii) Given, pointer moves an angle of 210° from its initial position in anti-clockwise direction.
Angle made by pointer with East direction

$= 30° + 210° = 240°$

Therefore, $\cos 240° = \cos(180° + 60°)$

$= -\cos 60° = -\frac{1}{2}$

(iii) sine and cosine both are negative, so quadrant is III and we know that $\sin 45° = \frac{1}{\sqrt{2}}$

Given that $\sin\theta = -\frac{1}{\sqrt{2}}$

$\Rightarrow \quad \sin\theta = \sin(180° + 45°)$

$\Rightarrow \quad \theta = 225°$

Hence, the pointer pointed in South-West direction

(iv) If $\tan\theta = -1$,
then $\theta = 135°$ or $315°$
Initially the pointer is at 30°.
So, angle moved by pointer is

$135° - 30° = 105°$

or $\quad 315° - 30° = 285°$

i.e. $\quad 105°$ or $285°$

27. **Hint** Given, $AB = 20$ cm, $AC = 12$ cm

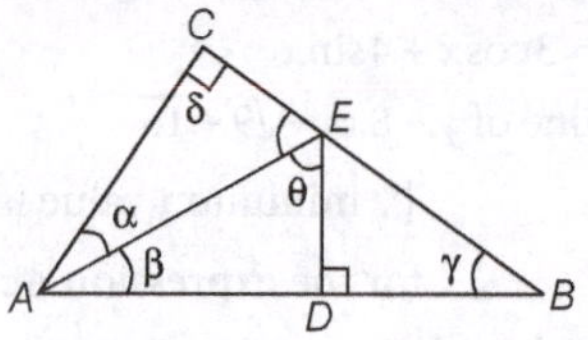

$\therefore \quad BC = \sqrt{20^2 - 12^2} = 16$ cm

Now, $\tan\gamma = \frac{AC}{BC} = \frac{12}{16} = \frac{3}{4}$

Since, $\Delta DEA \cong \Delta DEB$ by SAS property

Therefore, $\quad \gamma = \beta$

$\cos\gamma = \frac{BD}{BE} \Rightarrow BE = \frac{10}{4/5} = \frac{25}{2}$

$\tan\gamma = \frac{DE}{BD} \Rightarrow DE = \frac{10}{4/3} = \frac{15}{2}$

$CE = BC - BE = 16 - \frac{25}{2} = \frac{7}{2}$

$\tan\alpha = \frac{CE}{AC} = \frac{7}{24}, \cos\alpha = \frac{24}{25}, \sin\alpha = \frac{7}{25}$

$\tan\theta = \tan\left(\frac{\pi}{2} - \beta\right) = \cot\beta = \cot\gamma = \frac{4}{3},$

$\cos\theta = \frac{3}{5}, \sin\theta = \frac{4}{5}$

(i) (d) Use formula $\cos(A+B) = \cos A\cos B - \sin A\sin B$

(ii) (c) Use formula $\tan 2A = \frac{2\tan A}{1-\tan^2 A}$

(iii) (c) Since, $AE > CE \Rightarrow \delta > \alpha$

and $\quad AD > DE \Rightarrow \beta < \theta$

$\because \quad \gamma = \beta \Rightarrow \gamma < \theta$

$\therefore \quad \alpha + \beta < \theta + \delta$

(iv) (d) Use formula

$-2\sin A\sin B = \cos(A+B) - \cos(A-B)$

(v) (c) $\tan\beta + \tan\theta = \frac{3}{4} + \frac{4}{3} = \frac{9+16}{12} = \frac{25}{12}$

Subjective Questions

1. Here, radius of circle $r = 100$ cm

Length of arc, $l = 22$ cm

$\therefore \quad \theta = \frac{l}{r} = \frac{22}{100}$

$= \frac{22}{100} \times \frac{180°}{\pi}$

$= \frac{22}{100} \times 180° \times \frac{7}{22}$

$= \left(\frac{63}{5}\right)^\circ$

$= 12° + \left(\frac{3}{5} \times 60'\right)$

$= 12° \, 36'$.

2. We know that, radian measure $= \frac{\pi}{180} \times$ Degree measure

Since, $1' = \left(\frac{1}{60}\right)^\circ$

$\therefore \quad 30' = \left(\frac{30}{60}\right)^\circ = \left(\frac{1}{2}\right)^\circ$

Total degree to convert $= 125^\circ + \frac{1^\circ}{2} = \left(\frac{250+1}{2}\right)^\circ = \left(\frac{251}{2}\right)^\circ$

$\therefore$ Required radian measure $= \frac{\pi}{180} \times \frac{251}{2} = \frac{251}{360}\pi$

3. Degree measure $= \frac{180}{\pi} \times$ Radian measure

$\therefore \quad -2^c = \frac{180^\circ}{22} \times (-2) \times 7 \qquad \left[\because \pi = \frac{22}{7}\right]$

$= -\left[\frac{180^\circ \times 7}{11}\right] = -\left[114^\circ + \left(\frac{6}{11} \times 60\right)'\right]$

$= -\left[114^\circ + 32' + \left(\frac{8}{11} \times 60\right)''\right]$

$= -[114^\circ + 32' + 44'']$

$= -[114^\circ\, 32'\, 44'']$

4. (i) $\sin\left(\frac{-11}{3}\pi\right) = -\sin\left(\frac{11}{3}\pi\right) \qquad [\because \sin(-x) = -\sin x]$

$= -\left(-\frac{\sqrt{3}}{2}\right) = \frac{\sqrt{3}}{2}$

(ii) $\cot\left(\frac{-15}{4}\pi\right) = \cot\left(\frac{15}{4}\pi\right) \qquad [\because \cot(-x) = -\cot x]$

$= -(-1) = 1$

(iii) $\operatorname{cosec}(-1200^\circ) = -\operatorname{cosec} 1200^\circ$

$[\because \operatorname{cosec}(-\theta) = -\operatorname{cosec}\theta]$

$= -\operatorname{cosec}(1260 - 60^\circ) \qquad [\because 1170 < 1200 < 1260]$

$= -\operatorname{cosec}(90^\circ \times 14 - 60^\circ) = -\operatorname{cosec} 60^\circ$

$= -\left[\frac{2}{\sqrt{3}}\right] = -\frac{2}{\sqrt{3}}$

5. Hint Number of revolutions in one second

$= \frac{270}{60} = \frac{9}{2} = 4\frac{1}{2}$

Angle traced in one revlolution $= 2\pi$ radians

$\therefore$ Angle traced in one second $= \frac{9}{2} \times 2\pi = 9\pi$ radians

6. $\sin 78^\circ \cos 18^\circ - \cos 78^\circ \sin 18^\circ = \sin(78^\circ - 18^\circ)$

$[\because \sin A \cos B - \cos A \sin B = \sin(A - B)]$

$= \sin 60^\circ = \frac{\sqrt{3}}{2}$

7. $\tan\frac{19\pi}{3} = \tan\left(6\pi + \frac{\pi}{3}\right) = \tan\left(3 \times 2\pi + \frac{\pi}{3}\right)$

$= \tan\frac{\pi}{3} = \sqrt{3} \qquad [\because \tan(2n\pi + x) = \tan x]$

8. $\sin^2 24^\circ - \sin^2 6^\circ$

$= \sin(24^\circ + 6^\circ)\sin(24^\circ - 6^\circ)$

$[\because \sin^2 A - \sin^2 B = \sin(A+B)\sin(A-B)]$

$= \sin 30^\circ \sin 18^\circ$

$= \frac{1}{2} \times \frac{\sqrt{5}-1}{4} = \frac{\sqrt{5}-1}{8}$

9. Given, $\tan A = \frac{1 - \cos B}{\sin B}$

$\because \quad \cos 2x = 2\cos^2 x - 1 = 1 - 2\sin^2 x$

and $\quad \sin 2x = 2\sin x \cdot \cos x$

$\therefore \quad \tan A = \frac{1 - \left(1 - 2\sin^2\frac{B}{2}\right)}{2\sin\frac{B}{2}\cos\frac{B}{2}} = \frac{\sin\frac{B}{2}}{\cos\frac{B}{2}} = \tan\frac{B}{2}$

Now, $\because \tan 2A = \frac{2\tan A}{1 - \tan^2 A} \Rightarrow \tan 2A = \frac{2\tan\frac{B}{2}}{1 - \tan^2\frac{B}{2}} = \tan B$

$\therefore \quad \tan 2A = \tan B$

10. Hint Using formula

$\frac{\tan A + \tan B}{1 - \tan A \cdot \tan B} = \tan(A + B)$, we get

$\frac{\tan 55^\circ + \tan 35^\circ}{1 - \tan 55^\circ \tan 35^\circ} = \tan(55^\circ + 35^\circ) = \tan 90^\circ = \infty$

11. Angle traced out by minute hand in

$1 \text{ min} = \left(\frac{360}{60}\right)^\circ = 6^\circ$

$\therefore$ Angle traced by the minute hand in 30 min

$= (30 \times 6)^\circ = 180^\circ = \pi$ rad

We know that $\theta = \frac{l}{r}$.

$\therefore \quad \pi = \frac{l}{1.8} \qquad [\because \theta = \pi]$

$\Rightarrow \quad l = 1.8\,\pi = 5.65$ cm.

12. Hint $\theta = 30^\circ = 30^\circ \times \frac{\pi}{180^\circ} = \frac{\pi}{6}$

$\because \quad \theta = \frac{l}{r}$

$\therefore \quad r = \frac{l}{\theta} = \frac{88}{\frac{\pi}{6}} = \frac{88 \times 6}{\pi} = \frac{88 \times 6 \times 7}{22} = 168$ cm.

13. Given, $\cos\theta = -\frac{3}{5}$

Now, $\sin\theta = \sqrt{1 - \cos^2\theta} = \sqrt{1 - \left(\frac{-3}{5}\right)^2} = \sqrt{1 - \frac{9}{25}}$

$= \sqrt{\frac{16}{25}} = \frac{4}{5}$

$\therefore \quad \tan\theta = \dfrac{\sin\theta}{\cos\theta} = \dfrac{\frac{4}{5}}{\frac{-3}{5}} = -\dfrac{4}{3}$

Since, θ lies in the III quadrant.

$\therefore \quad \sin\theta = \dfrac{-4}{5}$ and $\tan\theta = \dfrac{4}{3}$.

14. $\tan 15° = \tan(60°-45°)$

$= \dfrac{\tan 60° - \tan 45°}{1+\tan 60° \tan 45°} \quad \left[\because \tan(A-B) = \dfrac{\tan A - \tan B}{1+\tan A \tan B}\right]$

$= \dfrac{\sqrt{3}-1}{1+\sqrt{3}\cdot 1} \quad [\because \tan 45° = 1 \text{ and } \tan 60° = \sqrt{3}]$

$= \dfrac{\sqrt{3}-1}{1+\sqrt{3}} = \dfrac{\sqrt{3}-1}{\sqrt{3}+1} = \dfrac{\sqrt{3}-1}{\sqrt{3}+1} \times \dfrac{\sqrt{3}-1}{\sqrt{3}-1}$ [on rationalisation]

$= \dfrac{(\sqrt{3}-1)^2}{(\sqrt{3})^2-(1)^2} \quad [\because (a-b)(a+b) = a^2-b^2]$

$= \dfrac{(\sqrt{3})^2+(1)^2-2\sqrt{3}}{3-1} [\because (a-b)^2 = a^2+b^2-2ab]$

$= \dfrac{3+1-2\sqrt{3}}{3-1} = \dfrac{4-2\sqrt{3}}{2} = \dfrac{2(2-\sqrt{3})}{2} = (2-\sqrt{3})$

15. Hint $\tan 2A = \tan[(A+B)+(A-B)]$

$= \dfrac{\tan(A+B)+\tan(A-B)}{1-\tan(A+B)\tan(A-B)}$

Now, put the given values $\tan(A+B) = p$ and $\tan(A-B) = q$ in above expression.

16. Hint Express $\dfrac{1-\cos\alpha+\sin\alpha}{1+\sin\alpha}$

$= \left(\dfrac{1-\cos\alpha+\sin\alpha}{1+\sin\alpha}\right)\left(\dfrac{1+\cos\alpha+\sin\alpha}{1+\cos\alpha+\sin\alpha}\right)$

17. Hint Simplify each trigonometric function individually.

18. Hint Use formula $\tan(A-B) = \dfrac{\tan A - \tan B}{1+\tan A \tan B}$ and put in it $\tan\theta = \dfrac{\sin\theta}{\cos\theta}$ **Ans.** $\tan(A-B) = \dfrac{16}{63}$

19. Given, $\sin A = \dfrac{3}{5}$ and $0 < A < \dfrac{\pi}{2}$

$\because \quad \cos^2 A = 1-\sin^2 A$

$\Rightarrow \quad \cos^2 A = 1-\left(\dfrac{3}{5}\right)^2 = 1-\dfrac{9}{25}$

$\Rightarrow \quad \cos^2 A = \dfrac{16}{25} \Rightarrow \cos A = \pm\dfrac{4}{5}$

$\Rightarrow \quad \cos A = \dfrac{4}{5} \quad \left[\because \cos A \text{ is + ve in } \left(0, \dfrac{\pi}{2}\right)\right]$

Again, given $\cos B = \dfrac{-12}{13}$, where $\pi < B < \dfrac{3\pi}{2}$

$\because \sin^2 B = 1-\cos^2 B, \sin^2 B = 1-\left(\dfrac{-12}{13}\right)^2 = \dfrac{25}{169}$

$\Rightarrow \sin B = \pm\dfrac{5}{13}$

$\Rightarrow \sin B = \dfrac{-5}{13} \quad \left[\sin B \text{ is } - \text{ ve in } \left(\pi, \dfrac{3\pi}{2}\right)\right]$

Now,

$\because \cos(A+B) = \cos A \cos B - \sin A \sin B$

$= \left(\dfrac{4}{5}\right)\left(\dfrac{-12}{13}\right) - \left(\dfrac{3}{5}\right)\left(\dfrac{-5}{13}\right) = \dfrac{-48+15}{65} = \dfrac{-33}{65}$

$\therefore \cos(A+B) = \dfrac{-33}{65}$

20. (i) $\dfrac{1+\cos 4x}{\cot x - \tan x} = \dfrac{1}{2}\sin 4x$

$\text{LHS} = \dfrac{1+\cos 4x}{\cot x - \tan x} = \dfrac{1+2\cos^2 2x - 1}{\dfrac{\cos x}{\sin x} - \dfrac{\sin x}{\cos x}}$

[using formula $\cos 2A = 2\cos^2 A - 1$]

$= \dfrac{2\cos^2 2x \cdot \cos x \cdot \sin x}{\cos^2 x - \sin^2 x}$

$= \dfrac{2\sin x \cos x \cdot \cos^2 2x}{\cos 2x}$

$= \sin 2x \cos 2x = \dfrac{1}{2}(2\sin 2x \cos 2x)$

$= \dfrac{1}{2}\sin 4x = \text{RHS}$ **Hence proved.**

(ii) $8\cos^3\dfrac{\pi}{9} - 6\cos\dfrac{\pi}{9} = 1$

$\text{LHS} = 8\cos^3\dfrac{\pi}{9} - 6\cos\dfrac{\pi}{9} = 2\left(4\cos^3\dfrac{\pi}{9} - 3\cos\dfrac{\pi}{9}\right)$

$= 2\cos\left(3 \times \dfrac{\pi}{9}\right)$

[using formula $\cos 3A = 4\cos^3 A - 3\cos A$]

$= 2\cos\dfrac{\pi}{3} = 2 \times \dfrac{1}{2} = 1 = \text{RHS}$ **Hence proved.**

(iii) **Hint** $\text{LHS} = 108\sin\dfrac{\pi}{18} - 144\sin^3\dfrac{\pi}{18}$

$= 36\left[3\sin\dfrac{\pi}{18} - 4\sin^3\dfrac{\pi}{18}\right] = 36\left[\sin\left(\dfrac{3\pi}{18}\right)\right]$

[using identity $\sin 3A = 3\sin A - 4\sin^3 A$]

$= 36\sin\dfrac{\pi}{6} = 18 = \text{RHS}$ **Hence proved.**

21. Hint $\sin 4A = 2\sin 2A \cos 2A$

$= 4\sin A \cos A(\cos^2 A - \sin^2 A)$

22. Hint We have, $\dfrac{\tan(\alpha+\theta)}{\tan(\alpha-\theta)} = \dfrac{n}{1}$

By componendo and dividendo rule, we get

$\dfrac{\tan(\alpha+\theta)+\tan(\alpha-\theta)}{\tan(\alpha+\theta)-\tan(\alpha-\theta)} = \dfrac{n+1}{n-1}$

$$\Rightarrow \frac{\frac{\sin(\alpha+\theta)}{\cos(\alpha+\theta)}+\frac{\sin(\alpha-\theta)}{\cos(\alpha-\theta)}}{\frac{\sin(\alpha+\theta)}{\cos(\alpha+\theta)}-\frac{\sin(\alpha-\theta)}{\cos(\alpha-\theta)}}=\frac{n+1}{n-1}$$

$$\Rightarrow \frac{\sin(\alpha+\theta)\cos(\alpha-\theta)+\sin(\alpha-\theta)\cos(\alpha+\theta)}{\sin(\alpha+\theta)\cos(\alpha-\theta)-\cos(\alpha+\theta)\sin(\alpha-\theta)}=\frac{n+1}{n-1}$$

$$\Rightarrow \frac{\sin(\alpha+\theta+\alpha-\theta)}{\sin(\alpha+\theta-\alpha+\theta)}=\frac{n+1}{n-1}\Rightarrow\frac{\sin2\alpha}{\sin2\theta}=\frac{n+1}{n-1}$$

23. Hint $\sin2\alpha\sin2\beta=\frac{1}{2}(2\sin2\alpha\sin2\beta)$

$$=\frac{1}{2}[\cos(2\alpha-2\beta)-\cos(2\alpha+2\beta)]$$

$$=\frac{1}{2}[\cos2(\alpha-\beta)-\cos90^\circ]=\frac{1}{2}(\cos2(\alpha-\beta)$$

$\because \quad -1\le\cos2(\alpha-\beta)\le1$

$\therefore \quad -\frac{1}{2}\le\frac{1}{2}\cos2(\alpha-\beta)\le\frac{1}{2}$

So, tha maximum value of $\sin2\alpha\sin2\beta$ is $\frac{1}{2}$ and the minimum value is $-\frac{1}{2}$.

24. LHS $=\cos\left(\frac{3\pi}{2}+x\right)\cos(2\pi+x)\left\{\cot\left(\frac{3\pi}{2}-x\right)+\cot(2\pi+x)\right\}$

$=\sin x\cos x(\tan x+\cot x)$

$\left[\because \cos\left(\frac{3\pi}{2}+x\right)=\sin x, \cos(2\pi+x)=\cos x.\right.$ and $\left.\cot\left(\frac{3\pi}{2}-x\right)=\tan x, \cot(2\pi+x)=\cot x.\right]$

$=\sin x\cos x\left(\frac{\sin x}{\cos x}+\frac{\cos x}{\sin x}\right)$

$=\sin x\cos x\left(\frac{\sin^2x+\cos^2x}{\sin x\cos x}\right)=1=\text{RHS}$

25. We have, LHS $=\cos^2 2x-\cos^2 6x$

$=1-\sin^2 2x-1+\sin^2 6x$

$=\sin^2 6x-\sin^2 2x$

$=\sin(6x+2x)\sin(6x-2x)$

$[\because \sin^2A-\sin^2B=\sin(A+B)\sin(A-B)]$

$=\sin8x\sin4x$

$=\sin4x\sin8x=\text{RHS}$

26. LHS $=\frac{\cos(90^\circ+\theta)\sec(-\theta)\tan(180^\circ-\theta)}{\sec(360^\circ-\theta)\sin(180^\circ+\theta)\cot(90^\circ-\theta)}$

$=\frac{-\sin\theta\sec\theta(-\tan\theta)}{\sec\theta(-\sin\theta)\tan\theta}$

$[\because \cos(90^\circ+\theta)=-\sin\theta, \sec(-\theta)=\sec\theta, \tan(180^\circ-\theta)=\tan\theta, \sec(360^\circ-\theta)=\sec\theta, \sin(180^\circ+\theta)=-\sin\theta, \cot(90^\circ-\theta)=\tan\theta]$

$=-1=\text{RHS}$ **Hence proved.**

27. Hint Let r be the radius of the circle and θ be the sector angle. Then,

Perimeter of the sector $=2r+\theta$ $\quad\left[\because \theta=\frac{l}{r}\right]$

Length of arc of semi-circle $=\pi r$ $\quad$ [$\because$ it is a semi-circle]

$\therefore \quad 2r+r\theta=\pi r$

$\Rightarrow \quad (2+\theta)=\pi$

$\Rightarrow \quad \theta=(\pi-2)$ rad

$\Rightarrow \quad \theta=(\pi-2)\times\frac{180^\circ}{\pi}$ $\quad\left[\because 1\text{ radian}=\frac{180^\circ}{\pi}\right]$

$\Rightarrow \quad \theta=\pi\times\frac{180^\circ}{\pi}-\frac{2\times180^\circ}{\pi}$

28. Let the three angles in AP be $a-d, a, a+d$.

$\therefore \quad (a-d)+a+(a+d)=180^\circ$

[$\because$ sum of all the angles of a triangle is 180°]

$\Rightarrow \quad 3a=180^\circ \Rightarrow a=60^\circ$

So, the angles are $(60-d)^\circ, 60^\circ, (60+d)^\circ$.

Measure of degrees of smallest angle $=(60-d)^\circ$

Measure of degrees in greatest angle $=(60+d)^\circ$

$\therefore$ Measure of radians in greatest angle

$=\frac{\pi}{180}\times(60+d)$ rad

According to the given condition,

$$\frac{(60-d)}{\frac{\pi}{180}\times(60+d)}=\frac{60}{\pi}\Rightarrow\frac{60-d}{60+d}\times\frac{180}{\pi}=\frac{60}{\pi}$$

$\Rightarrow \quad \frac{60-d}{60+d}=\frac{1}{3}$

$\Rightarrow \quad 180-3d=60+d$

$\Rightarrow \quad 4d=120\Rightarrow d=30$

Hence, the angles are (60°–30°), 60° and (60°+30°) i.e. 30°, 60° and 90°.

29. We have, $\tan\frac{\pi}{8}$

Let $\quad x=\frac{\pi}{4}\Rightarrow\frac{x}{2}=\frac{\pi}{8}$

$\therefore \quad \tan\frac{x}{2}=\frac{\sin\frac{x}{2}}{\cos\frac{x}{2}}=\frac{2\sin^2\frac{x}{2}}{2\sin\frac{x}{2}\cos\frac{x}{2}}$

$=\frac{1-\cos x}{\sin x}$ $\quad\left[\because \cos\theta=1-2\sin^2\frac{\theta}{2}\text{ and }\sin\theta=2\sin\frac{\theta}{2}\cos\frac{\theta}{2}\right]$ ...(i)

Now, on putting $x=\frac{\pi}{4}$ in Eq. (i), we get

$$\tan\frac{\pi}{8}=\frac{1-\cos\frac{\pi}{4}}{\sin\frac{\pi}{4}}=\frac{1-\frac{1}{\sqrt2}}{\frac{1}{\sqrt2}}=\sqrt2-1$$

30. Given, $\cos A = \frac{2}{3}$, where $0 < A < \frac{\pi}{2}$

$$\therefore \quad \sin A = \sqrt{1-\cos^2 A} = \sqrt{1-\frac{4}{9}} = \frac{\sqrt{5}}{3}$$

and $$\tan A = \frac{\sin A}{\cos A} = \frac{\sqrt{5}/3}{2/3} = \frac{\sqrt{5}}{2}$$

(i) $\sin 2A = 2\sin A\cos A = 2\times\frac{\sqrt{5}}{3}\times\frac{2}{3} = \frac{4\sqrt{5}}{9}$

(ii) $\cos 2A = 2\cos^2 A - 1$

$$= 2\left(\frac{2}{3}\right)^2 - 1 = 2\times\frac{4}{9} - 1$$

$$= \frac{8}{9} - 1 = -\frac{1}{9}$$

(iii) $$\tan 2A = \frac{2\tan A}{1-\tan^2 A} = \frac{2(\sqrt{5}/2)}{1-(\sqrt{5}/2)^2}$$

$$= \frac{\sqrt{5}}{1-5/4} = \frac{4\sqrt{5}}{4-5} = -4\sqrt{5}$$

31. $$\text{LHS} = \frac{\sin x}{\cos 3x} + \frac{\sin 3x}{\cos 9x} + \frac{\sin 9x}{\cos 27x}$$

$$= \frac{1}{2}\left[\frac{2\sin x\cos x}{\cos 3x\cos x} + \frac{2\sin 3x\cos 3x}{\cos 9x\cos 3x} + \frac{2\sin 9x\cos 9x}{\cos 27x\cos 9x}\right]$$

$$= \frac{1}{2}\left[\frac{\sin 2x}{\cos 3x\cos x} + \frac{\sin 6x}{\cos 9x\cos 3x} + \frac{\sin 18x}{\cos 27x\cos 9x}\right]$$

$$= \frac{1}{2}\left[\frac{\sin(3x-x)}{\cos 3x\cos x} + \frac{\sin(9x-3x)}{\cos 9x\cos 3x} + \frac{\sin(27x-9x)}{\cos 27x\cos 9x}\right]$$

$$= \frac{1}{2}\left[\frac{\sin 3x\cos x - \cos 3x\sin x}{\cos 3x\cos x} + \frac{\sin 9x\cos 3x - \cos 9x\sin 3x}{\cos 9x\cos 3x} + \frac{\sin 27x\cos 9x - \cos 27x\sin 9x}{\cos 27x\cos 9x}\right]$$

$$= \frac{1}{2}(\tan 3x - \tan x + \tan 9x - \tan 3x + \tan 27x - \tan 9x)$$

$$= \frac{1}{2}(\tan 27x - \tan x) = \text{RHS}$$

Hence proved.

32. Given, $\tan x = \frac{3}{4}$, $\pi < x < \frac{3\pi}{2}$

$$\because \quad \cos x = \pm\frac{1}{\sqrt{1+\tan^2 x}}$$

$$\therefore \quad \cos x = -\frac{1}{\sqrt{1+\tan^2 x}} \quad [\because x \text{ lies in III quadrant}]$$

$$= -\frac{1}{\sqrt{1+\frac{9}{16}}}$$

$$= -\frac{4}{\sqrt{25}} = -\frac{4}{5}$$

Now, $$\cos\frac{x}{2} = -\sqrt{\frac{1+\cos x}{2}}$$

$$= -\sqrt{\frac{1-\frac{4}{5}}{2}} = -\frac{1}{\sqrt{10}}$$

$$\left[\because \pi < x < \frac{3\pi}{2} \Rightarrow \frac{\pi}{2} < \frac{x}{2} < \frac{3\pi}{4} \Rightarrow \cos\frac{x}{2} < 0, \sin\frac{x}{2} > 0\right]$$

$$\because \quad \sin\frac{x}{2} = \sqrt{\frac{1-\cos x}{2}}$$

$$= \sqrt{\frac{1+4/5}{2}} = \sqrt{\frac{9}{10}} = \frac{3}{\sqrt{10}}$$

and $$\tan\frac{x}{2} = \frac{\sin x/2}{\cos x/2} = \frac{3/\sqrt{10}}{-1/\sqrt{10}} = -3$$

QR Code for Quick Test 1

QR Code for Quick Test 2

QR Code for Quick Test 3

QR Code for Quick Test 4

QR Code for Quick Test 5

Mind Map

Trigonometric Functions

Degree Measure

If the rotation from initial side to terminal side is $\left(\frac{1}{360}\right)$th of a complete revolution, then the angle is said to have a measure of one degree (1°).

$1° = 60'$; $1' = 60''$

Measure of an Angle

The amount of rotation from the initial side to the terminal side, is called the measure of an angle.

Radian Measure

Angle subtended at the centre by an arc of length one unit in a circle of radius 1 unit is said to have a measure of 1 radian.

$\theta = \frac{l}{r} = \frac{\textbf{Length of arc}}{\textbf{Radius}}$ or $l = r\theta$

The figures show the angles, whose measures are 1 radian (1^c) and – 1 radian (-1^c).

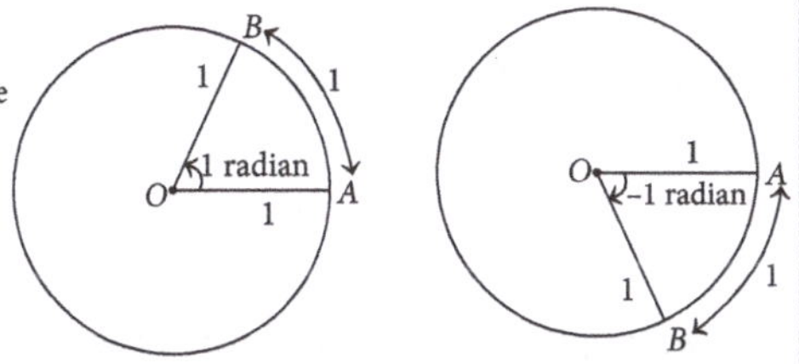

Transformation Formula

Transformation of Product into Sum or Difference

(i) $2\cos A \cos B = \cos(A + B) + \cos(A - B)$
(ii) $2\sin A \sin B = \cos(A - B) - \cos(A + B)$
(iii) $2\sin A \cos B = \sin(A + B) + \sin(A - B)$
(iv) $2\cos A \sin B = \sin(A + B) - \sin(A - B)$

Transformation of Sum or Difference into Product

- $\cos A + \cos B = 2\cos\left(\frac{A+B}{2}\right).\cos\left(\frac{A-B}{2}\right)$
- $\cos A - \cos B = -2\sin\left(\frac{A+B}{2}\right).\sin\left(\frac{A-B}{2}\right)$
- $\sin A + \sin B = 2\sin\left(\frac{A+B}{2}\right).\cos\left(\frac{A-B}{2}\right)$
- $\sin A - \sin B = 2\cos\left(\frac{A+B}{2}\right).\sin\left(\frac{A-B}{2}\right)$

Trigonometric Function of Any Angle

For any angle θ (shown in fig.) There are six circular function may be defined.

(i) $\cos\theta = x$
(ii) $\sin\theta = y$
(iii) $\tan\theta = \frac{y}{x}, x \neq 0$
(iv) $\cot\theta = \frac{x}{y}, y \neq 0$
(v) $\sec\theta = \frac{1}{x}, x \neq 0$
(vi) $\text{cosec}\,\theta = \frac{1}{y}, y \neq 0$

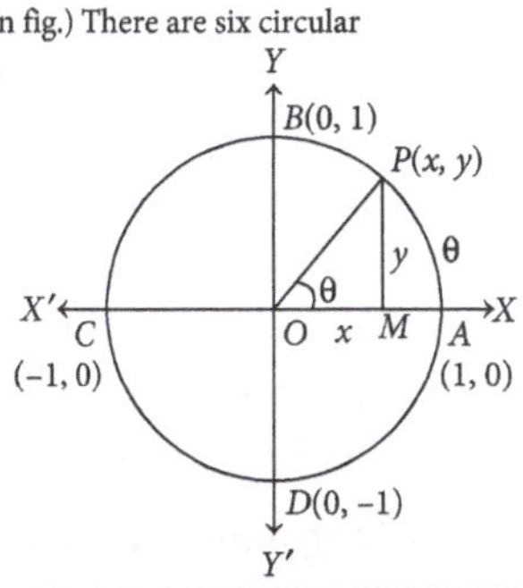

Domain and Range of Trigonometric Functions

T-function	Domain	Range
$\sin\theta$	R	$[-1, 1]$
$\cos\theta$	R	$[-1, 1]$
$\tan\theta$	$R - (2n+1)\frac{\pi}{2}, n \in I$	R
$\cot\theta$	$R - n\pi, n \in I$	R
$\text{cosec}\,\theta$	$R - n\pi, n \in I$	$R - (-1, 1)$
$\sec\theta$	$R - (2n+1)\frac{\pi}{2}, n \in I$	$R - (-1, 1)$

Trigonometric Identities

$\cos^2\theta + \sin^2\theta = 1$; $1 + \tan^2\theta = \sec^2\theta$; $1 + \cot^2\theta = \text{cosec}^2\theta$

Trigonometric Function of Multiple Angles

- $\cos 2x = \cos^2 x - \sin^2 x = 2\cos^2 x - 1$
  $= 1 - 2\sin^2 x = \frac{1 - \tan^2 x}{1 + \tan^2 x}$
- $\sin 2x = 2\sin x \cos x = \frac{2\tan x}{1 + \tan^2 x}$
- $\tan 2x = \frac{2\tan x}{1 - \tan^2 x}$
- $\cos 3x = 4\cos^3 x - 3\cos x$
- $\sin 3x = 3\sin x - 4\sin^3 x$
- $\tan 3x = \frac{3\tan x - \tan^3 x}{1 - 3\tan^2 x}$

Trigonometric Functions of Compound Angles

- $\cos(A + B) = \cos A \cos B - \sin A \sin B$
- $\cos(A - B) = \cos A \cos B + \sin A \sin B$
- $\sin(A + B) = \sin A \cos B + \cos A \sin B$
- $\sin(A - B) = \sin A \cos B - \cos A \sin B$
- $\tan(A + B) = \frac{\tan A + \tan B}{1 - \tan A \tan B}$

  where A, B and $(A + B)$ are not odd multiples of $\pi/2$.
- $\tan(A - B) = \frac{\tan A - \tan B}{1 + \tan A \tan B}$

  where A, B and $(A - B)$ are not odd multiples of $\pi/2$.
- $\cot(A + B) = \frac{\cot A \cot B - 1}{\cot B + \cot A}$

  where A, B and $(A + B)$ are not multiples of π.
- $\cot(A - B) = \frac{\cot A \cot B + 1}{\cot B - \tan A}$

  where A, B and $(A - B)$ are not multiples of π.

Sign of Trigonometric Functions

II Quadrant $\left(\frac{\pi}{2} < \theta < \pi\right)$

sin θ and cosec θ are positive.

(90°+θ) and (180° – θ)

I Quadrant $\left(0 < \theta < \frac{\pi}{2}\right)$

All trigonometric functions are positive.

(360°+θ) and (90° – θ)

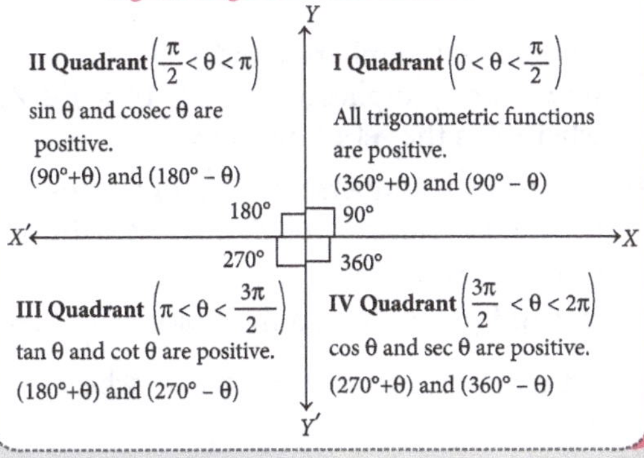

III Quadrant $\left(\pi < \theta < \frac{3\pi}{2}\right)$

tan θ and cot θ are positive.

(180°+θ) and (270° – θ)

IV Quadrant $\left(\frac{3\pi}{2} < \theta < 2\pi\right)$

cos θ and sec θ are positive.

(270°+θ) and (360° – θ)

Trigonometric Functions of Allied Angles

Two angles are said to be allied when their sum or difference is either zero or a multiple of $\pi/2$. The angles, $-\theta$, $\frac{\pi}{2} \pm \theta$, $\pi \pm \theta$, $2\pi \pm \theta$ etc., are angles allied to the angle θ, if θ is measured in radians.

Note If θ is measured in degrees, then the angles allied to θ are $-\theta$, $90° \pm \theta$, $180° \pm \theta$, $270° \pm \theta$, etc.

Topics to be Assessed Formatively

Trigonometric Equations

Equations which involve trigonometric functions of unknown angles (i.e. variable) are known as trigonometric equations.

e.g. $\sin x = \frac{3}{4}, \tan x + \sec x = -\sqrt{3}, \cos x + 4\sin x = 1$ etc.

Solution of a Trigonometric Equation

A solution of a trigonometric equation is the value of the unknown angle (i.e. variable) that satisfies the equation. We know that the values of $\sin x$ and $\cos x$ repeat after an interval of 2π and the values of $\tan x$ repeat after an interval of π.

Solutions of trigonometric equation is of two types

Principal Solution

The solution of a trigonometric equation for which the value of unknown angle say x lies between 0 and 2π, i.e. $0 \le x < 2\pi$, is called its **principal solution.**

Example 1 Find the principal solutions of the following equations.

(i) $\sin x = \frac{\sqrt{3}}{2}$ (ii) $\tan x = \frac{-1}{\sqrt{3}}$ (iii) $\sec x = -2$

Sol. (i) We have, $\sin x = \frac{\sqrt{3}}{2} \Rightarrow \sin x = \sin \frac{\pi}{3}$ $\left[\because \sin \frac{\pi}{3} = \frac{\sqrt{3}}{2}\right]$

$\therefore x = \frac{\pi}{3}$, which lies in I quadrant.

and $\sin x = \sin\left(\pi - \frac{\pi}{3}\right)$ $[\because \sin\theta = \sin(\pi - \theta)]$

$\therefore x = \frac{2\pi}{3}$, which lies in II quadrant.

Hence, the principal solutions are $x = \frac{\pi}{3}$ and $x = \frac{2\pi}{3}$.

(ii) We have, $\tan x = -\frac{1}{\sqrt{3}}$

Here, the value of $\tan x$ is negative. So, x lies in II and IV quadrants.

We know that $\tan \frac{\pi}{6} = \frac{1}{\sqrt{3}}$

Thus, $\tan\left(\pi - \frac{\pi}{6}\right) = -\frac{1}{\sqrt{3}} \Rightarrow \tan \frac{5\pi}{6} = \frac{-1}{\sqrt{3}}$

$\therefore x = \frac{5\pi}{6}$, which lies in II quadrant.

Also, $\tan\left(2\pi - \frac{\pi}{6}\right) = -\tan \frac{\pi}{6} \Rightarrow \tan \frac{11\pi}{6} = -\frac{1}{\sqrt{3}}$

$\therefore x = \frac{11\pi}{6}$, which lies in IV quadrant.

Hence, the principal solutions are $x = \frac{5\pi}{6}$ and $x = \frac{11\pi}{6}$.

(iii) We have, $\sec x = -2 \Rightarrow \cos x = -\frac{1}{2}$

Here, the value of $\cos x$ is negative. So, x lies in II and III quadrants.

We know that $\cos \frac{\pi}{3} = \frac{1}{2}$

Thus, $\cos\left(\pi - \frac{\pi}{3}\right) = -\frac{1}{2} \Rightarrow \cos \frac{2\pi}{3} = -\frac{1}{2}$

$\therefore x = \frac{2\pi}{3}$, which lies in II quadrant.

Also, $\cos\left(\pi + \frac{\pi}{3}\right) = -\frac{1}{2} \Rightarrow \cos \frac{4\pi}{3} = -\frac{1}{2}$

$\therefore x = \frac{4\pi}{3}$, which lies in III quadrant.

Hence, the principal solutions are $x = \frac{2\pi}{3}, \frac{4\pi}{3}$.

General Solution

The expression involving integer 'n' which gives all solutions of a trigonometric equation is called the general solution. The general solution of a trigonometric equation can be find out with the help of following theorems

Theorem 1. For any real numbers x and y, $\sin x = \sin y$, implies $x = n\pi + (-1)^n y, n \in Z$, where Z is the set of integers.

Theorem 2. For any real numbers x and y, $\cos x = \cos y$, implies $x = 2n\pi \pm y, n \in Z$, where Z is the set of integers.

Theorem 3. If x and y are not odd multiples of $\frac{\pi}{2}$, then $\tan x = \tan y$, implies $x = n\pi + y$, where $n \in Z$.

TEACHER TIPS

(i) The equation $\operatorname{cosec} x = \operatorname{cosec} y$ is equivalent to $\sin x = \sin y$. So, both have same general solution.
(ii) The equation $\sec x = \sec y$ is equivalent to $\cos x = \cos y$. So, both have same general solution.
(iii) The equation $\cot x = \cot y$ is equivalent to $\tan x = \tan y$. So, both have same general solution.
(iv) A trigonometric equation may have an infinite number of solutions.

Working Rule for Finding General Solution

The general solution of the given trigonometric equation can be find with the help of following steps

***Step* I** Simplify the given trigonometric equation by using factors or standard formulae to get the equations in the form of $\cos\theta =$ constant or $\sin\theta =$ constant or $\tan\theta =$ constant.

***Step* II** Find the value of θ which lies between 0 and 2π i.e. principal value of θ and say it is α, to get the equation of the form $\cos\theta = \cos\alpha$ or $\sin\theta = \sin\alpha$ or $\tan\theta = \tan\alpha$.

***Step* III** The general solution of the equation obtained in Step II can be find by using the following conditions.

- If $\sin\theta = \sin\alpha$, then $\theta = n\pi + (-1)^n\alpha,\ n \in Z$.
- If $\cos\theta = \cos\alpha$, then $\theta = 2n\pi \pm \alpha,\ n \in Z$.
- If $\tan\theta = \tan\alpha$, then $\theta = n\pi + \alpha,\ n \in Z$.

Thus, we get the required general solution of given trigonometric equation.

Example 2 **Find the general solution of the equation $2\cos^2 x + 3\sin x = 0$.**

Sol. Given equation is $2\cos^2 x + 3\sin x = 0$

$\Rightarrow\ 2(1-\sin^2 x) + 3\sin x = 0$ $\left[\because \sin^2 x + \cos^2 x = 1 \Rightarrow \cos^2 x = 1 - \sin^2 x\right]$

$\Rightarrow\ 2\sin^2 x - 3\sin x - 2 = 0$

$\Rightarrow\ 2\sin^2 x - 4\sin x + \sin x - 2 = 0$

$\Rightarrow\ 2\sin x(\sin x - 2) + 1(\sin x - 2) = 0$

$\Rightarrow\ (2\sin x + 1)(\sin x - 2) = 0$

$\Rightarrow\ \sin x = \dfrac{-1}{2}$

or $\sin x = 2$

Here, $\sin x = 2$ is not possible. $[\because \max(\sin\theta) = 1]$

$\therefore\ \sin x = -\dfrac{1}{2} = -\sin\dfrac{\pi}{6}$

$= \sin\left(\pi + \dfrac{\pi}{6}\right) = \sin\dfrac{7\pi}{6}$

Thus, $\sin x = \sin\dfrac{7\pi}{6}$, which is of the form $\sin\theta = \sin\alpha$.

So, $x = n\pi + (-1)^n\dfrac{7\pi}{6},\ n \in Z$

Some Important Results

Some standard forms of trigonometrical equations and their general solutions are given below

	Standard form	General solution
(i)	$\sin\theta = 0$	$\theta = n\pi,\ n \in Z$
(ii)	$\cos\theta = 0$	$\theta = (2n+1)\dfrac{\pi}{2}$, $n \in Z$
(iii)	$\tan\theta = 0$	$\theta = n\pi,\ n \in Z$
(iv)	$\sin^2\theta = \sin^2\alpha$, $\cos^2\theta = \cos^2\alpha$, $\tan^2\theta = \tan^2\alpha$	$\theta = n\pi \pm \alpha$, $n \in Z$
(v)	$\sin\theta = 1$	$\theta = (4n+1)\dfrac{\pi}{2}$, $n \in Z$
(vi)	$\cos\theta = 1$	$\theta = 2n\pi,\ n \in Z$
(vii)	$\cos\theta = -1$	$\theta = (2n+1)\pi$, $n \in Z$
(viii)	$\sin x = \sin\alpha$ and $\cos x = \cos\alpha$	$x = 2n\pi + \alpha$

Example 3 **Find the general solution for $\sec x = \sec(\pi + x)$.**

Sol. We have, $\sec x = \sec(\pi + x)$

$\Rightarrow\ \dfrac{1}{\cos x} = \dfrac{1}{\cos(\pi + x)}$ $\left[\because \sec x \cdot \cos x = 1 \Rightarrow \sec x = \dfrac{1}{\cos x}\right]$

$\Rightarrow\ \cos x = \cos(\pi + x)$

$\therefore$ The general solution is

$x = 2n\pi \pm (\pi + x)$ $\left[\because \cos x = \cos y \Rightarrow x = 2n\pi \pm y, \text{ where } n \in Z\right]$

$\Rightarrow\ x = 2n\pi + (\pi + x)$ or $x = 2n\pi - (\pi + x)$

$\Rightarrow\ x = 2n\pi + \pi + x$ or $x = 2n\pi - \pi - x$

$\Rightarrow x - x = \pi(2n+1)$ or $x + x = 2n\pi - \pi$

$\Rightarrow\ 0 = \pi(2n+1)$, which is absurd.

or $2x = \pi(2n-1)$

$\therefore\ x = \dfrac{\pi}{2}(2n-1)$,

But at $x = \dfrac{(2n-1)\pi}{2}$, given equation is not defined.

Hence, no general solution of given equation exist.

Example 4 Solve the equation $\sin\theta+\sin 3\theta+\sin 5\theta=0$.

Sol. We have, $\sin\theta+\sin 3\theta+\sin 5\theta=0$

$\Rightarrow (\sin 5\theta+\sin\theta)+\sin 3\theta=0$

$\Rightarrow 2\sin\left(\frac{5\theta+\theta}{2}\right)\cos\left(\frac{5\theta-\theta}{2}\right)+\sin 3\theta=0$

$\left[\because \sin A+\sin B=2\sin\left(\frac{A+B}{2}\right)\cos\left(\frac{A-B}{2}\right)\right]$

$\Rightarrow 2\sin 3\theta\cdot\cos 2\theta+\sin 3\theta=0$

$\Rightarrow \sin 3\theta(2\cos 2\theta+1)=0$

$\Rightarrow \sin 3\theta=0$ or $2\cos 2\theta+1=0$

$\Rightarrow \sin 3\theta=0$ or $\cos 2\theta=\frac{-1}{2}$

Now, if $\sin 3\theta=0 \Rightarrow 3\theta=n\pi, n\in Z$

$\Rightarrow \theta=\frac{n\pi}{3}, n\in Z$ ...(i)

and if $\cos 2\theta=\frac{-1}{2} \Rightarrow \cos 2\theta=\cos\frac{2\pi}{3}$

$\Rightarrow 2\theta=2m\pi\pm\frac{2\pi}{3}, m\in Z$

$\Rightarrow \theta=m\pi\pm\frac{\pi}{3}, m\in Z$ ...(ii)

From Eqs. (i) and (ii), we get general solution of the given equation, i.e.

$\theta=\frac{n\pi}{3}$ or $\theta=m\pi\pm\frac{\pi}{3}$, where $m, n\in Z$

Example 5 Solve the following equation.

$$\tan\theta+\tan 2\theta+\sqrt{3}\tan\theta\cdot\tan 2\theta=\sqrt{3}$$

Sol. We have, $\tan\theta+\tan 2\theta+\sqrt{3}\tan\theta\cdot\tan 2\theta=\sqrt{3}$

$\Rightarrow \tan\theta+\tan 2\theta=\sqrt{3}(1-\tan\theta\cdot\tan 2\theta)$

$\Rightarrow \frac{\tan\theta+\tan 2\theta}{1-\tan\theta\cdot\tan 2\theta}=\sqrt{3}$

$\left[\because \tan(x+y)==\frac{\tan x+\tan y}{1-\tan x\cdot\tan y}\right]$

$\Rightarrow \tan(\theta+2\theta)=\sqrt{3}\Rightarrow \tan 3\theta=\sqrt{3}$

$\Rightarrow \tan 3\theta=\tan\frac{\pi}{3}$ $\left[\because \tan\frac{\pi}{3}=\sqrt{3}\right]$

$\Rightarrow 3\theta=n\pi+\frac{\pi}{3}$

$[\because \tan x=\tan y\Rightarrow x=n\pi+y, n\in Z]$

$\therefore \theta=\frac{n\pi}{3}+\frac{\pi}{9}, n\in Z$

Example 6 Solve the following equations.

(i) $\tan^2\theta+(1-\sqrt{3})\tan\theta-\sqrt{3}=0$

(ii) $\sec^2 2x=1-\tan 2x$

Sol. (i) Given equation is

$\tan^2\theta+(1-\sqrt{3})\tan\theta-\sqrt{3}=0$

$\Rightarrow \tan^2\theta+\tan\theta-\sqrt{3}\tan\theta-\sqrt{3}=0$

$\Rightarrow \tan\theta(\tan\theta+1)-\sqrt{3}(\tan\theta+1)=0$

$\Rightarrow (\tan\theta+1)(\tan\theta-\sqrt{3})=0$

$\Rightarrow \tan\theta+1=0$ or $\tan\theta-\sqrt{3}=0$

$\Rightarrow \tan\theta=-1$ or $\tan\theta=\sqrt{3}$

$\Rightarrow \tan\theta=\tan\left(\frac{3\pi}{4}\right)$ or $\tan\theta=\tan\frac{\pi}{3}$

$\Rightarrow \theta=n\pi+\frac{3\pi}{4}$

or $\theta=m\pi+\frac{\pi}{3}$; $n, m\in Z$

(ii) Given equation is

$\sec^2 2x-1=-\tan 2x$

$\Rightarrow 1+\tan^2 2x-1=-\tan 2x$

$\Rightarrow \tan^2 2x=-\tan 2x$

$\Rightarrow \tan^2 2x+\tan 2x=0$

$\Rightarrow \tan 2x(\tan 2x+1)=0$

$\Rightarrow \tan 2x=0$ or $\tan 2x=-1$

$\Rightarrow 2x=n\pi$ or $\tan 2x=\tan\left(\frac{3\pi}{4}\right)$

$\Rightarrow 2x=n\pi$ or $2x=m\pi+\frac{3\pi}{4}$

$\Rightarrow x=\frac{n\pi}{2}$ or $x=\frac{m\pi}{2}+\frac{3\pi}{8}$; $m, n\in Z$

Example 7 Solve $\sin 3\alpha=4\sin\alpha\sin(x+\alpha)\sin(x-\alpha)$, where $\alpha\neq n\pi, n\in Z$.

Sol. We have, $\sin 3\alpha=4\sin\alpha\sin(x+\alpha)\sin(x-\alpha)$

$\Rightarrow \sin 3\alpha=4\sin\alpha(\sin^2 x-\sin^2\alpha)$

$[\because \sin(A+B)\sin(A-B)=\sin^2 A-\sin^2 B]$

$\Rightarrow 3\sin\alpha-4\sin^3\alpha=4\sin^2 x\sin\alpha-4\sin^3\alpha$

$\Rightarrow 3\sin\alpha=4\sin^2 x\sin\alpha$

$\Rightarrow \sin^2 x=\frac{3}{4}=\left(\frac{\sqrt{3}}{2}\right)^2$

$\Rightarrow \sin^2 x=\sin^2\frac{\pi}{3}$

$\therefore x=n\pi\pm\frac{\pi}{3}, n\in Z$

Example 8 Solve

$$\tan\theta+\tan\left(\theta+\frac{\pi}{3}\right)+\tan\left(\theta+\frac{2\pi}{3}\right)=3.$$

Sol. We have, $\tan\theta+\tan\left(\theta+\frac{\pi}{3}\right)+\tan\left(\theta+\frac{2\pi}{3}\right)=3$

$$\Rightarrow \quad \tan\theta + \frac{\tan\theta + \tan\frac{\pi}{3}}{1 - \tan\theta\tan\frac{\pi}{3}} + \frac{\tan\theta + \tan\frac{2\pi}{3}}{1 - \tan\theta\tan\frac{2\pi}{3}} = 3$$

$$\Rightarrow \quad \tan\theta + \frac{\tan\theta + \sqrt{3}}{1 - \sqrt{3}\tan\theta} + \frac{\tan\theta - \sqrt{3}}{1 + \sqrt{3}\tan\theta} = 3$$

$$\Rightarrow \quad \tan\theta + \frac{8\tan\theta}{1 - 3\tan^2\theta} = 3$$

$$\Rightarrow \quad \frac{\tan\theta - 3\tan^3\theta + 8\tan\theta}{1 - 3\tan^2\theta} = 3$$

$$\Rightarrow \quad \frac{9\tan\theta - 3\tan^3\theta}{1 - 3\tan^2\theta} = 3$$

$$\Rightarrow \quad \frac{3(3\tan\theta - \tan^3\theta)}{1 - 3\tan^2\theta} = 3$$

$$\Rightarrow \quad \frac{3\tan\theta - \tan^3\theta}{1 - 3\tan^2\theta} = 1 \Rightarrow \tan 3\theta = 1 = \tan\frac{\pi}{4}$$

$$\Rightarrow \quad 3\theta = n\pi + \frac{\pi}{4}, n \in Z \Rightarrow \theta = \frac{n\pi}{3} + \frac{\pi}{12}, n \in Z$$

Example 9 **Solve $4\sin x \sin 2x \sin 4x = \sin 3x$.**

Sol. We have, $4\sin x\sin 2x\sin 4x = \sin 3x$

$$\Rightarrow 4\sin x\sin(3x - x)\sin(3x + x) = \sin 3x$$

$$\Rightarrow \quad 4\sin x(\sin^2 3x - \sin^2 x) = \sin 3x$$

$$[\because \sin(A+B)\sin(A-B) = \sin^2 A - \sin^2 B]$$

$$\Rightarrow \quad 4\sin x\sin^2 3x - 4\sin^3 x = 3\sin x - 4\sin^3 x$$

$$\Rightarrow \quad 4\sin x\sin^2 3x - 3\sin x = 0$$

$$\Rightarrow \quad \sin x(4\sin^2 3x - 3) = 0$$

$$\Rightarrow \quad \sin x = 0 \text{ or } 4\sin^2 3x - 3 = 0$$

$$\Rightarrow \quad x = n\pi \text{ or } \sin^2 3x = \frac{3}{4}$$

$$\Rightarrow \quad x = n\pi \text{ or } \sin^2 3x = \left(\frac{\sqrt{3}}{2}\right)^2$$

$$\Rightarrow \quad x = n\pi \text{ or } \sin^2 3x = \sin^2\frac{\pi}{3}$$

$$\Rightarrow \quad x = n\pi \text{ or } 3x = m\pi \pm \frac{\pi}{3}, m, n \in Z$$

$$\Rightarrow \quad x = n\pi \text{ or } x = \frac{m\pi}{3} \pm \frac{\pi}{9}; m, n \in Z$$

Example 10 **Solve $\sqrt{3}\cos\theta + \sin\theta = 1$.**

Sol. Given equation is $\sqrt{3}\cos\theta + \sin\theta = 1$ ...(i)

which is of the form $a\cos\theta + b\sin\theta = c$.

Here, $a = \sqrt{3}, b = 1, c = 1$

On dividing both sides of Eq. (i) by

$\sqrt{a^2 + b^2} = \sqrt{(\sqrt{3})^2 + 1^2} = \sqrt{4} = 2$, we get

$$\frac{\sqrt{3}}{2}\cos\theta + \frac{1}{2}\sin\theta = \frac{1}{2}$$

$$\Rightarrow \quad \cos\frac{\pi}{6}\cos\theta + \sin\frac{\pi}{6}\sin\theta = \frac{1}{2}$$

$$\Rightarrow \quad \cos\left(\theta - \frac{\pi}{6}\right) = \frac{1}{2}$$

$$\Rightarrow \quad \cos\left(\theta - \frac{\pi}{6}\right) = \cos\frac{\pi}{3}$$

$$\Rightarrow \quad \theta - \frac{\pi}{6} = 2n\pi \pm \frac{\pi}{3}, n \in Z$$

$$\Rightarrow \quad \theta - \frac{\pi}{6} = 2n\pi + \frac{\pi}{3}, n \in Z$$

$$\text{or} \quad \theta - \frac{\pi}{6} = 2n\pi - \frac{\pi}{3}$$

$$\therefore \quad \theta = 2n\pi + \frac{\pi}{2} \text{ or } \theta = 2n\pi - \frac{\pi}{6}, n \in Z$$

Exercise

Multiple Choice Questions

1. The number of solutions of equation $\tan x + \sec x = 2\cos x$ lying in the interval $[0, 2\pi]$ is

(a) 0 (b) 1
(c) 2 (d) 3

2. The minimum value of $3\cos x + 4\sin x + 8$ is

(a) 5 (b) 9 (c) 7 (d) 3

3. If x and y are not odd multiples of $\frac{\pi}{2}$, then $\tan x = \tan y$ implies

(a) $x = n\pi + y$, where $n \in Z$ (b) $x = n\pi - y$, where $n \in Z$
(c) $x = n\pi \pm y$, where $n \in Z$ (d) None of the above

4. The solution of $\sin x = -\frac{\sqrt{3}}{2}$ is

(a) $x = n\pi + (-1)^n \frac{4\pi}{3}$, where $n \in Z$

(b) $x = n\pi + (-1)^n \frac{2\pi}{3}$, where $n \in Z$

(c) $x = n\pi + (-1)^n \frac{3\pi}{3}$, where $n \in Z$

(d) None of the above

5. The solution of $2\cos^2 x + 3\sin x = 0$ is

(a) $x = 2n\pi \pm \frac{2\pi}{3}, n \in Z$

(b) $x = 2n\pi \pm \frac{7\pi}{6}, n \in Z$

(c) $x = n\pi + (-1)^n \frac{7\pi}{6}, n \in Z$

(d) $x = n\pi + (-1)^n \frac{2\pi}{3}, n \in Z$

Very short answer Type Questions

Direction (Q. Nos 6-11) *Find the principal solution of the followings.*

6. $\tan x = \frac{1}{\sqrt{3}}$

7. $\operatorname{cosec} x = -2$

8. $\tan x = \sqrt{3}$

9. $\sec x = 2$

10. $\sin \frac{3x}{2} = 0$

11. $\sqrt{2} \cos x + 1 = 0$

12. Find the general solution for $\cos 3\theta = -\frac{1}{\sqrt{2}}$.

13. Find the general solution for $\sin 5\theta = \frac{1}{\sqrt{2}}$.

Short Answer Type I Questions

14. Solve $\cos 3x = \sin 2x$.

15. Solve $\sin 3x + \cos 2x = 0$.

16. If θ is the positive acute angle, then solve the equation $4 \cos^2 \theta - 4 \sin \theta = 1$.

17. Solve $7 \cos^2 \theta + 3 \sin^2 \theta = 4$.

Short Answer Type II Questions

18. Find the general solution of the equation
$$\sec^2 2x = 1 - \tan 2x.$$

19. Solve $\sin 2x - \sin 4x + \sin 6x = 0$.

20. Solve $\tan^3 x - 3 \tan x = 0$.

21. Solve $\cos \theta + \sqrt{3} \sin\theta = 1$

22. Solve $\sqrt{3} \cos \theta + \sin \theta = \sqrt{2}$.

23. Find the general solution of
$$\tan\theta + \tan 2\theta + \tan \theta \tan 2\theta = 1.$$

24. Find the general solution of
$$\cot x + \tan x = 2 \operatorname{cosec} x.$$

25. Find the number of values of x in the interval $[0, 5\pi]$ satisfying the following equation.
$$3 \sin^2 x - 7 \sin x + 2 = 0$$

26. Find the general solution $\sec^2 2x = 1 - \tan 2x$.

Long answer Type Questions

27. Find the general solution of the equation $\tan\theta + \tan 2\theta + \tan 3\theta = \tan\theta \tan 2\theta \tan 3\theta$.

28. Find the general solution of the equation $\sin x - 3 \sin 2x + \sin 3x = \cos x - 3 \cos 2x + \cos 3x$.

29. Find the general solution of the equation $\sin 2\theta + \sin 4\theta + \sin 6\theta = 0$.

Answers

1. (c) **2.** (d) **3.** (a) **4.** (a) **5.** (c)

6. $\frac{\pi}{6}, \frac{7\pi}{6}$ **7.** $\frac{7\pi}{6}, \frac{11\pi}{6}$ **8.** $n\pi + \frac{\pi}{3}$ **9.** $\frac{\pi}{3}, \frac{2\pi}{3}$

10. $\frac{2n\pi}{3}, n \in Z$ **11.** (i) $\frac{3\pi}{4}, \frac{5\pi}{4}$ (ii) $\frac{4\pi}{3}, \frac{5\pi}{3}$

12. $\frac{2n\pi}{3} \pm \frac{\pi}{4}$ **13.** $\frac{n\pi}{5} + (-1)^n \frac{\pi}{20}, n \in Z$

14. $\frac{2n\pi}{5} + \frac{\pi}{10}$ or $2n\pi - \frac{\pi}{2}, n \in I$

15. $\left(2n\pi - \frac{\pi}{2}\right)$ or $\left(\frac{2n\pi}{5} - \frac{\pi}{10}\right)$

16. 30°

17. $n\pi \pm \frac{\pi}{3}, n \in Z$

18. $\frac{n\pi}{2}$ or $\frac{n\pi}{2} + \frac{3\pi}{8}, n \in Z$

19. $\frac{n\pi}{4}$ or $n\pi \pm \frac{\pi}{6}$.

20. $n\pi$ or $n\pi \pm \frac{\pi}{3}$

21. $2n\pi + \frac{2\pi}{3}, 2n\pi$

22. $2n\pi + \frac{5\pi}{12}$ or $2n\pi - \frac{\pi}{12}, n \in Z$

23. $\frac{n\pi}{3} + \frac{\pi}{12}, n \in Z$

24. $2n\pi \pm \frac{\pi}{3}, n \in Z$

25. 6

26. $\frac{n\pi}{2}$ or x, $x = \frac{n\pi}{2} + \frac{3\pi}{8}, n \in Z.$

27. $\frac{n\pi}{3}, n \in Z$

28. $\frac{n\pi}{2} + \frac{\pi}{8}, n \in Z$

29. $\frac{n\pi}{4}$ or $n\pi \pm \frac{\pi}{3}$; $n, n \in Z$

04 CHAPTER

Without Mathematics, there's nothing you can do. Everything around you is Mathematics. Everything around you is numbers —*Shakuntala Devi*

Complex Numbers

YOUR STUDY GOALS

YOUR PRACTICE GOALS

- Objective Type Questions (29 Qs; 16 MCQs, 10 A-R, 3 Case Based)
- Subjective Type Questions (43 Qs; 14 VSA, 16 SA, 13 LA)

4.1 Complex Numbers

A number consisting of real number and imaginary number is called **complex number**. A complex number can be defined as a number of the form $\boldsymbol{a + ib}$, where a and b are real numbers.

e.g. $6 + 9i$, $-3 + 4i$ etc. are complex numbers.

Here, the symbol i is used to denote $\sqrt{-1}$ and it is called **iota.**

The complex number is generally denoted by z i.e. $\boldsymbol{z = a + ib}$.

Complex number z can be represented in the form of order pair i.e. z can be represented as (a, b).

Real and Imaginary Parts of a Complex Numbers

Let $z = a + ib$ be a complex number, then a is called the **real part** and b is called the **imaginary part** of z and they may be denoted as $\mathbf{Re}(z)$ and $\mathbf{Im}(z)$, respectively.

e.g. If $z = 2 + 3i$, then $\text{Re}(z) = 2$ and $\text{Im}(z) = 3$.

Purely Real and Purely Imaginary Complex Numbers

A complex number $z = a + ib$, is called **purely real**, if $b = 0$ i.e. $\text{Im}(z) = 0$ and is called **purely imaginary**, if $a = 0$ i.e. $\text{Re}(z) = 0$.

e.g. $z = 6$ and $z = \sqrt{5}$ are purely real and $z = 6i, z = -\frac{5}{2}i, z = \sqrt{3}i$ are purely imaginary.

Equality of Complex Numbers

Two complex numbers $z_1 = a + ib$ and $z_2 = c + id$ are said to be equal, if $a = c$ and $b = d$.

Some Important Results

(i) $\sqrt{a} \times \sqrt{b} \neq \sqrt{ab}$, if both a and b are negative real numbers.

(ii) $\sqrt{a} \times \sqrt{b} = \sqrt{ab}$, if atleast one of them is positive.

4.2 Integral Power of *i* (iota)

Positive Integral Powers of i

As we have seen, $i = \sqrt{-1}$. So, we can write the higher powers of i as follows

(i) $i^2 = -1$

(ii) $i^3 = i^2 \cdot i = (-1) \cdot i = -i$

(iii) $i^4 = (i^2)^2 = (-1)^2 = 1$

(iv) $i^5 = i^{4+1} = i^4 \cdot i = 1 \cdot i = i$

(v) $i^6 = i^{4+2} = i^4 \cdot i^2 = 1 \cdot i^2 = -1$

$\vdots \quad \vdots \quad \vdots \quad \vdots$

While evaluating i^n for $n > 4$, we are writing n as $4q + r$ for some $q, r \in N$ and $0 \le r \le 3$.

Then, $\quad i^n = i^{4q} \cdot i^r = (i^4)^q \cdot i^r = (1)^q \cdot i^r = i^r$

e.g. $\quad i^{17} = i^{4\times4+1} = i^{4\times4} \cdot i = (i^4)^4 \cdot i = 1 \cdot i = i$

Negative integral powers of i

Negative integral powers of i can be evaluated as follows

(i) $i^{-1} = \frac{1}{i} = \frac{1}{i} \times \frac{i}{i}$

[multiply numerator and denominator by i]

$= \frac{i}{i^2} = \frac{i}{(-1)} = -i \quad [\because i^2 = -1]$

(ii) $i^{-2} = \frac{1}{i^2} = \frac{1}{(-1)} = -1$

(iii) $i^{-3} = \frac{1}{i^3} = \frac{1}{i^3} \times \frac{i}{i}$

[multiply numerator and denominator by i]

$= \frac{i}{(i^4)} = \frac{i}{(1)} = i$

(iv) $i^{-4} = \frac{1}{i^4} = \frac{1}{1} = 1$

In order to compute i^{-n} for $n > 4$, first write $i^{-n} = \frac{1}{i^n} = \frac{1}{i^{4q+r}}$ for some $q, r \in N$ and $0 \le r \le 3$.

Then, evaluate i^{4q+r}.

Further, use above four negative integral powers of i.

e.g. $\quad i^{-15} = \frac{1}{i^{15}} = \frac{1}{i^{4\times3+3}} = \frac{1}{i^3} \quad [\because i^{4q+3} = i^3]$

$= \frac{1}{i^3} \times \frac{i}{i} = \frac{i}{i^4} = \frac{i}{1} = i \quad [\because i^4 = 1]$

4.3 Algebra of Complex Numbers

Addition of Two Complex Numbers

Let $z_1 = a_1 + ib_1$ and $z_2 = a_2 + ib_2$ be two complex numbers, then their addition is defined as

$$z = z_1 + z_2 = (a_1 + a_2) + i(b_1 + b_2)$$

e.g. (i) $(5 + 3i) + (-4 - i) = (5 - 4) + i(3 - 1) = 1 + 2i$

(ii) $(2 + 3i) + (-6 + 7i) = (2 - 6) + i(3 + 7) = -4 + 10i$

It can be observed that

(i) Real part of $(z_1 + z_2) = \text{Re}(z_1 + z_2) = \text{Re}(z_1) + \text{Re}(z_2)$

(ii) Imaginary part of $(z_1 + z_2) = \text{Im}(z_1 + z_2) = \text{Im}(z_1) + \text{Im}(z_2)$

Properties of Addition of Complex Numbers

If z_1 and z_2 are any two complex numbers, then

(i) Addition of complex numbers is closure i.e. $(z_1 + z_2)$ is also a complex number.

(ii) Addition of complex numbers is commutative i.e. $z_1 + z_2 = z_2 + z_1$

(iii) Addition of complex numbers is associative
i.e. $(z_1 + z_2) + z_3 = z_1 + (z_2 + z_3)$

(iv) **Existance of Additive Identity**

Complex number $0 = 0 + 0i$, is called additive identity of any complex number.

$z + 0 = z = 0 + z, \quad z \in C.$

(v) **Existence of Additive Inverse**

If z is a complex number, then the complex number $(-z)$ is called additive inverse of z.

i.e. $z + (-z) = 0 = (-z) + z$

Difference of Two Complex Numbers

Let $z_1 = a_1 + ib_1$ and $z_2 = a_2 + ib_2$ be two complex numbers. Then, their subtraction $z_1 - z_2$ is defined as the addition of z_1 and $(-z_2)$.

Thus, $z_1 - z_2 = z_1 + (-z_2) = (a_1 + ib_1) + (-a_2 - ib_2)$

$\Rightarrow \quad z_1 - z_2 = (a_1 - a_2) + i(b_1 - b_2)$

e.g.

(i) $(-4+7i)-(-11-23i) = (-4+7i)+(11+23i)$
$= (-4+11)+(7+23)i$
$= 7 + 30i$

(ii) $(6+5i) - (3+2i) = (6+5i) + (-3-2i)$
$= (6-3) + (5-2)\,i = 3 + 3i$

It can be observed that

(i) Real part of $(z_1 - z_2) = \text{Re}\,(z_1 - z_2)$
$= \text{Re}\,(z_1) - \text{Re}\,(z_2)$

(ii) Imaginary part of $(z_1 - z_2) = \text{Im}(z_1 - z_2)$
$= \text{Im}(z_1) - \text{Im}(z_2)$

Multiplication of Two Complex Numbers

The product of two complex numbers $z_1 = a + ib$ and $z_2 = c + id$ can be done as follows

$z_1 z_2 = (a + ib)(c + id) = ac + iad + ibc + i^2 bd$
$= ac + i(ad + bc) + (-1)bd \qquad [\because i^2 = -1]$
$z_1 z_2 = (ac - bd) + i(ad + bc)$

e.g. (i) $(-5+7i)(-13-3i)$
$= (-5)(-13)+(-5)(-3i)+(7i)(-13)+(7i)(-3i)$
$= 65+15i-91i-21i^2 = 65-76i+21 \qquad [\because i^2 = -1]$
$= 86-76i$

(ii) $(5i)\left(\frac{-2}{5}i\right) = \left[5 \times \left(\frac{-2}{5}\right)\right](i \times i)$
$= (-2)(i^2) = -2 \times (-1) = 2 \qquad [\because i^2 = -1]$

Properties of Multiplication of Complex Numbers

If z_1, z_2 and z_3 are any three complex number, then

(i) Multiplication of complex numbers is closure.
i.e. $z_1 \cdot z_2$ is also a complex number.

(ii) Multiplication of complex numbers is commutative. i.e. $z_1 \times z_2 = z_2 \times z_1$

(iii) Multiplication of complex numbers is associative.
i.e. $(z_1 \times z_2) \times z_3 = z_1 \times (z_2 \times z_3)$

(iv) **Existence of Multiplicative Identity**

Complex number $1 = 1 + 0i$ is called multiplicative. identity of any complex number.

$z \times 1 = z = 1 \times z, z \in C$

(v) **Existence of multiplicative Inverse or Reciprocal**

If z is a complex number, then the complex number $\frac{1}{z}$ is called multiplicative inverse of z.

i.e. $z \times \frac{1}{z} = 1 = \frac{1}{z} \times z$

(vi) Multiplication of complex number is distributive
i.e. $\quad z_1(z_2 + z_3) = z_1 z_2 + z_1 z_3$
and $\quad (z_1 + z_2) z_3 = z_1 z_3 + z_2 z_3$

Division of Two Complex Numbers

The division of a complex number z_1 by a non-zero complex number z_2 is defined as the multiplication of z_1 by the multiplicative inverse of z_2 and is denoted by $\frac{z_1}{z_2}$.

Therefore, $\quad \frac{z_1}{z_2} = z_1 \cdot z_2^{-1} = z_1 \cdot \left(\frac{1}{z_2}\right)$

Identities Related to Complex Numbers

Identity is an equation which is true for all values of the variable (complex number) involved in it. Here, we have the following identities.

(i) $(z_1 + z_2)^2 = z_1^2 + z_2^2 + 2z_1 z_2$,
for all complex numbers z_1 and z_2.

(ii) $(z_1 - z_2)^2 = z_1^2 - 2z_1 z_2 + z_2^2$

(iii) $(z_1+z_2)^3 = z_1^3 + 3z_1^2 z_2 + 3z_1 z_2^2 + z_2^3$

(iv) $(z_1-z_2)^3 = z_1^3 - 3z_1^2 z_2 + 3z_1 z_2^2 - z_2^3$

(v) $z_1^2 - z_2^2 = (z_1+z_2)(z_1-z_2)$

Example 1. **Write the real and imaginary parts of the complex number $\sqrt{37}+\sqrt{-19}$.**

Sol. Let $z = \sqrt{37}+\sqrt{-19}$

$= \sqrt{37}+\sqrt{19}\sqrt{-1}$

$\Rightarrow \quad z = \sqrt{37}+\sqrt{19}\,i \quad [\because \sqrt{-1}=i]$

$\therefore \quad \text{Re}(z) = \sqrt{37}$

and $\quad \text{Im}(z) = \sqrt{19}$

Example 2. **Find the real values of *a* and *b*, if $(3a-6)+2ib = -6b+(6+a)i$.**

Sol. We have, $(3a-6)+2ib = -6b+(6+a)i$

On equating real and imaginary parts, we get

$3a-6 = -6b \quad \text{...(i)}$

and $\quad 2b = 6+a \quad \text{...(ii)}$

Above equations can be rewritten as

$3a+6b = 6 \quad \text{...(iii)}$

and $\quad a-2b = -6 \quad \text{...(iv)}$

On solving Eqs. (iii) and (iv), we get

$a = -2$ and $b = 2$

Example 3. **Express the following in the form of $a+ib$, where $a, b \in R$.**

(i) i^{103} (ii) $(-\sqrt{-1})^{4x+3}$ (iii) $\left(i^{29}+\dfrac{1}{i^{29}}\right)$

(iv) i^{-41}

Sol. (i) $i^{103} = i^{25\times4+3} = (i^4)^{25}\cdot i^3 = (1)^{25}\cdot(-i)$

$[\because i^4 = 1 \text{ and } i^3 = -i]$

$= -i = 0 - i$

(ii) $(-\sqrt{-1})^{4x+3} = (-i)^{4x+3} = (-i)^{4x}\cdot(-i)^3$

$= i^{4x}(-i^3) = (i^4)^x(-(-i)) \quad [\because i^3 = -i]$

$= (1)^x(i) \quad [\because i^4 = 1]$

$= i = 0 + i$

(iii) $i^{29} + \dfrac{1}{i^{29}} = \dfrac{i^{29}\cdot i^{29}+1}{i^{29}} = \dfrac{(i^2)^{29}+1}{i^{29}}$

$= \dfrac{(-1)^{29}+1}{i^{29}} \quad [\because i^2 = -1]$

$= \dfrac{-1+1}{i^{29}} = 0 = 0 + 0i$

(iv) $i^{-41} = \dfrac{1}{i^{41}} = \dfrac{1}{i^{4\times10+1}}$

$= \dfrac{1}{(i^4)^{10}\cdot i} = \dfrac{1}{i} = \dfrac{i}{i^2} \quad [\because i^4 = 1]$

$= -i = 0 - i$

Example 4. **Evaluate $\dfrac{i^{592}+i^{590}+i^{588}+i^{586}+i^{584}}{i^{582}+i^{580}+i^{578}+i^{576}+i^{574}}$.**

Sol. Consider the given expression,

$\dfrac{i^{592}+i^{590}+i^{588}+i^{586}+i^{584}}{i^{582}+i^{580}+i^{578}+i^{576}+i^{574}}$

$= \dfrac{i^{584+8}+i^{584+6}+i^{584+4}+i^{584+2}+i^{584}}{i^{574+8}+i^{574+6}+i^{574+4}+i^{574+2}+i^{574}}$

$= \dfrac{i^{584}(i^8+i^6+i^4+i^2+1)}{i^{574}(i^8+i^6+i^4+i^2+1)}$

$= \dfrac{i^{584}}{i^{574}} = i^{584-574} = i^{10} = i^{4\times2+2}$

$= (i^4)^2\cdot i^2 = (1)^2\cdot i^2 \quad [\because i^4 = 1]$

$= -1 \quad [\because i^2 = -1]$

Example 5. **What is the value of $\dfrac{i^{4x+1}-i^{4x-1}}{2}$?**

NCERT Exemplar

Sol. Consider $\dfrac{i^{4x+1}-i^{4x-1}}{2} = \dfrac{i^{4x}\cdot i - i^{4x}\cdot i^{-1}}{2}$

$= \dfrac{i - \frac{1}{i}}{2} \quad [\because i^{4x} = 1]$

$= \dfrac{i^2-1}{2i} = \dfrac{-2}{2i} \quad [\because i^2 = -1]$

$= \dfrac{-1}{i} = \dfrac{-i}{i^2}$

$= \dfrac{-i}{-1} = i \quad [\because i^2 = -1]$

Example 6. **Express the following in the form of $a+ib$.**

$$\left[\left(\frac{1}{3}+\frac{7}{3}i\right)+\left(4+\frac{1}{3}i\right)\right]-\left(-\frac{4}{3}+i\right)$$

NCERT

Sol. Consider the given expression,

$\left[\left(\frac{1}{3}+\frac{7}{3}i\right)+\left(4+\frac{1}{3}i\right)\right]-\left(-\frac{4}{3}+i\right)$

$= \left[\left(\frac{1}{3}+4\right)+i\left(\frac{7}{3}+\frac{1}{3}\right)\right]-\left(-\frac{4}{3}+i\right)$

$$= \left(\frac{13}{3} + \frac{8}{3}i\right) + \left(\frac{4}{3} - i\right) = \left(\frac{13}{3} + \frac{4}{3}\right) + i\left(\frac{8}{3} - 1\right)$$

$= \frac{17}{3} + \frac{5}{3}i$, which is in the form of $a + ib$.

Example 7. **Express the following in the form $a + ib$.**

(i) $(-i)(3i)\left(-\frac{1}{6}i\right)^3$

(ii) $(-\sqrt{3} + \sqrt{-2})(-2 + \sqrt{-3})$

Sol. (i) $(-i)(3i)\left(-\frac{1}{6}i\right)^3 = (-3i^2)\left(-\frac{1}{216}i^3\right)$

$$= (-3 \times (-1))\left(-\frac{1}{216}(-i)\right)$$

$[\because i^2 = -1 \text{ and } i^3 = -i]$

$$= 3 \times \frac{1}{216} \times i = \frac{i}{72} = 0 + \frac{1}{72}i,$$

which is in the form of $a + ib$.

(ii) $(-\sqrt{3} + \sqrt{-2})(-2 + \sqrt{-3})$

$$= (-\sqrt{3} + i\sqrt{2})(-2 + i\sqrt{3})$$

$[\because \sqrt{-2} = \sqrt{2} \times \sqrt{-1} = \sqrt{2}\, i$, similarly $\sqrt{-3} = \sqrt{3}\, i\,]$

$$= 2\sqrt{3} - 3i - 2\sqrt{2}\, i + i^2\sqrt{6}$$

$$= 2\sqrt{3} - i(3 + 2\sqrt{2}) - \sqrt{6} \quad [\because i^2 = -1]$$

$$= (2\sqrt{3} - \sqrt{6}) - i(3 + 2\sqrt{2}),$$

which is in the form of $a + ib$.

Example 8. **If z_1 and z_2 are complex numbers, then prove that $\text{Re}(z_1 z_2) = \text{Re}(z_1)\,\text{Re}(z_2) - \text{Im}(z_1)\,\text{Im}(z_2)$.** NCERT

Sol. Let $z_1 = x_1 + iy_1$ and $z_2 = x_2 + iy_2$

Then, $z_1 z_2 = (x_1 x_2 - y_1 y_2) + i(x_1 y_2 + y_1 x_2)$

$\therefore \quad \text{Re}(z_1 z_2) = x_1 x_2 - y_1 y_2$

$= \text{Re}(z_1)\,\text{Re}(z_2) - \text{Im}(z_1)\,\text{Im}(z_2)$

Hence proved.

Example 9. **Simplify each of the following and express each in the form $a + ib$.**

(i) $(2 + \sqrt{-3})^2$ (ii) $\left[i^{18} + \left(\frac{1}{i}\right)^{25}\right]^3$

(iii) $(3 + \sqrt{-5})(3 - \sqrt{-5})$ NCERT

Sol. (i) $(2 + \sqrt{-3})^2 = (2 + \sqrt{3}i)^2 = 2^2 + 2\cdot(2)(\sqrt{3}i) + (\sqrt{3}i)^2$

$[\because (z_1 + z_2)^2 = z_1^2 + 2z_1 z_2 + z_2^2]$

$$= 4 + 4\sqrt{3}i + 3i^2 = 4 + 4\sqrt{3}i - 3$$

$$= 1 + 4\sqrt{3}\, i \quad [\because i^2 = -1]$$

(ii) $\left[i^{18} + \left(\frac{1}{i}\right)^{25}\right]^3 = \left[i^{4\times 4+2} + \left(\frac{i}{i^2}\right)^{25}\right]^3$

$$= \left[(i^4)^4 \cdot i^2 + \left(\frac{i}{-1}\right)^{25}\right]^3 \quad [\because i^2 = -1]$$

$$= \left[1\cdot(-1) + \frac{i^{25}}{(-1)}\right]^3$$

$[\because i^4 = 1 \text{ and } i^2 = -1]$

$$= [-1 - i^{4\times 6+1}]^3$$

$$= [-1 - (i^4)^6 \cdot i]^3 = [-1 - i]^3$$

$[\because i^4 = 1]$

$$= -[1 + i]^3 = -(1 + 3i + 3i^2 + i^3)$$

$[\because (z_1 + z_2)^3 = (z_1^3 + 3z_1^2 z_2 + 3z_1 z_2^2 + z_2^3]$

$$= -(1 + 3i - 3 - i)$$

$[\because i^2 = -1 \text{ and } i^3 = -i]$

$$= -(-2 + 2i) = 2 - 2i = 2(1 - i)$$

(iii) $(3 + \sqrt{-5})(3 - \sqrt{-5}) = (3 + \sqrt{5}i)(3 - \sqrt{5}i)$

$$= (3)^2 - (\sqrt{5}i)^2$$

$[\because (z_1 - z_2)(z_1 + z_2) = z_1^2 - z_2^2]$

$$= 9 - 5i^2 = 9 + 5 = 14 \quad [\because i^2 = -1]$$

$$= 14 + 0i$$

Example 10. **Reduce the following complex numbers in the standard form.**

(i) $\left(\frac{1}{1-4i} - \frac{2}{1+i}\right)\left(\frac{3-4i}{5+i}\right)$ NCERT

(ii) $\left(\frac{4i^3 - 1}{2i + 1}\right)^2$

Sol. (i) $\left(\frac{1}{1-4i} - \frac{2}{1+i}\right)\left(\frac{3-4i}{5+i}\right)$

$$= \left[\frac{1 + i - 2(1 - 4i)}{(1 - 4i)(1 + i)}\right]\left(\frac{3-4i}{5+i}\right)$$

$$= \left(\frac{1 + i - 2 + 8i}{1 + i - 4i - 4i^2}\right)\left(\frac{3-4i}{5+i}\right) = \left(\frac{-1 + 9i}{1 - 3i + 4}\right)\left(\frac{3-4i}{5+i}\right)$$

$[\because i^2 = -1]$

$$= \left(\frac{-1 + 9i}{5 - 3i}\right)\left(\frac{3-4i}{5+i}\right) = \frac{-3 + 4i + 27i - 36i^2}{25 + 5i - 15i - 3i^2}$$

$$= \frac{-3 + 31i + 36}{25 - 10i + 3} = \frac{33 + 31i}{28 - 10i} \quad [\because i^2 = -1]$$

$$= \frac{(33 + 31i)}{(28 - 10i)} \times \frac{(28 + 10i)}{(28 + 10i)}$$

[by rationalising the denominator]

$$= \frac{924 + 868i + 330i + 310i^2}{784 - 100i^2}$$

$$= \frac{924 + 1198i - 310}{784 + 100} = \frac{614 + 1198i}{884} = \frac{307}{442} + \frac{599}{442}i$$

(ii) $\left(\frac{4i^3 - 1}{2i + 1}\right)^2 = \left[\frac{4\cdot(-i) - 1}{2i + 1}\right]^2$ $[\because i^3 = -i]$

$$= \left(\frac{-4i - 1}{2i + 1}\right)^2 = \frac{(1 + 4i)^2}{(1 + 2i)^2} = \frac{1 + 16i^2 + 8i}{1 + 4i^2 + 4i}$$

$[\because (z_1 + z_2)^2 = z_1^2 + z_2^2 + 2z_1z_2]$

$$= \frac{1 - 16 + 8i}{1 - 4 + 4i} = \frac{-15 + 8i}{-3 + 4i}$$ $[\because i^2 = -1]$

$$= \frac{15 - 8i}{3 - 4i} = \frac{15 - 8i}{3 - 4i} \times \frac{3 + 4i}{3 + 4i}$$

[by rationalising the denominator]

$$= \frac{(15 - 8i)(3 + 4i)}{9 - 16i^2}$$

$[\because (z_1 - z_2)(z_1 + z_2) = z_1^2 - z_2^2]$

$$= \frac{45 + 60i - 24i - 32i^2}{9 + 16}$$

$$= \frac{45 + 36i + 32}{25} = \frac{77 + 36i}{25} = \frac{77}{25} + \frac{36}{25}i,$$

which is in the form of $(a + ib)$.

Example 11. **Express $(-2 + \sqrt{3}i)^{-1}$ in the form $(a + ib)$.**

Sol. We have, $(-2 + \sqrt{3}i)^{-1} = \frac{1}{-2 + \sqrt{3}i} \times \frac{-2 - \sqrt{3}i}{-2 - \sqrt{3}i}$

$$= \frac{-2 - \sqrt{3}i}{(-2)^2 - (\sqrt{3}i)^2} = \frac{-2 - \sqrt{3}i}{4 - 3i^2}$$

$[\because (z_1 + z_2)(z_1 - z_2) = z_1^2 - z_2^2]$

$$= \frac{-2 - \sqrt{3}i}{4 - 3(-1)}$$ $[\because i^2 = -1]$

$$= \frac{-2 - \sqrt{3}i}{7} = -\frac{2}{7} - \frac{\sqrt{3}}{7}i$$

Example 12. **Find the value of $2x^4 + 5x^3 + 7x^2 - x + 41$, when $x = -2 - \sqrt{3}i$.** *NCERT Exemplar, Competency Based Que*

Sol. We have, $x = -2 - \sqrt{3}i \Rightarrow x + 2 = -\sqrt{3}i$

On squaring both sides, we get

$$(x + 2)^2 = (-\sqrt{3}i)^2 \Rightarrow x^2 + 4 + 4x = 3i^2$$

$[\because (z_1 + z_2)^2 = z_1^2 + z_2^2 + 2z_1z_2]$

$\Rightarrow x^2 + 4x + 4 = -3$ $[\because i^2 = -1]$

$\Rightarrow x^2 + 4x + 7 = 0$

Now, divide $2x^4 + 5x^3 + 7x^2 - x + 41$ by $x^2 + 4x + 7$, we get

$$\begin{array}{r|l} & 2x^2 - 3x + 5 \\ x^2 + 4x + 7 & 2x^4 + 5x^3 + 7x^2 - x + 41 \\ & 2x^4 + 8x^3 + 14x^2 \\ & - \quad - \quad - \\ \hline & -3x^3 - 7x^2 - x + 41 \\ & -3x^3 - 12x^2 - 21x \\ & + \quad + \quad + \\ \hline & 5x^2 + 20x + 41 \\ & 5x^2 + 20x + 35 \\ & - \quad - \quad - \\ \hline & 6 \end{array}$$

Thus, $2x^4 + 5x^3 + 7x^2 - x + 41$

$$= (x^2 + 4x + 7)(2x^2 - 3x + 5) + 6$$

[$\because$ dividend = quotient × divisor + remainder]

$$= 0 \times (2x^2 - 3x + 5) + 6 = 6$$ $[\because x^2 + 4x + 7 = 0]$

Quick Test 1

1. Write the following as complex numbers.

(i) $5 - 7\sqrt{-21}$ (ii) $\sqrt{x}$; $x > 0$

(iii) $\frac{\sqrt{3}}{2} - \frac{\sqrt{-2}}{\sqrt{7}}$ (iv) $-b + \sqrt{-4ac}$; $a, c > 0$

(v) $\sqrt{-27}$ (vi) $\sqrt{-16} + 3\sqrt{-25} + \sqrt{-36} - \sqrt{-625}$

2. Write the real and imaginary parts of the following complex numbers.

(i) $z = \frac{\sqrt{17}}{2} + \frac{2}{\sqrt{70}}i$ (ii) $\frac{\sqrt{5}}{7}i$ (iii) $-\frac{1}{5} + \frac{i}{5}$

3. Find the values of x and y, if

(i) $x + i(3x - y) = 3 - 6i$

(ii) $(x - 2) + 3iy = -4y + (3 + x)i$

4. Express the following in the form of $a + ib$.

(i) i^{-35} (ii) i^{998}

(iii) $\left(i^{37} \times \frac{1}{i^{67}}\right)$ (iv) $i + \frac{1}{i}$

5. Find the value of the following expressions.

(i) $2i^2 + 6i^3 + 3i^{16} - 6i^{19} + 4i^{25}$

(ii) $1 + i^{10} + i^{110} + i^{1000}$

(iii) $(-i)^{4n+3}$, where n is a positive integer. *Competency Based Que*

(iv) $(2i)^3$

(v) $i^{12} + i^{13} + i^{14} + i^{15}$

6. Find the real value of a for which $3i^3 - 2ai^2 + (1 - a)i + 5$ is real.

NCERT Exemplar, Competency Based Que

7. Express the following in the form $a+ib$.

(i) $(1+\sqrt{-49})-(2+\sqrt{-25})$ NCERT

(ii) $(\sqrt{6}+5i)\left(\sqrt{6}-\frac{1}{5}i\right)$

(iii) $(1+i)^4$

(iv) $\frac{(1-i)^3}{1-i^3}$

(v) $\left[\left(\sqrt{5}+\frac{i}{2}\right)(\sqrt{5}-2i)\right]\div(6+5i)$

8. If $x+iy=\frac{a+i}{a-i}$, then prove that $ay-1=x$.

9. What is the smallest positive integer n, for which $(1+i)^{2n}=(1-i)^{2n}$? Competency Based Que

10. Find the value of the following.

(i) $x^4-12x^3+70x^2-204x+225$, when $x=3+4i$

(ii) $x^4+9x^3+35x^2-x+4$, when $x=-5+2\sqrt{-4}$

(iii) $2x^3+2x^2-7x+72$, when $x=\frac{3-5i}{2}$

Answers

1. (i) $5-7\sqrt{21}i$ (ii) $\sqrt{x}+0i$ (iii) $\frac{\sqrt{3}}{2}-\sqrt{\frac{14}{7}}i$ (iv) $-b+2\sqrt{ac}i$ (v) $0+3\sqrt{3}i$ (vi) $0+0i$

2. (i) $\text{Re}(z)=\frac{\sqrt{17}}{2}$, $\text{Im}(z)=\frac{2}{\sqrt{70}}$

(ii) $\text{Re}(z)=0$, $\text{Im}(z)=\frac{\sqrt{5}}{7}$

(iii) $\text{Re}(z)=\frac{-1}{5}$, $\text{Im}(z)=\frac{1}{5}$

3. (i) $x=3,\ y=15$ (ii) $x=\frac{-6}{7},\ y=\frac{5}{7}$

4. (i) $0+i$ (ii) $-1+0i$ (iii) $-1+0i$ (iv) $0+0i$

5. (i) $1+4i$ (ii) 0 (iii) i (iv) $-8i$ (v) 0

6. -2

7. (i) $-1+2i$ (ii) $7+\frac{24\sqrt{6}}{5}i$ (iii) $-4+0i$ (iv) $-2+0i$ (v) $\frac{72-15\sqrt{5}}{122}-\left(\frac{30+9\sqrt{5}}{61}\right)i$

9. 2

10. (i) 0 (ii) -160 (iii) 4

For Solutions, See QR Code on Pg. 117

4.4 Conjugate of a Complex Number

The conjugate of a complex number z, is the complex number, obtained by changing the sign of imaginary part of z. It is denoted by $\bar{z}$.

e.g. If $z=2+3i$, then $\bar{z}=2-3i$

and if $z=-4-3i$, then $\bar{z}=-4+3i$

- A pair of complex numbers z_1 and z_2 is said to be conjugate of each other, if $\bar{z}_1=z_2$ and $\bar{z}_2=z_1$.
- Conjugate of purely real complex number is same i.e. if $z=3$, then $\bar{z}=3$.

Properties of Conjugate of Complex Numbers

1. $(\bar{\bar{z}})=z$, where $\bar{z}$ is the conjugate of complex number z and $\bar{\bar{z}}$ is the conjugate of complex number $\bar{z}$.
2. $z\bar{z}=|z|^2=(\text{Re}(z))^2+(\text{Im}(z))^2$
3. $z+\bar{z}=2\,\text{Re}(z)$ 4. $z-\bar{z}=2i\,\text{Im}(z)$
5. $z=\bar{z}\Leftrightarrow z$ is purely real.
6. $z+\bar{z}=0\Leftrightarrow z$ is purely imaginary.
7. $\overline{z_1+z_2}=\bar{z}_1+\bar{z}_2$
8. $\overline{z_1-z_2}=\bar{z}_1-\bar{z}_2$
9. $\overline{z_1\cdot z_2}=\bar{z}_1\cdot\bar{z}_2$
10. $\overline{\left(\frac{z_1}{z_2}\right)}=\frac{\bar{z}_1}{\bar{z}_2}$, provided $z_2, \bar{z}_2\neq 0$
11. $(\bar{z})^n=\overline{(z^n)}$

4.5 Modulus (Absolute Value) of Complex Numbers

The modulus (or absolute value) of a complex number, $z=a+ib$ is defined as the non-negative real number $\sqrt{a^2+b^2}$. It is denoted by $|z|$ i.e. $|z|=\sqrt{a^2+b^2}$.

e.g. If $z=2+3i$, then $|z|=\sqrt{2^2+3^2}=\sqrt{4+9}=\sqrt{13}$

and if $z=1-i$, then $|z|=\sqrt{(1)^2+(-1)^2}=\sqrt{1+1}=\sqrt{2}$

Properties of Modulus of Complex Numbers

1. $|z|=|\bar{z}|=|-z|=|-\bar{z}|$
2. $|z|=0 \Leftrightarrow z=0$ i.e. $\text{Re}(z)=\text{Im}(z)=0$
3. $-|z| \le \text{Re}(z) \le |z|$
4. $-|z| \le \text{Im}(z) \le |z|$
5. $|z_1+z_2|^2=|z_1|^2+|z_2|^2+2\text{Re}(z_1\bar{z}_2)$
6. $|z_1-z_2|^2=|z_1|^2+|z_2|^2-2\text{Re}(z_1\bar{z}_2)$
7. $|z_1+z_2|^2+|z_1-z_2|^2=2(|z_1|^2+|z_2|^2)$
8. $|z_1z_2|=|z_1||z_2|$

   In general, if $z_1, z_2, \ldots, z_n$ are any complex numbers, then

   $$|z_1z_2\ldots z_n|=|z_1||z_2|\ldots|z_n| \quad \ldots\text{(i)}$$

   So, if $z_1=z_2=\ldots=z_n$, then from Eq. (i), we have $|z_1^n|=|z_1|^n$. Thus, we have $|z^n|=|z|^n$.
9. $\left|\dfrac{z_1}{z_2}\right|=\dfrac{|z_1|}{|z_2|}$, provided $z_2 \ne 0$
10. $|z_1+z_2| \le |z_1|+|z_2|$
11. $|z_1-z_2| \ge |z_1|-|z_2|$

Note *Properties 10 and 11 are called triangle inequality.*

4.6 Argand Plane

A complex number $z=a+ib$ can be represented by a unique point $P(a, b)$ in the cartesian plane referred to a pair of rectangular axes. A purely real number a, i.e. $(a+0i)$ is represented by the point $(a, 0)$ on X-axis.

Therefore, X-axis is called **real** axis.

A purely imaginary number ib i.e. $(0+ib)$ is represented by the point $(0, b)$ on Y-axis. Therefore, Y-axis is called **imaginary axis.** The intersection (common) of two axes is called zero complex number i.e. $z=0+0i$.

Similarly, the representation of complex numbers as points in the plane is known as **Argand diagram.** The plane representing complex numbers as points, is called **Complex plane or Argand plane or Gaussian plane.**

If two complex numbers z_1 and z_2 are represented by the points P and Q in the complex plane, then

$|z_1-z_2|=PQ=$ Distance between P and Q

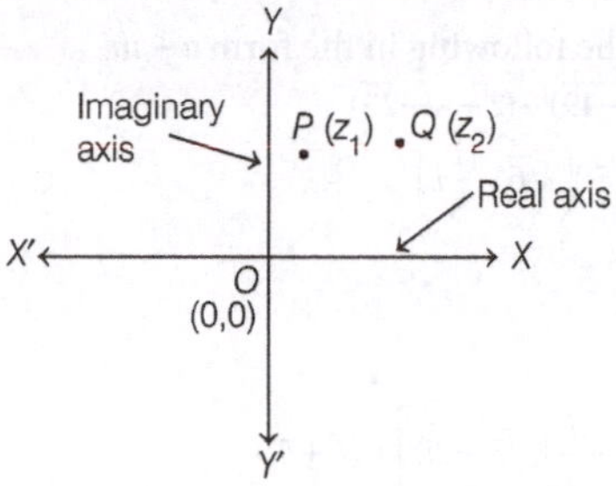

e.g. The complex numbers such as $2+3i$, $-2+3i$, $-1+0i$, $0+4i$ and $-4-2i$, which correspond to the ordered pairs $(2, 3)$, $(-2, 3)$, $(-1, 0)$, $(0, 4)$ and $(-4, -2)$ respectively, can be represented geometrically by the points A, B, C, D and E respectively, in the cartesian plane, as shown in the figure.

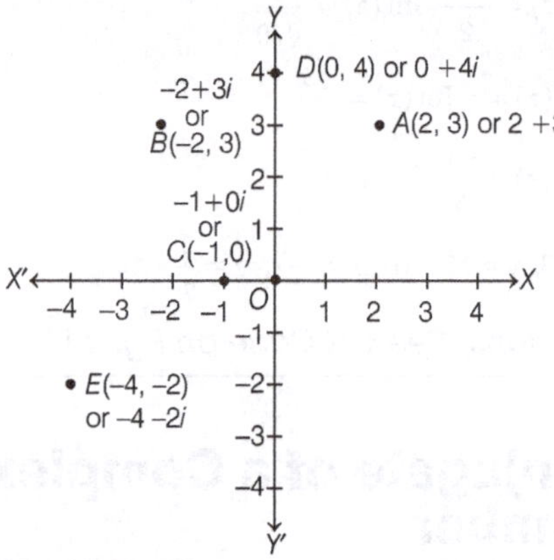

Representation of Conjugate of z on Argand Plane

Geometrically, the mirror image of the complex number $z=a+ib$ (represented by the ordered pair (a, b) about the X-axis) is called **conjugate of z** which is represented by the ordered pair $(a, -b)$.

If $z=a+ib$, then $\bar{z}=a-ib$

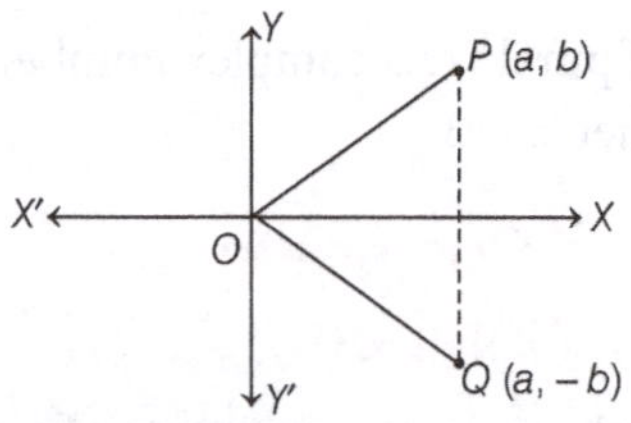

Representation of Modulus of z on Argand Plane

Geometrically, the distance of the complex number $z=a+ib$ [represented by the ordered pair (a, b)] from origin, is called the modulus of z.

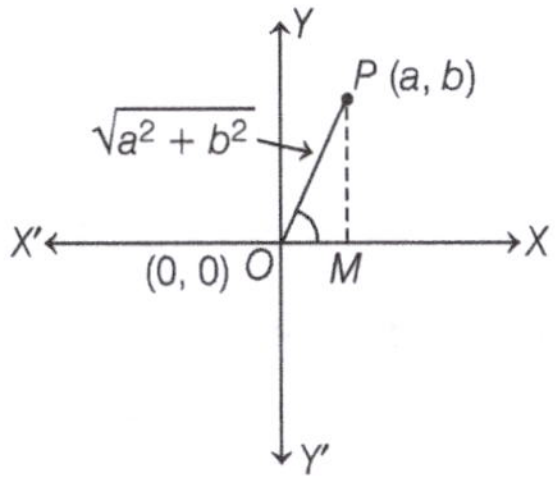

$$\therefore \quad OP = \sqrt{(a-0)^2+(b-0)^2} = \sqrt{a^2+b^2}$$
$$= \sqrt{\{\text{Re}(z)\}^2+\{\text{Im}(z)\}^2} = |a+ib|$$

Point to Remember

(i) $\sqrt{i^2} = \pm i = \pm\sqrt{-1}$

(ii) Multiplicative inverse of z is
$$\frac{1}{z} = z^{-1} = \frac{\bar{z}}{|z|^2}$$

(iii) $z\bar{z} = |z|^2$

(iv) In the set of complex numbers, $z_1 > z_2$ or $z_1 < z_2$ are meaningless but $|z_1| > |z_2|$ meaningful because $|z_1|$ and $|z_2|$ are real numbers.

Example 1. Find the conjugate of complex number $3+i$.

Sol. Let $z = 3+i$

$\therefore \quad \bar{z} = 3-i$

[$\because$ the conjugate of complex number z, is the complex number, obtained by changing the sign of imaginary part of z]

Example 2. Simplify the following complex number.
$\overline{9-i}+\overline{6+i^3}-\overline{9+i^2}$

Sol. $\overline{9-i}+\overline{6+i^3}-\overline{9+i^2}$

$= (9+i)+\overline{6-i}-\overline{9-1}$ $\quad$ [$\because i^3 = -i$ and $i^2 = -1$]

$= (9+i)+(6+i)-\bar{8} = 15+2i-8 = 7+2i$

Example 3. Find the real numbers x and y, if $(x-iy)(3+5i)$ is the conjugate of $-6-24i$.

Sol. We have, $(x-iy)(3+5i)$ is the conjugate of $-6-24i$.

$$(x-iy)(3+5i) = -6+24i$$
[$\because$ conjugate of $-6-24i = -6+24i$]

$\Rightarrow \quad 3x-3iy+5ix-5i^2y = -6+24i$

$\Rightarrow \quad (3x+5y)+i(5x-3y) = -6+24i$ [$\because i^2 = -1$] ...(i)

On equating real and imaginary parts from both sides of Eq. (i), we get

$$3x+5y = -6 \quad \text{...(ii)}$$

and $\quad 5x-3y = 24 \quad$ ...(iii)

On multiplying Eq. (i) by 3 and Eq. (ii) by 5 and then adding the result, we get

$$9x+15y+25x-15y = -18+120$$

$\Rightarrow \quad 34x = 102$

$\Rightarrow \quad x = 3$

On substituting $x = 3$ in Eq. (i), we get

$$9+5y = -6$$

$\Rightarrow \quad 5y = -15$

$\Rightarrow \quad y = -3$

Hence, the required values of x and y are 3 and -3, respectively.

Example 4. Solve the equation $z^2 = \bar{z}$, where $z = x+iy$. *NCERT Exemplar*

Sol. We have,

$$z^2 = \bar{z} \Rightarrow (x+iy)^2 = x-iy$$

$\Rightarrow \quad x^2+(iy)^2+2xyi = x-iy$

[$\because (z_1+z_2)^2 = z_1^2+z_2^2+2z_1z_2$]

$\Rightarrow \quad x^2-y^2+2xyi = x-iy \quad$ [$\because i^2 = -1$]

On equating real and imaginary parts, we get

$$x^2-y^2 = x \quad \text{...(i)}$$

and $\quad 2xy = -y \quad$ ...(ii)

From Eq. (ii), we get

$$2xy+y = 0 \Rightarrow y(2x+1) = 0$$

$\Rightarrow \quad y = 0$ or $x = -\frac{1}{2}$

Case I When $y = 0$.

In this case, we have $x^2 = x$ $\quad$ [from Eq. (i)]

$\Rightarrow \quad x^2-x = 0$

$\Rightarrow \quad x(x-1) = 0$

$\Rightarrow \quad x = 0$ or $x = 1$

$\therefore \quad z = 0+0i$ or $z = 1+0i$

Case II When $x = -\frac{1}{2}$.

In this case, we get

$$\frac{1}{4}-y^2 = -\frac{1}{2} \quad \text{[from Eq. (i)]}$$

$\Rightarrow \quad y^2 = \frac{1}{4}+\frac{1}{2} = \frac{3}{4}$

$\Rightarrow \quad y = \pm\frac{\sqrt{3}}{2}$

$\therefore \quad z = -\frac{1}{2}\pm i\frac{\sqrt{3}}{2}$

Hence, the solutions of given equation are $0+0i, 1+0i, -\frac{1}{2}+i\frac{\sqrt{3}}{2}$ and $-\frac{1}{2}-i\frac{\sqrt{3}}{2}$.

Example 5. If $z_1 = 3+2i$ and $z_2 = 2-i$, then verify that

(i) $\overline{z_1+z_2} = \bar{z}_1+\bar{z}_2$

(ii) $\overline{z_1 z_2} = \bar{z}_1\,\bar{z}_2$

Sol. Given that $z_1 = 3+2i$ and $z_2 = 2-i$

(i) Now, $z_1 + z_2 = (3+2i)+(2-i) = 5+i$

$\Rightarrow \quad \overline{z_1 + z_2} = \overline{5+i} = 5-i \quad$ ...(i)

Now, consider $\bar{z}_1 + \bar{z}_2 = (\overline{3+2i})+(\overline{2-i})$

$= 3-2i+2+i = 5-i \quad$ ...(ii)

From Eqs. (i) and (ii), we get

$\overline{z_1 + z_2} = \bar{z}_1 + \bar{z}_2$

(ii) Now, $z_1 z_2 = (3+2i)(2-i)$

$= 6-3i+4i-2i^2$

$= 6+i-2(-1) = 8+i \quad [\because i^2 = -1]$

$\Rightarrow \quad \overline{z_1 z_2} = \overline{8+i} = 8-i \quad$ ...(i)

Now, consider $\bar{z}_1\,\bar{z}_2 = (3-2i)(2+i)$

$= 6+3i-4i-2i^2$

$= 6-i-2(-1) = 8-i \quad$ ...(ii) $\quad [\because i^2 = -1]$

From Eqs. (i) and (ii), we get

$\overline{z_1 z_2} = \bar{z}_1\,\bar{z}_2$

Example 6. If $x+iy = \dfrac{(a^2+1)^2}{2a-i}$, then what is the value of x^2+y^2?

Sol. We have, $\quad x+iy = \dfrac{(a^2+1)^2}{2a-i} \quad$ ...(i)

$\therefore \quad \overline{x+iy} = \overline{\left\{\dfrac{(a^2+1)^2}{2a-i}\right\}}$

$\Rightarrow \quad x-iy = \dfrac{\overline{(a^2+1)^2}}{\overline{2a-i}} \quad \left[\because \overline{\left(\dfrac{z_1}{z_2}\right)} = \dfrac{\bar{z}_1}{\bar{z}_2}\right]$

$\Rightarrow \quad x-iy = \dfrac{(a^2+1)^2}{2a+i} \quad$ ...(ii)

On multiplying Eqs. (i) and (ii), we get

$(x+iy)(x-iy) = \dfrac{(a^2+1)^2(a^2+1)^2}{(2a-i)(2a+i)}$

$\Rightarrow \quad x^2-(iy)^2 = \dfrac{(a^2+1)^4}{(2a)^2-i^2}$

$[\because (z_1+z_2)(z_1-z_2) = z_1^2-z_2^2]$

$\Rightarrow \quad x^2-i^2y^2 = \dfrac{(a^2+1)^4}{4a^2+1} \quad [\because i^2 = -1]$

$\therefore \quad x^2+y^2 = \dfrac{(a^2+1)^4}{4a^2+1}$

Example 7. Find the modulus of the following complex numbers.

(i) $4+3i^7$ $\qquad$ (ii) $\dfrac{\sqrt{3}-i\sqrt{2}}{2\sqrt{3}-i\sqrt{2}}$

Sol. (i) We have, $4+3i^7 = 4+3(i^4)(i^2)i$

$= 4+3(1)(-1)i \quad [\because i^4 = 1,\ i^2 = -1]$

$= 4-3i$

$\therefore$ Modulus $= |4+3i^7| = |4-3i|$

$= \sqrt{4^2+(-3)^2} = \sqrt{16+9} = \sqrt{25} = 5$

(ii) Let $z = \dfrac{\sqrt{3}-i\sqrt{2}}{2\sqrt{3}-i\sqrt{2}} = \dfrac{\sqrt{3}-i\sqrt{2}}{2\sqrt{3}-i\sqrt{2}} \times \dfrac{2\sqrt{3}+i\sqrt{2}}{2\sqrt{3}+i\sqrt{2}}$

[by rationalising the denominator]

$= \dfrac{6+i\sqrt{6}-2\sqrt{6}\,i-2i^2}{(2\sqrt{3})^2-(\sqrt{2}\,i)^2}$

$[\because (z_1-z_2)(z_1+z_2) = z_1^2-z_2^2]$

$= \dfrac{6-\sqrt{6}\,i+2}{12+2} \quad [\because i^2 = -1]$

$= \dfrac{8-\sqrt{6}\,i}{14} = \dfrac{8}{14}-\dfrac{\sqrt{6}}{14}i$

$\Rightarrow \quad z = \dfrac{4}{7}-\dfrac{\sqrt{6}}{14}i$

Now, modulus of z, $|z| = \sqrt{\left(\dfrac{4}{7}\right)^2+\left(\dfrac{-\sqrt{6}}{14}\right)^2}$

$= \sqrt{\dfrac{16}{49}+\dfrac{6}{196}}$

$= \sqrt{\dfrac{64+6}{196}} = \sqrt{\dfrac{70}{196}} = \sqrt{\dfrac{5}{14}}$

Example 8. If $|z| = 1$, then find the value of $\dfrac{1+z}{1+\bar{z}}$.

Sol. Given, $|z| = 1 \Rightarrow |z|^2 = 1$

$\Rightarrow \quad z\bar{z} = 1 \quad [\because |z|^2 = z\bar{z}]$

Now, $\dfrac{1+z}{1+\bar{z}} = \dfrac{z\bar{z}+z}{1+\bar{z}} = \dfrac{z(\bar{z}+1)}{(\bar{z}+1)} = z \quad [\because 1 = z\bar{z}]$

Example 9. Find the multiplicative inverse of the following complex numbers.

(i) $5+3i$ $\qquad$ (ii) $(3+\sqrt{2}i)^2$

Sol. (i) Let $z = 5+3i$, then $\bar{z} = 5-3i$

and $|z| = \sqrt{(5)^2+(3)^2} = \sqrt{25+9} = \sqrt{34}$

Therefore, the multiplicative inverse of $(5+3i)$ is given by

$z^{-1} = \dfrac{\bar{z}}{|z|^2} = \dfrac{5-3i}{(\sqrt{34})^2} = \dfrac{5-3i}{34} = \dfrac{5}{34}-\dfrac{3}{34}i$

Multiplicative inverse of a complex number can also be find in the following manner

$$z^{-1}=\frac{1}{z}=\frac{1}{5+3i}\times\frac{5-3i}{5-3i}$$
$$=\frac{5-3i}{(5)^2-(3i)^2}=\frac{5-3i}{25+9}$$
$$=\frac{5}{34}-\frac{3}{34}i$$

(ii) Let $z=(3+\sqrt{2}i)^2=9+2i^2+6\sqrt{2}i$

$$=7+6\sqrt{2}i \qquad [\because i^2=-1]$$

Then, $\bar{z}=7-6\sqrt{2}i$

and $|z|=\sqrt{(7)^2+(6\sqrt{2})^2}=\sqrt{49+72}=\sqrt{121}=11$

Therefore, the multiplicative inverse of $(3+\sqrt{2}i)^2$ is given by

$$z^{-1}=\frac{\bar{z}}{|z|^2}=\frac{7-6\sqrt{2}i}{(11)^2}$$
$$=\frac{7-6\sqrt{2}i}{121}=\frac{7}{121}-\frac{6\sqrt{2}}{121}i$$

Example 10. If $|z|=1$, then prove that $\frac{z-1}{z+1}$; $(z\neq 1)$ is a purely imaginary number. What will you conclude, if $z=1$?

Competency Based Que

Sol. Let $z=a+ib$ such that $|z|=\sqrt{a^2+b^2}=1$

$$\Rightarrow \quad a^2+b^2=1$$

Now, consider $\left(\frac{z-1}{z+1}\right)=\left(\frac{a+ib-1}{a+ib+1}\right)$

$$=\frac{(a-1+ib)}{(a+1+ib)}\times\frac{(a+1-ib)}{(a+1-ib)}$$

[by rationalising the denominator]

$$=\frac{[(a-1)+ib][(a+1)-ib]}{(a+1)^2-(ib)^2}$$

$$[\because (z_1+z_2)(z_1-z_2)=z_1^2-z_2^2]$$

$$=\frac{a^2-1-iab+ib+iab+ib-i^2b^2}{(a+1)^2-i^2b^2}$$

$$=\frac{(a^2+b^2-1)+2bi}{(a+1)^2+b^2} \qquad [\because i^2=-1]$$

$$=\frac{(1-1)+2bi}{a^2+1+2a+b^2} \qquad [\because a^2+b^2=1]$$

$$=\frac{0+2bi}{(a^2+b^2)+1+2a}=\frac{2bi}{2+2a} \qquad [\because a^2+b^2=1]$$

$$=0+\frac{bi}{1+a}$$

Clearly, real part of $\left(\frac{z-1}{z+1}\right)$ is zero and imaginary part of $\left(\frac{z-1}{z+1}\right)$ is $\frac{b}{1+a}$.

$\therefore \quad \left(\frac{z-1}{z+1}\right)=\frac{ib}{1+a}$ is purely imaginary.

Again, when $z=1$, then

$$\left(\frac{z-1}{z+1}\right)=\frac{1-1}{1+1}=0,$$

which is both purely real and purely imaginary.

Example 11. Find the complex number satisfying the equation $z+\sqrt{2}|(z+1)|+i=0$.

NCERT Exemplar, Competency Based Que

Sol. We have, $z+\sqrt{2}|(z+1)|+i=0$

Let $z=x+iy$.

Then, $\quad (x+iy)+\sqrt{2}|(x+iy+1)|+i=0$

$\Rightarrow \quad x+i(y+1)+\sqrt{2}|(x+1)+iy|=0$

$\Rightarrow \quad x+i(y+1)+\sqrt{2}\sqrt{(x+1)^2+y^2}=0$

[if $z=a+ib$, then $|z|=\sqrt{a^2+b^2}$]

$\Rightarrow x+\sqrt{2}\sqrt{x^2+1+2x+y^2}+i(y+1)=0+0i$

On equating real and imaginary parts, we get

$$x+\sqrt{2}\sqrt{x^2+1+2x+y^2}=0 \qquad \text{...(i)}$$

and $\quad y+1=0 \qquad$ ...(ii)

From Eq. (ii), we get

$$y=-1$$

Now, on substituting $y=-1$ in Eq. (i), we get

$$x+\sqrt{2}\sqrt{x^2+1+2x+1}=0$$

$\Rightarrow \quad x=-\sqrt{2}\sqrt{x^2+2x+2}$

On squaring both sides, we get

$$x^2=2(x^2+2x+2)$$

$\Rightarrow \quad x^2=2x^2+4x+4$

$\Rightarrow \quad x^2+4x+4=0$

$\Rightarrow \quad (x+2)^2=0$

$\Rightarrow \quad x+2=0 \Rightarrow x=-2$

Hence, $\quad z=x+iy=-2-i$

Example 12. If $(2+i)(2+2i)(2+3i)\ldots(2+ni)=x+iy$, then prove that $5\cdot 8\cdot 13\ldots(4+n^2)=x^2+y^2$.

NCERT Exemplar, Competency Based Que

Sol. We have, $(2+i)(2+2i)(2+3i)\ldots(2+ni)=x+iy$

On taking modulus both sides, we get

$$|(2+i)(2+2i)(2+3i)\ldots(2+ni)|=|x+iy|$$

$\Rightarrow \quad |2+i||2+2i|\dots|2+ni| = |x+iy|$

$[\because |z_1 z_2 \dots z_n| = |z_1||z_2|\dots|z_n|]$

$\Rightarrow (\sqrt{4+1})(\sqrt{4+4})\dots(\sqrt{4+n^2}) = \sqrt{x^2+y^2}$

On squaring both sides, we get

$5 \cdot 8 \dots (4+n^2) = x^2+y^2$ **Hence proved.**

Example 13. **If $z_1 = \sqrt{3}+i\sqrt{3}$ and $z_2 = \sqrt{3}+i$, then find the quadrant in which $\left(\dfrac{z_1}{z_2}\right)$ lies.** *NCERT Exemplar*

Sol. We have, $z_1 = \sqrt{3}+i\sqrt{3}$ and $z_2 = \sqrt{3}+i$.

$\therefore \quad \dfrac{z_1}{z_2} = \dfrac{\sqrt{3}(1+i)}{\sqrt{3}+i} = \dfrac{\sqrt{3}(1+i)}{(\sqrt{3}+i)} \times \dfrac{\sqrt{3}-i}{\sqrt{3}-i}$

[by rationalising the denominator]

$= \dfrac{\sqrt{3}(1+i)(\sqrt{3}-i)}{(\sqrt{3})^2-(i)^2}$

$[\because (z_1+z_2)(z_1-z_2) = z_1^2 - z_2^2]$

$= \dfrac{\sqrt{3}(\sqrt{3}-i+i\sqrt{3}-i^2)}{3-i^2}$

$= \dfrac{\sqrt{3}[\sqrt{3}+i(\sqrt{3}-1)+1]}{3+1}$ $[\because i^2 = -1]$

$= \dfrac{\sqrt{3}}{4}[(\sqrt{3}+1)+i(\sqrt{3}-1)]$

$= \dfrac{\sqrt{3}(\sqrt{3}+1)}{4} + \dfrac{i\sqrt{3}(\sqrt{3}-1)}{4}$,

which is represented by a point in I quadrant.

Quick Test 2

1. Find the conjugate of the following complex numbers.

(i) $-i\sqrt{5}$

(ii) $\sqrt{3}$

(iii) $(6+5i)^2$

(iv) $-6-i7$

(v) $\dfrac{(3-2i)(2+3i)}{(1+2i)(2-i)}$

NCERT Exemplar

2. (i) Find the real numbers x and y, if $(x-iy)(3+5i)$ is the conjugate of $(-1-3i)$. *Competency Based Que*

(ii) If $z = 12+5i$, then verify that $z+\bar{z} = 2\text{Re}(z)$.

3. Find the modulus of the following complex numbers.

(i) $6+2i^{19}$

(ii) $\dfrac{1+i}{1-i} - \dfrac{1-i}{1+i}$

4. Find the conjugate and modulus of the following complex numbers. *Competency Based Que*

(i) $(3-2i)(3+2i)(1+i)$ (ii) $\dfrac{2+3i}{3+2i}$

5. Find the multiplicative inverse of the following complex numbers.

(i) $\sqrt{5}+3i$ (ii) $2-3i$ *NCERT*

(iii) $(1+i\sqrt{3})^2$

6. If $a+ib = \dfrac{(x+i)^2}{2x^2+1}$, then prove that $a^2+b^2 = \dfrac{(x^2+1)^2}{(2x^2+1)^2}$. *NCERT*

7. If $z = 12-5i$, then verify that

(i) $-|z| \le \text{Re}(z) \le |z|$

(ii) $-|z| \le \text{Im}(z) \le |z|$

8. If $f(z) = \dfrac{7-z}{1-z^2}$, where $z = 1+2i$, then find $|f(z)|$.

NCERT Exemplar, Competency Based Que

9. If $\dfrac{z-1}{z+1}$ is a purely imaginary number $(z \ne -1)$, then find the value of $|z|$. *NCERT Exemplar, Competency Based Que*

10. If $z_1 = \sqrt{2}-3i$ and $z_2 = 5-i\sqrt{2}$, then find the quadrant in which $\dfrac{z_1}{z_2}$ lies.

Answers

1. (i) $i\sqrt{5}$ (ii) $\sqrt{3}$ (iii) $11-60i$ (iv) $-6+7i$ (v) $\dfrac{63}{25}+\dfrac{16}{25}i$

2. (i) $x = \dfrac{6}{17}$ and $y = -\dfrac{7}{17}$

3. (i) $2\sqrt{10}$ (ii) 2

4. (i) $\bar{z} = 13-13i$ and $|z| = 13\sqrt{2}$ (ii) $\bar{z} = \dfrac{12}{13}-\dfrac{5}{13}i$ and $|z| = 1$

5. (i) $\dfrac{\sqrt{5}}{14}-\dfrac{3}{14}i$ (ii) $\dfrac{2}{13}+\dfrac{3}{13}i$ (iii) $\dfrac{-1}{8}-\dfrac{\sqrt{3}}{8}i$

8. $\dfrac{\sqrt{5}}{2}$ **9.** $|z| = 1$ **10.** IV quadrant.

For Solutions, See QR Code on Pg. 117

Exam Pattern Questions

Objective Type Questions

Multiple Choice Questions

1. If $x = \sqrt{-16}$, then

(a) $x = 4i$ (b) $x = 4$
(c) $x = -4$ (d) All of these

2. Write the real and imaginary part of $1 - 2i^2$.

(a) 1, 2 (b) 1, −2 (c) 3, 0 (d) 0, 3

3. Find the real and imaginary parts of the following complex number $-13 - \sqrt{-36}$.

(a) $\text{Re}(z) = -13, \text{Im}(z) = -36$
(b) $\text{Re}(z) = -13, \text{Im}(z) = -6$
(c) $\text{Re}(z) = 13, \text{Im}(z) = -36$
(d) $\text{Re}(z) = -13, \text{Im}(z) = 6$

4. Which of the pair is purely real and purely imaginary?

(a) $7 - \sqrt{30}i$ and $6i$ (b) $\sqrt{8} + i$ and $\sqrt{-7}$
(c) $30 + \sqrt{3}$ and $\sqrt{7}i$ (d) $2\sqrt{5}$ and $\sqrt{7}(1+2i)$

5. If $z \neq 0$ is a complex number, then

(a) $\text{Re}(z) = 0 \Rightarrow \text{Im}(z^2) = 0$
(b) $\text{Re}(z^2) = 0 \Rightarrow \text{Im}(z^2) = 0$
(c) $\text{Re}(z) = 0 \Rightarrow \text{Re}(z^2) = 0$
(d) None of the above

6. Two complex numbers are equal, if and only if

(a) their real and imaginary parts are separately equal.
(b) their real parts are only equal.
(c) their imaginary parts are only equal.
(d) None of the above

7. Evaluate : $i^n + i^{n+1} + i^{n+2} + i^{n+3}$

(a) 1 (b) −1 (c) 2 (d) 0

8. If $z = i^{91} + i^{92} + i^{93} + i^{94}$, then z is equal to

Competency Based Que

(a) 1 (b) i (c) 0 (d) −1

9. The value of $(1+i)(1+i^2)(1+i^3)(1+i^4)$ is

Competency Based Que

(a) 1 (b) 2 (c) i (d) 0

10. If $z + 0 = z$, where $z = x + iy$ and $0 = 0 + i0$, then 0 is called

(a) additive identity (b) additive inverse
(c) closure (d) None of these

11. If $z = 5i\left(-\frac{3}{5}i\right)$, then z is equal to

(a) $0 + 3i$ (b) $3 + 0i$ (c) $0 - 3i$ (d) $-3 + 0i$

12. Multiplicative inverse of complex number $\sqrt{10} + \sqrt{7}i$ is

(a) $\frac{\sqrt{10}}{17} + \frac{\sqrt{7}}{17}i$ (b) $\frac{\sqrt{10}}{17} - \frac{\sqrt{7}}{17}i$
(c) $\sqrt{10} + \frac{\sqrt{7}}{17}i$ (d) $\sqrt{10} - \frac{\sqrt{7}}{17}i$

13. If $z_1 = 6 + 3i$ and $z_2 = 2 - i$, then $\frac{z_1}{z_2}$ is equal to

(a) $\frac{1}{5}(9 + 12i)$ (b) $9 + 12i$
(c) $3 + 2i$ (d) $\frac{1}{5}(12 + 9i)$

14. The modulus of $z = 1 + \frac{4}{3}i$ is

(a) $\frac{8}{3}$ (b) $\frac{5}{3}$
(c) $\frac{-5}{3}$ (d) $\frac{5}{4}$

15. The conjugate of a complex number $\sqrt{5} - \sqrt{-18}$ is

Competency Based Que

(a) $\sqrt{5} + 3\sqrt{2}$ (b) $-\sqrt{5} + \sqrt{18}i$
(c) $-5 - 3\sqrt{2}i$ (d) $\sqrt{5} + 3\sqrt{2}i$

16. The value of $(z + 4)(\bar{z} + 4)$ is equivalent to

Competency Based Que

(a) $|z + 4|^2$ (b) $|z - 4|$
(c) $z^2 + 4$ (d) None of these

Assertion-Reason Based Questions

Directions (Q. Nos. 17-26) *In the questions given below are two statements labelled as Assertion (A) and Reason (R). In the context of the two statements, which one of the following is correct?*

(a) Both A and R are correct; R is the correct explanation of A.
(b) Both A and R are correct; R is not the correct explanation of A.
(c) A is correct; R is incorrect.
(d) R is correct; A is incorrect.

17. **Assertion** (A) If x and y are real numbers and $\frac{(1+i)x - 2i}{3+i} + \frac{(2-3i)y + i}{3-i} = i$, then $x = 3$ and $y = -1$.

Reason (R) If $2a + i4b$ and $2i$ represent the same complex number, then $a = 0$ and $b = \frac{1}{2}$.

Competency Based Que

18. Assertion (A) Simplest form of i^{-35} is $-i$.

Reason (R) Additive inverse of $(1-i)$ is equal to $-1+i$.

Competency Based Que

19. Assertion (A) If $z_1 = 2+3i$ and $z_2 = 3-2i$, then $z_1 - z_2 = -1+5i$.

Reason (R) If $z_1 = a+bi$ and $z_2 = c+di$, then $z_1 - z_2 = (a-c)+i(b-d)$.

20. Assertion (A) $(2+3i)[(3+2i)+(2+i)] = 1+21i$.

Reason (R) $z_1(z_2+z_3) = z_1z_2 + z_1z_3$.

21. Assertion (A) Simplest form of $\dfrac{5+\sqrt{2}i}{1-\sqrt{2}i}$ is $1-2\sqrt{2}i$.

Reason (R) The value of $(1+i)^5(1-i)^5 = 32$.

Competency Based Que

22. Assertion (A) If $(1+i)^6 = a+ib$, then $b = -8$.

Reason (R) If $(1-i)^3 = a+ib$, then $\dfrac{a}{b} = 1$.

23 Assertion (A) If $\sqrt{a+ib} = x+iy$, then $\sqrt{a-ib} = x-iy$.

Reason (R) A complex number z is said to be purely imaginary, if $\text{Re}(z) = 0$.

24. Assertion (A) Multiplicative inverse of $2-3i$ is $2+3i$.

Reason (R) If $z = 3+4i$, then $\bar{z} = 3-4i$.

25. Assertion (A) If $z = \dfrac{1+2i}{1-3i}$, then $|z| = \dfrac{1}{\sqrt{2}}$.

Reason (R) If $z = a+ib$, then $|z| = \sqrt{a^2+b^2}$.

Competency Based Que

26. Assertion (A) If z is a complex number, then $(\bar{z}^{-1})(\bar{z})$ is equal to 4.

Reason (R) The region of the complex plane for which $\left|\dfrac{z-a}{z+\bar{a}}\right| = 1\,[\text{Re}(a) \neq 0]$ is Y-axis.

Competency Based Que

Case Study Based Questions

27. Two complex numbers $z_1 = a+ib$ and $z_2 = c+id$ are said to be equal, if $a = c$ and $b = d$.

On the basis of above information, answer the following questions.

Competency Based Que

(i) If $(3a-6)+2ib = -6b+(6+a)i$, then the real values of a and b are respectively

(a) $-2, 2$ (b) $2, -2$ (c) $3, -3$ (d) $4, 2$

(ii) If $(2a+2b)+i(b-a) = -4i$, then the real values of a and b are respectively

(a) 2, 3 (b) $2, -2$
(c) 3, 1 (d) $-2, 2$

(iii) If $\left(\dfrac{1-i}{1+i}\right)^{100} = a+ib$, then the values of a and b are respectively

(a) 1, 0 (b) 0, 1 (c) 1, 2 (d) 2, 1

(iv) If $\dfrac{(1+i)^2}{2-i} = x+iy$, then the value of $x+y$ is

(a) $\dfrac{1}{5}$ (b) $\dfrac{3}{5}$ (c) $\dfrac{4}{5}$ (d) $\dfrac{2}{5}$

(v) If $(x+y)+i(x-y) = 4+6i$, then xy is equal to

(a) 5 (b) -5 (c) 4 (d) -4

28. We have, $i = \sqrt{-1}$. So, we can write the higher powers of i as follows

(i) $i^2 = -1$

(ii) $i^3 = i^2 \cdot i = (-1)\cdot i = -i$

(iii) $i^4 = (i^2)^2 = (-1)^2 = 1$

(iv) $i^5 = i^{4+1} = i^4 \cdot i = 1\cdot i = i$

(v) $i^6 = i^{4+2} = i^4 \cdot i^2 = 1\cdot i^2 = -1$

$\vdots \quad \vdots \quad \vdots \quad \vdots$

In order of compute i^n for $n > 4$, write $i^n = i^{4q+r}$ for some $q, r \in N$ and $0 \le r \le 3$.

Then, $i^n = i^{4q}\cdot i^r = (i^4)^q \cdot i^r = (1)^q \cdot i^r = i^r$.

In general, for any integer k,

$i^{4k} = 1, i^{4k+1} = i, i^{4k+2} = -1$ and $i^{4k+3} = -i$.

On the basis of above information, answer the following questions.

(i) The value of i^{37} is equal to

(a) i (b) $-i$ (c) 1 (d) -1

(ii) The value of i^{-30} is equal to

(a) i (b) 1 (c) -1 (d) $-i$

(iii) If $z = i^9 + i^{19}$, then z is equal to

(a) $0+0i$ (b) $1+0i$
(c) $0+i$ (d) $1+2i$

(iv) The value of $\left[i^{19} + \left(\dfrac{1}{i}\right)^{25}\right]^2$ is equal to

(a) -4 (b) 4 (c) i (d) 1

(v) If $z = i^{-39}$, then simplest form of z is equal to

(a) $1+0i$ (b) $0+i$
(c) $0+0i$ (d) $1+i$

29. The conjugate of a complex number z, is the complex number, obtained by changing the sign of imaginary part of z. It is denoted by $\bar{z}$.

The modulus (or absolute value) of a complex number, $z = a+ib$ is defined as the non-negative real number $\sqrt{a^2+b^2}$. It is denoted by $|z|$ i.e. $|z| = \sqrt{a^2+b^2}$.

Multiplicative inverse of z is $\frac{\bar{z}}{|z|^2}$. It is also called reciprocal of z i.e. $z\bar{z} = |z|^2$.

On the basis of above information, answer the following questions. **Competency Based Que**

(i) If $(x - iy)(3 + 5i)$ is the conjugate of $-6 - 24i$, then the value of $x + y$ is equal to

(a) 0 (b) 1 (c) 2 (d) 3

(ii) The value of $(z + 3)(\bar{z} + 3)$ is equivalent to

(a) $|z + 3|^2$ (b) $|z - 3|$
(c) $z^2 + 3$ (d) None of these

(iii) If $f(z) = \frac{7 - z}{1 - z^2}$, where $z = 1 + 2i$, then $|f(z)|$ is equal to

(a) $\frac{|z|}{2}$ (b) $|z|$
(c) $2|z|$ (d) None of these

(iv) If $z_1 = 1 - 3i$ and $z_2 = -2 + 4i$, then $|z_1 + z_2|$ is equal to

(a) $\sqrt{2}$ (b) 2 (c) $\sqrt{3}$ (d) 1

(v) If $z = 3 + 4i$, then $\frac{z + \bar{z}}{2}$ is equal to

(a) 1 (b) 2 (c) 3 (d) 4

Subjective Questions

Very Short Answer Type Questions

1. Find the value of $\sqrt{-25} + 3\sqrt{-4} + 2\sqrt{-9}$.

2. Find the values of x and y, if $x + 4iy = ix + y + 3$.

3. Find the value of i^{-1097}.

4. Find the value of $1 + i^2 + i^4 + i^6 + \ldots + i^{20}$.

NCERT Exemplar, Competency Based Que

5. Show that $1 + i^{10} + i^{20} + i^{30}$ is a real number.

6. Find the real values of x and y, if $(1 + i)(x + iy) = 2 - 5i$. **Competency Based Que**

7. Prove that $\left(\frac{2 + 3i}{3 + 4i}\right)\left(\frac{2 - 3i}{3 - 4i}\right)$ is purely real.

8. Express $\frac{1}{-2 + \sqrt{-3}}$ in the form $a + ib$.

9. Express $\frac{2 - \sqrt{-25}}{1 - \sqrt{-16}}$ in the form $a + ib$.

10. Express $\left(-2 - \frac{1}{3}i\right)^3$ in the form $a + ib$.

11. Find the real and imaginary parts of the conjugate of the complex number $-5i^{-15} - 6i^{-8}$. **Competency Based Que**

12. If $(1 + i)z = (1 - i)\bar{z}$, then show that $z = -i\bar{z}$.

13. If $z_1 = 2 - i$ and $z_2 = 1 + i$, then find $\left|\frac{z_1 + z_2 + 1}{z_1 - z_2 + 1}\right|$.

NCERT Exemplar

14. If $z_1 = 3 + i$ and $z_2 = 1 + 4i$, then verify that $|z_1 + z_2| < |z_1| + |z_2|$.

Short Answer Type Questions

15. Find the real values of x and y, if
$(x^4 + 2xi) - (3x^2 + iy) = (3 - 5i) + (1 + 2iy)$.

16. Let $z_1 = 2 - i$ and $z_2 = -2 + i$, then find $\text{Re}\left(\frac{z_1 z_2}{\bar{z}_1}\right)$.

17. Find the multiplicative inverse of $1 + i$.

18. What is the reciprocal of $3 + \sqrt{7}i$?

19. Express $\frac{1}{1 - \cos\theta + 2i\sin\theta}$ in the form $a + ib$. **Competency Based Que**

20. Find the smallest positive integral value of n for which $\frac{(1 + i)^n}{(1 - i)^{n-2}}$ is a real number.

Competency Based Que

21. If $z = 2 - 3i$, then show that $z^2 - 4z + 13 = 0$ and hence find the value of $4z^3 - 3z^2 + 169$.

22. What is the conjugate of $\frac{2 - i}{(1 - 2i)^2}$?

NCERT Exemplar, Competency Based Que

23. Find $\left|(1 + i)\frac{(2 + i)}{(3 + i)}\right|$.

NCERT Exemplar

24. If $z_1 = 3 + 2i$ and $z_2 = 2 - 4i$, then verify that
$|z_1 + z_2|^2 + |z_1 - z_2|^2 = 2(|z_1|^2 + |z_2|^2)$.

25. If $z_1 = 3 + 2i$ and $z_2 = 1 - 3i$, then find the modulus of z_1 and z_2. Also, verify that $|z_1 z_2| = |z_1||z_2|$.

26. Solve the equation $|z| = z + 1 + 2i$. **Competency Based Que**

27. If $|z_1|=|z_2|=\ldots=|z_n|=1$, then show that

$$|z_1+z_2+\ldots+z_n|=\left|\frac{1}{z_1}+\frac{1}{z_2}+\ldots+\frac{1}{z_n}\right|.$$

Competency Based Que

28. If $(1+i)(1+2i)(1+3i)\ldots(1+ni)=(x+iy)$, then show that $2\cdot5\cdot10\ldots(1+n^2)=x^2+y^2$.

29. If z_1 and z_2 are complex numbers such that $\frac{4z_1}{5z_2}$ is purely imaginary number, then find $\left|\frac{z_1-z_2}{z_1+z_2}\right|$.

Competency Based Que

30. Find the quadrant in which conjugate of $\frac{1+2i}{1-i}$ lies.

Long Answer Type Questions

31. Express the following complex numbers in the form $a+ib$.

(i) $(-2-5i)\div(3-6i)$ (ii) $\frac{(3+\sqrt{5}i)(3-\sqrt{5}i)}{(\sqrt{3}+\sqrt{2}i)-(\sqrt{3}-\sqrt{2}i)}$ **NCERT**

32. If $\frac{(1+i)^2}{2-i}=x+iy$, then find the value of $2x-y$.

33. If $x+iy=\sqrt{\frac{1+i}{1-i}}$, then prove that $x^2+y^2=1$.

34. Find real θ such that $\frac{3+2i\sin\theta}{1-2i\sin\theta}$ is purely real.

Competency Based Que

35. Find the value of $(4+3\sqrt{-20})^{1/2}+(4-3\sqrt{-20})^{1/2}$.

36. Find the value of $x^4+4x^3+6x^2+4x+9$, when $x=\sqrt{-2}-1$.

37. Find the conjugate and modulus of the following complex numbers.

(i) $(1-i)^{-2}+(1+i)^{-2}$ (ii) $\frac{3+2i}{2-5i}+\frac{3-2i}{2+5i}$

38. If z is a complex number such that $|z-1|=|z+1|$, then show that $\text{Re}(z)=0$.

39. If $z=x+iy$, then prove that $|x|+|y|\le\sqrt{2}|z|$.

40. If $|z+1|=z+2(1+i)$, then find the value of z.

NCERT Exemplar

41. If $z_1=3+5i$ and $z_2=2-3i$, then verify that $\overline{\left(\frac{z_1}{z_2}\right)}=\frac{\bar{z}_1}{\bar{z}_2}$.

42. Let z be a complex number satisfying $|z|=3|z-1|$. Then, prove that $\left|z-\frac{9}{8}\right|=\frac{3}{8}$.

43. If $z=x+iy$, then show that $z\bar{z}+2(z+\bar{z})+a=0$, where $a\in R$, represents a circle. **NCERT Exemplar**

Hints & Solutions

1. (a) Given, $x=\sqrt{-16}=\sqrt{16}\sqrt{-1}$

$x=4i$ $[\because\sqrt{-1}=i]$

2. (c) Let $z=1-2i^2=1-2\times(-1)$ $[\because i^2=-1]$

$=1+2=3$

$\Rightarrow$ $z=3+0.i$

Hence, Re $(z)=3$, and Im $(z)=0$.

3. (b) Let $z=-13-\sqrt{-36}=-13-\sqrt{36}\sqrt{-1}$

$=-13-6i$ $[\because\sqrt{-1}=i]$

Here, $\text{Re}(z)=-13$ and $\text{Im}(z)=-6$

4. (c) In a complex number $z=a+ib$

if $b=0$, then z is purely real and if $a=0$, then z is purely imaginary.

So, pair $(30+\sqrt{3})$ and $\sqrt{7}i$ is purely real and purely imaginary.

5. (a) We know that $z=a+ib$ but $z\ne0$.

Let $a=0\Rightarrow z=ib$

$\Rightarrow$ $z^2=(ib)^2=i^2b^2$

$\Rightarrow$ $z^2=-b^2$ $[\because i^2=-1]$

$\text{Im}(z^2)=0$

So, $\text{Re}(z)=0$

$\Rightarrow$ $\text{Im}(z^2)=0$

6. (a) Two complex numbers $z_1=a+ib$ and $z_2=c+id$ are equal, if $a=c$ and $b=d$ i.e. if their real and imaginary parts are separately equal.

7. (d) Given, $i^n+i^{n+1}+i^{n+2}+i^{n+3}$

$=i^n(1+i+i^2+i^3)$

$=i^n(1+i-1-i)$ $[\because i^2=-1, i^3=-i]$

$=i^n\times0$

$=0$

8. (c) We have, $z=i^{91}+i^{92}+i^{93}+i^{94}$

$=i^{91}[1+i+i^2+i^3]$

$=i^{91}[1+i+(-1)+(-i)]$ $[\because i^2=-1, i^3=-i]$

$=i^{91}[1+i-1-i]=i^{91}\times0=0$

9. (d) Let $x = (1+i)(1+i^2)(1+i^3)(1+i^4)$

$= (1+i)(1-1)(1-i)(1+1)$ $[\because i^2 = -1, i^3 = -i]$

$= 0$

10. (a) If $z + 0 = z = 0 + z$, where z is a complex number, then 0 is additive identity of z.

11. (b) Given, $z = 5i\left(\frac{-3}{5}i\right)$

$= 5 \times \left(\frac{-3}{5}\right) i^2$

$= \frac{-15}{5}(-1)$ $[\because i^2 = -1]$

$= 3$

$= 3 + 0i$

12. (b) Let $z = \sqrt{10} + \sqrt{7}i$

$z^{-1} = \frac{1}{z} = \frac{1}{\sqrt{10} + \sqrt{7}i}$

On multiplying and dividing by $(\sqrt{10} - \sqrt{7}i)$, we get

$z^{-1} = \frac{1}{\sqrt{10} + \sqrt{7}i}\left(\frac{\sqrt{10} - \sqrt{7}i}{\sqrt{10} - \sqrt{7}i}\right)$

$z^{-1} = \frac{\sqrt{10} - \sqrt{7}i}{\sqrt{100} - \sqrt{49}i^2} = \frac{\sqrt{10} - \sqrt{7}i}{10 - 7(-1)}$ $[\because i^2 = -1]$

$= \frac{\sqrt{10} - \sqrt{7}i}{17} = \frac{\sqrt{10}}{17} - \frac{\sqrt{7}}{17}i$

13. (a) We have, $z_1 = 6 + 3i$ and $z_2 = 2 - i$

$\therefore \quad \frac{z_1}{z_2} = \frac{6+3i}{2-i}$

On multiplying by $(2+i)$ in numerator and denominator, we get

$\frac{z_1}{z_2} = \frac{6+3i}{2-i}\left(\frac{2+i}{2+i}\right)$

$\frac{z_1}{z_2} = \frac{9+12i}{5} = \frac{1}{5}(9+12i)$

14. (b) We have, $z = 1 + \frac{4}{3}i$

Modulus of z,

$|z| = \sqrt{1^2 + \left(\frac{4}{3}\right)^2}$ $[\because |z| = \sqrt{a^2 + b^2}]$

$= \sqrt{1 + \frac{16}{9}}$

$= \sqrt{\frac{9+16}{9}} = \frac{5}{3}$

15. (d) Let $z = \sqrt{5} - \sqrt{-18}$

$z = \sqrt{5} - \sqrt{18}\sqrt{-1}$

$z = \sqrt{5} - 3\sqrt{2}i$ $[\because \sqrt{-1} = i]$

$\bar{z} = \sqrt{5} + 3\sqrt{2}i$ $[\because \overline{a + bi} = a - bi]$

16. (a) Let $z = a + ib$

$\bar{z} = a - ib$

$|z + 4| = |a + ib + 4| = |(a+4) + ib|$

$= \sqrt{(a+4)^2 + b^2}$

Now, $(z+4)(\bar{z}+4) = (a + ib + 4)(a - ib + 4)$

$= (a + 4 + ib)(a + 4 - ib)$

$= (a+4)^2 - (i^2b^2) = (a+4)^2 + b^2$

$= (\sqrt{(a+4)^2 + b^2})^2 = |z + 4|^2$

17. (b) **Assertion** We have, $\frac{(1+i)x - 2i}{3+i} + \frac{(2-3i)y + i}{3-i} = i$

$\Rightarrow \quad i = \frac{x + (x-2)i}{3+i} + \frac{2y + i(1-3y)}{3-i}$

$\Rightarrow \quad i = \frac{[x + (x-2)i](3-i) + [2y + i(1-3y)](3+i)}{(3+i)(3-i)}$

$\Rightarrow \quad i = \frac{3x + 3i(x-2) - xi - (x-2)i^2 + 6y + 3i(1-3y) + 2yi + (1-3y)i^2}{9 - i^2}$

$\Rightarrow \quad i = \frac{3x + i(3x - 6 - x) + (x-2) + 6y + i(3 - 9y + 2y) - (1-3y)}{10}$ $[\because i^2 = -1]$

$\Rightarrow \quad i = \frac{3x + i(2x-6) + (x-2) + 6y + i(3-7y) - 1 + 3y}{10}$

$\Rightarrow \quad 10i = (4x + 9y - 3) + i(2x - 7y - 3)$

On comparing real and imaginary parts, we get

$4x + 9y - 3 = 0$

$\Rightarrow \quad 4x + 9y = 3$ ...(i)

and $\quad 2x - 7y - 3 = 10$

$\Rightarrow \quad 2x - 7y = 13$ ...(ii)

On multiplying by 2 in Eq. (ii), we get

$4x - 14y = 26$ ...(iii)

On solving Eqs. (i) and (iii), we get

$x = 3$ and $y = -1$

So, Assertion is correct.

Reason Given, $2a + i4b = 2i$

On comparing real and imaginary parts, we get

$a = 0$

$b = \frac{1}{2}$

So, Reason is also correct.

Hence, Both A and R are correct and R is not the correct explanation of A.

18. (d) **Assertion** $i^{-35} = \frac{1}{i^{35}} = \frac{1}{(i^2)^{17} i} = \frac{1}{-i} \times \frac{i}{i} = \frac{i}{-i^2} = i$

Reason Additive inverse of z is $-z$.

$\therefore$ Additive inverse of $(1-i)$ is

$$-(1-i)=-1+i$$

Hence, Assertion is incorrect and Reason is correct.

19. (a) **Assertion** Given, $z_1=2+3i, z_2=3-2i$

$$\therefore \quad z_1-z_2=(2+3i)-(3-2i)$$
$$=(2-3)+i[3-(-2)]=-1+5i$$

Hence, Assertion and Reason both are correct and Reason is the correct explanation of Assertion.

20. (a) **Assertion** For any three complex numbers z_1, z_2 and z_3, distributive law is

$$z_1(z_2+z_3)=z_1z_2+z_1z_3$$

and $\quad (z_1+z_2)z_3=z_1z_3+z_2z_3.$

$$\therefore \quad (2+3i)[(3+2i)+(2+i)]$$
$$=(2+3i)(3+2i)+(2+3i)(2+i)$$
$$=(6-6)+13i+(4-3)+8i$$
$$=1+21i$$

Hence, Assertion and Reason both are correct and Reason is the correct explanation of Assertion.

21. (d) **Assertion** We have,

$$\frac{5+\sqrt{2}i}{1-\sqrt{2}i}=\frac{5+\sqrt{2}i}{1-\sqrt{2}i}\times\frac{1+\sqrt{2}i}{1+\sqrt{2}i}$$
$$=\frac{5+5\sqrt{2}i+\sqrt{2}i-2}{1-(\sqrt{2}i)^2}$$
$$=\frac{3+6\sqrt{2}i}{1+2}=\frac{3(1+2\sqrt{2}i)}{3}$$
$$=1+2\sqrt{2}i$$

Reason $(1+i)^5(1-i)^5=(1-i^2)^5=2^5=32$

Hence, Assertion is incorrect and Reason is correct.

22. (b) **Assertion** We have,

$$(1+i)^6=[(1+i)^2]^3$$
$$=(1+i^2+2i)^3$$
$$[\because (z_1+z_2)^2=z_1^2+z_2^2+2z_1z_2]$$
$$=(1-1+2i)^3 \quad [\because i^2=-1]$$
$$\Rightarrow \quad (1+i)^6=(2i)^3=8i^3=-8i \quad [\because i^3=-i]$$
$$=a+ib$$
$$\therefore \quad b=-8$$

Reason $(1-i)^3=1^3-i^3-3(1)^2i+3(1)(i)^2$

$$[\because (z_1-z_2)^3=z_1^3-3z_1^2z_2+3z_1z_2^2-z_2^3]$$
$$=1-(-i)-3i-3 \quad [\because i^3=-i \text{ and } i^2=-1]$$
$$\Rightarrow \quad (1-i)^3=-2-2i=a+ib$$
$$\Rightarrow \quad a=-2$$

and $\quad b=-2$

$$\therefore \quad \frac{a}{b}=\frac{-2}{-2}=1$$

Hence, Assertion and Reason both are correct and Reason is not the correct explanation of Assertion.

23. (b) **Assertion** We have, $\sqrt{a+ib}=x+iy$

On taking square both sides, we get

$$a+ib=(x+iy)^2$$
$$\Rightarrow \quad a+ib=x^2-y^2+2xyi$$

On comparing real and imaginary parts, we get

$$a=x^2-y^2, b=2xy$$

Now, $\quad a-ib=x^2-y^2-2xyi$

$$\Rightarrow \quad a-ib=(x-iy)^2$$

On taking square root both sides, we get

$$\sqrt{a-ib}=\sqrt{(x-iy)^2}$$
$$\sqrt{a-ib}=(x-iy)$$

So, Assertion is correct.

Reason In a purely imaginary number, real part is always 0. So, Reason is also correct.

Hence, both Assertion and Reason are correct and Reason is not the correct explanation of Assertion.

24. (d) **Assertion** Let $z=2-3i$

Then, $\bar{z}=2+3i$ and $|z|^2=2^2+(-3)^2=13$

Therefore, the multiplicative inverse of $2-3i$ is

$$z^{-1}=\frac{\bar{z}}{|z|^2}=\frac{2+3i}{13}=\frac{2}{13}+\frac{3}{13}i$$

Reason If $z=a+ib$, then conjugate of z

i.e. $\quad \bar{z}=a-ib$

$$\therefore \quad z=3+4i \Rightarrow \bar{z}=3-4i$$

Hence, Assertion is false and Reason is true.

25. (a) **Assertion** Let $z=\dfrac{1+2i}{1-3i}$

$$\therefore \quad z=\frac{1+2i}{1-3i}\times\frac{1+3i}{1+3i}=\frac{1+3i+2i+6i^2}{1^2-(3i)^2}$$
$$[\because (a+b)(a-b)=a^2-b^2]$$
$$=\frac{1+5i+6(-1)}{1-9i^2} \quad [\because i^2=-1]$$
$$=\frac{1+5i-6}{1+9}=\frac{-5+5i}{10}=\frac{-1+i}{2}$$
$$\Rightarrow \quad z=-\frac{1}{2}+\frac{1}{2}i$$
$$\therefore \quad |z|=\sqrt{\left(-\frac{1}{2}\right)^2+\left(\frac{1}{2}\right)^2} \quad [\because |a+ib|=\sqrt{a^2+b^2}]$$
$$=\sqrt{\frac{1}{4}+\frac{1}{4}}=\sqrt{\frac{2}{4}}=\sqrt{\frac{1}{2}}=\frac{1}{\sqrt{2}}$$

Hence, Assertion and Reason both are true and Reason is the correct explanation of Assertion.

26. (d) **Assertion** Let $z = a + ib$

$$\bar{z} = a - ib$$

and $$\bar{z}^{-1} = \frac{1}{a - ib} = \frac{a + ib}{a^2 + b^2}$$

Now, $$(\bar{z}^{-1})(\bar{z}) = \left(\frac{a + ib}{a^2 + b^2}\right)(a - ib)$$

$$= \frac{(a + ib)(a - ib)}{a^2 + b^2}$$

$$= \frac{a^2 - (i^2 b^2)}{a^2 + b^2} = \frac{a^2 + b^2}{a^2 + b^2} = 1 \quad [\because i^2 = -1]$$

So, Assertion is incorrect.

Reason Let $z = x + iy$

and $a = p + iq, \quad p \neq 0$

$$\bar{a} = p - iq$$

$$\because \quad \left|\frac{z - a}{z + \bar{a}}\right| = 1$$

$$\Rightarrow \quad |z - a| = |z + \bar{a}|$$

$$\Rightarrow \quad |(x - p) + i(y - q)| = |(x + p) + i(y - q)|$$

$$\Rightarrow \sqrt{(x - p)^2 + (y - q)^2} = \sqrt{(x + p)^2 + (y - q)^2}$$

$$\Rightarrow \quad (x - p)^2 + (y - q)^2 = (x + p)^2 + (y - q)^2$$

$$\Rightarrow \quad (x - p)^2 = (x + p)^2$$

$$\Rightarrow \quad x^2 + p^2 - 2xp = x^2 + p^2 + 2xp$$

$$\Rightarrow \quad 4xp = 0$$

$$\Rightarrow \quad x = 0$$

So, Reason is correct.

Hence, Reason is correct but Assertion is incorrect.

27. (i) (a) We have, $(3a - 6) + 2ib = -6b + (6 + a)i$

On equating real and imaginary parts, we get

$$3a - 6 = -6b \quad \ldots(\text{i})$$

and $$2b = 6 + a \quad \ldots(\text{ii})$$

Above equations can be rewritten as

$$3a + 6b = 6 \quad \ldots(\text{iii})$$

and $$a - 2b = -6 \quad \ldots(\text{iv})$$

On multiplying Eq. (iv) by 3 and then adding with Eq. (iii), we get

$$3a + 6b + 3a - 6b = 6 - 18$$

$$\Rightarrow \quad 6a = -12 \Rightarrow a = -2$$

On substituting $a = -2$ in Eq. (iv), we get

$$-2 - 2b = -6$$

$$\Rightarrow \quad -2b = -6 + 2 \Rightarrow b = \frac{-4}{-2} = 2$$

$$\therefore \quad a = -2 \text{ and } b = 2$$

(ii) (b) We have, $(2a + 2b) + i(b - a) = -4i$, which can be rewritten as

$$(2a + 2b) + i(b - a) = 0 - 4i$$

On equating real and imaginary parts, we get

$$2a + 2b = 0$$

$$\Rightarrow \quad a + b = 0 \quad [\because 2 \neq 0] \ldots(\text{i})$$

and $$b - a = -4 \quad \ldots(\text{ii})$$

On adding Eqs. (i) and (ii), we get

$$a + b + b - a = 0 - 4$$

$$\Rightarrow \quad 2b = -4 \Rightarrow b = -2$$

On substituting $b = -2$ in Eq. (i), we get

$$a - 2 = 0 \Rightarrow a = 2$$

$$\therefore \quad a = 2 \text{ and } b = -2$$

(iii) (a) Given that $$\left(\frac{1 - i}{1 + i}\right)^{100} = a + ib$$

$$\Rightarrow \quad \left[\frac{(1 - i)}{(1 + i)} \cdot \frac{(1 - i)}{(1 - i)}\right]^{100} = a + ib$$

$$\Rightarrow \quad \left(\frac{1 + i^2 - 2i}{1 - i^2}\right)^{100} = a + ib$$

$$\Rightarrow \quad \left(\frac{-2i}{2}\right)^{100} = a + ib \quad [\because i^2 = -1]$$

$$\Rightarrow \quad (i^4)^{25} = a + ib$$

$$\Rightarrow \quad 1 = a + ib \quad [\because i^4 = 1]$$

On comparing real and imaginary parts both sides, we get

$$a = 1 \text{ and } b = 0$$

$$\therefore \quad (a, b) = (1, 0)$$

(iv) (d) Given that $$\frac{(1 + i)^2}{2 - i} = x + iy$$

$$\Rightarrow \quad \frac{(1 + i^2 + 2i)}{2 - i} = x + iy$$

$$\Rightarrow \quad \frac{2i}{2 - i} = x + iy$$

$$\Rightarrow \quad \frac{2i(2 + i)}{(2 - i)(2 + i)} = x + iy$$

$$\Rightarrow \quad \frac{4i + 2i^2}{4 - i^2} = x + iy$$

$$\Rightarrow \quad \frac{4i - 2}{4 + 1} = x + iy \Rightarrow \frac{-2}{5} + \frac{4i}{5} = x + iy$$

On comparing both sides, we get

$$x = -\frac{2}{5} \text{ and } y = \frac{4}{5}$$

$$\therefore \quad x + y = \frac{-2}{5} + \frac{4}{5} = \frac{2}{5}$$

(v) (b) Given that $(x + y) + i(x - y) = 4 + 6i$

On comparing both sides, we get

$$x + y = 4 \quad \ldots(\text{i})$$

and $$x - y = 6 \quad \ldots(\text{ii})$$

On adding Eqs. (i) and (ii), we get

$$2x = 10 \Rightarrow x = 5$$

$\therefore$ From Eq. (i), we get

$$5 + y = 4$$

$\Rightarrow$ $y = 4 - 5 = -1$

$\therefore$ $xy = 5(-1) = -5$

28. (i) (a) We have, $i^{37} = (i)^{36+1} = (i)^{4\times 9}\, i$

$= (i^4)^9 \cdot i = (1)^9 \cdot i = i$ $\quad [\because i^4 = 1]$

(ii) (c) We have, $i^{-30} = \dfrac{1}{i^{30}}$

Now, $i^{30} = (i)^{4\times 7+2}$

$= (i^{4\times 7})\, i^2 = (i^4)^7\,(-1)$ $\quad [\because i^2 = -1]$

$= (1)^7\,(-1) = -1$ $\quad [\because i^4 = 1]$

$\Rightarrow$ $i^{-30} = \dfrac{1}{(-1)} = -1$

(iii) (a) $i^9 + i^{19} = i^9\,(1 + i^{10}) = i^9\,[1 + (i^2)^5]$

[taking i^9 common]

$= i^9\,[(1 + (-1)^5] = i^9\,(1-1) = 0$ $\quad [\because i^2 = -1]$

$= 0 + 0i$

(iv) (a) $\left[i^{19} + \left(\dfrac{1}{i}\right)^{25}\right]^2 = \left[i^{4\times 4+3} + \dfrac{1}{i^{4\times 6+1}}\right]^2$

$= \left[(i^4)^4\,(i)^3 + \dfrac{1}{(i^4)^6\, i}\right]^2 = \left[(1)^4\,(i)^3 + \dfrac{1}{(1)^6 i}\right]^2$

$[\because i^4 = 1]$

$= \left(-i + \dfrac{1}{i}\right)^2 = \left(-i + \dfrac{i}{i^2}\right)^2$ $\quad [\because i^3 = -i]$

$= \left(-i + \dfrac{i}{-1}\right)^2$ $\quad [\because i^2 = -1]$

$= (-i - i)^2 = (-2i)^2 = 4i^2 = -4$

(v) (b) $i^{-39} = \dfrac{1}{i^{39}}$

On multiplying and dividing by i, we get

$= \dfrac{i}{i^{40}} = \dfrac{i}{(i^4)^{10}} = \dfrac{i}{(1)^{10}} = \dfrac{i}{1} = i$ $\quad [\because i^4 = 1]$

$= 0 + i$

29. (i) (a) We have, $(x - iy)(3 + 5i)$ is the conjugate of $-6 - 24\,i$.

$\Rightarrow$ $(x - iy)(3 + 5i) = -6 + 24i$

$[\because$ conjugate of $-6 - 24\,i = -6 + 24\,i]$

$\Rightarrow$ $3x - 3iy + 5ix - 5i^2 y = -6 + 24i$

$\Rightarrow$ $(3x + 5y) + i(5x - 3y) = -6 + 24\,i$ $\quad$ ...(i)

$[\because i^2 = -1]$

On equating real and imaginary parts both sides of Eq. (i), we get

$3x + 5y = -6$ $\quad$ ...(ii)

and $5x - 3y = 24$ $\quad$ ...(iii)

On multiplying Eq. (ii) by 3 and Eq. (iii) by 5 and then adding the result, we get

$$9x + 15y + 25x - 15y = -18 + 120$$

$\Rightarrow$ $34x = 102 \Rightarrow x = 3$

On substituting $x = 3$ in Eq. (ii), we get

$$9 + 5y = -6$$

$\Rightarrow$ $5y = -15$

$\Rightarrow$ $y = -3$

Now, $x + y = 3 + (-3) = 0$

(ii) (a) Given that $(z + 3)(\bar{z} + 3)$

Let $z = x + iy$

$\Rightarrow$ $(z + 3)(\bar{z} + 3) = (x + iy + 3)(x + 3 - iy)$

$= (x + 3)^2 - (iy)^2 = (x + 3)^2 + y^2$

$= |x + 3 + iy|^2 = |z + 3|^2$

(iii) (a) Let $z = 1 + 2i$

$\Rightarrow$ $|z| = \sqrt{1 + 4} = \sqrt{5}$

Now, $f(z) = \dfrac{7 - z}{1 - z^2} = \dfrac{7 - 1 - 2i}{1 - (1 + 2i)^2}$

$= \dfrac{6 - 2i}{1 - 1 - 4i^2 - 4i} = \dfrac{6 - 2i}{4 - 4i}$

$= \dfrac{3 - i}{2 - 2i} = \dfrac{(3 - i)(2 + 2i)}{(2 - 2i)(2 + 2i)}$

$= \dfrac{6 - 2i + 6i - 2i^2}{4 - 4i^2}$

$= \dfrac{6 + 4i + 2}{4 + 4}$

$= \dfrac{8 + 4i}{8} = 1 + \dfrac{1}{2} i$

$f(z) = 1 + \dfrac{1}{2} i$

$\therefore$ $|f(z)| = \sqrt{1 + \dfrac{1}{4}} = \dfrac{\sqrt{4 + 1}}{4} = \dfrac{\sqrt{5}}{2} = \dfrac{|z|}{2}$

(iv) (a) Given, $z_1 = 1 - 3i$ and $z_2 = -2 + 4i$

$\therefore$ $z_1 + z_2 = (1 - 3i) + (-2 + 4i) = -1 + i$

$|z_1 + z_2| = \sqrt{(-1)^2 + (1)^2} = \sqrt{1 + 1} = \sqrt{2}$

(v) (c) Given, $z = 3 + 4i$

$\therefore$ $\bar{z} = 3 - 4i$

$\Rightarrow$ $z + \bar{z} = (3 + 4i) + (3 - 4i) = 6$

Now, $\dfrac{z + \bar{z}}{2} = \dfrac{6}{2} = 3$

Subjective Questions

1. We have, $\sqrt{-25} + 3\sqrt{-4} + 2\sqrt{-9}$

$= \sqrt{25}\,\sqrt{-1} + 3\sqrt{4}\,\sqrt{-1} + 2\sqrt{9}\,\sqrt{-1}$

$= 5 \times i + 3 \times 2 \times i + 2 \times 3 \times i$ $\quad [\because \sqrt{-1} = i]$

$= 5i + 6i + 6i = 17i$

2. We have, $x+4iy=ix+y+3$

$\Rightarrow \quad x+4yi=(y+3)+ix$

On comparing real and imaginary parts, we get

$$x=y+3 \quad \ldots(\text{i})$$

and $$4y=x \quad \ldots(\text{ii})$$

On solving Eqs. (i) and (ii), we get

$$y=1 \text{ and } x=4$$

Hence, the values of x and y are 4 and 1, respectively.

3. We have, $i^{-1097}=\dfrac{1}{i^{1097}}$

$$=\frac{1}{i^{4\times 274}\cdot i}=\frac{1}{i} \quad \left[\because i^4=1 \; \therefore (i^4)^{274}=1\right]$$

$$=\frac{i}{i^2}=-i \quad [\because i^2=-1]$$

Hence, the value of i^{-1097} is $(-i)$.

4. We have, $1+i^2+i^4+i^6+\ldots\ldots i^{20}$

$=1+(i^2+i^4)+(i^6+i^8)+(i^{10}+i^{12})+(i^{14}+i^{16})+(i^{18}+i^{20})$

$=1+[i^2+(i^2)^2]+[(i^2)^3+(i^2)^4]+[(i^2)^5+(i^2)^6]+[(i^2)^7+(i^2)^8]+[(i^2)^9+(i^2)^{10}]$

$=1+(-1+1)+(-1+1)+(-1+1)+(-1+1)+(-1+1) \quad [\because i^2=-1]$

$=1+0+0+0+0+0=1$

5. Given, $1+i^{10}+i^{20}+i^{30}$

$=1+i^{4\times 2+2}+i^{4\times 5}+i^{4\times 7+2}$

$=1+(i^4)^2\cdot i^2+(i^4)^5+(i^4)^7\cdot i^2$

$=1+(1)^2\cdot(-1)+(1)^5+(1)^7\cdot(-1) \quad \left[\because i^4=1 \text{ and } i^2=-1\right]$

$=1-1+1-1$

$=0$, which is a real number.

6. We have, $(1+i)(x+iy)=2-5i$

$\Rightarrow \quad x+iy+ix+i^2y=2-5i$

$\Rightarrow \quad x+i(y+x)-y=2-5i \quad [\because i^2=-1]$

$\Rightarrow \quad (x-y)+i(x+y)=2-5i$

On equating real and imaginary parts from both sides, we get

$$x-y=2 \quad \ldots(\text{i})$$

and $$x+y=-5 \quad \ldots(\text{ii})$$

On solving Eqs. (i) and (ii), we get

$$x=\frac{-3}{2} \text{ and } y=\frac{-7}{2}$$

7. We have, $\left(\dfrac{2+3i}{3+4i}\right)\left(\dfrac{2-3i}{3-4i}\right)=\dfrac{(2+3i)(2-3i)}{(3+4i)(3-4i)}$

$$=\frac{2^2-(3i)^2}{3^2-(4i)^2}=\frac{4-(-9)}{9-(-16)} \quad [\because i^2=-1]$$

$=\dfrac{13}{25}$, which is purely real.

8. Let $z=\dfrac{1}{-2+\sqrt{-3}}$

$\Rightarrow \quad z=\dfrac{1}{-2+\sqrt{3}i} \quad [\because i=\sqrt{-1}]$

On multiplying by $(-2-\sqrt{3}i)$ in numerator and denominator, we get

$$z=\frac{-2-\sqrt{3}i}{(-2+\sqrt{3}i)(-2-\sqrt{3}i)}$$

$\Rightarrow \quad z=\dfrac{-2-\sqrt{3}i}{(-2)^2-(\sqrt{3}i)^2}=\dfrac{-2-\sqrt{3}i}{4+3}$

$\Rightarrow \quad z=\dfrac{-2-\sqrt{3}i}{7}$

$\therefore \quad z=\dfrac{-2}{7}-\dfrac{\sqrt{3}}{7}i$

9. Let $z=\dfrac{2-\sqrt{-25}}{1-\sqrt{-16}}\Rightarrow z=\dfrac{2-5i}{1-4i}$

On multiplying by $(1+4i)$ in numerator and denominator, we get

$$z=\frac{(2-5i)(1+4i)}{(1-4i)(1+4i)}\Rightarrow z=\frac{2-5i+8i-20i^2}{1^2-(4i)^2}$$

$\Rightarrow \quad z=\dfrac{22+3i}{17} \quad [\because i^2=-1]$

$\therefore \quad z=\dfrac{22}{17}+\dfrac{3}{17}i$

10. Let $z=\left(-2-\dfrac{1}{3}i\right)^3=-\left(2+\dfrac{i}{3}\right)^3 \quad [\because (-1)^3=-1]$

$=-\left[8+\dfrac{i^3}{27}+3\times 2\times\dfrac{i}{3}\left(2+\dfrac{i}{3}\right)\right]$

$=-\left[8-\dfrac{i}{27}+4i+\dfrac{2i^2}{3}\right] \quad [\because i^3=-i]$

$=-\left[8-\dfrac{2}{3}+i\left(\dfrac{-1}{27}+4\right)\right] \quad [\because i^2=-1]$

$=\dfrac{-22}{3}-\dfrac{107}{27}i$

11. Let $z=-5i^{-15}-6i^{-8}$

$=\dfrac{-5}{i^{15}}-\dfrac{6}{i^8}=\dfrac{-5}{(i^4)^3\cdot i^3}-\dfrac{6}{(i^4)^2} \quad [\because i^{15}=i^{4\times 3+3}]$

$=\dfrac{-5}{(1)^3\cdot(-i)}-\dfrac{6}{(1)^2} \quad [\because i^4=1 \text{ and } i^3=-i]$

$=\dfrac{-5}{-i}-6=\dfrac{5}{i}-6=\dfrac{5-6i}{i}=\dfrac{(5-6i)i}{i\cdot i}$

[by rationalising the denominator]

$=\dfrac{5i-6i^2}{i^2}=\dfrac{5i+6}{-1}=-6-5i \quad [\because i^2=-1]$

$\therefore \quad \bar{z}=-6+5i$

Hence, $\text{Re}(\bar{z})=-6$ and $\text{Im}(\bar{z})=5$

12. We have, $(1+i)z = (1-i)\bar{z}$

$$\Rightarrow \quad \frac{z}{\bar{z}} = \frac{1-i}{1+i} \times \frac{1-i}{1-i} = \frac{1-1-2i}{1+1}$$

$$\Rightarrow \quad \frac{z}{\bar{z}} = \frac{-2i}{2} = -i$$

$\therefore \quad z = -i\bar{z}$ **Hence proved.**

13. We have, $z_1 = 2 - i$ and $z_2 = 1 + i$

$$\therefore \left|\frac{z_1 + z_2 + 1}{z_1 - z_2 + 1}\right| = \left|\frac{2 - i + 1 + i + 1}{2 - i - 1 - i + 1}\right| = \left|\frac{4}{2 - 2i}\right| = \left|\frac{2}{1-i}\right|$$

$$= \frac{2}{|1-i|} \qquad \left[\because \left|\frac{z_1}{z_2}\right| = \frac{|z_1|}{|z_2|}\right]$$

$$= \frac{2}{\sqrt{(1)^2 + (-1)^2}} = \frac{2}{\sqrt{1+1}} = \frac{2}{\sqrt{2}} \times \frac{\sqrt{2}}{\sqrt{2}} = \sqrt{2}$$

14. We have, $z_1 = 3 + i$ and $z_2 = 1 + 4i$

$$\therefore \quad |z_1| = \sqrt{3^2 + 1^2} = \sqrt{9+1} = \sqrt{10}$$

and $\quad |z_2| = \sqrt{1^2 + 4^2} = \sqrt{1+16} = \sqrt{17}$

$$\therefore \quad |z_1| + |z_2| = \sqrt{10} + \sqrt{17} = 3.16 + 4.12 = 7.28 \quad \text{...(i)}$$

Now, $\quad z_1 + z_2 = 3 + i + 1 + 4i = 4 + 5i$

$$\therefore \quad |z_1 + z_2| = \sqrt{4^2 + 5^2}$$

$$= \sqrt{16 + 25}$$

$$= \sqrt{41} = 6.40 \quad \text{...(ii)}$$

From Eqs. (i) and (ii), we get

$$|z_1 + z_2| < |z_1| + |z_2|$$

15. We have, $(x^4 + 2xi) - (3x^2 + iy) = (3 - 5i) + (1 + 2iy)$

$$\Rightarrow \quad (x^4 - 3x^2) + (2x - y)i = 4 + (-5 + 2y)i$$

On equating real and imaginary parts from both sides, we get

$$x^4 - 3x^2 = 4 \quad \text{... (i)}$$

and $\quad 2x - y = -5 + 2y$

$$\Rightarrow \quad 2x - 3y = -5 \quad \text{...(ii)}$$

On solving Eq. (i), we get

$$x^4 - 3x^2 = 4$$

$$\Rightarrow \quad x^4 - 3x^2 - 4 = 0$$

$$\Rightarrow \quad x^4 - 4x^2 + x^2 - 4 = 0$$

$$\Rightarrow \quad (x^2 - 4)(x^2 + 1) = 0$$

$$\Rightarrow \quad x^2 - 4 = 0$$

$$[\because x^2 + 1 \neq 0, \text{ for any real value of } x]$$

$$\therefore \quad x = \pm 2$$

On putting $x = \pm 2$ in Eq. (ii), we get

$$y = 3, \text{ when } x = 2 \text{ and } y = \frac{1}{3}, \text{ when } x = -2$$

Thus, $x = -2, y = \frac{1}{3}$ or $x = 2, y = 3$.

16. We have, $z_1 = 2 - i$ and $z_2 = -2 + i$

Now, $\quad \dfrac{z_1 z_2}{\bar{z}_1} = \dfrac{(2 - i)(-2 + i)}{\overline{(2 - i)}} = \dfrac{-(2 - i)(2 - i)}{2 + i}$

$$= -\frac{(4 + i^2 - 4i)}{2 + i} = -\frac{(4 - 1 - 4i)}{2 + i} = -\frac{(3 - 4i)}{2 + i} \times \frac{2 - i}{2 - i}$$

[by rationalising the denominator]

$$= -\frac{(6 - 3i - 8i + 4i^2)}{4 - i^2} \qquad [\because (z_1 + z_2)(z_1 - z_2) = z_1^2 - z_2^2]$$

$$= -\frac{(6 - 11i - 4)}{5} \qquad [\because i^2 = -1]$$

$$= -\frac{(2 - 11i)}{5} = \frac{-2}{5} + \frac{11}{5}i$$

$$\therefore \quad \text{Re}\left(\frac{z_1 z_2}{\bar{z}_1}\right) = \text{Re}\left(\frac{-2}{5} + \frac{11i}{5}\right) = \frac{-2}{5}$$

17. Let $z = 1 + i$

Multiplicative inverse of $z = \dfrac{1}{z} = \dfrac{1}{1+i}$

On multiplying by $(1-i)$ in numerator and denominator, we get

$$\frac{1}{z} = \frac{1-i}{(1+i)(1-i)}$$

$$\Rightarrow \quad \frac{1}{z} = \frac{1-i}{1-i^2}$$

$$\Rightarrow \quad \frac{1}{z} = \frac{1}{2} - \frac{i}{2} \qquad [\because i^2 = -1]$$

18. Let $z = 3 + \sqrt{7}i$

Reciprocal of $z = \dfrac{1}{z} = \dfrac{1}{3 + \sqrt{7}i}$

On multiplying by $(3 - \sqrt{7}i)$ in numerator and denominator, we get

$$\frac{1}{z} = \frac{3 - \sqrt{7}i}{(3 + \sqrt{7}i)(3 - \sqrt{7}i)}$$

$$\Rightarrow \quad \frac{1}{z} = \frac{3 - \sqrt{7}i}{9 - 7i^2}$$

$$\Rightarrow \quad \frac{1}{z} = \frac{3 - \sqrt{7}i}{16} \qquad [\because i^2 = -1]$$

$$\therefore \quad \frac{1}{z} = \frac{3}{16} - \frac{\sqrt{7}}{16}i$$

19. Let $z = \dfrac{1}{1 - \cos\theta + 2i\sin\theta}$

On multiplying by $(1 - \cos\theta - 2i\sin\theta)$ in numerator and denominator, we get.

$$z = \frac{[(1 - \cos\theta) - 2i\sin\theta]}{[(1 - \cos\theta) + 2i\sin\theta][(1 - \cos\theta) - 2i\sin\theta]}$$

$$\Rightarrow \quad z = \frac{1 - \cos\theta - 2i\sin\theta}{(1 - \cos\theta)^2 - 4i^2\sin^2\theta}$$

$$\Rightarrow \quad z=\frac{1-\cos\theta-2i\sin\theta}{1+\cos^2\theta-2\cos\theta+4\sin^2\theta} \qquad [\because i^2=-1]$$

$$\Rightarrow \quad z=\frac{1-\cos\theta-2i\sin\theta}{1+1-2\cos\theta+3\sin^2\theta} \qquad [\because \sin^2 A+\cos^2 A=1]$$

$$\therefore \quad z=\frac{1-\cos\theta}{2-2\cos\theta+3\sin^2\theta}+i\left[\frac{-2\sin\theta}{2-2\cos\theta+3\sin^2\theta}\right]$$

20. Let $z=\frac{(1+i)^n}{(1-i)^{n-2}}$

On multiplying by $(1-i)^2$ in numerator and denominator, we get

$$z=\frac{(1+i)^n\,(1-i)^2}{(1-i)^{n-2}(1-i)^2}=\left(\frac{1+i}{1-i}\right)^n(1-i)^2$$

On multiplying by $(1+i)^n$ in numerator and denominator, we get.

$$z=\left[\frac{(1+i)\,(1+i)}{(1-i)\,(1+i)}\right]^n[1+i^2-2i]$$

$$\Rightarrow \quad z=\left[\frac{(1+i)^2}{1^2-i^2}\right]^n[1-1-2i] \qquad [\because i^2=-1]$$

$$\Rightarrow \quad z=\left[\frac{1+i^2+2i}{1-(-1)}\right]^n(-2i)$$

$$\Rightarrow \quad z=\left(\frac{1-1+2i}{2}\right)^n(-2i)$$

$$\Rightarrow \quad z=i^n(-2i)$$

$$\Rightarrow \quad z=-2(i)^{n+1}$$

Hence, at $n=1$, $\frac{(1+i)^n}{(1-i)^{n-2}}$ is a real number.

21. We have, $z=2-3i$

$\Rightarrow \quad z-2=-3i$

On squaring both sides, we get

$$(z-2)^2=(-3i)^2$$

$$\Rightarrow \quad z^2+4-4z=9i^2$$

$$\Rightarrow \quad z^2-4z+4=-9 \qquad [\because i^2=-1]$$

$$\Rightarrow \quad z^2-4z+13=0$$ **Hence proved.**

Now, on dividing $4z^3-3z^2+169$ by $z^2-4z+13$.

$$\begin{array}{r} 4z+13 \\ z^2-4z+13\;\big)\;4z^3-3z^2+169 \\ 4z^3-16z^2+52z \\ -\quad+\quad- \\ \hline 13z^2-52z+169 \\ 13z^2-52z+169 \\ -\quad+\quad- \\ \hline 0 \end{array}$$

Thus, $(4z^3-3z^2+169)=(z^2-4z+13)\,(4z+13)$
$=0\,(4z+13)=0$

22. Let $z=\frac{2-i}{(1-2i)^2}=\frac{2-i}{[1^2-2(1)(2i)+(2i)^2]}$

$[\because (z_1-z_2)^2=z_1^2-2z_1z_2+z_2^2]$

$$\Rightarrow \quad z=\frac{2-i}{(1-4i-4)} \qquad [\because i^2=-1]$$

$$\Rightarrow \quad z=\frac{2-i}{-3-4i} \Rightarrow z=\frac{2-i}{-3-4i}\times\frac{-3+4i}{-3+4i}$$

[by rationalising the denominator]

$$\Rightarrow \quad z=\frac{(2-i)(-3+4i)}{(-3)^2-(4i)^2} \qquad [\because (z_1+z_2)(z_1-z_2)=z_1^2-z_2^2]$$

$$\Rightarrow \quad z=\frac{-6+8i+3i-4i^2}{9-16i^2}$$

$$\Rightarrow \quad z=\frac{-6+11i+4}{9+16} \qquad [\because i^2=-1]$$

$$\Rightarrow \quad z=\frac{-2+11i}{25} \Rightarrow z=-\frac{2}{25}+\frac{11}{25}i$$

Hence, $\bar{z}=-\frac{2}{25}-\frac{11}{25}i$

23. Let $z=\frac{(1+i)(2+i)}{(3+i)}=\frac{2+i+2i+i^2}{3+i}=\frac{2+3i-1}{3+i}$

$$\Rightarrow \quad z=\frac{1+3i}{3+i} \qquad [\because i^2=-1]$$

Now, $|z|=\left|\frac{1+3i}{3+i}\right|=\frac{|1+3i|}{|3+i|}$ $\quad\left[\because \left|\frac{z_1}{z_2}\right|=\frac{|z_1|}{|z_2|}\right]$

$$=\frac{\sqrt{1^2+3^2}}{\sqrt{3^2+1^2}}=1$$

Hence, $\left|(1+i)\frac{(2+i)}{(3+i)}\right|=1$

24. We have, $z_1=3+2i$ and $z_2=2-4i$

Now, $\quad \text{LHS}=|z_1+z_2|^2+|z_1-z_2|^2$

On substituting the values of z_1 and z_2, we get

$$\text{LHS}=|3+2i+2-4i|^2+|3+2i-2+4i|^2$$
$$=|5-2i|^2+|1+6i|^2$$
$$=(5)^2+(-2)^2+(1)^2+(6)^2$$

[if $z=a+ib$, then $|z|^2=a^2+b^2$]

$$=25+4+1+36$$

$\therefore \quad \text{LHS}=66 \qquad \ldots\text{(i)}$

and $\text{RHS}=2(|z_1|^2+|z_2|^2)=2(|3+2i|^2+|2-4i|^2)$

$$=2[(3)^2+(2)^2+(2)^2+(-4)^2]$$
$$=2(9+4+4+16)=2\times 33$$

$\therefore \quad \text{RHS}=66 \qquad \ldots\text{(ii)}$

From Eqs. (i) and (ii), we get

LHS = RHS **Hence proved.**

25. Given, $z_1 = 3+2i$ and $z_2 = 1-3i$

Clearly, $|z_1| = \sqrt{(3)^2+(2)^2} = \sqrt{9+4} = \sqrt{13}$

and $|z_2| = \sqrt{1^2+(-3)^2} = \sqrt{1+9} = \sqrt{10}$

$\therefore \quad |z_1||z_2| = \sqrt{13}\sqrt{10} = \sqrt{130}$ ...(i)

Now, consider $z_1 z_2 = (3+2i)(1-3i) = 3-9i+2i-6(i^2)$

$= 3-7i+6 = 9-7i \quad [\because i^2 = -1]$

$\therefore \quad |z_1 z_2| = \sqrt{9^2+(-7)^2}$

$= \sqrt{81+49} = \sqrt{130}$ ...(ii)

From Eqs. (i) and (ii), we get

$$|z_1 z_2| = |z_1||z_2|$$

26. We have, $|z| = z+1+2i$

Let $z = x+iy$

$\therefore \quad |x+iy| = x+iy+1+2i$

$\Rightarrow \quad \sqrt{x^2+y^2} = (x+1)+i(y+2)$

On comparing both sides, we get

$\sqrt{x^2+y^2} = x+1$ ...(i)

and $y+2 = 0$

$\Rightarrow \quad y = -2$ ...(ii)

On putting $y = -2$ in Eq. (i), we get

$$\sqrt{x^2+4} = x+1$$

On squaring both sides, we get

$$x^2+4 = (x+1)^2$$

$\Rightarrow \quad x^2+4 = x^2+1+2x$

$\Rightarrow \quad x = \frac{3}{2}$ ...(iii)

From Eqs. (ii) and (iii), we get

$$x = \frac{3}{2} \text{ and } y = -2$$

Now, $z = \frac{3}{2}+(-2)i = \frac{3}{2}-2i$

27. Given, $|z_1| = |z_2| = ... = |z_n| = 1$

$\Rightarrow \quad |z_1|^2 = |z_2|^2 = ... = |z_n|^2 = 1$

$\therefore \quad z_1\bar{z}_1 = z_2\bar{z}_2 = ... = z_n\bar{z}_n = 1 \quad [\because |z|^2 = z\bar{z}]$

$\therefore \quad z_1 = \frac{1}{\bar{z}_1}, z_2 = \frac{1}{\bar{z}_2}, ..., z_n = \frac{1}{\bar{z}_n}$

$\therefore \quad |z_1+z_2+...+z_n| = |\overline{z_1+z_2+...+z_n}|$

$= |\bar{z}_1+\bar{z}_2+...+\bar{z}_n|$

$= \left|\frac{1}{z_1}+\frac{1}{z_2}+...+\frac{1}{z_n}\right|$

28. Given, $(1+i)(1+2i)(1+3i)...(1+ni) = x+iy$

On taking modulus both sides, we get

$|1+i||1+2i||1+3i|...|1+ni| = |x+iy| \quad [\because |a \cdot b| = |a| \cdot |b|]$

$\Rightarrow \sqrt{1+1}\sqrt{1+4}\sqrt{1+9}...\sqrt{1+n^2} = \sqrt{x^2+y^2}$

$\Rightarrow \quad \sqrt{2}\sqrt{5}\sqrt{10}...\sqrt{1+n^2} = \sqrt{x^2+y^2}$

On squaring both sides, we get

$2 \cdot 5 \cdot 10...(1+n^2) = x^2+y^2$ **Hence proved.**

29. Since, $\frac{4z_1}{5z_2}$ is purely imaginary number.

$\therefore \quad \frac{4z_1}{5z_2} = \lambda i$, for some $\lambda \in R$

$\Rightarrow \quad \frac{z_1}{z_2} = \frac{5\lambda}{4}i$ ...(i)

Now, consider $\left|\frac{z_1-z_2}{z_1+z_2}\right| = \left|\frac{\frac{z_1}{z_2}-1}{\frac{z_1}{z_2}+1}\right|$

[divide numerator and denominator by z_2]

$= \left|\frac{\frac{5}{4}\lambda i-1}{\frac{5}{4}\lambda i+1}\right| = \left|\frac{5\lambda i-4}{5\lambda i+4}\right|$ [using Eq. (i)]

$= \frac{|5\lambda i-4|}{|5\lambda i+4|} = \frac{|-4+5\lambda i|}{|4+5\lambda i|} = \frac{\sqrt{(-4)^2+(5\lambda)^2}}{\sqrt{(4)^2+(5\lambda)^2}} = 1$

Hence, $\left|\frac{z_1-z_2}{z_1+z_2}\right| = 1$

30. Let $z = \frac{1+2i}{1-i}$

$\Rightarrow \quad z = \frac{1+2i}{(1-i)} \times \frac{(1+i)}{(1+i)}$

$\Rightarrow \quad z = \frac{1+2i+i+2i^2}{1^2-i^2} \Rightarrow z = \frac{1+3i-2}{2} \quad [\because i^2 = -1]$

$\Rightarrow \quad z = \frac{-1}{2}+\frac{3i}{2}$

Conjugate of $z = \bar{z} = \frac{-1}{2}-\frac{3}{2}i$,

which is represented by a point in III quadrant.

31. (i) $(-2-5i) \div (3-6i) = \frac{-2-5i}{3-6i} = \frac{-(2+5i)}{3-6i} \times \frac{(3+6i)}{(3+6i)}$

[by rationalising the denominator]

$= -\frac{[6+12i+15i+30i^2]}{(3)^2-(6i)^2}$

$[\because (z_1+z_2)(z_1-z_2) = z_1^2-z_2^2]$

$= \frac{-[6+27i-30]}{9+36} \quad [\because i^2 = -1]$

$= \frac{-(-24+27i)}{45} = \frac{24}{45}-\frac{27}{45}i$

$= \frac{8}{15}-\frac{3}{5}i = \frac{8}{15}+i\left(\frac{-3}{5}\right)$,

which is the form of $(a+ib)$.

(ii) $\dfrac{(3+\sqrt{5}i)(3-\sqrt{5}i)}{(\sqrt{3}+\sqrt{2}i)-(\sqrt{3}-\sqrt{2}i)} = \dfrac{(3)^2-(\sqrt{5}i)^2}{\sqrt{3}+\sqrt{2}i-\sqrt{3}+\sqrt{2}i}$

$[\because (z_1+z_2)(z_1-z_2) = z_1^2 - z_2^2]$

$= \dfrac{9+5}{2\sqrt{2}i} = \dfrac{14}{2\sqrt{2}i} = \dfrac{7}{\sqrt{2}i} \times \dfrac{\sqrt{2}i}{\sqrt{2}i}$

[by rationalising the denominator]

$= \dfrac{7\sqrt{2}i}{2i^2} = \dfrac{7\sqrt{2}i}{-2} = 0 - i\dfrac{7\sqrt{2}}{2}$ $[\because i^2 = -1]$

$= 0 + i\left(\dfrac{-7\sqrt{2}}{2}\right)$, which is the form of $(a+ib)$.

32. We have, $\dfrac{(1+i)^2}{2-i} = x+iy$

$\Rightarrow \quad x+iy = \dfrac{(1+i)^2(2+i)}{(2-i)(2+i)}$

[by rationalising the denominator]

$\Rightarrow \quad x+iy = \dfrac{(1-1+2i)(2+i)}{4-i^2}$

$\Rightarrow \quad x+iy = \dfrac{2i(2+i)}{4-(-1)}$

$\Rightarrow \quad x+iy = \dfrac{4i+2i^2}{5}$

$\Rightarrow \quad x+iy = \dfrac{-2+4i}{5}$

$\Rightarrow \quad x+iy = \dfrac{-2}{5} + \dfrac{i4}{5}$

On comparing both sides, we get

$x = \dfrac{-2}{5}$ and $y = \dfrac{4}{5}$

Now, $\quad 2x-y = \dfrac{-4}{5} - \dfrac{4}{5} = \dfrac{-8}{5}$

33. We have, $x+iy = \sqrt{\dfrac{1+i}{1-i}}$

$\Rightarrow \quad x+iy = \sqrt{\dfrac{1+i}{1-i} \times \dfrac{1+i}{1+i}}$

[by rationalising the denominator]

$\Rightarrow \quad x+iy = \sqrt{\dfrac{(1+i)^2}{1-i^2}}$ $[\because (z_1-z_2)(z_1+z_2) = z_1^2 - z_2^2]$

$\Rightarrow \quad x+iy = \dfrac{1+i}{\sqrt{1+1}} = \dfrac{1+i}{\sqrt{2}}$ $[\because i^2 = -1]$

$= \dfrac{1}{\sqrt{2}} + \dfrac{i}{\sqrt{2}}$ ...(i)

Now, on taking conjugate both sides, we get

$\overline{x+iy} = \overline{\left(\dfrac{1}{\sqrt{2}} + \dfrac{i}{\sqrt{2}}\right)}$

$\Rightarrow \quad x-iy = \dfrac{1}{\sqrt{2}} - \dfrac{i}{\sqrt{2}}$ ...(ii)

On multiplying Eqs. (i) and (ii), we get

$(x+iy)(x-iy) = \left(\dfrac{1}{\sqrt{2}} + \dfrac{i}{\sqrt{2}}\right)\left(\dfrac{1}{\sqrt{2}} - \dfrac{i}{\sqrt{2}}\right)$

$\Rightarrow \quad x^2 - (iy)^2 = \left(\dfrac{1}{\sqrt{2}}\right)^2 - \left(\dfrac{i}{\sqrt{2}}\right)^2$

$[\because (z_1+z_2)(z_1-z_2) = z_1^2 - z_2^2]$

$\Rightarrow \quad x^2 - i^2y^2 = \dfrac{1}{2} - \dfrac{i^2}{2}$

$\Rightarrow \quad x^2 + y^2 = \dfrac{1}{2} + \dfrac{1}{2} = 1$ $[\because i^2 = -1]$

Hence proved.

34. Let $\quad z = \dfrac{3+2i\sin\theta}{1-2i\sin\theta}$

On multiplying by $(1+2i\sin\theta)$ in numerator and denominator, we get

$z = \dfrac{3+2i\sin\theta}{1-2i\sin\theta}\left(\dfrac{1+2i\sin\theta}{1+2i\sin\theta}\right)$

$\Rightarrow \quad z = \dfrac{(3+2i\sin\theta)+6i\sin\theta+4i^2\sin^2\theta}{1-4i^2\sin^2\theta}$

$\Rightarrow \quad z = \dfrac{3+8i\sin\theta-4\sin^2\theta}{1+4\sin^2\theta}$

$\Rightarrow \quad z = \dfrac{3-4\sin^2\theta}{1+4\sin^2\theta} + \dfrac{8\sin\theta}{1+4\sin^2\theta}i$

For z to be real, Im $(z) = 0$.

$\therefore \quad \dfrac{8\sin\theta}{1+4\sin^2\theta} = 0$

$\Rightarrow \quad \sin\theta = 0$

$\Rightarrow \quad \theta = n\pi, n \in I$

35. $(4+3\sqrt{-20}) = 4+6i\sqrt{5}$

Let $(4+3\sqrt{-20})^{1/2} = x+iy$

$\Rightarrow \quad 4+6i\sqrt{5} = x^2 - y^2 + (2xy)i$

$\Rightarrow \quad x^2 - y^2 = 4$

and $\quad 2xy = 6\sqrt{5}$

$\therefore \quad x^2 + y^2 = \sqrt{(x^2-y^2)^2 + 4x^2y^2}$

$= \sqrt{16+180} = \sqrt{196} = 14$

On solving the equations $x^2 + y^2 = 14$ and $x^2 - y^2 = 4$, we get

$x^2 = 9$ and $y^2 = 5$

$\Rightarrow \quad x = \pm 3$ and $y = \pm\sqrt{5}$

Since, $xy > 0$, it follows that x and y are of the same sign.

$\therefore \quad x = 3, y = \sqrt{5}$

or $\quad x = -3, y = -\sqrt{5}$

So, $(4+3\sqrt{-20})^{1/2} = (4+6i\sqrt{5})^{1/2}$

$= \pm(3+\sqrt{5}i)$ ...(i)

$\therefore\ (4-3\sqrt{-20})^{1/2} = \pm(3-\sqrt{5}i)$ ...(ii)

Hence, $(4+3\sqrt{-20})^{1/2} + (4-3\sqrt{-20})^{1/2} = \pm 6$

36. Given, $x = \sqrt{-2} - 1$

$\Rightarrow\ x + 1 = \sqrt{2}i$

On squaring both sides, we get

$x^2 + 1 + 2x = -2$

$\Rightarrow\ x^2 + 2x + 3 = 0$

Now, divide $x^4 + 4x^3 + 6x^2 + 4x + 9$ by $x^2 + 2x + 3$.

$$\begin{array}{r|l} & x^2 + 2x - 1 \\ x^2+2x+3 & x^4 + 4x^3 + 6x^2 + 4x + 9 \\ & x^4 + 2x^3 + 3x^2 \\ & - \quad - \quad - \\ \hline & 2x^3 + 3x^2 + 4x + 9 \\ & 2x^3 + 4x^2 + 6x \\ & - \quad - \quad - \\ \hline & -x^2 - 2x + 9 \\ & -x^2 - 2x - 3 \\ & + \quad + \quad + \\ \hline & 12 \end{array}$$

Thus, $(x^4 + 4x^3 + 6x^2 + 4x + 9)$

$= (x^2 + 2x + 3)(x^2 + 2x - 1) + 12$

$= 0\,(x^2 + 2x - 1) + 12 = 12$

37. (i) Let $z = (1-i)^{-2} + (1+i)^{-2}$

$$= \frac{1}{(1-i)^2} + \frac{1}{(1+i)^2} = \frac{(1+i)^2 + (1-i)^2}{(1-i)^2\,(1+i)^2}$$

$$= \frac{1 + i^2 + 2i + 1 + i^2 - 2i}{(1-i^2)^2} = \frac{1-1+1-1}{(1+1)^2} = \frac{0}{4}$$

$[\because i^2 = -1]$

$= 0 = 0 + 0i$

$\therefore\ \bar{z} = \overline{0+0i} = 0$ and $|z| = \sqrt{0+0} = 0$

(ii) Let $z = \dfrac{3+2i}{2-5i} + \dfrac{3-2i}{2+5i}$

$$= \frac{(3+2i)(2+5i) + (3-2i)(2-5i)}{(2-5i)(2+5i)}$$

$$= \frac{6 + 15i + 4i + 10i^2 + 6 - 15i - 4i + 10i^2}{(2)^2 - (5i)^2}$$

$[\because (z_1 - z_2)(z_1 + z_2) = z_1^2 - z_2^2]$

$$= \frac{6 + 20i^2 + 6}{4 - 25i^2}$$

$$= \frac{12 - 20}{4 + 25} = \frac{-8}{29} = \frac{-8}{29} + 0i \quad [\because i^2 = -1]$$

Now, $\bar{z} = -\dfrac{8}{29} - 0i = \dfrac{-8}{29}$

and $|z| = \sqrt{\left(\dfrac{-8}{29}\right)^2 + 0^2} = \sqrt{\dfrac{64}{841}} = \dfrac{8}{29}$

38. Let $z = x + iy$, then $|z - 1| = |z + 1|$

$|x + iy - 1| = |x + iy + 1|$

$\Rightarrow\ |(x-1) + iy| = |(x+1) + iy|$

$\Rightarrow\ \sqrt{(x-1)^2 + y^2} = \sqrt{(x+1)^2 + y^2}$

$\Rightarrow\ (x-1)^2 + y^2 = (x+1)^2 + y^2$

$\Rightarrow\ x^2 + 1 - 2x = x^2 + 1 + 2x$

$\Rightarrow\ 4x = 0$

$\Rightarrow\ x = 0$

$\therefore\ \text{Re}(z) = 0$

39. Given, $z = x + iy$

Since, $(|x| + |y|)^2 = |x|^2 + |y|^2 + 2|x||y|$

$= x^2 + y^2 + 2|x||y|$

$= |z|^2 + 2|x||y|$ ...(i)

and $(|x| - |y|)^2 = |x|^2 + |y|^2 - 2|x||y|$

$= x^2 + y^2 - 2|x||y|$

$= |z|^2 - 2|x||y|$ ...(ii)

On adding Eqs. (i) and (ii), we get

$(|x| + |y|)^2 + (|x| - |y|)^2 = 2|z|^2$

$\Rightarrow\ (|x| + |y|)^2 \le 2|z|^2$

$\therefore\ |x| + |y| \le \sqrt{2}|z|$

40. Given, $|z + 1| = z + 2(1+i)$ ...(i)

$\Rightarrow\ |x + iy + 1| = x + iy + 2(1+i)$ [let $z = x + iy$]

$\Rightarrow\ \sqrt{(x+1)^2 + y^2} = (x+2) + i(y+2)$

On squaring both sides, we get

$(x+1)^2 + y^2 = (x+2)^2 + i^2(y+2)^2 + 2i(x+2)(y+2)$

$\Rightarrow\ x^2 + y^2 + 2x + 1 = x^2 + 4x + 4 - y^2 - 4y - 4 + 2i(x+2)(y+2)$

On equating real and imaginary parts, we get

$x^2 + y^2 + 2x + 1 = x^2 - y^2 + 4x - 4y$

$\Rightarrow\ 2y^2 - 2x + 4y + 1 = 0$ ...(ii)

and $2(x+2)(y+2) = 0$

$\Rightarrow\ x + 2 = 0$ or $y + 2 = 0$

$\Rightarrow\ x = -2$ or $y = -2$

For $x = -2$,

$2y^2 + 4 + 4y + 1 = 0$

$\Rightarrow\ 2y^2 + 4y + 5 = 0$

$\Rightarrow\ b^2 - 4ac < 0$ $[\because a = 2, b = 4, c = 5]$

$\Rightarrow 2y^2 + 4y + 5$ has no real roots.

For $y = -2$,

$$2(-2)^2 - 2x + 4(-2) + 1 = 0$$

$$\Rightarrow \quad 8 - 2x - 8 + 1 = 0$$

$$\Rightarrow \quad x = \frac{1}{2}$$

$$\therefore \quad z = x + iy = \frac{1}{2} - 2i$$

41. Given, $z_1 = 3 + 5i$ and $z_2 = 2 - 3i$

Now, $\frac{z_1}{z_2} = \frac{3+5i}{2-3i} = \frac{3+5i}{2-3i} \times \frac{2+3i}{2+3i}$

[by rationalising the denominator]

$$= \frac{6 + 9i + 10i + 15i^2}{4 - 9i^2} = \frac{6 + 19i - 15}{4 + 9} \qquad [\because i^2 = -1]$$

$$= \frac{-9 + 19i}{13} = \frac{-9}{13} + \frac{19}{13}i$$

$$\therefore \text{LHS} = \overline{\left(\frac{z_1}{z_2}\right)} = \overline{\left(\frac{-9}{13} + \frac{19}{13}i\right)} = \frac{-9}{13} - \frac{19}{13}i \qquad \ldots\text{(i)}$$

Now, consider RHS $= \frac{\bar{z}_1}{\bar{z}_2} = \frac{\overline{3+5i}}{\overline{2-3i}} = \frac{3-5i}{2+3i}$

$$= \frac{3-5i}{2+3i} \times \frac{2-3i}{2-3i}$$

[by rationalising the denominator]

$$= \frac{6 - 9i - 10i + 15i^2}{4 - 9i^2}$$

$$= \frac{6 - 19i - 15}{4 + 9} \qquad [\because i^2 = -1]$$

$$= \frac{-9 - 19i}{13} = \frac{-9}{13} - \frac{19}{13}i \qquad \ldots\text{(ii)}$$

From Eqs. (i) and (ii), we get

$$\overline{\left(\frac{z_1}{z_2}\right)} = \frac{\bar{z}_1}{\bar{z}_2}$$

42. We have, $|z| = 3|z - 1|$

Let $z = x + iy$

$$\therefore \quad |x + iy| = 3|x + iy - 1|$$

$$\Rightarrow \quad \sqrt{x^2 + y^2} = 3\sqrt{(x-1)^2 + y^2}$$

On squaring both sides, we get

$$\Rightarrow \quad x^2 + y^2 = 9(x^2 + 1 - 2x + y^2)$$

$$\Rightarrow \quad 8x^2 + 8y^2 - 18x + 9 = 0$$

$$\Rightarrow \quad x^2 + y^2 - \frac{9}{4}x + \frac{9}{8} = 0$$

$$\Rightarrow \quad x^2 + \frac{81}{64} - \frac{9x}{4} + y^2 = \frac{81}{64} - \frac{9}{8}$$

$$\Rightarrow \quad \left(x - \frac{9}{8}\right)^2 + y^2 = \frac{9}{64}$$

$$[\because a^2 - 2ab + b^2 = (a-b)^2]$$

$$\Rightarrow \quad \left|\left(x - \frac{9}{8}\right) + iy\right| = \frac{3}{8}$$

$$\Rightarrow \quad \left|z - \frac{9}{8}\right| = \frac{3}{8}$$

43. We have, $z = x + iy \Rightarrow \bar{z} = x - iy$

$$\therefore \quad z\bar{z} + 2(z + \bar{z}) + a = 0$$

$$\Rightarrow \quad (x + iy)(x - iy) + 2(x + iy + x - iy) + a = 0$$

$$\Rightarrow \quad x^2 + y^2 + 4x + a = 0$$

$$\Rightarrow \quad (x+2)^2 + (y-0)^2 = (\sqrt{4-a})^2, a \le 4,$$

which represents a circle.

QR Code for Quick Test 1

QR Code for Quick Test 2

Mind Map

Complex Numbers

Properties of Addition of Complex Numbers

(i) Addition of complex numbers is closure.
(ii) Addition of complex numbers is commutative.
(iii) Addition of complex numbers is associative .
(iv) **Existence of Additive Identity**
Complex number $0 = 0 + 0\,i$, is called additive identity of any complex number.
(v) **Existence of Additive Inverse**
If z is a complex number, then the complex number $(-z)$ is called additive

Difference of Two Complex Numbers

If $z_1 = a_1 + ib_1$ and $z_2 = a_2 + ib_2$ are two complex numbers. Then, their subtraction $z_1 - z_2$ is defined as the addition of z_1 and $(-z_2)$.

Thus, $z_1 - z_2 = z_1 + (-z_2) = (a_1 + ib_1) + (-a_2 - ib_2)$

$\therefore\ z_1 - z_2 = (a_1 - a_2) + i(b_1 - b_2)$

Multiplication of Two Complex Numbers

The product of two complex numbers $z_1 = a + ib$ and $z_2 = c + id$ can be done as follows

$$z_1z_2 = (a + ib)(c + id) = ac + iad + ibc + i^2bd$$
$$= ac + i(ad + bc) + (-1)bd \quad [\because i^2 = -1]$$
$$\therefore\ z_1z_2 = (ac - bd) + i(ad + bc)$$

Properties of Multiplication of Complex Numbers

(i) Multiplication of complex numbers is closure.
(ii) Multiplication of complex numbers is commutative.
(iii) Multiplication of complex numbers is associative.
(iv) **Existence of Multiplicative Identity**
Complex number $1 = 1 + 0i$ is called multiplicative identity of any complex number.
$$z \times 1 = z = 1 \times z,\ z \in C$$
(v) **Existence of multiplicative Inverse or Reciprocal**
If z is a complex number, then the complex number $\frac{1}{z}$ is called multiplicative inverse of z.
(vi) Multiplication of complex number is distributive.

Addition of Two Complex Numbers

If $z_1 = a_1 + i\,b_1$ and $z_2 = a_2 + i\,b_2$ are two complex numbers, then their addition is defined as

$z = z_1 + z_2 = (a_1 + a_2) + i\,(b_1 + b_2)$

Equality of Complex Numbers

Two complex numbers $z_1 = a + ib$ and $z_2 = c + id$ are said to be equal, if $a = c$ and $b = d$.

Complex Number

A number consisting of real number and imaginary number is called **complex number**. A complex number can be defined as a number of the form $\boldsymbol{a + ib}$, where a and b are real numbers.

Division of Two Complex Numbers

The division of a complex number z_1 by a non-zero complex number z_2 is defined as the multiplication of z_1 by the multiplicative inverse of z_2 and is denoted by $\frac{z_1}{z_2}$.

Modulus (Absolute Value) of Complex Numbers

The modulus (or absolute value) of a complex number, $z = a + ib$ is defined as the non-negative real number $\sqrt{a^2 + b^2}$.

It is denoted by $|z|$ i.e. $|z| = \sqrt{a^2 + b^2}$.

Argand Plane

The plane representing complex numbers as points, is called **Complex plane or Argand plane or Gaussian plane.**

Identities Related to Complex Numbers

Identity is an equation which is true for all values of the variable (complex number) involved in it. Here, we have the following identities.

(i) $(z_1 + z_2)^2 = z_1^{\ 2} + z_2^2 + 2z_1z_2$, for all complex numbers z_1 and z_2.
(ii) $(z_1 - z_2)^2 = z_1^2 - 2\,z_1z_2 + z_2^2$
(iii) $(z_1 + z_2)^3 = z_1^{\ 3} + 3z_1^2\,z_2 + 3z_1z_2^2 + z_2^3$
(iv) $(z_1 - z_2)^3 = z_1^3 - 3z_1^2\,z_2 + 3z_1z_2^2 - z_2^3$
(v) $z_1^2 - z_2^2 = (z_1 + z_2)(z_1 - z_2)$

Real and Imaginary Parts of a Complex Numbers

If $z = a + ib$ is a complex number, then a is called the **real part** and b is called the **imaginary part** of z and they may be denoted as **Re(z)** and **Im(z)**, respectively.
e.g. If $z = 2 + 3\,i$, then Re $(z) = 2$ and Im $(z) = 3$.

Integral Power of i (iota) Positive Integral Powers of i

As we have seen, $i = \sqrt{-1}$. So, we can write the higher powers of i as follows
(i) $i^2 = -1$
(ii) $i^3 = i^2 \times i = (-1)\,i = -i$
(iii) $i^4 = (i^2)^2 = (-1)^2 = 1$
(iv) $i^5 = i^{4+1} = i^4 \times i = 1 \times i = i$
(v) $i^6 = i^{4+2} = i^4 \times i^2 = 1 \times i^2 = -1$
⋮
While evaluating i^n for $n > 4$, we are writing n as $4q + r$ for some $q, r \in N$ and $0 \le r \le 3$.
Then, $i^n = i^{4q} . i^r = (i^4)^q . i^r = (1)^q . i^r = i^r$
e.g. $i^{17} = i^{4 \times 4 + 1} = i^{4\times4} . i = (i^4)^4 . i = 1 . i = i$

Conjugate of a Complex Number

The conjugate of a complex number z, is the complex number, obtained by changing the sign of imaginary part of z. It is denoted by $\bar{z}$.

Topics to be Assessed Formatively

Polar Form of a Complex Number

Let O be the origin and OX and OY be the real axis and imaginary axis, respectively.

Let $z = x + iy$ be represented by a point $P(x, y)$.

Draw $PQ \perp OX$. Then, $OQ = x$ and $PQ = y$.

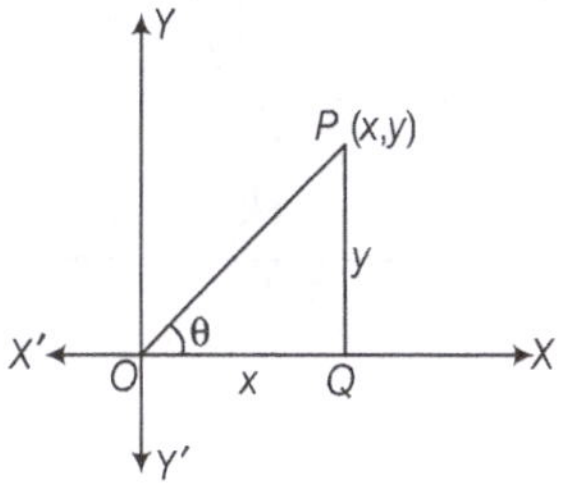

Now, join OP.

Let $\quad OP = r$ and $\angle POX = \theta$

Then, in ΔPQO, $\cos\theta = \dfrac{x}{r}$

$$\Rightarrow \quad x = r\cos\theta \quad \text{...(i)}$$

and $$\sin\theta = \frac{y}{r}$$

$$\Rightarrow \quad y = r\sin\theta \quad \text{...(ii)}$$

On squaring Eqs. (i) and (ii) and then adding, we get

$$r^2\cos^2\theta + r^2\sin^2\theta = x^2 + y^2$$

$$\Rightarrow \quad r^2(\cos^2\theta + \sin^2\theta) = x^2 + y^2$$

$$\Rightarrow \quad r^2 = x^2 + y^2 \quad [\because \cos^2\theta + \sin^2\theta = 1]$$

$$\Rightarrow \quad r = \sqrt{x^2 + y^2} = |z|$$

On dividing Eq. (ii) by Eq. (i), we get

$$\tan\theta = \frac{y}{x}$$

$$\Rightarrow \quad \theta = \tan^{-1}\left(\frac{y}{x}\right) = \arg(z)$$

Thus, the **polar form** of a complex number z is $z = r(\cos\theta + i\sin\theta)$ and (r, θ) is called **polar coordinates** of the point P.

Here, r is called the modulus of complex number z.

Angle θ is known as **argument** or **amplitude** of $z = x + iy$ and is written as 'arg(z)'. The **argument** of a complex number is not defined **uniquely**.

Thus, $$\theta = \tan^{-1}\left(\frac{y}{x}\right) = \tan^{-1}\left(\frac{\mathrm{Im}(z)}{\mathrm{Re}(z)}\right)$$

This angle θ has infinitely many values differing by multiples of 2π.

e.g. Let $\quad z = 1 + i\sqrt{3}$

Here, $\quad x = 1,\ y = \sqrt{3}$

$$\therefore \quad \arg(z) = \tan^{-1}\left(\frac{\sqrt{3}}{1}\right) = \tan^{-1}\left\{\tan\frac{\pi}{3}\right\}$$

$$\Rightarrow \quad \arg(z) = \frac{\pi}{3}$$

Principal Argument

The unique value of θ such that $-\pi < \theta \le \pi$ is called the **principal value of the argument** or **amplitude** or **principal argument.**

The argument of a complex number depends upon the quadrant in which the point P lies.

In different quadrant, the signs of real and imaginary parts of a complex number $z = x + iy$, its argument and graph are given in the tables on the next page.

I General Argument, when $0 \le \theta \le 2\pi$

Quadrant	Signs of x and y	Argument	Graph
In I quadrant	$x > 0, y > 0$	$0 < \theta < \frac{\pi}{2}$	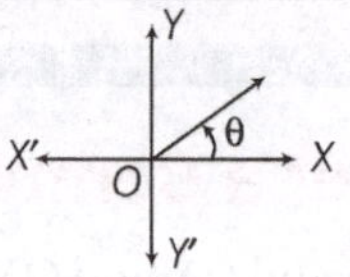
In II quadrant	$x < 0, y > 0$	$\frac{\pi}{2} < \theta < \pi$	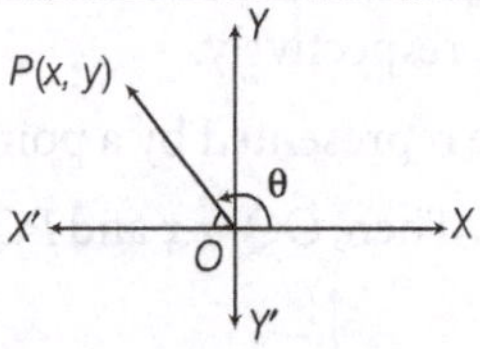
In III quadrant	$x < 0, y < 0$	$\pi < \theta < \frac{3\pi}{2}$	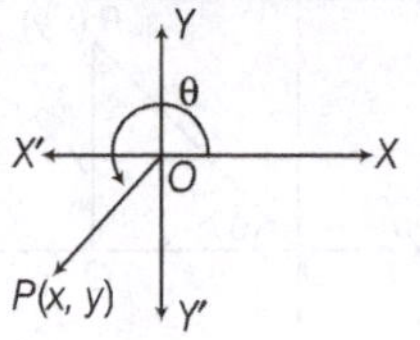
In IV quadrant	$x > 0, y < 0$	$\frac{3\pi}{2} < \theta < 2\pi$	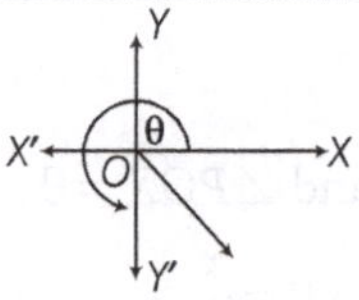

II Principal Argument, when $-\pi < \theta \le \pi$

Quadrant	Signs of x and y	Argument	Graph
In I quadrant	$x > 0, y > 0$	$\theta = \alpha$ and $0 < \theta < \frac{\pi}{2}$	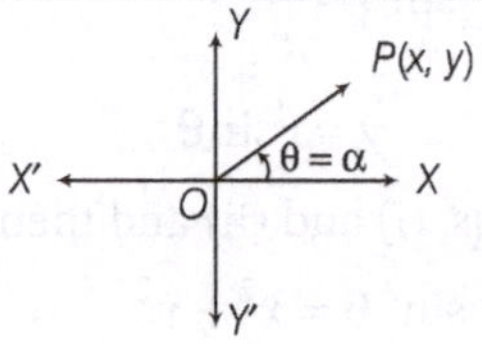
In II quadrant	$x < 0, y > 0$	$\theta = \pi - \alpha$ and $\frac{\pi}{2} < \theta < \pi$	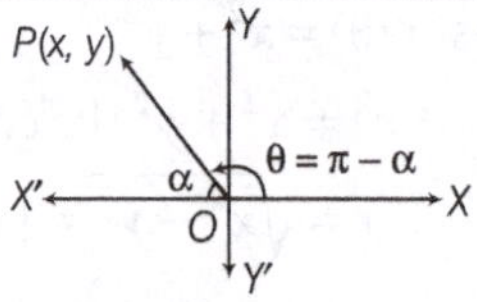
In III quadrant	$x < 0, y < 0$	$\theta = -(\pi - \alpha) = \alpha - \pi$ and $-\pi < \theta < -\frac{\pi}{2}$	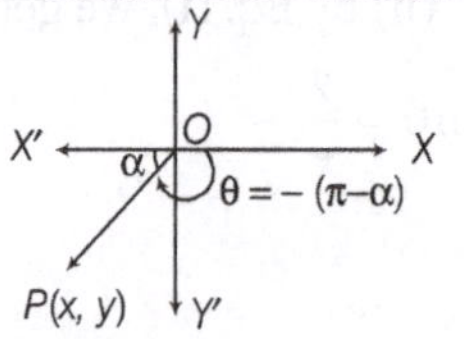
In IV quadrant	$x > 0, y < 0$	$\theta = -\alpha$ and $-\frac{\pi}{2} < \theta < 0$	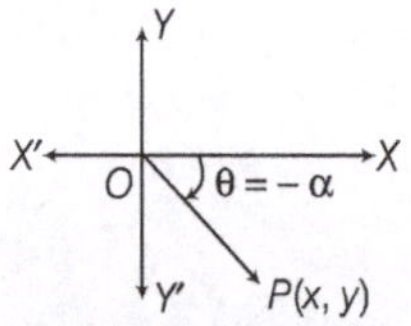

Properties of Argument of Complex Numbers

If z_1, z_2 and z_3 are three complex numbers, then

(i) $\arg(z_1z_2)=\arg(z_1)+\arg(z_2)$
In general, $\arg(z_1z_2z_3\ldots z_n)=\arg(z_1)+\arg(z_2)+\arg(z_3)+\ldots+\arg(z_n)$

(ii) $\arg\left(\dfrac{z_1}{z_2}\right)=\arg(z_1)-\arg(z_2)$

(iii) $\arg\left(\dfrac{z}{\bar{z}}\right)=2\arg(z)$

(iv) $\arg(z^n)=n\arg(z)$

(v) If $\arg\left(\dfrac{z_2}{z_1}\right)=\theta$, then $\arg\left(\dfrac{z_1}{z_2}\right)=-\theta$

(vi) $\arg(\bar{z})=-\arg(z)$

(vii) If $\arg(z)=0\Rightarrow z$ is real.

(viii) $\arg(z_1\bar{z}_2)=\arg(z_1)-\arg(z_2)$

(ix) $|z_1+z_2|=|z_1-z_2|\Rightarrow \arg(z_1)-\arg(z_2)=\dfrac{\pi}{2}$

(x) $|z_1+z_2|=|z_1|+|z_2|\Rightarrow \arg(z_1)=\arg(z_2)$

(xi) $|z_1+z_2|^2=|z_1|^2+|z_2|^2\Rightarrow \dfrac{z_1}{z_2}$ is purely imaginary.

Example 1 Find the argument of the complex number $1-i$.

Sol. Let $z=1-i$ and $\arg(z)=\theta$.

Here, $\text{Re}(z)=1$ and $\text{Im}(z)=-1$

Then, $\tan\theta=\dfrac{\text{Im}(z)}{\text{Re}(z)}=\dfrac{-1}{1}=-1$

Now, let $\tan\alpha=|\tan\theta|$, where $0<\alpha<\dfrac{\pi}{2}$.

Then, $\tan\alpha=|-1|=1$

$\Rightarrow \tan\alpha=\tan\dfrac{\pi}{4} \Rightarrow \alpha=\dfrac{\pi}{4}$

Since, z lies in IV quadrant.

$\therefore \arg(z)=-\alpha=-\dfrac{\pi}{4}$

Example 2 Find z, if $|r|=2$ and $\arg(z)=\dfrac{\pi}{4}$.

Sol. Here, we have $r=2$ and $\theta=\dfrac{\pi}{4}$

$$\therefore z=r(\cos\theta+i\sin\theta)$$
$$=2\left(\cos\frac{\pi}{4}+i\sin\frac{\pi}{4}\right)=2\left(\frac{1}{\sqrt{2}}+i\frac{1}{\sqrt{2}}\right)$$
$$=\sqrt{2}+i\sqrt{2}$$

Example 3 Find the modulus and argument of the complex number $\dfrac{1+2i}{1-3i}$. [NCERT]

Sol. Let $z=\dfrac{1+2i}{1-3i}$

Then, $z=\dfrac{1+2i}{1-3i}\times\dfrac{1+3i}{1+3i}$ [by rationalising the denominator]

$$=\frac{(1+2i)(1+3i)}{1^2-(3i)^2}\quad [\because (z_1+z_2)(z_1-z_2)=z_1^2-z_2^2]$$
$$=\frac{1+3i+2i+6i^2}{1-9i^2}$$
$$=\frac{1+5i-6}{1+9}=\frac{-5+5i}{10}=-\frac{1}{2}+\frac{i}{2}\quad [\because i^2=-1]$$

$\Rightarrow z=-\dfrac{1}{2}+\dfrac{i}{2}$

$$\therefore |z|=\sqrt{\left(-\frac{1}{2}\right)^2+\left(\frac{1}{2}\right)^2}=\sqrt{\left(\frac{1}{2}\right)^2+\left(\frac{1}{2}\right)^2}$$
$$=\sqrt{2\cdot\left(\frac{1}{2}\right)^2}=\frac{1}{2}\times\sqrt{2}=\frac{1}{\sqrt{2}}$$

Let $\arg(z)=\theta$.

Then, $\tan\theta=\dfrac{\frac{1}{2}}{\left(-\frac{1}{2}\right)}=-1$

Now, let $\tan\alpha=|\tan\theta|$, where $0<\alpha<\dfrac{\pi}{2}$

$\Rightarrow \tan\alpha=|-1|=1=\tan\dfrac{\pi}{4} \Rightarrow \alpha=\dfrac{\pi}{4}$

Since, z lies is II quadrant, therefore

$$\theta=\pi-\alpha=\pi-\frac{\pi}{4}=\frac{3\pi}{4}$$

Hence, $|z|=\dfrac{1}{\sqrt{2}}$ and $\arg(z)=\dfrac{3\pi}{4}$

Method for Writing Polar Form of a Complex Number

For writing polar form of a complex number, *we use the following steps*

***Step* I** Firstly, write the given complex in standard form $z=a+ib$ (if not given in standard form).

***Step* II** Put the standard form $z=a+ib$ equal to $r(\cos\theta+i\sin\theta)$ and equate the real and imaginary part from both sides.

***Step* III** Find the value of r and θ by using suitable method.

***Step* IV** Put the values of r and θ in $z = r[\cos\theta + i\sin\theta]$ which gives the required polar form of given complex number.

Example 4 Change the complex number $2 - 2i$ into polar form.

Sol. Let $z = 2 - 2i = r(\cos\theta + i\sin\theta)$

On equating real and imaginary parts, we get

$$r\cos\theta = 2 \quad \text{...(i)}$$

and $$r\sin\theta = -2 \quad \text{...(ii)}$$

On squaring and adding Eqs. (i) and (ii), we get

$$r^2\cos^2\theta + r^2\sin^2\theta = 2^2 + (-2)^2$$

$$\Rightarrow \quad r^2(\cos^2\theta + \sin^2\theta) = 4 + 4$$

$$[\because \cos^2\theta + \sin^2\theta = 1]$$

$$\Rightarrow \quad r^2 = 8$$

$$\Rightarrow \quad r = 2\sqrt{2} \quad [\because r > 0]$$

Now, from Eqs. (i) and (ii), we get

$r\cos\theta = 2$ and $r\sin\theta = -2$

$$\Rightarrow \quad \tan\theta = -1 \quad \left[\because \tan\theta = \frac{\sin\theta}{\cos\theta}\right]$$

Now, let $\tan\alpha = |\tan\theta|$, where $0 < \alpha < \frac{\pi}{2}$

$$\Rightarrow \quad \tan\alpha = 1 = \tan\frac{\pi}{4}$$

$$\Rightarrow \quad \alpha = \frac{\pi}{4}$$

Since, z lies in IV quadrant.

$$\therefore \quad \arg(z) = -\alpha = -\frac{\pi}{4}$$

Hence, polar form of z is $2\sqrt{2}\left[\cos\left(-\frac{\pi}{4}\right) + \sin\left(-\frac{\pi}{4}\right)\right]$.

Quadratic Equation

An equation of the form $ax^2 + bx + c, a \neq 0$ is called quadratic equation in variable x, where a, b and c are numbers (real or complex).

Nature of Roots of Quadratic Equation

The roots of quadratic equation $ax^2 + bx + c = 0, a \neq 0$ are

$$\alpha = \frac{-b + \sqrt{b^2 - 4ac}}{2a}$$

and $$\beta = \frac{-b - \sqrt{b^2 - 4ac}}{2a}.$$

Now, if we look at these roots of quadratic equation $ax^2 + bx + c = 0; a \neq 0$, we observe that the roots depend upon the value of the quantity $b^2 - 4ac$. This quantity is known as the **discriminant** of the quadratic equation and denoted by D.

There are following cases :

***Case* I** **If $b^2 - 4ac = 0$** i.e. $D = 0$, then $\alpha = \beta = -\frac{b}{2a}$.

Thus, if $b^2 - 4ac = 0$, then the quadratic equation has **real and equal roots** and each equal to $-b/2a$.

***Case* II** If a, b and c are rational numbers and $b^2 - 4ac > 0$ and it is a perfect square, then $D = \sqrt{b^2 - 4ac}$ is a rational number and hence α and β are **rational and unequal.**

***Case* III** If $b^2 - 4ac > 0$ and it is not a perfect square, then roots are irrational and unequal.

***Case* IV** If $b^2 - 4ac < 0$, then the roots are **complex conjugate of each other.**

Quadratic Equations with Real Coefficients

Let us consider the following quadratic equation $ax^2 + bx + c = 0$ with real coefficients a, b, c and $a \neq 0$. Also, let us assume that $b^2 - 4ac < 0$. Now, we can find the square root of negative real numbers in the set of complex numbers. Therefore, the solutions of the above equation are available in the set of complex numbers which are given by

$$x = \frac{-b \pm \sqrt{b^2 - 4ac}}{2a} = \frac{-b \pm \sqrt{-(4ac - b^2)}}{2a}$$

$$= \frac{-b \pm \sqrt{4ac - b^2}\, i}{2a}$$

Example 5 Solve $2x^2 - 2\sqrt{3}\,x + \frac{21}{8} = 0$.

Sol. We have, $$2x^2 - 2\sqrt{3}\,x + \frac{21}{8} = 0 \quad \text{...(i)}$$

On comparing Eq. (i) with $ax^2 + bx + c = 0$, we get

$$a = 2,\ b = -2\sqrt{3} \text{ and } c = \frac{21}{8}$$

$$\because \quad x = \frac{-b \pm \sqrt{b^2 - 4ac}}{2a}$$

$$\therefore \quad x = \frac{-(-2\sqrt{3}) \pm \sqrt{(-2\sqrt{3})^2 - 4 \times 2 \times \frac{21}{8}}}{2 \times 2}$$

$$= \frac{2\sqrt{3} \pm \sqrt{12-21}}{4} = \frac{2\sqrt{3} \pm \sqrt{-9}}{4}$$

$$= \frac{2\sqrt{3} \pm 3i}{4} = \frac{\sqrt{3}}{2} \pm \frac{3}{4}i \quad [\because \sqrt{-1} = i]$$

Hence, the roots are $\frac{\sqrt{3}}{2} + \frac{3}{4}i$ and $\frac{\sqrt{3}}{2} - \frac{3}{4}i$.

Example 6 **Solve $\sqrt{3}\,x^2 - \sqrt{2}\,x + 3\sqrt{3} = 0$.**

Sol. We have, $\sqrt{3}x^2 - \sqrt{2}x + 3\sqrt{3} = 0$ ...(i)

On comparing Eq. (i) with $ax^2 + bx + c = 0$, we get

$$a = \sqrt{3},\ b = -\sqrt{2} \text{ and } c = 3\sqrt{3}$$

On putting these values in $x = \frac{-b \pm \sqrt{b^2 - 4ac}}{2a}$, we get

$$x = \frac{-(-\sqrt{2}) \pm \sqrt{(-\sqrt{2})^2 - 4 \times \sqrt{3} \times 3\sqrt{3}}}{2\sqrt{3}}$$

$$= \frac{\sqrt{2} \pm \sqrt{2-36}}{2\sqrt{3}} = \frac{\sqrt{2} \pm \sqrt{-34}}{2\sqrt{3}} = \frac{\sqrt{2} \pm i\sqrt{34}}{2\sqrt{3}}$$

$[\because \sqrt{-1} = i]$

Hence, the roots are $\frac{\sqrt{2}}{2\sqrt{3}} + \frac{i\sqrt{34}}{2\sqrt{3}}$ and $\frac{\sqrt{2}}{2\sqrt{3}} - i\frac{\sqrt{34}}{2\sqrt{3}}$.

Example 7 **Solve $(y+1)(y-3)+7=0$.**

Sol. We have, $(y+1)(y-3)+7=0$

$$\Rightarrow \quad y^2 - 2y - 3 + 7 = 0 \Rightarrow y^2 - 2y + 4 = 0$$

On comparing with $ay^2 + by + c = 0$, we get

$$a = 1,\ b = -2 \text{ and } c = 4$$

$$\because \quad y = \frac{-b \pm \sqrt{b^2 - 4ac}}{2a}$$

$$\therefore \quad y = \frac{2 \pm \sqrt{(-2)^2 - 4 \times 1 \times 4}}{2 \times 1} = \frac{2 \pm \sqrt{4-16}}{2}$$

$$= \frac{2 \pm \sqrt{-12}}{2} = \frac{2 \pm 2\sqrt{3}\,i}{2} = 1 \pm \sqrt{3}\,i \quad [\because \sqrt{-1} = i]$$

$$\therefore \quad y = 1 + \sqrt{3}\,i \text{ or } y = 1 - \sqrt{3}\,i$$

Hence, the roots of the given equation are $1 + \sqrt{3}\,i$ and $1 - i\sqrt{3}$.

Quadratic Equations with Complex Coefficients

Let a, b and c be complex numbers, such that $a \neq 0$. Then, the equation $ax^2 + bx + c = 0$ is called quadratic equation with complex coefficients and its roots are given by

$$\alpha = \frac{-b + \sqrt{b^2 - 4ac}}{2a} \text{ and } \beta = \frac{-b - \sqrt{b^2 - 4ac}}{2a}$$

Here, α and β are complex numbers as a, b and c are complex numbers. Since, complex numbers cannot be compared, so the sign of the discriminant $b^2 - 4ac$ cannot be determined. While the roots are equal, if $b^2 - 4ac = 0$ and unequal, if $b^2 - 4ac \neq 0$.

Example 8 **Solve $x^2 - 5ix - 6 = 0$.**

Sol. Given, $x^2 - 5ix - 6 = 0$ ... (i)

On comparing Eq. (i) with $ax^2 + bx + c = 0$, we get

$a = 1$, $b = -5i$ and $c = -b$

$$\because \quad \alpha = \frac{-b + \sqrt{b^2 - 4ac}}{2a} \text{ and } \beta = \frac{-b - \sqrt{b^2 - 4ac}}{2a}$$

$$\therefore \quad \alpha = \frac{5i + \sqrt{(-5i)^2 - 4 \times 1 \times (-6)}}{2 \times 1}$$

$$\text{and} \quad \beta = \frac{5i - \sqrt{(-5i)^2 - 4 \times 1 \times (-6)}}{2 \times 1}$$

$$\Rightarrow \quad \alpha = \frac{5i + \sqrt{-25+24}}{2} \text{ and } \beta = \frac{5i - \sqrt{-25+24}}{2}$$

$$\Rightarrow \quad \alpha = \frac{5i + \sqrt{-1}}{2} \text{ and } \beta = \frac{5i - \sqrt{-1}}{2}$$

$$\Rightarrow \quad \alpha = \frac{5i + i}{2} \text{ and } \beta = \frac{5i - i}{2} \Rightarrow \alpha = \frac{6i}{2} \text{ and } \beta = \frac{4i}{2}$$

$$\therefore \quad \alpha = 3i \text{ and } \beta = 2i$$

Exercise

Objective Type Questions

1 Representation of $z = x + iy$ in terms of r and θ is called

(a) polar form of the complex number
(b) cartesian form of the complex number
(c) complex form of the complex number
(d) real form of the complex number

2 If $x = r\cos\theta$, $y = r\sin\theta$ such that $x, y > 0$ and $z = r(\cos\theta + i\sin\theta)$, then

I. $|z| = r = \sqrt{x^2 - y^2}$ II. $|z| = r = \sqrt{x^2 + y^2}$

III. $\arg(z) = \theta$

(a) I and III are correct (b) II and III are correct
(c) All are correct (d) None of these

3 Polar form of the complex number $z = 1 + i\sqrt{3}$, is

(a) $2\left(\cos\frac{\pi}{3} - i\sin\frac{\pi}{3}\right)$ (b) $2\left(\cos\frac{\pi}{3} + i\sin\frac{\pi}{3}\right)$
(c) $2\left(\cos\frac{2\pi}{3} + i\sin\frac{2\pi}{3}\right)$ (d) $2\left(\cos\frac{4\pi}{3} + i\sin\frac{4\pi}{3}\right)$

4 The value of arg (x), when $x < 0$ is

(a) 0 (b) $\frac{\pi}{2}$ (c) π (d) None of these

Very Short Answer Type Questions

5 Find the principal argument of the following complex numbers.

(i) $1+i\sqrt{3}$ (ii) $-1-i\sqrt{3}$

(iii) $-\sqrt{3}+i$ (iv) $\frac{1}{1+i}$

(v) $\frac{1+2i}{1-3i}$

6 Find z, if $|r| = 4$ and $\arg(z) = \frac{5\pi}{6}$.

Short Answer Type Questions

7 If $z_1 = \sqrt{2} - 3i$ and $z_2 = 5 - i\sqrt{2}$, then find the quadrant in which $\frac{z_1}{z_2}$ lies.

8 Find the modulus and principal argument of $-\sqrt{3} + i$.

9 Find the modulus and argument of complex number $\frac{1+i}{1-i}$.

10 Find the argument of the complex number $\frac{1+3i}{1-2i}$.

Long Answer Type Questions

11 Convert each of the following complex numbers in polar form.

(i) $1-i$ (ii) $-1+i$

(iii) $\frac{1-i}{1+i}$

12 Write the complex number $\sqrt{3} + i$ into polar form and determine the modulus and the principal value of the argument.

13 Change the complex number $-4 + i4\sqrt{3}$ in polar form.

14 Convert the complex number $\frac{-16}{1+i\sqrt{3}}$ into polar form.

15 Express the complex number $\frac{2+6\sqrt{3}i}{5+\sqrt{3}i}$ into polar form.

16 Convert the complex number $z = \frac{i-1}{\cos\frac{\pi}{3} + i\sin\frac{\pi}{3}}$ into polar form.

Directions (Q. Nos. 18-23) *Solve the following equations.*

17 $x^2 + 4 = 0$

18 $x^2 + 2 = 0$

19 $\sqrt{2}x^2 + x + \sqrt{2} = 0$ [NCERT]

20 $x^2 + 5x + \frac{25}{2} = 0$

21 $x^2 + x + \frac{1}{\sqrt{2}} = 0$

22 $21x^2 - 28x + 10 = 0$

23 $5x^2 - 4ix + 9 = 0$

Answers

1. (a) **2.** (b) **3.** (b) **4.** (c)

5. (i) $\arg(z) = \frac{\pi}{3}$ (ii) $\frac{-2\pi}{3}$ (iii) $\frac{5\pi}{3}$

(iv) $\frac{-\pi}{4}$ (v) $\frac{3\pi}{4}$

6. $-2\sqrt{3} + 2i$ **7.** IVth quadrant

8. $2, \frac{5\pi}{6}$ **9.** $1, \frac{\pi}{2}$ **10.** $\frac{3\pi}{4}$

11.

(i) $z = \sqrt{2}\left[\cos\left(\frac{-\pi}{4}\right) + i\sin\left(\frac{-\pi}{4}\right)\right]$

(ii) $z = \sqrt{2}\left[\cos\frac{3\pi}{4} + i\sin\frac{3\pi}{4}\right]$

(iii) $z = 1\left[\cos\left(\frac{-\pi}{2}\right) + i\sin\left(\frac{-\pi}{2}\right)\right]$

12. $2\left(\cos\frac{\pi}{6} + i\sin\frac{\pi}{6}\right); 2, \frac{\pi}{6}$

13. $8\left(\cos\frac{2\pi}{3} + i\sin\frac{2\pi}{3}\right)$ **14.** $8\left(\cos\frac{2\pi}{3} + i\sin\frac{2\pi}{3}\right)$

15. $2\left(\cos\frac{\pi}{3} + i\sin\frac{\pi}{3}\right)$ **16.** $\sqrt{2}\left(\cos\frac{5\pi}{12} + i\sin\frac{5\pi}{12}\right)$

17. $x = \pm 2i$ **18.** $x = \pm\sqrt{2}i$

19. $x = \frac{-1 \pm \sqrt{7}\,i}{2\sqrt{2}}$ **20.** $x = -\frac{5}{2} \pm i\frac{5}{2}$

21. $x = \frac{-1 \pm i\sqrt{\sqrt{4-\sqrt{2}}}}{2}$

22. $x = \frac{2}{3} + i\frac{\sqrt{14}}{21}$ **23.** $\frac{4i \pm 14i}{10}$

05

CHAPTER

> Mathematics is the most beautiful and most powerful creation of the human being. — *Stefan Banach*

Linear Inequalities

YOUR STUDY GOALS

YOUR PRACTICE GOALS

- Objective Type Questions (27 Qs; 17 MCQs, 8 A-R, 2 Case Based)
- Subjective Type Questions (25 Qs; 14 VSA, 6 SA, 5 LAs)

5.1 Inequality

A statement involving the symbols $>, <, \leq$ or $\geq$ is called an **inequality** or **inequation**.

Here, the symbols $<$ (less than), $>$ (greater than), $\leq$ (less than or equal to) and $\geq$ (greater than or equal to) are known as symbols of **inequalities**.

e.g. $5 < 7, x \leq 2, x + y \geq 11$

Linear Inequality

An inequality is said to be linear, if the variable(s) occurs in first degree only and there is no term involving the product of the variables.

e.g. $ax + b \leq 0,\ ax + by + c > 0,\ ax \leq 4.$

Linear Inequality in One Variable

A linear inequality which has only one variable, is called linear inequality in one variable.

e.g. $ax + b < 0$, where $a \neq 0$

Linear Inequality in Two Variables

A linear inequality which have only two variables, is called linear inequality in two variables.

e.g. $3x + 11y \leq 0,\ 4x + 3y > 0$

5.2 Intervals on a Number Line

Closed Interval

If a and b are real numbers such that $a < b$, then the set of all real numbers x such that $a \leq x \leq b$, is called a closed interval and is denoted by $[a, b]$.

$\therefore \quad [a, b] = \{x : a \leq x \leq b, x \in R\}$

On the number line, $[a, b]$ may be represented as follows

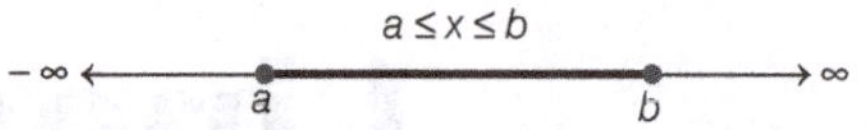

Here, end points of the interval i.e. a and b are included in the interval. So, on number line, draw filled circle (•) at a and b.

Open Interval

If a and b are real numbers such that $a < b$, then the set of all real numbers x such that $a < x < b$, is called an open interval and is denoted by (a, b) or $]a, b[$.

$\therefore \quad (a, b) = \{x : a < x < b, x \in R\}$

On the number line, (a, b) may be represented as follows

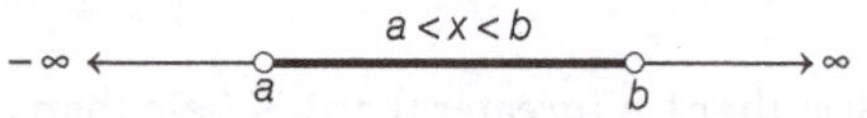

Here, end points of the interval i.e. a and b are not included in the interval. So, on number line, draw open circle (o) at a and b.

Semi-open or Semi-closed Intervals

If a and b are real numbers such that $a < b$,

then $(a, b] = \{x : a < x \leq b, x \in R\}$

and $[a, b) = \{x : a \leq x < b, x \in R\}$

are known as semi-open or semi-closed intervals.

On the number line, these intervals may be represented as follows

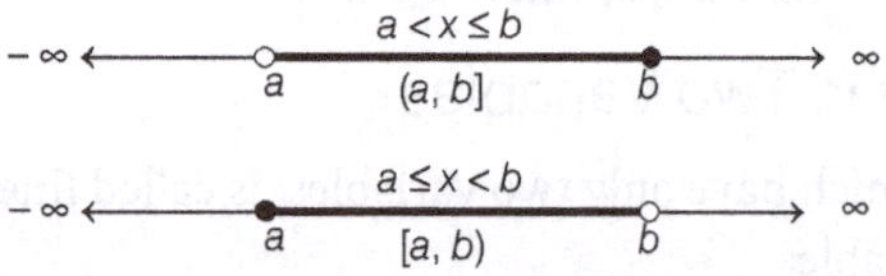

5.3 Solution of an Inequality

Any solution of an inequality is the value(s) of variable(s) which makes it a true statement. We solve inequalities with systematic technique.

Some properties or rules which are used to solve the inequalities, are given below.

Addition or Subtraction

Some number may be added (or subtracted) to (from) both sides of an inequality i.e. if $a > b$, then for any number c,

$$a + c > b + c \text{ or } a - c > b - c$$

e.g. (i) $10 > 5 \Rightarrow 10 + 7 > 5 + 7$ [adding 7 on both sides]

$\Rightarrow 17 > 12$, which is true.

(ii) $-8 > -13 \Rightarrow -8 - 2 > -13 - 2$

[subtracting 2 from both sides]

$\Rightarrow -10 > -15$, which is true.

Multiplication or Division

If both sides of an inequality are multiplied (or divided) by the same positive number, then the sign of inequality remains the same. But when both sides are multiplied (or divided) by the same negative number, then the sign of inequality is reversed.

Let a, b and c be three real numbers such that $a > b$ and $c \neq 0$.

- If $c > 0$, then $\frac{a}{c} > \frac{b}{c}$ and $ac > bc$.
- If $a > b$ and $c < 0$, then $\frac{a}{c} < \frac{b}{c}$ and $ac < bc$.

e.g. (i) $6 > 4 \Rightarrow 6 \times 2 > 4 \times 2$

[multiplying both sides by 2]

$\Rightarrow \quad 12 > 8$, which is true.

(ii) $-32 < -24 \Rightarrow \frac{-32}{-8} > \frac{-24}{-8}$

[dividing both sides by -8]

$\Rightarrow 4 > 3$, which is true.

Solution Set

The set of all solutions of an inequality is called the solution set of the inequality.

Example 1. **Solve the linear inequality $3x - 5 < x + 7$, when** *NCERT Exemplar*

(i) x is a natural number. (ii) x is a whole number.
(iii) x is an integer. (iv) x is a real number.

Sol. We have, $3x - 5 < x + 7$

$\Rightarrow \quad 3x - 5 + 5 < x + 7 + 5$ [adding 5 both sides]

$\Rightarrow \quad 3x < x + 12$

$\Rightarrow \quad 3x - x < x + 12 - x$ [subtracting x from both sides]

$\Rightarrow \quad 2x < 12$

$\Rightarrow \quad \frac{2x}{2} < \frac{12}{2}$ [dividing both sides by 2]

$\Rightarrow \quad x < 6$

Now, if

(i) x is a natural number, then the solution set is $\{1, 2, 3, 4, 5\}$.

(ii) x is a whole number, then the solution set is $\{0, 1, 2, 3, 4, 5\}$.

(iii) x is an integer, then the solution set is $\{..., -3, -2, -1, 0, 1, 2, 3, 4, 5\}$

(iv) x is a real number, then the solution set is $\{x : x \in R \text{ and } x < 6\}$ i.e. any real number less than 6. This can also be written as $(-\infty, 6)$.

Example 2. **Solve the linear inequality $-5x + 25 > 0$.**

Sol. We have, $-5x + 25 > 0$

On adding $5x$ to both sides, we get

$-5x + 25 + 5x > 0 + 5x$

$\Rightarrow \quad 25 > 5x \Rightarrow 5x < 25$

On dividing both sides by 5, we get

$\Rightarrow \quad \frac{5x}{5} < \frac{25}{5} \Rightarrow x < 5$

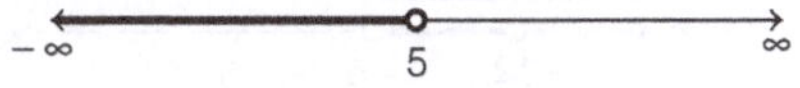

Hence, the required solution set is $(-\infty, 5)$.

Example 3. **Solve the inequality $\frac{5 - 3x}{3} \le \frac{x}{6} - 5$.**

Sol. We have, $\frac{5 - 3x}{3} \le \frac{x}{6} - 5$

$\Rightarrow \quad \frac{5 - 3x}{3} \le \frac{x - 30}{6}$

$\Rightarrow \quad \frac{5 - 3x}{3} \times 6 \le \frac{x - 30}{6} \times 6$ [multiplying both sides by 6]

$\Rightarrow \quad 2(5 - 3x) \le x - 30$

$\Rightarrow \quad 10 - 6x \le x - 30$

$\Rightarrow \quad 10 - 6x - 10 \le x - 30 - 10$ [subtracting 10 from both sides]

$\Rightarrow \quad -6x \le x - 40$

$\Rightarrow \quad -6x - x \le x - 40 - x$ [subtracting x from both sides]

$\Rightarrow \quad -7x \le -40$

$\Rightarrow \quad 7x \ge 40$ [multiplying by -1 on both sides]

$\Rightarrow \quad \frac{7x}{7} \ge \frac{40}{7}$ [dividing by 7 on both sides]

$\Rightarrow \quad x \ge \frac{40}{7}$

i.e. $\quad x \in \left[\frac{40}{7}, \infty\right)$

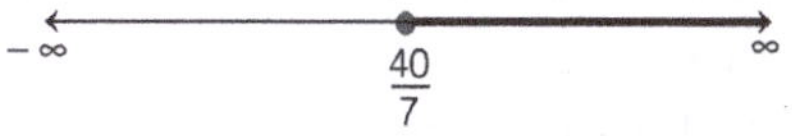

Hence, the required solution set is $\left[\frac{40}{7}, \infty\right)$.

Example 4. **Solve the following inequalities and show that the graph of the solution in each case on number line.**

(i) $\frac{1}{x - 4} < 0$ (ii) $\frac{x + 1}{x + 3} \ge 1$

(iii) $\frac{x}{4} < \frac{5x - 2}{3} - \frac{7x - 3}{5}$

(iv) $\frac{(2x - 1)}{3} \le \frac{(3x - 2)}{4} - \frac{(2 - x)}{5}$

Sol. (i) We have, $\frac{1}{x - 4} < 0$

$\Rightarrow \quad x - 4 < 0 \quad \left[\because \frac{a}{b} < 0 \text{ and } a > 0 \Rightarrow b < 0\right]$

$\Rightarrow \quad x < 4$

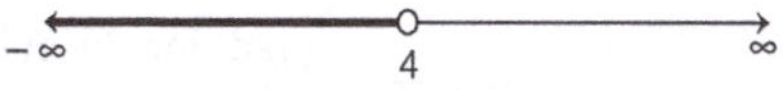

$\therefore \quad x \in (-\infty, 4)$

(ii) We have, $\frac{x + 1}{x + 3} \ge 1 \Rightarrow \frac{x + 1}{x + 3} - 1 \ge 1 - 1$ [subtracting 1 from both sides]

$\Rightarrow \quad \frac{x + 1 - x - 3}{x + 3} \ge 0 \Rightarrow \frac{-2}{x + 3} \ge 0$

$\Rightarrow \quad \frac{2}{x + 3} \le 0$ [multiplying both sides by -1]

$\Rightarrow \quad x + 3 < 0 \quad \left[\because \frac{a}{b} \le 0 \text{ and } a > 0 \Rightarrow b < 0\right]$

$\Rightarrow \quad x < -3$

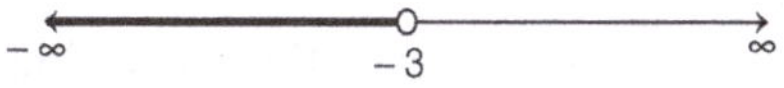

$\therefore \quad x \in (-\infty, -3)$

(iii) We have, $\frac{x}{4} < \frac{5x-2}{3} - \frac{7x-3}{5}$

$\Rightarrow \quad \frac{x}{4} < \frac{5(5x-2)-3(7x-3)}{15}$

$\Rightarrow \quad \frac{x}{4} < \frac{25x-10-21x+9}{15}$

$\Rightarrow \quad 60\left(\frac{x}{4}\right) < \frac{(4x-1)}{15} \times 60$

[multiplying both sides by LCM (4, 15) = 60]

$\Rightarrow \quad 15x < 4(4x-1)$

$\Rightarrow \quad 15x < 16x - 4$

$\Rightarrow \quad 15x + 4 < 16x - 4 + 4$ [adding 4 on both sides]

$\Rightarrow \quad 15x + 4 < 16x$

$\Rightarrow 15x + 4 - 15x < 16x - 15x$

[subtracting $15x$ from both sides]

$\Rightarrow \quad 4 < x$ or $x > 4$

$\therefore \quad x \in (4, \infty)$

On number line, it can be represented as

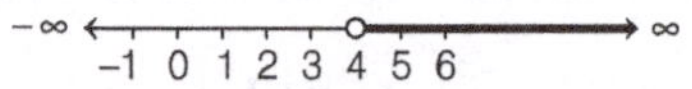

Here, the dark portion on the number line represents the solution of inequality.

(iv) We have, $\frac{2x-1}{3} \le \frac{3x-2}{4} - \frac{2-x}{5}$

On multiplying both sides by LCM of 3, 4 and 5 i.e. 60, we get

$\frac{2x-1}{3} \times 60 \le \frac{3x-2}{4} \times 60 - \frac{2-x}{5} \times 60$

$\Rightarrow \quad 20(2x-1) \le 15(3x-2) - 12(2-x)$

$\Rightarrow \quad 40x - 20 \le 45x - 30 - 24 + 12x$

$\Rightarrow \quad 40x - 20 \le 57x - 54$

$\Rightarrow \quad 40x - 20 + 20 \le 57x - 54 + 20$

[adding 20 on both sides]

$\Rightarrow \quad 40x \le 57x - 34$

$\Rightarrow \quad 40x - 57x \le 57x - 34 - 57x$

[subtracting $57x$ from both sides]

$\Rightarrow \quad -17x \le -34$

$\Rightarrow \quad \frac{-17x}{17} \le \frac{-34}{17}$ [dividing both sides by 17]

$\Rightarrow \quad -x \le -2 \Rightarrow x \ge 2$

[multiplying both sides by –1]

Thus, the solution set is $[2, \infty)$.

On number line, it can be represented as

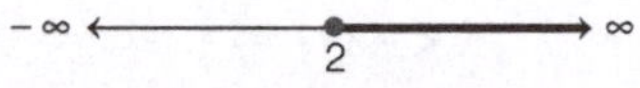

Here, the dark portion on the number line represents the solution of inequality.

Example 5. Solve the inequality $\frac{x-2}{x+5} > 2$.

Also, represent the solution on the number line.

Sol. We have, $\frac{x-2}{x+5} > 2$

On subtracting 2 from both sides, we get

$\frac{x-2}{x+5} - 2 > 0$

$\Rightarrow \quad \frac{x-2-2(x+5)}{x+5} > 0$

$\Rightarrow \quad \frac{x-2-2x-10}{x+5} > 0$

$\Rightarrow \quad \frac{-(x+12)}{x+5} > 0 \Rightarrow \frac{x+12}{x+5} < 0$

[multiplying both sides by –1]

On multiplying both sides by $(x+5)^2$, we get

$\frac{(x+12)}{(x+5)} \times (x+5)^2 < 0 \times (x+5)^2$

$\Rightarrow \quad (x+12)(x+5) < 0$

So, the factors $(x+12)$ and $(x+5)$ have opposite signs.

$\therefore \quad (x+12) > 0$ and $(x+5) < 0$ ...(i)

or $\quad (x+12) < 0$ and $(x+5) > 0$ ...(ii)

From inequality (i), we get

$x+12 > 0$ and $x+5 < 0$

$\Rightarrow \quad x+12-12 > -12$

and $\quad x+5-5 < 0-5$

[subtracting 12 from first inequality and 5 from second]

$\Rightarrow \quad x > -12$ and $x < -5$ ...(iii)

On number line, it can be represented as

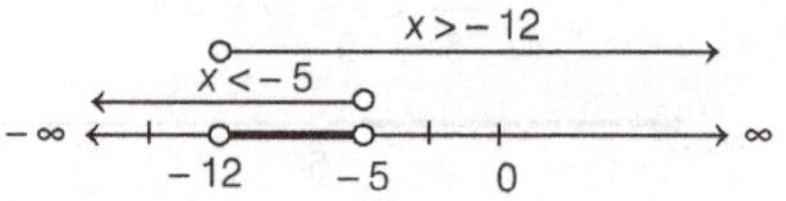

$\therefore \quad -12 < x < -5$ ...(iv)

From inequality (ii), we get

$(x+12) < 0$ and $(x+5) > 0$

$\Rightarrow \quad x+12-12 < 0-12$ and $x+5-5 > -5$

[subtracting 12 from first inequality and 5 from second]

$\Rightarrow \quad x < -12$ and $x > -5$

On number line, it can be represented as

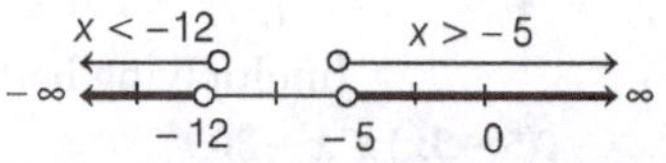

Since, no common point occurs, so it has no solution.

Hence, the solution set of the given inequality is $-12 < x < -5$ i.e. $(-12, -5)$.

Example 6. **Solve the inequality $\dfrac{x-4}{x+2} \le 2$. Also, represents the solution on number line.**

Sol. We have, $\dfrac{x-4}{x+2} \le 2 \Rightarrow \dfrac{x-4}{x+2} - 2 \le 2 - 2$

[subtracting 2 from both sides]

$\Rightarrow \quad \dfrac{x-4}{x+2} - 2 \le 0 \Rightarrow \dfrac{x-4-2(x+2)}{x+2} \le 0$

$\Rightarrow \quad \dfrac{x-4-2x-4}{x+2} \le 0$

$\Rightarrow \quad \dfrac{-x-8}{x+2} \le 0$, here $x+2 \ne 0$ i.e. $x \ne -2$

$\Rightarrow \quad \dfrac{x+8}{x+2} \ge 0$ [multiplying both sides by -1]

$\Rightarrow \quad \dfrac{x+8}{x+2} \times (x+2)^2 \ge 0 \times (x+2)^2$

[multiplying both sides by $(x+2)^2$]

$\Rightarrow \quad (x+8)(x+2) \ge 0$

So, the factors $(x+8)$ and $(x+2)$ have the same signs.

Case I When both are non-negative

In this case, we have

$x+8 \ge 0$ and $x+2 > 0$

[$\because x+2 \ne 0$ because for $x=-2$, denominator of given inequality becomes 0]

$\Rightarrow \quad x \ge -8$ and $x > -2$

On number line, these can be represented as

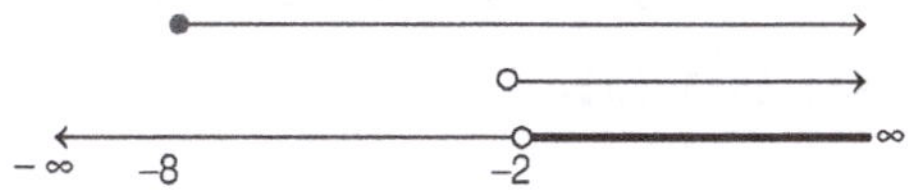

Clearly, the common values of x are lying in the interval $(-2, \infty)$.

$\therefore$ The solution set is $(-2, \infty)$.

Case II When both are non-positive

In this case, we have

$x+8 \le 0$ and $x+2 < 0$

$\Rightarrow \quad x \le -8$ and $x < -2$

On number line, these can be represented as

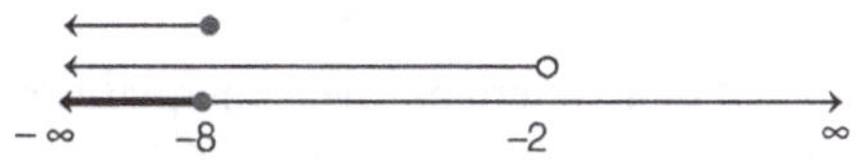

Clearly, the common values of x are lying in the interval $(-\infty, -8]$.

$\therefore$ The solution set is $(-\infty, -8]$.

Hence, the required solution set is $(-\infty, -8] \cup (-2, \infty)$ and this can be represented on the number line as

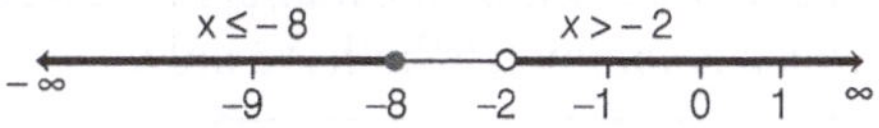

Quick Test 1

1. Solve $-12x > 30$, when

(i) x is a natural number. (ii) x is an integer. NCERT

2. Solve the given inequality for x is a real number, $3(x-1) \le 2(x-3)$

3. Solve for real x, $2(2x+3) - 10 < 6(x-2)$ NCERT

Directions (Q. Nos. 4-6) *Solve the inequalities for real x and also represent the graph of solution on number line.*

4. $\dfrac{x}{3} > \dfrac{x}{2} + 1$ **5.** $37 - (3x+5) \ge 9x - 8(x-3)$ NCERT

6. $\dfrac{x}{2} \ge \dfrac{(5x-2)}{3} - \dfrac{(7x-3)}{5}$ NCERT

7. Solve inequality $\dfrac{x+1}{x+2} \ge 1$

8. Solve inequality $\dfrac{x+8}{x-2} \ge 0$

9. Solve inequality $\dfrac{2x+5}{x-1} > 5$

10. Solve inequality $\dfrac{x}{x-5} > \dfrac{1}{2}$

Answers

1. (i) No solution (ii) $\{\ldots, -4, -3\}$
2. $(-\infty, -3]$
3. $(4, \infty)$
4. $(-\infty, -6)$ 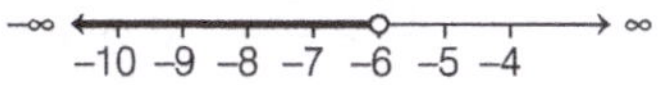

5. $(-\infty, 2]$ 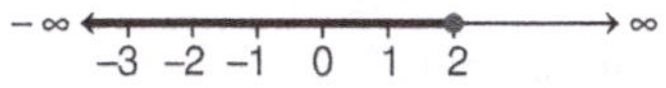

6. $\left[-\dfrac{2}{7}, \infty\right)$ 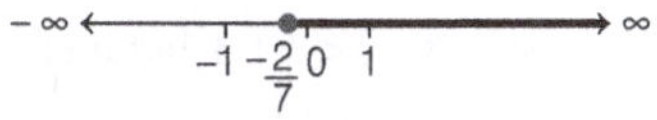

7. $(-\infty, -2)$
8. $(-\infty, -8] \cup (2, \infty)$
9. $\left(1, \dfrac{10}{3}\right)$
10. $(-\infty, -5) \cup (5, \infty)$

For Solutions, See QR Code on Pg. 145

5.4 System of Inequalities in One Variable and Their Solutions

Two or more inequalities taken together comprise a system of inequalities and the solution of the system of inequalities are the solutions common to all the inequalities comprising the system.

e.g. $x = 10$ is the solution of the system of inequalities

$$4x + 3 \le 91 \text{ and } 2x \ge x + 8$$

Solution of System of Linear Inequalities in One Variable

The solution set of a system of linear inequalities in one variable is defined as the **intersection of the solution set** of the linear inequalities in the system.

e.g. If the solution sets of linear inequalities in the system are $(-\infty, 5]$ and $[5, \infty)$, then the solution of system of linear inequalities in one variable is 5 only. Because if we represent the solution sets on the number line, we see that the value which is common to both is 5 only.

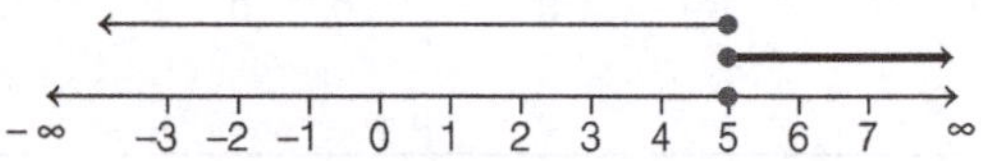

Type 1 When two separate linear inequalities are given.

Type 2 When inequalities of the form $a \le \frac{cx+d}{e} \le b$, where $a, b, c, d \in R$ are given.

Example 1. **Solve the following system of inequalities $2x - 3 < 7$ and $2x > -4$. Also, represent the solution graphically on the number line.**

Sol. We have, the following inequalities

$$2x - 3 < 7 \quad \text{...(i)}$$

and $$2x > -4 \quad \text{...(ii)}$$

From inequality (i), we get

$$2x - 3 < 7$$

$\Rightarrow$ $2x - 3 + 3 < 7 + 3$ [adding 3 on both sides]

$\Rightarrow$ $2x < 10$

$\Rightarrow$ $\frac{2x}{2} < \frac{10}{2}$ [dividing both sides by 2]

$\Rightarrow$ $x < 5$

$\therefore$ The solution set is $(-\infty, 5)$.

From inequality (ii), we get

$$2x > -4$$

$\Rightarrow$ $\frac{2x}{2} > \frac{-4}{2}$ [dividing both sides by 2]

$\Rightarrow$ $x > -2$

$\therefore$ The solution set is $(-2, \infty)$.

Now, let us draw the graphs of the solutions of both inequalities on number line.

Graph for Solution set $(-\infty, 5)$.

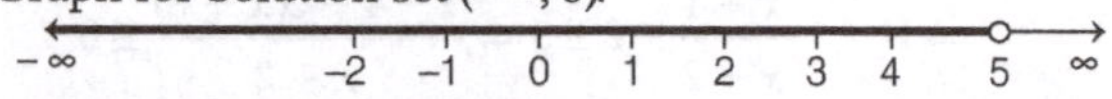

Graph for solution set $(-2, \infty)$

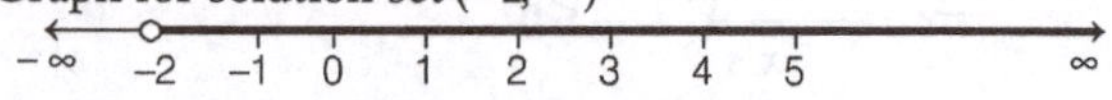

It can be seen that the values of x, which are common to both are lying in the interval $(-2, 5)$.

Hence, the solution set of given system of inequations is $(-2, 5)$ and this can be represented graphically on the number line as

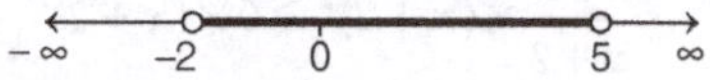

Example 2. **Solve the following system of inequalities and represent the solution graphically on the number line.**

$$5(2x - 7) - 3(2x + 3) \le 0, 2x + 19 \le 6x + 47$$ NCERT

Sol. We have, the following inequalities

$$5(2x - 7) - 3(2x + 3) \le 0 \quad \text{...(i)}$$

and $$2x + 19 \le 6x + 47 \quad \text{...(ii)}$$

From inequality (i), we get

$$5(2x - 7) - 3(2x + 3) \le 0$$

$\Rightarrow$ $10x - 35 - 6x - 9 \le 0 \Rightarrow 4x - 44 \le 0$

$\Rightarrow$ $4x - 44 + 44 \le 44$ [adding 44 on both sides]

$\Rightarrow$ $4x \le 44 \Rightarrow x \le 11$ ...(iii) [dividing both sides by 4]

$\therefore$ The solution set is $(-\infty, 11]$.

From inequality (ii), we get

$$2x + 19 \le 6x + 47$$

$\Rightarrow$ $2x + 19 - 2x \le 6x + 47 - 2x$ [subtracting $2x$ from both sides]

$\Rightarrow$ $19 \le 4x + 47$

$\Rightarrow$ $19 - 47 \le 4x + 47 - 47$ [subtracting 47 from both sides]

$\Rightarrow$ $-28 \le 4x$

or $4x \ge -28$

$\Rightarrow$ $\frac{4x}{4} \ge \frac{-28}{4}$ [dividing both sides by 4]

$\Rightarrow$ $x \ge -7$

$\therefore$ The solution set is $[-7, \infty)$.

Now, let us draw the graphs of the solutions of both inequalities on number line.

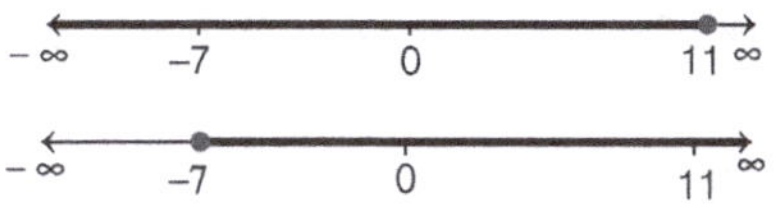

It can be seen that the values of x, which are common to both are lying in the interval $[-7, 11]$.

Hence, the solution set of given system of inequations is $[-7, 11]$ and this can be represented graphically on the number line as

Example 3. **Solve the inequalities $-3 \le 4 - \frac{7x}{2} \le 18$.** NCERT

Sol. We have, $-3 \le 4 - \frac{7x}{2} \le 18$

On subtracting 4 from each term, we get

$$-3-4 \le 4 - \frac{7x}{2} - 4 \le 18 - 4 \Rightarrow -7 \le -\frac{7x}{2} \le 14$$

On multiplying each term by $\left(\frac{-2}{7}\right)$, we get

$$-7\left(-\frac{2}{7}\right) \ge \frac{-7}{2} x \times \left(\frac{-2}{7}\right) \ge 14 \times \left(\frac{-2}{7}\right)$$

[multiplying each term by the same negative number, the sign of inequalities will get changed]

$\Rightarrow \quad 2 \ge x \ge -4$ or $-4 \le x \le 2$ or $x \in [-4, 2]$

Hence, the solution set of given system of inequations is $[-4, 2]$.

Example 4. **Find all pairs of consecutive even positive integers, both of which are larger than 5 such that their sum is less than 23.**

Sol. Let x be the smaller of the two positive even integers so that the other one is $x + 2$. Then, we should have

$$x > 5 \quad \text{...(i)}$$

$$x + 2 > 5 \quad \text{...(ii)}$$

and their sum, $x + (x + 2) < 23$ ...(iii)

From inequalities (i) and (ii), we get

$$x > 5 \quad \text{...(iv)}$$

From inequality (iii), we get

$$2x + 2 < 23$$

$\Rightarrow \quad 2x + 2 - 2 < 23 - 2$

[subtracting 2 from both sides]

$\Rightarrow \quad 2x < 21$

$\Rightarrow \quad x < 10.5$ [dividing both sides by 2]...(v)

Now, from inequalities (iv) and (v), we get $5 < x < 10.5$

Since, x is an even number.

So, x can take the values 6, 8 and 10.

Hence, the required possible pairs will be (6, 8), (8, 10) and (10, 12).

Example 5. **A solution is to be kept between 40°C and 45°C. What is the range of temperature in degree Fahrenheit, if the conversion formula is $F = \frac{9}{5}C + 32$?**

Sol. We have, $40 < C < 45$

On multiplying each term by $\frac{9}{5}$, we get

$$40 \times \frac{9}{5} < \frac{9}{5}C < \frac{9}{5} \times 45$$

$$\Rightarrow \quad 8 \times 9 < \frac{9}{5}C < 9 \times 9$$

$$\Rightarrow \quad 72 < \frac{9}{5}C < 81$$

On adding 32 in each term, we get

$$72 + 32 < \frac{9}{5}C + 32 < 81 + 32$$

$$\Rightarrow \quad 104 < F < 113 \quad \left[\because F = \frac{9}{5}C + 32\right]$$

Hence, the required range of temperature is between 104° F and 113° F.

Example 6. **A man wants to cut three lengths from a single piece of board of length 91 cm, the second length is to be 3 cm longer than the shortest and third length is to be twice as long as the shortest. What are the possible lengths for the shortest board, if third piece is to be atleast 5 cm longer than the second?** NCERT

Sol. Let the length of the shortest piece be x cm so that the lengths of second and third piece are $(x + 3)$ cm and $2x$ cm, respectively.

Then, $x + (x + 3) + 2x \le 91$ ...(i)

and $2x \ge (x + 3) + 5$ ...(ii)

From inequality (i), we get

$$4x + 3 \le 91 \Rightarrow 4x + 3 - 3 \le 91 - 3$$

[subtracting 3 from both sides]

$\Rightarrow \quad 4x \le 88$

$\Rightarrow \quad \frac{4x}{4} \le \frac{88}{4}$ [dividing both sides by 4]

$\Rightarrow \quad x \le 22$ ...(iii)

From inequality (ii), we get

$$2x \ge x + 8 \Rightarrow 2x - x \ge x + 8 - x$$

[subtracting x from both sides]

$\Rightarrow \quad x \ge 8$ ...(iv)

From inequalities (iii) and (iv), we get

$$8 \le x \le 22$$

Hence, the shortest piece must be atleast 8 cm long but not more than 22 cm long.

Example 7. A solution of 8% boric acid is to be diluted by adding a 2% boric acid solution to it. The resulting mixture is to be more than 4% but less than 6% boric acid. If we have 640 L of the 8% solution, then how many litres of the 2% solution will have to be added? NCERT

Sol. Let x be the number of litres of 2% boric acid solution.
Then, total mixture = $(640 + x)$ L

According to the question,

$$2\% \text{ of } x + 8\% \text{ of } 640 > 4\% \text{ of } (640 + x)$$

$$\Rightarrow \frac{2x}{100} + \frac{8}{100} \times 640 > \frac{4}{100}(640 + x)$$

$$\Rightarrow 2x + 5120 > 2560 + 4x$$

[multiplying both sides by 100]

$$\Rightarrow 2x + 5120 - 2x > 2560 + 4x - 2x$$

[subtracting $2x$ from both sides]

$$\Rightarrow 5120 > 2560 + 2x$$

$$\Rightarrow 5120 - 2560 > 2560 + 2x - 2560$$

[subtracting 2560 from both sides]

$$\Rightarrow 2560 > 2x \Rightarrow 2x < 2560 \Rightarrow x < 1280$$

[dividing both sides by 2] ...(i)

Also, 2% of x + 8% of 640 < 6% of $(640 + x)$

$$\Rightarrow \frac{2}{100} \times x + \frac{8}{100} \times 640 < \frac{6}{100} \times (640 + x)$$

$$\Rightarrow 2x + 5120 < 3840 + 6x$$

[multiplying both sides by 100]

$$\Rightarrow 2x + 5120 - 2x < 3840 + 6x - 2x$$

[subtracting $2x$ from both sides]

$$\Rightarrow 5120 < 3840 + 4x$$

$$\Rightarrow 5120 - 3840 < 3840 + 4x - 3840$$

[subtracting 3840 from both sides]

$$\Rightarrow 1280 < 4x \text{ or } 4x > 1280$$

$$\Rightarrow \frac{4x}{4} > \frac{1280}{4}$$ [dividing both sides by 4]

$$\Rightarrow x > 320 \qquad \text{...(ii)}$$

Now, from inequalities (i) and (ii), we get

$$320 < x < 1280$$

Thus, the number of litres of the 2% boric acid solution will have to be greater than 320 L and less than 1280 L.

Quick Test 2

Directions (Q. Nos. 1-2) *Solve the following system of inequalities.*

1. (i) $3x - 7 > 2(x - 6)$ and $6 - x > 11 - 2x$
(ii) $7(2 - 3x) > 18 - 19x$ and $5 + 3x < 5x + 6$
(iii) $\frac{2x + 1}{7x - 1} > 5$ and $\frac{x + 7}{x - 8} > 2$ Competency Based Que

2. (i) $7 \le \frac{(3x + 11)}{2} \le 11$ NCERT
(ii) $\frac{4}{x + 1} \le 3 \le \frac{6}{x + 1}, x > 0$ NCERT Exemplar

3. Anu needs a minimum of 450 marks in five tests in statistics course to obtain B grade. In her first four tests, she scored 90, 94, 96 and 88 marks. What should her score be in the 5th test so that she can make B grade?

4. Find all pairs of consecutive odd positive integers, both of which are smaller than 18 such that their sum is more than 20. Competency Based Que

5. IQ (Intelligence Quotient) of a person is given by formula, $IQ = \frac{MA}{CA} \times 100$, where MA stands for the mental age of the person and CA for the chronological age. If $80 \le IQ \le 140$ for a group of 12 yr children, then find the range of their mental age. NCERT

6. A company manufactures cassettes and its cost equation for a week is $C = 300 + 1.5x$ and its revenue equation is $R = 2x$, where x is the number of cassettes sold in a week. How many cassettes must be sold for the company to realise a profit? Competency Based Que

7. The longest side of a triangle is 3 times the shortest side and the third side is 2 cm shorter than the longest side. If the perimeter of the triangle is atleast 61 cm, then find the minimum length of the shortest side. NCERT

8. A solution of 9% acid is to be diluted by adding 3% acid solution to it. The resulting mixture is to be more than 5% but less than 7% acid. If there is 460 L of the 9% solution, then how many litres of 3% solution will have to be added? Competency Based Que

Answers

1. (i) $(5, \infty)$ 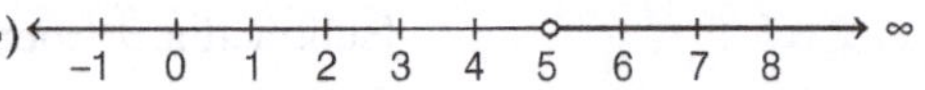

(ii) No solution exists (iii) No solution exists

2. (i) $\left[1, \frac{11}{3}\right]$ (ii) $\left[\frac{1}{3}, 1\right]$

3. 82 marks

4. (11, 13), (13, 15) and (15, 17)

5. $9.6 \le MA \le 16.8$

6. More than 600 cassettes.

7. 9 cm

8. More than 230 litres but less than 920 litres.

For Solutions, See QR Code on Pg. 145

Exam Pattern Questions

Objective Type Questions

Multiple Choice Questions

1. Two real numbers or two algebraic expressions related by the symbol '<', '>', '≤' or '≥' forms an

(a) equation (b) inequality
(c) set (d) None of these

2. If $a < b$ and $c < 0$, then

(a) $\frac{a}{c} = \frac{b}{c}$ (b) $\frac{a}{c} > \frac{b}{c}$
(c) $\frac{a}{c} < \frac{b}{c}$ (d) None of these

3. The graph of the solution of inequality $\frac{3x-4}{2} \geq \frac{x+1}{4} - 1$ on number line is

(a) 

(b) 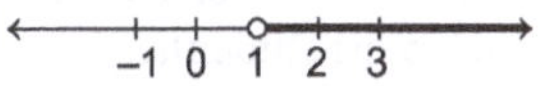

(c) 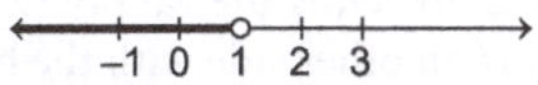

(d) 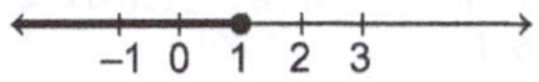

4. If x is a real number and $|x| < 3$, then x lies between

(a) $x \geq 3$ (b) $x \leq -3$
(c) $-3 \leq x \leq 3$ (d) $-3 < x < 3$

5. The solution set for inequality $2x + 1 > 3x - 2$ is

(a) $(-\infty, 3)$ (b) $(-\infty, 2)$
(c) $(-\infty, -3)$ (d) $(-\infty, 1)$

6. If $-3x + 17 < -13$, then

(a) $x \in (10, \infty)$ (b) $x \in [10, \infty)$
(c) $x \in (-\infty, 10]$ (d) $x \in [-10, 10)$

7. If $x + \frac{x}{2} < 11 - \frac{x}{3}$, then

(a) $x \in (-\infty, 3)$ (b) $x \in (-\infty, 2)$
(c) $(-\infty, 0)$ (d) $x \in (-\infty, 6)$

8. The solution set of $\frac{2x-1}{3} \geq \left(\frac{3x-2}{4}\right) - \left(\frac{2-x}{5}\right)$ is

(a) $(-\infty, 3)$ (b) $(-\infty, 2]$
(c) $\left(-\infty, -\frac{1}{2}\right)$ (d) $\left(-\infty, -\frac{1}{2}\right]$

9. The inequality $\frac{2}{x} < 3$ is true, when x belongs to

(a) $\left[\frac{2}{3}, \infty\right)$ (b) $\left(-\infty, \frac{2}{3}\right]$
(c) $(-\infty, 0) \cup \left(\frac{2}{3}, \infty\right)$ (d) None of these

10. The solution set of the inequality $4x + 3 < 6x + 7$ is

(a) $[-2, \infty)$ (b) $(-\infty, -2)$
(c) $(-2, \infty)$ (d) None of these

11. Solution set for inequality $2(x-3) \geq 3(x-1)$

(a) $(-\infty, 3]$ (b) $(-\infty, 2]$ (c) $(-\infty, -3]$ (d) $(-\infty, 1]$

12. The solution set for inequality $\frac{x}{2} \geq \frac{4x-1}{15}$ is

(a) $\left[\frac{2}{7}, \infty\right)$ (b) $\left[\frac{-2}{7}, \infty\right)$ (c) $\left[\frac{2}{15}, \infty\right)$ (d) $\left[\frac{-2}{15}, \infty\right)$

13. What is the solution set of $\left|\frac{2}{x-6}\right| \geq 1$, where $x \neq 6$?

(a) $[4, 6) \cup (6, 8]$ (b) $(6, 8)$
(c) $(-\infty, 4)$ (d) $(6, \infty)$

14. Let $\frac{C}{5} = \frac{F-32}{9}$. If C lies between 10 and 20, then

(a) $50 < F < 78$ (b) $50 < F < 68$
(c) $49 < F < 68$ (d) $49 < F < 78$

15. The solution of the system of inequalities $3x - 7 < 5 + x$ and $11 - 5x \leq 1$ on the number line is

(a) 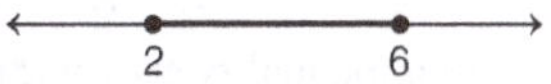

(b) 

(c) 

(d) None of the above

16. The solution set of the inequalities $3x - 7 > 2(x - 6)$ and $6 - x > 11 - 2x$, is

(a) $(-5, \infty)$ (b) $[5, \infty)$ (c) $(5, \infty)$ (d) $[-5, \infty)$

17. The solution set of the inequalities $6 \leq -3(2x - 4) < 12$ is

(a) $(-\infty, 1]$ (b) $(0, 1]$
(c) $(0, 1] \cup [1, \infty)$ (d) $[1, \infty)$

Assertion-Reason Based Questions

Directions (Q. Nos. 18-25) *In the questions given below are two statements labelled as Assertion (A) and Reason (R). In the context of the two statements, which one of the following is correct?*

(a) Both A and R are correct; R is the correct explanation of A.
(b) Both A and R are correct; R is not the correct explanation of A.
(c) A is correct; R is incorrect.
(d) R is correct; A is incorrect.

18. **Assertion** (A) If $-3x+12<0$, then $x \in (4, \infty)$.

Reason (R) If $a<b$, then $ka>kb$, if $k<0$.

19. **Assertion** (A) The inequality $ax+by<0$ is strict inequality.

Reason (R) The inequality $ax+b \geq 0$ is slack inequality.

20. **Assertion** (A) The inequality $3x+2y \geq 5$ is the linear inequality.

Reason (R) The solution of $5x-3<7$, when x is a real number, is $(-\infty, 2)$.

21. **Assertion** (A) If $3x+8>2$, then $x \in \{-1, 0, 1, 2, \ldots\}$, when x is an integer.

Reason (R) The solution set of the inequality $4x+3<5x+7, \forall\ x \in R$ is $[4, \infty)$.

22. **Assertion** (A) The solution set of $4x-2 \leq 2x+10, x \in W$ is $\{0, 1, 2, 3, 4, 5\}$.

Reason (R) In number system, the symbol W denotes the set of whole numbers i.e. $W=\{0, 1, 2, 3, 4, \ldots\}$.

23. **Assertion** (A) The solution set of the inequality $-4x>30$, when $x \in N$ is a null set i.e. ϕ.

Reason (R) Since, $x \in N, x \geq 1$.

24. **Assertion** (A) If $P>Q$ and $R>0$, then $\frac{P}{R}<\frac{Q}{R}$.

Reason (R) If both sides of a inequality are divided by the same negative number, then the sign at inequality is reversed.

25. **Assertion** (A) If $2x-7>5-x$ and $11-5x \leq 1$, then $x \in (4, \infty)$.

Reason (R) The solution set of system of linear inequations is the intersection of solution set of individual inequations.

Case Study Based Questions

26. A manufacturing company produces certain goods. The company manager used to make a data record on daily basis about the cost and revenue of these goods separately. The cost and revenue functions of a product are given by $C(x)=20x+4000$ and $R(x)=60x+2000$ respectively, where x is the number of goods produced and sold.

Based on above information, answer the following questions.

(i) How many goods must be sold to realise some profit?

(a) $x<50$ (b) $x>50$ (c) $x \geq 50$ (d) $x \leq 50$

(ii) If the cost and revenue functions of a product are given by $C(x)=3x+400$ and $R(x)=5x+20$ respectively, where x is the number of items produced by the manufacturer, then how many items must be sold to realise some profit?

(a) $x \leq 190$ (b) $x<190$ (c) $x \geq 190$ (d) $x>190$

(iii) Let x and b be the real numbers. If $b>0$ and $x<b$, then

(a) x is always positive (b) x is always negative
(c) x is real number (d) None of these

(iv) The solution set of $20x+4000 \leq 60x+2000$, when x is real number is given by

(a) $(-50, \infty)$ (b) $[50, \infty)$
(c) $(50, \infty)$ (d) $[-50, \infty)$

27. Himanshu was teaching 'method to solve a linear inequality in one variable' to his students.

Step I Collect all terms involving the variable x on one side and constant terms on other side with the help of above rules and then reduce it in the form $ax<b$ or $ax \leq b$ or $ax>b$ or $ax \geq b$.

Step II Divide the inequality by the coefficient of variable x. This gives the solution set of given inequality.

Step III Write the solution set.

Based on above information, answer the following questions.

(i) The solution set of $24x<100$, when x is a natural number is

(a) $\{1, 2, 3, 4\}$ (b) $(1, 4)$
(c) $[1, 4)$ (d) $(1, 4]$

(ii) The solution set of $24x<100$, when x is an integer is

(a) $\{\ldots -4, -3, -2, -1, 0, 1, 2, 3, 4\}$
(b) $(-\infty, 4)$
(c) $[4, \infty)$
(d) $(-\infty, 4]$

(iii) The solution set of $-5x+25>0$ is

(a) $[5, \infty)$ (b) $(-\infty, 5]$ (c) $(5, \infty)$ (d) $(-\infty, 5)$

(iv) The solution set of $x+\frac{x}{2}+\frac{x}{3}<22$ is

(a) $(-\infty, 12)$ (b) $(-\infty, 12]$
(c) $[12, \infty)$ (d) $[-6, \infty)$

Subjective Questions

Very Short Answer Type Questions

1. Solve $12 + 1\frac{5}{6}x \le 5 + 3x$, when

(i) $x \in N$ (ii) $x \in R$

2. Solve $24x < 100$, when

(i) x is a natural number. (ii) x is an integer. **NCERT**

3. Solve the inequality $3x + \frac{2x}{3} < 22$.

4. Solve the inequality $\frac{1}{2}\left(\frac{3}{5}x + 4\right) \ge \frac{1}{3}(x - 6)$.

5. Solve the inequality $3(2 - x) \ge 2(1 - x)$ and show the graph of solution on number line. **NCERT**

6. Solve the inequality $\frac{2x - 3}{4} + 9 \ge 3 + \frac{4x}{3}$ and show the graph of the solution on number line.

7. Solve the following inequalities and show the graph of the solution in each case on number line.

(i) $\frac{x}{4} < \frac{5x - 2}{3} - \frac{7x - 3}{5}$ (ii) $\frac{(2x - 1)}{3} \le \frac{(3x - 2)}{4} - \frac{(2 - x)}{5}$

8. The sum of three consecutive integers must not be more than 12. What are the integers?

9. Solve the inequalities $6 \le -3(2x - 4) < 12$. **NCERT**

10. Solve the linear inequalities $-15 < \frac{3(x - 2)}{5} \le 0$.

11. Solve for x, $4x + 3 \ge 2x + 17$, $3x - 5 < -2$.

12. Solve the following system of linear inequalities.

$\frac{4x}{3} - \frac{9}{4} < x + \frac{3}{4}$ and $\frac{7x - 1}{3} - \frac{7x + 2}{6} > x$

13. Solve for x, $-5 \le \frac{2 - 3x}{4} \le 9$. **NCERT Exemplar**

14. Solve $2x - 1 > x + \frac{7 - x}{3} > 2$, $x \in R$.

Short Answer Type Questions

15. Ram needs a minimum of 360 marks in four tests in a Mathematics course to obtain A grade. In his first three tests, he scored 88, 96 and 79 marks. What should his score be in the fourth test so that he can make A grade?

16. Anshu obtained 73, 67 and 72 marks in the Mathematics test. How many marks should he get in his fourth test, so as to have an average of atleast 75?

17. In the first four examinations, each of 100 marks, Mohan got 94, 73, 72 and 84 marks. If a final average greater than or equal to 80 and less than 90 is needed to obtain a final grade B in a course, then what range of marks in the fifth (last) examination will result, if Mohan receiving grade B in the course?

18. Find all the pairs of consecutive odd positive integers, both of which are smaller than 10 such that their sum is more than 11. **NCERT**

19. A solution is to be kept between 68°F and 77°F. What is the range of temperature in degree Celsius (C), if the Celsius/Fahrenheit (F) conversion formula is given by $F = \frac{9}{5}C + 32$?

20. The longest side of a triangle is twice the shortest side and the third side is 2 cm longer than the shortest side. If the perimeter of the triangle is more than 166 cm, then find the minimum length of the shortest side. **NCERT Exemplar**

Long Answer Type Questions

21. Solve the inequality $\frac{-3x + 10}{x + 1} > 0$. Also, represent the solution on the number line.

22. Solve $\frac{5}{x - 2} > 3$ and represent the solution set on number line.

23. Solve graphically for x, $|x - 1| \le 5$, $|x| \ge 2$.

24. Solve the following system of linear inequalities and represent the solution graphically on the number line.

$2(2x + 3) - 10 < 6(x - 2)$ and $\frac{2x - 3}{4} + 6 \ge 2 + \frac{4x}{3}$

25. How many litres of water will have to be added to 1125 L of the 45% solution of acid so that the resulting mixture will contain more than 25% but less than 30% acid content? **NCERT**

Hints & Solutions

Objective Type Questions

1. (b) A statement involving the symbols $>, <, \leq$ or $\geq$ is called an inequality or inequation.

2. (b) We have, $a < b$ and $c < 0$

On dividing both sides of $a < b$ by c,

$$\frac{a}{c} > \frac{b}{c}$$

[$\because$ c is a negative number, sign at inequality will get reversed]

3. (a) We have,

$$\frac{3x-4}{2} \geq \frac{x+1}{4} - 1$$

$$\Rightarrow \frac{3x-4}{2} \geq \frac{x+1-4}{4}$$

$$\Rightarrow \frac{3x-4}{2} \geq \frac{x-3}{4}$$

On multiplying by 4 both sides, we get

$$6x - 8 \geq x - 3$$

On adding 8 both sides, we get

$$6x \geq x + 5$$

On subtracting x from both sides, we get

$$5x \geq 5$$

On dividing by 5 both sides, we get

$$x \geq 1$$

i.e. $x \in [1, \infty)$

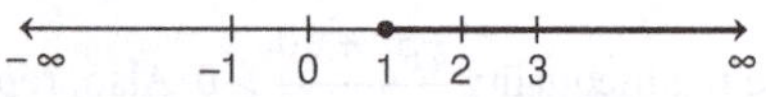

4. (d) Given that $|x| < 3$

For real values of x, it lies between -3 and 3

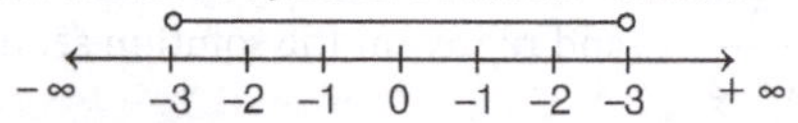

$\Rightarrow -3 < x < 3$

i.e $(-3, 3)$.

5. (a) We have,

$$2x + 1 > 3x - 2$$

On transferring the term $2x$ to RHS and the term (-2) to LHS, we get

$$1 + 2 > 3x - 2x$$

$$\Rightarrow 3 > x$$

$$\Rightarrow x < 3$$

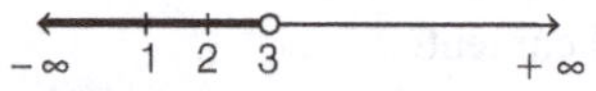

Since, all the numbers on the left side of 3 will be less than it.

$\therefore$ Solution set $= (-\infty, 3)$

6. (a) We have, $-3x + 17 < -13$

On adding $(3x + 13)$ both sides, we get

$$30 < 3x$$

On dividing both sides by 3, we get

$$10 < x \Rightarrow x > 10$$

i.e. $x \in (10, \infty)$

7. (d) We have,

$$x + \frac{x}{2} < 11 - \frac{x}{3}$$

$$\Rightarrow x + \frac{x}{2} + \frac{x}{3} < 11$$

$$\Rightarrow \frac{6x + 3x + 2x}{6} < 11 \Rightarrow \frac{11x}{6} < 11$$

$$\Rightarrow \frac{11x}{6} \times \frac{6}{11} < 11 \times \frac{6}{11} \quad \left[\text{multiplying both sides by } \frac{6}{11}\right]$$

$$\Rightarrow x < 6$$

$\therefore x \in (-\infty, 6)$

8. (b) We have,

$$\frac{2x-1}{3} \geq \left(\frac{3x-2}{4}\right) - \left(\frac{2-x}{5}\right)$$

On multiplying by 60 both sides, we get

$$20(2x - 1) \geq 15(3x - 2) - 12(2 - x)$$

$$\Rightarrow 40x - 20 \geq 45x - 30 - 24 + 12x$$

$$\Rightarrow 40x - 20 \geq 57x - 54$$

On subtracting $40x$ from both sides, we get

$$-20 \geq 17x - 54$$

On adding 54 both sides, we get

$$34 \geq 17x$$

$$\Rightarrow 17x \leq 34$$

On dividing by 17 both sides, we get

$$x \leq 2$$

i.e. $x \in (-\infty, 2]$

9. (c) Given inequality,

$$\frac{2}{x} < 3$$

$$\Rightarrow \frac{2}{x} - 3 < 3 - 3 \quad \text{[subtracting 3 from both sides]}$$

$$\Rightarrow \frac{2 - 3x}{x} < 0$$

$$\Rightarrow \frac{(2 - 3x)}{x} \times x^2 < 0 \times x^2 \quad \text{[multiplying both sides by } x^2\text{]}$$

$$\Rightarrow (2 - 3x)\,x < 0$$

So, both the factors have opposite sign.

Therefore, $2 - 3x < 0$ and $x > 0$ or $2 - 3x > 0$ and $x < 0$

$\Rightarrow 3x-2>0$ and $x>0$ or $3x-2<0$ and $x<0$

$\Rightarrow x>\frac{2}{3}$ and $x>0$ or $x<\frac{2}{3}$ and $x<0$

$\Rightarrow x\in\left(\frac{2}{3},\infty\right)$ or $x\in(-\infty,0)$

Therefore, the common solution for x is

$(-\infty,0)\cup\left(\frac{2}{3},\infty\right)$.

10. (c) We have, $4x+3<6x+7$

$\Rightarrow \quad 4x-6x<6x+4-6x$

$\Rightarrow \quad -2x<4 \Rightarrow x>-2$

i.e. all the real numbers which are greater than -2, are the solutions of the given inequality. Hence, the solution set is $(-2,\infty)$.

11. (c) We have, $2(x-3)\ge 3(x-1)$

$\Rightarrow \quad 2x-6\ge 3x-3$

$\Rightarrow \quad 2x-6-3x+6\ge 3x-3-3x+6$

[subtracting $3x$ from both sides and adding 6 on both sides]

$\Rightarrow \quad -x\ge 3 \Rightarrow x\le -3$

Thus, the solution set is $(-\infty,-3]$.

12. (b) We have,

$$\frac{x}{2}\ge\frac{4x-1}{15}$$

$\Rightarrow \quad \frac{x}{2}\times 30\ge\frac{(4x-1)}{15}\times 30$

[on multiplying by 30 on both sides of inequality]

$\Rightarrow \quad 15x\ge 2(4x-1)$

$\Rightarrow \quad 15x\ge 8x-2$

$\Rightarrow \quad 15x-8x\ge -2$

$\Rightarrow \quad 7x\ge -2$ [transferring $8x$ to LHS]

$\Rightarrow \quad \frac{7x}{7}\ge\frac{-2}{7}$

[dividing by 7 on both sides of inequality]

$\Rightarrow \quad x\ge\frac{-2}{7}$

$\therefore$ The solution set $=\left[\frac{-2}{7},\infty\right)$

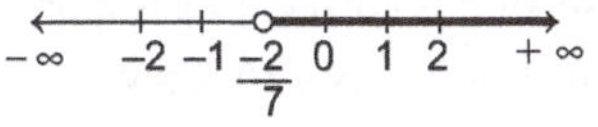

13. (a) Given that

$$\left|\frac{2}{x-6}\right|\ge 1, \text{ where } x\ne 6$$

$\Rightarrow \quad -\left(\frac{2}{x-6}\right)\le 1$ or $\frac{2}{x-6}\ge 1$

$\Rightarrow \quad \frac{2}{x-6}\ge -1$ or $\frac{x-6}{2}\le 1$

[taking inverse of second inequality]

$\Rightarrow \quad \frac{x-6}{2}\ge -1$ or $\frac{x-6}{2}\le 1$

[taking inverse of first inequality]

$\Rightarrow \quad x-6\ge -2$ or $x-6\le 2$

[multiplying by 2 in both inequalities]

$\Rightarrow \quad x\ge 4$ or $x\le 8$

[adding 6 in both inequalities]

Therefore, solution set $=[4,6)\cup(6,8]$.

14. (b) Given, $\frac{C}{5}=\frac{F-32}{9}$ and $10<C<20$

$\Rightarrow \quad C=\frac{5F-(32)5}{9}$

Since, $\quad 10<C<20$

$\Rightarrow \quad 10<\frac{5F-160}{9}<20$

$\Rightarrow \quad 90<5F-160<180$

$\Rightarrow \quad 90+160<5F<180+160$

$\Rightarrow \quad 250<5F<340$

$\Rightarrow \quad \frac{250}{5}<F<\frac{340}{5}$

$\therefore \quad 50<F<68$

15. (b) Given inequalities,

$3x-7<5+x$ ...(i)

and $11-5x\le 1$ ...(ii)

From inequality (i), we get

$3x-7<5+x$

or $x<6$ ...(iii)

Also, from inequality (ii), we get

$11-5x\le 1$

or $-5x\le -10$ i.e. $x\ge 2$ ...(iv)

If we draw the graph of inequalities (iii) and (iv) on the number line, we see that the values of x, which are common to both, are shown by bold line in figure.

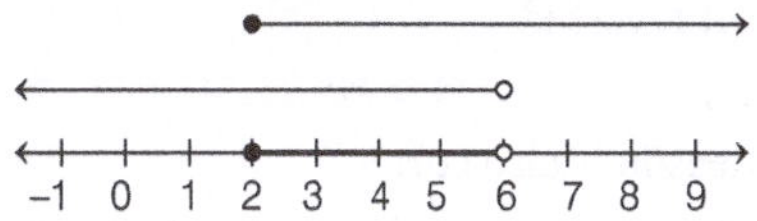

16. (c) We have, $3x-7>2(x-6)$

$\Rightarrow \quad 3x-7>2x-12$

On transferring the term $2x$ to LHS and the term (-7) to RHS, we get

$3x-2x>-12+7$

$\Rightarrow \quad x>-5$ ...(i)

and $6-x>11-2x$

On transferring the term $(-2x)$ to LHS and the term 6 to RHS, we get

$$-x+2x>11-6$$

$\Rightarrow$ $x>5$ ...(ii)

Draw the graph of inequations (i) and (ii) on the number line,

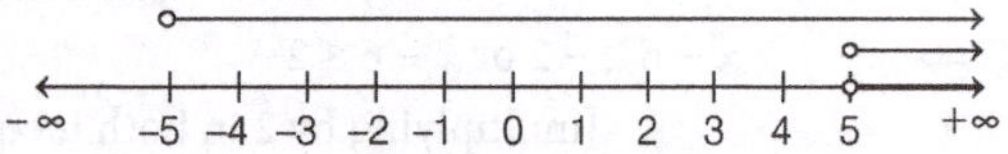

Hence, solution set of the equations are real numbers, x lying on greater than 5 excluding 5.

i.e. $x>5$

$\therefore$ Solution set is $(5, \infty)$.

17. (b) The given inequality $6 \le -3(2x-4) < 12$

$$6 \le -6x+12 < 12$$

On adding (-12) to each term, we get

$$6-12 \le -6x+12-12 < 12-12$$

$\Rightarrow$ $-6 \le -6x < 0$

On dividing by (-6) to each term, we get

$$\frac{-6}{-6} \ge \frac{-6x}{-6} > \frac{0}{-6}$$

$\Rightarrow$ $1 \ge x > 0 \Rightarrow 0 < x \le 1$

$\therefore$ Solution set is $(0, 1]$.

18. (a) **Assertion** We have,

$$-3x+12<0$$

On subtracting 12 from both sides, we get

$$-3x<-12$$

On multiplying both sides by $\frac{-1}{3}$, we get

$$x>4$$

i.e. $x \in (4, \infty)$

$\therefore$ Assertion is correct.

Reason We have, $a<b$ and $k<0$

$\because$ $k<0$

$\therefore$ k is negative number.

$\therefore$ On multiplying both sides $a<b$ by k, sign of inequality will get reversed.

Hence, $ka>kb$

So, Reason is also correct.

Hence, Assertion and Reason both are correct; Reason is the correct explanation of Assertion.

19. (b) Let us consider some inequalities

$ax+b<0$ ...(i)

$ax+b>0$ ...(ii)

$ax+b \le 0$ ...(iii)

$ax+b \ge 0$ ...(iv)

$ax+by>c$ ...(v)

$ax+by \le c$ ...(vi)

$ax^2+bx+c<0$ ...(vii)

$ax^2+bx+c \le 0$ ...(viii)

Inequalities (i), (ii), (v) and (vii) are strict inequalities, while inequalities (iii), (iv), (vi) and (viii) are slack inequalities.

$\therefore$ Assertion and Reason both are correct but Reason is not correct explaination of Assertion.

20. (b) **Assertion** An inequality is said to be linear, if the variables occur in first degree only and there is no term involving the product of the variables.

$\therefore$ Inequality $3x+2y \ge 5$ is a linear inequality.

$\therefore$ Assertion is correct.

Reason We have, $5x-3<7, x \in R$

On adding 3 both sides, we get

$$5x<10$$

On dividing both sides by 5, we get

$$x<2$$

i.e. $x \in (-\infty, 2)$

$\therefore$ Reason is correct.

Hence, Assertion and Reason both are correct ; Reason is not the correct explanation of Assertion.

21. (c) **Assertion** $3x+8>2, x \in I$

On subtracting 8 from both sides, we get

$$3x>-6$$

On dividing both sides by 3, we get

$$x>-2$$

i.e. $x \in \{-1, 0, 1, 2, \ldots\}$

$\therefore$ Assertion is correct.

Reason We have,

$$4x+3<5x+7, \forall x \in R$$

On subtracting $(4x+7)$ from both sides, we get

$$-4<x$$

$\Rightarrow$ $x>-4$

i.e. $x \in (-4, \infty)$

$\therefore$ Reason is incorrect.

Hence, Assertion is correct ; Reason is incorrect.

22. (d) We have, $4x-2 \le 2x+10, x \in W$

$\Rightarrow$ $2x-2 \le 10$ [subtracting $2x$ from both sides]

$\Rightarrow$ $2x \le 12$ [adding 2 in both sides]

$\Rightarrow$ $x \le 6$ [dividing both sides by 2]

i.e. $x \in (-\infty, 6]$

$\Rightarrow$ $x \in \{-\infty, \ldots, -2-1, 0, 1, 2, 3, 4, 5, 6\}$

Hence, Reason is correct; Assertion is incorrect.

23. (a) **Assertion** We have,

$$-4x > 30$$

On multiplying by $\frac{-1}{4}$ in both sides, we get

$$x < \frac{-30}{4}$$

$$\Rightarrow \quad x < \frac{-15}{2}$$

$$\Rightarrow \quad x \in \left(-\infty, \frac{-15}{2}\right)$$

$$\because \quad x \in N$$

$\therefore$ x cannot be less than 1.

$\therefore$ The solution set of the inequality $-4x > 30$ is null set i.e. ϕ.

So, Assertion is correct.

Reason Since, $x \in N$

and natural number $(N) = \{1, 2, 3, 4, 5, ...\}$

$$\therefore \quad x \geq 1$$

$$\Rightarrow \quad x \in [1, \infty)$$

$\therefore$ Reason is correct.

Hence, Assertion and Reason both are correct; Reason is correct explanation of Assertion.

24. (d) **Assertion** We have, $P > Q$ and $R > 0$

$$\because \quad R > 0$$

$\therefore$ R is a positive number.

So, $$\frac{P}{R} > \frac{Q}{R}$$

Therefore, Assertion is incorrect.

Reason is correct.

Hence, Reason is correct; Assertion is incorrect.

25. (a) **Assertion** We have,

$$2x - 7 > 5 - x \quad \ldots(i)$$

and $$11 - 5x \leq 1 \quad \ldots(ii)$$

From inequality (i), we get

$$2x - 7 > 5 - x$$

On adding $(x + 7)$ both sides, we get

$$3x > 12$$

On dividing each side by 3, we get

$$x > 4$$

i.e. $$x \in (4, \infty)$$

Now, from inequality (ii), we get

$$11 - 5x \leq 1$$

On adding $5x$ in each side, we get

$$11 \leq 1 + 5x$$

On subtracting 1 from each side, we get

$$10 \leq 5x$$

On dividing each side by 5, we get

$$2 \leq x$$

$$\Rightarrow \quad x \geq 2$$

i.e. $$x \in [2, \infty)$$

The values of x, which are common to both are lying in the interval $(4, \infty)$.

$\therefore$ Assertion is correct and Reason is also correct.

Hence, Assertion and Reason both are correct; Reason is the correct explanation of Assertion.

26. (i) (b) We know that Profit = Revenue − Cost

$\therefore$ In order to realise some profit and revenue should be greater than the cost.

Thus, we should have $R(x) > C(x)$

$$\Rightarrow \quad 60x + 2000 > 20x + 4000$$

On subtracting $20x$ from both sides, we get

$$40x + 2000 > 4000$$

On subtracting 2000 from both sides, we get

$$40x > 2000$$

On dividing both sides by 40, we get

$$x > 50$$

Hence, the manufacturer must sell more than 50 items to realise some profit.

(ii) (d) In order to realise some profit, we should have

$$R(x) > C(x)$$

$$\Rightarrow \quad 5x + 20 > 3x + 400$$

$$\Rightarrow \quad 2x + 20 > 400$$

[subtracting $3x$ from both sides]

$$\Rightarrow \quad 2x > 380$$

$$\Rightarrow \quad x > 190$$

Hence, the manufacturer must sell more than 190 items to realise some profit.

(iii) (c) We have, $b > 0$ and $x < b$

It means x is always less than the positive number.

$\therefore$ x is real number.

(iv) (b) We have, $20x + 4000 \leq 60x + 2000$

$$\Rightarrow \quad 4000 \leq 40x + 2000$$

$$\Rightarrow \quad 2000 \leq 40x$$

$$\Rightarrow \quad 40x \geq 2000$$

$$\Rightarrow \quad x \geq 50$$

$$\therefore \quad x \in [50, \infty)$$

27. (i) (a) We have, $24x < 100$

$$\Rightarrow \quad x < \frac{100}{24}$$

[dividing by 24 on both sides]

$$\Rightarrow \quad x < \frac{25}{6}$$

Since, x is a natural number.

$\therefore$ Solutions of the inequality $x < \frac{25}{6}$ are all natural numbers, which are less than $\frac{25}{6}$.

$\therefore \quad x = \{1, 2, 3\ 4\}$

(ii) (a) Given inequality is

$$24x < 100$$

$$\Rightarrow \quad x < \frac{100}{24} \Rightarrow x < \frac{25}{6}$$

Since, x is an integer.

$\therefore$ The solution set of the inequality $x < \frac{25}{6}$ is $\{\ldots, -2, -1, 0, 1, 2, 3, 4\}$.

(iii) (d) We have, $-5x + 25 > 0$

On adding $5x$ both sides, we get

$$25 > 5x$$

On dividing by 5 both sides, we get

$$5 > x \Rightarrow x < 5$$

i.e. $\quad x \in (-\infty, 5)$

(iv) (a) We have, $x + \frac{x}{2} + \frac{x}{3} < 22$

On multiplying both sides by 6, we get

$$6x + 3x + 2x < 22 \times 6$$

$$\Rightarrow \quad 11x < 132 \Rightarrow x < 12$$

i.e. $\quad x \in (-\infty, 12)$

Subjective Questions

1. We have, $\quad 12 + 1\frac{5}{6}x \le 5 + 3x$

$$\Rightarrow \quad 12 + \frac{11}{6}x \le 5 + 3x$$

$$\Rightarrow \quad 72 + 11x \le 30 + 18x$$

$$\Rightarrow \quad 42 \le 7x \Rightarrow x \ge 6$$

i.e. $\quad x \in [6, \infty)$

(i) $x \in N$,

$x = \{6, 7, 8, 9 \ldots\}$

(ii) $x \in R$,

$x \in [6, \infty)$

2. We have, $\quad 24x < 100$

On dividing both sides by 24, we get

$$\frac{24x}{24} < \frac{100}{24}$$

$$\Rightarrow \quad x < \frac{25}{6}$$

(i) When x is a natural number, then solutions of the inequality $x < \frac{25}{6}$, are all natural numbers, which are less than $\frac{25}{6}$. In this case, the following values of x make the statement true.

$$x = 1, 2, 3, 4$$

Hence, the solution set of inequality is $\{1, 2, 3, 4\}$.

(ii) When x is an integer.

In this case, solutions of given inequality are

$$\ldots, -4, -3, -2, -1, 0, 1, 2, 3, 4$$

Hence, the solution set of inequality is

$$\{\ldots, -4, -3, -2, -1, 0, 1, 2, 3, 4\}$$

3. We have, $\quad 3x + \frac{2x}{3} < 22$

$$\Rightarrow \quad \frac{9x + 2x}{3} < 22$$

$$\Rightarrow \quad \frac{11x}{3} < 22$$

$$\Rightarrow \quad \frac{11x}{6} < 11 \quad \text{[dividing both sides by 2]}$$

On multiplying both sides by $\frac{6}{11}$, we get

$$\frac{11x}{6} \times \frac{6}{11} < 11 \times \frac{6}{11} \Rightarrow x < 6$$

$$\therefore \quad x \in (-\infty, 6)$$

This can be represented on number line as

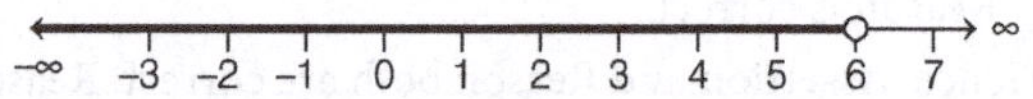

Here, the dark portion on the number line represents the solution of inequality.

4. We have, $\quad \frac{1}{2}\left(\frac{3}{5}x + 4\right) \ge \frac{1}{3}(x - 6)$

$$\Rightarrow \quad \frac{1}{2}\left(\frac{3x + 20}{5}\right) \ge \frac{1}{3}(x - 6)$$

$$\Rightarrow \quad \frac{1}{10}(3x + 20) \ge \frac{1}{3}(x - 6)$$

$$\Rightarrow \quad \frac{30}{10}(3x + 20) \ge \frac{30}{3}(x - 6)$$

[multiplying both sides by LCM (10, 3) = 30]

$$\Rightarrow \quad 3(3x + 20) \ge 10(x - 6)$$

$$\Rightarrow \quad 9x + 60 \ge 10x - 60$$

$$\Rightarrow \quad 9x + 60 - 60 \ge 10x - 60 - 60$$

[subtracting 60 from both sides]

$$\Rightarrow \quad 9x \ge 10x - 120$$

$$\Rightarrow \quad 9x - 10x \ge 10x - 120 - 10x$$

[subtracting $10x$ from both sides]

$$\Rightarrow \quad -x \ge -120 \Rightarrow x \le 120$$

[multiplying both sides by -1]

$$\therefore \quad x \in (-\infty, 120]$$

On number line, it can be represented as

Here, the dark portion on the number line represents the solution of inequality.

5. We have, $3(2-x) \geq 2(1-x)$

$\Rightarrow \quad 6-3x \geq 2-2x$

$\Rightarrow \quad 4 \geq x \Rightarrow x \leq 4$

i.e. $\quad x \in (-\infty, 4]$

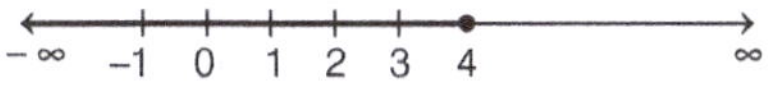

6. We have, $\dfrac{2x-3}{4}+9 \geq 3+\dfrac{4x}{3}$

$\Rightarrow \quad \dfrac{2x-3+36}{4} \geq \dfrac{9+4x}{3} \Rightarrow \dfrac{2x+33}{4} \geq \dfrac{9+4x}{3}$

$\Rightarrow \quad 6x+99 \geq 36+16x$

$\Rightarrow \quad 63 \geq 10x \Rightarrow x \leq 6.3$

i.e. $\quad x \in (-\infty, 6.3]$

7. (i) We have, $\dfrac{x}{4} < \dfrac{5x-2}{3} - \dfrac{7x-3}{5}$

$\Rightarrow \quad \dfrac{x}{4} < \dfrac{5(5x-2)-3(7x-3)}{15}$

$\Rightarrow \quad \dfrac{x}{4} < \dfrac{25x-10-21x+9}{15}$

$\Rightarrow \quad 60\left(\dfrac{x}{4}\right) < \dfrac{(4x-1)}{15} \times 60$

[multiplying both sides by LCM (4, 15) = 60]

$\Rightarrow \quad 15x < 4(4x-1) \Rightarrow 15x < 16x-4$

$\Rightarrow \quad 15x+4 < 16x-4+4$

[adding 4 on both sides]

$\Rightarrow \quad 15x+4 < 16x$

$\Rightarrow \quad 15x+4-15x < 16x-15x$

[subtracting $15x$ from both sides]

$\Rightarrow \quad 4 < x$ or $x > 4$

$\therefore \quad x \in (4, \infty)$

On number line, it can be represented as

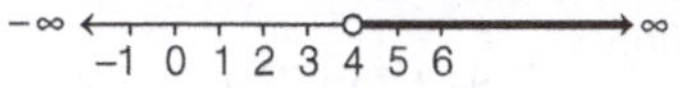

Here, the dark portion on the number line represents the solution of inequality.

(ii) We have, $\dfrac{2x-1}{3} \leq \dfrac{3x-2}{4} - \dfrac{2-x}{5}$

On multiplying both sides by LCM of 3, 4 and 5 i.e. 60, we get

$\dfrac{2x-1}{3} \times 60 \leq \dfrac{3x-2}{4} \times 60 - \dfrac{2-x}{5} \times 60$

$\Rightarrow \quad 20(2x-1) \leq 15(3x-2)-12(2-x)$

$\Rightarrow \quad 40x-20 \leq 45x-30-24+12x$

$\Rightarrow \quad 40x-20 \leq 57x-54$

$\Rightarrow \quad 40x-20+20 \leq 57x-54+20$

[adding 20 on both sides]

$\Rightarrow \quad 40x \leq 57x-34$

$\Rightarrow \quad 40x-57x \leq 57x-34-57x$

[subtracting $57x$ from both sides]

$\Rightarrow \quad -17x \leq -34$

$\Rightarrow \quad \dfrac{-17x}{17} \leq \dfrac{-34}{17}$ [dividing both sides by 17]

$\Rightarrow \quad -x \leq -2$

$\Rightarrow \quad x \geq 2$ [multiplying both sides by –1]

Thus, the solution set is $[2, \infty)$.

On number line, it can be represented as

Here, the dark portion on the number line represents the solution of inequality.

8. Let the three consecutive integers be x, $(x+1)$ and $(x+2)$. Since, sum of these integers must not be more than 12 i.e. $x+(x+1)+(x+2)$ cannot exceed 12. Therefore,

$x+(x+1)+(x+2) \leq 12$

$\Rightarrow \quad 3x+3 \leq 12$

$\Rightarrow \quad 3x+3-3 \leq 12-3$ [subtracting 3 from both sides]

$\Rightarrow \quad 3x \leq 9$

$\therefore \quad x \leq 3$ [dividing both sides by 3]

9. We have, $6 \leq -3(2x-4) < 12$ or $6 \leq -6x+12 < 12$

On subtracting 12 from each term, we get

$6-12 \leq -6x+12-12 < 12-12$

$\Rightarrow \quad -6 \leq -6x < 0$

On dividing each term by – 6, we get

$\dfrac{-6}{-6} \geq \dfrac{-6x}{-6} > \dfrac{0}{-6}$

[dividing each term by the same negative number, then sign of inequalities will get changed]

$\Rightarrow \quad 1 \geq x > 0$

which can be written as $0 < x \leq 1$.

Hence, the solution set of given system of inequations is $(0, 1]$.

10. We have, $-15 < \dfrac{3(x-2)}{5} \leq 0$

$\Rightarrow \quad -15 \times 5 < \dfrac{3(x-2)}{5} \times 5 \leq 0 \times 5$

[multiplying each term by 5]

$\Rightarrow \quad -75 < 3(x-2) \leq 0$

$\Rightarrow \quad \dfrac{-75}{3} < \dfrac{3(x-2)}{3} \leq \dfrac{0}{3}$ [dividing each term by 3]

$\Rightarrow \quad -25 < x-2 \leq 0$

$\Rightarrow \quad -25+2 < x-2+2 \leq 0+2$ [adding 2 on all sides]

$\Rightarrow \quad -23 < x \leq 2$

$\therefore \quad x \in (-23, 2]$

Hence, the solution set of the given system of inequations is $(-23, 2]$.

11. We have, $4x+3 \geq 2x+17$ ...(i)

and $3x-5 < -2$ ...(ii)

From inequality (i), we get

$$2x \geq 14 \Rightarrow x \geq 7$$

i.e. $x \in [7, \infty)$

From inequality (ii), we get

$$3x-5 < -2$$

$\Rightarrow$ $3x < 3 \Rightarrow x < 1$

i.e. $x \in (-\infty, 1)$

There is no common value of x, which satisfies both inequalities.

Hence, the given system of inequalities has no solution set.

12. We have, $\frac{4x}{3} - \frac{9}{4} < x + \frac{3}{4}$ ...(i)

and $\frac{7x-1}{3} - \frac{7x+2}{6} > x$ ...(ii)

From inequality (i), we get

$$16x - 27 < 12x + 9$$

[multiplying both sides by 12]

$\Rightarrow$ $4x - 27 < 9$

[subtracting $12x$ from both sides]

$\Rightarrow$ $4x < 36$ [adding 27 both sides]

$\Rightarrow$ $x < 9$

$\Rightarrow$ $x \in (-\infty, 9)$

From inequality (ii), we get

$$\frac{7x-1}{3} - \frac{7x+2}{6} > x$$

$\Rightarrow$ $14x - 2 - 7x - 2 > 6x$

[multiplying by 6 both sides]

$\Rightarrow$ $x > 4$ i.e. $x \in (4, \infty)$

The values of x, which are common to both are lying in the interval $(4, 9)$. Hence, solution set is $(4, 9)$.

13. We have, $-5 \leq \frac{2-3x}{4} \leq 9$

$\Rightarrow$ $-20 \leq 2 - 3x \leq 36$

[multiplying by 4 in each term]

$\Rightarrow$ $-22 \leq -3x \leq 34$

[subtracting 2 from each term]

$\Rightarrow$ $22 \geq 3x \geq -34$

[multiplying by -1 in each term]

$\Rightarrow$ $\frac{22}{3} \geq x \geq \frac{-34}{3}$ [dividing by 3 in each term]

$\Rightarrow$ $\frac{-34}{3} \leq x \leq \frac{22}{3}$

i.e. $x \in \left[\frac{-34}{3}, \frac{22}{3}\right]$

14. We have, $2x - 1 > x + \frac{7-x}{3} > 2, x \in R$

$\Rightarrow$ $2x - 1 > \frac{7+2x}{3} > 2$

$\Rightarrow$ $6x - 3 > 7 + 2x > 6$

[multiplying by 3 in each term]

$\Rightarrow$ $6x - 3 > 7 + 2x$ and $7 + 2x > 6$

$\Rightarrow$ $4x > 10$ and $2x > -1$

$\Rightarrow$ $x > \frac{5}{2}$ and $x > \frac{-1}{2}$

$\Rightarrow$ $x \in \left(\frac{5}{2}, \infty\right)$ and $x \in \left(\frac{-1}{2}, \infty\right)$

$\therefore$ $x \in \left(\frac{5}{2}, \infty\right)$

15. Let Ram score x marks in the fourth test.

Then, the sum of Ram's test scores should be greater than or equal to 360

i.e. $88 + 96 + 79 + x \geq 360$

$\Rightarrow$ $263 + x \geq 360$

$\Rightarrow$ $263 + x - 263 \geq 360 - 263$

[subtracting 263 from both sides]

$\therefore$ $x \geq 97$

Hence, Ram should score 97 or greater than 97 marks in fourth test to obtain A grade.

16. Let Anshu score x marks in the fourth test.

Then, $\frac{73 + 67 + 72 + x}{4} \geq 75$

$\Rightarrow$ $212 + x \geq 300$

[multiplying by 4 in each side]

$\Rightarrow$ $x \geq 88$

[subtracting 212 from each side]

Hence, Anshu should obtain greater than or equal to 88 marks to get average of atleast 75.

17. Hint Let x be the score obtained by Mohan in the last examination.

Then, $80 \leq \frac{94 + 73 + 72 + 84 + x}{5} < 90$

$\Rightarrow$ $80 \leq \frac{323 + x}{5} < 90$

$\Rightarrow$ $400 \leq 323 + x < 450$

[multiplying both sides by 5]

$\Rightarrow$ $400 - 323 \leq 323 + x - 323 < 450 - 323$

[subtracting 323 from each term]

$\Rightarrow$ $77 \leq x < 127$

Since, the upper limit is 100, therefore the required range is $77 \leq x \leq 100$.

18. Let x be the smaller of two positive odd integers so that other one is $x + 2$.

We should have, $x < 10$ ...(i)

$$x+2<10 \Rightarrow x<8 \qquad \ldots(ii)$$

and $\quad x+(x+2)>11 \qquad \ldots(iii)$

From inequalities (i) and (ii), we get

$$x<8 \qquad \ldots(iv)$$

From inequality (iii), we get

$$2x+2>11$$

$\Rightarrow \quad 2x>9$

$\Rightarrow \quad x>4.5 \qquad \ldots(v)$

From inequalities (iv) and (v), we get

$$4.5<x<8$$

Since, x is an odd number, so x can take the values 5 and 7.

Hence, the required possible pairs will be (5, 7) and (7, 9).

19. It is given that $68<F<77$

On putting $F=\frac{9}{5}C+32$, we get

$$68<\frac{9}{5}C+32<77$$

$\Rightarrow \quad 68-32<\frac{9}{5}C<77-32$

[subtracting 32 from each term]

$\Rightarrow \quad 36<\frac{9}{5}C<45$

$\Rightarrow \quad 36\times\frac{5}{9}<C<45\times\frac{5}{9}$

$\left[\text{multiplying by } \frac{5}{9} \text{ in each term}\right]$

$\Rightarrow \quad 4\times5<C<5\times5$

$\Rightarrow \quad 20<C<25$

Hence, the required range of temperature is between 20° C and 25° C.

20. Let the length of shortest side be x cm.

Then, according to question, we have

length of longest side $=2x$ cm

and length of third side $=(x+2)$ cm

Since, the perimeter of the triangle is more than 166 cm.

$\therefore \quad 2x+x+(x+2)>166$

$\Rightarrow \quad 4x+2>166$

$\Rightarrow \quad 4x>164$ [subtracting 2 from both sides]

$\Rightarrow \quad x>41$ [dividing both sides by 4]

Hence, the length of the shortest side should be greater than 41 cm.

21. We have, $\quad \frac{-3x+10}{x+1}>0$

$\Rightarrow \quad \frac{-3x+10}{x+1}\times(x+1)^2>0\times(x+1)^2$

[multiplying both sides by $(x+1)^2$]

$\Rightarrow \quad (-3x+10)(x+1)>0$

The factors $(-3x+10)$ and $(x+1)$ have the same signs.

Case I When both are positive

In this case, we have

$$-3x+10>0 \text{ and } x+1>0$$

$\Rightarrow \quad 10>3x$ and $x>-1$

$\Rightarrow \quad 3x<10$ and $x>-1$

$\Rightarrow \quad x<\frac{10}{3}$ and $x>-1$

$\Rightarrow \quad -1<x<\frac{10}{3}$

$\therefore$ The solution set is $\left(-1,\frac{10}{3}\right)$.

Case II When both are negative

In this case, we have

$$-3x+10<0 \text{ and } x+1<0$$

$\Rightarrow \quad 3x>10$ and $x<-1$

$\Rightarrow \quad x>\frac{10}{3}$ and $x<-1$, which is impossible.

[$\because$ system of inequalities have no common solution]

Hence, the required solution set is $\left(-1,\frac{10}{3}\right)$ and this can be represented on the number line as

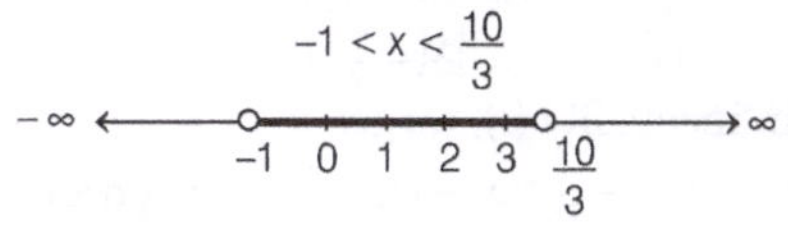

22. We have, $\frac{5}{x-2}>3$

On subtracting 3 from both sides, we get

$$\frac{5}{x-2}-3>0$$

$\Rightarrow \quad \frac{5-3x+6}{x-2}>0$

$\Rightarrow \quad \frac{-3x+11}{x-2}>0$

On multiplying by $[-(x-2)^2]$, we get

$$(3x-11)(x-2)<0$$

So, the factors $(3x-11)$ and $(x-2)$ have opposite signs.

$\therefore \quad (3x-11)>0$ and $(x-2)<0 \qquad \ldots(i)$

or $\quad (3x-11)<0$ and $(x-2)>0 \qquad \ldots(ii)$

From inequality (i), we get

$$3x-11>0 \text{ and } (x-2)<0$$

$\Rightarrow \quad x>\frac{11}{3}$ and $x<2$

i.e. $\quad x\in\left(\frac{11}{3},\infty\right)$ and $x\in(-\infty,2) \qquad \ldots(iii)$

On number line, it can be represented as

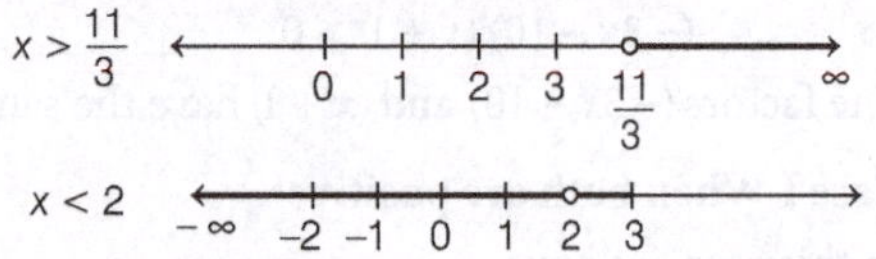

Since, no common point, so it has no solution.

From inequality (ii), we get

$$3x - 11 < 0 \text{ and } x - 2 > 0$$

$\Rightarrow$ $$x < \frac{11}{3} \text{ and } x > 2$$

$$x \in \left(-\infty, \frac{11}{3}\right) \text{ and } x \in (2, \infty) \quad \ldots\text{(iv)}$$

On number line, it can be represented as

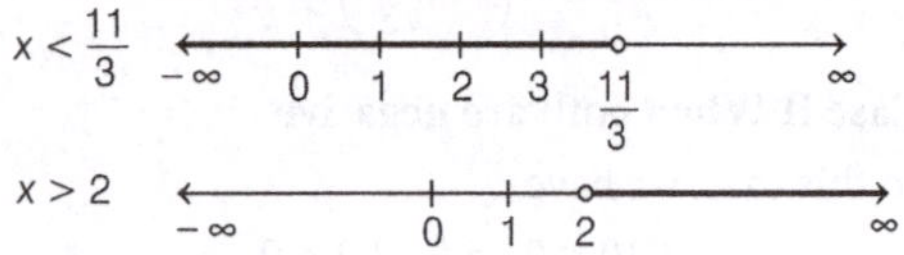

$\therefore$ $$2 < x < \frac{11}{3}$$

Hence, the solution set of the given inequality is $2 < x < \frac{11}{3}$ i.e. $\left(2, \frac{11}{3}\right)$.

23. We have, the following inequalities

$$|x - 1| \le 5 \quad \ldots\text{(i)}$$

and $$|x| \ge 2 \quad \ldots\text{(ii)}$$

From inequality (i), we get

$$-5 \le x - 1 \le 5 \quad [\because |x| \le a \Rightarrow -a \le x \le a]$$

$\Rightarrow$ $$-5 + 1 \le x \le 5 + 1 \quad \text{[adding 1 on each term]}$$

$\Rightarrow$ $$-4 \le x \le 6 \Rightarrow x \in [-4, 6]$$

$\therefore$ The solution set is $[-4, 6]$.

From inequality (ii), we get

$$|x| \ge 2$$

$\Rightarrow$ $$x \ge 2 \text{ or } x \le -2 \quad [\text{for } |x| \ge a, x \ge a \text{ or } x \le -a]$$

$\Rightarrow$ $$x \in [2, \infty) \text{ or } (-\infty, -2]$$

$\Rightarrow$ $$x \in (-\infty, -2] \cup [2, \infty)$$

$\therefore$ The solution set is $(-\infty, -2] \cup [2, \infty)$.

Now, let us draw the graphs of solutions of both inequalities.

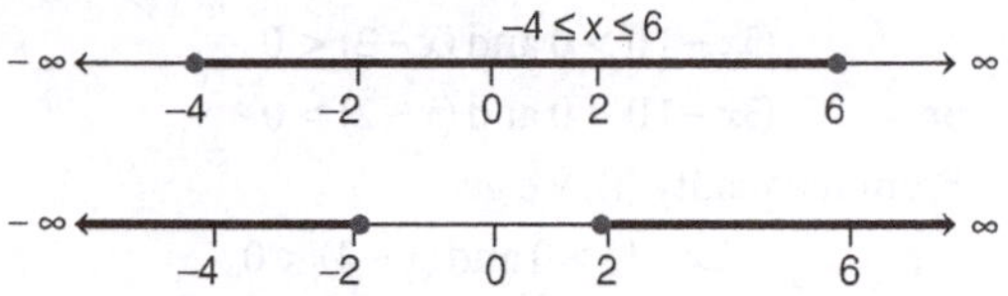

Clearly, the common values of x are lying in the interval $[-4, -2]$ or $[2, 6]$.

Hence, the required solution set is $[-4, -2] \cup [2, 6]$.

24. Given system of linear inequalities is

$$2(2x + 3) - 10 < 6(x - 2) \quad \ldots\text{(i)}$$

and $$\frac{2x - 3}{4} + 6 \ge 2 + \frac{4x}{3} \quad \ldots\text{(ii)}$$

From inequality (i), we get

$$2(2x + 3) - 10 < 6(x - 2)$$

$\Rightarrow$ $$4x + 6 - 10 < 6x - 12$$

$\Rightarrow$ $$4x - 4 < 6x - 12$$

$\Rightarrow$ $$4x - 4 + 4 < 6x - 12 + 4 \quad \text{[adding 4 on both sides]}$$

$\Rightarrow$ $$4x < 6x - 8$$

$\Rightarrow$ $$4x - 6x < 6x - 8 - 6x$$

[subtracting $6x$ from both sides]

$\Rightarrow$ $$-2x < -8$$

$\Rightarrow$ $$2x > 8$$ [dividing both sides by -1, then inequality sign will also change]

$\Rightarrow$ $$\frac{2x}{2} > \frac{8}{2} \quad \text{[dividing both sides by 2]}$$

$\therefore$ $$x > 4 \quad \ldots\text{(iii)}$$

Thus, any value of x greater than 4 satisfies the inequality.

$\therefore$ The solution set is $(4, \infty)$.

The representation of solution of inequality (i) is

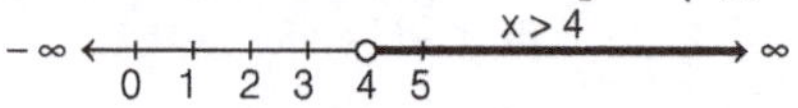

From inequality (ii), we get

$$\frac{2x - 3}{4} + 6 \ge 2 + \frac{4x}{3}$$

$\Rightarrow$ $$\frac{2x - 3 + 24}{4} \ge \frac{6 + 4x}{3} \Rightarrow \frac{2x + 21}{4} \ge \frac{6 + 4x}{3}$$

$\Rightarrow$ $$3(2x + 21) \ge 4(6 + 4x)$$

[multiplying both sides by LCM (4, 3) = 12]

$\Rightarrow$ $$6x + 63 \ge 24 + 16x$$

$\Rightarrow$ $$6x + 63 - 63 \ge 24 + 16x - 63$$

[subtracting 63 from both sides]

$\Rightarrow$ $$6x \ge 16x - 39$$

$\Rightarrow$ $$6x - 16x \ge 16x - 39 - 16x$$

[subtracting $16x$ from both sides]

$\Rightarrow$ $$-10x \ge -39$$

$\Rightarrow$ $$10x \le 39$$

[multiplying both sides by -1, the inequality sign will also change]

$\Rightarrow$ $$\frac{10x}{10} \le \frac{39}{10} \quad \text{[dividing both sides by 10]}$$

$\Rightarrow$ $$x \le 3.9 \quad \ldots\text{(iv)}$$

Thus, any value of x less than or equal to 3.9 satisfies the inequality.

$\therefore$ The solution set is $(-\infty, 3.9]$.

Its representation on number line is

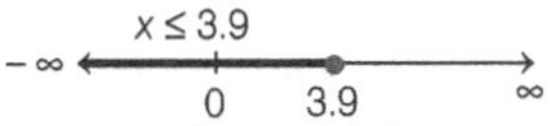

We see that there is no common value of x, which satisfies both inequalities (iii) and (iv).

Hence, the given system of inequalities has no solution set.

25. Let x L of water be required to be added.

Then, total mixture $= (x + 1125)$ L.

Clearly, the amount of acid in the resulting mixture is 45% of 1125 L. Since, the resulting mixture contains more than 25% but less than 30% acid content.

Therefore, we have

$$45\% \text{ of } 1125 > 25\% \text{ of } (x + 1125) \quad \ldots\text{(i)}$$

and

$$45\% \text{ of } 1125 < 30\% \text{ of } (x + 1125) \quad \ldots\text{(ii)}$$

From Eq. (i), we get

$$\frac{45}{100} \times 1125 > \frac{25}{100} \times (x + 1125)$$

$$\Rightarrow \quad 45 \times 1125 > 25 \times (x + 1125)$$

[multiplying both sides by 100]

$$\Rightarrow \quad 45 \times 1125 > 25x + (25 \times 1125)$$

$$\Rightarrow \quad 45 \times 1125 - (25 \times 1125) > 25x$$

[subtracting 25×1125 from both sides]

$$\Rightarrow \quad 25x < 1125(45 - 25)$$

$$\Rightarrow \quad 25x < 1125 \times 20$$

$$\Rightarrow \quad x < \frac{1125 \times 20}{25} = 225 \times 4$$

$$= 900$$

[dividing both sides by 25]

$$\Rightarrow \quad x < 900 \quad \ldots\text{(iii)}$$

From Eq. (ii), we get

$$\frac{45}{100} \times 1125 < \frac{30}{100} \times (x + 1125)$$

$$\Rightarrow \quad 45 \times 1125 < 30x + (30 \times 1125)$$

[multiplying both sides by 100]

$$\Rightarrow \quad 45 \times 1125 - (30 \times 1125) < 30x$$

[subtracting 30×1125 from both sides]

$$\Rightarrow \quad 30x > 1125(45 - 30)$$

$$\Rightarrow \quad x > \frac{1125 \times 15}{30} = \frac{1125}{2}$$

$$= 562.5$$

$$\Rightarrow \quad x > 562.5 \quad \ldots\text{(iv)}$$

From Eqs. (iii) and (iv), we get

$$562.5 < x < 900$$

Thus, the required number of litres of water that is to be added will have to be more than 562.5 L but less than 900 L.

QR Code for Quick Test 1

QR Code for Quick Test 2

Mind Map

Closed Interval

If a and b are real numbers such that $a < b$, then the set of all real numbers x such that $a \leq x \leq b$ is called a closed interval and is denoted by $[a, b]$.

$$[a, b]=\{x : a \leq x \leq b, x \in R\}$$

On the number line, $[a, b]$ may be represented as follows

$a \leq x \leq b$

$-\infty$ ← a —— b → ∞

Open Interval

If a and b are real numbers such that $a < b$, then the set of all real numbers x such that $a < x < b$, is called an open interval and is denoted by (a, b) or $]a, b[$.

$\therefore \quad (a, b)=\{x : a < x < b, x \in R\}$

On the number line, (a, b) may be represented as follows

$a < x < b$

$-\infty$ ← a —— b → ∞

Semi-open or Semi-closed Intervals

If a and b are real numbers such that $a < b$, then $(a, b] =\{x : a < x \leq b, x \in R\}$ and $[a, b) =\{x : a \leq x < b, x \in R\}$ are known as semi-open or semi-closed intervals. On the number line, these intervals may be represented as follows

$a < x \leq b$

$-\infty$ ← a —— b → ∞

$(a, b]$

$a \leq x < b$

$-\infty$ ← a —— b → ∞

$[a, b)$

Intervals on a Number Line

Inequality

A statement involving the symbols $>, <, \leq$ or $\geq$ is called an inequality or inequation.

e.g. $x < 2, x + y \geq 11$

Linear Inequalities

An inequality is said to be linear, if the variable(s) occurs in first degree only and there is no term involving the product of the variables.

e.g. $ax + b \leq 0, ax + by + c > 0$

Linear Inequality in Two Variables

A linear inequality which have only two variables, is called linear inequality in two variables.

e.g. $3x + 11y \leq 0, 4x + 3y > 0$

Linear Inequality in One Variable

A linear inequality which has only one variable, is called linear inequality in one variable.

e.g. $ax + b < 0$, where $a \neq 0$

Solution of an Inequality

Any solution of an inequality is the value(s) of variable(s) which makes it a true statement.

Solution of System of Linear Inequalities in One Variable

The solution set of a system of linear inequalities in one variable is defined as the **intersection of the solution set** of the liner inequalities in the system.

Multiplication or Division

If both sides of an inequality are multiplied (or divided) by the same positive number, then the sign of inequality remains the same. But when both sides are multiplied (or divided) by the same negative number, then the sign of inequality is reversed.

Let a, b and c be three real numbers such that $a > b$ and $c \neq 0$. Then,

- If $c > 0$, then $\frac{a}{c} > \frac{b}{c}$ and $ac > bc$.
- If $c < 0$, then $\frac{a}{c} < \frac{b}{c}$ and $ac < bc$

Addition or Subtraction

Some number may be added (or subtracted) to (from) both sides of an inequality

i.e. if $a > b$, then for any number c,

$a + c > b + c$ or $a - c > b - c$

Topics to be Assessed Formatively

Linear Inequalities in Two Variables and their Graphical Solutions

An inequality of the form $ax + by + c > 0$ or $ax + by + c < 0$ or $ax + by + c \geq 0$ or $ax + by + c \leq 0$, where $a \neq 0$ and $b \neq 0$, is called a linear inequality in two variables x and y.

The region containing all the solutions of an inequality, is called the **solution region.**

Concept of Half Planes

The graph of $ax + by + c = 0$ is a straight line which divides the **cartesian plane or XY-plane** into two parts. Each part is known as half plane.

Types of Half Planes

1. **Left and right half planes** A vertical line will divide the XY-plane in two parts, left half plane and right half plane.

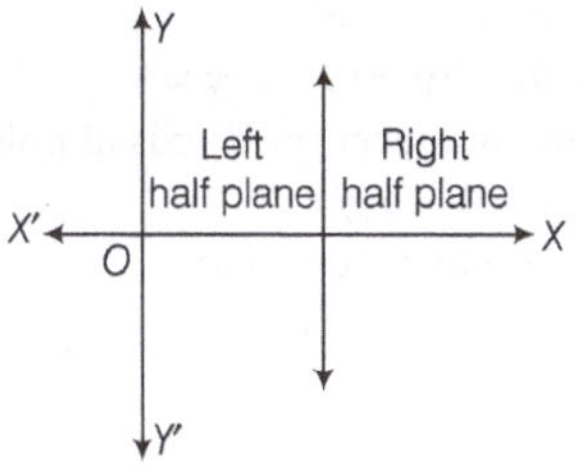

2. **Lower and upper half planes** A non-vertical line will divide the XY-plane into two parts, lower half plane and upper half plane.

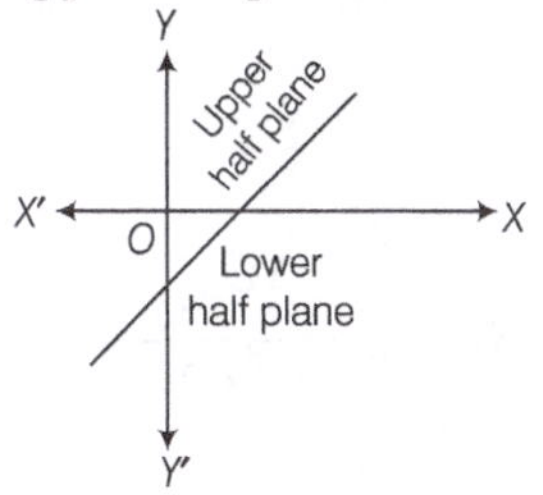

3. **Closed half plane** A half plane in XY-plane is called a closed half plane, if the line separating the plane is also included in the half plane.

   Therefore, the graph of a linear inequality involving sign $\leq$ or $\geq$ is always closed half plane.

4. **Open half plane** A half plane in XY-plane is called an open half plane, if the line separating the plane is not included in the half plane.

   Therefore, the graph of a linear inequality involving sign $<$ or $>$ is always an open half plane.

TEACHER TIPS

Many practical situations involves system of inequalities as number of units produced, number of articles purchased, number of hours worked etc. The variables x and y often represent quantities that cannot have negative values. In such cases, $x \geq 0$, $y \geq 0$ and the solution region lies only in the first quadrant.

Graph of Linear Inequality in Two Variables

Working rule for drawing the graph of linear inequalities in two variables are discussed below

***Step* I** Consider the equation $ax + by = c$, in place of given inequality $ax + by \leq c$ or $ax + by \geq c$ or $ax + by < c$ or $ax + by > c$, which represents a straight line in XY-plane.

***Step* II** Put $x = 0$ in the equation obtained in step I to get the point, where the line meets Y-axis. Similarly, put $y = 0$ to obtain a point, where the line meets the X-axis.

***Step* III** Draw a line joining the points obtained in step II. If the inequality is of the form of $<$ or $>$, then draw dotted line to indicate that the points on the line are excluded. Otherwise, draw a thick or dark line to indicate that the points on this line are included.

***Step* IV** Take any point (preferable origin, i.e. 0, 0), not lying on the line, and check whether this satisfies the given linear inequality or not.

***Step* V** If the inequality is satisfied, by this point, then shade that portion of the plane which contains the chosen point. The shaded region obtained in step V, represents the graph of linear inequality in two variable.

Note The gaph of linear inequality in two variable always represent a half plane (closed or open).

Example 1 Draw the graph of the inequality $y+8\geq 2x$.

Sol. Given inequality is $y+8\geq 2x$.

In equation form, it can be written as

$$y+8=2x \text{ or } 2x-y=8$$

On putting $y=0$, we get $2x-0=8 \Rightarrow x=4$

$\therefore$ Line $2x-y=8$ cuts the X-axis at $A(4, 0)$.

On putting $x=0$, we get

$$2\times 0-y=8 \Rightarrow y=-8$$

$\therefore$ Line $2x-y=8$ cuts the Y-axis at $B(0, -8)$.

On joining the points $A(4, 0)$ and $B(0, -8)$ by a dark line, we get the line AB.

[$\because$ given inequality has sign $\geq$, so we draw dark line]

Let us check whether $O(0, 0)$ satisfy the inequality or not and shade the suitable region.

On putting $x=y=0$ in $y+8\geq 2x$, we get

$$0+8\geq 2\cdot 0$$

$\Rightarrow$ $8\geq 0$, which is true.

So, shade the region containing $(0, 0)$ i.e. the region above the line AB.

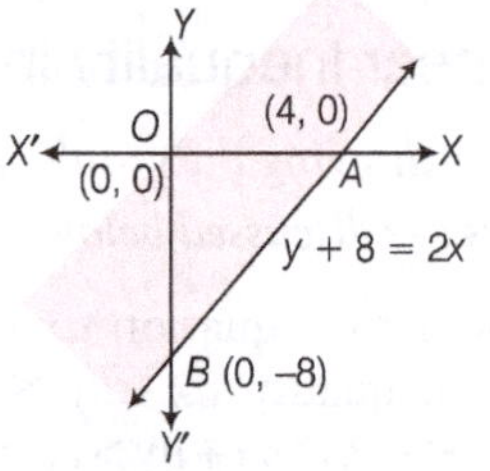

Here, the shaded region including the points on the line represent the graph of the inequality.

Example 2 Check whether the half plane $x+2y\geq 4$ contains origin.

Sol. We have, $x+2y\geq 4$

On putting $x=0$ and $y=0$, we get

$$0+2(0)\geq 4 \Rightarrow 0\geq 4$$

This is a false statement. Therefore, the given half plane does not contain the origin.

Example 3 Solve the inequality $2x+y>3$ graphically.

Sol. We have, $2x+y>3$

In equation form, given inequality can be written as

$$2x+y=3 \quad \text{...(i)}$$

On putting $x=0$ in Eq. (i), we get

$$2(0)+y=3 \Rightarrow y=3$$

$\therefore$ Line meets Y-axis at point $A(0, 3)$.

On putting $y=0$ in Eq. (i), we get

$$2x+0=3 \Rightarrow x=\frac{3}{2}$$

$\therefore$ Line meets X-axis at point $B\left(\frac{3}{2}, 0\right)$.

On joining the points $A(0, 3)$ and $B\left(\frac{3}{2}, 0\right)$ by dotted line, we get the line AB. [$\because$ given inequality has sign $>$, so we join the points by a dotted line]

Now, take a point not lying on the line say $(0, 0)$ to check whether it satisfies the given inequality or not.

On putting $x=0$ and $y=0$ in given inequality, we get

$2(0)+0>3 \Rightarrow 0>3$, which is false.

So, we shade the portion which does not contain $(0, 0)$ i.e. the region above the dotted line AB.

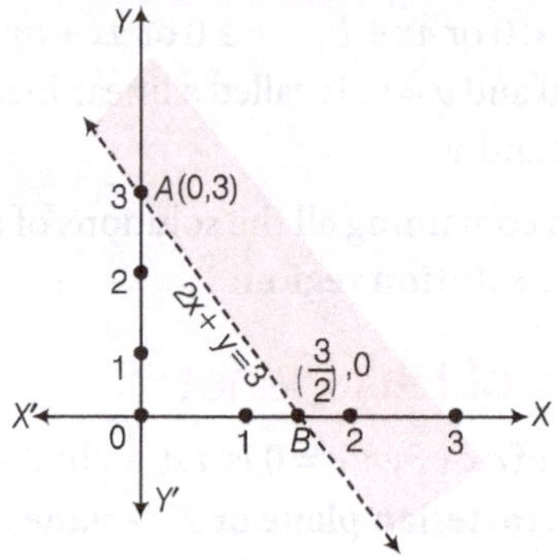

Here, the shaded region is the required solution region.

Example 4 Solve the inequality $5x+2y\leq 10$ graphically.

Sol. Given inequality is $5x+2y\leq 10$.

Corresponding equation of line is

$$5x+2y=10$$

On putting $x=0$, we get

$$5(0)+2y=10 \Rightarrow y=5$$

Thus, the line intersect the Y-axis at point $(0, 5)$.

On putting $y=0$, we get

$$5x+2(0)=10 \Rightarrow x=2$$

x	0	2
y	5	0

Thus, the line intersect the X-axis at point $(2, 0)$.

Join the points $A(0, 5)$ and $B(2, 0)$ with a dark line

[$\because$ given inequality has sign $\leq$]

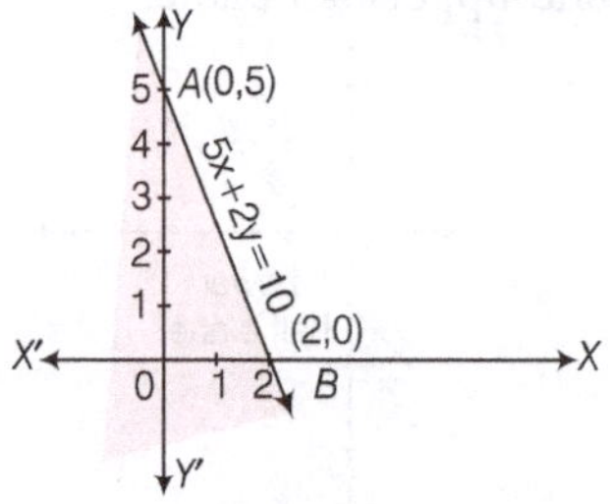

On putting $(0, 0)$ in given inequality, we get

$$5\times 0+2\times 0\leq 10 \Rightarrow 0\leq 10, \text{ which is true.}$$

So, the half plane of $5x+2y\leq 10$ contains origin.

Now, shade the half plane which contain $(0, 0)$.

Here, the shaded region including the points on the line represent the solution region of given inequality.

Graphical Solutions of System of Linear Inequalities in Two Variables

A system of linear inequalities in two variables can be solved by graphical method.

For finding the solution, we use the following steps

***Step* I** Draw the graph of all the given inequalities.

***Step* II** Find the common shaded region, which satisfies all the given linear inequalities.

***Step* III** This common region is the required solution region of the system of given inequalities. If there is no common region, then the system of inequalities has no solution.

Example 5 Solve the system of inequalities graphically.

$$x+y>6;\ 2x-y<0$$

Sol. We have, $x+y>6$ ...(i)

and $2x-y<0$ ...(ii)

Take inequality (i), $x+y>6$

In equation form, it can be written as $x+y=6$

Now, let us construct the following table

x	0	6
y	6	0

Thus, the line $x+y=6$ passes through points A (0,6) and B (6, 0).

Now, on joining the points A and B by a dotted line, we get the line AB.

On putting $x=0$ and $y=0$ in inequality (i), we get

$0+0>6 \Rightarrow 0>6$, which is false.

So, for inequality $x+y>6$, shade the half plane which does not contain origin.

Take inequality (ii), $2x-y<0$

In equation form it can be written as $2x-y=0$.

Now, let us construct the following table

x	0	2
y	0	4

Thus, the line $2x-y=0$ passes through points O (0,0) and C (2, 4).

Now, on joining the points O and C by a dotted line, we get the line OC.

On putting $x=0$ and $y=6$ in inequality (ii), we get

$$2(0)-6<0$$

$\Rightarrow$ $-6<0$, which is true.

So, for inequality $2x-y<0$, shade the half plane region which contain (0, 6) excluding all points on line.

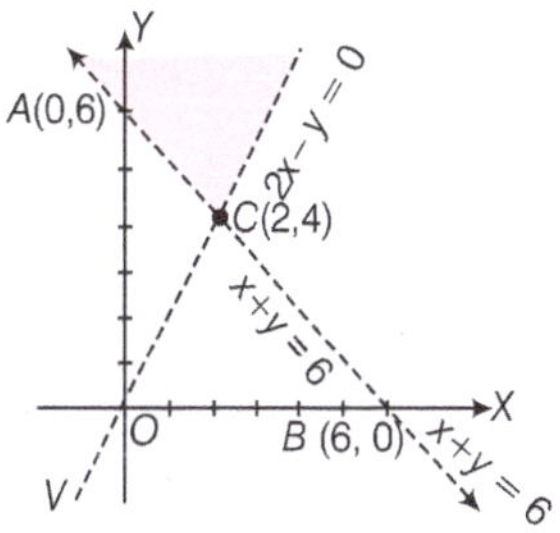

The common shaded region represent the solution region for system of inequality.

Example 6 Solve the inequalities graphically $3x+4y\le 60$, $x+3y\le 30$, $x\ge 0$, $y\ge 0$.

Sol. We have the following inequalities

$3x+4y\le 60$ ...(i)

$x+3y\le 30$ ...(ii)

$x\ge 0$ ...(iii)

and $y\ge 0$ ...(iv)

Take inequality (i),

$$3x+4y\le 60$$

In equation form, it can be written as $3x+4y=60$.

Now, let us construct the following table

x	0	20
y	15	0

Thus, the line intersect X-axis at $A(20, 0)$ and Y-axis at $B(0, 15)$. Now, joining the points A and B by a dark line, we get the line AB.

On putting (0, 0) in the given inequality, we get

$0+0\le 60 \Rightarrow 0\le 60$, which is true.

So, for $3x+4y\le 60$, shade the half plane which contain (0, 0).

Take inequality (ii), $x+3y\le 30$

In equation form, it can be written as $x+3y=30$

Now, let us construct the following table

x	0	30
y	10	0

Thus, the line intersect X-axis at $C(30, 0)$ and Y-axis at $D(0, 10)$.

Now, joining the points C and D by a dark line, we get the line CD.

On putting (0, 0) in the given inequality, we get

$0+0\le 30 \Rightarrow 0\le 30$, which is true.

So, for $x+3y\le 30$, shade the half plane which contain (0, 0).

Take inequalities (iii) and (iv), $x\ge 0$ and $y\ge 0$

These inequalities represent the region of first quadrant including the points on the axes. So, shade the first quadrant.

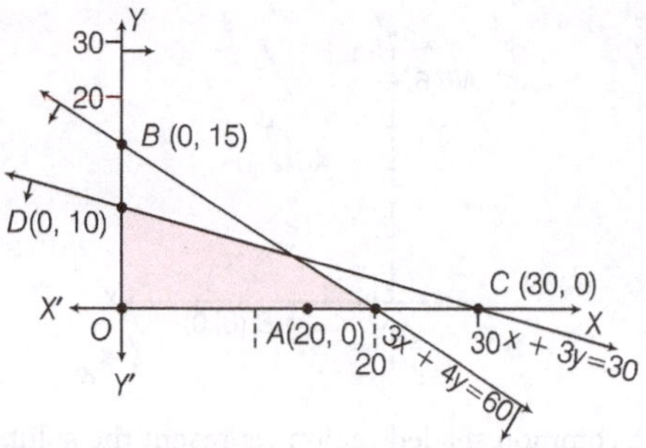

Here, common shaded region represent the solution region for system of inequalities.

Inequality between Coordinate Axes

The inequality $x \geq 0$ consist of Y-axis and the plane on the right side of Y-axis. Also, $y \geq 0$ consist of X-axis and the plane above the X-axis.

$\therefore$ The inequalities $x \geq 0$ and $y \geq 0$, together represent the first quadrant including the point on the axes.

Example 7 **Solve the following inequalities graphically.** $3x+2y \leq 12, x+y \geq 8, -x+y \geq 4, 5x \leq 10; x, y \geq 0.$

Sol. We have, inequalities

$$3x + 2y \leq 12 \quad \ldots(i)$$
$$x + y \geq 8 \quad \ldots(ii)$$
$$-x + y \geq 4 \quad \ldots(iii)$$
$$5x \leq 10 \quad \ldots(iv)$$
and $$x, y \geq 0 \quad \ldots(v)$$

Take inequality (i), $3x + 2y \leq 12$

Corresponding equation of line is $3x + 2y = 12$.

Consider the table of values, satisfying the equation $3x + 2y = 12$.

x	0	4
y	6	0

Thus, the line passing through the points $A(0, 6)$ and $B(4, 0)$.

Join the points A and B by a dark line.

On putting (0, 0) in the given inequality (i), we get

$$3(0) + 2(0) \leq 12 \Rightarrow 0 \leq 12, \text{ which is true.}$$

So, half plane corresponding to $3x + 2y \leq 12$ contains the origin.

Take inequality (ii), $x + y \geq 8$

Corresponding equation of line is $x + y = 8$.

Consider the table of values, satisfying the equation $x + y = 8$.

x	0	8
y	8	0

Thus, the line passing through the points $C(0, 8)$ and $D(8, 0)$.

Join the points C and D by a dark line.

On putting (0, 0) in the given inequality (ii), we get

$$0 + 0 \geq 8 \Rightarrow 0 \geq 8, \text{ which is false.}$$

So, half plane corresponding to $x + y \geq 8$ does not contain the origin.

Take inequality (iii), $-x + y \geq 4$

Corresponding equation of line is $-x + y = 4$.

Consider the table of values, satisfying the equation $-x + y = 4$.

x	0	-4
y	4	0

Thus, the line passing through the points $E(0, 4)$ and $F(-4, 0)$.

Join the points E and F by a dark line.

On putting (0, 0) in the given inequality (iii), we get

$$-0 + 0 \geq 4 \Rightarrow 0 \geq 4, \text{ which is false.}$$

So, half plane cooresponding to $-x + y \geq 4$ does not contain the origin.

Take inequality (iv), $5x \leq 10$

Corresponding equation of line is $5x = 10 \Rightarrow x = 2$.

This is a line parallel to Y-axis at a distance of 2 units to the right of Y-axis.

On putting (0, 0) in the given inequality (iv), we get

$$5(0) \leq 10 \Rightarrow 0 \leq 10, \text{ which is true.}$$

So, half plane corresponding to $5x \leq 10$ contains the origin.

Take inequality (v), $x, y \geq 0$

These inequalities represent the region of first quadrant including the points on the axes.

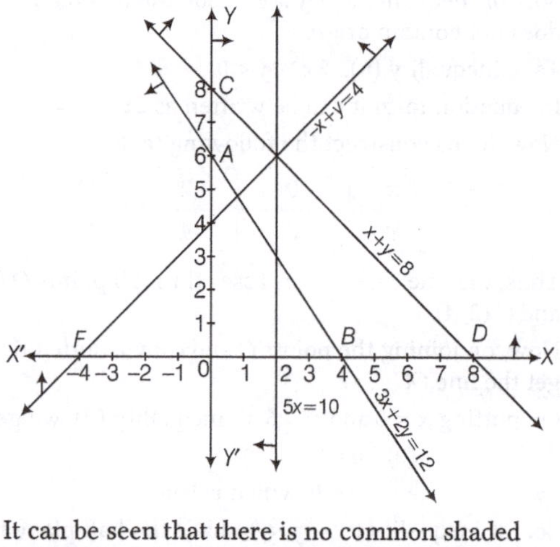

It can be seen that there is no common shaded region, therefore given system of inequalities has no solution.

Example 8 Solve the inequalities $|y - x| \leq 3$, graphically.

Sol. We have, $|y - x| \leq 3$

$\Rightarrow$ $-3 \leq y - x \leq 3$

$\Rightarrow$ $y - x \geq -3$ ...(i)

and $y - x \leq 3$ ...(ii)

Take inequality (i),

$$y - x \geq -3$$

In equation form, it can be written as

$$y - x = -3$$

Now, let us construct the table of values, which satisfy the above equation.

x	0	3
y	-3	0

Thus, the line passing through the points $A(3, 0)$ and $B(0, -3)$.

Now, join the points A and B by a dark line.

On putting (0, 0) in inequality (i), we get $0 \geq -3$, which is true.

So, for $y - x \geq -3$, shade the half plane which contain origin.

Take inequality (ii),

$$y - x \leq 3$$

In equation form, it can be written as

$$y - x = 3$$

Now, let us construct the table of values, which satisfy the above equation.

x	0	-3
y	3	0

Thus, the line passing through the points $C(-3, 0)$ and $D(0, 3)$. Now, join these points by a dark line.

On putting (0, 0) in inequality (ii), we get $0 \leq 3$, which is true.

So, for $y - x \leq 3$, shade the half plane which contain origin.

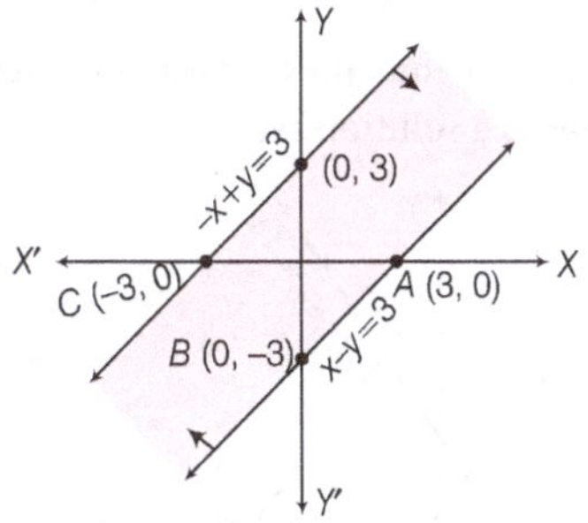

The common shaded region represent the solution region for system of inequalities.

Method of Finding the Linear Inequalities when Their Solution region is Given

When solution region is given, then *we can find the linear inequalities with the help of following steps*

***Step* I** Choose a line given in graphs.

***Step* II** Check whether the shaded region corresponding to the line and the origin lie on the same side or on the opposite side of the line.

***Step* III** ***Case* I** If the shaded region and origin lie on the same side, then corresponding inequality has sign $\leq$.

***Case* II** If the shaded region and origin lie on the opposite side, then corresponding inequality has sign $\geq$.

***Step* IV** Repeat these steps for each line to get all inequalities.

Example 9 Find the linear inequalities for which the shaded region in given figure is the solution set.

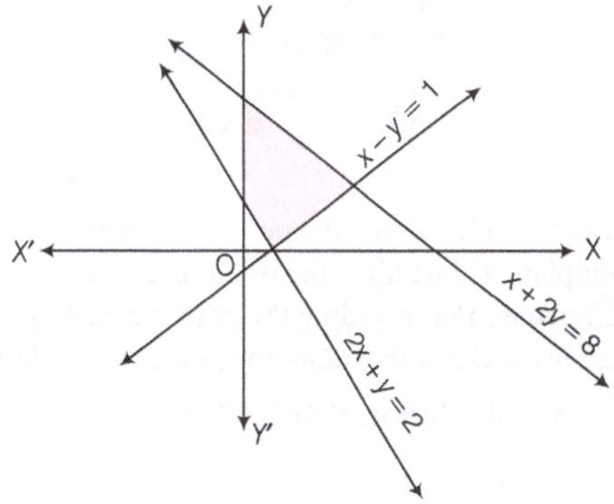

Sol. Consider the equation of the line $x + 2y = 8$.

We observe that the shaded region and the origin lie on the same side of the line. So, the corresponding inequality is $x + 2y < 8$.

Now, consider the equation of the line $x - y = 1$.

We observe that the shaded region and origin both lie on the same side. So, corresponding inequality is $x - y < 1$.

Consider the equation of the line $2x + y = 2$. We observe that the shaded region and origin lie on opposite side of the line. So, corresponding inequality is $2x + y > 2$.

We also observe that the shaded region is above X-axis and is on the right side of Y-axis. So, $x \geq 0$ and $y \geq 0$.

Thus, the linear inequalities corresponding to the given solution set are

$$x + 2y < 8,$$
$$2x + y > 2$$
$$x - y > 1$$
$$x \geq 0,\ y \geq 0$$

Exercise

Multiple Choice Questions

1. Which of the following is/are true?

(a) A vertical line will divide the plane in left and right half planes

(b) A non-vertical line will divide the plane into left and right half planes

(c) Both (a) and (b)

(d) None of the above

2. If a point $P(\alpha, \beta)$ on the line $ax + by = c$, then

(a) $a\alpha + b\beta > c$ (b) $a\alpha + b\beta < c$

(c) $a\alpha + b\beta = c$ (d) None of these

3. The graph of the inequality $ax + by > c$ is represented by shading in the ...

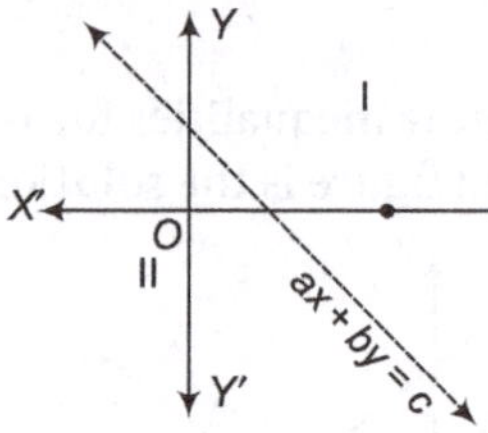

(a) half plane I including the point on $ax + by = c$

(b) half plane II including the point on $ax + by = c$

(c) half plane I not including the point on $ax + by = c$

(d) half plane II not including the point on $ax + by = c$

4. The graphical solution of $3x - 6 \geq 0$ is

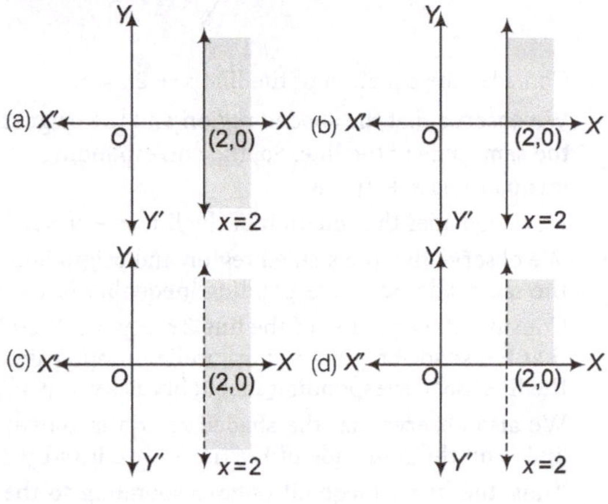

5. Consider the inequality $40x + 20y \leq 120$, where x and y are whole numbers. Then, its solution set is

(a) (0, 0), (5, 5), (1, 1), (2, 2), (3, 0)

(b) (0, 0), (0, 1), (0, 2), (0, 3), (0, 4), (0, 5), (0, 6)

(c) (1, 0), (2, 0), (3, 0), (4, 0), (5, 0)

(d) None of the above

Very Short Answer Type Questions

6. Check whether the plane $5x + 2y \leq 5$ contains origin or not.

7. Check whether the given plane $3x - 6y \leq 0$ contains the point (3, 1).

8. Solve graphically $4x + 3y \geq 12$ and $4x - 5y \geq -20$

Short Answer Type Questions

9. Solve the following inequalities $2x + y \geq 6$ and $3x + 4y \leq 12$ graphically.

10. Solve the following system of inequalities graphically.

$x + y \leq 4, 2x - y > 0, x \geq 0$ and $y \geq 0$.

11. Solve graphically $x - y \leq 2$; $x + 2y \leq 8$; $x, y \geq 0$.

Long Answer Type Questions

12. Solve the inequality $3y - 5x < 30$ graphically.

13. Solve the inequality $x + y \geq 5$ graphically.

14. Solve the following inequalities $2x + y \geq 4$, $x + y \leq 3$ and $2x - 3y \leq 6$ graphically.

15. Solve the system of inequalities graphically.

$$x + y \leq 4,$$
$$3x + y \geq 3,$$
$$x + 4y \geq 4,$$
$$x \leq 3, y \leq 2$$

16. Solve graphically the following system of inequalities.

$$x + 2y \leq 3; 3x + 4y \geq 12, \ x \geq 0 \text{ and } y \geq 1$$

17. Solve the following system of inequalities graphically.

$$3x + 2y \leq 150,$$
$$x + 4y \leq 80,$$
$$x \leq 15, y \geq 0, x \geq 0$$

18. Find the linear inequalities for which shaded area in figure below is a solution set.

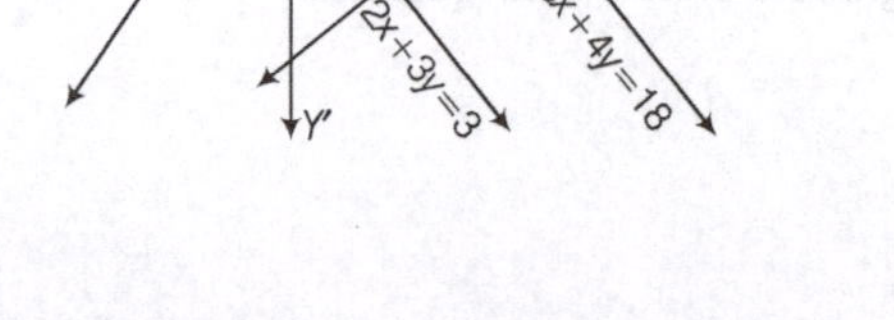

Answers

1. (a)
2. (c)
3. (c)
4. (a)
5. (b)
6. Given plane contains origin.
7. Given plane does not contain the point (3, 1).
8.

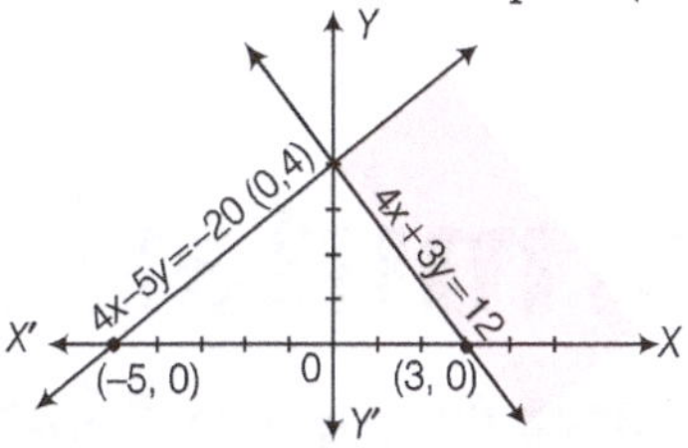

9. Solve as Example 3.

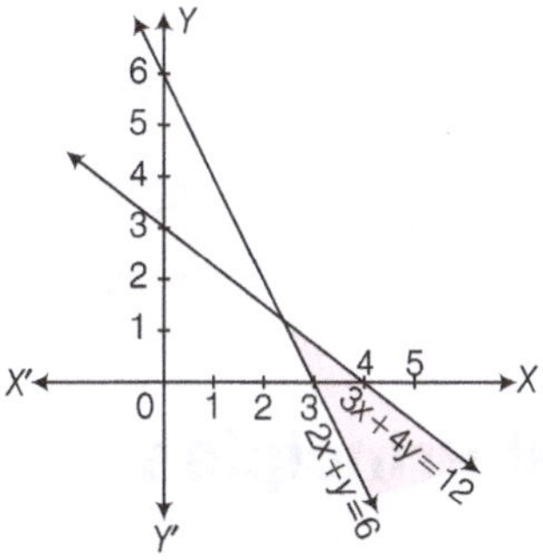

10.

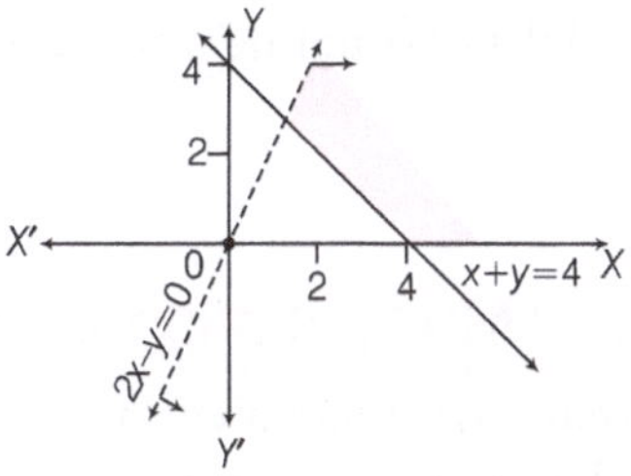

11.

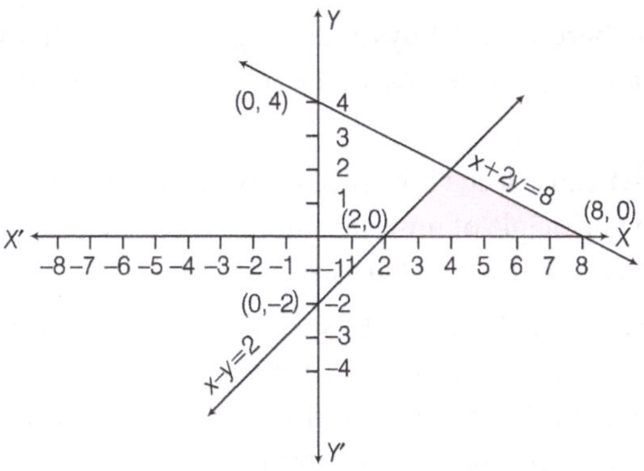

12. Firstly consider the given inequation as a equation.

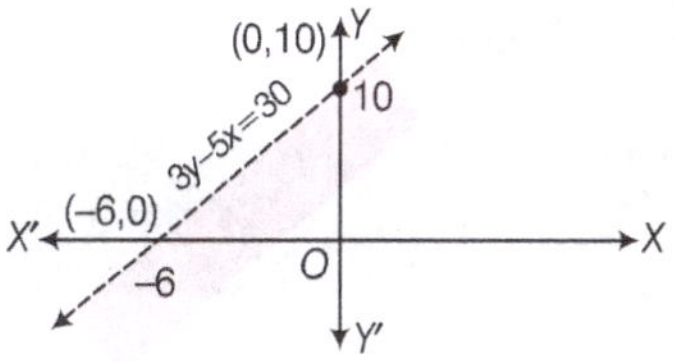

13.

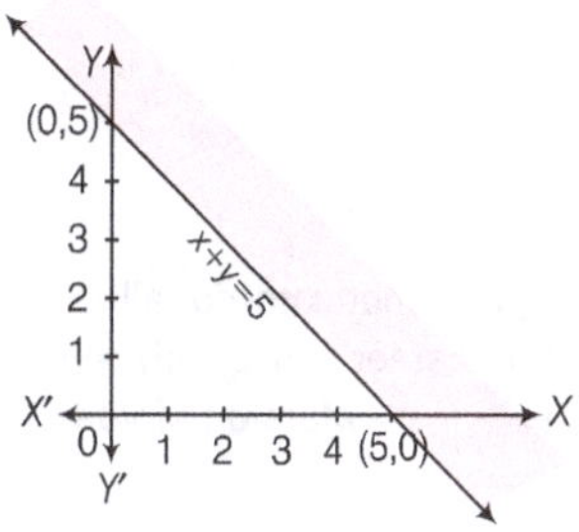

14.

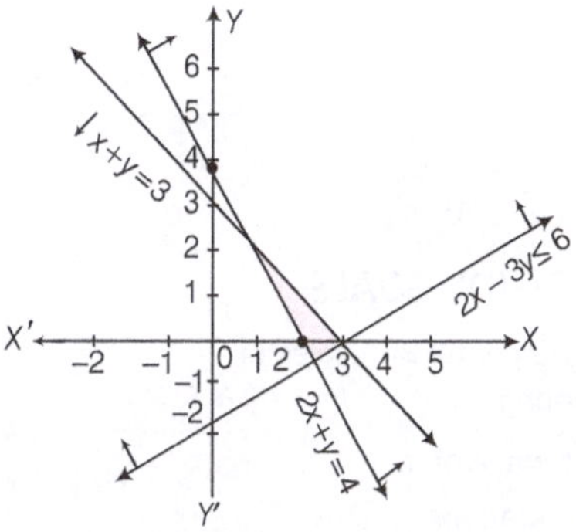

15.

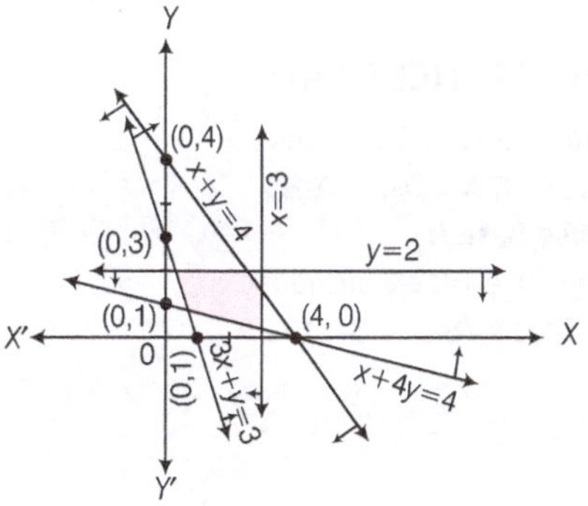

16. Do yourself.
17. Do yourself.
18.

$$3x + 4y \le 18$$
$$-7x + 4y \le 14$$
$$2x + 3y \ge 3$$
$$x - 6y \le 3,$$
$$x \ge 0,$$
$$y \ge 0$$

06

CHAPTER

Mathematics is a great motivator for all humans because its career starts with zero and it never ends. —*Marston Morse*

Permutations and Combinations

▼ **YOUR STUDY GOALS**

6.1 Fundamental Principles of Counting

6.2 Factorial Notation

6.3 Permutations

6.4 Combinations

▼ **YOUR PRACTICE GOALS**

- Objective Type Questions (29 Qs ; 17 MCQs, 10 A-R, 2 Case Based)
- Subjective Type Questions (43 Qs; 16 VSA, 17 SA, 10 LA)

6.1 Fundamental Principles of Counting

There are two fundamental principles of counting which are the base of permutations a

nd combinations.

Fundamental Principle of Multiplication (FPM)

If an event can occur in m different ways, another event can occur in n different ways, then the total number of occurrence of the events in the given order is $m \times n$.

e.g. In a school, there are 200 boys and 150 girls. The teacher wants to select a boy and a girl to represent the school in inter school competition.

Here, the teacher can select a boy in 200 ways and a girl in 150 ways. So, by principle of multiplication, the teacher can select a boy and a girl in $200 \times 150 = 30000$ ways.

If there are n events, say $E_1, E_2, \dots, E_n$ such that E_1 can occur in m_1 ways, E_2 can occur in m_2 ways, E_3 can occur in m_3 ways and so on, then the total number of ways in which all the events can occur in the stated order is $m_1 \times m_2 \times m_3 \times \dots \times m_n$.

Fundamental Principle of Addition (FPA)

If there are two events such that they can be performed independently in m and n ways respectively, then either of the two events can occur in $(m+n)$ ways.

e.g. In a school, there are 200 boys and 150 girls. The teacher wants to select either a boy or a girl to represent the school in inter school competition.

Here, the teacher can select a boy in 200 ways and a girl in 150 ways.

So, by fundamental principle of addition, the teacher can select a boy or a girl in $(200+150)=350$ ways.

Note *The above principle can be extended to any finite number of operation.*

6.2 Factorial Notation

Let n be a positive integer. Then, the continued product of first n natural numbers is called the **n factorial** and it is denoted by $n!$ or $\underline{|n}$.

Thus, $n! = n(n-1)(n-2)(n-3)\ldots\ldots 3\cdot 2\cdot 1.$

Zero Factorial

It does not make any sense to define 0! as product of the integers from 1 to 0. So, we define $0! = 1$.

Factorial of proper fractions or negative integers are not defined. $n!$ is defined only for whole numbers i.e. for non-negative integers.

Example 1. A room has 7 doors. In how many ways can a man enter the room through one door and come out through a different door?

Sol. Here, we need to perform two operations.

(i) Selecting a door to enter.

(ii) Selecting a door to come out.

Clearly, the man can enter the room through anyone of the seven doors.

So, there are seven ways of entering into the room. Note that the man can come out through anyone of the remaining six doors. So, he can come out through a different door in 6 ways.

Hence, by fundamental principle of multiplication, required number of ways $= 7\times 6 = 42$

Example 2. In an examination, there are three multiple choice questions and each question has 4 choices. Find the number of ways in which a student can fail to get all answer correct.

Sol. Since, each question can be answered in 4 ways.

So, the total number of ways answering 3 questions is

$$4\times 4\times 4 = 64.$$

Out of these possible ways, only one way will be correct and hence the number of ways in which a student can fail to get all correct answer is $64-1=63$.

Example 3. How many 3-digit numbers can be formed from the digits 1, 2, 3, 4 and 5, assuming

(i) repetition of digits allowed? *NCERT*

(ii) repetition of digits not allowed?

Sol. (i) There are 5 digits *viz.* 1, 2, 3, 4 and 5.

Since, every digit can be repeated any number of times and we have to form 3-digit numbers.

So, hundred's place can be fill in 5 ways, ten's place can be fill in 5 ways and unit's place can be fill in 5 ways.

Hundred's place	Ten's place	Unit's place
↓ 5 ways	↓ 5 ways	↓ 5 ways

Thus, by the fundamental principle of multiplication, the number of ways in which the 3-digit number can be formed $= 5\times 5\times 5 = 125$

Hence, the number of 3-digit numbers is 125.

(ii) When repetition of digits is not allowed.

Hundred's place can be fill in 5 ways.

Ten's place can be fill in 4 ways.

Unit's place can be fill in 3 ways.

Hundred's place	Ten's place	Unit's place
↓ 5 ways	↓ 4 ways	↓ 3 ways

According to FPM, total number of ways

$$= 5\times 4\times 3 = 60$$

Hence, number of 3-digit numbers is 60.

Example 4. Find the number of positive integers greater than 6000 and less than 7000, which are divisible by 5, provided that no digit is to be repeated.

Sol. We know, a number will be divisible by 5, if either 0 or 5 is at unit place.

$\therefore$ Number of ways to fill the unit's place = 2

Also, note that for number to be greater than 6000 and less than 7000, digit 6 should be fixed at thousand's place.

$\therefore$ Number of ways to fill the thousand's place = 1

Now, we are left with 8 digits. So, hundred's place can be filled in 8 ways and ten's place can be filled in 7 ways.

Thousand's place	Hundred's place	Ten's place	Unit's place
6			0 or 5
↓ 1 way	↓ 8 ways	↓ 7 ways	↓ 2 ways

$\therefore$ According to FPM, total number of numbers

$$=1\times 8\times 7\times 2=112$$

Example 5. How many 4-digit numbers are there, when a digit may be repeated any number of times?

Sol. The thousand's place of a 4-digit number cannot be 0. So, thousand's place can be filled with any digit from 1 to 9. Thus, thousand's place can be filled in 9 ways.

Since, repetition of digits is allowed, therefore each of the remaining 3 places can be filled in 10 ways by using the digits from 0 to 9.

Hence, by the fundamental principle of multiplication, the required number of numbers

$$=9\times 10\times 10\times 10=9000$$

Example 6. By using the digits 0, 1, 2, 3, 4 and 5 (repetition not allowed) numbers are formed using any number of digits. Find the total number of non-zero numbers that can be formed.

Sol. Required number of numbers

= Number of 1-digit number + Number of 2-digit numbers + ... + Number of 6-digit numbers

$$=5+5\times 5+5\times 5\times 4+5\times 5\times 4\times 3+5\times 5\times 4\times 3\times 2+5\times 5\times 4\times 3\times 2\times 1$$

$$=5+25+100+300+600+600=1630$$

Example 7. How many numbers are there between 100 and 1000, which have exactly one of their digits as 7? *Competency Based*

Sol. We have to find numbers between 100 and 1000 i.e. 3-digit numbers which have exactly one of their digits as 7.

Now, consider the following cases

Case I When 7 is at unit's place

In this case, ten's and hundred's place can be filled using the digits 0, 1, 2, 3, 4, 5, 6, 8 and 9.

Clearly, the number of ways to fill ten's place = 9

[using anyone of the digits 0, 1, 2, 3, 4, 5, 6, 8, 9]

and the number of ways to fill the hundred's place = 8

[$\because$ hundred's place cannot filled by zero]

$\therefore$ Number of such numbers $=1\times 9\times 8=72$

Case II When 7 is at ten's place

In this case, unit's and hundred's place can be filled using the digits 0, 1, 2, 3, 4, 5, 6, 8 and 9.

Clearly, the number of ways to fill unit's place = 9

[using anyone of digits 0, 1, 2, 3, 4, 5, 6, 8, 9]

and the number of ways to fill hundred's place = 8

[$\because$ hundred's place cannot be filled by zero]

$\therefore$ Number of such numbers $=9\times 1\times 8=72$

Case III When 7 is at hundred's place

In this case, unit's and ten's place can be filled using the digits 0, 1, 2, 3, 4, 5, 6, 8 and 9.

Clearly, the number of ways to fill unit's place = 9

[using anyone of the digits 0, 1, 2, 3, 4, 5, 6, 8, 9]

and the number of ways to fill the ten's place = 9

[using anyone of the digits 0, 1, 2, 3, 4, 5, 6, 8, 9]

$\therefore$ Number of such numbers $=1\times 9\times 9=81$

Hence, the total number of required numbers

$$=72+72+81=225$$

Example 8. Compute $\frac{8!}{4!}$.

Or Is $\frac{8!}{4!}=2!$?

Sol. We have, $\frac{8!}{4!}=\frac{8\times 7\times 6\times 5\times 4!}{4!}$

$[\because n!=n(n-1)(n-2)\ldots 3\cdot 2\cdot 1]$

$$=8\times 7\times 6\times 5=1680$$

Also, $2!=2\times 1=2\neq 1680$

$\therefore$ $\frac{8!}{4!}\neq 2!$

Example 9. Evaluate $\frac{n!}{(n-r)!}$ when $r=3$.

Sol. We have, $\frac{n!}{(n-r)!}=\frac{n!}{(n-3)!}$ $[\because r=3, \text{ given}]$

$$=\frac{n(n-1)(n-2)(n-3)!}{(n-3)!}$$

$[\because n!=n(n-1)(n-2)(n-3)!]$

$$=n(n-1)(n-2)$$

Example 10. Prove that $\frac{(2n)!}{n!}=\{1\cdot 3\cdot 5\ldots(2n-1)\}2^n$.

Sol. LHS $=\frac{(2n)!}{n!}=\frac{1\cdot 2\cdot 3\cdot 4\cdot 5\ldots(2n-2)(2n-1)(2n)}{n!}$

$$= \frac{\{1\cdot3\cdot5\ldots(2n-1)\}\{2\cdot4\cdot6\ldots(2n-2)(2n)\}}{n!}$$

$$= \frac{\{1\cdot3\cdot5\ldots(2n-1)\}\cdot2^n\{1\cdot2\cdot3\ldots(n-1)n\}}{n!}$$

$$= \frac{\{1\cdot3\cdot5\ldots(2n-1)\}\cdot2^n\cdot n!}{n!} [\because 1\cdot2\cdot3\ldots(n-1)\,n = n!]$$

$$= \{1\cdot3\cdot5\ldots(2n-1)\}2^n$$

= RHS **Hence proved.**

Example 11. What is the largest integer n such that 33! is divisible by 2^n? *Competency Based*

Sol. Clearly, $33! = 1\times2\times3\times\ldots\times30\times31\times32\times33$

$$= [2\cdot4\cdot6\cdot8\ldots30\cdot32](1\cdot3\cdot5\cdot7\cdot9\ldots31\cdot33)$$

$$= [2^1\cdot2^2\cdot(2\times3)\cdot2^3\cdot(2\times5)\cdot(2^2\times3)\cdot(2\times7)\cdot2^4 (2\times9)\cdot(2^2\times5)\cdot(2\times11)\cdot(2^3\times3)\cdot(2\times13) (2^2\times7)\cdot(2\times15)\cdot2^5]\cdot(1\cdot3\cdot5\ldots31\cdot33)$$

$$= [2^{1+2+1+3+1+2+1+4+1+2+1+3+1+2+1+5} \cdot(3\cdot5\cdot3\cdot7\cdot9\cdot5\cdot11\cdot3\cdot13\cdot7\cdot15)](1\cdot3\cdot5\ldots31\cdot33)$$

$$= 2^{31}[3\cdot5\cdot3\cdot7\cdot9\cdot5\cdot11\cdot3\cdot13\cdot7\cdot15](1\cdot3\cdot5\ldots31\cdot33)$$

Thus, 33! is divisible by 2^{31}.

Hence, the maximum power of 2 in 33! is 31.

Quick Test 1

1. Pratiksha wants to go abroad by ship and to return by airline. She has a choice of 6 different ships to go and 4 airlines to return. In how many ways can she perform her journey?
2. A telegraph has 5 arms and each arm is capable of 4 distinct positions, including the position of rest, then find the total number of signals that can be made. *Competency Based Que*
3. How many words (with or without meaning) of three distinct letters of the English alphabet are there?
4. A sequence is a ternary sequence, if it contains digits 0, 1 and 2, then find the total number of ternary sequences of length 9 which either begin with 210 or end with 210. *Competency Based Que*
5. A team consists of 6 boys and 4 girls and other has 5 boys and 3 girls. How many single matches can be arranged between the two teams when a boy plays against a boy and a girl play against a girl?
6. A gentleman has 6 friends to invite. In how many ways can he send invitation cards to them, if he has three servants to carry the cards? *Competency Based Que*
7. How many 3-digit numbers more than 600 can be formed using the digits 2, 3, 4, 6 and 7?
8. Five persons entered in the lift cabin on the ground floor of 8-floor house. Suppose each of them can leave the cabin independently at any floor beginning with the first. Find the total number of ways in which each of the five persons can leave the cabin *Competency Based Que*
   (i) at anyone of the 7 floors.
   (ii) at different floors.
9. Find the value of $1\cdot1!+2\cdot2!+3\cdot3!+\ldots+n\cdot n!$. *Competency Based Que*
10. If $(n+1)! = 12\times(n-1)!$, then find the value of n. *Competency Based Que*

Answers

1. 24 2. 1023 3. 15600 4. 1431 5. 42 6. 729 7. 50 8. (i) 7^5 (ii) 2520 9. $(n+1)!-1$ 10. 3

For Solutions, See QR Code on Pg. 177

6.3 Permutation

A permutation is an arrangement of objects in a definite order. Arrangement can be made by taking some or all objects at a time.

e.g. If there are three objects say A, B and C, then the permutations of these three objects taking two at a time are AB, BA, AC, CA, BC, CB and the permutations of these three objects taking all at a time are ABC, ACB, BAC, BCA, CAB, CBA.

Here, in each case, number of permutations is 6.

Note that the order of arrangement is important. Because when the order is changed, then different permutation is obtained.

Permutations, when all the Objects are Distinct

When all given objects are distinct, then we can find the number of permutations with the help of following theorem.

Theorem 1

The number of permutations of n different objects taken r at a time, where $0 < r \le n$ and the objects do not repeat, is $n(n-1)(n-2)\dots(n-r+1)$, which is denoted by ^nP_r or $P(n, r)$.

i.e. $P(n, r) = {}^nP_r = \dfrac{n!}{(n-r)!}, 0 \le r \le n$

- When $r = 0$, then $^nP_0 = 0! = \dfrac{n!}{(n-0)!} = \dfrac{n!}{n!} = 1$
- When $r = n$, then $^nP_n = \dfrac{n!}{(n-n)!} = \dfrac{n!}{0!} = n!$ $[\because 0! = 1]$
- Number of permutations of n different things taken all at a time $= n!$

Permutations with Repetitions

When repetition of objects is allowed, then number of permutations can be obtained with the help of following theorem.

Theorem 2

The number of permutations of n different objects taken r at a time, when each may be repeated any number of times in each arrangement, is n^r (permutation with repetitions).

- The number of permutations of n different objects, all at a time, when each may be repeated any number of times in each arrangement, is n^n.

Permutations when all the Objects are not Distinct Objects

When all the objects are not distinct i.e. some objects are of same kind, then we can find the number of permutations with the help of following theorems (without proof).

Theorem 3

The number of permutations of n objects, where p objects are of the same kind or identical and other are distinct, is given by $\dfrac{n!}{p!}$.

Theorem 4

The number of permutations of n objects, where p_1 objects are of one kind, p_2 objects are of second kind, ... p_k objects are of kth kind and the rest if any, are of different kind is $\dfrac{n!}{p_1!\,p_2!\dots p_k!}$.

Some Important Results

- If r particular things out of n different things are taken together, then we count these r particular thing as one thing and remaining $(n-r)$ things as separate thing.
  Then, total number of things $= (n-r)+1 = n-r+1$
  Since, these are different things, therefore number of permutations of these things $= (n-r+1)!$.
  But r particular things can also be arranged among themselves in $r!$ ways.
  $\therefore$ Required number of permutations $= (n-r+1)!r!$
- If r particular things are identical, then required number of permutations $= (n-r+1)!$
- The number of permutations of n different objects taken r at a time, when a particular object is taken in each arrangement, is $r \cdot {}^{n-1}P_{r-1}$.
- The number of permutations of n objects taken r at a time, when a particular object is never taken in each arrangement, is $^{n-1}P_r$.
- The number of permutations of n different objects taken r at a time in which two specific objects always occur together, is $2!(r-1) \cdot {}^{n-2}P_{r-2}$.

Example 1. If $P(n, 5) = 20P(n, 3)$, then find n.

Sol. We have, $P(n, 5) = 20\,P(n, 3)$

$$\Rightarrow \frac{n!}{(n-5)!} = 20 \times \frac{n!}{(n-3)!}$$
$$\Rightarrow (n-3)! = 20 \times (n-5)!$$
$$\Rightarrow (n-3)(n-4)(n-5)! = 20 \times (n-5)!$$
$$\Rightarrow (n-3)(n-4) = 20 = 5 \times 4$$
$$\Rightarrow (n-3)(n-4) = (8-3)(8-4)$$

On comparing, we get

$$n = 8$$

Example 2. If $P(9, r) = 3024$, then find r.

Sol. We have, $P(9, r) = 3024 \Rightarrow {}^9P_r = 3024$

$$\Rightarrow \frac{9!}{(9-r)!} = 3024 \Rightarrow \frac{9!}{(9-r)!} = 9 \times 336$$
$$\Rightarrow \frac{9!}{(9-r)!} = 9 \times 8 \times 7 \times 6$$
$$\Rightarrow \frac{9!}{(9-r)!} = \frac{9 \times 8 \times 7 \times 6 \times 5!}{5!}$$

[multiplying numerator and denominator by 5!]

$$\Rightarrow \frac{9!}{(9-r)!} = \frac{9!}{5!}$$
$$\Rightarrow (9-r)! = 5!$$

On comparing, we get

$$9 - r = 5 \Rightarrow r = 4$$

Example 3. **If $^9P_5 + 5 \cdot {}^9P_4 = {}^{10}P_r$, then find r.**

Sol. We have, $^9P_5 + 5 \cdot {}^9P_4 = {}^{10}P_r$

$$\Rightarrow \frac{9!}{(9-5)!} + 5 \cdot \frac{9!}{(9-4)!} = \frac{10!}{(10-r)!}$$

$$\Rightarrow \frac{9!}{4!} + 5 \cdot \frac{9!}{5!} = \frac{10!}{(10-r)!}$$

$$\Rightarrow \frac{9!}{4!} + 5 \cdot \frac{9!}{5 \times 4!} = \frac{10 \times 9!}{(10-r)!}$$

$$\Rightarrow \frac{1}{4!} + \frac{1}{4!} = \frac{10}{(10-r)!}$$

[dividing both sides by 9!]

$$\Rightarrow \frac{2}{4!} = \frac{10}{(10-r)!}$$

$$\Rightarrow (10-r)! = 5 \times 4! = 5!$$

$$\Rightarrow (10-r)! = 5!$$

On comparing, we get

$$10 - r = 5 \Rightarrow r = 5$$

Example 4. **If $P(2n-1, n) : P(2n+1, n-1) = 22 : 7$, then find n.**

Sol. We have, $^{2n-1}P_n : {}^{2n+1}P_{n-1} = 22 : 7$

$$\therefore \frac{\dfrac{(2n-1)!}{(2n-1-n)!}}{\dfrac{(2n+1)!}{(2n+1-n+1)!}} = \frac{22}{7}$$

$$\frac{(2n-1)!}{(n-1)!} \times \frac{(n+2)!}{(2n+1)!} = \frac{22}{7}$$

$$\Rightarrow \frac{(2n-1)!}{(n-1)!} \times \frac{(n+2)(n+1)n(n-1)!}{(2n+1)(2n)(2n-1)!} = \frac{22}{7}$$

$$\Rightarrow \frac{(n+2)(n+1)n}{(2n+1)(2n)} = \frac{22}{7}$$

$$\Rightarrow \frac{(n+1)(n+2)}{2(2n+1)} = \frac{22}{7}$$

$$\Rightarrow 7(n^2 + 3n + 2) = 44(2n+1)$$

$$\Rightarrow 7n^2 + 21n + 14 = 88n + 44$$

$$\Rightarrow 7n^2 - 67n - 30 = 0$$

$$\Rightarrow 7n^2 - 70n + 3n - 30 = 0$$

$$\Rightarrow 7n(n-10) + 3(n-10) = 0$$

$$\Rightarrow (n-10)(7n+3) = 0$$

$$\Rightarrow n = 10 \text{ or } n = \frac{-3}{7}$$

[$\because$ n can not be negative]

$$\therefore n = 10$$

Example 5. **In how many ways two different rings can be worn in four fingers with atmost one in each finger?**

Competency Based Que

Sol. Clearly, required number of ways is same as number of permutations (or arrangements) of 4 different things taken 2 at a time.

So, required number of ways = $^4P_2$

$$= \frac{4!}{2!} = \frac{4 \times 3 \times 2!}{2!} = 12$$

Example 6. **How many words, with or without meaning can be made from the letters of the word 'MONDAY,' assuming that no letter is repeated, if**

(i) 4 letters are used at a time?

(ii) all letters are used at a time? *NCERT*

Sol. Here, the word 'MONDAY' contains 6 distinct letters.

(i) Required number of words = Number of permutations of 6 different things taken 4 at a time

$$= {}^6P_4 = \frac{6!}{2!} = \frac{6 \times 5 \times 4 \times 3 \times 2!}{2!} = 360$$

(ii) Required number of words = Number of permutations of 6 different things taken all at a time

$$= {}^6P_6 = 6! = 720$$

Example 7. **Three men have 4 coats, 5 waistcoats and 6 caps. In how many ways, can they wear them?**

Sol. Here, all the 4 coats are distinct, all 5 waistcoats are distinct and all 6 caps are also distinct.

So, the number of ways in which three men can wear 4 coats will be equal to the number of arrangements of 4 different things taken 3 at a time.

$\therefore$ Number of ways in which three men can wear 4 coats

$$= {}^4P_3 = \frac{4!}{(4-3)!} = 4 \times 3 \times 2 = 24$$

Similarly, number of ways in which three men can wear 5 waistcoats

$$= {}^5P_3 = \frac{5!}{(5-3)!} = \frac{5 \times 4 \times 3 \times 2!}{2!} = 5 \times 4 \times 3 = 60$$

and number of ways in which three men can wear 6 caps

$$= {}^6P_3 = \frac{6!}{(6-3)!} = \frac{6 \times 5 \times 4 \times 3!}{3!} = 6 \times 5 \times 4 = 120$$

Since, all three men wear coat, waistcoat and cap. So, by the fundamental principle of multiplication,

required number of ways = $^4P_3 \times {}^5P_3 \times {}^6P_3$

$$= 24 \times 60 \times 120 = 172800$$

Example 8. Find the number of 5-digit telephone numbers having atleast one of their digits repeated. *NCERT Exemplar*

Sol. Using the digits 0, 1, 2, ..., 9, the number of 5-digit telephone numbers, which can be formed is 10^5 (since, repetition is allowed).

The number of 5-digit telephone numbers, which have none of the digits repeated $= {}^{10}P_5 = \frac{10!}{5!} = 30240$

Hence, the required number of telephone numbers having atleast one of their digits repeated

$= 10^5 - {}^{10}P_5 = 100000 - 30240 = 69760$

Example 9. How many permutations of the letters of the words 'INDIA' are there?

Sol. Here, we have 5 letters, two of which are of the same kind and others are different.

∴ Required number of permutations (or arrangements)

$= \frac{5!}{2!} = 5 \times 4 \times 3 = 60$

Example 10. Find the number of different words that can be formed from the letters of the word 'TRIANGLE' so that

(i) all vowels occur together.

(ii) all vowels do not occur together.

Sol. There are 8 distinct letters in the word TRIANGLE, out of which 3 are vowels, namely A, E, I and 5 are consonants, namely T, R, N, G, L.

(i) Since, the vowels have to occur together, so let us assume them as a single object (AEI).

$\boxed{\text{AEI}}$ T, R, N, G, L

Now, this single object together with 5 remaining letters will be counted as 6 objects and these can be arrange in ${}^6P_6 = 6!$ ways.

Corresponding to each of these permutations, we have

$3! = 6$

Permutations of the three vowels A, E and I taken all at a time.

Hence, by fundamental principle of multiplication, the required number of words $= 6! \times 3! = 4320$

(ii) Clearly, required number of words = Number of all possible arrangements of 8 letters taken all at a time − Number of permutation in which the vowels are always together

$= {}^8P_8 - 6! \times 3! = 8! - 6! \times 3!$

$= 8 \times 7 \times 6! - 6! \times 3! = 6!(56 - 6)$

$= 720 \times 50 = 36000$

Example 11. The principal wants to arrange 5 students on the platform such that the boy SALIM occupies the second position and such that the girl SITA is always adjacent to the girl RITA. How many such arrangements are possible? *Competency Based Que*

Sol. Since, SALIM occupies the second position and the two girls RITA and SITA are always adjacent to each other. So, none of these two girls can occupy the first seat. Thus, first seat can be occupied by any one of the remaining two students in 2 ways. Second seat can be occupied by SALIM in only one way.

Now, in the remaining three seats SITA and RITA can be seated in the following four ways.

	I	II	III	IV	V
1.	X	SALIM	SITA	RITA	X
2.	X	SALIM	RITA	SITA	X
3.	X	SALIM	X	SITA	RITA
4.	X	SALIM	X	RITA	SITA

Now, only one seat is left which can be occupied by the 5th student in one way.

Hence, the total number of required type of arrangements $= 2 \times 4 \times 1 = 8$

Example 12. How many different words can be formed using all the letters of the word 'ALLAHABAD'? In how many of them, vowels occupy the even position?

Sol. There are 9 letters in the word ALLAHABAD.

In these letters, A occurs 4 times, L occurs twice and rest all are different.

∴ Total number of words $= \frac{9!}{4!\,2!} = 7560$

Now, consider the case, where vowels occupy the even position.

Since, the word ALLAHABAD contains 4 vowels, A, A, A and A and there are 4 even places, therefore 4 vowels can be arranged in 4 even places.

This can be done in ${}^4P_4 \times \frac{1}{4!} = 4! \times \frac{1}{4!} = 1$ way

Now, we are left with 5 places and 5 letters, which can be arranged in ${}^5P_5 \times \frac{1}{2!} = \frac{5!}{2!}$ ways

[∵ out of 5 letters, 2 are alike]

Hence, the number of words in which vowels occupy the even position $= \frac{5!}{2!} = 60$

Quick Test 2

1. If ${}^nP_4 : {}^nP_5 = 1:2$, find n.
2. Prove that following.
   (i) ${}^nP_n = 2 \cdot {}^nP_{n-2}$
   (ii) ${}^nP_n = {}^nP_{n-1}$
   (iii) ${}^nP_r = n \cdot {}^{n-1}P_{r-1}$
3. In a class of 10 students, there are 3 girls A, B, C. Find the number of different ways, they can be arranged in a row such that no two of three girls are consecutive. *Competency Based Que*
4. Find the total number of 9-digit numbers which have all different digits. *NCERT Exemplar*
5. There are 6 items in column A and 6 items in column B. A student is asked to match each item in column A with an item in column B. How many possible, correct or incorrect answers are there to this question?
6. In how many ways can 4 red, 3 yellow and 2 green discs be arranged in a row if the discs of the same colour are indistinguishable?
7. How many 3-digit even numbers can be made using the digits 1, 2, 3, 4, 5, 6, 7, if no digit is repeated?
8. How many words can be formed out of the letters of the word 'ORIENTAL' so that the vowels always occupy the odd places? *Competency Based Que*
9. How many different words can be formed with the letters of the word 'HARYANA'? How many of these
   (i) have H and N together?
   (ii) begin with H and end with N?
   (iii) have three vowels together?
10. How many numbers greater than 1000000 can be formed using the digits 2, 3, 0, 3, 4, 2, 3? *Competency Based Que*

Answers

1. 6 3. $7! \times 336$ 4. $9 \times 9!$ 5. $6!$ ways 6. 1260 7. 90 8. 576 9. 840 (i) 240 (ii) 20 (iii) 120 10. 360

For Solutions, See QR Code on Pg. 177

6.4 Combinations

The word combination means selection. Each of the different selection, which is made by taking some or all of a number of different objects at a time, irrespective of their arrangements is called a combination.

e.g. The different combinations formed from three letters A, B, C taking two at a time are AB, AC, BC.

In a combination, only selection is made and the order in which the objects are arranged is immaterial, while in a permutation, not only the selection is formed but also an arrangement in a definite order is considered.

Meaning of ${}^nC_r$

The number of combinations of n distinct objects taken r at a time is given by ${}^nC_r = \dfrac{n!}{r!(n-r)!}$.

In particular,

(i) When $r = 0$, then ${}^nC_0 = 1$

(ii) When $r = n$, then ${}^nC_n = 1$

Some Important Theorems

Theorem 1 ${}^nP_r = {}^nC_r \cdot r!, 0 < r \le n$

Proof The number of combinations of n distinct objects taken r at a time is ${}^nC_r$. In these combinations, r things can be arranged among themselves in $r!$ ways. So, we have $r!$ permutations.

Therefore, the total number of permutations of n different things taken r at a time is ${}^nC_r \times r!$, which is equal to ${}^nP_r$.

$\therefore$ $\quad {}^nP_r = r! \times {}^nC_r, 0 < r \le n$ **Hence proved.**

Theorem 2 ${}^nC_r = {}^nC_{n-r}, 0 \le r \le n$

Theorem 3 ${}^nC_r + {}^nC_{r-1} = {}^{n+1}C_r$

Proof LHS $= {}^nC_r + {}^nC_{r-1}$

$$= \frac{n!}{r!(n-r)!} + \frac{n!}{(r-1)!(n-r+1)!}$$

$$= \frac{n!(n-r+1)}{r!(n-r+1)(n-r)!} + \frac{n!r}{r(r-1)!(n-r+1)!}$$

$$= \frac{n!(n-r+1)}{r!(n-r+1)!} + \frac{n!r}{r!(n-r+1)!}$$

$$= n!\left[\frac{n-r+1+r}{r!(n-r+1)!}\right] = \frac{(n+1)\,n!}{r!(n+1-r)!}$$

$$= \frac{(n+1)!}{r!(n+1-r)!} = {}^{n+1}C_r = \text{RHS} \quad \textbf{Hence proved.}$$

Theorem 4

If ${}^nC_x = {}^nC_y$, then either $x = y$ or $x + y = n$.

Theorem 5

If n is an even natural number, then the greatest of the values ${}^nC_0, {}^nC_1, {}^nC_2, \ldots, {}^nC_n$ is ${}^nC_{n/2}$.

If n is an odd natural number, then the greatest of the values ${}^nC_0, {}^nC_1, {}^nC_2, \ldots, {}^nC_n$ is ${}^nC_{\frac{n-1}{2}} = {}^nC_{\frac{n+1}{2}}$.

Example 1. If ${}^nP_r = 840$ and ${}^nC_r = 35$, then find r.
NCERT Exemplar, Competency Based Que

Sol. We know that ${}^nP_r = r! \times {}^nC_r$

$\therefore \quad 840 = r! \times 35$

$\Rightarrow \quad r! = \frac{840}{35} \Rightarrow r! = 24 = 4!$

$\therefore \quad r = 4$

Example 2. If ${}^{n+2}C_8 : {}^{n-2}P_4 = 57:16$, then find the value of n.
Competency Based Que

Sol. We have, ${}^{n+2}C_8 : {}^{n-2}P_4 = 57:16$

$$\Rightarrow \frac{{}^{n+2}C_8}{{}^{n-2}P_4} = \frac{57}{16}$$

$$\Rightarrow \frac{\frac{(n+2)!}{8!(n+2-8)!}}{\frac{(n-2)!}{(n-2-4)!}} = \frac{57}{16}$$

$$\Rightarrow \frac{(n+2)!}{8!(n-6)!} \times \frac{(n-6)!}{(n-2)!} = \frac{57}{16}$$

$$\Rightarrow \frac{(n+2)(n+1)n(n-1)(n-2)!}{8!} \times \frac{1}{(n-2)!} = \frac{57}{16}$$

$$\Rightarrow (n+2)(n+1)n(n-1) = \frac{57}{16} \times 8!$$

$$= \frac{19 \times 3}{16} \times 8 \times 7 \times 6 \times 5 \times 4 \times 3 \times 2 \times 1$$

$$\Rightarrow (n-1)n(n+1)(n+2) = 19 \times 3 \times 7 \times 6 \times 5 \times 4 \times 3$$

$$\Rightarrow (n-1)n(n+1)(n+2) = 19 \times (3 \times 7) \times (6 \times 3) \times (5 \times 4)$$

$$\Rightarrow (n-1)n(n+1)(n+2) = 18 \times 19 \times 20 \times 21$$

On comparing both sides, we get

$$n - 1 = 18$$

$$\therefore \quad n = 19$$

Example 3. Prove that ${}^{2n}C_n = \frac{2^n\{1 \cdot 3 \cdot 5 \ldots (2n-1)\}}{n!}$.

Sol. We have, ${}^{2n}C_n = \frac{2n!}{(2n-n)!\,n!} = \frac{(2n)!}{n!\,n!}$

$$= \frac{(2n)(2n-1)(2n-2)\ldots 5 \cdot 4 \cdot 3 \cdot 2 \cdot 1}{n!\,n!}$$

$$= \frac{\{1 \cdot 3 \cdot 5 \ldots (2n-1)\}\{2 \cdot 4 \cdot 6 \ldots 2n\}}{n!\,n!}$$

$$= \frac{\{1 \cdot 3 \cdot 5 \ldots (2n-1)\} \times 2^n\{1 \cdot 2 \cdot 3 \ldots n\}}{n!\,n!}$$

$$= \frac{\{1 \cdot 3 \cdot 5 \ldots (2n-1)\} \times 2^n \times n!\}}{n!\,n!}$$

$$= \frac{2^n\{1 \cdot 3 \cdot 5 \ldots (2n-1)\}}{n!} \quad \textbf{Hence proved.}$$

Example 4. Prove that the product of r consecutive positive integers is divisible by $r!$.

Sol. Let the r consecutive positive integers be $(n+1), (n+2), (n+3), \ldots, (n+r)$.

Then, product $= (n+1)(n+2)(n+3)\ldots(n+r)$

$$= \frac{n!(n+1)(n+2)(n+3)\ldots(n+r)}{n!}$$

$$= \frac{1 \cdot 2 \cdot 3 \ldots n(n+1)(n+2)\ldots(n+r)}{n!}$$

$$= \frac{(n+r)!}{n!} = \frac{(n+r)!}{r!\{(n+r)-r\}!}(r!)$$

$$= ({}^{n+r}C_r)\,r!$$

which is divisible by $r!$. **Hence proved.**

Example 5. If ${}^nC_{10} = {}^nC_{12}$, then find the value of ${}^{23}C_n$.
Competency Based Que

Sol. We know that ${}^nC_x = {}^nC_y$

$\Rightarrow \quad x + y = n$ or $x = y$

Here, $x \neq y$, so $x + y = n$

$\Rightarrow \quad n = 10 + 12 = 22$

Now, ${}^{23}C_n = {}^{23}C_{22} = {}^{23}C_{23-22} = {}^{23}C_1 = 23$

Example 6. Find the value of the expression ${}^{47}C_4 + \sum_{j=1}^{5} {}^{52-j}C_3$.
Competency Based Que

Sol. We have, ${}^{47}C_4 + \sum_{j=1}^{5} {}^{52-j}C_3$

$$= {}^{47}C_4 + {}^{52-1}C_3 + {}^{52-2}C_3 + {}^{52-3}C_3 + {}^{52-4}C_3 + {}^{52-5}C_3$$

$$= {}^{47}C_4 + {}^{51}C_3 + {}^{50}C_3 + {}^{49}C_3 + {}^{48}C_3 + {}^{47}C_3$$

$= ({}^{47}C_3 + {}^{47}C_4) + {}^{48}C_3 + {}^{49}C_3 + {}^{50}C_3 + {}^{51}C_3$

$= {}^{48}C_4 + {}^{48}C_3 + {}^{49}C_3 + {}^{50}C_3 + {}^{51}C_3$

$[\because {}^nC_{r-1} + {}^nC_r = {}^{n+1}C_r]$

$= {}^{49}C_4 + {}^{49}C_3 + {}^{50}C_3 + {}^{51}C_3$

$= {}^{50}C_4 + {}^{50}C_3 + {}^{51}C_3 = {}^{51}C_4 + {}^{51}C_3 = {}^{52}C_4$

$= \frac{52 \times 51 \times 50 \times 49}{4 \times 3 \times 2 \times 1} = 13 \times 17 \times 25 \times 49 = 270725$

Example 7. In how many ways, can 5 sportsmen be selected from a group of 10?

Sol. Required number of ways $= {}^{10}C_5 = \frac{10 \times 9 \times 8 \times 7 \times 6}{5 \times 4 \times 3 \times 2 \times 1}$

$= 2 \times 3 \times 7 \times 6 = 252$

Example 8. If there are 15 persons in a party and if each two of them shake hands with each other. How many hand-shakes happen in the party?

Sol. The total number of hand-shakes is same as the number of ways of selecting 2 persons among 15 persons $= {}^{15}C_2 = \frac{15 \times 14}{2 \times 1} = 15 \times 7 = 105$

Example 9. From a class of 25 students, 10 are to be chosen for an excursion party. There are 3 students who decide that either all of them will join or none of them will join. In how many ways can the excursion party be chosen? NCERT

Sol. Here, we have to choose 10 students out of a class of 25 students such that the given condition satisfied.
Let us make the following cases and find the number of possible choices in each case.

Case I When 3 particular students join the party
In this case, we have to choose 7 students out of 22 students. This can be done in ${}^{22}C_7$ ways.

Case II When 3 particular students do not join the party In this case, we have to choose 10 students out of 22 students. This can be done in ${}^{22}C_{10}$ ways.

Hence, the total number of ways in which excursion party can be chosen $= {}^{22}C_7 + {}^{22}C_{10}$

Example 10. We wish to select 6 persons from 8 but if the person A is chosen, then B must be chosen. In how many ways, can the selection be made?

Competency Based Que

Sol. Let us make the following cases and find the number of possible selection in each case.

Case I When A is chosen
In this case, B must be chosen also. So, we have to choose 4 persons out of 6 persons.

This can be done in ${}^6C_4 = {}^6C_2 = \frac{6 \times 5}{2 \times 1} = 15$ ways

Case II When A is not chosen
In this case, we have to choose 6 persons out of 7 persons. This can be done in ${}^7C_6 = {}^7C_1 = 7$ ways

Hence, the total number of ways of selecting 6 persons $= 15 + 7 = 22$

Example 11. There are 10 professors and 20 students out of whom a committee of 2 professors and 3 students is to be formed. Find the number of ways in which this can be done. Further find in how many of these committees Competency Based Que

(i) a particular professors is included?

(ii) a particular student is included?

(iii) a particular student is excluded?

Sol. Clearly, for forming a committee of 2 professors and 3 students, we have to select 2 professors out of 10 professors and 3 students out of 20 students.
This can be done in ${}^{10}C_2 \times {}^{20}C_3$ ways.

Thus, required number of ways

$= \frac{10 \times 9}{2 \times 1} \times \frac{20 \times 19 \times 18}{3 \times 2 \times 1} = 45 \times 60 \times 19 = 51300$

(i) When a particular professor is included, then we have to select 1 professor out of 9 professors and 3 students out of 20 students.
This can be done in ${}^9C_1 \times {}^{20}C_3$ ways.
Thus, required number of ways $= 9 \times \frac{20 \times 19 \times 18}{3 \times 2 \times 1}$

$= 9 \times 60 \times 19 = 10260$

(ii) When a particular student is included, then we have to select 2 professors out of 10 professors and 2 students out of 19 students.
This can be done in ${}^{10}C_2 \times {}^{19}C_2$.
Thus, required number of ways $= \frac{10 \times 9}{2 \times 1} \times \frac{19 \times 18}{2 \times 1}$

$= 45 \times 171 = 7695$

(iii) When a particular student is excluded, then we have to select 2 professors out of 10 professors and 3 students out of 19 students. This can be done in ${}^{10}C_2 \times {}^{19}C_3$ ways.
Thus, required number of ways

$= \frac{10 \times 9}{2 \times 1} \times \frac{19 \times 18 \times 17}{3 \times 2 \times 1} = 45 \times 969 = 43605$

Example 12. A committee of 7 members has to be formed from 9 boys and 4 girls. In how many ways, can this be done when the committee consists of

(i) exactly 3 girls? **(ii) atleast 3 girls?**

(iii) atmost 3 girls? NCERT

Sol. There are 9 boys and 4 girls. We have to form a committee of 7 members.

(i) Since, the committee has exactly 3 girls. Therefore, we have to choose 3 girls out of 4 girls and the rest 4 members out of 9 boys.

This can be done in ${}^4C_3 \times {}^9C_4$ ways.

Hence, the number of ways of forming a committee

$$= {}^4C_3 \times {}^9C_4$$
$$= {}^4C_1 \times \frac{9\times8\times7\times6}{4\times3\times2\times1}$$
$$= 4\times9\times7\times2 = 504 \text{ ways} \quad [\because {}^nC_1 = n]$$

(ii) Since, the committee has atleast 3 girls. Therefore, we can form the committee in the following ways.

(a) Selecting 3 girls and 4 boys

(b) Selecting 4 girls and 3 boys

Hence, the number of ways of forming a committee

$$= {}^4C_3 \times {}^9C_4 + {}^4C_4 \times {}^9C_3$$
$$= 504 + 1\times\frac{9\times8\times7}{3\times2\times1} \quad [\because {}^nC_n = 1]$$
$$= 504 + 84 = 588 \text{ ways}$$

(iii) Since, the committee has atmost 3 girls, therefore we can form committee in the following ways.

(a) Selecting 7 boys and no girl

(b) Selecting 6 boys and 1 girl

(c) Selecting 5 boys and 2 girls

(d) Selecting 4 boys and 3 girls

Hence, the number of ways of forming a committee

$$= {}^9C_7 \times {}^4C_0 + {}^9C_6 \times {}^4C_1 + {}^9C_5 \times {}^4C_2 + {}^9C_4 \times {}^4C_3$$
$$= {}^9C_2 + {}^9C_3 \times 4 + {}^9C_4 \times {}^4C_2 + {}^9C_4 \times {}^4C_1$$
$$[\because {}^nC_r = {}^nC_{n-r},\ {}^nC_1 = n \text{ and } {}^nC_0 = 1]$$
$$= 36 + 84\times4 + 126\times6 + 126\times4$$
$$= 36 + 336 + 756 + 504 = 1632$$

Example 13. A sports team of 11 students is to be constituted, choosing atleast 5 from class XI and atleast 5 from class XII. If there are 20 students in each of these classes, then how many ways can the team be constituted? *NCERT Exemplar*

Sol. To constitute a sports team, there are two cases.

Case I Choosing 5 students from class XI and 6 students from class XII.

$\therefore$ In this case, number of ways

$$= {}^{20}C_5 \times {}^{20}C_6$$

Case II Choosing 6 students from class XI and 5 students from class XII.

In this case, number of ways $= {}^{20}C_6 \times {}^{20}C_5$

$$\therefore \text{Required number of ways} = ({}^{20}C_5 \times {}^{20}C_6) + ({}^{20}C_6 \times {}^{20}C_5)$$
$$= 2({}^{20}C_5 \times {}^{20}C_6)$$

Hence, the total number of ways of constituting a sports team of 11 students is $2({}^{20}C_5 \times {}^{20}C_6)$ ways.

Example 14. If a polygon has 44 diagonals, then find the number of its sides. *NCERT Exemplar, Competency Based Que*

Sol. Let n be the number of sides of a polygon.

Then, number of vertices of polygon $= n$

Now, number of diagonals $= {}^nC_2 - n = 44$

[$\because$ diagonal will be formed by joining any two non-adjacent vertices of the polygon, therefore total number of diagonals is ${}^nC_2 - n$]

$$\Rightarrow \frac{n(n-1)}{2} - n = 44$$
$$\Rightarrow n^2 - 3n - 88 = 0$$
$$\Rightarrow n = -8 \text{ or } 11$$
$$\therefore n = 11 \quad [\because n \text{ cannot be negative}]$$

Example 15. Find the number of triangles that are formed by choosing the vertices from a set of 12 points, seven of which lie on the same line. *NCERT Exemplar, Competency Based Que*

Sol. We know that a triangle will be formed by taking three non-collinear points at a time.

$$\therefore \text{Required number of triangles} = {}^{12}C_3 - {}^7C_3$$
$$= \frac{12\times11\times10}{3\times2\times1} - \frac{7\times6\times5}{3\times2}$$
$$= 220 - 35 = 185$$

Hence, the total number of such triangles is 185.

Example 16. The English alphabet has 5 vowels and 21 consonants. How many words with 2 different vowels and 2 different consonants can be formed from the alphabet?

Sol. Here, we have 5 vowels and 21 consonants. We have to select 2 vowels out of 5 vowels and 2 consonants out of 21 consonants. Also, we have to arrange these 4 letters which can be done in 4! ways.

$$\therefore \text{Required number of words} = {}^5C_2 \times {}^{21}C_2 \times 4!$$
$$= \frac{5\times4}{2} \times \frac{21\times20}{2} \times 24$$
$$= 10\times210\times24 = 50400$$

Hence, the total number of words having 2 consonants and 2 vowels is 50400.

Quick Test 3

1. Find the value of ${}^{15}C_8 + {}^{15}C_9 - {}^{15}C_6 - {}^{15}C_7$.

2. Determine n, if ${}^{2n}C_3 : {}^{n}C_3 = 12:1$. *NCERT*

3. If ${}^{28}C_{2r} : {}^{24}C_{2r-4} = 225:11$, then find r.

4. A person wishes to make up as many different parties as he can out of his 36 friends such that each party consists of the same number of persons. How many friends should he invite?

5. Find the number of ways in which 2 black and 3 red balls can be selected from a bag containing 5 black and 6 red balls.

6. Find the number of ways in which a mixed double game can be arranged from amongst 9 married couples, if no husband and wife play in the same game. *Competency Based Que*

7. In how many ways can a student choose a programme of 5 courses, if 9 courses are available and 2 specific courses are compulsory for every student?

8. A boy has 3 library tickets and 8 books of his interest in the library. Of these 8, he does not want to borrow Mathematics Part II, unless Mathematics Part I is also borrowed. In how many ways can he choose the three books to be borrowed? *NCERT Exemplar, Competency Based Que*

9. In how many ways can a cricket team be selected from a group of 25 players containing 10 batsmen, 8 bowlers, 5 all-rounders and 2 wicket keepers, assuming that the team of 11 players requires 5 batsmen, 3 all-rounders, 2 bowlers and 1 wicket keeper?

10. What is the number of ways of choosing 4 cards from a pack of 52 playing cards? In how many of these
(i) four cards are of the same suit?
(ii) four cards belong to four different suits?
(iii) are face cards?
(iv) two are red and two are black cards?
(v) cards are of the same colour? *NCERT*

Answers

1. 0
2. 5
3. 7
4. 18
5. 200 ways
6. 1512
7. 35
8. 41
9. 141120
10. (i) 2860 (ii) 13^4 (iii) 495 (iv) 105625 (v) 29900

For Solutions, See QR Code on Pg. 177

Exam Pattern Questions

Objective Type Questions

Multiple Choice Questions

1. If repetition of the digits is allowed, then the number of even natural number having 3 digits is

(a) 250 (b) 350
(c) 450 (d) 550

2. The number of ways in which 8 distinct toys can be distributed among 5 children is

(a) 5^8 (b) 8^5
(c) $^8P_5$ (d) $^5P_5$

3. Suppose you have a suitcase with a number lock. The number lock has 4 wheels each labelled with 10 digits from 0 to 9. The lock can be opened, if 4 specific digits are arranged in a particular sequence with no repetition. Some how, you have forgotten this specific sequence of digits. You remember only the first digit which is 7. In order to open the lock, how many sequences of 3 digits you may have to check with? **Competency Based Que**

(a) 720 (b) 504
(c) 500 (d) None of these

4. The number of natural numbers smaller than 10^4 in the decimal notation of which all the digits are distinct is **Competency Based Que**

(a) 5274 (b) 5275
(c) 5276 (d) 5277

5. The sum of the digits in unit place of all the numbers formed with the help of 3, 4, 5 and 6 taken all at a time is **NCERT Exemplar, Competency Based Que**

(a) 432 (b) 108
(c) 36 (d) 18

6. The value of $P(n, n-1)$ is

(a) n (b) $2n$
(c) $n!$ (d) $2n!$

7. If $^{15}P_r = 2730$, then ^5P_r

(a) 3 (b) 30
(c) 15 (d) 60

8. 10 students are participating in a competition. In how many different ways can the first 3 prizes be won?

(a) 920 (b) 720
(c) 30 (d) 120

9. How many 3-letter words with or without meaning, can be formed out of the letters of the word, 'LOGARITHMS'. If repetition of letter is not allowed.

(a) 720 (b) 420
(c) 5040 (d) None of these

10. In how many different ways can be letters of the word 'CYCLE' be arranged?

(a) 120 (b) 720 (c) 80 (d) 60

11. The number of permutations of letters of the word 'INSTITUTE' is

(a) 30240 (b) 60480
(c) 15120 (d) 24000

12. The number of ways in which three girls and nine boys can be seated in two vans, each having numbered seats, 3 in the front and 4 at the back such that 3 girls always sit together in a back row on adjacent seats, is **Competency Based Que**

(a) $^{11}P_9 \times 4! \times 3$ (b) $^{11}P_9 \times 4 \times 3!$
(c) $^{14}P_{12} \times 4! \times 3$ (d) $^{14}P_{12} \times 4 \times 3!$

13. All the letters of the word 'EAMCOT' are arranged in different possible ways. The number of such arrangements in which no two vowels are adjacent to each other is **Competency Based Que**

(a) 360 (b) 144
(c) 72 (d) 54

14. If $^nC_{12} = {}^nC_8$, then n is equal to

(a) 20 (b) 12
(c) 6 (d) 30

15. Total number of words formed by 2 vowels and 3 consonants taken from 4 vowels and 5 consonants is equal to

(a) 60 (b) 120 (c) 7200 (d) 720

16. The number of combinations of 4 different objects A, B, C, D taken 2 at a time is

(a) 4 (b) 6 (c) 7 (d) 8

17. The number of triangles that are formed by choosing the vertices from a set of 12 points, seven of which lie on the same line is **Competency Based Que**

(a) 105 (b) 15
(c) 175 (d) 185

Assertion-Reason Based Questions

Directions (Q. Nos. 18-27) *In the questions given below are two statements labelled as Assertion (A) and Reason (R). In the context of the two statements, which one of the following is correct?*

(a) Both A and R are correct; R is the correct explanation of A.
(b) Both A and R are correct; R is not the correct explanation of A.
(c) A is correct; R is incorrect.
(d) R is correct; A is incorrect.

18. **Assertion** (A) $6! = 720$

Reason (R) $n! = n(n-1)(n-2)(n-3)...3\cdot 2\cdot 1$

19. **Assertion** (A) If $(n+2)! = 2550 \times n!$, then $n = 49$

Reason (R) $(n+3)! = (n+2)(n+1)...3\cdot 2\cdot 1$

Competency Based Que

20. **Assertion** (A) The number of ways of arranging n boys and n girls in a circle such that no two boys are consecutive, is $(n-1)^2!$.

Reason (R) The number of ways of arranging n distinct objects in a circle is $(n-1)!$. Competency Based Que

21. **Assertion** (A) The number of permutations of letters of word 'ROOT' is 10.

Reason (R) The number of permutations of letters of word 'INSTITUTE' is $\frac{9!}{2!3!}$.

22. **Assertion** (A) If $P(n, 5) = 20\, P(n, 3)$, then value of n is 8.

Reason (R) $P(n, r) = {}^nP_r = \frac{n!}{(n-r)!}, 0 \le r \le n.$

23. **Assertion** (A) If the letters W, I, F, E are arranged in a row in all possible ways and the words (with or without meaning) so formed are written as in a dictionary, then the word WIFE occurs in the 24th position.

Reason (R) The number of ways of arranging four distinct objects taken all at a time is $C(4,4)$.

Competency Based Que

24. **Assertion** (A) From a group of students, if we have to select few for a competition, then situation relates to combination.

Reason (R) Situations where arrangement or order is required are related to permutations.

25. **Assertion** (A) If ${}^{n-1}C_r : {}^nC_r : {}^{n+1}C_r = 6:9:13$, then values of n and r are 12 and 4, respectively.

Reason (R) ${}^nC_r = \frac{n!}{r!\,(n-r)!}$.

26. **Assertion** (A) If ${}^nC_9 = {}^nC_4$, then $n = 13$

Reason (R) If x and y are non-negative integers such that ${}^nC_x = {}^nC_y \Rightarrow x = y$ or $x + y = n$

27. **Assertion** (A) If a polygon has 54 diagonals, then the number of its sides is 13.

Reason (R) When diagonals are formed by joining any two non-adjacent vertices of a polygon, then the total number of diagonals is ${}^nC_2 - n$. Competency Based Que

Case Study Based Questions

28. Five students Ajay, Shyam, Yojana, Rahul and Akansha are sitting in a playground in a line.

(i) Total number of ways of sitting arrangement of five students is

(a) 120 (b) 60
(c) 24 (d) 30

(ii) Total number of arrangement of sitting, if Ajay and Yojana sit together, is

(a) 60 (b) 48
(c) 72 (d) 120

(iii) Total number of arrangements 'Yojana and Rahul sitting at extreme position' is

(a) 24 (b) 36
(c) 48 (d) 12

(iv) Total number of arrangement, if Shyam is sitting in the middle is

(a) 24 (b) 12
(c) 6 (d) 36

29. Republic day is a national holiday of India. It honours the date on which the constitution of India come into effect on 26 January 1950, turning the nation into a newly formed republic.

Answer the following questions, which are based on the word 'REPUBLIC'.

(i) Find the number of arrangement of the letters of the word 'REPUBLIC'.

(a) 40300 (b) 30420
(c) 40320 (d) 40400

(ii) How many arrangements start with a vowel?

(a) 12015 (b) 15120
(c) 12018 (d) 15100

(iii) Which concept is used for finding the arrangements start with a vowel?

(a) FPA (b) FPM
(c) Combination (d) None of these

(iv) How many arrangement start with a consonants?

(a) 25200 (b) 25000
(c) 15120 (d) 12015

Subjective Type Questions

Very Short Answer Type Questions

1. In how many ways, two books of different languages can be selected from 10 Hindi, 5 English and 7 Sanskrit books? **Competency Based Que**

2. Prove that $(n!)^2 \le (n!) \times n^n < (2n)!$, for all $n \in N$.

3. Find n, if $(n+1)! = 12 \times (n-1)!$.

4. Find the value of n, if $\dfrac{n!}{(n-2)!} = 930, n \ge 2$.

5. If $P(n, 4) = 12 \cdot P(n, 2)$, then find n.

6. $P(n, 4) : P(n-1, 4) = 5 : 3, n > 4$

7. If ${}^{56}P_{r+6} : {}^{54}P_{r+3} = 30800 : 1$, then find r.

8. If ${}^{n}C_8 = {}^{n}C_6$, then find ${}^{n}C_2$.

9. If ${}^{18}C_x = {}^{18}C_{x+2}$, then find x.

10. If ${}^{2n}C_3 : {}^{n}C_2 = 44 : 3$, then find n.

11. Evaluate ${}^{n}C_r + 2 \cdot {}^{n}C_{r-1} + {}^{n}C_{r-2}$. **Competency Based Que**

12. Evaluate $\sum_{r=1}^{5} {}^{5}C_r$. **Competency Based Que**

13. Everybody in a room shakes hands with everybody else. The total number of hand-shakes is 66.
Find the total number of persons in the room. **NCERT Exemplar, Competency Based Que**

14. In how many ways a committee of 3 men and 2 women can be chosen from 7 men and 5 women?

15. 7 points lie on a circle. How many chords can be drawn by joining these points? **Competency Based Que**

16. If 20 lines are drawn in a plane such that no two of them are parallel and no three are concurrent, in how many points will they intersect each other? **NCERT Exemplar, Competency Based Que**

Short Answer Type Questions

17. 25 buses are running between Delhi and Noida. In how many ways can a person go from Delhi to Noida and return by a different bus?

18. There are 6 multiple choice questions in an examination. How many sequences of answers are possible, if the first three questions have 4 choices each and remaining three have 3 each?

19. In how many ways, can 50 voters vote for the 4 candidates contesting the election for the post of secretary of their association?

20. How many numbers are there between 100 and 1000 such that every digit is either 2 or 9?

21. How many 4-digit numbers can be formed using the digits 1 to 9, if repetition of digits is not allowed? **NCERT**

22. How many different words can be formed with the letters of the word 'TRIANGLE' so that the word begin with T and end with E?

23. In how many ways can the letters of the word 'PENCIL' be arranged so that N and E are always together?

24. In a certain city, all telephone numbers have six digits, the first two digits always being 41 or 42 or 46 or 62 or 64. How many telephone numbers have all six digits distinct? **Competency Based Que**

25. How many 3 letter words can be made using the letters of the word 'ORIENTAL'?

26. Find how many arrangements can be made with the letters of the word 'MATHEMATICS'?

27. How many words can be formed using the letter A thrice, the letter B twice and the letter C thrice?

28. If $r < s \le n$, then prove that ${}^{n}P_s$ is divisible by ${}^{n}P_r$.

29. From a committee of 8 persons, in how many ways can we choose a chairman and a vice-chairman assuming one person cannot hold more than one position? **NCERT**

30. How many numbers between 400 and 1000 can be made with the digits 2, 3, 4, 5, 6 and 0, when repetition of digits is not allowed? **Competency Based Que**

31. Eighteen guests have to be seated, half on each side of a long table. Four particular guests desire to sit on one particular side and three others on the other side. Find the number of ways in which the seating arrangement can be made. **Competency Based Que**

32. In how many ways, can a team of 3 boys and 3 girls be selected from 5 boys and 4 girls?

33. Determine the number of 5 cards combinations out of a deck of 52 cards, if atleast one of the 5 cards has to be a king. **Competency Based Que**

Long Answer Type Questions

34. Find the number of different signals that can be generated by arranging atleast 2 flags in order (one below the other) on a vertical staff, if five different flags are available. **NCERT**

35. Find the number of 4 letter words, with or without meaning which can be formed out of the letters of the word 'POSE', when

(i) the repetition of the letters is not allowed.

(ii) the repetition of the letters is allowed.

36. A code word is to consist of two distinct English alphabets followed by two distinct numbers between 1 and 9. e.g. CA23 is a code word. **Competency Based Que**

(i) How many such code words are there?

(ii) How many of them end with an even integer?

37. In how many ways can 3 prizes be distributed among 4 boys, when

(i) no boy gets more than one prize?

(ii) a boy may get any number of prizes?

(iii) no boy gets all the prizes?

38. How many numbers are there between 100 and 1000 which have exactly one of their digits as 7? **Competency Based Que**

39. How many different signals can be made using any number of flags from 6 flags of different colours? **NCERT Exemplar, Competency Based Que**

40. How many four-digit numbers are there with distinct digits?

41. Find the number of ways in which 5 boys and 5 girls be seated in a row so that

(i) no two girls sit together.

(ii) boys and girls sit alternatively. **NCERT Exemplar**

42. In how many ways can the letters of the word 'PERMUTATIONS' be arranged, if

(i) words start with P and end with S?

(ii) vowels are all together? **NCERT**

Hints & Solutions

Objective Type Questions

1. (c) In the even natural numbers, number of ways by unit place of that number can be filled = 5 ways

$[0, 2, 4, 6, 8]$

Number of ways by 10th place can be filled = 10 ways

$[0, 1, 2, ..., 9]$

and number of ways by 100th place can be filled = 9 ways

$[1, 2, 3, ..., 9]$

Therefore, total number of even natural numbers having 3 digits $= 5 \times 10 \times 9 = 450$

2. (a) Total number of toys = 8

Total number of students = 5

$\because$ Each toy can be distributed in 5 different ways.

$\therefore$ Total number of ways in which 8 toys can be distributed among 5 children is 5^8.

3. (b) There are 10 digits out of which digit 7 is filled in first place.

$\because$ Repetition of digits is not allowed.

$\therefore$ Second place can be fill in 9 ways, third place can be fill in 8 ways and last place can be fill in 7 ways.

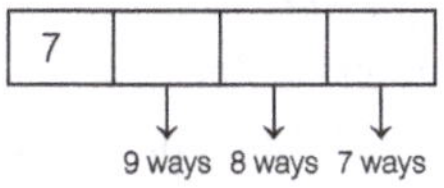

According to FPM,

Total numbers of ways $= 9 \times 8 \times 7 = 504$ ways

4. (a) $\because 10^4 = 10000$

$\therefore$ Numbers smaller than 10^4 are 1 to 9999.

Numbers of ways, 4 digit number to be formed without repetition of digits $= 9 \times 9 \times 8 \times 7 = 4536$

Number of ways, 3 digit number to be formed without repetition of digit $= 9 \times 9 \times 8 = 648$

Number of ways, 2 digit number to be formed without repetition of digit $= 9 \times 9 = 81$

Number of ways, 1 digit number to be formed = 9

Therefore, required ways to form a number smaller than $10^4 = 4536 + 648 + 81 + 9 = 5274$

5. (b) Total number of digits = 4

First we form a four digit number with 3 in unit place.

Total number of ways to form a four digit number with 3 in unit place = 3!.

Similarly, total number of ways to form a four digit number with 4 in unit place = 3!.

Total number of ways to form a four digit number with 5 in unit place = 3!

and total number of ways to form a four digit number with 6 in unit place = 3!.

The sum of digits in unit place of all the numbers formed by 3, 4, 5 and 6 taken all at a time

$= 3 \times 3! + 4 \times 3! + 5 \times 3! + 6 \times 3!$

$= 3! \times [3 + 4 + 5 + 6]$

$= 3! \times 18 = 108$ $\quad [\because 3! = 3 \times 2 \times 1 = 6]$

6. (c) We have, $P(n, n-1)$

$$= {}^nP_{n-1} = \frac{n!}{[n-(n-1)]!}$$

$$= \frac{n!}{(n-n+1)!} = \frac{n!}{1} = n!$$

7. (d) We have, ${}^{15}P_r = 2730 = 15 \times 182$

$\Rightarrow$ ${}^{15}P_r = 15 \times 14 \times 13$

On multiplying and dividing by $(12 \times 11 \times 10 \times ... \times 3 \times 2 \times 1)$, we get

$${}^{15}P_r = \frac{15 \times 14 \times 13 \times 12 \times 11 \times ... \times 3 \times 2 \times 1}{12 \times 11 \times 10 \times ... \times 3 \times 2 \times 1}$$

$$\Rightarrow \quad \frac{15!}{(15-r)!} = \frac{15!}{12!}$$

$$\Rightarrow \quad \frac{1}{(15-r)!} = \frac{1}{12!}$$

On comparing both sides, we get

$$r = 3$$

Now, $\quad {}^5P_r = {}^5P_3 = \frac{5!}{2!} = 5 \times 4 \times 3 = 60$

8. (b) Total number of participants = 10

Total number of prizes = 3

Number of different ways to win first three prizes

$$= {}^{10}P_3$$

$$= \frac{10!}{(10-3)!} = \frac{10!}{7!} = \frac{10 \times 9 \times 8 \times 7!}{7!}$$

$$= 10 \times 9 \times 8$$

$$= 720 \text{ ways}$$

Hence, 10 students have 720 ways to win first 3 prizes.

9. (a) Total number of letters in 'LOGARITHMS' = 10

So, the number of 3 letters word = Number of arrangements of 10 letters taken 3 at a time

$$= {}^{10}P_3$$

$$= 10 \times 9 \times 8$$

$$= 720$$

10. (d) Total number of letters in word 'CYCLE' = 5

In the word CYCLE, the letter 'C' comes twice.

$\therefore$ Number of ways that the letters of the word 'CYCLE' be arranged $= \frac{5!}{2!} = \frac{120}{2} = 60.$

11. (a) Total number of letters in the word 'INSTITUTE' is 9.

In which 'I' and 'T' come twice and thrice, respectively.

$\therefore$ Number of permutations $= \frac{9!}{2!\,3!}$

$$= \frac{9 \times 8 \times 7 \times 6 \times 5 \times 4}{2}$$

$$= 30240$$

12. (b) We have 14 seats in two vans and there are 9 boys and 3 girls.

Now, the number of ways of choosing back seats = 2

The number of ways of arranging 3 girls on adjacent seats $= 2 \times 3!$

and the number of ways of arranging 9 boys on the remaining 11 seats $= {}^{11}P_9$.

Therefore, the required number of ways

$$= 2 \times 2\,(3!) \times {}^{11}P_9$$

$$= {}^{11}P_9 \times 4 \times 3!$$

13. (b) In the word, 'EAMCOT', there are 3 consonants M, C, T and 3 vowels E, A, O.

Since, no two vowels have to be together, so the possible choices for vowels are the blank spaces.

_M_C_T

These vowels can be arranged in ${}^4P_3$ ways.

3 consonants can be arranged in 3! ways.

Hence, required number of ways $= 3! \times {}^4P_3$

$$= 3! \times \frac{4!}{1} = 144$$

14. (a) Given, ${}^nC_{12} = {}^nC_8$

$\Rightarrow$ ${}^nC_{n-12} = {}^nC_8$ $\quad [\because {}^nC_r = {}^nC_{n-r}]$

$\therefore$ $n - 12 = 8$

$\Rightarrow$ $n = 12 + 8 = 20$

15. (c) Total number of vowels = 4

and total number of consonants = 5

Number of ways selecting 2 vowels from 4 vowels $= {}^4C_2$

and number of ways selecting 3 consonants from 5 consonants $= {}^5C_3$

Number of ways to form a word using 5 letters = 5!

Total number of words formed by 2 vowels and 3 consonants taken from 4 vowels and 5 consonants

$$= {}^4C_2 \times {}^5C_3 \times 5!$$

$$= \frac{4!}{2!\,2!} \times \frac{5!}{3!\,2!} \times 5!$$

$$= \frac{4 \times 3}{2} \times \frac{5 \times 4}{2} \times 5 \times 4 \times 3 \times 2 \times 1$$

$$= 60 \times 120$$

$$= 7200$$

16. (b) Total number of objects = 4

Combination of 4 different objects taken 2 at a time

$$= {}^4C_2 = \frac{4!}{2!\,2!}$$

$$= \frac{4 \times 3 \times 2!}{2! \times 2 \times 1} = 6$$

17. (d) We have a set of 12 points.

Number of triangles formed with 12 points = $^{12}C_3$

But according to the given condition, 7 points lie on the same straight line. Thus, the selection of 3 points out of 7 collinear points = $^7C_3$.

Thus, the required number of triangles

$$= {}^{12}C_3 - {}^7C_3 = \frac{12!}{3!\,9!} - \frac{7!}{3!\,4!}$$
$$= \frac{12\times 11\times 10}{3\times 2} - \frac{7\times 6\times 5}{3\times 2}$$
$$= 220 - 35 = 185$$

18. (a) We know that

$$n! = n(n-1)(n-2)(n-3)\ldots 3\cdot 2\cdot 1$$
$$\therefore\quad 6! = 6(6-1)(6-2)(6-3)(6-4)(6-5)$$
$$= 6\times 5\times 4\times 3\times 2\times 1 = 720$$

19. (c) We have,

$$(n+2)! = 2550\times n!$$
$$\Rightarrow\quad (n+2)(n+1)n! = 2550\times n!$$
$$\Rightarrow\quad (n+2)(n+1) = 2550 = 51\times 50$$
$$\Rightarrow\quad (n+1) = 50$$
$$\therefore\quad n = 49$$

20. (d) Reason is correct as on fixing one object anywhere in the circle, the remaining $(n-1)$ objects can be arranged in $(n-1)!$ ways.

Assertion is incorrect, as after arranging boys on the circle in $(n-1)!$ ways, girls can be arranged in between the boys in $n!$ ways (for any arrangement of boys).

Hence, number of arrangements is $n!\,(n-1)!$.

21. (d) There are 4 letters in word 'ROOT' of which there are 2 O's and rest are different. Therefore, the required number of arrangements

$$= \frac{4!}{2!} = 12$$

There are 9 letters in 'INSTITUTE' of which there are 2 I's, 3 T's and rest are different.

Therefore, required number of arrangements = $\dfrac{9!}{2!\,3!}$

Hence, Assertion is incorrect and Reason is correct.

22. (a) Given, $P(n,5) = 20\,P(n,3)$

$$\Rightarrow\quad \frac{n!}{(n-5)!} = 20\times\frac{n!}{(n-3)!}$$
$$\Rightarrow\quad \frac{1}{(n-5)!} = \frac{20}{(n-3)(n-4)(n-5)!}$$
$$\Rightarrow\quad \frac{(n-3)(n-4)(n-5)!}{(n-5)!} = 20$$
$$\Rightarrow\quad (n-3)(n-4) = 20$$
$$\Rightarrow\quad n^2 - 4n - 3n + 12 = 20$$
$$\Rightarrow\quad n^2 - 8n + n - 8 = 0$$
$$\Rightarrow\quad (n-8)(n+1) = 0 \qquad [\because n \neq -1]$$
$$\Rightarrow\quad n - 8 = 0$$
$$\therefore\quad n = 8$$

23. (c) Number of ways of arranging four distinct objects in a line is $^4P_4 = 4! = 24$. Hence, Reason is incorrect. Again, when W, I, F, E are arranged in all possible ways, then number of words formed is $4! = 24$ and WIFE occurs last of all as its letters are against alphabetical order.

24. (b) A combination is a grouping or selection of different objects taken some or all at a time.

A permutation represents an arrangement or order or a sequence of different objects taken some or all at a time.

25. (a) Given, $^{n-1}C_r : {}^nC_r : {}^{n+1}C_r = 6:9:13$

$$\Rightarrow\quad \frac{(n-1)!}{r!\,(n-1-r)!} : \frac{n!}{r!(n-r)!} : \frac{(n+1)!}{r!\,(n+1-r)!} = 6:9:13$$
$$\Rightarrow\quad \frac{(n-1)!}{(n-1-r)!} : \frac{n(n-1)!}{(n-r)!} : \frac{(n+1)n\,(n-1)!}{(n+1-r)!} = 6:9:13$$

We have,
$$\frac{\dfrac{(n-1)!}{(n-1-r)!}}{\dfrac{n(n-1)!}{(n-r)!}} = \frac{6}{9}$$
$$\Rightarrow\quad \frac{(n-1)!\,(n-r)!}{(n-1-r)!\,n(n-1)!} = \frac{6}{9}$$
$$\Rightarrow\quad \frac{n-r}{n} = \frac{6}{9}$$
$$\Rightarrow\quad 6n = 9n - 9r$$
$$\Rightarrow\quad 3n = 9r$$
$$\Rightarrow\quad n = 3r$$

Also,
$$\frac{\dfrac{n(n-1)!}{(n-r)!}}{\dfrac{(n+1)n\,(n-1)!}{(n+1-r)!}} = \frac{9}{13}$$
$$\Rightarrow\quad \frac{n!(n+1-r)!}{(n+1)!\,(n-r)!} = \frac{9}{13}$$
$$\Rightarrow\quad \frac{n+1-r}{n+1} = \frac{9}{13}$$

On substituting $n = 3r$ in the above equation,

$$\frac{3r+1-r}{3r+1} = \frac{9}{13}$$
$$\Rightarrow\quad 13(2r+1) = 9(3r+1)$$
$$\Rightarrow\quad 26r + 13 = 27r + 9$$
$$\Rightarrow\quad r = 4$$
$$\Rightarrow\quad n = 3\times 4 = 12$$

26. (a) Given, $^nC_9 = {}^nC_4$

$\Rightarrow \quad 9+4=n \quad [\because 9 \neq 4]$

$\therefore \quad n=13$

27. (d) We know that total number of diagonals in a polygon is $^nC_2 - n$.

Given polygon has 54 diagonals.

$\therefore \quad {}^nC_2 - n = 54$

$\Rightarrow \quad \frac{n!}{(n-2)!2!} - n = 54$

$\Rightarrow \quad \frac{n(n-1)}{2} - n = 54$

$\Rightarrow \quad \frac{n(n-3)}{2} = 54$

$\Rightarrow \quad n(n-3) = 108$

$\Rightarrow \quad n^2 - 3n - 108 = 0$

$\Rightarrow \quad (n-12)(n+9) = 0$

$\Rightarrow \quad n - 12 = 0$

$[\because n \neq -9$ (number of sides cannot be negative)$]$

$\Rightarrow \quad n = 12$

Thus, the polygon has 12 sides.

Hence, Assertion is incorrect and Reason is correct.

28. (i) (a) We have five students.

$\therefore$ Total number of arrangements is $5! = 120$.

(ii) (b) Ajay and Yojana sit together.

Total number of arrangements $= 4! \times 2!$

$= 24 \times 2 = 48$

(iii) (d) Total number of arrangements of Yojana and Rahul sitting in extreme position is

$2! \times 3! = 2 \times 6 = 12$

(iv) (a) Number of arrangements of Shyam in middle is

$4! = 24$

29. (i) (c) The letters in the word 'REPUBLIC' are all distinct. There are 8 letters in given word.

$\therefore$ Number of arrangements of the word 'REPUBLIC'

$= 8!$

$= 40320$

(ii) (b) Vowels in a given word are 'E, I, U' if we start a word from vowel, we can choose 1 vowel from 3 vowels in $^3C_1$ ways.

Further, remaining 7 letters can be arranged in 7! ways.

$\therefore$ Total number of arrangements start with a vowel

$= {}^3C_1 \times 7!$

$= 3 \times 5040$

$= 15120$

(iii) (c) Combination

(iv) (a) The consonants in given word are 'R, P, B, L, C'.

If we start a word from consonants, we can choose 1 consonant from 5 consonants in $^5C_1$ ways.

Further, remaining 7 letters can be arranged in 7! ways.

$\therefore$ Total number of arrangements start with a consonants $= {}^5C_1 \times 7!$

$= \frac{5!}{1!\,4!} \times 7 \times 6 \times 5 \times 4 \times 3 \times 2 \times 1$

$= 5 \times 5040 = 25200$

Subjective Questions

1. Clearly, a book of Hindi and a book of English can be selected in $10 \times 5 = 50$ ways.

A book of English and a book of Sanskrit can be selected in $5 \times 7 = 35$ ways.

A book of Hindi and a book of Sanskrit can be selected in $10 \times 7 = 70$ ways.

Now, by fundamental principle of addition, two books of different languages can be selected in $50 + 35 + 70 = 155$ ways.

2. We have, $(n!)^2 = (n!) \times (n!) = n! \times (1 \times 2 \times 3 \times \ldots \times n)$

$\leq (n!) \times n^n \quad [\because r \leq n$ for each $r = 1, 2, 3, \ldots, n]$

$\therefore \quad (n!)^2 \leq (n!) \times n^n \quad \ldots$(i)

Again, we have

$(2n!) = (1 \times 2 \times 3 \times \ldots \times n) \times (n+1) \times (n+2) \times \ldots \times 2n$

$> (n!) \times n^n \quad [\because (n+r) > n$ for each $r = 1, 2, \ldots, n]$

Thus, $(n!) \times n^n < (2n)! \quad \ldots$(ii)

From Eqs. (i) and (ii), we get

$(n!)^2 \leq (n!) \times n^n < (2n)!$ **Hence proved.**

3. We have, $(n+1)! = 12 \times (n-1)!$

$\Rightarrow \quad (n+1)n\,(n-1)! = 12 \times (n-1)!$

$\Rightarrow \quad n^2 + n = 12$

$\Rightarrow \quad n^2 + n - 12 = 0$

$\Rightarrow \quad n^2 + 4n - 3n - 12 = 0$

$\Rightarrow \quad (n+4)(n-3) = 0$

$\Rightarrow \quad (n+4) = 0$ or $(n-3) = 0$

$\Rightarrow \quad n = -4$ or $n = 3$

$\therefore \quad n = 3 \quad [\because n \neq -4]$

4. We have, $\frac{n!}{(n-2)!} = 930, n \geq 2$

$\Rightarrow \quad \frac{n(n-1)\,(n-2)!}{(n-2)!} = 31 \times 30$

$\Rightarrow \quad n(n-1) = 31 \times 30$

$\therefore \quad n = 31$

5. We have, $^nP_4 = 12 \cdot {^nP_2} \Rightarrow \frac{n!}{(n-4)!} = 12 \times \frac{n!}{(n-2)!}$

$\Rightarrow \quad 1 = 12 \times \frac{1}{(n-2)(n-3)}$

$\Rightarrow \quad (n-2)(n-3) = 12$

$\Rightarrow \quad (n-2)(n-3) = 4 \times 3$

On comparing, $n = 6$

6. We have, $^nP_4 : {^{n-1}P_4} = 5:3$

$\Rightarrow \quad \frac{n!}{(n-4)!} : \frac{(n-1)!}{(n-1-4)!} = 5:3$

$\Rightarrow \quad \frac{\frac{n(n-1)!}{(n-4)(n-5)!}}{\frac{(n-1)!}{(n-5)!}} = \frac{5}{3}$

$\Rightarrow \quad \frac{n}{n-4} = \frac{5}{3}$

$\Rightarrow \quad 3n = 5n - 20$

$\Rightarrow \quad 2n = 20 \Rightarrow n = 10$

7. We have, $^{56}P_{(r+6)} : {^{54}P_{(r+3)}} = 30800:1$

$\Rightarrow \quad \frac{\frac{56!}{(56-r-6)!}}{\frac{54!}{(54-r-3)!}} = \frac{30800}{1}$

$\Rightarrow \quad \frac{\frac{56 \times 55 \times 54!}{(50-r)!}}{\frac{54!}{(51-r)!}} = \frac{30800}{1}$

$\Rightarrow \quad 56 \times 55\,(51-r) = 56 \times 550$

$\Rightarrow \quad 56 \times 55\,(51-r) = 56 \times 55 \times 10$

On comparing both sides, we get

$51 - r = 10 \Rightarrow r = 41$

8. We have, $^nC_8 = {^nC_6}$

$\Rightarrow \quad ^nC_{(n-8)} = {^nC_6} \quad [\because {^nC_r} = {^nC_{(n-r)}}]$

$\Rightarrow \quad n - 8 = 6$

$\Rightarrow \quad n = 14$

Now, $^nC_2 = {^{14}C_2}$

$= \frac{14!}{2!\,12!} \quad \left[\because {^nC_r} = \frac{n!}{r!\,(n-r)!}\right]$

$= \frac{14 \times 13}{2 \times 1} = 91$

9. We have, $^{18}C_x = {^{18}C_{x+2}}$

$\Rightarrow \quad ^{18}C_{(18-x)} = {^{18}C_{x+2}} \quad [\because {^nC_r} = {^nC_{(n-r)}}]$

$\Rightarrow \quad 18 - x = x + 2$

$\Rightarrow \quad 2x = 16$

$\therefore \quad x = 8$

10. We have, $^{2n}C_3 : {^nC_2} = 44:3$

$\Rightarrow \quad \frac{2n!}{3!\,(2n-3)!} : \frac{n!}{2!\,(n-2)!} = 44:3$

$\Rightarrow \quad \frac{2n(2n-1)(2n-2)}{3 \cdot 2!} : \frac{n(n-1)}{2!} = 44:3$

$\Rightarrow \quad \frac{2n(2n-1)2(n-1)}{3 \cdot 2!} : \frac{n(n-1)}{2!} = 44:3$

$\Rightarrow \quad \frac{4(2n-1)}{3} : 1 = 44:3$

$\Rightarrow \quad \frac{\frac{4}{3}(2n-1)}{1} = \frac{44}{3}$

$\Rightarrow \quad 2n - 1 = 11$

$\therefore \quad n = 6$

11. Consider, $^nC_r + 2 \cdot {^nC_{r-1}} + {^nC_{r-2}}$

$= {^nC_r} + {^nC_{r-1}} + {^nC_{r-1}} + {^nC_{r-2}}$

$= ({^nC_r} + {^nC_{r-1}}) + ({^nC_{r-1}} + {^nC_{r-2}})$

$= {^{n+1}C_r} + {^{n+1}C_{r-1}} = {^{n+2}C_r}$

$[\because {^nC_r} + {^nC_{r-1}} = {^{n+1}C_r}]$

12. Consider, $\sum_{r=1}^{5} {^5C_r} = {^5C_1} + {^5C_2} + {^5C_3} + {^5C_4} + {^5C_5}$

$= {^5C_1} + {^5C_2} + {^5C_2} + {^5C_1} + {^5C_0} \quad [\because {^nC_r} = {^nC_{n-r}}]$

$= 5 + 2 \times {^5C_2} + 5 + 1 \quad [\because {^nC_1} = n \text{ and } {^nC_0} = 1]$

$= 11 + 2 \times \frac{5 \times 4}{2 \times 1}$

$= 11 + 20 = 31$

13. Let total number of persons be n.

Since, total number of hand-shakes = 66

$\therefore \quad {^nC_2} = 66$

$\Rightarrow \quad \frac{n(n-1)}{2} = 66$

$\Rightarrow \quad n^2 - n - 132 = 0$

$\Rightarrow \quad (n-12)(n+11) = 0$

$\therefore \quad n = 12 \quad [\because n \text{ cannot be negative}]$

14. Total number of men = 7

and total number of women = 5

Number of ways selecting 3 men form 7 men = $^7C_3$

and number of ways selecting 2 women from 5 women $= {^5C_2}$

$\therefore$ Total number of ways to form a committee of 3 men and 2 women = $^7C_3 \times {^5C_2}$

$= \frac{7!}{3!\,4!} \times \frac{5!}{2!\,3!}$

$= \frac{7 \times 6 \times 5}{3 \times 2} \times \frac{5 \times 4}{2} = 35 \times 10 = 350$

15. Clearly, number of chords that can be drawn from 7 points $= {}^7C_2 = \dfrac{7\times6}{2\times1} = 21$

Hence, the total number of chords that can be drawn through 7 points on a circle is 21.

16. It is given that no two lines are parallel means all lines are intersecting and no three lines are concurrent means no three lines intersect at a point.

$\therefore$ Number of points of intersection $= {}^{20}C_2$

$$= \frac{20\times19}{2} = 190$$

17. Here, we need to perform two operations.

(i) Selecting a bus to go from Delhi to Noida.

(ii) Selecting a bus to return from Noida to Delhi.

Clearly, a person can travel anyone of the 25 buses.

So, there are 25 ways to travel Delhi to Noida.

$\because$ A person can return anyone of the remaining 24 buses. So, he can return through a different bus in 24 ways.

Hence, by fundamental principle of multiplication, required number of ways $= 25\times24$

$$= 600$$

18. Since, each of first three questions can be answered in 4 ways.

So, the total number of ways to answering the first 3 questions $= 4\times4\times4 = 64$

Now, each of remaining 3 questions can be answered in 3 ways.

So, the total number of ways to answering the remaining 3 questions is $= 3\times3\times3 = 27$.

The number of possible ways to answer $= 64\times27 = 1728$

19. $\because$ Each voter can be voted to anyone of the four candidates.

$\therefore$ Each of voters can be voted in 4 ways.

Hence, the total number of ways in which all 50 voters can be voted is 4^{50}.

20. Between 100 to 1000, there are only 3 digit numbers exist.

$\because$ Each digit of the number can be 2 or 9.

$\therefore$ Number of ways to fill unit place = 2,

number of ways to fill 10th place = 2

and number of ways to fill 100th place = 2

Therefore, total number of ways $= 2\times2\times2 = 8$

21. Total number of digits = 9

$\because$ Repetition of digits are not allowed.

$\therefore$ Number of ways to fill unit place = 9,

number of ways to fill 10th place = 8,

number of ways fill 100th place = 7

and number of ways to fill 1000th place = 6

6 ways 7 ways 8 ways 9 ways

Hence, total number of 4 digit numbers formed using the digits 1 to 9 $= 9\times8\times7\times6 = 3024$

22. Total number of letters in the word 'TRIANGLE' is 8. and each letter is distinct.

$\because$ Word begin with T and end with E.

$\therefore$ Number of letters remains = 6

T _ _ _ _ _ _ E

6 ways 5 ways 4 ways 3 ways 2 ways 1 way

Number of ways to fill the remaining 6 places = 6!

$$= 720$$

Hence, 720 different words can be formed with the letters of word 'TRIANGLE' so that the word begin with 'T' and end with 'E'.

23. Total number of letters in the word 'PENCIL' = 6

$\because$ N and E are always together.

$\therefore$ Number of ways that N and E come together = 2!

N and E always come together. So, N and E are consider as one letter and remaining four letters are P, C, I, L.

Number of words made by 5 letters = 5!

Total number of words made by the letters of word 'PENCIL' so that N and E are always together

$$= 2!\times5!$$
$$= 2\times120 = 240$$

24. Suppose first two digits are 4 and 1. Then, we have 8 digits, namely 0, 2, 3, 5, 6, 7, 8, 9 which can be used in constructing the telephone numbers.

Clearly, third digit can be selected in 8 ways.

Fourth digit can be selected in $8-1=7$ ways.

Fifth digit can be selected in 6 ways.

and sixth digit can be selected in 5 ways.

Now, according to fundamental principle of multiplication, number of telephone starting with 41

$$= 8\times7\times6\times5 = 1680$$

Similarly, if the telephone number start with 42 or 46 or 62 or 64, the remaining 4 digits can be selected in 1680 ways.

Hence, the total number of required telephone numbers

$$= 1680\times5 = 8400$$

25. Total number of letters in the word 'ORIENTAL' = 8 and each letter is distinct.

Number of 3 letter words made using the letters of the word "ORIENTAL' = $^8P_3$

$$= \frac{8!}{(8-3)!} = \frac{8!}{5!} = 8 \times 7 \times 6 = 336$$

26. Total number of letters in the word 'MATHEMATICS' = 11 and M, A, T come twice.

Total number of arrangements can be made with the letters of the word 'MATHEMATICS' = $\frac{11!}{2!\,2!\,2!}$

$$= 4989600$$

27. Given, number of letter A = 3

number of letter B = 2

and number of letter C = 3

Total number of letters = 3 + 2 + 3 = 8

Number of words can be formed using the letter A thrice, letter B twice and letter C thrice = $\frac{8!}{3!\,2!\,3!}$

$$= \frac{8 \times 7 \times 6 \times 5 \times 4}{3 \times 2 \times 1 \times 2 \times 1} = 560$$

28. We have,

$$^nP_s = n(n-1)(n-2)\ldots(n-r+1)(n-r)\ldots(n-s+1)$$
$$= \{n(n-1)(n-2)\ldots(n-r+1)\}$$
$$\times \{(n-r)(n-r-1)\ldots(n-s+1)\}$$
$$= {}^nP_r \times \{(n-r)(n-r-1)\ldots(n-s+1)\},$$

which is divisible by ^nP_r.

Hence, ^nP_s is divisible by ^nP_r.

29. Since, a person cannot hold more than one position, therefore required number of ways is same as number of permutations of 8 different things taken 2 at a time.

Hence, required number of ways = $^8P_2$

$$= \frac{8!}{(8-2)!} = \frac{8!}{6!} = \frac{8 \times 7 \times 6!}{6!} = 56$$

30. Given digits are 2, 3, 4, 5, 6 and 0.

Any number between 400 and 1000 will be of three digits.

Since, the number should be greater than 400.

Therefore, hundred's place can be filled up by anyone of the digits 4, 5 and 6 in 3 ways.

$\therefore \quad k = 3$

Now, remaining two places can be filled up by the remaining five digits in $^5P_2$ ways.

Hence, required number = $3 \times {}^5P_2 = 3 \times \frac{5!}{3!} = 60$

31. Since, four particular guests want to sit on a particular side A (say) and three others on the other side B (say).

So, we are left with 11 guests. Out of which, we choose 5 for side A in $^{11}C_5$ ways and the remaining 6 for side B in $^6C_6$ ways.

Hence, the number of selections for the two sides is $^{11}C_5 \times {}^6C_6$.

Now, 9 persons on each side of the table can be arranged among themselves in 9! ways.

Hence, the total number of arrangements

$$= {}^{11}C_5 \times {}^6C_6 \times 9! \times 9! = \frac{11!}{6!\,5!} \times 9! \times 9!$$

32. There are 5 boys and 4 girls. We have to select 3 boys out of 5 boys and 3 girls out of 4 girls.

Clearly, 3 boys out of 5 boys can be selected in $^5C_3$ ways, and 3 girls out of 4 girls can be selected in $^4C_3$ ways.

Hence, by fundamental principle of multiplication, number of ways of selecting 3 boys and 3 girls

$$= {}^5C_3 \times {}^4C_3 = {}^5C_2 \times {}^4C_1 \quad [\because {}^nC_r = {}^nC_{n-r}]$$
$$= \frac{5 \times 4}{2 \times 1} \times 4 \quad [\because {}^nC_1 = n]$$
$$= 10 \times 4 = 40$$

33. Selection of 5 cards can be done in the following ways.

Case	I	II	III	IV
King	1	2	3	4
Other cards	4	3	2	1

The number of combinations in case I, II, III and IV are respectively,

$^4C_1 \times {}^{48}C_4$; $^4C_2 \times {}^{48}C_3$; $^4C_3 \times {}^{48}C_2$ and $^4C_4 \times {}^{48}C_1$

Hence, the required number of combinations

$$= {}^4C_1 \times {}^{48}C_4 + {}^4C_2 \times {}^{48}C_3 + {}^4C_3 \times {}^{48}C_2 + {}^4C_4 \times {}^{48}C_1$$
$$= 4 \times \frac{48 \times 47 \times 46 \times 45}{4 \times 3 \times 2 \times 1} + \frac{4 \times 3}{2 \times 1} \times \frac{48 \times 47 \times 46}{3 \times 2 \times 1} + 4 \times \frac{48 \times 47}{2 \times 1} + 1 \times 48$$
$$= 8 \times 47 \times 46 \times 45 + 48 \times 47 \times 46 + 2 \times 48 \times 47 + 48$$
$$= 778320 + 103776 + 4512 + 48 = 886656$$

34. Here, we have to find number of different signals that can be generated by arranging atleast 2 flags in order. So, a signal can consists of either 2 flags, 3 flags, 4 flags or 5 flags. If a signal consists of 2 flags, then two vacant places are available.

$\therefore$ Number of ways of filling first vacant place = 5 and number of ways of filling second vacant place = 4

Thus, total number of signals consisting 2 flags

$$= 5 \times 4 = 20$$

Now, if a signal consists of 3 flags, then three vacant places are available.

$\therefore$ Number of ways of filling first vacant place = 5, number of ways of filling second vacant place = 4 and number of ways of filling third vacant place = 3

Thus, total number of signals consisting 3 flags

$$= 5 \times 4 \times 3 = 60$$

Similarly, total number of signals consisting 4 flags

$$= 5 \times 4 \times 3 \times 2 = 120$$

and total number of signals consisting 5 flags

$$= 5 \times 4 \times 3 \times 2 \times 1 = 120$$

Since, different signals can be generated by arranging either 2 flags or 3 flags or 4 flags or 5 flags.

$\therefore$ Total number of signals = 20 + 60 + 120 + 120 = 320

35. (i) When the repetition of the letters is not allowed. In this case, the total number of words is the same as the number of ways of filling 4 places □□□□ by 4 different letters chosen from P, O, S, E.

The first place can be filled by any of the 4 letters.

Thus, there are 4 ways of filling first place. Since, the repetition of letters is not allowed, so the second place can be filled by any of the remaining 3 letters in 3 different ways.

Following it, the third place can be filled by any of the two remaining letters in 2 different ways and the fourth place can be filled by the remaining one letter in 1 way.

So, by the fundamental principle of multiplication, the required number of 4 letter words

$$= 4 \times 3 \times 2 \times 1 = 24$$

Hence, the required number of words = 24

(ii) When the repetition of the letters is allowed.

In this case, each of the four different places can be filled in succession in 4 different ways.

So, the required number of 4 letter words

$$= 4 \times 4 \times 4 \times 4 = 4^4 = 256$$

36. (i) There are in all 26 English alphabets, we have to choose 2 distinct alphabets.

1st alphabet can be selected in 26 ways.

2nd alphabet can be selected in 25 ways.

Again, out of 9 digits (1 to 9),

1st number can be selected in 9 ways.

2nd number can be selected in 8 ways.

Thus, by the fundamental principle of multiplication, the number of distinct codes $= 26 \times 25 \times 9 \times 8$

$$= 46800$$

(ii) As above, two distinct alphabets can be selected in 26×25 ways.

We have, in all 1, 2, 3, 4, 5, 6, 7, 8 and 9 digits, unit place can be filled up in 4 ways (by 2, 4, 6, 8).

Ten's place can be filled up in 8 ways.

[since, one of the digits is already used]

Thus, the number of such codes $= 26 \times 25 \times 8 \times 4$

$$= 20800$$

37. (i) The first prize can be given in 4 ways. The 2nd prize can be given in 3 ways because the boy who got the first prize cannot receive the prize.

The 3rd prize can be given to anyone of the remaining 2 boys in 2 ways.

$\therefore$ Total number of ways in which all the prizes can be given $= 4 \times 3 \times 2 = 24$

(ii) The first prize can be given in 4 ways.

The 2nd prize can also be given in 4 ways because it may be obtained by the boy who has already received a prize.

Similarly, 3rd prize can be given in 4 ways.

Hence, the number of ways in which all the prizes can be given $= 4 \times 4 \times 4 = 4^3 = 64$

(iii) If anyone of the 4 boys may get all the prizes, then the number of ways in which all the 3 prizes given to a boy = 4.

So, the number of ways in which no boy get all the prizes = 64 − 4 = 60

38. We have to find numbers between 100 and 1000, i.e. 3-digit numbers which have exactly one of their digits as 7.

Now, consider the following cases

Case I When 7 is at unit's place.

In this case, ten's and hundred's places can be filled using the digits 0, 1, 2, 3, 4, 5, 6, 8 and 9.

Clearly, the number of ways to fill ten's place = 9

[using anyone of the digits 0, 1, 2, 3, 4, 5, 6, 8, 9]

and the number of ways to fill the hundred's place = 8

$\therefore$ Number of such numbers $= 1 \times 9 \times 8 = 72$

Case II When 7 is at ten's place.

In this case, unit's and hundred's places can be filled using the digits 0, 1, 2, 3, 4, 5, 6, 8 and 9.

Clearly, the number of ways to fill unit's place = 9

[using anyone of digits 0, 1, 2, 3, 4, 5, 6, 8, 9]

and the number of ways to fill hundred's place = 8

[$\because$ hundred's place cannot be filled by zero]

$\therefore$ Number of such numbers $= 9 \times 1 \times 8 = 72$

Case III When 7 is at hundred's place.

In this case, unit's and ten's places can be filled using the digits 0, 1, 2, 3, 4, 5, 6, 8 and 9.

Clearly, the number of ways to fill unit's place = 9

[using anyone of the digits 0, 1, 2, 3, 4, 5, 6, 8, 9]

and the number of ways to fill the ten's place = 9

[using anyone of the digits 0, 1, 2, 3, 4, 5, 6, 8, 9]

$\therefore$ Number of such numbers $= 1 \times 9 \times 9 = 81$

Hence, the total number of required numbers

$$= 72 + 72 + 81 = 225$$

39. Clearly, the signals can be made using one or two or three or four or five or six flags.

Now, number of signals when one flag is used at a time

$$= {}^6P_1$$

Number of signals when two flags are used at a time

$$= {}^6P_2$$

Similarly, number of signals when three, four, five, six flags are used at a time are ${}^6P_3$, ${}^6P_4$, ${}^6P_5$, ${}^6P_6$, respectively.

Hence, by fundamental principle of addition, total number of signals formed from 6 flags of different colours $= {}^6P_1 + {}^6P_2 + {}^6P_3 + {}^6P_4 + {}^6P_5 + {}^6P_6$

$$= \frac{6!}{(6-1)!} + \frac{6!}{(6-2)!} + \frac{6!}{(6-3)!} + \frac{6!}{(6-4)!} + \frac{6!}{(6-5)!} + \frac{6!}{(6-6)!}$$

$$= \frac{6!}{5!} + \frac{6!}{4!} + \frac{6!}{3!} + \frac{6!}{2!} + \frac{6!}{1!} + \frac{6!}{0!}$$

$$= 6 + 6 \times 5 + 6 \times 5 \times 4 + 6 \times 5 \times 4 \times 3 + 6 \times 5 \times 4 \times 3 \times 2 + 6!$$

$$= 6 + 30 + 120 + 360 + 720 + 720 = 1956$$

40. The total number of arrangements of ten digits 0, 1, 2, 3, 4, 5, 6, 7, 8, 9 taking 4 at a time is ${}^{10}P_4$. But these arrangements also include those numbers which have 0 at thousand's place. That type of numbers are not four-digit numbers. When 0 is fixed at thousand's place, we have to arrange remaining 9 digits by taking 3 at a time. The number of such arrangements is ${}^9P_3$.

So, the total number of numbers having 0 at thousand's place $= {}^9P_3$

Hence, total numbers of four-digit numbers

$$= {}^{10}P_4 - {}^9P_3 = 5040 - 504 = 4536$$

41. We have, 5 boys and 5 girls.

(i) Since, no two girls sit together, therefore the possible choices for girls are the places marked as '×'.

$$\times B_1 \times B_2 \times B_3 \times B_4 \times B_5 \times$$

Clearly, the girls can be arranged in ${}^6P_5$ ways and the boys can be arranged in 5! ways.

Hence, by fundamental principle of multiplication, required number of ways $= {}^6P_5 \times 5!$

$$= \frac{6!}{(6-5)!} \times 5! = 6! \times 5! = 86400$$

(ii) Let us first seat the 5 girls (or 5 boys). This can be done in 5! ways.

Now, for each such arrangement, 5 boys can be seated only at the cross marked places as shown below.

(I) $\times G_1 \times G_2 \times G_3 \times G_4 \times G_5$

(II) $G_1 \times G_2 \times G_3 \times G_4 \times G_5 \times$

In Case I, 5 boys can be seated in ${}^5P_5 = 5!$ ways

Thus, number of ways of seating $= 5! \times 5! = (5!)^2$

Similarly, in Case II, number of ways of seating $= (5!)^2$

Hence, the required number of ways $= (5!)^2 + (5!)^2$

$$= 2 \cdot (5!)^2 = 28800$$

42. There are 12 letters in the word 'PERMUTATIONS'.

In this letter, T occurs twice and rest all are different.

Also, the word 'PERMUTATIONS' contains all the five vowels, namely, A, E, I, O and U.

(i) Since, the words start with P and end with S, so let us fix P at first place and S at last place.

P	I	II	III	IV	V	VI	VII	VIII	IX	X	S

Now, we are left with 10 letters (E, R, M, U, T, A, T, I, O, N) which can be arranged in $\frac{10!}{2!}$ ways.

[$\because$ T occurs twice]

Hence, the required number of ways $= \frac{10!}{2!} = 1814400$

(ii) Since, the vowels are together, so let us take all vowels together and consider it as one letter.

[A E I O U], P, R, M, T, T, N, S

Now, we have 8 letters, which can be arranged in $\frac{8!}{2!}$ ways. Corresponding to each such arrangement, all the five vowels can be arranged among themselves in ${}^5P_5 = 5!$ ways.

Hence, the required number of ways $= \frac{8!}{2!} \times 5!$

$$= 2419200$$

QR Code for Quick Test 1

QR Code for Quick Test 2

QR Code for Quick Test 3

Mind Map

Permutations and Combinations

Fundamental Principles of Counting

Fundamental Principle of Addition (FPA)
If there are two events such that they can be performed independently in m and n ways respectively, then either of the two events can occur in $(m + n)$ ways.

Fundamental Principle of Multiplication (FPM)
If an event can occur in m different ways, another event can occur in n different ways, then the total number of occurrence of the events in the given order is $m \times n$.

Permutation
A permutation is an arrangement of objects in a definite order. Arrangement can be made by taking some or all objects at a time.

Permutations, when all the Objects are Distinct
The number of permutations of n different objects taken r at a time, where $0 < r \leq n$ and the objects do not repeat, is $n(n-1)(n-2)\ldots(n-r+1)$, which is denoted by ${}^nP_r$ or $P(n, r)$.

i.e. $P(n, r) = {}^nP_r = \dfrac{n!}{(n-r)!}, 0 \leq r \leq n$

Permutation with Repetition
The number of permutations of n different objects taken r at a time, when each may be repeated any number of times in each arrangement, is n^r (permutation with repetitions).
The number of permutations of n different objects, all at a time, when each may be repeated any number of times in each arrangement, is n^n.

Permutations when all the Objects are not Distinct Objects
(i) The number of permutations of n objects, where p objects are of the same kind or identical and other are distinct, is given by $\dfrac{n!}{p!}$.
(ii) The number of permutations of n objects, where p_1 objects are of one kind, ...p_2 objects are of second kind, ...p_k objects are of kth kind and the rest if any, are of different kind is $\dfrac{n!}{p_1!\,p_2!\ldots p_k!}$.

Factorial Notation
Let n be a positive integer. Then, the continued product of first n natural numbers is called the ***n* factorial** and it is denoted by $n!$ or $\underline{|n}$.
Thus, $n! = n(n-1)(n-2)(n-3) \ldots\ldots 3.2.1.$

Zero Factorial
It does not make any sense to define 0! as product of the integers from 1 to 0. So, we define $0! = 1$.

Combinations
The word combination means selection. Each of the different selection, which is made by taking some or all of a number of different objects at a time, irrespective of their arrangements is called a combination.

Meaning of ${}^nC_r$
The number of combinations of n distinct objects taken at a time is given by

$${}^nC_r = \frac{n!}{r!(n-r)!}$$

In particular,
(i) When $r = 0$, then ${}^nC_r = 1$
(ii) When $r = n$, then ${}^nC_r = 1$

Some Important Theorems
(i) ${}^nP_r = {}^nC_r \times r!, 0 < r \leq n$
(ii) ${}^nC_r = {}^nC_{n-r}, 0 \leq r \leq n$
(iii) ${}^nC_r + {}^nC_{r-1} = {}^{n+1}C_r$
(iv) If ${}^nC_x = {}^nC_y$, then either $x = y$ or $x + y = n$.
(v) If n is an even natural number, then the greatest of the values ${}^nC_0, {}^nC_1, {}^nC_2, \ldots, {}^nC_n$ is ${}^nC_{n/2}$.
If n is an odd natural number, then the greatest of the values ${}^nC_0, {}^nC_1, {}^nC_2, \ldots, {}^nC_n$ is ${}^nC_{n-1/2} = {}^nC_{n+1/2}$.

07
CHAPTER

Mathematics is a great motivator for all humans because its career starts with zero and it never ends. —*Marston Morse*

Binomial Theorem

YOUR STUDY GOALS

YOUR PRACTICE GOALS

- Objective Type Questions (32 Qs ; 20 MCQs, 10 A-R, 2 Case Based)
- Subjective Type Questions (22 Qs; 5 VSA, 8 SA, 9 LA)

7.1 Binomial Expression

An algebraic expression consisting of two terms with +ve or −ve sign between them, is called binomial expression.

e.g. $(a+2b)$, $\left(\frac{p}{x^2}-\frac{q}{x^4}\right)$, $\left(3x-\frac{2}{y}\right)$

Expansion of Binomial Expression

In earlier classes, we have already studied that

- $(a+b)^0=1$
- $(a+b)^1=a+b$
- $(a+b)^2=a^2+2ab+b^2$
- $(a+b)^3=a^3+3a^2b+3ab^2+b^3$
- $(a+b)^4=a^4+4a^3b+6a^2b^2+4ab^3+b^4$

From the above expressions, it can be observed that

- number of terms in expansion is one more than the index.
- power of first quantity a goes on decreasing by 1, while power of second quantity b goes on increasing by 1.
- in each term of the expansion, the sum of indices a and b is same and equal to the index of $(a+b)$.

The coefficient of variables in the expansion of $(a+b)^n$, follow a particular pattern, which is given below.

Index	Coefficients	Combinations
0	1	${}^0C_0$ (=1)
1	1 ∇ 1	${}^1C_0$ (=1), ${}^1C_1$ (=1)
2	1 ∇ 2 ∇ 1	${}^2C_0$ (=1), ${}^2C_1$ (=2), ${}^2C_2$ (=1)
3	1 ∇ 3 ∇ 3 ∇ 1	${}^3C_0$ (=1), ${}^3C_1$ (=3), ${}^3C_2$ (=3), ${}^3C_3$ (=1)
4	1 ∇ 4 ∇ 6 ∇ 4 ∇ 1	${}^4C_0$ (=1), ${}^4C_1$ (=4), ${}^4C_2$ (=6), ${}^4C_3$ (=4), ${}^4C_4$ (=1)
5	1 ∇ 5 ∇ 10 ∇ 10 ∇ 5 ∇ 1	${}^5C_0$ (=1), ${}^5C_1$ (=5), ${}^5C_2$ (=10), ${}^5C_3$ (=10), ${}^5C_4$ (=5), ${}^5C_5$ (=1)

It can be seen that the addition of 1's in the row for index 1 gives rise to 2 in the row for index 2. The addition of 1, 2 and 2, 1 in the row for index 2 gives rise to 3 and 3 in the row for index 3 and so on. This structure or array of numbers, is known as **Pascal's triangle.**

In this structure table, we can also use a rule of combination

i.e. $${}^nC_r = \frac{n!}{r!(n-r)!} \quad (0 \le r \le n)$$

and n is a non-negative integer.

Clearly, with the help of combination pattern, row of the Pascal's triangle for any index can be written without writing the earlier rows.

e.g. For $n=7$, the row would be

$${}^7C_0, {}^7C_1, {}^7C_2, {}^7C_3, {}^7C_4, {}^7C_5, {}^7C_6, {}^7C_7$$

7.2 Binomial Theorem for any Positive Integer *n*

If a and b are two real numbers, then for any positive integer n, we have

$$(a+b)^n = {}^nC_0a^nb^0 + {}^nC_1a^{n-1}b^1 + {}^nC_2a^{n-2}b^2 + \dots + {}^nC_na^0b^n$$

$$(a+b)^n = {}^nC_0a^n + {}^nC_1a^{n-1}b + {}^nC_2a^{n-2}b^2 + \dots + {}^nC_nb^n$$

$$(a+b)^n = \sum_{r=0}^{n} {}^nC_ra^{n-r}b^r$$

Where, ${}^nC_0, {}^nC_1, \dots, {}^nC_n$ are called **binomial coefficients.** These binomial coefficients can also be written as $C_0, C_1, \dots, C_n$.

Some Important Observations

- The total number of terms in the expansion is one more than the index i.e. the total number of terms in $(a+b)^n$ is $n+1$.
- In the successive terms of the expansion, powers of the first quantity a go on decreasing by 1 i.e. it is n in the first term, $(n-1)$ in the second term and so on ending with zero in the last term whereas the powers of the second quantity b increase by 1 i.e. it is zero in first term, 1 in the second term and so on ending with n in the last term.
- In each term of the expansion, the sum of the indices of a and b is the same and is equal to the index of $a+b$ i.e. n.
- The binomial coefficients of terms equidistant from the beginning and end are equal.

Particular Cases of Binomial Theorem

(i) **Expansion of $(a-b)^n$** Replacing b by $(-b)$ in the expansion of $(a+b)^n$, we get

$$(a-b)^n = [a+(-b)]^n$$
$$= {}^nC_0a^n - {}^nC_1a^{n-1}b + {}^nC_2a^{n-2}b^2 + \dots + (-1)^n\,{}^nC_nb^n$$

$$\therefore (a-b)^n = \sum_{r=0}^{n}(-1)^r\,{}^nC_ra^{n-r}b^r$$

(ii) **Expansion of $(1+x)^n$** Replacing a by 1 and b by x in the expansion of $(a+b)^n$, we get

$$(1+x)^n = {}^nC_01^n + {}^nC_11^{n-1}x + {}^nC_21^{n-2}x^2 + \dots + {}^nC_nx^n$$
$$= 1 + {}^nC_1x + {}^nC_2x^2 + \dots + {}^nC_nx^n$$
$$= 1 + nx + \frac{n(n-1)}{2!}x^2 + \dots + x^n$$

$$\therefore (1+x)^n = \sum_{r=0}^{n} {}^nC_rx^r$$

(iii) **Expansion of $(1-x)^n$** Replacing a by 1 and b by $(-x)$ in the expansion of $(a+b)^n$, we get

$$(1-x)^n = {}^nC_0 - {}^nC_1x + {}^nC_2x^2 + \dots + (-1)^n\,{}^nC_n\,x^n$$

$$= 1 - nx + \frac{n(n-1)}{2!}x^2 - \ldots + (-1)^n x^n$$

$$\therefore (1-x)^n = \sum_{r=0}^{n} (-1)^r \, {}^nC_r x^r$$

(iv) $(x+a)^n + (x-a)^n$

$$= 2[{}^nC_0 x^n + {}^nC_2 x^{n-2} a^2 + \ldots]$$

$$= 2 \text{ (Sum of terms at odd places)}$$

Here, the last term will be ${}^nC_n a^n$ or ${}^nC_{n-1} x a^{n-1}$ according as n is even or odd, respectively.

- If n is odd, then number of terms is $\left(\frac{n+1}{2}\right)$.
- If n is even, then number of terms is $\left(\frac{n}{2}+1\right)$.

(v) $(x+a)^n - (x-a)^n = 2[{}^nC_1 x^{n-1} a^1 + {}^nC_3 x^{n-3} a^3 + \ldots]$

$$= 2 \text{ (Sum of terms at even places)}$$

Here, last term will be ${}^nC_{n-1} x a^{n-1}$ or ${}^nC_n a^n$ as n is even or odd, respectively.

- If n is odd, then number of terms is $\left(\frac{n+1}{2}\right)$.
- If n is even, then number of terms is $\left(\frac{n}{2}\right)$.

Example 1. Using binomial theorem, expand the following expressions.

(i) $(x^2+2y)^5$ **(ii)** $(2x-3)^6$

(iii) $\left(\frac{2x}{3} - \frac{3}{2x}\right)^4$ **(iv)** $(x^2-2x+1)^3$

(v) $(1-x+x^2)^4$ **(vi)** $\left(1+\frac{x}{2}-\frac{2}{x}\right)^4, x \neq 0$

Sol. (i) By binomial theorem, we have

$$(a+b)^n = {}^nC_0 a^n + {}^nC_1 a^{n-1} b + {}^nC_2 a^{n-2} b^2 + \ldots + {}^nC_n (b)^n$$

Here, $a = x^2$, $b = 2y$ and $n = 5$

$$\therefore (x^2+2y)^5 = {}^5C_0 (x^2)^5 + {}^5C_1 (x^2)^4 (2y) + {}^5C_2 (x^2)^3 (2y)^2 + {}^5C_3 (x^2)^2 (2y)^3 + {}^5C_4 (x^2)(2y)^4 + {}^5C_5 (2y)^5$$

$$= x^{10} + 5x^8(2y) + 10x^6(4y^2) + 10x^4(8y^3) + 5x^2(16y^4) + 32y^5$$

$$\left[\because {}^5C_0 = {}^5C_5 = 1, {}^5C_1 = {}^5C_4 = 5 \text{ and } {}^5C_2 = {}^5C_3 = \frac{5!}{3!\,2!} = 10\right]$$

$$= x^{10} + 10x^8 y + 40x^6 y^2 + 80x^4 y^3 + 80x^2 y^4 + 32y^5$$

(ii) We have, $(2x-3)^6 = {}^6C_0 (2x)^6 - {}^6C_1 (2x)^5 \times 3 + {}^6C_2 (2x)^4 (3)^2 - {}^6C_3 (2x)^3 (3)^3 + {}^6C_4 (2x)^2 (3)^4 - {}^6C_5 (2x) 3^5 + {}^6C_6 3^6$

$$= {}^6C_0\, 64x^6 - {}^6C_1 32x^5 \times 3 + {}^6C_2 16x^4 \times 9 - {}^6C_3 8x^3 \times 27 + {}^6C_2\, 4x^2 \times 81 - {}^6C_1\, 2x \times 243 + {}^6C_0 729$$

$$[\because {}^nC_r = {}^nC_{n-r}]$$

$$= 64x^6 - 6 \times 32 \times 3 \times x^5 + \frac{6 \times 5}{2} \times 16x^4 \times 9 - \frac{6 \times 5 \times 4 \times 8x^3 \times 27}{6} + \frac{6 \times 5}{2} \times 4x^2 \times 81 - 6 \times 2x \times 243 + 729$$

$$= 64x^6 - 576x^5 + 2160x^4 - 4320x^3 + 4860x^2 - 2916x + 729$$

(iii) We have, $\left(\frac{2x}{3} - \frac{3}{2x}\right)^4$

$$= {}^4C_0\left(\frac{2x}{3}\right)^4 - {}^4C_1\left(\frac{2x}{3}\right)^3\left(\frac{3}{2x}\right) + {}^4C_2\left(\frac{2x}{3}\right)^2\left(\frac{3}{2x}\right)^2 - {}^4C_3\left(\frac{2x}{3}\right)\left(\frac{3}{2x}\right)^3 + {}^4C_4\left(\frac{3}{2x}\right)^4$$

$$= 1 \times \frac{16x^4}{81} - 4 \times \frac{8x^3}{27}\left(\frac{3}{2x}\right) + 6 \cdot \frac{4x^2}{9}\left(\frac{9}{4x^2}\right) - 4\left(\frac{2x}{3}\right)\left(\frac{27}{8x^3}\right) + 1 \times \left(\frac{81}{16x^4}\right)$$

$$\left[\because {}^4C_0 = {}^4C_4 = 1, {}^4C_3 = {}^4C_1 = 4 \text{ and } {}^4C_2 = \frac{4!}{2!2!} = \frac{4 \times 3 \times 2!}{2 \times 1 \times 2!} = 6\right]$$

$$= \frac{16}{81}x^4 - \frac{16}{9}x^2 + 6 - \frac{9}{x^2} + \frac{81}{16x^4}$$

(iv) We have, $(x^2 - 2x + 1)^3 = [(x-1)^2]^3 = (x-1)^6$

$$= {}^6C_0 (x)^6 - {}^6C_1 (x)^5 (1)^1 + {}^6C_2 (x)^4 (1)^2 - {}^6C_3 (x)^3 (1)^3 + {}^6C_4 (x)^2 (1)^4 - {}^6C_5 (x)^1 (1)^5 + {}^6C_6 (1)^6$$

$$= x^6 - 6x^5 + 15x^4 - 20x^3 + 15x^2 - 6x + 1$$

(v) We have, $(1 - x + x^2)^4 = [(1-x) + x^2]^4$

$$= {}^4C_0 (1-x)^4 + {}^4C_1 (1-x)^3 (x^2) + {}^4C_2 (1-x)^2 (x^2)^2 + {}^4C_3 (1-x)(x^2)^3 + {}^4C_4 (x^2)^4$$

[on taking $(1-x)$ as a and x^2 as b]

$$= (1-x)^4 + 4x^2(1-x)^3 + 6x^4(1-x)^2 + 4x^6(1-x) + 1\cdot x^8$$

$$= (1-4x+6x^2-4x^3+x^4) + 4x^2(1-3x+3x^2-x^3) + 6x^4(1-2x+x^2) + 4(1-x)x^6 + x^8$$

$$= 1-4x+6x^2-4x^3+x^4+4x^2-12x^3+12x^4-4x^5 + 6x^4-12x^5+6x^6+4x^6-4x^7+x^8$$

$$= 1-4x+10x^2-16x^3+19x^4-16x^5+10x^6-4x^7+x^8$$

(vi) We have, $\left(1+\frac{x}{2}-\frac{2}{x}\right)^4 = \left[\left(1+\frac{x}{2}\right)-\frac{2}{x}\right]^4$

$$= {}^4C_0\left(1+\frac{x}{2}\right)^4 - {}^4C_1\left(1+\frac{x}{2}\right)^3\left(\frac{2}{x}\right) + {}^4C_2\left(1+\frac{x}{2}\right)^2\left(\frac{2}{x}\right)^2 - {}^4C_3\left(1+\frac{x}{2}\right)\left(\frac{2}{x}\right)^3 + {}^4C_4\left(\frac{2}{x}\right)^4$$

$$= \left(1+\frac{x}{2}\right)^4 - 4\left(1+\frac{x}{2}\right)^3\frac{2}{x} + \frac{4\times 3}{2}\left(1+\frac{x}{2}\right)^2\frac{4}{x^2} - 4\left(1+\frac{x}{2}\right)\times\frac{8}{x^3} + \frac{16}{x^4}$$

$$= \left(1+\frac{x}{2}\right)^4 - \frac{8}{x}\left(1+\frac{x}{2}\right)^3 + \frac{24}{x^2}\left(1+\frac{x}{2}\right)^2 - \frac{32}{x^3}\left(1+\frac{x}{2}\right) + \frac{16}{x^4}$$

Now, on expanding $\left(1+\frac{x}{2}\right)^4, \left(1+\frac{x}{2}\right)^3, \left(1+\frac{x}{2}\right)^2$, we get

$$\left(1+\frac{x}{2}-\frac{2}{x}\right)^4 = \left(1+4\cdot\frac{x}{2}+6\cdot\frac{x^2}{4}+4\cdot\frac{x^3}{8}+\frac{x^4}{16}\right) - 8\cdot\frac{1}{x}\left(1+3\cdot\frac{x}{2}+3\cdot\frac{x^2}{4}+\frac{x^3}{8}\right)$$

$$+ 24\cdot\frac{1}{x^2}\left(1+x+\frac{x^2}{4}\right) - 32\times\frac{1}{x^3}\left(1+\frac{x}{2}\right) + \frac{16}{x^4}$$

$$= \left(1+2x+\frac{3x^2}{2}+\frac{x^3}{2}+\frac{x^4}{16}\right) - \left(\frac{8}{x}+12+6x+x^2\right) + \left(\frac{24}{x^2}+\frac{24}{x}+6\right) - \left(\frac{32}{x^3}+\frac{16}{x^2}\right) + \frac{16}{x^4}$$

$$= \frac{x^4}{16}+\frac{x^3}{2}+x^2\left(\frac{3}{2}-1\right)+x(2-6)+(1-12+6) + (24-8)\frac{1}{x}+(24-16)\frac{1}{x^2}-\frac{32}{x^3}+\frac{16}{x^4}$$

$$= \frac{x^4}{16}+\frac{x^3}{2}+\frac{x^2}{2}-4x-5+\frac{16}{x}+\frac{8}{x^2}-\frac{32}{x^3}+\frac{16}{x^4}$$

Example 2. Prove that $\sum_{r=0}^{n} 3^r\,{}^nC_r = 4^n$.

NCERT, Competency Based

Sol. We have, $\sum_{r=0}^{n} {}^nC_r \times 3^r$

$$= {}^nC_0 3^0 + {}^nC_1 3 + {}^nC_2 3^2 + {}^nC_3 3^3 + \dots + {}^nC_n 3^n$$

[on putting $r = 0, 1, 2, \dots, n$]

$$= {}^nC_0 + {}^nC_1 3 + {}^nC_2 3^2 + {}^nC_3 3^3 + \dots + {}^nC_n 3^n$$

$$= (1+3)^n$$

$[\because (1+x)^n = {}^nC_0 + {}^nC_1 x + {}^nC_2 x^2 + \dots + {}^nC_n x^n]$

$$= 4^n$$

Hence proved.

Example 3. Find $(x+1)^6 + (x-1)^6$. Hence, evaluate $(\sqrt{2}+1)^6 + (\sqrt{2}-1)^6$.

NCERT

Sol. We have,

$$(x+1)^6 = {}^6C_0 x^6 + {}^6C_1 x^5 \times 1 + {}^6C_2 x^4 \times (1)^2 + {}^6C_3 x^3 \times (1)^3 + {}^6C_4 x^2 \times (1)^4 + {}^6C_5 x \times (1)^5 + {}^6C_6 (1)^6$$

$$\Rightarrow (x+1)^6 = {}^6C_0 x^6 + {}^6C_1 x^5 + {}^6C_2 x^4 + {}^6C_3 x^3 + {}^6C_2 x^2 + {}^6C_1 x + {}^6C_0 \quad [\because {}^nC_r = {}^nC_{n-r}]$$

$$\Rightarrow (x+1)^6 = x^6 + 6x^5 + \frac{6\times 5}{2}x^4 + \frac{6\times 5\times 4}{6}x^3 + \frac{6\times 5}{2}x^2 + 6x + 1$$

$$\Rightarrow (x+1)^6 = x^6 + 6x^5 + 15x^4 + 20x^3 + 15x^2 + 6x + 1 \quad \text{...(i)}$$

Similarly,

$$\Rightarrow (x-1)^6 = x^6 - 6x^5 + 15x^4 - 20x^3 + 15x^2 - 6x + 1 \quad \text{...(ii)}$$

On adding Eqs. (i) and (ii), we get

$$(x+1)^6 + (x-1)^6 = 2[x^6 + 15x^4 + 15x^2 + 1]$$

Now, on putting $x = \sqrt{2}$, we get

$$(\sqrt{2}+1)^6 + (\sqrt{2}-1)^6 = 2[(\sqrt{2})^6 + 15(\sqrt{2})^4 + 15(\sqrt{2})^2 + 1]$$

$$= 2(2^3 + 15\times 2^2 + 15\times 2 + 1)$$

$$= 2(8 + 15\times 4 + 30 + 1)$$

$$= 2(8 + 60 + 30 + 1) = 2\times 99 = 198$$

Example 4. Evaluate $(x^2 - \sqrt{1-x^2})^4 + (x^2 + \sqrt{1-x^2})^4$.

NCERT Exemplar

Sol. Let $E = (x^2 - \sqrt{1-x^2})^4 + (x^2 + \sqrt{1-x^2})^4$

On putting $\sqrt{1-x^2} = y$, we get

$$E = (x^2 - y)^4 + (x^2 + y)^4$$

$= 2[{}^4C_0(x^2)^4\, y^0 + {}^4C_2(x^2)^2\, y^2 + {}^4C_4(x^2)^0\, y^4]$

$\left[\begin{array}{l}\text{if } n \text{ is even, then } (x-a)^n + (x+a)^n \\ = 2\{{}^nC_0\, x^n a^0 + {}^nC_2\, x^{n-2}a^2 + {}^nC_4 x^{n-4}\, a^4 + \dots\}\end{array}\right]$

$= 2[1 \times x^8 \times 1 + 6 \times x^4 y^2 + 1 \times 1 \times y^4]$

$= 2[x^8 + 6x^4(1-x^2) + (1-x^2)^2]$ [put $y = \sqrt{1-x^2}$]

$= 2(x^8 + 6x^4 - 6x^6 + 1 + x^4 - 2x^2)$

$= 2x^8 - 12x^6 + 14x^4 - 4x^2 + 2$

Example 5. Using binomial theorem, expand the following expressions.

(i) $(96)^3$ **(ii) $(102)^5$**

Sol. (i) Here, given number without index is 96. Since, it is less than 100, so it can be written as $100 - 4$.

Now, $(96)^3 = (100-4)^3$

$= {}^3C_0(100)^3 - {}^3C_1(100)^2\, 4 + {}^3C_2(100)\, 4^2 - {}^3C_3 4^3$ [by binomial theorem]

$= 1000000 - 3(10000)4 + 3(100)(16) - 64$

$= 1000000 - 120000 + 4800 - 64$

$= 884736$

(ii) Here, given number without index is 102. Since, it is more than 100, so it can be written as $(100+2)$.

We have, $(102)^5 = (100+2)^5$

$= {}^5C_0(100)^5 + {}^5C_1(100)^4(2)^1 + {}^5C_2(100)^3(2)^2 + {}^5C_3(100)^2(2)^3 + {}^5C_4(100)(2)^4 + {}^5C_5(2)^5$

$= {}^5C_0(10)^{10} + {}^5C_1(10)^8\, 2 + {}^5C_2(10)^6 \times 4 + {}^5C_2(10)^4 \times 8 + {}^5C_1(10)^2 \times 16 + {}^5C_0 \times 32$

$[\because {}^nC_r = {}^nC_{n-r}]$

$= 1 \times 10^{10} + 5 \times 10^8 \times 2 + \dfrac{5\times 4}{2} \times 10^6 \times 4 + \dfrac{5 \times 4}{2} \times 10^4 \times 8 + 5 \times 100 \times 16 + 32$

$= 10000000000 + 1000000000 + 40000000 + 800000 + 8000 + 32$

$= 11040808032$

Example 6. Find an approximation of $(0.99)^5$, using the first three terms of its expansion. NCERT

Sol. We have, $(0.99)^5 = (1-0.01)^5$

$= 1 - {}^5C_1 \times (0.01) + {}^5C_2 \times (0.01)^2 - \dots$

$= 1 - 0.05 + 10 \times 0.0001 - \dots$

$\approx 1.001 - 0.05$

≈ 0.951

Example 7. Which number is larger, $(1.1)^{10000}$ or 1000?

Sol. We have, $(1.1)^{10000} = [1 + (0.1)]^{10000}$ NCERT

$= {}^{10000}C_0 + {}^{10000}C_1(0.1) + {}^{10000}C_2(0.1)^2 + \dots$

[expanding by binomial theorem]

$= 1 + 10000(0.1) +$ Other positive terms

$= 1 + 1000 +$ Other positive terms

$= 1001 +$ Other positive terms > 1000

Hence, $(1.1)^{10000} > 1000$

Example 8. Find the remainder when $6^n - 5n$ is divided by 25. Competency Based Que

Sol. We have, $6^n - 5n = (1+5)^n - 5n$

$= {}^nC_0 + {}^nC_1(5) + {}^nC_2(5)^2 + \dots + {}^nC_n 5^n - 5n$

$= 1 + n(5) + \dfrac{n(n-1)}{2} \times (25) + \dots + 1(5^n) - 5n$

$= 1 + 25\left[\dfrac{n(n-1)}{2} + \dots + 5^{n-2}\right]$

$= 25 \times$ an integer $+ 1$

Hence, $6^n - 5n$ leaves the remainder 1, when divided by 25.

Example 9. Using binomial theorem, show that the expression $7^9 + 9^7$ is divisible by 64. Competency Based Que

Sol. We have, $7^9 + 9^7 = (1+8)^7 - (1-8)^9$

$= [{}^7C_0 + {}^7C_1(8) + {}^7C_2(8)^2 + {}^7C_3(8)^3 + \dots + {}^7C_7(8)^7]$
$- [{}^9C_0 - {}^9C_1(8)^1 + {}^9C_2(8)^2 - \dots - {}^9C_9(8)^9]$

$= [1 + 7 \times (8)^1 + 21 \times (8)^2 + 35(8)^3 \dots + (8)^7]$
$- [1 - 9 \times (8)^1 + 36(8)^2 - \dots - (8)^9]$

$= (7+9) \times (8)^1 + (21-36) \times (8)^2 + \dots$

$= 2 \times 64 - 15 \times 64 + \dots = 64(2 - 15 + \dots)$

Hence, it is clear that it is divisible by 64.

Hence proved.

Example 10. If a and b are distinct integers, then prove that $(a-b)$ is a factor of $a^n - b^n$, whenever n is a positive integer. NCERT

Sol. We can write, $a^n = (a - b + b)^n$, then using expansion of $(x+y)^n$, we get

$a^n = {}^nC_0(a-b)^n + {}^nC_1(a-b)^{n-1}b + {}^nC_2(a-b)^{n-2}b^2 + \dots + {}^nC_n b^n$

[taking $(a-b)$ as x and b as y]

$\Rightarrow \quad a^n = (a-b)[{}^nC_0(a-b)^{n-1} + {}^nC_1(a-b)^{n-2}b + {}^nC_2(a-b)^{n-3}b^2 + \dots] + b^n$

$\therefore \quad a^n - b^n = (a-b)[{}^nC_0(a-b)^{n-1} + \dots]$

Hence, $(a-b)$ is a factor of $a^n - b^n$. **Hence proved.**

Exam Pattern Questions

Objective Type Questions

Multiple Choice Questions

1. In every term, the sum of indices of a and b in the expansion of $(a+b)^n$ is

(a) n (b) $n+1$
(c) $n+2$ (d) $n-1$

2. $(100)^{50}+(99)^{50}$ NCERT Exemplar

(a) $<(101)^{50}$ (b) $=101$
(c) $>(101)^{50}$ (d) >101

3. The coefficient of x^n in the expansion of $(1+x)(1-x)^n$ is Competency Based Que

(a) $1-n$ (b) $(-1)^n(1-n)$
(c) $n-1$ (d) $(-1)^n(n-1)$

4. If the coefficients of 2nd, 3rd and the 4th terms in the expansion of $(1+x)^n$ are in AP, then value of n is Competency Based Que

(a) 2 (b) 7
(c) 11 (d) 14

5. The coefficient of x^3y^4 in $(2x+3y^2)^5$ is

(a) 360 (b) 720 (c) 240 (d) 1080

6. Total number of terms in the expansion of $(x+y)^{99}$ is DIKSHA APP

(a) 98 (b) 99 (c) 101 (d) 100

7. The fourth term in the expansion of $(x-2y)^{12}$ is

(a) $-1760x^9\times y^3$ (b) $-1670\,x^9\times y^3$
(c) $-7160\,x^9\times y^3$ (d) $-1607\,x^9\times y^3$

8. If $(a+b)^4-(a-b)^4=8ab\,(a^2+b^2)$, then the value of $(\sqrt{5}+\sqrt{2})^4-(\sqrt{5}-\sqrt{2})^4$ is Competency Based Que

(a) 56 (b) $56\sqrt{10}$
(c) $56\sqrt{2}$ (d) $56\sqrt{5}$

9. The coefficients of x^p and x^q (p and q are positive integers) in the expansion of $(1+x)^{p+q}$ are

(a) equal
(b) equal with opposite signs
(c) reciprocal of each other
(d) None of the above

10. The value of $(\sqrt{2}+1)^5-(\sqrt{2}-1)^5$ is

(a) 80 (b) 85 (c) 82 (d) 72

11. If the fourth term of the binomial expansion of $\left(px+\frac{1}{x}\right)^n$ is $\frac{5}{2}$, then Competency Based Que

(a) $n=6, p=6$ (b) $n=8, p=6$
(c) $n=8, p=\frac{1}{2}$ (d) $n=6, p=\frac{1}{2}$

12. The fourth root of 10001 upto 7 decimal palces is

(a) 10.0222499 (b) 10.0024499
(c) 10.0002499 (d) 10.0224499

13. What is the sum of the coefficients of first and last terms in the expansion of $(1+x)^{2n}$, where n is a natural number? Competency Based Que

(a) 1 (b) 2 (c) n (d) $2n$

14. In the expansion of $\left(\sqrt{x}+\frac{1}{3x^2}\right)^{10}$, the value of third term independent of x is Competency Based Que

(a) 5 (b) 8 (c) 45 (d) 90

15. In the expansion of $(1+x)^{50}$, the sum of the coefficients of odd powers of x is Competency Based Que

(a) 2^{26} (b) 2^{49}
(c) 2^{50} (d) 2^{51}

16. What is the value of n, if the coefficients of the second term of $(x-y)^3$ is equal to the third term of the expansion $(x+y)^n$? Competency Based Que

(a) -2 (b) 3 (c) 4 (d) 5

17. What is the fourth term of $(x-5y)^{96}$?

(a) $-125\;{}^{96}C_3\,x^{93}y^3$ (b) $625\;{}^{96}C_3\,x^{93}y^4$
(c) $625\;{}^{96}C_4\,x^{92}y^4$ (d) $-125\;{}^{96}C_4\,x^{92}\,y^4$

18. Number of terms in $(1+14x+49x^2)^{60}$ is Competency Based Que

(a) 101 (b) 22 (c) 121 (d) 300

19. If $x=99^{50}+100^{50}$ and $y=(101)^{50}$, then

(a) $x=y$ (b) $x<y$
(c) $x>y$ (d) None of these

20. If the 5th term in the expansion of $(2x-1)^8$ is 1120, what is the value of x?

(a) 5 (b) -1
(c) 10 (d) -10

Assertion-Reason Based Questions

Directions (Q. Nos. 21-30) *In the question given below are two statements labelled as Assertion (A) and Reason (R). In the context of the two statements, which one of the following is correct?*

(a) Both A and R are correct; R is the correct explanation of A.
(b) Both A and R are correct; R is not the correct explanation of A.
(c) A is correct; R is incorrect.
(d) R is correct; A is incorrect.

21. **Assertion** (A) Binomial expansion of $(1+x)^n$ in descending powers of x is $^nC_0x^n + {}^nC_1x^{n-1} + {}^nC_2x^{n-2}...+1$.

Reason (R) Binomial expansion of $(x+y)^n = {}^nC_0x^n + {}^nC_1x^{n-1}y + {}^nC_2x^{n-2}y^2 +...+ {}^nC_ny^n$

22. **Assertion** (A) Total number of terms in the expansion of $\left(x-\frac{5}{x}\right)^6$ are 7.

Reason (R) Total number of terms in binomial expansion of $(a+b)^n$ are $(n+1)$.

23. **Assertion** (A) If 25^{15} is divided by 13, then the remainder is 12.

Reason (R) In the expansion of $(a+b)^n$, the sum of the indices of a and b is n.

24. **Assertion** (A) $(x^2+a)^5 = x^{10}+5x^8a+10x^6a^2 +10x^4a^3+5x^2a^4+a^5$

Reason (R) $(x+a)^n = {}^nC_0x^n + {}^nC_1x^{n-2}a + ... + {}^nC_{n-1}xa^{n-1} + {}^nC_na^n$

25. **Assertion** (A) In the expansion of $\left(x-\frac{1}{x}\right)^6$ 4th term is independent from x.

Reason (R) Coefficient of x^r in $(1+x)^n$ is ^nC_r.

26. **Assertion** (A) The coefficient of 5th term in the expansion of $\left(\frac{x}{3}-3y\right)^7$ is 115.

Reason (R) The expansion of $(a-b)^n = {}^nC_0\,a^n - {}^nC_1\,a^{n-1}\,b + {}^nC_2a^{n-2}\,b^2 + ... + (-1)^n\,{}^nC_n\,b^n$

27. **Assertion** (A) The coefficients of the expansions are arranged in an array. This array is called Pascal's triangle.

Reason (R) There are twelve terms in the expression of $(4x+7y)^{10}$.

28. **Assertion** (A) In the binomial expansion of $(a+b)^n$, the powers of a decreases while the powers of b increases.

Reason (R) Each term of the expansion is of the form $a^{n-k}\,b^k$, where k increases from 0 to n.

29. If $(1+x)^n = \sum_{r=0}^{n} C_r\,x^r$, then

Assertion (A) $\left(1+\frac{C_1}{C_0}\right)\left(1+\frac{C_2}{C_1}\right)...\left(1+\frac{C_n}{C_{n-1}}\right) = \frac{(n+1)^n}{n!}$

Reason (R) $^nC_r = \frac{n(n-1)\dots(n-r+1)}{r(r-1)\dots1}$

30. **Assertion** (A) The binomial theorem provides an expansion for the expression $(a+b)^n$, where $a, b, n \in R$.

Reason (R) All coefficients in a binomial expansion may be obtained by Pascal's triangle.

Case Study Based Questions

31. During preparation for an examination, a group of friends are studying and discussing different aspects of different concepts of binomial theorem. One of the friend suddenly wrote

$(a+2b)\,(a+2b)\,(a+2b)\,(a+2b)\,(a+2b)$

and asked some questions to his friends based on it and asked them to answer the following questions.

(i) How many terms are in the above expansion?
(ii) What is the third term in the expansion in decreasing order of a?
(iii) What is Pascal's coefficient of 4th term in the expansion $(a+2b)^5$?
(iv) How many terms are there in the expansion of $(a-2b)\,(a-2b)\,(a-2b)\,(a-2b)$?

32. Four friends applied the knowledge of Binomial Theorem while playing a game to make the equations by observing some conditions they make some equations.

(i) Expand, $(1-x+x^2)^4$.
(ii) Expand the expression, $(1-3x)^7$.
(iii) Show that 11^9+9^{11} is divisible by 10.

Subjective Questions

Very Short Answer Type Questions

1. Using the binomial theorem, expand $(2x-3)^4$.

2. Expand $\left(x+\frac{1}{x}\right)^5$.

3. Expand the expression $(1-x)^6$.

4. Expand the expression $\left(\frac{2}{x}+\frac{x}{2}\right)^5$.

5. Evaluate $(\sqrt{3}+\sqrt{2})^6+(\sqrt{3}-\sqrt{2})^6$

Short Answer Type Questions

6. Prove that $\sum_{r=0}^{n} 5^r \cdot {}^nC_r = 6^n$.

Competency Based Que

7. Using binomial theorem, expand the following expressions.

(i) $(2+\sqrt{3})^7+(2-\sqrt{3})^7$

(ii) $(\sqrt{2}+1)^6+(\sqrt{2}-1)^6$

8. Using binomial theorem, expand the following expansions.

(i) $(1+x+x^2)^3$

(ii) $(1-x-x^2)^3$

(iii) $(x^2+3+2\sqrt{3}x)^3$

9. Using binomial theorem, evaluate each of the following.

(i) $(0.98)^4$ (ii) $(101)^4$ (iii) $(90)^5$

10. If the coefficient of $(2r+4)$th and $(r-2)$th terms in the expansion of $(1+x)^{18}$ are equal, then the value of r.

DIKSHA APP

11. Find the remainder, when 5^{99} is divided by 13.

12. Find the coefficient of x^5 in the expansion of $(1+x)^3\ (1-x)^6$.

Competency Based Que

13. If P and Q are the sum of odd an even terms in the expansion $(y+b)^n$, then prove that $(y+b)^{2n}+(y-b)^{2n}=2(P^2+Q^2)$

Competency Based Que

Long Answer Type Questions

14. Using binomial theorem, expand $(x+y)^6-(x-y)^6$. Hence, find the value of $(\sqrt{2}+1)^6-(\sqrt{2}-1)^6$.

15. Using binomial theorem, prove that $2^{3n}-7n-1$ is divisible by 49, where $n \in N$.

Competency Based Que

16. Show that $9^{n+1}-8n-9$ is divisible by 64, whenever n is a positive integer.

Competency Based Que

17. Find the coefficient of a^4 in the product $(1+2a)^4\ (2-a)^5$ using binomial theorem.

NCERT

18. The sum of the coefficients of the first three terms in the expansion of $\left(x-\frac{3}{x^2}\right)^m$, $x \neq 0$, m being a natural number, is 559. Find the term of the expansion containing x^3.

NCERT

19. Find $(a+b)^4-(a-b)^4$.

Hence, evaluate $(\sqrt{3}+\sqrt{2})^4-(\sqrt{3}-\sqrt{2})^4$.

NCERT, Competency Based Que

20. Find the value of $(a^2+\sqrt{a^2-1})^4+(a^2-\sqrt{a^2-1})^4$.

21. The first three terms in the expansion of $(x+y)^n$ are 1, 56 and 1372, respectively. Find the values of x and y.

22. Using binomial theorem, determine which number is larger $(1.2)^{4000}$ or 800?

NCERT Exemplar

Hints & Solutions

Objective Type Questions

1. (a) The general formula of a binomial expansion.

$$(a+b)^n = {}^nC_0\, a^n\, b^0 + {}^nC_1 a^{n-1} b^1 + \ldots + {}^nC_n a^0 b^n$$

Take any two terms and look at the indices of a and b.

For first term, we get $n + 0 = n$

For second term, we get $(n-1)+1 = n$

For last term, we get $0 + n = n$

2. (a) Since, $(101)^{50} = (100+1)^{50}$

$$= 100^{50} + {}^{50}C_1 100^{49} + {}^{50}C_2 100^{48} + \ldots + 1 \quad \ldots\text{(i)}$$

and $(99)^{50} = (100-1)^{50}$

$$= 100^{50} - {}^{50}C_1 100^{49} + {}^{50}C_2 100^{48} - \ldots + 1 \quad \ldots\text{(ii)}$$

On subtracting Eq. (ii) from Eq. (i), we get

$$(101)^{50} - (99)^{50} = 2\{{}^{50}C_1\, 100^{49} + {}^{50}C_3 100^{47} + \ldots\}$$
$$= 2 \times {}^{50}C_1 100^{49} + (2 \times {}^{50}C_3 \times 100^{47} + \ldots)$$
$$= 100 \times (100)^{49} + \text{a positive number} > (100)^{50}$$
$$\Rightarrow \quad (100)^{50} + (99)^{50} < (101)^{50}$$

3. (b) $(1+x)(1-x)^n = (1+x)({}^nC_0 - {}^nC_1 x + \ldots + (-1)^{n-1} \times {}^nC_{n-1} x^{n-1} + (-1)^n \times {}^nC_n x^n)$

Coefficient of $x^n = (-1)^n\, {}^nC_n + (-1)^{n-1} \times {}^nC_{n-1}$
$$= (-1)^n + (-1)^{n-1} n$$
$$= (-1)^n - (-1)^n n$$
$$= (-1)^n (1-n)$$

4. (b) Coefficients of 2nd, 3rd and 4th term in the given expansion are ${}^nC_1$, ${}^nC_2$ and ${}^nC_3$.

We have, $\quad 2 \times {}^nC_2 = {}^nC_1 + {}^nC_3$

On dividing both sides by ${}^nC_2$, we get

$$2 = \frac{{}^nC_1}{{}^nC_2} + \frac{{}^nC_3}{{}^nC_2}$$
$$\Rightarrow \quad 2 = \frac{2}{n-1} + \frac{n-2}{3}$$
$$\Rightarrow \quad 6n - 6 = 6 + n^2 + 2 - 3n$$
$$\Rightarrow \quad n^2 - 9n + 14 = 0$$
$$\therefore \quad n = 7 \quad [\because n \neq 2 \text{ as } 2 > 3 \text{ in the 4th term}]$$

5. (b) Given, $(2x + 3y^2)^5$

Therefore, by the binomial theorem

$$(2x+3y^2)^5 = {}^5C_0\, (2x)^5\, (3y^2)^0 + {}^5C_1\, (2x)^4\, (3y^2)^1 + {}^5C_2 (2x)^3\, (3y^2)^2 + {}^5C_3\, (2x)^2\, (3y^2)^3 + {}^5C_4\, (2x)^1\, (3y^2)^4 + {}^5C_5\, (3y^2)^5$$
$$= 32x^5 + 240x^4y^2 + 720\, x^3y^4 + 1080x^2y^6 + 810xy^8 + 243y^{10}$$

$\therefore$ Coefficient of x^3y^4 is 720.

6. (d) We know that there are $(n+1)$ terms in the expansion of $(a+b)^n$.

There are 100 terms in the expansion of $(x+y)^{99}$.

7. (a) Given, $(x-2y)^{12}$

Expand upto fourth term.

$$(x-2y)^{12} = {}^{12}C_0\, (x)^{12} - {}^{12}C_1\, (x)^{11}\, (2y) + {}^{12}C_2\, (x)^{10}\, (2y)^2 - {}^{12}C_3\, (x)^9\, (2y)^3$$
$$= x^{12} - 24x^{11}y + 264x^{10}y^2 - 1760\, x^9\, y^3$$

The fourth term in the expansion $(x-2y)^{12}$ is $-1760x^9y^3$.

8. (b) Given that $(a+b)^4 - (a-b)^4 = 8ab\,(a^2 + b^2)$

$$\therefore \quad (\sqrt{5}+\sqrt{2})^4 - (\sqrt{5}-\sqrt{2})^4 = 8\sqrt{5}\sqrt{2}[(\sqrt{5})^2 + (\sqrt{2})^2]$$
$$= 8\sqrt{10}(5+2)$$
$$= 56\sqrt{10}$$

9. (a) Coefficient of x^p and x^q in the expansion of $(1+x)^{p+q}$ are ${}^{p+q}C_p$ and ${}^{p+q}C_q$.

$$\because \quad {}^{p+q}C_q = \frac{(p+q)!}{p!\,q!}$$
$$ {}^{p+q}C_q = \frac{(p+q)!}{q!\,p!} = {}^{p+q}C_p$$

10. (c) $(\sqrt{2}+1)^5 - (\sqrt{2}-1)^5$

$$= 2\,[{}^5C_1(\sqrt{2})^4 + {}^5C_3(\sqrt{2})^2 + {}^5C_5(\sqrt{2})^0]$$
$$= 2\,[5 \times 4 + 10 \times 2 + 1]$$
$$= 2 \times 41 = 82$$

11. (d) Given, $\left(px + \dfrac{1}{x}\right)^n$

Expand upto fourth term

$$\left(px + \frac{1}{x}\right)^n = {}^nC_0\, (px)^n + {}^nC_1\, (px)^{n-1}\left(\frac{1}{x}\right) + {}^nC_2 (px)^{n-2}\left(\frac{1}{x}\right)^2 + {}^nC_3\, (px)^{n-3}\left(\frac{1}{x}\right)^3$$

Fourth term is ${}^nC_3\, (px)^{n-3}\left(\dfrac{1}{x}\right)^3$.

Now, $\quad {}^nC_3 p^{n-3}\, x^{n-3}\, x^{-3} = \dfrac{5}{2} x^0$

$$ {}^nC_3\, p^{n-3}\, x^{n-3-3} = \frac{5}{2} x^0$$

On comparing the power of x, we get $n = 6$.

Then, $\quad {}^6C_3\, p^3\, x^0 = \dfrac{5}{2}$

$$\frac{6 \times 4 \times 5}{1 \times 2 \times 3} p^3 = \frac{5}{2}$$
$$p^3 = \frac{1}{8}$$

$$p = \left(\frac{1}{2}\right)^3$$

$$p = \frac{1}{2} \text{ and } n = 6.$$

12. (c) Given $(10001)^{1/4} = (10000 + 1)^{1/4}$

The binomial expansion of $(a + b)^n$ is

$$(a + b)^n = a^n \left(1 + \frac{b}{a}\right)^n = 10\left(1 + \frac{1}{10000}\right)^{1/4}$$

$$= 10\left(1 + \frac{1}{4}\cdot\frac{1}{10000} + \frac{\frac{1}{4}\left(\frac{1}{4} - 1\right)}{2!}\cdot\frac{1}{(10000)^2} + \ldots\right)$$

$$= 10\,[1 + 0.000025 - 0.000000001 + \ldots]$$

$$= 10 + 0.00025 - 0.0000001 = 10.0002499$$

13. (b) Given expansion is $(1 + x)^{2n}$.

$$\Rightarrow {}^{2n}C_0 \times 1^{(2n-0)} \times x^0 + {}^{2n}C_1 \times 1^{(2n-1)} \times x^1 + \ldots + {}^{2n}C_{2n} \times 1^{(2n-2n)} \times x^{2n}$$

Here, first term $= {}^{2n}C_0 \times 1 \times 1 = 1$

and last term $= {}^{2n}C_{2n} \times 1 \times x^{2n} = 1 \times x^{2n} = x^{2n}$

Coefficient of first term = 1 and coefficient of last term = 1

$\therefore$ Sum of the coefficient of first and last term $= 1 + 1 = 2$.

14. (a) Given, expression is

$$\left(\sqrt{x} + \frac{1}{3x^2}\right)^{10}$$

Expansion of given expression upto third term is

$$\left(\sqrt{x} + \frac{1}{3x^2}\right)^{10} = {}^{10}C_0\,(\sqrt{x})^{10} + {}^{10}C_1\,(\sqrt{x})^9\left(\frac{1}{3x^2}\right) + {}^{10}C_2\,(\sqrt{x})^8\left(\frac{1}{3x^2}\right)^2$$

$$= x^5 + \frac{10}{3}x^{5/2} + 5$$

Hence, third term independent of x is 5.

15. (b) $(1 + x)^{50} = {}^{50}C_0 + {}^{50}C_1\,x + {}^{50}C_2\,x^2 + \ldots + {}^{50}C_n\,x^{50}$ ...(i)

Here, $n = 50$

Using the above formula, the sum of coefficient of odd powers of x is

$$S = {}^{50}C_1 + {}^{50}C_3 + {}^{50}C_5 + \ldots + {}^{50}C_{49}$$

$$\Rightarrow S = 2^{50-1} = 2^{49}$$

$\therefore$ Sum of the coefficient of odd powers of $x = 2^{49}$.

16. (b) Given, coefficient of the second term of $(x - y)^3$ is ${}^3C_1$ and the coefficient of the third term of the expansion $(x + y)^n$ is ${}^nC_2$

$$\therefore \quad {}^3C_1 = {}^nC_2$$

$$\Rightarrow \quad 3 = \frac{n!}{2!\,(n-2)!}$$

$$\Rightarrow \quad 3 = \frac{n(n-1)\,(n-2)!}{2 \times (n-2)!}$$

$$\Rightarrow \quad 6 = n^2 - n$$

$$\Rightarrow \quad n^2 - n - 6 = 0$$

$$\Rightarrow \quad n^2 - 3n + 2n - 6 = 0$$

$$\Rightarrow \quad (n-3)(n+2) = 0 \Rightarrow n = 3, -2$$

Since, n cannot be negative,

$$\therefore \quad n = 3$$

17. (a) Given, $(x - 5y)^{96}$

Expand upto fourth term, we get

$$(x - 5y)^{96} = {}^{96}C_0\,(x)^{96} - {}^{96}C_1(x)^{95}(5y) + {}^{96}C_2\,(x)^{94}\,(5y)^2 - {}^{96}C_3\,(x)^{93}\,(5y)^3$$

$$= {}^{96}C_0\,x^{96} - {}^{96}C_1\,x^{95}\,(5y) + {}^{96}C_2\,x^{94}\,25y^2 - {}^{96}C_3\,x^{93}\cdot 125y^3$$

Hence, fourth term is $-125\;{}^{96}C_3\,x^{93}y^3$.

18. (c) We have,

$$(1 + 14x + 49x^2)^{60} = ((1 + 7x)^2)^{60} = (1 + 7x)^{120}$$

$$[\because (1 + 7x)^2 = 1 + 49x + 14x]$$

$\therefore$ The number of terms $= 120 + 1 = 121$

19. (b) Using binomial theorem, we have

$$101^{50} - 99^{50} = (100 + 1)^{50} - (100 - 1)^{50}$$

$$= ({}^{50}C_0 100^{50} + {}^{50}C_1\,100^{49} + {}^{50}C_2\,100^{48} + \ldots) - ({}^{50}C_0\,100^{50} - {}^{50}C_1\,100^{49} + {}^{50}C_2\,100^{48} \ldots)$$

$$= 2\,({}^{50}C_1\,100^{49} + {}^{50}C_3 100^{47} + {}^{50}C_5\,100^{45} + \ldots)$$

$$= 2\,(50\cdot 100^{49} + {}^{50}C_3\,100^{47} + {}^{50}C_5\,100^{45} + \ldots)$$

$$= 100\cdot 100^{49} + 2({}^{50}C_3 100^{47} + {}^{50}C_5\,100^{45} + \ldots)$$

$$\Rightarrow 101^{50} - 99^{50} = 100^{50} + 2({}^{50}C_3\,100^{47} + {}^{50}C_5\,100^{45} + \ldots)$$

$$\Rightarrow \quad 101^{50} - 99^{50} > 100^{50}$$

$$\Rightarrow \quad 101^{50} - 99^{50} > 100^{50}$$

$$101^{50} > 99^{50} + 100^{50}$$

$$\therefore \quad x < y$$

20. (b) Given, expression is $(2x - 1)^8$.

Expand upto 5th term, we get

$$(2x - 1)^8 = {}^8C_0\,(2x)^8 - {}^8C_1\,(2x)^7 + {}^8C_2\,(2x)^6\,(1)^2 - {}^8C_3\,(2x)^5\,(1)^3 + {}^8C_4\,(2x)^4\,(1)^4$$

It is given that fifth term is 1120.

$$\therefore \quad {}^8C_4\,(2x)^4 = 1120$$

$$70 \times 16x^4 = 1120$$

$$x^4 = 1 \Rightarrow x = \pm 1$$

21. (a) We know that

$$(x + y)^n = {}^nC_0\,x^n + {}^nC_1\,x^{n-1}y + {}^nC_2x^{n-2}y^2 + \ldots + {}^nC_ny^n$$

On substituting $y = 1$ in above equation, we get

$$(x + 1)^n = {}^nC_0\,x^n + {}^nC_1x^{n-1} + {}^nC_2\,x^{n-2} + \ldots\ldots + {}^nC_n\,1^n$$

$$= {}^nC_0\,x^n + {}^nC_1\,x^{n-1} + {}^nC_2\,x^{n-2} + \ldots + 1$$

22. (a) We know that total number of terms in binomial expansion of $(a+b)^n$ are $(n+1)$.

$\therefore$ Total number of terms in the expansion of $\left(x-\dfrac{5}{x}\right)^6$ are $(6+1)$ i.e. 7.

23. (b) As, $25^{15} = (26-1)^{15}$

$$= {}^{15}C_0\, 26^{15} - {}^{15}C_1\, 26^{14} + \dots - {}^{15}C_{15}$$
$$= {}^{15}C_0 26^{15} - {}^{15}C_1\, 26^{14} + \dots - 1$$

Since, 25^{15} is divided by 13. So, remainder $= 13-1 = 12$

24. (a) We know that

$$(x+a)^n = {}^nC_0 x^n + {}^nC_1 x^{n-1} a + {}^nC_n a^n \dots + {}^nC_{n-1} x\, a^{n-1} + {}^nC_n\, a^n$$

$$\therefore\ (x^2+a)^5 = {}^5C_0\,(x^2)^5\,(a)^0 + {}^5C_1\,(x^2)^4\,(a) + {}^5C_2 (x^2)^3 (a)^2 + {}^5C_3\,(x^2)^2\,(a)^3 + {}^5C_4\,(x^2)\,(a)^4 + {}^5C_5\,(x^2)^0\,(a)^5$$
$$= x^{10} + 5x^8 a + 10x^6 a^2 + 10x^4 a^3 + 5x^2 a^4 + a^5$$

25. (b) $\left(x-\dfrac{1}{x}\right)^6 = {}^6C_0 x^6 \left(\dfrac{1}{x}\right)^0 + {}^6C_1\, x^5 \left(\dfrac{1}{x}\right) + {}^6C_2\, x^4 \left(\dfrac{1}{x}\right)^2 + {}^6C_3\, x^3 \left(\dfrac{1}{x}\right)^3 + {}^6C_4\, x^2 \left(\dfrac{1}{x}\right)^4 + {}^6C_5 x \left(\dfrac{1}{x}\right)^5 + {}^6C_6\, x^0 \left(\dfrac{1}{x}\right)^6$

$$= x^6 + {}^6C_1\, x^4 + {}^6C_2\, x^2 + {}^6C_3 + {}^6C_4 \frac{1}{x^2} + {}^6C_5 \frac{1}{x^4} + \frac{1}{x^6}$$

Here, the 4th term is independent of x.

$\therefore$ Assertion and Reason both are correct but Reason is not the correct explanation of Assertion.

26. (d) $\left(\dfrac{x}{3} - 3y\right)^7$

$$= {}^7C_0 \left(\frac{x}{3}\right)^7 - {}^7C_1 \left(\frac{x}{3}\right)^6 \times 3y + {}^7C_2 \left(\frac{x}{3}\right)^5 (3y)^2 - {}^7C_3 \left(\frac{x}{3}\right)^4 (3y)^3 + {}^7C_4 \left(\frac{x}{3}\right)^3 (3y)^4 - {}^7C_5 \left(\frac{x}{3}\right)^2 (3y)^5 + {}^7C_6 \left(\frac{x}{3}\right)(3y)^6 + {}^7C_7\,(3y)^7$$

$$= {}^7C_0 \frac{x^7}{2187} - {}^7C_1 \frac{x^6}{729} \times 3y + {}^7C_2 \frac{x^5}{243} 9y^2 - {}^7C_3 \frac{x^4}{81} 27y^3 + {}^7C_4 \frac{x^3}{27} 81y^4 - {}^7C_5 \frac{x^2}{9} 243y^5 + {}^7C_6 \frac{x}{3} \times 729y^6 + 2187y^7$$

$$= \frac{x^7}{2187} - \frac{7x^6 y}{243} + \frac{7}{9} x^5 y^2 - \frac{7}{3} x^4 y^3 + 105x^3 y^4 - 567x^2 y^5 + 8505xy^6 + 2187\, y^7$$

The coefficient of 5th term is 105.

Thus, Assertion is incorrect but Reason is correct.

27. (c) Assertion is correct but Reason is incorrect as the number of terms in $(4x+7y)^{10}$ is 11.

28. (a) In the expansion, the exponents of $a (n-k)$ decrease as k increases, while the exponents of b (k) increase.

Since, both Assertion and Reason are true, and Reason is the correct explanation of A.

29. (b) We have, $\left(1+\dfrac{C_1}{C_0}\right)\left(1+\dfrac{C_2}{C_1}\right)\dots\left(1+\dfrac{C_n}{C_{n-1}}\right)$

$$= \left(1+\frac{n}{1}\right)\left[1+\frac{\frac{n(n-1)}{2!}}{n}\right]\dots\left(1+\frac{1}{n}\right)$$

$$= \left(\frac{1+n}{1}\right)\cdot\frac{(1+n)}{2}\cdot\frac{(1+n)}{3}\dots\frac{(1+n)}{n} = \frac{(1+n)^n}{n!}$$

and $\quad {}^nC_r = \dfrac{n(n-1)\dots(n-r+1)}{r\,(r-1)\dots 1}$

$\therefore$ Assertion and Reason both are correct but Reason is not the correct explanation of Assertion.

30. (d) Assertion is incorrect but Reason is correct.

31. (i) We have,

$(a+2b)\,(a+2b)\,(a+2b)\,(a+2b)\,(a+2b) = (a+2b)^5$

We know that there are $(n+1)$ terms in the expansion of $(a+b)^n$.

$\therefore$ $(a+2b)^5$ has $5+1$ i.e. 6 terms in its expansion.

(ii) $(a+2b)^5 = {}^5C_0\, a^5\, (2b)^0 + {}^5C_1\, a^4\, (2b)^1 + {}^5C_2\, a^3\, (2b)^2 + {}^5C_3\, a^2\, (2b)^3 + {}^5C_4\, a (2b)^4 + {}^5C_5\, a^0\, (2b)^5$

$\therefore$ The third term in the expansion of $(a+2b)^5$ is ${}^5C_2\, a^3\, 4b^2$ i.e. $40\, a^3 b^2$.

(iii) From part (ii) the 4th term in the expansion of $(a+2b)^5$ is ${}^5C_3\, a^2 (8b^3)$.

$\therefore$ The Pascal's coefficient of 4th term is ${}^5C_3$.

(iv) We have, $(a-2b)\,(a-2b)\,(a-2b)\,(a-2b) = (a-2b)^4$

We know that there are $(n+1)$ terms in the expansion of $(a+b)^n$.

$\therefore$ There are $(4+1)$ i.e. 5 terms in the expansion of $(a-2b)^4$.

32. (i) We have, $(1-x+x^2)^4 = [(1-x)+x^2]^4$

$$= {}^4C_0\,(1-x)^4 + {}^4C_1 (1-x)^3\,(x^2) + {}^4C_2 (1-x)^2 (x^2)^2 + {}^4C_3\,(1-x)\,(x^2)^3 + {}^4C_4 (x^2)^4$$
$$= (1-x)^4 + 4x^2 (1-x)^3 + 6x^4\,(1-x)^2 + 4x^6 (1-x) + 1\cdot x^8$$
$$= (1-4x+6x^2-4x^3+x^4) + 4x^2\,(1-3x+3x^2-x^3) + 6x^4\,(1-2x+x^2) + 4(1-x)\,x^6 + x^8$$
$$= 1-4x+6x^2-4x^3+x^4+4x^2-12x^3+12x^4 - 4x^5+6x^4-12x^5+6x^6+4x^6-4x^7+x^8$$
$$= 1-4x+10x^2-16x^3+19x^4-16x^5+10x^6 - 4x^7+x^8$$

(ii) Here, $a = 1, b = 3x$ and $n = 7$

Given, $(1-3x)^7 = {}^7C_0(1)^7 - {}^7C_1(1)^6(3x)^1 + {}^7C_2(1)^5(3x)^2 - {}^7C_3(1)^4(3x)^3 + {}^7C_4(1)^3(3x)^4 - {}^7C_5(1)^2(3x)^5 + {}^7C_6(1)^1(3x)^6 - {}^7C_7(1)^0(3x)^7$

$= 1 - 21x + 189x^2 - 945x^3 + 2835x^4 - 5103x^5 + 5103x^6 - 2187x^7$

(iii) $11^9 + 9^{11} = (10+1)^9 + (10-1)^{11}$

$= ({}^9C_0 \cdot 10^9 + {}^9C_1 \cdot 10^8 + \ldots {}^9C_9) + ({}^{11}C_0 \cdot 10^{11} - {}^{11}C_1 \cdot 10^{10} + \ldots - {}^{11}C_{11})$

$= {}^9C_0 \cdot 10^9 + {}^9C_1 \cdot 10^8 + \ldots + {}^9C_8 \cdot 10 + 1 + 10^{11} - {}^{11}C_1 \cdot 10^{10} + \ldots + {}^{11}C_{10} \cdot 10 - 1$

$= 10[{}^9C_0 \cdot 10^8 + {}^9C_1 \cdot 10^7 + \ldots + {}^9C_8 + {}^{11}C_0 \cdot 10^{10} - {}^{11}C_1 \cdot 10^9 + \ldots + {}^{11}C_{10}]$

$= 10K$, which is divisible by 10.

Subjective Questions

1. $(2x-3)^4 = {}^4C_0(2x)^4(-3)^0 + {}^4C_1(2x)^3(-3)^1 + {}^4C_2(2x)^2(-3)^2 + {}^4C_3(2x)(-3)^3 + {}^4C_4(2x)^0(-3)^4$

$= 16x^4 - 96x^3 + 216x^2 - 216x + 81$

2. $\left(x+\frac{1}{x}\right)^5 = {}^5C_0 x^5\left(\frac{1}{x}\right)^0 + {}^5C_1 x^4\left(\frac{1}{x}\right)^1 + {}^5C_2 x^3\left(\frac{1}{x}\right)^2 + {}^5C_3 x^2\left(\frac{1}{x}\right)^3 + {}^5C_4 x\left(\frac{1}{x}\right)^4 + {}^5C_5 x^0\left(\frac{1}{x}\right)^5$

$= x^5 + 5x^3 + 10x + \frac{10}{x} + \frac{5}{x^3} + \frac{1}{x^5}$

3. $(1-x)^6 = {}^6C_0 1^6(-x)^0 + {}^6C_1 1^5(-x)^1 + {}^6C_2 1^4(-x)^2 + {}^6C_3 1^3(-x)^3 + {}^6C_4 1^2(-x)^4 + {}^6C_5 1(-x)^5 + {}^6C_6 1^0(-x)^6$

$= 1 - 6x + 15x^2 - 20x^3 + 15x^4 - 6x^5 + x^6$

4. $\left(\frac{2}{x}+\frac{x}{2}\right)^5 = {}^5C_0\left(\frac{2}{x}\right)^5\left(\frac{x}{2}\right)^0 + {}^5C_1\left(\frac{2}{x}\right)^4\left(\frac{x}{2}\right)^1 + {}^5C_2\left(\frac{2}{x}\right)^3\left(\frac{x}{2}\right)^2 + {}^5C_3\left(\frac{2}{x}\right)^2\left(\frac{x}{2}\right)^3 + {}^5C_4\left(\frac{2}{x}\right)\left(\frac{x}{2}\right)^4 + {}^5C_5\left(\frac{2}{x}\right)^0\left(\frac{x}{2}\right)^5$

$= \left(\frac{2}{x}\right)^5 + \frac{40}{x^3} + \frac{20}{x} + 5x + \frac{5}{8}x^3 + \left(\frac{x}{2}\right)^5$

5. Consider $(\sqrt{3}+\sqrt{2})^6 + (\sqrt{3}-\sqrt{2})^6$

We know, $(x+y)^n + (x-y)^n = 2[{}^nC_0 x^n + {}^nC_2 x^{n-2}y^2 + {}^nC_4 x^{n-4}y^4 + \ldots]$

Here $x = \sqrt{3}, y = \sqrt{2}, n = 6$

$\therefore \quad (\sqrt{3}+\sqrt{2})^6 + (\sqrt{3}-\sqrt{2})^6$

$= 2[{}^6C_0(\sqrt{3})^6 + {}^6C_2(\sqrt{3})^4(\sqrt{2})^2 + {}^6C_4(\sqrt{3})^2(\sqrt{2})^4 + {}^6C_6(\sqrt{2})^6]$

$= 2[1 \times 27 + 15 \times 9 \times 2 + 15 \times 3 \times 4 + 1 \times 8]$

$= 2[27 + 270 + 180 + 8] = 970$

6. Binomial theorem can be written as

$$\sum_{r=0}^{n} {}^nC_r \cdot (1)^{n-r} x^r = (1+x)^n$$

$$\sum_{r=0}^{n} {}^nC_r x^r = (1+x)^n$$

On substituting $x = 5$ in the above equation,

$$\sum_{r=0}^{n} {}^nC_r 5^r = (1+5)^n$$

$$\Rightarrow \sum_{r=0}^{n} 5^r . {}^nC_r = 6^n$$

7. (i) $(2+\sqrt{3})^7$

$= {}^7C_0(2)^7 \times (\sqrt{3})^0 + {}^7C_1(2)^6 \times (\sqrt{3})^1 + {}^7C_2(2)^5(\sqrt{3})^2 + {}^7C_3(2)^4(\sqrt{3})^3 + {}^7C_4(2)^3(\sqrt{3})^4 + {}^7C_5(2)^2(\sqrt{3})^5 + {}^7C_6(2)^1(\sqrt{3})^6 + {}^7C_7 \times 2^0 \times (\sqrt{3})^7 \quad \ldots(i)$

and $(2-\sqrt{3})^7$

$= {}^7C_0(2)^7(\sqrt{3})^0 - {}^7C_1(2)^6(\sqrt{3})^1 + {}^7C_2(2)^5(\sqrt{3})^2 - {}^7C_3(2)^4(\sqrt{3})^3 + {}^7C_4(2)^3(\sqrt{3})^4 - {}^7C_5(2)^2(\sqrt{3})^5 + {}^7C_6(2)^1(\sqrt{3})^6 - {}^7C_7 2^0(\sqrt{3})^7 \quad \ldots(ii)$

Now, on adding Eqs. (i) and (ii), we get

$(2+\sqrt{3})^7 + (2-\sqrt{3})^7$

$= 2[{}^7C_0(2)^7(\sqrt{3})^0 + {}^7C_2(2)^5(\sqrt{3})^2 + {}^7C_4(2)^3(\sqrt{3})^4 + {}^7C_6(2)^1(\sqrt{3})^6]$

$= 10084$

8. (i) $(1+x+x^2)^3 = [(1+x)+x^2]^3$

$= {}^3C_0(1+x)^3 + {}^3C_1(1+x)^2(x^2)^1 + {}^3C_2(1+x)(x^2)^2 + {}^3C_3(x^2)^3$

$= x^6 + 3x^5 + 6x^4 + 7x^3 + 6x^2 + 3x + 1$

(ii) $(1-x-x^2)^3 = [(1-x)-x^2]^3$

$= {}^3C_0(1-x)^3 - {}^3C_1(1-x)^2(x^2)^1 + {}^3C_2(1-x)^1(x^2)^2 - {}^3C_3(x^2)^3$

$= -x^6 - 3x^5 + 5x^3 - 3x + 1$

(iii) $(x^2 + 3 + 2\sqrt{3}x)^3 = (x+\sqrt{3})^6$

$= {}^6C_0(x)^6 \times (\sqrt{3})^0 + {}^6C_1 x^5(\sqrt{3})^1 + {}^6C_2 x^4(\sqrt{3})^2 + {}^6C_3 x^3(\sqrt{3})^3 + {}^6C_4 x^2(\sqrt{3})^4 + {}^6C_5 x^1(\sqrt{3})^5 + {}^6C_6(\sqrt{3})^6$

$= x^6 + 6\sqrt{3}x^5 + 45x^4 + 60\sqrt{3}x^3 + 135x^2 + 54\sqrt{3}x + 27$

9. (i) $(0.98)^{14} = (1-0.02)^{14}$

$= 1 + {}^{14}C_1(-0.02)^1 + {}^{14}C_2(-0.02)^2 + {}^{14}C_3(-0.02)^3$

Neglecting higher powers of (0.01)

$= 1 - 14(0.02) + 91(0.0004) - 364(0.000008)$

$= 1 - 0.28 + 0.0364 - 0.002912$

$= 0.753488$

(ii) $(101)^4 = (100+1)^4$

$= {}^4C_0(100)^4(1)^0 + {}^4C_1(100)^3(1) + {}^4C_2(100)^2(1)^2 + {}^4C_3(100)(1)^3 + {}^4C_4(100)^0(1)^4$

$= 100000000 + 4 \times 1000000 + 6 \times 10000 + 4 \times 100 + 1$

$= 104060401$

(iii) $(90)^5 = (100-10)^5$

$= {}^5C_0(100)^5(-10)^0 + {}^5C_1(100)^4(-10)^1 + {}^5C_2(100)^3(-10)^2 + {}^5C_3(100)^2(-10)^3 + {}^5C_4(100)(-10)^4 + {}^5C_5(100)^0(-10)^5$

$= (100)^5 - 5 \times (100)^4(10) + 10 \times (100)^3(10)^2 - 10 \times (100)^2 \times (10)^3 + 5 \times 100 \times (10)^4 - (10)^5$

$= 5904900000$

10. The coefficient of the $(2r+4)$th and $(r-2)$th terms in the expansion of $(1+x)^{18}$ are ${}^{18}C_{2r+3}$ and ${}^{18}C_{r-3}$.

Given, ${}^{18}C_{2r+3} = {}^{18}C_{r-3}$

$\Rightarrow \quad 2r+3 = r-3$ or $2r+3+r-3 = 18$

$\Rightarrow \quad r = -6$ or $3r = 18$

$r = 6$

But r cannot be negative.

$\therefore r = 6$

11. We know that $5^4 = 625 = 13 \times 48 + 1$

$\Rightarrow \quad 5^4 = 13\lambda + 1,$

where λ is a positive integer.

$\Rightarrow (5^4)^{24} = (13\lambda+1)^{24}$

$= {}^{24}C_0(13\lambda)^{24} + {}^{24}C_1(13\lambda)^{23} + {}^{24}C_2(13\lambda)^{22} + \cdots + {}^{24}C_{23}(13\lambda) + {}^{24}C_{24}$

[by binomial theorem]

$\Rightarrow 5^{96} = 13[{}^{24}C_0 13^{23}\lambda^{24} + {}^{24}C_1 13^{22}\lambda^{23} + \ldots + {}^{24}C_{23}\lambda] + 1$

$=$ (a multiple of 13) + 1

On multiplying both sides by 5^3, we get

$5^{96} \cdot 5^3 = 5^3 \cdot$ (a multiple of 13) $+ 5^3$

$\Rightarrow \quad 5^{99} =$ [(a multiple of 13) $+ [13 \times 9 + 8]$

$[\because 5^3 = 125 = 13 \times 9 + 8]$

Hence, the required remainder is 8.

12. We have, $(1+x)^3(1-x)^6$

$= (1 + x^3 + 3x + 3x^2)(1 - {}^6C_1x + {}^6C_2x^2 - {}^6C_3x^3 + {}^6C_4x^4 - {}^6C_5x^5 + {}^6C_6x^6)$

$\therefore$ Coefficient of $x^5 = -{}^6C_5 + {}^6C_2 + 3\,{}^6C_4 - 3\,{}^6C_3$

$= -6 + 15 + 45 - 60$

$= -6$

13. $(y+b)^n = {}^nC_0y^n + {}^nC_1y^{n-1}b + {}^nC_2y^{n-2}b^2 + \ldots + {}^nC_nb^n$

$= P + Q \quad \ldots$(i)

$(y-b)^n = {}^nC_0y^n - {}^nC_1y^{n-1}b + {}^nC_2y^{n-2}b^2 - \ldots + (-1)^n\,{}^nC_nb^n$

$= P - Q \quad \ldots$(ii)

On squaring Eqs. (i) and (ii) and then adding them, we get

$(y+b)^{2n} + (y-b)^{2n} = (P+Q)^2 + (P-Q)^2$

$= P^2 + Q^2 + 2PQ + P^2 + Q^2 - 2PQ$

$= 2(P^2 + Q^2)$

14. $(x+y)^6 - (x-y)^6$

$= {}^6C_0x^6 + {}^6C_1x^5y + {}^6C_2x^4y^2 + {}^6C_3x^3y^3 + {}^6C_4x^2y^4 + {}^6C_5xy^5 + {}^6C_6y^6 - [{}^6C_0x^6 + {}^6C_1x^5(-y)^1 + {}^6C_2x^4(-y)^2 + {}^6C_3x^3(-y)^3 + {}^6C_4x^2(-y)^4 + {}^6C_5x(-y)^5 + {}^6C_6(-y)^6]$

$= 2(6x^5y + 20x^3y^3 + 6xy^5)$

$= 4xy(3x^4 + 10x^2y^2 + 3y^4)$

On substituting $x = \sqrt{2}$ and $y = 1$, we get

$(\sqrt{2}+1)^6 - (\sqrt{2}-1)^6 = 4(\sqrt{2})(1)[3 \times (\sqrt{2})^4 + 10(\sqrt{2})^2(1)^2 + 3(1)^4]$

$= 4\sqrt{2}(3 \times 4 + 10 \times 2 + 3 \times 1)$

$= 4\sqrt{2}(12 + 20 + 3) = 140\sqrt{2}$

15. We have, $\quad (2^3)^n - 7n - 1 = (8)^n - 7n - 1$

Here, 8 can be written as $1+7$. $\quad [\because 7$ is a factor of 49]

$\therefore \quad 2^{3n} - 7n - 1 = (1+7)^n - 7n - 1$

Using binomial theorem, we get

$2^{3n} - 7n - 1 = ({}^nC_0 + {}^nC_17 + {}^nC_27^2 + \ldots + {}^nC_n7^n) - 7n - 1$

$= \left(1 + \frac{n!}{1!(n-1)!} \cdot 7 + \frac{n!}{2!(n-2)!} 7^2 + \ldots + 7^n\right) - 7n - 1$

$= 1 + 7n + \frac{n(n-1)}{2} 7^2 + \ldots + 7^n - 7n - 1$

$= \frac{n(n-1)}{2} 7^2 + \ldots + 7^n$

Take common number outside, we get

$2^{3n} - 7n - 1 = 7^2\left[\frac{n(n-1)}{2} + \ldots + 7^{n-2}\right]$

$= 49\left[\frac{n(n-1)}{2} + \ldots + 7^{n-2}\right]$

Here, we see that above series is a multiple of 49.

Therefore, $2^{3n} - 7n - 1$ is divisible by 49. **Hence proved.**

16. We have, $9^{n+1} = (1+8)^{n+1}$

$= {}^{n+1}C_0 + {}^{n+1}C_1(8) + {}^{n+1}C_2(8)^2 + {}^{n+1}C_3(8)^3 + \ldots + {}^{n+1}C_{n+1}(8)^{n+1}$

$= 1 + (n+1) \times 8 + {}^{n+1}C_2(8)^2 + {}^{n+1}C_3(8)^3 + \ldots + {}^{n+1}C_{n+1}(8)^{n+1}$

$= 1 + 8n + 8 + {}^{n+1}C_2(8)^2 + {}^{n+1}C_3(8)^3 + \ldots + {}^{n+1}C_{n+1}(8)^{n+1}$

$\Rightarrow 9^{n+1} - 8n - 9 = {}^{n+1}C_2(8)^2 + {}^{n+1}C_3(8)^3 + \ldots + {}^{n+1}C_{n+1}(8)^{n+1}$

$= 64\,[{}^{n+1}C_2 + {}^{n+1}C_3 \cdot 8 + \ldots + {}^{n+1}C_{n+1} \cdot 8^{n-1}]$

which shows that $9^{n+1} - 8n - 9$ is divisible by 64.

17. Coefficient of a^4 in the product $(1+2a)^4\,(2-a)^5$

= The coefficient of a^4 in $[{}^4C_0 + {}^4C_1(2a) + {}^4C_2(2a)^2 + {}^4C_3(2a)^3 + {}^4C_4(2a)^4] \times [{}^5C_0(2)^5 - {}^5C_1(2)^4 a + {}^5C_2(2)^3(a)^2 - {}^5C_3(2)^2(a)^3 + {}^5C_4(2)(a)^4 - {}^5C_5(a)^5]$

= Coefficient of a^4 in $[1 + 8a + 24a^2 + 32a^3 + 16a^4] \times [32 - 80a + 80a^2 - 40a^3 + 10a^4 - a^5]$

$= [(1 \times 10) + (8 \times (-40)) + (24 \times 80) + (32 \times (-80)) + (16 \times 32)]$

$= 10 - 320 + 1920 - 2560 + 512 = -438$

18. The coefficient of first three terms of $\left(x - \dfrac{3}{x^2}\right)^m$ are ${}^mC_0, (-3)\cdot{}^mC_1$ and $9\cdot{}^mC_2$.

Then, we have ${}^mC_0 - 3\cdot{}^mC_1 + 9\cdot{}^mC_2 = 559$

$\Rightarrow \quad 1 - 3m + 9m\dfrac{(m-1)}{2} = 559$

$\Rightarrow \quad m = 12$

[$\because m$ being a natural number]

$\therefore$ The required term $= -5940x^3$

19. $(a+b)^4 - (a-b)^4 = 2[{}^4C_1\,a^3b^1 + {}^4C_3\,ab^3]$

$= 2\,[4a^3b + 4ab^3]$

$= 8\,(a^3b + ab^3)$

On putting $a = \sqrt{3}, b = \sqrt{2}$, we get

$(\sqrt{3}+\sqrt{2})^4 - (\sqrt{3}-\sqrt{2})^4 = [8\,(\sqrt{3})^3(\sqrt{2}) + (\sqrt{3})(\sqrt{2})^3]$

$= 8\,(3\sqrt{6} + 2\sqrt{6}) = 40\sqrt{6}$

20. Using binomial theorem for positive integral index, we have,

$(a^2 + \sqrt{a^2-1})^4 + (a^2 - \sqrt{a^2-1})^4 = (x+y)^4 + (x-y)^4$,

where $x = a^2, y = \sqrt{a^2-1}$

$= \{{}^4C_0\,x^4 + {}^4C_1\,x^3y + {}^4C_2\,x^2y^2 + {}^4C_3\,xy^3 + {}^4C_4y^4\} + \{{}^4C_0\,x^4 - {}^4C_1\,x^3y + {}^4C_2\,x^2y^2 - {}^4C_3\,xy^3 + {}^4C_4y^4\}$

$= 2\,\{{}^4C_0\,x^4 + {}^4C_2x^2\,y^2 + {}^4C_4y^4\}$

$= 2\,\{1(a^2)^4 + 6(a^2)^2(\sqrt{a^2-1})^2 + 1(\sqrt{a^2-1})^4\}$

$= 2\{a^8 + 6a^4(a^2-1) + (a^2-1)^2\}$

$= 2\,\{a^8 + 6a^6 - 5a^4 - 2a^2 + 1\}$

21. $(x+y)^n = {}^nC_0\,x^ny^0 + {}^nC_1\,x^{n-1}\,y + {}^nC_2\,x^{n-2}\,y^2 + {}^nC_3\,x^{n-3}\,y^3 + \ldots$

Given, ${}^nC_0\,x^ny^0 = 1 \Rightarrow x^n = 1$ ...(i)

${}^nC_1\,x^{n-1}y = 56 \Rightarrow nx^{n-1}\,y = 56$ ...(ii)

${}^nC_2\,x^{n-2}\,y^2 = 1372$

$\Rightarrow \quad \dfrac{n\,(n-1)}{2}\cdot x^{n-2}\,y^2 = 1372$ ...(iii)

On dividing Eq. (i) by (ii), we get

$\dfrac{x^n}{nx^{n-1}y} = \dfrac{1}{56}$

$\Rightarrow \quad \dfrac{x}{ny} = \dfrac{1}{56}$

$\Rightarrow \quad 56x = ny$ ...(iv)

On dividing Eq. (ii) by (iii), we get

$\dfrac{nx^{n-1}\,y}{\dfrac{n\,(n-1)}{2}x^{n-2}\,y^2} = \dfrac{56}{1372}$

$\Rightarrow \quad \dfrac{2x}{(n-1)y} = \dfrac{56}{1372}$

$\Rightarrow \quad 2744\,x = 56\,(n-1)y$ ...(v)

On dividing Eq. (i) by (iii), we get

$\dfrac{x^n}{\dfrac{n\,(n-1)}{2}x^{n-2}\,y^2} = \dfrac{1}{1372}$

$\Rightarrow \quad \dfrac{2x^2}{n(n-1)y^2} = \dfrac{1}{1372}$

$\Rightarrow \quad 2744x^2 = n\,(n-1)\,y^2$ ...(vi)

On solving Eqs. (iv), (v) and (vi). **Ans.** $x = 1, y = 7$

22. $(1.2)^{4000} = (1+0.2)^{4000}$

$= {}^{4000}C_0 + {}^{4000}C_1\,(0.2)$ + Sum of positive terms

$= 1 + 4000\,(0.2)$ + Positive number

$= 1 + 800$ + Positive number > 800

Mind Map

Binomial Theorem

Binomial Expression

An algebraic expression consisting of two terms with +ve or –ve sign between them, is called binomial expression.

Binomial Theorem for any Positive Integer n

If a and b are two real numbers, then for any positive integer n, we have

$$(a+b)^n = {}^nC_0 a^n b^0 + {}^nC_1 a^{n-1} b^1 + {}^nC_2 a^{n-2} b^2 + \ldots + {}^nC_n a^0 b^n$$

$$(a+b)^n = {}^nC_0 a^n + {}^nC_1 a^{n-1} b + {}^nC_2 a^{n-2} b^2 + \ldots + {}^nC_n b^n$$

$$(a+b)^n = \sum_{r=0}^{n} {}^nC_r a^{n-r} b^r$$

where, ${}^nC_0, {}^nC_1, \ldots, {}^nC_n$ are called **binomial coefficients.** These binomial coefficients can also be written as C_0, C_1,

Some Important Observations

- The total number of terms in the expansion is one more than the index i.e. the total number of terms in $(a + b)^n$ is $n + 1$.
- In the successive terms of the expansion, powers of the first quantity a go on decreasing by 1 i.e. it is n in the first term, $(n - 1)$ in the second term and so on ending with zero in the last term whereas the powers of the second quantity b increase by 1 i.e. it is zero in first term, 1 in the second term and so on ending with n in the last term.
- In each term of the expansion, the sum of the indices of a and b is the same and is equal to the index of $a + b$ i.e. n.
- The binomial coefficients of terms equidistant from the beginning and end are equal.

Particular Cases of Binomial Theorem

(i) **Expansion of $(a - b)^n$** Replacing b by $(-b)$ in the expansion of $(a + b)^n$, we get

$$(a-b)^n = [a + (-b)]^n = {}^nC_0 a^n - {}^nC_1 a^{n-1} b + {}^nC_2 a^{n-2} b^2 + \ldots + (-1)^n \, {}^nC_n b^n$$

(ii) **Expansion of $(1 + x)^n$** Replacing a by 1 and b by x in the expansion of $(a + b)^n$, we get

$$(1+x)^n = {}^nC_0 1^n + {}^nC_1 1^{n-1} x + {}^nC_2 1^{n-2} x^2 + \ldots + {}^nC_n x^n$$

(iii) **Expansion of $(1 - x)^n$** Replacing a by 1 and b by $(-x)$ in the expansion of $(a + b)^n$, we get

$$(1-x)^n = {}^nC_0 - {}^nC_1 x + {}^nC_2 x^2 + \ldots + (-1)^n \, {}^nC_n x^n$$

(iv) $(x + a)^n + (x - a)^n = 2\,[{}^nC_0 x^n + {}^nC_2 x^{n-2} a^2 + \ldots]$

(v) $(x + a)^n - (x - a)^n = 2\,[{}^nC_1 x^{n-1} a^1 + {}^nC_3 x^{n-3} a^3 + \ldots]$

Topics to be Assessed Formatively

General Term in a Binomial Expansion

Consider the binomial expansion of $(a+b)^n$, i.e.

$(a+b)^n = {}^nC_0a^n + {}^nC_1a^{n-1}b + {}^nC_2a^{n-2}b^2 + \ldots + {}^nC_nb^n$

Here, we observed that

$$\text{First term} = {}^nC_0a^n$$
$$\text{Second term} = {}^nC_1a^{n-1}b^1$$
$$\text{Third term} = {}^nC_2a^{n-2}b^2$$
$$\text{Fourth term} = {}^nC_3a^{n-3}b^3$$
$$\vdots \quad \vdots$$
$$(r+1)\text{th term} = {}^nC_ra^{n-r}b^r$$

Here, $(r+1)$th term is called the **general term of the expansion $(a+b)^n$** and it is denoted by T_{r+1}.

i.e. $$T_{r+1} = {}^nC_ra^{n-r}b^r$$

Thus, $(n+1)$th term, $T_{n+1} = {}^nC_na^{n-n}b^n = {}^nC_nb^n$

Note The pth term from the end in $(a+b)^n$ is $[(n+1)-(p-1) = (n-p+2)$th term from the beginning.

General term of particular cases of Binomial Theorem

(i) In the expansion of $(a-b)^n$

$$T_{r+1} = (-1)^r\ {}^nC_r\ x^{n-r}\ y^r$$

(ii) In the expansion of $(1+x)^n$

$$T_{r+1} = {}^nC_rx^r$$

(iii) In the expansion of $(1-x)^n$

$$T_{r+1} = (-1)^r\ {}^nC_r\ x^r$$

Example 1 **Evaluate the following terms.**

(i) General term in the expansion of $(3x^2+4y)^{10}$.

(ii) 6th term in the expansion of $\left(y-\frac{1}{y}\right)^{20}$.

(iii) 7th term in the expansion of $\left(2x+\frac{y}{3}\right)^{15}$.

(iv) 5th term from the end in the expansion of $\left(x-\frac{1}{x^2}\right)^{12}$.

Sol. (i) The general term in the expansion of $(3x^2+4y)^{10}$ is

$$T_{r+1} = {}^{10}C_r\ (3x^2)^{10-r}\ (4y)^r \quad [\because T_{r+1} = {}^nC_r\ a^{n-r}\ b^r]$$
$$= {}^{10}C_r \times 3^{10-r} \times 4^r \times x^{20-2r}\ y^r$$

(ii) The general term in the expansion of $\left(y-\frac{1}{y}\right)^{20}$ is

$$T_{r+1} = {}^{20}C_r\ (y)^{20-r}\left(-\frac{1}{y}\right)^r \quad [\because T_{r+1} = {}^nC_r\ a^{n-r}\ b^r]$$
$$= {}^{20}C_r\ y^{20-2r}\ (-1)^r$$

For determining 6th term, put $r=5$, we get

$$T_{5+1} = {}^{20}C_5\ y^{20-10}\ (-1)^5$$
$$= -{}^{20}C_5\ y^{10}$$

(iii) The general term in the expansion of $\left(2x+\frac{y}{3}\right)^{15}$ is

$$T_{r+1} = {}^{15}C_r\ (2x)^{15-r}\left(\frac{y}{3}\right)^r \quad [\because T_{r+1} = {}^nC_r\ a^{n-r}\ b^r]$$
$$= {}^{15}C_r\ 2^{15-r} \times 3^{-r} \times x^{15-r}\ y^r$$

For determining 7th term, put $r=6$, we get

$$T_{6+1} = {}^{15}C_6\ 2^{15-6} \times 3^{-6} \times x^{15-6}y^6$$
$$= {}^{15}C_6\ 2^9 \times 3^{-6} \times x^9y^6$$

(iv) The 5th term from the end in the expansion of $\left(x-\frac{1}{x^2}\right)^{12}$ = The $(12-5+2)$th i.e. 9th term from the beginning in the expansion of $\left(x-\frac{1}{x^2}\right)^{12}$.

[$\because$ pth term from the end $=(n-p+2)$th term from the beginning]

Now, $$T_{8+1} = {}^{12}C_8\ (x)^{12-8}\left(-\frac{1}{x^2}\right)^8$$
$$= {}^{12}C_4x^4\ (-1)^8 \cdot \frac{1}{x^{16}}$$
$$= \frac{12\times11\times10\times9}{4\times3\times2\times1} \times x^{4-16}$$
$$= 495\ x^{-12}$$
$$= \frac{495}{x^{12}}$$

Example 2 **Find the coefficient of 5th term in the expansion of $\left(\frac{x}{3}-3y\right)^7$.**

Sol. Given expression is $\left(\frac{x}{3}-3y\right)^7$.

General term, $T_{r+1}={}^7C_r\left(\frac{x}{3}\right)^{7-r}\cdot(-3y)^r$

Put $r+1=5\Rightarrow r=4$

Then, $T_5=T_{4+1}={}^7C_4\left(\frac{x}{3}\right)^3(-3y)^4$

$$=\frac{7\cdot6\cdot5}{3\cdot2\cdot1}\cdot\frac{x^3}{27}\times81\cdot y^4=35\cdot3\,x^3y^4=105\,x^3y^4$$

Thus, the coefficient of 5th term is 105.

Example 3 **If the third term in the expansion of $\left(\frac{1}{x}+x^{\log_{10}x}\right)^5$ is 1000, then find x.**

Sol. Given, $T_3=1000\Rightarrow{}^5C_2\left(\frac{1}{x}\right)^{5-2}(x^{\log_{10}x})^2=1000$

$\Rightarrow\quad 10\,(x^{\log_{10}x})^2\times x^{-3}=1000\Rightarrow x^{2\log_{10}x}\times x^{-3}=100$

$\Rightarrow\quad x^{2\log_{10}x-3}=10^2\Rightarrow 2\log_{10}x-3=\log_x 10^2$

$\Rightarrow\quad 2\log_{10}x-3=\frac{2}{\log_{10}x}$

$\Rightarrow\quad 2y-3=\frac{2}{y}$, where $y=\log_{10}x$

$\Rightarrow\quad 2y^2-3y-2=0\Rightarrow(2y+1)(y-2)=0$

$\Rightarrow\quad y=2$ or $y=-\frac{1}{2}\Rightarrow\log_{10}x=2$ or $\log_{10}x=-\frac{1}{2}$

$\Rightarrow\quad x=10^2$ or $x=10^{-1/2}$ $\therefore$ $x=100$ or $x=\frac{1}{\sqrt{10}}$

Middle Terms in a Binomial Expansion

We know that the binomial expansion $(a+b)^n$ has $(n+1)$ terms. So for middle terms, there are two cases.

***Case* I** If n is even, then total number of terms (i.e. $n+1$) is odd and the middle term is $\left(\frac{n}{2}+1\right)$th term.

e.g. In the expansion of $(a+b)^6$, the middle term is $\left(\frac{6}{2}+1\right)$ th term i.e. 4th term.

***Case* II** If n is odd, then total number of terms i.e. $(n+1)$ is even and there are two middle terms which are given by $\left(\frac{n+1}{2}\right)$th and $\left(\frac{n+1}{2}+1\right)$th terms

i.e. by $\left(\frac{n+1}{2}\right)$th and $\left(\frac{n+3}{2}\right)$th terms.

e.g. In the expansion of $(a+b)^9$, the middle terms are $\left(\frac{9+1}{2}\right)$th and $\left(\frac{9+1}{2}+1\right)$th terms

i.e. 5th and 6th terms.

Example 4 **Find the middle term(s) in the expansion of $(2x+3y)^9$.**

Sol. The index of binomial expansion $(2x+3y)^9$ is $n=9$, which is an odd number. So there are two middle terms which are given by $\left(\frac{9+1}{2}\right)$th and $\left(\frac{9+1}{2}+1\right)$th terms

i.e. T_5 and T_6 terms.

In the expansion of $(2x+3y)^9$,

$$T_5={}^9C_4(2x)^{9-4}(3y)^4$$

$$\Rightarrow\quad T_5=\frac{9\times8\times7\times6}{4\times3\times2\times1}(2x)^5(3y)^4$$

$$=126\times32\times81x^5y^4=326592x^5y^4$$

and $\quad T_6={}^9C_5(2x)^{9-5}(3y)^5=\frac{9\times8\times7\times6}{4\times3\times2\times1}(2x)^4(3y)^5$

$$=126\times16\times243x^4y^5=489888x^4y^5$$

Hence, middle terms in the expansion of $(2x+3y)^9$ are $489888\,x^4y^5$ and $326592\,x^5y^4$.

Example 5 **Find the middle term in the expansion of $\left(\frac{2x^2}{3}+\frac{3}{2x^2}\right)^{10}$.**

Sol. Here, value of index n is 10 (even), therefore there is only one middle term given by $T_{\frac{10}{2}+1}$ i.e. T_6.

Now, $T_6=T_{5+1}={}^{10}C_5\left(\frac{2x^2}{3}\right)^{10-5}\left(\frac{3}{2x^2}\right)^5$

$$[\because T_{r+1}={}^nC_ra^{n-r}b^r]$$

$$={}^{10}C_5\left(\frac{2x^2}{3}\right)^5\left(\frac{3}{2x^2}\right)^5$$

$$=\frac{10\cdot9\cdot8\cdot7\cdot6}{5\cdot4\cdot3\cdot2\cdot1}\cdot\frac{2^5\cdot x^{10}}{3^5}\cdot\frac{3^5}{2^5\cdot x^{10}}=252$$

which is independent of x.

Example 6 **Find the middle term in the expansion of $\left(3-\frac{x^3}{6}\right)^7$.**

Sol. Here, $n=7$ (odd)

$\therefore$ Middle terms are given by $\left(\frac{n+1}{2}\right)$th and $\left(\frac{n+3}{2}\right)$th terms, i.e. $\left(\frac{7+1}{2}\right)$th term and $\left(\frac{7+3}{2}\right)$th term,

i.e. 4th and 5th terms.

Now, $T_4 = T_{3+1} = {}^7C_3(3)^{7-3}\left(-\frac{x^3}{6}\right)^3$ $[\because T_{r+1} = {}^nC_r a^{n-r} b^r]$

$$= \frac{7\times6\times5}{6}\times\frac{3^4(-1)^3\times x^9}{6^3} = \frac{35\times81\times(-1)\times x^9}{6\times6\times6}$$

$$= -\frac{35\times27\times x^9}{2\times6\times6} = -\frac{105}{8}x^9$$

and $T_5 = T_{4+1} = {}^7C_4 3^{7-4}\left(-\frac{x^3}{6}\right)^4$

$$= {}^7C_3\frac{3^3(-1)^4(x^3)^4}{6^4} \quad [\because {}^nC_r = {}^nC_{r-1}]$$

$$= \frac{7\times6\times5}{6}\times\frac{3\times3\times3\times x^{12}}{6\times6\times6\times6} = \frac{35}{48}x^{12}$$

Some More Problems Related to Binomial Theorem

There are many other types of problems related to binomial theorem along with its general term, middle term, coefficients of consecutive terms, etc.

Some of these types of problems are discussed below

|TYPE I|

FINDING THE COEFFICIENT FOR A GIVEN INDEX OF THE VARIABLE

In this type of problem firstly find the general term, then equate the index of variable obtained in general term to the given index to find the value of r and then use the value of r to calculate required coefficient.

Note Sometimes we simply expand the given expression to get the coefficient of required term.

Example 7 Find the coefficient of x^4 in $(1+2x)^4(2-x)$.

Sol. We have, $(1+2x)^4(2-x)$

$= [1+{}^4C_1(2x)^1+{}^4C_2(2x)^2+{}^4C_3(2x)^3+{}^4C_4(2x)^4](2-x)$

$= (1+4\times2x+6\times4x^2+4\times8x^3+1\times16x^4)(2-x)$

$= (1+8x+24x^2+32x^3+16x^4)(2-x)$

$= 2+16x+48x^2+64x^3+32x^4-x-8x^2-24x^3-32x^4-16x^5$

$= 2+15x+40x^2+40x^3-16x^5$

Since, there is no term containing x^4.

So, coefficient of x^4 is zero.

Example 8 Find the coefficients of x^{32} and x^{-17} in the expansion of $\left(x^4-\frac{1}{x^3}\right)^{15}$.

Sol. Let general term in the expansion of $\left(x^4-\frac{1}{x^3}\right)^{15}$ be T_{r+1}.

Then, $T_{r+1} = {}^{15}C_r(x^4)^{15-r}\left(-\frac{1}{x^3}\right)^r$

$\Rightarrow \quad T_{r+1} = (-1)^r \times {}^{15}C_r x^{60-7r}$ ...(i)

This term will contain x^{32}, if $60-7r=32$

$\Rightarrow \quad 7r = 60-32 = 28 \Rightarrow r = \frac{28}{7} = 4$

Thus, the term containing x^{32} is T_{4+1} i.e. T_5.

$\therefore$ Coefficient of $x^{32} = (-1)^4\times{}^{15}C_4 = 1365$ [using Eq. (i)]

Now, for the term containing x^{-17}, put

$$60-7r=-17 \Rightarrow 7r=77 \Rightarrow r=11$$

Thus, the term containing x^{-17} is T_{11+1} i.e. T_{12}.

$\therefore$ Coefficient of $x^{-17} = (-1)^{11}\times{}^{15}C_{11} = -1365$

[using Eq. (i)]

|TYPE II|

FINDING THE TERM INDEPENDENT OF THE VARIABLE

Independent Term (Constant term) The term which is free from variable (say x), i.e. which have x^0 is known as independent term.

e.g. The $(n+1)$th term in the expansion of $\left(x+\frac{1}{x}\right)^{2n}$, where $x\neq0$, is

$$T_{n+1} = {}^{2n}C_n x^n\left(\frac{1}{x}\right)^n = {}^{2n}C_n$$

which is an independent term.

In this type of problem, we put the index of variable x obtained in general term, equal to zero and find the value of r to calculate the required term.

Example 9 If the term free from x in the expansion of $\left(\sqrt{x}-\frac{k}{x^2}\right)^{10}$ is 405, then find the value of k.

Sol. Given expansion is $\left(\sqrt{x}-\frac{k}{x^2}\right)^{10}$.

Let T_{r+1} be the general term.

Then, $T_{r+1} = {}^{10}C_r(\sqrt{x})^{10-r}\left(\frac{-k}{x^2}\right)^r$

$= {}^{10}C_r(x)^{\frac{1}{2}(10-r)}(-k)^r \cdot x^{-2r}$

$= {}^{10}C_r x^{5-\frac{r}{2}}(-k)^r \cdot x^{-2r}$

$= {}^{10}C_r\, x^{5-\frac{r}{2}-2r}(-k)^r$

$= {}^{10}C_r\, x^{\frac{10-5r}{2}}(-k)^r$

For this term to be free from x, $\frac{10-5r}{2}=0$

$\Rightarrow \quad 10-5r=0 \Rightarrow r=2$

Thus, $T_{2+1}=T_3$ is free from x.

Now, $T_3 = T_{2+1} = {}^{10}C_2(-k)^2 = 405$

[$\because$ term free from $x = 405$]

$\Rightarrow \quad \frac{10\times 9\times 8!}{2!\times 8!}(-k)^2 = 405$

$\Rightarrow \quad 45k^2 = 405$

$\Rightarrow \quad k^2 = \frac{405}{45} = 9$

$\therefore \quad k = \pm 3$

| TYPE III |

PROBLEMS BASED ON THREE CONSECUTIVE TERMS

In this type of problem, relation between consecutive terms or coefficient of consecutive terms of the expansion will be given. For finding the required values, we use the given relation and the following results.

$$\frac{{}^nC_r}{{}^nC_{r-1}} = \frac{n-r+1}{r} \text{ and } \frac{{}^nC_{r+1}}{{}^nC_r} = \frac{n-r}{r+1}$$

Example 10 **If the coefficient of second, third and fourth terms in the expansion of $(1+x)^{2n}$ are in AP, then show that $2n^2-9n=-7$.**

Sol. The general term of $(1+x)^{2n}$ is $T_{r+1} = {}^{2n}C_r x^r$

Then, $T_2 = {}^{2n}C_1x^1, T_3 = {}^{2n}C_2x^2$ and $T_4 = {}^{2n}C_3x^3$

Given, coefficients of T_2, T_3 and T_4 are in AP.

i.e. ${}^{2n}C_1, {}^{2n}C_2$ and ${}^{2n}C_3$ are in AP.

$\therefore\ 2\times {}^{2n}C_2 = {}^{2n}C_1 + {}^{2n}C_3 \Rightarrow 2 = \frac{{}^{2n}C_1}{{}^{2n}C_2} + \frac{{}^{2n}C_3}{{}^{2n}C_2}$

$\Rightarrow \quad 2 = \frac{(2n)!}{1!(2n-1)!}\times\frac{2!(2n-2)!}{(2n)!} + \frac{(2n)!}{3!(2n-3)!}\times\frac{2!(2n-2)!}{(2n)!}$

$\Rightarrow \quad 2 = \frac{2(2n-2)!}{(2n-1)(2n-2)!} + \frac{2(2n-2)(2n-3)!}{6(2n-3)!}$

$\Rightarrow \quad 2 = \frac{2}{2n-1} + \frac{2n-2}{3} \Rightarrow 2 = \frac{6+(2n-2)(2n-1)}{3(2n-1)}$

$\Rightarrow \quad 6(2n-1) = 6+4n^2-2n-4n+2$

$\Rightarrow \quad 12n-6 = 8+4n^2-6n$

$\Rightarrow \quad 4n^2-6n-12n = -6-8 \Rightarrow 4n^2-18n=-14$

$\therefore \quad 2n^2-9n=-7$ [dividing by 2]

Hence proved.

Exercise

Multiple Choice Questions

1. The coefficient of x^{11} in the expansion $\left(x^3-\frac{2}{x^2}\right)^{12}$ is

(a) ${}^{12}C_{11}(-1)^{11}2^{11}$ (b) ${}^{12}C_6(-1)^6 2^6$
(c) ${}^{12}C_5(-1)^5 2^5$ (d) None of these

2. The coefficient of 11th term from the end in the expansion of $\left(2x-\frac{1}{x^2}\right)^{25}$ is

(a) ${}^{25}C_{15}\times 2^{10}$ (b) $-{}^{25}C_{15}\times 2^{10}$
(c) ${}^{25}C_{20}\times 2^{10}$ (d) $-{}^{25}C_{20}\times 2^{10}$

3. If the coefficient of x^7 and x^8 in $\left(2+\frac{x}{3}\right)^n$ are equal, then n is equal to

(a) 15 (b) 45
(c) 55 (d) 56

4. The coefficient of x in the expansion of $(1-3x+7x^2)(1-x)^{16}$ is

(a) 17 (b) 19 (c) -17 (d) -19

5. If the term free from x in the expansion of $\left(\sqrt{x}-\frac{k}{x^2}\right)^{15}$ is 455, then the values of k are

(a) 1 (b) 2 (c) 3 (d) -1

Very Short Answer Type Questions

6. Find the general term in the expansion of the following expressions.

(i) $(3y+6z)^9$ (ii) $(5x-3)^{12}$
(iii) $\left(x-\frac{4}{z}\right)^{50}$ (iv) $\left(z^2-\frac{3}{z^2}\right)^{35}$

7. Evaluate the following terms in the given expansion.

(i) 7th term in the binomial expansion $(x+3y)^8$.

(ii) 5th term in the binomial expansion $\left(2x-\frac{1}{x}\right)^{10}$.

(iii) 6th term in the binomial expansion $\left(z^2+\frac{2z}{3}+\frac{1}{9}\right)^5$.

(iv) 4th term in the expansion $(x-2y)^{12}$.

8. (i) Find the rth term from the end in the expansion of $(x+a)^n$.

(ii) Find the 5th term from the end in the expansion $\left(\frac{4x}{5}-\frac{5}{2x}\right)^9$.

Short Answer Type I Questions

9. Find the coefficient of

(i) x^3 in the expansion of $\left(3x-\frac{1}{x}\right)^7$.

(ii) y^5 in the expansion of $\left(3y^2+\frac{1}{3y}\right)^{10}$.

(iii) x^4 in the expansion of $(2x+5)^8$.

(iv) x^{11} in the expansion of $\left(x^3-\frac{2}{x^2}\right)^{12}$.

10. Find the middle term(s) in the given expansion.

(i) $\left(\frac{x}{3}+9y\right)^{10}$ (ii) $\left(3x-\frac{x^3}{6}\right)^9$

(iii) $\left(2ax-\frac{b}{x^2}\right)^{12}$ (iv) $\left(x^2-\frac{2}{x}\right)^{10}$

11. Find the term independent of x (constant term) in the expansion of

(i) $\left(\sqrt{x}+\frac{1}{3x^2}\right)^{10}$ (ii) $\left(2x-\frac{1}{x}\right)^9$

(iii) $\left(3x^2+\frac{1}{x^5}\right)^7$ (iv) $\left(x+\frac{1}{x}\right)^{10}$

12. Find the coefficient of x^6y^3 in the expansion of $(x+2y)^9$.

13. In the binomial expansion $(x+y)^n$, the coefficients of 4th and 13th terms are equal. Find the value of n.

14. If p is a real number and the middle term in the expansion of $\left(\frac{p}{2}+2\right)^8$ is 1120, then find the value of p.

15. Show that the term containing x^3 does not exist in the expansion of $\left(3x+\frac{1}{2x}\right)^8$.

16. Find the coefficient of x^4 in the expansion of $(1+x+x^2+x^3)^{11}$.

17. Find the coefficient of x in the expansion of $(1-3x+7x^2)(1-x)^{16}$.

18. If the coefficients of $(r-5)$th and $(2r-1)$th terms in the expansion $(1+x)^{34}$ are equal, then find the value of r.

Short Answer Type II Questions

19. Find a, if the 17th and 18th terms of the expansion $(2+a)^{50}$ are equal.

20. Find the two successive terms in the expansion of $(1+x)^{24}$, whose coefficients are in the ratio $1:4$.

21. If the coefficients of 2nd, 3rd and 4th terms in the expansion of $(1+x)^n$ are in AP, then find the value of n.

22. Find the coefficient of a^4 in the product $(1+2a)^4(2-a)^5$ using binomial theorem.

23. The sum of the coefficients of the first three terms in the expansion of $\left(x-\frac{3}{x^2}\right)^m$, $x\neq 0$, m being a natural number, is 559. Find the term of the expansion containing x^3.

Long Answer Type Questions

24. The coefficients of three consecutive terms in the expansion of $(1+a)^n$ are in the ratio $1:7:42$. Find the value of n.

25. The 2nd, 3rd and 4th terms in the binomial expansion $(x+a)^n$ are 240, 720 and 1080, respectively. Find the values of x, a and n.

26. Find the term independent of x in the expansion of $(1+x+2x^3)\left(\frac{3}{2}x^2-\frac{1}{3x}\right)^9$.

27. Show that the middle term in the expansion of $(1+x)^{2n}$ is $\frac{1\cdot3\cdot5\ldots(2n-1)}{n!}2^n\cdot x^n$, where n is a positive integer.

28. Show that the middle term in the expansion of $\left(x^2+\frac{1}{x^2}+2\right)^n$ is $\frac{1\cdot3\cdot5\ldots(2n-1)}{n!}\cdot2^n$.

Answers

1. (c) **2.** (b) **3.** (a) **4.** (d)

5. (d)

6. (i) $^{9}C_r\,(3y)^{9-r}\,(6z)^r$ (ii) $^{12}C_r\,(5x)^{12-r}\,(-3)^r$

(iii) $^{50}C_r\,(x)^{50-r}\left(-\frac{4}{z}\right)^r$ (iv) $^{35}C_r\;z^{70-4r}\,(-3)^r$

7. (i) $20412x^2\,y^6$ (ii) $13440x^2$

(iii) $\frac{252}{243}z^2$ (iv) $-1760x^9y^3$

8. (i) $^{n}C_{n-r+1}\;x^{r-1}\;a^{n-r+1}$ (ii) $\frac{-5040}{x}$

9. (i) 5103 (ii) 252

(iii) 700000 (iv) −25344

10. (i) $61236\,x^5y^5$ (ii) $\frac{189}{8}x^{17}$ and $\frac{-21}{16}\,x^{19}$

(iii) $\frac{59136\,a^6\,b^6}{x^6}$ (iv) $-8064\,x^5$

11. (i) 5

(ii) No independent term.

(iii) 5103

(iv) 252

12. 672 **13.** 15 **14.** ± 2 **16.** 990

17. −19 **18.** 14 **19.** 1 **20.** 5th and 6th terms

21. 7 **22.** − 438 **23.** $-5940x^3$ **24.** 55

25. $x = 2$, $a = 3$ and $n = 5$ **26.** $\frac{17}{54}$

08

CHAPTER

Doing Mathematics should always mean finding patterns and crafting beautiful explanations. —*Paul Lockhart*

Sequences and Series

YOUR STUDY GOALS

- 8.1 Sequence
- 8.2 Geometric Progression (GP)
- 8.3 Sum of n terms of a GP
- 8.4 Geometric Mean (GM)

YOUR PRACTICE GOALS

- Objective Type Questions (32 Qs; 20 MCQs, 10 A-R, 2 Case Based)
- Subjective Type Questions (31 Qs; 8 VSA, 12 SA, 10 LA)

8.1 Sequence

A sequence is an arrangement of numbers according to some rules.

e.g. 1, 4, 9, 16, ... is a sequence as each number follow a rule of n^2, where $n = 1, 2, 3$.

The various numbers occurring in a sequence are called its **terms**. The terms of a sequence are generally denoted by $a_1, a_2, a_3, \ldots, a_n, \ldots$ etc. The number occurring at the nth place of a sequence is called its ***n*th term** and it is denoted by a_n.

The nth term is also called the **general term** of the sequence.

e.g. 2, 6, 10, 14, ... is a sequence. Here, $a_1 = 2, a_2 = 6, a_3 = 10$ and $a_4 = 14$.

Types of Sequence

1. **Finite Sequence** A sequence containing finite number of terms, is called a finite sequence.

   e.g. 1, 3, 5, 7 is a finite sequence as it contains only 4 terms.

2. **Infinite Sequence** A sequence which is not a finite sequence is known as infinite sequence i.e. in an infinite sequence, **number of terms never ends.**

e.g.

- 1, 3, 5, 7,... is an infinite sequence, as it contains infinite number of terms.
- The sequence of successive quotients obtained in the division of 10 by 3 i.e. 3, 3.3, 3.33, 3.333, ...is an infinite sequence.

Real Sequence

A sequence of real numbers is a function whose domain is the set of natural numbers and whose range is a subset of real numbers.

e.g. Let $a: N \rightarrow R$ given by $a(n) = 4n - 3$, then

$$a(1) = a_1 = 4(1) - 3 = 1$$
$$a(2) = a_2 = 4(2) - 3 = 5$$
$$a(3) = a_3 = 4(3) - 3 = 9$$
$$\vdots \quad \vdots \quad \vdots \quad \vdots$$

i.e. 1, 5, 9, ... is the required sequence.

Fibonacci Sequence

Sometimes an arrangement of numbers has no visible pattern but the sequence can be represented by the recurrence relation.

The sequence 1, 1, 2, 3, 5, 8, ... has no visible pattern but its recurrence relation is

$$a_1 = a_2 = 1 \text{ and } a_{n+1} = a_n + a_{n-1}, n \geq 2$$

This sequence is called Fibonacci sequence.

8.2 Geometric Progression (GP)

A sequence of non-zero numbers is said to be a geometric progression if the ratio of each term, except the first one, by its preceding term is always constant.

In other words, we can say that a sequence $a_1, a_2, \ldots, a_n$ is called geometric progression (geometric sequence) if each term is non-zero and it follows the relation

$$\frac{a_{k+1}}{a_k} = r \text{ (constant) for all } k \in N.$$

The constant ratio is called **common ratio** of the GP and it is denoted by r.

In a GP, we usually denote the first term by a, the **nth term** by T_n or a_n.

Thus, a GP can be written as $a, ar, ar^2, ar^3, \ldots$ and so on.

General Term of a GP

If a is the first term of a GP and its common ratio is r, then general term or nth term,

$$T_n = ar^{n-1} \text{ or } l = ar^{n-1},$$

where l is the last term.

mth Term of a Finite GP from the End

Let a be the first term and r be the common ratio of a GP having n terms.

Then, mth term from the end is $(n - m + 1)$th term from the beginning.

$\therefore$ **mth term from the end** $= ar^{n-m+1-1}$

$$= ar^{n-m}, \text{ where } n > m.$$

Also, **mth term from the end** $= l\left(\frac{1}{r}\right)^{n-1}$,

where l is the last term of the finite GP.

Selection of Terms in GP

Sometimes we have to select certain number of terms in GP. The convenient method of selecting terms is given below.

Number of Terms	Terms	Common ratio
3	$\frac{a}{r}, a, ar$	r
4	$\frac{a}{r^3}, \frac{a}{r}, ar, ar^3$	r^2
5	$\frac{a}{r^2}, \frac{a}{r}, a, ar, ar^2$	r

Properties of Geometric Progression

- If all the terms of a GP are multiplied by the same quantity, then the resulting sequence is also a GP with the same common ratio.
- The reciprocals of the terms of a given GP also form a GP.
- If each term of a GP is raised to same power, then the resulting sequence is also a GP.
- In a finite GP, the product of the terms equidistant from the beginning and from the end is always same and is equal to the product of the first and the last terms.

Example 1. **Write first five terms of sequence**
(i) $a_n = (-1)^{n-1} 5^{n+1}$ (ii) $a_n = 2n^2 - n + 1$ NCERT

Sol. (i) We have, $a_n = (-1)^{n-1} 5^{n+1}$

On putting $n = 1$, we get

$$a_1 = (-1)^{1-1} 5^{1+1} = (-1)^0 5^2 = 25$$

On putting $n = 2$, we get

$$a_2 = (-1)^{2-1} 5^{2+1} = (-1)^1 5^3 = -125$$

On putting $n = 3$, we get

$$a_3 = (-1)^{3-1} 5^{3+1} = (-1)^2 5^4 = 625$$

On putting $n = 4$, we get

$$a_4 = (-1)^{4-1} 5^{4+1} = (-1)^3 5^5 = -3125$$

On putting $n = 5$, we get

$$a_5 = (-1)^{5-1} 5^{5+1} = (-1)^4 5^6 = 15625$$

Hence, the first five terms of the given sequence are 25, – 125, 625, – 3125, 15625.

(ii) We have, $a_n = 2n^2 - n + 1$

On putting $n = 1$, we get

$$a_1 = 2(1)^2 - 1 + 1 = 2 - 1 + 1 = 2$$

On putting $n = 2$, we get

$$a_2 = 2(2)^2 - 2 + 1 = 8 - 2 + 1 = 7$$

On putting $n = 3$, we get

$$a_3 = 2(3)^2 - 3 + 1 = 18 - 3 + 1 = 16$$

On putting $n = 4$, we get

$$a_4 = 2(4)^2 - 4 + 1 = 32 - 4 + 1 = 29$$

On putting $n = 5$, we get

$$a_5 = 2(5)^2 - 5 + 1 = 50 - 5 + 1 = 46$$

Hence, the first five terms of the given sequence are 2, 7, 16, 29 and 46.

Example 2. **Let the sequence a_n is defined as follows $a_1 = 2, a_n = a_{n-1} + 3$ for $n \ge 2$. Find first five terms and write corresponding series.** Competency Based Que

Sol. We have, $a_1 = 2$ and $a_n = a_{n-1} + 3$, for $n \ge 2$

On putting $n = 2$, we get

$$a_2 = a_1 + 3 = 2 + 3 = 5$$

On putting $n = 3$, we get

$$a_3 = a_2 + 3 = 5 + 3 = 8$$

On putting $n = 4$, we get

$$a_4 = a_3 + 3 = 8 + 3 = 11$$

On putting $n = 5$, we get

$$a_5 = a_4 + 3 = 11 + 3 = 14$$

Thus, first five terms of given sequence are 2, 5, 8, 11 and 14. Also, the corresponding series is
$2 + 5 + 8 + 11 + 14 + \dots$

Example 3. **For what values of k, the numbers $-\frac{2}{7}, k, -\frac{7}{2}$ are in GP?** NCERT, Competency Based Que

Sol. Given, $-\frac{2}{7}, k, -\frac{7}{2}$ are in GP.

Then, $\frac{a_2}{a_1} = \frac{a_3}{a_2}$

$$\left[\because \text{common ratio } (r) = \frac{a_2}{a_1} = \frac{a_3}{a_2} = \frac{a_4}{a_3} = \dots\right]$$

$$\therefore \quad \frac{k}{-2/7} = \frac{-7/2}{k} \Rightarrow \frac{7}{-2}k = \frac{-7}{2} \times \frac{1}{k}$$

$$\Rightarrow \quad 7k \times 2k = -7 \times (-2)$$

$$\Rightarrow \quad 14k^2 = 14 \Rightarrow k^2 = 1$$

$$\therefore \quad k = \pm 1$$

Example 4. **If a, b, c and d are in GP, then show that $(a^2 + b^2 + c^2)(b^2 + c^2 + d^2) = (ab + bc + cd)^2$.** NCERT, Competency Based Que

Sol. Given, a, b, c, d are in GP.

$$\therefore \quad \frac{b}{a} = \frac{c}{b} = \frac{d}{c} = r \text{ (let)}$$

$$\Rightarrow \quad b = ar,\ c = br,\ d = cr$$

$$\Rightarrow \quad b = ar,\ c = (ar)r,\ d = (br)r$$

$$\Rightarrow \quad b = ar,\ c = ar^2,\ d = br^2$$

$$\Rightarrow \quad b = ar,\ c = ar^2,\ d = (ar)r^2 = ar^3 \quad \dots(i)$$

Now, we have to prove that

$$(a^2 + b^2 + c^2)(b^2 + c^2 + d^2) = (ab + bc + cd)^2$$

$$\text{LHS} = (a^2 + b^2 + c^2)(b^2 + c^2 + d^2)$$

$$= (a^2 + a^2r^2 + a^2r^4)(a^2r^2 + a^2r^4 + a^2r^6)$$

$$= a^2(1 + r^2 + r^4)a^2r^2(1 + r^2 + r^4)$$

$$= a^4r^2(1 + r^2 + r^4)^2 = [a^2r(1 + r^2 + r^4)]^2$$

$$= [a^2r + a^2r^3 + a^2r^5]^2$$

$$= [a \cdot ar + ar \cdot ar^2 + ar^2 \cdot ar^3]^2$$

$$= [ab + bc + cd]^2 \quad \text{[from Eq. (i)]}$$

$$= \text{RHS}$$

Hence proved.

Example 5. **Find the 12th term of the sequence – 6, 18, – 54, ...**

Sol. Given, sequence is – 6, 18, – 54, ...

Clearly, the successive ratio of the terms is same.

So, the given sequence forms a GP with first term, $a=-6$ and common ratio, $r=\frac{18}{-6}=-3$.

$\therefore \quad T_{12}=(-6)(-3)^{12-1} \qquad [\because T_n=ar^{n-1}]$

$=(-1)^{12}6\cdot 3^{12-1}=(1)\cdot 6\cdot 3^{11}$

$=2\cdot 3\cdot 3^{11}=2\cdot 3^{12}$

Hence, the nth term of given GP is $(-1)^n 6\cdot 3^{n-1}$ and 12th term is $2\cdot 3^{12}$.

Example 6. Which term of the GP 5, 10, 20, 40, ... is 5120?

Sol. Given, GP is 5, 10, 20, 40, ...

Here, $a=5$ and $r=\frac{10}{5}=2$

Let nth term of given GP $=5120$ i.e. $T_n=5120$

Now, $T_n=ar^{n-1}=5120$

$\Rightarrow \quad 5(2)^{n-1}=5120 \qquad [\because a=5 \text{ and } r=2]$

$\Rightarrow \quad 2^{n-1}=\frac{5120}{5}=1024$

$\Rightarrow \quad 2^{n-1}=1024$

$\Rightarrow \quad 2^{n-1}=2^{10}$

On equating the powers, we get

$n-1=10$

$\Rightarrow \quad n=10+1=11$

Hence, 11th term of given GP is 5120.

Example 7. If the 4th and 9th term of a GP are 54 and 13122 respectively, then find the GP.

Sol. Let a be the first term and r be the common ratio of GP.

Given, 4th term, $T_4=54$

$\Rightarrow \quad ar^{4-1}=54$

$\Rightarrow \quad ar^3=54 \qquad \text{...(i)}$

and 9th term, $T_9=13122$

$\Rightarrow \quad ar^{9-1}=13122$

$\Rightarrow \quad ar^8=13122 \qquad \text{...(ii)}$

On dividing Eq. (ii) by Eq. (i), we get

$$\frac{ar^8}{ar^3}=\frac{13122}{54}$$

$\Rightarrow \quad r^5=243$

$\Rightarrow \quad r^5=(3)^5$

$\therefore \quad r=3$

On putting the value of r in Eq. (i), we get

$a(3)^3=54$

$\Rightarrow \quad 27a=54$

$\Rightarrow \quad a=\frac{54}{27}=2$

$\therefore$ Required GP is $a, ar, ar^2, ar^3, \ldots$ i.e. 2, 6, 18, 54, ...

Example 8. The first term of a GP is 1. The sum of the third and fifth terms is 90. Find the common ratio of the GP.

Sol. Let r be the common ratio of the GP.

It is given that first term of GP is 1.

$\therefore$ Third term of GP i.e. $a_3=ar^2=r^2$

and fifth term of GP i.e. $a_5=ar^4=r^4$

Given, $a_3+a_5=90$

$\therefore \quad r^2+r^4=90$

$\Rightarrow \quad r^4+r^2-90=0$

$\Rightarrow \quad r^4+10r^2-9r^2-90=0$

$\Rightarrow \quad (r^2+10)(r^2-9)=0$

$\Rightarrow \quad r^2=9,\ r^2\neq -10$

$\Rightarrow \quad r=\pm 3$

Hence, the common ratio of the given GP is -3 or 3.

Example 9. Find four numbers forming a GP in which the third term is greater than the first term by 9 and the second term is greater than 4th by 18.

Sol. Let the GP be $a, ar, ar^2, ar^3, \ldots$

Given, third term = First term + 9

$\Rightarrow \quad T_3=a+9$

$\Rightarrow \quad ar^2=a+9$

$\Rightarrow \quad ar^2-a=9 \qquad \text{...(i)}$

Again, second term = Fourth term + 18

$T_2=T_4+18$

$\Rightarrow \quad ar=ar^3+18$

$ar-ar^3=18 \qquad \text{...(ii)}$

On dividing Eq. (i) by Eq. (ii), we get

$$\frac{ar^2-a}{ar-ar^3}=\frac{9}{18}$$

$$\Rightarrow \quad \frac{a(r^2-1)}{ar(1-r^2)}=\frac{1}{2}$$

$$\Rightarrow \quad \frac{-1}{r}\frac{(1-r^2)}{(1-r^2)}=\frac{1}{2}$$

$$\Rightarrow \quad -\frac{1}{r}=\frac{1}{2}$$

$\Rightarrow \quad r=-2$

On putting $r = -2$ in Eq. (ii), we get

$$a(-2) - a(-2)^3 = 18$$

$\Rightarrow \quad -2a + 8a = 18$

$\Rightarrow \quad 6a = 18$

$\Rightarrow \quad a = 3$

$\therefore$ GP is $3, 3(-2), 3(-2)^2, 3(-2)^3, \dots$

i.e. $\quad 3, -6, 12, -24, \dots$

Example 10. **The $(m+n)$th and $(m-n)$th terms of a GP are p and q, respectively. Show that the mth and nth terms are $\sqrt{pq}$ and $p\left(\frac{q}{p}\right)^{m/2n}$, respectively.**

Competency Based Que

Sol. Let a be the first term and r be the common ratio.

Then, $\quad a_{m+n} = p$ and $a_{m-n} = q$

$\Rightarrow \quad ar^{m+n-1} = p$ and $ar^{m-n-1} = q$

$\Rightarrow \quad \frac{ar^{m+n-1}}{ar^{m-n-1}} = \frac{p}{q} \Rightarrow r^{2n} = \frac{p}{q}$

$\Rightarrow \quad r = \left(\frac{p}{q}\right)^{1/2n} \Rightarrow \frac{1}{r} = \left(\frac{q}{p}\right)^{1/2n}$

Now, $\quad a_m = ar^{m-1}$

$\Rightarrow \quad a_m = ar^{(m+n-1)}\left(\frac{1}{r}\right)^n$

$\Rightarrow \quad a_m = a_{m+n}\left(\frac{1}{r}\right)^n \quad [\because a_{m+n} = ar^{m+n-1}]$

$\Rightarrow \quad a_m = p\left(\frac{q}{p}\right)^{n/2n} \quad \left[\because a_{m+n} = p \text{ and } \frac{1}{r} = \left(\frac{q}{p}\right)^{1/2n}\right]$

$\Rightarrow \quad a_m = p\left(\frac{q}{p}\right)^{1/2} = \sqrt{pq}$

and $\quad a_n = ar^{n-1}$

$\Rightarrow \quad a_n = ar^{m+n-1}\left(\frac{1}{r}\right)^m = a_{m+n}\left(\frac{1}{r}\right)^m$

$[\because a_{m+n} = ar^{m+n-1}]$

$\therefore \quad a_n = p\left(\frac{q}{p}\right)^{m/2n} \quad \left[\because a_{m+n} = p \text{ and } \frac{1}{r} = \left(\frac{q}{p}\right)^{1/2n}\right]$

Example 11. **Find the 8th term from the end of the sequence 3, 6, 12, ..., 25th term.**

Competency Based Que

Sol. We know that if a sequence has n terms, then mth term from end is equal to $(n-m+1)$th term from the beginning.

Here, $a = 3$, $r = \frac{6}{3} = 2$, $m = 8$ and $n = 25$.

Now, 8th term from the end of sequence is equal to the $(25-8+1)$ i.e. 18th term from the beginning.

$\because \quad T_n = ar^{n-1}$

$\therefore \quad T_{18} = ar^{18-1} = 3(2)^{18-1} \quad [\because a = 3 \text{ and } r = 2]$

$= 3(2)^{17} = 3 \times 131072 = 393216$

Hence, the 8th term from the end is 393216.

Example 12. **Find four numbers in GP, whose sum is 85 and product is 4096.**

Sol. Let the four numbers in GP be

$$\frac{a}{r^3}, \frac{a}{r}, ar, ar^3 \quad \dots(i)$$

Product of four numbers = 4096 [given]

$\Rightarrow \quad \left(\frac{a}{r^3}\right)\left(\frac{a}{r}\right)(ar)(ar^3) = 4096$

$\Rightarrow \quad a^4 = 4096 \Rightarrow a^4 = 8^4$

On comparing the base of the power 4, we get

$$a = 8$$

and sum of four numbers = 85 [given]

$\Rightarrow \quad \frac{a}{r^3} + \frac{a}{r} + ar + ar^3 = 85$

$\Rightarrow \quad a\left[\frac{1}{r^3} + \frac{1}{r} + r + r^3\right] = 85$

$\Rightarrow \quad 8\left[r^3 + \frac{1}{r^3}\right] + 8\left[r + \frac{1}{r}\right] = 85 \quad [\because a = 8]$

$\Rightarrow \quad 8\left[\left(r + \frac{1}{r}\right)^3 - 3\left(r + \frac{1}{r}\right)\right] + 8\left(r + \frac{1}{r}\right) = 85$

$[\because a^3 + b^3 = (a+b)^3 - 3ab(a+b)]$

$\Rightarrow \quad 8\left(r + \frac{1}{r}\right)^3 - 16\left(r + \frac{1}{r}\right) - 85 = 0 \quad \dots(ii)$

On putting $\left(r + \frac{1}{r}\right) = x$ in Eq. (ii), we get

$$8x^3 - 16x - 85 = 0$$

$\Rightarrow (2x-5)(4x^2 + 10x + 17) = 0 \Rightarrow \quad 2x - 5 = 0$

$[\because 4x^2 + 10x + 17 = 0 \text{ has imaginary roots}]$

$\Rightarrow \quad x = \frac{5}{2} \Rightarrow r + \frac{1}{r} = \frac{5}{2} \quad \left[\because x = r + \frac{1}{r}\right]$

$\Rightarrow \quad 2r^2 - 5r + 2 = 0$

$\Rightarrow \quad (r-2)(2r-1) = 0 \Rightarrow r = 2 \text{ or } r = \frac{1}{2}$

On putting $a = 8$ and $r = 2$ or $r = \frac{1}{2}$ in Eq. (i), we obtain four numbers as

$$\frac{8}{2^3}, \frac{8}{2}, 8 \times 2, 8 \times 2^3 \text{ or } \frac{8}{(1/2)^3}, \frac{8}{(1/2)}, 8 \times \frac{1}{2}, 8 \times \left(\frac{1}{2}\right)^3$$

i.e. 1, 4, 16, 64 or 64, 16, 4, 1.

Quick Test 1

1. Find the 7th term of the sequence, whose nth term is $a_n = \frac{n^2}{2^n}$.

2. Write the first five terms of each of the following sequence whose nth terms are given below. NCERT, Competency Based Que

(i) $a_n = \frac{n}{n+1}$ (ii) $a_n = (-1)^{n+1} 3^{n+2}$

3. The Fibonacci sequence is defined by $1 = a_1 = a_2$ and $a_n = a_{n-1} + a_{n-2}, n > 2$. Find $\frac{a_{n+1}}{a_n}$, for $n = 1, 2, 3, 4, 5$. NCERT, Competency Based Que

4. For what values of x are the numbers $(x+9), (x-6)$ and 4 are in GP?

5. If a, b, c and d are in GP, prove that $(a^2 - b^2), (b^2 - c^2)$ and $(c^2 - d^2)$ are also in GP. Competency Based Que

6. Find the following

(i) 11th term of the GP 3, 6, 12, 24, ...

(ii) 12th term of the GP $\frac{1}{a^3 x^3}, ax, a^5 x^5, \ldots$

7. Find 12th term of a GP, whose 8th term is 192 and common ratio is 2.

8. Find the geometric series whose 5th and 8th terms are 80 and 640, respectively.

9. Find the 6th term from the end of the GP $8, 4, 2, \ldots, \frac{1}{1024}$. Competency Based Que

10. The sum of first three terms of a GP is $\frac{39}{10}$ and their product is 1. Find the common ratio and the terms. Competency Based Que

Answers

1. $\frac{49}{128}$ 2. (i) $\frac{1}{2}, \frac{2}{3}, \frac{3}{4}, \frac{4}{5}, \frac{5}{6}$ (ii) 27, −81, 243, −729, 2187

3. The terms are $1, 2, \frac{3}{2}, \frac{5}{3}$ and $\frac{8}{5}$. 4. 0 or 16

6. (i) 3072 (ii) $(ax)^{41}$ 7. 3072

8. $5 + 10 + 20 + 40 + \ldots$

9. $\frac{1}{32}$ 10. $\frac{2}{5}$ or $\frac{5}{2}$; $\frac{5}{2}, 1, \frac{2}{5}$ or $\frac{2}{5}, 1, \frac{5}{2}$

For Solutions, See QR Code on Pg. 223

8.3 Sum of *n* Terms of a GP

If a and r are the first term and common ratio of a GP respectively, then sum of n terms of this GP is given by

$$S_n = \frac{a(1-r^n)}{1-r}, \text{ where } r < 1$$

or

$$S_n = \frac{a(r^n - 1)}{r-1}, \text{ where } r > 1 \, (r \neq 1)$$

Suppose a GP contains n terms with first term = a, common ratio = r and last term = l.

The sum of GP

$$S_n = \frac{a - lr}{1-r}, \text{ where } r < 1 \text{ or } S_n = \frac{lr - a}{r-1}, \text{ where } r > 1$$

Sum of an Infinite GP

If a is the first term and r is the common ratio of a GP such that $|r| < 1$, then sum of an infinite GP is given by

$$S = \frac{a}{1-r}$$

8.4 Geometric Mean (GM)

Let a and b be two positive numbers. If we insert a single number G between them so that a, G, b is a GP, then G is called **geometric mean**. Thus, a, G, b are in GP.

$$\therefore \quad \frac{G}{a} = \frac{b}{G} \Rightarrow G^2 = ab$$

$$\Rightarrow \quad G = \sqrt{ab}$$

Inserting *n* GM's between Two Positive Numbers

Let a and b be two positive numbers and $G_1, G_2, G_3, \ldots, G_n$ be the n GM's between a and b.

Then, $a, G_1, G_2, G_3, \ldots, G_n, b$ are in GP with common ratio r and number of terms is $n+2$.

Here, b is $(n+2)$th term

i.e. $b = T_{n+2} = ar^{n+1} \Rightarrow r = \left(\frac{b}{a}\right)^{\frac{1}{n+1}}$

Thus, the n geometric means are given below.

$$G_1 = ar = a\left(\frac{b}{a}\right)^{\frac{1}{n+1}}$$

$$G_2 = ar^2 = a\left(\frac{b}{a}\right)^{\frac{2}{n+1}}$$

....

....

$$G_n = ar^n = a\left(\frac{b}{a}\right)^{\frac{n}{n+1}}$$

Relation between AM and GM

Let A and G be the AM and GM of two positive real numbers a and b, respectively.

Then, $A = \frac{a+b}{2}$ and $G = \sqrt{ab}$

Now, $A - G = \frac{a+b}{2} - \sqrt{ab} = \frac{a+b-2\sqrt{ab}}{2}$

$$= \frac{(\sqrt{a}-\sqrt{b})^2}{2} \geq 0$$

$\therefore \quad A \geq G$

TEACHER TIPS

- Three numbers p, q, r are said to be in geometric progression if $\frac{q}{p} = \frac{r}{q} \Rightarrow q^2 = pr$ i.e. square of the middle term is equal to product of the extremes.
- In a GP, if number of terms is odd, then assume the middle term as a and common difference as r.
- If number of terms is even, then assume the middle terms as $\frac{a}{r}$, ar and the common ratio as r^2.
- If the terms of the GP are not given, then the terms are chosen as $a, ar, ar^2, ar^3, \ldots$.
- If the ratio of the AM and GM of two positive numbers a and b is $m:n$, then $a:b = (m+\sqrt{m^2-n^2}):(m-\sqrt{m^2-n^2})$

Example 1. Find the sum of the following sequences

(i) $\frac{1}{2}, \frac{3}{2}, \frac{9}{2}, \ldots$, upto 10 terms.

(ii) $2, -\frac{1}{2}, \frac{1}{8}, \ldots$, upto 12 terms.

Sol. (i) We have, sequence $\frac{1}{2}, \frac{3}{2}, \frac{9}{2}, \ldots$

Clearly, the successive ratio of the terms is same.

So, the above sequence forms a GP, with first term, $a = \frac{1}{2}$ and common ratio, $r = \frac{3}{2} \div \frac{1}{2} = 3$.

$$\therefore \quad S_{10} = \frac{\frac{1}{2}[3^{10}-1]}{3-1} \quad \left[\because S_n = \frac{a(r^n-1)}{r-1} \text{ as } r>1\right]$$

$$= \frac{1}{2}\frac{(3^{10}-1)}{2} = \frac{1}{4}[3^{10}-1] = 14762$$

(ii) We have, sequence $2, -\frac{1}{2}, \frac{1}{8}, \ldots$

Clearly, the successive ratio of the terms is same.

So, the above sequence forms a GP with first term, $a = 2$ and common ratio, $r = -\frac{1}{2} \div 2 = \frac{-1}{4}$.

$$\therefore S_{12} = \frac{2\left[1-\left(\frac{-1}{4}\right)^{12}\right]}{1-\left(\frac{-1}{4}\right)} \quad \left[\because S_n = \frac{a(1-r^n)}{1-r} \text{ as } r<1\right]$$

$$= \frac{2\left[1-\frac{1}{(4)^{12}}\right]}{1+\frac{1}{4}} = \frac{2\left[1-\left(\frac{1}{4}\right)^{12}\right]}{\left(\frac{5}{4}\right)} = \frac{8}{5}\left[1-\left(\frac{1}{4}\right)^{12}\right]$$

Example 2. Find the sum of the series $2 + 6 + 18 + 54 + \ldots + 4374$.

Competency Based Que

Sol. Given series is $2+6+18+54+\ldots+4374$.

Clearly, the successive ratio of the terms is same.

So, the above series forms a GP with first term, $a = 2$, common ratio, $r = \frac{6}{2} = 3 > 1$ and last term, $l = 4374$.

$$\therefore \text{Required sum} = \frac{(lr-a)}{(r-1)} = \frac{(4374\times 3-2)}{(3-1)}$$

$$= \frac{13120}{2} = 6560$$

Hence, the sum of the given series is 6560.

Example 3. How many terms of GP $3, \frac{3}{2}, \frac{3}{4}, \ldots$ are needed to give the sum $\frac{3069}{512}$?

NCERT

Sol. Given GP is $3, \frac{3}{2}, \frac{3}{4}, \ldots$

Here, $a = 3$ and $r = \frac{3}{2} \div 3 = \frac{1}{2}$

Let n be the number of terms needed.

Then, $S_n = \frac{3069}{512}$

$\Rightarrow \frac{a(1-r^n)}{1-r} = \frac{3069}{512}$ $[\because r<1]$

$\Rightarrow \frac{3\left\{1-\frac{1}{2^n}\right\}}{1-\frac{1}{2}} = \frac{3069}{512}$

$\Rightarrow 6\left(1-\frac{1}{2^n}\right) = \frac{3069}{512} \Rightarrow 1-\frac{1}{2^n} = \frac{3069}{3072}$

$\Rightarrow \frac{1}{2^n} = 1-\frac{3069}{3072} = \frac{3072-3069}{3072}$

$\Rightarrow \frac{1}{2^n} = \frac{3}{3072} = \frac{1}{1024}$

$\Rightarrow 2^n = 1024 \Rightarrow 2^n = 2^{10}$

On comparing the powers from both sides, we get

$n = 10$

Hence, 10 terms are needed to give the sum $\frac{3069}{512}$.

Example 4. Find the sum of the series $4+44+444+\ldots+n$ terms.

Sol. Given series is $4+44+444+\ldots+n$ terms.

On taking common 4, we get

$S_n = 4(1+11+111+\ldots+n \text{ terms})$

On multiplying and dividing each term of RHS by 9, we get

$S_n = \frac{4}{9}(9+99+999+\ldots+n \text{ terms})$

$\Rightarrow S_n = \frac{4}{9}[(10-1)+(100-1)+(1000-1)+\ldots n \text{ terms}]$

$\Rightarrow S_n = \frac{4}{9}[(10+100+1000+\ldots+n \text{ terms}) - (1+1+1+\ldots+n \text{ terms})]$ ...(i)

For first series, $10+100+1000+\ldots+n$ terms

$r = \frac{100}{10} = 10 > 1$

$\therefore$ Sum $= 10(1+10+10^2+\ldots+n \text{ terms})$

$= 10\left(\frac{10^n-1}{10-1}\right)$ $\left[\because \text{sum} = \frac{a(r^n-1)}{r-1}\right]$

$= \frac{10}{9}(10^n-1)$

Now, from Eq. (i), we get

$S_n = \frac{4}{9}\left[\frac{10}{9}(10^n-1)-n\right] = \frac{40}{81}(10^n-1)-\frac{4}{9}n$

Example 5. Find the sum of the series $0.7+0.77+0.777+\ldots$ upto n terms.

Sol. Given series is $0.7+0.77+0.777+\ldots$ upto n terms

Let $S_n = 0.7+0.77+0.777+\ldots$ upto n terms

$\Rightarrow S_n = 7[0.1+0.11+0.111+0.1111+\ldots \text{ upto } n \text{ terms}]$

$\Rightarrow S_n = \frac{7}{9}[0.9+0.99+0.999+\ldots \text{ upto } n \text{ terms}]$

$\Rightarrow S_n = \frac{7}{9}\left[\frac{9}{10}+\frac{99}{100}+\frac{999}{1000}+\ldots \text{upto } n \text{ terms}\right]$

$\Rightarrow S_n = \frac{7}{9}\left[1-\frac{1}{10}+1-\frac{1}{100}+1-\frac{1}{1000}+\ldots \text{upto } n \text{ terms}\right]$

$\Rightarrow S_n = \frac{7}{9}\left[1+1+\ldots \text{upto } n \text{ terms} - \left(\frac{1}{10}+\frac{1}{10^2}+\frac{1}{10^3}+\ldots \text{upto } n \text{ terms}\right)\right]$

$\Rightarrow S_n = \frac{7}{9}\left[n-\frac{1}{10}\frac{\left(1-\frac{1}{10^n}\right)}{\left(1-\frac{1}{10}\right)}\right]$

$\Rightarrow S_n = \frac{7}{9}\left[n-\frac{(1-10^{-n})}{9}\right]$

$\Rightarrow S_n = \frac{7}{9}\left[\frac{9n-1+10^{-n}}{9}\right]$

$\therefore S_n = \frac{7}{81}[9n-1+10^{-n}]$

Example 6. Find the sum to n terms of the sequence given by $a_n = 2^n+3n, n \in N$. *Competency Based Que*

Sol. Let S_n denote the sum to n terms of the given sequence. Then, $S_n = a_1+a_2+a_3+\ldots+a_n$

$\Rightarrow S_n = (2^1+3\times 1)+(2^2+3\times 2)+(2^3+3\times 3)+\ldots+(2^n+3\times n)$

$\Rightarrow S_n = (2^1+2^2+2^3+\ldots+2^n)+(3\times 1+3\times 2+3\times 3+\ldots+3\times n)$

$= (2^1+2^2+2^3+\ldots+2^n)+3(1+2+3+\ldots+n)$

$= 2\left(\frac{2^n-1}{2-1}\right)+3\left\{\frac{n}{2}(1+n)\right\}$

$\left[\because 2+2^2+2^3+2^n \text{ is a GP and } \Sigma n = \frac{n(n+1)}{2}\right]$

$= 2(2^n-1)+\frac{3n}{2}(n+1)$

Example 7. **Find the sum of series $\frac{3}{5}+\frac{4}{5^2}+\frac{3}{5^3}+\frac{4}{5^4}+...$ + 2n terms.**

Sol. We have, $S=\frac{3}{5}+\frac{4}{5^2}+\frac{3}{5^3}+\frac{4}{5^4}+...+2n$ terms

$$=\left[\frac{3}{5}+\frac{3}{5^3}+...+n\text{ terms}\right]+\left[\frac{4}{5^2}+\frac{4}{5^4}+...+n\text{ terms}\right]$$

$$=\frac{\frac{3}{5}\left[1-\left(\frac{1}{5^2}\right)^n\right]}{\left[1-\left(\frac{1}{5^2}\right)\right]}+\frac{\frac{4}{5^2}\left[1-\left(\frac{1}{5^2}\right)^n\right]}{\left[1-\left(\frac{1}{5^2}\right)\right]}$$

$$\left[\because \text{sum of GP}=\frac{a(1-r^n)}{1-r}, |r|<1\right]$$

$$=\frac{\frac{3}{5}\left[1-\frac{1}{5^{2n}}\right]}{1-\frac{1}{25}}+\frac{\frac{4}{25}\left[1-\frac{1}{5^{2n}}\right]}{1-\frac{1}{25}}$$

$$=\frac{\frac{3}{5}\left[1-\frac{1}{5^{2n}}\right]}{\frac{24}{25}}+\frac{\frac{4}{25}\left[1-\frac{1}{5^{2n}}\right]}{\frac{24}{25}}$$

$$=\frac{5}{8}\left[1-\frac{1}{5^{2n}}\right]+\frac{1}{6}\left[1-\frac{1}{5^{2n}}\right]$$

$$=\left[1-\frac{1}{5^{2n}}\right]\left[\frac{5}{8}+\frac{1}{6}\right]=\left[1-\frac{1}{5^{2n}}\right]\left[\frac{15+4}{24}\right]$$

$$=\frac{19}{24}\left[1-\frac{1}{5^{2n}}\right]$$

Example 8. **A GP consists of an even number of terms. If the sum of all the terms is 5 times the sum of terms occupying odd places, then find its common ratio.** Competency Based Que

Sol. Let the GP be $a, ar, ar^2, ar^3, ar^4, ..., ar^{2n-2}, ar^{2n-1}$, where $a, ar^2, ar^4, ar^6, ...$ occupy odd places and $ar, ar^3, ar^5, ar^7 ...$ occupy even places.

According to the question,

Sum of all terms = 5 × Sum of terms occupying odd places

i.e. $a+ar+ar^2+...+ar^{2n-1}$

$$=5\times(a+ar^2+ar^4+...+ar^{2n-2})$$

$$\Rightarrow \frac{a(r^{2n}-1)}{r-1}=\frac{5a[(r^2)^n-1]}{r^2-1}\quad\left[\because S_n=\frac{a(r^n-1)}{r-1}, r>1\right]$$

$$\Rightarrow \frac{r^{2n}-1}{r-1}=\frac{5(r^{2n}-1)}{(r-1)(r+1)}$$

$$\Rightarrow 1=\frac{5}{r+1}$$

$$\Rightarrow r+1=5$$

$$\therefore r=4$$

Example 9. **The sum of the first three terms of a GP is 16 and the sum of next three terms is 128. Determine the first term, the common ratio and the sum to n terms of the GP.** NCERT

Sol. Let the GP be $a, ar, ar^2, ar^3, ...$

According to the given condition,

sum of first three terms $=a+ar+ar^2=16$ ...(i)

and sum of next three terms

$=ar^3+ar^4+ar^5=128$ ...(ii)

On dividing Eq. (i) by Eq. (ii), we get

$$\frac{a+ar+ar^2}{ar^3+ar^4+ar^5}=\frac{16}{128}$$

$$\Rightarrow \frac{a(1+r+r^2)}{ar^3(1+r+r^2)}=\frac{1}{8}$$

$$\Rightarrow \left(\frac{1}{r}\right)^3=\left(\frac{1}{2}\right)^3$$

On comparing the base of the power 3 from both sides, we get

$$\frac{1}{r}=\frac{1}{2}\Rightarrow r=2$$

On putting $r=2$ in Eq. (i), we get

$$a+2a+4a=16$$

$$\Rightarrow 7a=16\Rightarrow a=\frac{16}{7}$$

Now, sum of n terms, $S_n=\frac{a(r^n-1)}{r-1}$ $[\because r=2>1]$

$$=\frac{\frac{16}{7}(2^n-1)}{2-1}=\frac{16}{7}(2^n-1)$$

Hence, $a=\frac{16}{7}$, $r=2$ and $S_n=\frac{16}{7}(2^n-1)$

Example 10. **If S_1, S_2 and S_3 are respectively the sum of n, $2n$ and $3n$ terms of a GP, then prove that $S_1(S_3-S_2)=(S_2-S_1)^2$.** Competency Based Que

Sol. Let a be the first term and r be the common ratio of the given GP. Then,

$$S_1(S_3-S_2)=\frac{a(1-r^n)}{(1-r)}\cdot\left\{\frac{a(1-r^{3n})}{(1-r)}-\frac{a(1-r^{2n})}{(1-r)}\right\}$$

$$=\frac{a(1-r^n)}{(1-r)}\cdot\frac{(a-ar^{3n}-a+ar^{2n})}{(1-r)}$$

$$= \frac{a(1-r^n)}{(1-r)} \cdot \frac{ar^{2n}(1-r^n)}{(1-r)} = \frac{a^2r^{2n}(1-r^n)^2}{(1-r)^2} \quad ...(i)$$

And $$(S_2 - S_1)^2 = \left\{\frac{a(1-r^{2n})}{(1-r)} - \frac{a(1-r^n)}{(1-r)}\right\}^2$$

$$= \frac{(a - ar^{2n} - a + ar^n)^2}{(1-r)^2}$$

$$= \frac{\{(ar^n(1-r^n)\}^2}{(1-r)^2} = \frac{a^2r^{2n}(1-r^n)^2}{(1-r)^2} \quad ...(ii)$$

Hence, from Eqs. (i) and (ii), we get

$$S_1(S_3 - S_2) = (S_2 - S_1)^2 \quad \textbf{Hence proved.}$$

Example 11. Find the sum of the series $9^{1/3} \cdot 9^{1/9} \cdot 9^{1/27} \ldots \infty$. *Competency Based Que*

Sol. Let $E = 9^{\frac{1}{3}} \cdot 9^{\frac{1}{9}} \cdot 9^{\frac{1}{27}} \ldots \infty = 9^{\frac{1}{3}+\frac{1}{9}+\frac{1}{27}+\ldots\infty}$ ...(i)

Again, let $S_\infty = \frac{1}{3} + \frac{1}{9} + \frac{1}{27} + \ldots + \infty$

Clearly, the successive ratio of the terms is same. So, the above obtained series forms a GP, with first term, $a = \frac{1}{3}$ and common ratio, $\frac{1}{9} \div \frac{1}{3} = \frac{1}{3}$

Then, $$S_\infty = \frac{\frac{1}{3}}{1 - \frac{1}{3}} = \frac{1}{2} \quad \left[\because S_\infty = \frac{a}{1-r}\right]$$

Then, from Eq. (i), we get

$$E = 9^{\frac{1}{2}} = 3$$

Example 12. If $|x| < 1$ and $|y| < 1$, then find the sum of infinity of the following series $(x+y)+(x^2+xy+y^2)+(x^3+x^2y+xy^2+y^3)+\ldots$ upto ∞. *Competency Based Que*

Sol. We have,

$$(x+y)+(x^2+xy+y^2)+(x^3+x^2y+xy^2+y^3) + \ldots \text{ upto } \infty$$

$$= \frac{1}{x-y}[(x^2-y^2)+(x^3-y^3)+(x^4-y^4) + \ldots \text{ upto } \infty]$$

$$\left[\because \frac{x^n - a^n}{x-a} = x^{n-1} \cdot a^0 + x^{n-2} \cdot a + x^{n-3} \cdot a^2 + \ldots + a^{n-1}, n \in N\right]$$

$$= \frac{1}{x-y}\{(x^2+x^3+x^4+\ldots \text{ upto } \infty) - (y^2+y^3+y^4+\ldots \text{ upto } \infty)\}$$

$$= \frac{1}{x-y}\left\{\frac{x^2}{1-x} - \frac{y^2}{1-y}\right\}$$

$$= \frac{1}{x-y} \frac{\{x^2 - x^2y - y^2 + y^2x\}}{(1-x)(1-y)}$$

$$= \frac{1}{(x-y)} \frac{\{(x^2-y^2) - xy(x-y)\}}{(1-x)(1-y)}$$

$$= \frac{x+y-xy}{(1-x)(1-y)}$$

Example 13. At the end of each year, the value of a certain machine has depreciated by 20% of its value at the beginning of that year. If its initial value was ₹ 1250, then find the value at the end of 5 yr. *NCERT Exemplar, Competency Based Que*

Sol. Given, depreciation in value of machine = 20%

After each year, the value of the machine is 80% (100 − 20)% of its value of the previous year, so at the end of 5 yr, the machine will depreciate as many times as 5.

Hence, we have to find the 6th terms of the GP whose first term a_1 is 1250 and common ratio r is 0.8.

Hence, value at the end 5 yr $= t_6 = a_1r^5 \quad [\because T_n = ar^{n-1}]$

$$= 1250\,(0.8)^5$$

$$= 409.6$$

Example 14. The lengths of three unequal edges of a rectangular solid block are in GP. If the volume of the block is 216 cm^3 and the total surface area is 252 cm^2, then find the length of its edges. *NCERT Exemplar*

Sol. Let the length, breadth and height of rectangular solid block be $\frac{a}{r}$, a and ar, respectively.

$\therefore$ Volume $= \frac{a}{r} \times a \times ar = 216 \text{ cm}^3$

$\Rightarrow \quad a^3 = 216$

$\Rightarrow \quad a^3 = 6^3$

$\therefore \quad a = 6$

Surface area $= 2\left(\frac{a^2}{r} + a^2 r + a^2\right) = 252$

$\Rightarrow \quad 2a^2\left(\frac{1}{r} + r + 1\right) = 252$

$\Rightarrow \quad 2 \times 36\left(\frac{1 + r^2 + r}{r}\right) = 252$

$\Rightarrow \quad \frac{1 + r^2 + r}{r} = \frac{252}{2 \times 36}$

$$\Rightarrow \quad 1+r^2+r=\frac{126}{36}r$$

$$\Rightarrow \quad 1+r^2+r=\frac{21}{6}r$$

$$\Rightarrow \quad 6+6r^2+6r=21r$$

$$\Rightarrow \quad 6r^2-15r+6=0$$

$$\Rightarrow \quad 2r^2-5r+2=0$$

$$\Rightarrow \quad (2r-1)(r-2)=0$$

$$\Rightarrow \quad r=\frac{1}{2} \text{ or } 2$$

For $r=\frac{1}{2}$, length $=\frac{a}{r}=\frac{6\times 2}{1}=12$ cm,

breadth $=a=6$ cm

and height $=ar=6\times\frac{1}{2}=3$ cm

For $r=2$, length $=\frac{a}{r}=\frac{6}{2}=3$ cm,

breadth $=a=6$ cm

and height $=ar=6\times 2=12$ cm

Example 15. Find a geometric mean of 4 and 16.

Sol. Geometric mean between 4 and 16 $=\sqrt{4\times 16}$

$$=\sqrt{64}=8$$

Example 16. Insert 4 GM's between 3 and 96.

Sol. Let G_1, G_2, G_3 and G_4 be the required GM's.

Then, $3, G_1, G_2, G_3, G_4, 96$ are in GP.

Let r be the common ratio. Here, 96 is the 6th term.

$$\therefore \quad 96=ar^{6-1}=3r^5$$

$$\Rightarrow \quad 32=r^5 \Rightarrow (2)^5=r^5 \Rightarrow r=2$$

$$\therefore \quad G_1=ar=3\cdot 2=6$$

$$G_2=ar^2=3\cdot 2^2=12$$

$$G_3=ar^3=3\cdot 2^3=24$$

and $G_4=ar^4=3\cdot 2^4=48$

Example 17. Find two positive numbers whose difference is 12 and whose AM exceeds the GM by 2. *Competency Based Que*

Sol. Let the two numbers be a and b such that $a>b$.

Given, $a-b=12$ ...(i)

and AM − GM = 2

$$\Rightarrow \quad \frac{a+b}{2}-\sqrt{ab}=2 \quad \left[\because \text{AM}=\frac{a+b}{2} \text{ and GM}=\sqrt{ab}\right]$$

$$\Rightarrow \quad a+b-2\sqrt{ab}=4 \Rightarrow (\sqrt{a}-\sqrt{b})^2=4$$

$$\Rightarrow \quad \sqrt{a}-\sqrt{b}=2 \quad \text{...(ii)}$$

Now, $a-b=12 \Rightarrow (\sqrt{a}+\sqrt{b})(\sqrt{a}-\sqrt{b})=12$

$[\because x^2-y^2=(x-y)(x+y)]$

$$\Rightarrow \quad (\sqrt{a}+\sqrt{b})\times(2)=12 \quad \text{[using Eq. (ii)]}$$

$$\Rightarrow \quad \sqrt{a}+\sqrt{b}=6 \quad \text{...(iii)}$$

On adding Eqs. (ii) and (iii), we get

$$2\sqrt{a}=8 \Rightarrow \sqrt{a}=4$$

$$\Rightarrow \quad a=(4)^2 \Rightarrow a=16$$

Then, from Eq. (i), we get

$$16-b=12 \Rightarrow b=16-12=4$$

Hence, the required numbers are 16 and 4.

Example 18. The sum of two numbers is 6 times their geometric mean, show that the numbers are in the ratio $(3+2\sqrt{2}):(3-2\sqrt{2})$. *NCERT*

Sol. Let the numbers be a and b.

According to the question, $a+b=6\sqrt{ab}$

$$\Rightarrow \quad \frac{a+b}{2\sqrt{ab}}=\frac{3}{1}$$

Now, on applying componendo and dividendo rule, we get

$$\frac{a+b+2\sqrt{ab}}{a+b-2\sqrt{ab}}=\frac{3+1}{3-1}$$

$$\Rightarrow \quad \frac{(\sqrt{a})^2+(\sqrt{b})^2+2\sqrt{ab}}{(\sqrt{a})^2+(\sqrt{b})^2-2\sqrt{ab}}=\frac{4}{2}$$

$$\Rightarrow \quad \frac{(\sqrt{a}+\sqrt{b})^2}{(\sqrt{a}-\sqrt{b})^2}=\frac{2}{1}$$

$$\Rightarrow \quad \frac{\sqrt{a}+\sqrt{b}}{\sqrt{a}-\sqrt{b}}=\frac{\sqrt{2}}{1}$$

Again, on applying componendo and dividendo rule, we get

$$\frac{\sqrt{a}+\sqrt{b}+\sqrt{a}-\sqrt{b}}{\sqrt{a}+\sqrt{b}-(\sqrt{a}-\sqrt{b})}=\frac{\sqrt{2}+1}{\sqrt{2}-1}$$

$$\Rightarrow \quad \frac{2\sqrt{a}}{2\sqrt{b}}=\frac{\sqrt{2}+1}{\sqrt{2}-1} \Rightarrow \frac{\sqrt{a}}{\sqrt{b}}=\frac{\sqrt{2}+1}{\sqrt{2}-1}$$

Now, on squaring both sides, we get

$$\frac{a}{b}=\frac{(\sqrt{2}+1)^2}{(\sqrt{2}-1)^2}$$

$$\Rightarrow \quad \frac{a}{b}=\frac{2+1+2\sqrt{2}}{2+1-2\sqrt{2}} \quad \left[\begin{array}{l}\because (a+b)^2=a^2+b^2+2ab \text{ and}\\ (a-b)^2=a^2+b^2-2ab\end{array}\right]$$

$$\Rightarrow \quad \frac{a}{b}=\frac{3+2\sqrt{2}}{3-2\sqrt{2}}$$

$\therefore \quad a:b=(3+2\sqrt{2}):(3-2\sqrt{2})$ **Hence proved.**

Example 19. **If A and G are AM and GM respectively between two positive numbers, then prove that the numbers are $A \pm \sqrt{(A+G)(A-G)}$.** Competency Based Que

Sol. Let the numbers be α and β.

Then, $\dfrac{\alpha+\beta}{2} = A$ [arithmetic mean]

$\Rightarrow \quad \alpha+\beta = 2A$

and $\sqrt{\alpha\beta} = G$ [geometric mean]

$\Rightarrow \quad \alpha\beta = G^2$

Now, quadratic equation having roots α and β is

$$x^2 - (\alpha+\beta)x + \alpha\beta = 0$$

$$\Rightarrow \quad x^2 - 2Ax + G^2 = 0$$

$$\Rightarrow \quad x = \frac{2A \pm \sqrt{4A^2 - 4\times1\times G^2}}{2\times1}$$

$$\left[\because x = \frac{-b \pm \sqrt{b^2-4ac}}{2a}\right]$$

$$= \frac{2A \pm 2\sqrt{A^2-G^2}}{2}$$

$$= A \pm \sqrt{(A+G)(A-G)}$$

$[\because a^2 - b^2 = (a+b)(a-b)]$

Hence proved.

Quick Test 2

1. Find the sum of following geometric progressions.

(i) $\sqrt{7}, \sqrt{21}, 3\sqrt{7}, \ldots n$ terms

(ii) $1, -a, a^2, -a^3, \ldots n$ terms (if $a \neq -1$)

(iii) $1, \sqrt{3}, 3, 3\sqrt{3}, \ldots$ upto 10 terms

2. Given a GP with $a = 729$ and 7th term 64, determine S_7. NCERT

3. The sum of the some terms of a GP is 315, whose first term and the common ratio are 5 and 2, respectively. Find the last term and the number of terms. NCERT

4. The common ratio of a GP is $-\dfrac{4}{5}$ and the sum to infinity is $\dfrac{80}{9}$. Find the first term. Competency Based Que

5. One side of an equilateral triangle is 18 cm. The mid-points of its sides are joined to form another triangle whose mid-points in term are joined to form further another triangle and so on upto infinity. Find the sum of Competency Based Que

(i) perimeters of all the triangles.

(ii) areas of all the triangles.

6. Insert three GMs between 1 and 256. NCERT

7. Find the minimum value of $4^x + 4^{1-x}, x \in R$. NCERT Exemplar

8. If x, y, z are distinct positive integers, then prove that $(x+y)(y+z)(z+x) > 8xyz$. NCERT Exemplar

9. Find the value of n so that $\dfrac{a^{n+1}+b^{n+1}}{a^n+b^n}$ may be the geometric mean between a and b. NCERT, Competency Based Que

10. If arithmetic mean and geometric mean between two positive numbers are 5 and 4 respectively, then find the two numbers. Competency Based Que

Answers

1. (i) $\dfrac{\sqrt{7}}{2}(\sqrt{3}+1)(3^{n/2}-1)$ (ii) $\dfrac{1-(-a)^n}{1+a}$ (iii) $121(\sqrt{3}+1)$

2. 2059

3. $n = 6$, $T_6 = 160$

4. 16

5. (i) 108 cm (ii) $108\sqrt{3}$ cm^2

6. 4, 16, 64 or −4, 16, 64 are the three GMs.

7. 4

9. $\dfrac{-1}{2}$

10. 2 and 8

For Solutions, See QR Code on Pg. 223

Exam Pattern Questions

Objective Type Questions

Multiple Choice Questions

1. If the sequence $a_n = 3\,(2^n), \forall\, n \in N$ is GP, then its common ratio is

(a) −2 (b) 2 (c) 3 (d) $\frac{1}{2}$

2. The third term of GP is 4, the product of the first five terms is

(a) 1024 (b) 512
(c) 256 (d) None of these

3. The 9th term of the GP $2, 2\sqrt{2}, 4, 8\sqrt{2}, \ldots$ is

(a) 16 (b) 64 (c) 32 (d) 128

4. 8th term from the end of the sequence $12, 4, \frac{4}{3}, \frac{4}{9}, \ldots,$ is

(a) $12 \times \left(\frac{1}{3}\right)^{13}$ (b) $12 \times \left(\frac{1}{3}\right)^{12}$
(c) $3 \times \left(\frac{1}{12}\right)^{12}$ (d) 0

5. Which term of the sequence 3, 6, 12, …… is 768?

(a) 10 (b) 8
(c) 9 (d) None of these

6. Which of the following is a GP?

(a) 2, 4, 8, 16, ...
(b) $\frac{1}{9}, \frac{-1}{27}, \frac{1}{81}, \frac{-1}{243}, \ldots$
(c) 0.01, 0.0001, 0.000001,...
(d) All of the above

7. The nth and 10th terms of a GP 5, 25, 125, ..., are respectively

(a) $5^{n-1}, 5^9$ (b) $5^{n-2}, 5^8$
(c) $5^n, 5^{10}$ (d) None of these

8. The third term of a geometric progression is 4. The product of the first five terms is

(a) 4^3 (b) 4^5
(c) 4^4 (d) None of these

9. If the sum of three numbers in GP is 38 and their product is 1728, then numbers are

(a) 8, 12, 18 (b) 8, 16, 32
(c) 18, 12, 6 (d) None of these

10. The sum of the GP $\frac{1}{3}, \frac{1}{9}, \frac{1}{27}, \frac{1}{81}, \ldots$ to n terms is **Competency Based Que**

(a) $\frac{2}{3}\left[\left(\frac{1}{3}\right)^n - 1\right]$ (b) $\frac{1}{2}\left[1 - \left(\frac{1}{3}\right)^n\right]$
(c) $\frac{1}{2}\left[\left(\frac{1}{3}\right)^n - 1\right]$ (d) None of these

11. The sum of the GP 3, 6, 12, …… upto 10 terms is

(a) 1024 (b) 2048 (c) 3069 (d) 1023

12. If the first and fourth terms of a GP are 2 and 54 respectively, then the sum of first seven terms is

(a) 2186 (b) 2187
(c) 2086 (d) None of these

13. The length of three unequal edges of a rectangular solid block are in GP. The volume of the Hock is 216 cm^3 and the total surface area is 252 cm^2. The length of the longest edge is **Competency Based Que**

(a) 6 cm (b) 14 cm (c) 10 cm (d) 12 cm

14. In a GP of even number of terms, the sum of all terms is 8 times the sum of the odd terms, then the common ration of GP is

(a) $\frac{1}{5}$ (b) 7 (c) $\frac{-4}{5}$ (d) $\frac{1}{4}$

15. Let S be the sum, P be the product and R be the sum of the reciprocals of 3 terms of a GP. Then, $P^2R^3 : S^3$ is equal to **Competency Based Que**

(a) 2 : 1 (b) 1 : 2 (c) 2 : 3 (d) 1 : 1

16. The sum to infinity of the GP $\frac{-5}{4}, \frac{5}{16}, \frac{-5}{64}, \ldots,$ is

(a) 1 (b) 0
(c) −1 (d) None of these

17. The product 32, $(32)^{1/6}$, $(32)^{1/36}$, … upto ∞ is equal to

(a) 64 (b) 16 (c) 32 (d) 0

18. If $b = a + a^2 + a^3 + \ldots + \infty$, then a is equal to

(a) $\frac{b}{1+b}$ (b) $\frac{b}{1-b}$
(c) $\frac{1}{1-b}$ (d) $\frac{1}{1+b}$

19. The geometric mean of 9 and 81 is

(a) 18 (b) 27 (c) 54 (d) 36

20. The difference between the AM and GM of two positive real numbers a and b is **Competency Based Que**

(a) $\frac{a+b-\sqrt{ab}}{2}$

(b) $\frac{a+b+\sqrt{ab}}{-2a}$

(c) $\frac{a+b-2\sqrt{ab}}{2}$

(d) None of the above

Assertion-Reason Based Questions

Directions (Q. Nos. 21-30) *In the questions given below are two statements labelled as Assertion (A) and Reason (R). In the context of the two statements, which one of the following is correct?*

(a) Both A and R are correct; R is the correct explanation of A

(b) Both A and R are correct; R is not the correct explanation of A

(c) A is correct; R is incorrect

(d) R is correct; A is incorrect

21. Assertion (A) 1, 2, 4, 8, ... is a GP, 4, 8, 16, 32, ... is a GP and $1+4, 2+8, 4+16, 8+32, \ldots$ is also a GP.

Reason (R) Let general term of a GP with common ratio r be T_{k+1} and general term of another GP with common ratio r be T_{k+1}, then the series whose general term $T_{k+1} = T_{k+1} + T_{k+1}$ is also a GP with common ratio r.

22. Assertion (A) The 4th term from end of the GP 3, 6, 12, 24, ... 3072 is 384.

Reason (R) The nth term from end of a GP with last term l and common ratio r is given by $a_n = l\left(\frac{1}{r}\right)^{n-1}$.

23. Assertion (A) If the numbers $\frac{-3}{7}, k, \frac{-7}{3}$ are in GP, then $k = \pm 1$.

Reason (R) If a_1, a_2, a_3 are in GP, then $\frac{a_2}{a_1} = \frac{a_3}{a_2}$.

24. Assertion (A) Sum of 10 terms of GP $3+9+27+\ldots$ is $\frac{3}{2}(3^{10}-1)$.

Reason (R) 20th term of the GP $\frac{5}{2}, \frac{5}{4}, \frac{5}{8}, \ldots$ is $\frac{5}{2^{20}}$.

Competency Based Que

25. Assertion (A) The sum of the series $1+\frac{1}{\sqrt{5}}+\frac{1}{5}+\frac{1}{5\sqrt{5}}+\ldots$ upto 16 terms is 25.

Reason (R) If 27, x, 3 are in GP, then $x = \pm 9$.

26. Assertion (A) Sum of n terms of the GP $x^3, x^5, x^7, \ldots (x \neq \pm 1)$ is $\frac{x^3(1-x^{2n})}{(1-x^2)}$.

Reason (R) If a is the first term and r is common ratio of a GP, then sum to n terms is given as $S_n = \frac{a(r^n-1)}{r-1}$ or $\frac{a(1-r^n)}{1-r}$ if $r \neq 1$.

27. Assertion (A) The GM between 2 and 8 is 4.

Reason (R) GM between a and b is $\sqrt{ab}$.

28. Assertion (A) The minimum value of $4^{\sin^2 x} + 4^{\cos^2 x}$ is 2.

Reason (R) AM $\geq$ GM **Competency Based Que**

29. Assertion (A) The arithmetic mean (AM) between two numbers is 34 and their geometric mean is 16. Then, the numbers are 4 and 64.

Reason (R) For two numbers a and b, AM $= \frac{a+b}{2}$ and GM $= \sqrt{ab}$.

30. Assertion (A) GM between 0.008 and 0.2 is 0.04.

Reason (R) If x, y, z are in GP, then y is GM between x and z.

Case Study Based Questions

31. A sequence of non-zero numbers is said to be a geometric progression, if the ratio of each term except the first one by its preceding term is always constant.

Rahul being a plant lover decides to open a nursery and he bought few plants with pots. He wants to place pots in such a way that number of pots in first row is 2, in second row is 4 and in third row is 8 and so on.

Based on the above information, answer the following questions.

(i) The constant multiple by which the number of pots is increasing in every row is

(a) 2 (b) 4

(c) 8 (d) 1

(ii) The number of pots in 8th row is

(a) 156 (b) 256

(c) 300 (d) 456

(iii) The difference in number of pots placed in 7th row and 5th row is

(a) 86 (b) 50 (b) 90 (d) 96

(iv) If Rahul wants to place 510 pots in total, then the total number of rows formed in this arrangement is

(a) 7 (b) 8

(c) 9 (d) 5

32. Each side of an equilateral triangle is 24 cm. The mid-point of its are joined to form another triangle. This process is going continuously. **Competency Based Que**

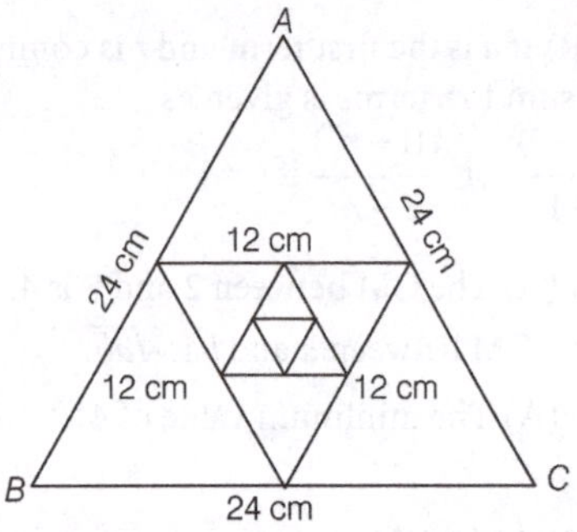

Based on the above information, answer the following questions.

(i) The side of the 5th triangle is (in cm)

(a) 8 (b) 3 (c) 1.5 (d) 1.25

(ii) The sum of perimeters of first 6 triangles is (in cm)

(a) $\frac{569}{4}$ (b) $\frac{567}{4}$ (c) 150 (d) 124

(iii) The sum of area of all triangles is (in cm)

(a) $192\sqrt{3}$ (b) $112\sqrt{3}$ (c) $34\sqrt{3}$ (d) $162\sqrt{3}$

(iv) The perimeter of 7th triangle is (in cm)

(a) $\frac{7}{8}$ (b) $\frac{9}{8}$ (c) $\frac{5}{8}$ (d) $\frac{3}{4}$

Subjective Questions

Very Short Answer Type Questions

1. Find first 5 terms of the sequence defined by $a_1 = -1$, $a_n = \frac{a_{n-1}}{n}$, for $n \geq 2$. **Competency Based Que**

2. Which term of the GP 3, 6, 12, 24, ... is 3072?

3. The 5th, 8th and 11th terms of a GP are x, y, z respectively. Show that $y^2 = xz$.

4. Find the sum of geometric series $3 + 6 + 12 + \ldots + 1536$.

5. Find the sum of n terms of the GP series $p, q, \frac{q^2}{p}, \frac{q^3}{p^2}, \ldots\ldots$ for $\frac{q}{p} > 1$.

6. Evaluate $3^{1/2} \times 3^{1/4} \times 3^{1/8} \times \ldots \infty$.

7. Insert four numbers between 1 and 1024 so that the resulting sequence is GP.

8. Find three numbers in GP in which the third term is greater than the first by 9 and the second term is greater than the fourth by 36.

Short Answer Type Questions

9. If $a^x = b^y = c^z$ and x, y, z are in GP, then show that $\log_b a = \log_c b$. **Competency Based Que**

10. If 5th and 8th terms of a GP are 48 and 384 respectively, then find the GP, if terms of GP are real numbers.

11. If the 4th, 10th and 16th terms of a GP are x, y and z respectively, then prove that x, y, z are in GP.

12. The fifth term of a GP is 1875. If the first term is 3, then find the common ratio.

13. Find three numbers in GP such that their sum is $\frac{13}{3}$ and sum of their squares is $\frac{91}{9}$.

14. Find three numbers in GP whose sum is 19 and product is 216.

15. Evaluate $\sum_{k=1}^{11} (2 + 3^k)$.

16. Evaluate $\sum_{r=1}^{n} (3^r - 2^r)$. **Competency Based Que**

17. Find the sum of the series upto n terms, $0.5 + 0.55 + 0.555 + \ldots$ to n terms.

18. The first term and seventh term of a GP are 729 and 64. Determine S_9. **Competency Based Que**

19. The ratio of the sum of the first three terms and the sum of the first six terms of a GP is 125 : 152. Find the common ratio. **Competency Based Que**

20. Find two positive numbers a and b whose AM and GM are 34 and 16, respectively.

Long Answer Type Questions

21. If a, b, c, d and p are different real numbers such that $(a^2 + b^2 + c^2)p^2 - 2(ab + bc + cd) + (b^2 + c^2 + d^2) \leq 0$, then show that a, b, c, d are in GP.

22. If a, b are the roots of $x^2 - 3x + p = 0$ and c, d are the roots of $x^2 - 12x + q = 0$, where a, b, c, d form a GP, then prove that $(q + p) : (q - p) = 17 : 15$. **Competency Based Que**

23. If a, b and c are in GP, then prove that

$$\frac{(a+b+c)^2}{a^2+b^2+c^2} = \frac{a+b+c}{a-b+c}.$$

24. For a sequence, if $S_n = \left(\frac{4}{3}\right)^n - 1$, find nth term. Hence, show that it is a GP. Also, find the common ratio.

25. Let S be the sum, P be the product and R be the sum of reciprocals of n terms in a GP. Prove that $P^2R^n = S^n$.

26. If the sum of the terms of an infinite GP is 15 and sum of the squares of these terms is 45, then find the GP.

27. Write the rational number corresponding to the decimal expansion $0.3\overline{56}$. **Competency Based Que**

28. The sum of an infinite GP is 57 and the sum of their cubes is 9747. Find the GP. **Competency Based Que**

29. After striking a floor a certain ball rebounds $\left(\frac{4}{5}\right)$th of the height from which it has fallen. Find the total distance that is travels before coming to rest, if it is gently dropped from a height of 120 m.

30. A trophy is to be made out of waste material in the form of equilateral triangle as base, on this base another equilateral triangle is kept so that its vertices are mid-point of sides of the triangle, again another equilateral triangle is kept on the second equilateral triangle obtained in the same manner and process continues. Now, sides of each equilateral triangle are decorated with green ribbon to give natural environmental look. If side of equilateral triangle at the base is 30 cm. Find the total length of ribbon required to decorate the trophy. **Competency Based Que**

31. The ratio of the AM and GM of two positive numbers a and b is $m: n$, show that $a:b = (m+\sqrt{m^2-n^2}):(m-\sqrt{m^2-n^2})$. **Competency Based Que**

Hints & Solutions

Objective Type Questions

1. (b) We have,

$$a_n = 3(2^n)$$

and $$a_{n+1} = 3(2^{n+1})$$

Now, common ratio $= \frac{a_n+1}{a_n} = \frac{3(2^{n+1})}{3(2^n)} = 2$

2. (a) Let the first five term of GP be

$\frac{a}{r^2}, \frac{a}{r}, a, ar, ar^2$, where r is common ratio.

Given, third term = 4

i.e. $a = 4$ ...(i)

$\therefore$ Product of first five term

$$= \frac{a}{r^2}\cdot\frac{a}{r}\cdot a\cdot ar\cdot ar^2$$

$$= a^5 = (4)^5 \quad [\because \text{using Eq. (i)}]$$

$$= 1024$$

3. (c) **Hint** Given GP is $2, 2\sqrt{2}, 4, 8\sqrt{2}, \ldots$

$\therefore$ 9th term, $T_9 = a\cdot r^{n-1}$

$$= 2(\sqrt{2})^8 \quad \left[\because r = \frac{2\sqrt{2}}{2} = \sqrt{2}\right]$$

$$= 32$$

4. (b) Given sequence is in GP.

Here, $a = 12, r = \frac{1}{3}, n = 20, m = 8$

We know that if a sequence has n terms, then mth term from end is equal to $(n-m+1)$th term from the beginning.

$\therefore$ mth term from end $= ar^{n-m+1-1}$

$$= ar^{n-m}$$

$$= 12\cdot\left(\frac{1}{3}\right)^{20-8} = 12\times\left(\frac{1}{3}\right)^{12}$$

5. (c) **Hint** We have, 3, 6, 12, ...

Let nth term be 768.

$$T_n = a\cdot r^n = 768$$

Here, $a = 3$ and $r = 2$

$\therefore$ $3\cdot(2)^n = 768 \Rightarrow n = 9$

6. (d) (a) We have,

$a_1 = 2, \frac{a_2}{a_1} = 2, \frac{a_3}{a_2} = 2, \frac{a_4}{a_3} = 2$ and so on

(b) We observe, $a_1 = \frac{1}{9}, \frac{a_2}{a_1} = \frac{-1}{3}$,

$\frac{a_3}{a_2} = \frac{-1}{3}, \frac{a_4}{a_3} = \frac{-1}{3}$ and so on.

(c) We have, $a_1 = 0.01, \frac{a_2}{a_1} = 0.01,$

$\frac{a_3}{a_2} = 0.01, \frac{a_4}{a_3} = 0.01$ and so on.

It is observed that in each case, every term except the first term bears a constant ratio to the term immediately preceding it.

Thus, options (a), (b) and (c) all are in GP.

7. (c) Here, $a = 5$ and $r = 5$. Thus, the nth term of the given GP is given by $a_n = ar^{n-1}$

$\Rightarrow \quad a_n = 5(5)^{n-1} = 5^n$

Now, for $n = 10$, we have $a_{10} = 5^{10}$

8. (b) Here, $T_3 = 4 \Rightarrow ar^2 = 4$

$\therefore$ Product of first five terms

$= a \cdot ar \cdot ar^2 \cdot ar^3 \cdot ar^4$

$= a^5 r^{10} = (ar^2)^5 = 4^5$

9. (a) Let the three numbers be $\frac{a}{r}$, a and ar.

Then, $\quad$ product = 1728

$\Rightarrow \quad \frac{a}{r} \cdot a \cdot ar = 1728$

$\Rightarrow \quad a^3 = 1728 \Rightarrow a = 12$

$\because \quad$ Sum = 38

$\therefore \quad \frac{a}{r} + a + ar = 38$

$\Rightarrow \quad a\left(\frac{1}{r} + 1 + r\right) = 38$

$\Rightarrow 12\left(\frac{1+r+r^2}{r}\right) = 38$

$\Rightarrow \quad 6 + 6r + 6r^2 = 19r$

$\Rightarrow \quad 6r^2 - 13r + 6 = 0$

$\Rightarrow (3r-2)(2r-3) = 0$

$\Rightarrow \quad r = 3/2$ and $2/3$

Hence, on putting the values of a and r, the required numbers are 8, 12, 18.

10. (b) **Hint** Let $S_n = \frac{1}{3}, \frac{1}{9}, \frac{1}{27}, \frac{1}{81}, \ldots$ to n terms

Here, $a = \frac{1}{3}$ and $r = \frac{1}{3}$

We know that $\quad S_n = \frac{a(1-r^n)}{1-r} \quad [\because r < 1]$

$\therefore \quad S_n = \frac{\frac{1}{3}\left[1-\left(\frac{1}{3}\right)^n\right]}{1-\frac{1}{3}}$

$= \frac{1}{2}\left[1-\left(\frac{1}{3}\right)^n\right]$

11. (c) **Hint** Given GP is 3, 6, 12, ... upto 10 terms

Here, $a = 3, r = 2, n = 10$

Using $\quad S_n = \frac{a(r^n - 1)}{r-1}, \quad r > 1$

12. (a) Given, $a = 2$

$\therefore \quad T_4 = ar^3 = 54 \Rightarrow r = 3$

Now, using $S_n = \frac{a(r^n - 1)}{r-1}$, for $r > 1$

13. (d) **Hint** Let the length, breadth and height of the rectangular block be $\frac{a}{r}$, a and ar.

Since, they are in GP.

$\therefore \quad$ Volume = lbh

$\Rightarrow \quad 216 = \frac{a}{r} \cdot a \cdot ar$

$\Rightarrow \quad 216 = a^3$

$\Rightarrow \quad a = 6 \quad \ldots$(i)

Now, total surface area = $2(lb + bh + lh)$

$\Rightarrow \quad 252 = 2\left[\frac{a}{r} \cdot a + a \cdot ar + \frac{a}{r} \cdot ar\right]$

$\Rightarrow \quad 252 = 2\left[\frac{a^2}{r} + a^2 r + a^2\right] \quad \ldots$(ii)

On solving Eqs. (i) and (ii), we get

$r = 2$ or $\frac{1}{2}$

$\therefore$ The length of the longest edge = 12 cm

14. (b) **Hint** Let there be $2n$ terms in a GP, a be the first term and r be the common ratio.

According to the question,

$S_{2n} = 8 \times$ [sum of odd terms]

$\Rightarrow \quad \frac{a(r^{2n} - 1)}{r-1} = 8a\left(\frac{(r^2)^n - 1}{r^2 - 1}\right)$

On solving, we get

$r = 7$

15. (d) **Hint** Let us take a GP with three terms $\frac{a}{r}$, a, ar.

Then, $\quad S = \frac{a}{r} + a + ar = \frac{a(r^2 + r + 1)}{r}$,

$P = \frac{a}{r} \cdot a \cdot ar = a^3$

and $\quad R = \frac{r}{a} + \frac{1}{a} + \frac{1}{ar} = \frac{1}{a}\left(\frac{r^2 + r + 1}{r}\right)$

On putting the values of S, P and R in $\frac{P^2 R^3}{S^3}$, we get

$\frac{P^2 R^3}{S^3} = \frac{1}{1}$

16. (c) The given GP has first term $(a) = \frac{-5}{4}$ and the common ratio $(r) = \frac{-1}{4}$. Also, $|r| < 1$

Hence, the sum to infinity is given by

$$S = \frac{a}{1-r} = \frac{\frac{-5}{4}}{1-\left(\frac{-1}{4}\right)} = -1$$

17. (a) We have, $32, (32)^{1/6}, (32)^{1/36}, \ldots$ upto ∞

Here, we can write above

$$32^{\left[1+\frac{1}{6}+\frac{1}{36}+\ldots+\infty\right]}$$

Here, $1+\frac{1}{6}+\frac{1}{36}+\ldots\infty$ is an infinite GP. First term $a = 1$, common ratio, $r = \frac{1}{6}$

$$\because \quad S_\infty = \frac{a}{1-r}$$

$$\Rightarrow \quad 1+\frac{1}{6}+\frac{1}{36}+\ldots\infty = \frac{1}{1-\frac{1}{6}} = \frac{6}{5}$$

Thus, $$32^{\left[1+\frac{1}{6}+\frac{1}{36}+\ldots\infty\right]} = [32]^{6/5}$$

$$= 2^{5\times\frac{6}{5}} = 2^6 = 64$$

18. (a) **Hint** First term $a = a$, common ratio $r = a$

Then, $$b = \frac{a}{1-a} \qquad \left[\because S_\infty = \frac{a}{1-r}\right]$$

$$\Rightarrow \quad a = b(1-a)$$

$$\Rightarrow \quad a = b - ab \Rightarrow a + ab = b$$

$$\therefore \quad a = \frac{b}{1+b}$$

19. (b) Geometric mean of 9 and 81

$$= \sqrt{9\times 81} = 27$$

20. (c) **Hint** AM $= \frac{a+b}{2}$ and GM $= \sqrt{ab}$

$$\therefore \quad \text{AM} - \text{GM} = \frac{a+b}{2} - \sqrt{ab} = \frac{a+b-2\sqrt{ab}}{2}$$

21. (a) **Hint** Sequences $a_1, a_2, \ldots, a_n$ are in GP.

The common ratio r we take as $\frac{a_{k+1}}{a_k}$.

Thus, the general term of a GP with common ratio r be take T_{k+1} and general term of another GP with common ratio r be T_{k+1}, then the series whose general term $T_{k+1} = T_{k+1} + T_{k+1}$ is also a GP with common ratio r.

22. (a) **Hint** $\because \quad a_n = l\left(\frac{1}{r}\right)^{n-1}$

Given, GP is 3, 6, 12, 24, …, 3072.

Here, $l = 3072, r = 2$

$\therefore$ 4th term from end, $a_4 = 3072\left(\frac{1}{2}\right)^{4-1} = 384$

23. (a) **Hint** $\frac{-3}{7}, k, \frac{-7}{3}$ are in GP.

$\because a_1, a_2, a_3$ are in GP, then

$$\frac{a_2}{a_1} = \frac{a_3}{a_2}$$

$$\therefore \quad \frac{k}{-3/7} = \frac{-7/3}{k}$$

$$\Rightarrow \quad k = \pm 1$$

24. (b) **Hint** Given GP is $3+9+27+\ldots$

Here, $a = 3$ and $r = 3 > 1$

$$\because \quad S_n = \frac{a(r^n-1)}{r-1} \text{ for } r > 1$$

$$\therefore \quad S_{10} = \frac{3(3^{10}-1)}{3-1}$$

$$= \frac{3}{2}(3^{10}-1)$$

Hence, Assertion is correct.

For Reason, $a = \frac{5}{2}$ and $r = \frac{1}{2}$

$\therefore$ 20th term, $T_{20} = a\cdot r^{20-1}$

$$= \frac{5}{2}\left(\frac{1}{2}\right)^{19} = \frac{5}{2^{20}}$$

25. (d) **Hint** $1+\frac{1}{\sqrt{5}}+\frac{1}{5}+\frac{1}{5\sqrt{5}}+\ldots$ upto 16th terms.

Above series is GP.

Here, $a = 1, r = \frac{1}{\sqrt{5}}, n = 16$

$$\because \quad S_n = \frac{a(1-r^n)}{1-r} \text{ [for } r < 1]$$

$$\Rightarrow \quad S_{16} = \frac{1\left[1-\left(\frac{1}{\sqrt{5}}\right)^{16}\right]}{1-\frac{1}{\sqrt{5}}} \neq 25$$

Now, 27, x, 3 in GP.

$$\Rightarrow \quad \frac{x}{27} = \frac{3}{x}$$

$$\Rightarrow \quad x^2 = 81$$

$$\Rightarrow \quad x = \pm 9$$

Hence, Reason is correct but Assertion is incorrect.

26. (a) **Hint** Sum of GP, $S_n = \dfrac{a(r^n - 1)}{r-1}$ for $r > 1$

and if $r < 1$, then $S_n = \dfrac{a(1-r^n)}{1-r}$

So, sum of the GP $x^3, x^5, x^7, \ldots$

$$= \frac{x^3(1-x^{2n})}{1-x^2}$$

27. (a) **Hint** GM between a and $b = \sqrt{ab}$

GM between 2 and 8 $= \sqrt{2\times 8} = \sqrt{16} = 4$

28. (a) **Hint** Let two terms be $4^{\sin^2 x}$ and $4^{\cos^2 x}$.

Using AM-GM relation,

$$\text{AM} \geq \text{GM}$$

$$\Rightarrow \frac{4^{\sin^2 x} + 4^{\cos^2 x}}{2} \geq \sqrt{4^{\sin^2 x} \cdot 4^{\cos^2 x}}$$

$$\Rightarrow \frac{4^{\sin^2 x} + 4^{\cos^2 x}}{2} \geq \sqrt{4^{(\sin^2 x + \cos^2 x)}}$$

$$\Rightarrow \frac{4^{\sin^2 x} + 4^{\cos^2 x}}{2} \geq \sqrt{(4)^1}$$

$$\Rightarrow 4^{\sin^2 x} + 4^{\cos^2 x} \geq 4$$

The minimum value of given expression is 4.

29. (a) **Hint** Let two numbers be a and b.

$$\text{AM} = 34 \Rightarrow \frac{a+b}{2} = 34 \quad \ldots\text{(i)}$$

and $\quad \text{GM} = 16 \Rightarrow \sqrt{ab} = 16 \quad \ldots\text{(ii)}$

On solving Eqs. (i) and (ii), we get

$a = 4$ and $b = 64$

30. (a) GM between 0.008 and 0.2

$$= \sqrt{0.008 \times 0.2} = \sqrt{0.0016} = 0.04$$

Hence, 0.008, 0.04, 0.2 are in GP.

Thus, we can say x, y, z are in GP, then y is GM between x and z.

31. (i) (a) Number of pots in each row is 2, 4, 8, …

This forms a geometric progression,

where $\quad a = 2, r = \dfrac{4}{2} = 2$

Hence, the constant multiple by which the number of pots is increasing in every row is 2.

(ii) (b) Number of pots in 8th row

$$a_8 = ar^{8-1} = 2(2)^7 = 2^8 = 256$$

(iii) (d) Number of pots in 7th row,

$$a_7 = 2(2)^{7-1} = 2 \cdot 2^6 = 2^7 = 128$$

Number of pots in 5th row,

$$a_5 = 2(2)^{5-1} = 2 \cdot 2^4 = 2^5 = 32$$

$\therefore$ Required difference $= 128 - 32 = 96$

(iv) (b) Let there be n number of rows.

$$\therefore \quad S_n = 510 = \frac{2(2^n - 1)}{2-1}$$

$$\Rightarrow \quad \frac{510}{2} = 2^n - 1$$

$$\Rightarrow \quad 255 = 2^n - 1 \Rightarrow 256 = 2^n$$

$$\Rightarrow \quad 2^8 = 2^n \Rightarrow n = 8$$

32. (i) (c) Side of first triangle is 24.

Side of second triangle is $\dfrac{24}{2} = 12$

Similarly, side of third triangle is $\dfrac{12}{2} = 6$

$$\therefore \quad a = 24, r = \frac{12}{24} = \frac{1}{2}$$

$\therefore$ Side of the fifth triangle,

$$a_5 = ar^4 = 24 \times \left(\frac{1}{2}\right)^4$$

$$= \frac{24}{16} = \frac{3}{2} = 1.5 \text{ cm}$$

(ii) (b) Perimeter of first triangle $= 24 \times 3 = 72$

Perimeter of second triangle $= \dfrac{72}{2} = 36$

Similarly, perimeter of third triangle $= \dfrac{36}{2} = 18$

$$\therefore \quad a = 72, r = \frac{36}{72} = \frac{1}{2}$$

$\therefore$ Sum of perimeters of first 6 triangles,

$$S_6 = \frac{a(1-r^6)}{1-r} = \frac{72\left(1-\left(\frac{1}{2}\right)^6\right)}{1-\frac{1}{2}}$$

$$= \frac{72 \times 63 \times 2}{2^6} = \frac{567}{4} \text{ cm}$$

(iii) (a) Area of first triangle $= \dfrac{\sqrt{3}}{4}(24)^2$

Area of second triangle $= \dfrac{\sqrt{3}}{4}\left(\dfrac{24}{2}\right)^2$

$$\therefore \quad a = \frac{\sqrt{3}}{4}(24)^2, r = \frac{1}{4}$$

$\therefore$ Sum of area of all triangles

$$= \frac{a}{1-r} = \frac{\sqrt{3}}{4}\frac{(24)^2}{1-\frac{1}{4}}$$

$$= \frac{\sqrt{3} \times (24)^2}{3} = 192\sqrt{3} \text{ cm}^2$$

(iv) (b) Here, $a = 72, r = \dfrac{1}{2}$

$$a_7 = (72)\left(\frac{1}{2}\right)^6 = \frac{72}{64} = \frac{9}{8} \text{ cm}$$

Subjective Questions

1. **Hint** Put $n = 2, 3, 4, 5$ in nth term.

Ans. $-1, -\frac{1}{2}, -\frac{1}{6}, -\frac{1}{24}, -\frac{1}{120}$

2. Given GP is 3, 6, 12, 24, ...

Let nth term be 3072.

Then, $T_n = 3072$

$\because$ First term, $a = 3$ and common ratio, $r = \frac{6}{3} = 2$

Now, $T_n = ar^{n-1}$

and put equal to 3072.

Ans. 11th term is 3072.

3. **Hint** Given, $ar^4 = x, ar^7 = y$ and $ar^{10} = z$

$\Rightarrow \quad y^2 = (ar^7)^2 = a^2r^{14} = (ar^4)(ar^{10})$

4. **Hint** Given GP is $3 + 6 + 12 + \ldots + 1536$

Here, $a = 3, r = \frac{6}{3} = 2 > 1$

Now, use $S_n = \frac{a(r^n - 1)}{r - 1}, r > 1$

Ans. 3069

5. **Hint** First term $= p$ and common ratio, $r = \frac{q}{p}$.

$\therefore \quad S_n = \frac{a(r^n - 1)}{r - 1}$, for $r > 1$

Ans. $\frac{p^2}{p-q}\left[1-\left(\frac{q}{p}\right)^n\right]$

6. Let $\quad y = 3^{\frac{1}{2}} \times 3^{\frac{1}{4}} \times 3^{\frac{1}{8}} \times \ldots \infty$

$\Rightarrow \quad y = 3^{\frac{1}{2}+\frac{1}{4}+\frac{1}{8}+\ldots+\infty}$

$\Rightarrow \quad y = 3^{\frac{1}{2}\left[1+\frac{1}{2}+\left(\frac{1}{2}\right)^2+\ldots+\infty\right]}$

$\Rightarrow \quad y = 3^{\frac{2}{2}} = 3^1 = 3$

7. **Hint** Let G_1, G_2, G_3 and G_4 be the required three GM's.

Then, $1, G_1, G_2, G_3, G_4$, 1024 are in GP.

Ans. 4, 16, 64 and 256

8. Let the four numbers be a, ar, ar^2 and ar^3.

It is given that

$$ar^2 - a = 9 \text{ and } ar - ar^3 = 36$$

$\Rightarrow \quad a(r^2 - 1) = 9$ and $ar(1 - r^2) = 36$

$\Rightarrow \quad \frac{ar(1-r^2)}{a(r^2-1)} = \frac{36}{9}$

$\Rightarrow \quad r = -4$

On putting $r = -4$ in $a(r^2 - 1) = 9$, we get

$$a = \frac{3}{5}$$

Hence, the three numbers are $\frac{3}{5}, \frac{-12}{5}, \frac{48}{5}$.

9. Let $a^x = b^y = c^z = k$

$\therefore \quad x = \log_a k, y = \log_b k, z = \log_c k$

Since, x, y, z are in GP.

$\therefore \quad \frac{\log_b k}{\log_a k} = \frac{\log_c k}{\log_b k}$

$\Rightarrow \quad \frac{\log_k a}{\log_k b} = \frac{\log_k b}{\log_k c}$

$\therefore \quad \log_b a = \log_c b$ **Hence proved.**

10. Let a be the first term and r be the common ratio of the given GP.

According to the question,

$T_5 = 48 \Rightarrow ar^4 = 48$ ...(i)

and $T_8 = 384 \Rightarrow ar^7 = 384$ ...(ii)

On dividing Eq. (ii) by Eq. (i), we get

$$\frac{ar^7}{ar^4} = \frac{384}{48}$$

$\Rightarrow \quad r^3 = 8 \Rightarrow r = 2$

On putting $r = 2$ in Eq. (i), we get $a = 3$

$\therefore$ The required GP is 3, 6, 12, ...

11. Let a be the first term and r be the common ratio of GP.

Given, $a_4 = ar^3 = x$ ...(i)

$a_{10} = ar^9 = y$ ...(ii)

$a_{16} = ar^{15} = z$ ...(iii)

On dividing Eq. (ii) by Eq. (i) and Eq. (iii) by Eq. (ii), we get

$$\frac{y}{x} = a^6 \text{ and } \frac{z}{y} = a^6$$

Therefore, $\frac{y}{x} = \frac{z}{y}$

Thus, x, y, z are in GP.

12. **Hint** Given, $T_5 = ar^4 = 1875$ and first term, $a = 3$

$\therefore \quad 3 \cdot r^4 = 1875$ **Ans.** 5

13. **Hint** Let the three numbers in GP be a, ar and ar^2.

Given, sum of numbers $= \frac{13}{3}$

$\Rightarrow \quad a + ar + ar^2 = \frac{13}{3}$ ...(i)

and sum of the square of numbers $= \frac{91}{9}$

$\Rightarrow \quad a^2 + (ar)^2 + a^2r^4 = \frac{91}{9}$ ...(ii)

Now, on solving Eqs. (i) and (ii) and find the values of a and r.

Ans. $\frac{1}{3}, 1, 3$ or $3, 1, \frac{1}{3}$

14. Let three numbers in GP be $\frac{a}{r}, a, ar$.

Then, according to the question,

$$\frac{a}{r} + a + ar = 19 \quad \ldots(\text{i})$$

and $$\frac{a}{r} \cdot a \cdot ar = 216$$

$$\Rightarrow \quad a^3 = 216 \Rightarrow a = 6$$

On putting the value $a = 6$ in Eq. (i), we get

$$r = \frac{2}{3} \text{ or } \frac{3}{2}$$

Ans. 4, 6, 9 or 9, 6, 4

15. Hint $\sum_{k=1}^{11}(2+3^k) = \sum_{k=1}^{11} 2 + \sum_{k=1}^{11} 3^k$

$$= 11 \times 2 + 3^1 + 3^2 + \ldots + 3^{11}.$$

Ans. 265741

16. Hint Let $S_n = \sum_{r=1}^{n} (3^r - 2^r)$

$$\Rightarrow \quad S_n = (3^1 + 3^2 + 3^3 + \ldots \text{upto } n \text{ terms}) + (2^1 + 2^2 + 2^3 + \ldots \text{ upto } n \text{ terms})$$

$$\Rightarrow \quad S_n = \frac{3(3^n - 1)}{3-1} + \frac{2(2^n - 1)}{2-1} \quad \left[\because S_n = \frac{a(r^n - 1)}{r-1}, r > 1\right]$$

$$= \frac{3^{n+1} - 3}{2} + \frac{2^{n+1} - 2}{1} = \frac{3^{n+1} - 3 + 2^{n+2} - 4}{2}$$

Ans. $\frac{1}{2}(3^{n+1} + 2^{n+2} - 7)$

17. Hint Let $S_n = 0.5 + 0.55 + 0.555 + \ldots$ upto n terms.

$$\Rightarrow \quad S_n = 5\,[0.1 + 0.11 + 0.111 + \ldots \text{ upto } n \text{ terms}$$

$$\Rightarrow \quad S_n = \frac{5}{9}\left[\frac{9}{10} + \frac{99}{100} + \frac{999}{1000} + \ldots \text{ upto } n \text{ terms}\right]$$

$$\Rightarrow \quad S_n = \frac{5}{9}\left[1 + 1 + 1 + \ldots \text{ upto } n \text{ terms} - \left(\frac{1}{10} + \frac{1}{10^2} + \frac{1}{10^3}\right) + \ldots \text{ upto } n \text{ terms}\right]$$

$$\Rightarrow \quad S_n = \frac{5}{9}\left[n - \frac{1}{10}\frac{\left(1 - \frac{1}{10^n}\right)}{\left(1 - \frac{1}{10}\right)}\right]$$

Ans. $\frac{5}{9}\left[n - \frac{1}{9}(1 - 10^{-n})\right]$

18. Hint Given, $ar^6 = 64 \Rightarrow r = \frac{2}{3}$

$$\because \quad S_n = \frac{a(1 - r^n)}{1 - r} \quad [\because r < 1]$$

$$\therefore \quad S_9 = \frac{729\left(1 - \left(\frac{2}{3}\right)^9\right)}{1 - \frac{2}{3}}$$

Ans. 2130.1

19. Let a be the first term and r be the common ratio of given GP.

Given, $S_3 : S_6 = 125 : 152$

$$\Rightarrow \quad \frac{\frac{a(r^3 - 1)}{r-1}}{\frac{a(r^6 - 1)}{r-1}} = \frac{125}{152} \Rightarrow \frac{r^3 - 1}{r^6 - 1} = \frac{125}{152}$$

$$\Rightarrow \quad \frac{r^3 - 1}{(r^3 - 1)(r^3 + 1)} = \frac{125}{152}$$

$$\Rightarrow \quad \frac{1}{r^3 + 1} = \frac{125}{152}$$

$$\Rightarrow \quad r^3 + 1 = \frac{152}{125}$$

$$\Rightarrow \quad r^3 = \frac{152}{125} - 1$$

$$\Rightarrow \quad r^3 = \frac{27}{125}$$

$$\Rightarrow \quad r = \frac{3}{5}$$

Hence, the common ratio is $\frac{3}{5}$.

20. Hint Given, a and b are two position numbers.

$$\therefore \quad \text{AM} = \frac{a+b}{2} \text{ and GM} = \sqrt{ab}$$

According to the question,

$$\frac{a+b}{2} = 34 \text{ and } \sqrt{ab} = 16$$

$$\Rightarrow \quad a + b = 68 \quad \ldots(\text{i})$$

and $$ab = 256 \quad \ldots(\text{ii})$$

On solving Eqs. (i) and (ii), we get

$$a = 4 \text{ and } b = 64$$

Ans. 4, 64

21. Given, $(a^2 + b^2 + c^2)p^2 - 2(ab + bc + cd)\,p + (b^2 + c^2 + d^2) \le 0$

$$\Rightarrow (a^2p^2 - 2abp + b^2) + (b^2p^2 - 2bcp + c^2) + (c^2p^2 - 2cdp + d^2) \le 0$$

$$\Rightarrow \quad (ap-b)^2+(bp-c)^2+(cp-d)^2 \le 0 \quad \dots(i)$$

Now, if a, b, c, d and p are real numbers, then

$$(ap-b)^2 \ge 0, (bp-c)^2 \ge 0 \text{ and } (cp-d)^2 \ge 0 \quad \dots(ii)$$

From Eqs. (i) and (ii), we get

$$(ap-b)^2+(bp-c)^2+(cp-d)^2 = 0$$

$$\Rightarrow \quad ap-b=0,\ bp-c=0,\ cp-d=0$$

$$\Rightarrow \quad p=\frac{b}{a}, p=\frac{c}{a}, p=\frac{d}{a}$$

$$\Rightarrow \quad \frac{b}{a}=\frac{c}{a}=\frac{d}{a}$$

Then, a, b, c, d are in GP. **Hence proved.**

22. $\because a$ and b are roots of $x^2-3x+p=0$.

$$\therefore \quad a+b=3 \quad \dots (i)$$

and $$ab=p \quad \dots (ii)$$

$\because c$ and d are roots of $x^2-12x+q=0$.

$$\therefore \quad c+d=12 \quad \dots (iii)$$

and $$cd=q \quad \dots (iv)$$

Given, a, b, c and d are in GP.

Let r be the common ratio of given GP, then

$$a=a, b=ar, c=ar^2 \text{ and } d=ar^3$$

From Eqs. (i) and (iii), we get

$$c+d=4(a+b)$$

$$\Rightarrow \quad ar^2+ar^3=4(a+ar)$$

$$\Rightarrow \quad r^2=4$$

From Eqs. (ii) and (iv), we get

$$\frac{q}{p}=\frac{cd}{ab}$$

$$\Rightarrow \quad \frac{q}{p}=\frac{ar^2 \cdot ar^3}{a \cdot ar}=r^4=16$$

$$\Rightarrow \quad \frac{q}{p}=\frac{16}{1} \Rightarrow \frac{q+p}{q-p}=\frac{16+1}{16-1}$$

[$\because$ using componendo and dividendo rule]

$$\therefore \quad (q+p):(q-p)=17:15$$ **Hence proved.**

23. Let numbers be $a, b=ar$ and $c=ar^2$.

We have, $$\frac{(a+b+c)^2}{a^2+b^2+c^2}=\frac{a+b+c}{a-b+c}$$

$$\text{LHS}=\frac{(a+b+c)^2}{a^2+b^2+c^2}=\frac{(a+ar+ar^2)^2}{a^2+(ar)^2+(ar^2)^2}$$

$$=\frac{a^2(1+r+r^2)^2}{a^2(1+r^2+r^4)}=\frac{(1+r+r^2)^2}{\{(1+r^2)^2-r^2\}}$$

$$=\frac{(1+r+r^2)^2}{(1+r+r^2)(1-r+r^2)}$$

$$=\frac{1+r+r^2}{1-r+r^2}=\frac{a+ar+ar^2}{a-ar+ar^2}$$

$$=\frac{a+b+c}{a-b+c}=\text{RHS}$$ **Hence proved.**

24. Hint $S_n=\left(\frac{4}{3}\right)^n-1$

$$\therefore \quad S_{n-1}=\left(\frac{4}{3}\right)^{n-1}-1$$

Now, $$t_n=S_n-S_{n-1}$$

$$=\left[\left(\frac{4}{3}\right)^n-1\right]-\left[\left(\frac{4}{3}\right)^{n-1}-1\right]$$

$$=\left(\frac{4}{3}\right)^n-1-\left(\frac{4}{3}\right)^{n-1}+1$$

$$=\left(\frac{4}{3}\right)^n-\left(\frac{4}{3}\right)^n\left(\frac{4}{3}\right)^{-1}=\left(\frac{4}{3}\right)^n\left[1-\frac{3}{4}\right]$$

$$\Rightarrow \quad t_n=\frac{1}{4}\left(\frac{4}{3}\right)^n$$

$$\therefore \quad t_{n-1}=\frac{1}{4}\left(\frac{4}{3}\right)^{n-1}$$

We know that common ratio $=\dfrac{t_n}{t_{n-1}}$

Ans. $t_n=\frac{1}{4}\left(\frac{4}{3}\right)^n$ and $r=\frac{4}{3}$

25. Let the GP be $a, ar, ar^2, ar^3, \dots, ar^{n-1}$.

Given, $S=$ Sum of n terms

$$=a+ar+ar^2+ar^3+\dots+ar^{n-1}$$

$$=\frac{a(r^n-1)}{r-1} \quad [\text{let } r>1] \quad \dots(i)$$

and $R=$ Sum of the reciprocals of n terms

$$=\frac{1}{a}+\frac{1}{ar}+\frac{1}{ar^2}+\dots+\frac{1}{ar^{n-1}} \quad \left[\because \frac{1}{r}<1\right]$$

$$=\frac{\frac{1}{a}\left[1-\left(\frac{1}{r}\right)^n\right]}{1-\frac{1}{r}}=\frac{1}{a}\left[\frac{1}{1}-\frac{1}{r^n}\right]\times\frac{1}{\frac{r-1}{r}}$$

$$=\frac{1}{a}\left[\frac{r^n-1}{r^n}\right]\times\frac{r}{r-1} \Rightarrow R=\frac{(r^n-1)r}{ar^n(r-1)} \quad \dots(ii)$$

and $P=$ Product of n terms

$$=a\times ar\times ar^2\times ar^3\times\dots\times ar^{n-1}$$

$$=a^{1+1+1+\dots+n\text{ terms}}\, r^{1+2+3+\dots+(n-1)\text{terms}}$$

$$=a^n r^{\frac{n(n-1)}{2}} \quad \left[\because \Sigma n=\frac{n(n+1)}{2}\right]$$

$\Rightarrow \quad P^2 = a^{2n} r^{n(n-1)}$ ...(iii)

Now, we have to prove $P^2R^n = S^n$

or $\quad P^2 = \dfrac{S^n}{R^n}$ or $P^2 = \left(\dfrac{S}{R}\right)^n$

$$\text{RHS} = \left(\frac{S}{R}\right)^n = \left[\frac{a(r^n-1)}{r-1} \times \frac{ar^n(r-1)}{(r^n-1)r}\right]^n$$

[using Eqs. (i) and (ii)]

$= [a^2 r^n r^{-1}]^n = (a^2 r^{n-1})^n = [a^{2n} r^{n(n-1)}]$

$= P^2 =$ LHS [from Eq. (iii)] **Hence proved.**

26. Hint We have, $\dfrac{a}{1-r} = 15$ ...(i)

and $\dfrac{a^2}{1-r^2} = 45$ ...(ii)

On dividing the square of Eq. (i) by Eq. (ii), we get

$$\frac{a^2}{(1-r)^2} \times \frac{1-r^2}{a^2} = \frac{225}{45}$$

$$\Rightarrow \quad \frac{1+r}{1-r} = 5 \Rightarrow r = \frac{2}{3}$$

Ans. $5, \dfrac{10}{3}, \dfrac{20}{9}, \dfrac{40}{27}, \ldots$

27. Given, decimal expansion is $0.3\overline{56}$.

In series form, it can be written as

$$0.3\overline{56} = 0.3 + 0.056 + 0.00056 + 0.0000056 + \ldots + \infty$$

$$\Rightarrow \quad 0.3\overline{56} = 0.3 \times \frac{10}{10} + 0.056 \times \frac{10^3}{10^3} + 0.00056 \times \frac{10^5}{10^5} + \ldots + \infty$$

$$= \frac{3}{10} + \frac{56}{10^3} + \frac{56}{10^5} + \ldots + \infty$$

$$\Rightarrow \quad 0.3\overline{56} = \frac{3}{10} + \left[\frac{56}{10^3} + \frac{56}{10^5} + \ldots + \infty\right]$$

Here, first term $a = \dfrac{56}{10^3}$

and common ratio $= \dfrac{56}{10^5} \div \dfrac{56}{10^3} = \dfrac{1}{10^2} < 1$

$$\therefore \quad 0.3\overline{56} = \frac{3}{10} + \frac{\frac{56}{10^3}}{1 - \frac{1}{10^2}} \quad \left[\because S_\infty = \frac{a}{1-r}, \text{if } r < 1\right]$$

$$= \frac{3}{10} + \frac{56}{10^3} \times \frac{10^2}{99}$$

$$= \frac{3}{10} + \frac{56}{990}$$

$$\therefore \quad 0.3\overline{56} = \frac{353}{990}$$

28. Let a be the first term and r be the common ratio of an infinite GP.

Given, sum = 57 $\Rightarrow \dfrac{a}{1-r} = 57$ ...(i)

and sum of the cubes = 9747

$\Rightarrow \quad a^3 + a^3r^3 + a^3r^6 + \ldots = 9747$

$\Rightarrow \quad \dfrac{a^3}{1-r^3} = 9747$ ...(ii)

On dividing the cube of Eq. (i) by Eq. (ii), we get

$$\frac{a^3}{(1-r)^3} \cdot \frac{(1-r^3)}{a^3} = \frac{(57)^3}{9747}$$

$$\Rightarrow \quad \frac{1-r^3}{(1-r)^3} = 19$$

$$\Rightarrow \quad \frac{1+r+r^2}{(1-r)^2} = 19$$

$[\because a^3 - b^3 = (a-b)(a^2+b^2+ab)]$

$$\Rightarrow \quad 1 + r + r^2 = 19(1 + r^2 - 2r)$$

$[\because (a-b)^2 = a^2 + b^2 - 2ab]$

$$\Rightarrow \quad 18r^2 - 39r + 18 = 0$$

$$\Rightarrow \quad (3r-2)(6r-9) = 0$$

$$\Rightarrow \quad r = 2/3 \text{ or } r = 3/2$$

$$\Rightarrow \quad r = 2/3$$

$[\because r \neq 3/2$, because $-1 < r < 1$ for an infinite GP$]$

On putting $r = \dfrac{2}{3}$ in Eq. (i), we get

$$\frac{a}{1-(2/3)} = 57$$

$$\Rightarrow \quad 3a = 57$$

$$\Rightarrow \quad a = 19$$

Hence, the required GP is $19, 19 \times \dfrac{2}{3}, 19 \times \left(\dfrac{2}{3}\right)^2, \ldots$

i.e. $19, \dfrac{38}{3}, \dfrac{76}{9}, \ldots$

29. Given, a ball dropped from a height of 120 m.

After dropping 120 m, height after first rebound

$$= \frac{4}{5} \times 120 = 96 \text{ m}$$

After dropping 96 m, height after second rebound

$$= \frac{4}{5} \times 96 = 76.8 \text{ m}$$

After dropping 76.8 m, height after third rebound

$$= \frac{4}{5} \times 76.8 = 61.44$$

Continuing this way, we get a geometric series $96 + 76.8 + 61.44 + \ldots,$

where first term $a = 96$ and $r = \frac{4}{5}$.

The sum of an infinite geometric series is given by,

$$S = \frac{a}{1-r}$$

So,
$$S = \frac{96}{1-\frac{4}{5}} = \frac{96}{\frac{1}{5}} = 480$$

Thus, total distance $= 120 + 2 \times 480 = 1080$ m

Hence, the total distance travelled by the ball before coming to rest is 1080 m.

30. **Hint** By the given condition, we have

perimeter of first triangle = 90 cm,

perimeter of second triangle = 45 cm,

perimeter of third triangle $= \frac{45}{2}$ cm

and so on.

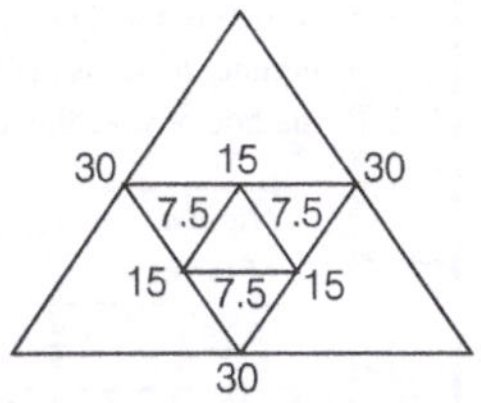

$\therefore$ Sum of perimeters of triangles form an GP

$$90 + \frac{90}{2} + \frac{90}{4} + \frac{90}{8} + \ldots + \infty$$

Here, $a = 90$ and $r = \frac{1}{2}$

Now, use sum of an infinite GP,

$$S = \frac{a}{1-r}$$

Ans. 180 cm

31. $\text{AM} = \frac{a+b}{2}, \text{GM} = \sqrt{ab}$

$$\therefore \quad m : n = \frac{a+b}{2\sqrt{ab}}$$

$$\text{RHS} = \frac{m + \sqrt{m^2 - n^2}}{m - \sqrt{m^2 - n^2}} = \frac{\frac{m}{n} + \sqrt{\frac{m^2}{n^2} - 1}}{\frac{m}{n} - \sqrt{\frac{m^2}{n^2} - 1}}$$

$$= \frac{\frac{a+b}{2\sqrt{ab}} + \sqrt{\frac{a^2 + b^2 + 2ab}{4ab} - 1}}{\frac{a+b}{2\sqrt{ab}} - \sqrt{\frac{a^2 + b^2 + 2ab}{4ab} - 1}}$$

$$= \frac{\frac{a+b}{2\sqrt{ab}} + \sqrt{\frac{a^2 + b^2 - 2ab}{4ab}}}{\frac{a+b}{2\sqrt{ab}} - \sqrt{\frac{a^2 + b^2 - 2ab}{4ab}}}$$

$$= \frac{a + b + (a - b)}{a + b - (a - b)} = \frac{a}{b}$$

$= \text{LHS}$ **Hence proved.**

QR Code for Quick Test 1

QR Code for Quick Test 2

Mind Map

Sequences and Series

Types of Sequence

Finite Sequence
A sequence containing finite number of terms, is called a finite sequence.
e.g. 1, 3, 5, 7 is a finite sequence as it contains only 4 terms.

Infinite Sequence
A sequence which is not a finite sequence is known as infinite sequence i.e. in an infinite sequence, **number of terms never ends.**
e.g. 1, 3, 5, 7,... is an infinite sequence, as it contains infinite number of terms.

Real Sequence
A sequence of real numbers is a function whose domain is the set of natural numbers and whose range is a subset of real numbers.
e.g. Let $a : N \to R$ given by $a(n) = 4n - 3$, then
$a(1) = a_1 = 4(1) - 3 = 1$
$a(2) = a_2 = 4(2) - 3 = 5$
$a(3) = a_3 = 4(3) - 3 = 9$
⋮ ⋮
i.e. 1, 5, 9, ... is required sequence.

Fibonacci Sequence
The sequence 1, 1, 2, 3, 5, 8, … has no visible pattern but its recurrence relation is $a_1 = a_2 = 1$ and $a_{n+1} = a_n + a_{n-1}, n \geq 2$
This sequence is called Fibonacci sequence.

Sequence
A sequence is an arrangement of numbers according to some rules.
e.g. 1, 4, 9, 16, … is a sequence as each number follow a rule of n^2, where n=1, 2, 3.

Geometric Progression (GP)
A sequence of non-zero numbers is said to be a geometric progression if the ratio of each term, except the first one, by its preceding term is always constant.
e.g. Sequence $a_1, a_2, a_3, \ldots, a_n$

Here, $\frac{a_{k+1}}{a_k} = r$ (constant) for all $k \in N$

i.e. if 2, 4, 8, 16, 32, ...

Then, $r = \frac{4}{2} = 2$

The constant ratio is called **common ratio** of the GP and it is denoted by ***r***.

Sum of *n* Terms of a GP
If a and r are the first term and common ratio of a GP respectively, then the sum of n terms of this GP is given by

$$S_n = \frac{a(1 - r^n)}{1 - r}, \text{ where } r < 1$$

or $S_n = \frac{a(r^n - 1)}{r - 1}$, where $r > 1$ $(r \neq 1)$

Sum of an Infinite GP
If a is the first term and r is the common ratio of a GP such that $|r| < 1$, then sum of an infinite GP is given by

$$S = \frac{a}{1 - r}$$

Geometric Mean (GM)
Let a and b be two positive numbers. If we insert a single number G between them so that a, G, b is a GP, then G is called geometric mean. Thus, a, G, b are in GP.
Then, geometric mean, $G = \sqrt{ab}$

General Term of a GP
If a is the first term of a GP and its common ratio is r, then general term or nth term, $T_n = ar^{n-1}$ or $l = ar^{n-1}$, where l is the last term.

*m*th Term of a Finite GP from the End
Let a be the first term and r be the common ratio of a GP having n terms. Then, mth term from the end is $(n - m + 1)$th term from the beginning.
∴ *m*th term from the end = $ar^{n-m+1-1}$ = ar^{n-m}, where $n > m$,
Also, mth term from the end $= l\left(\frac{1}{r}\right)^{n-1}$,
where l is the last term of finite GP.

Selection of Terms in GP
Sometimes we have to select certain number of terms in GP. The convenient method of selecting terms is given below.

Number of terms	Terms	Common ratio
3	$\frac{a}{r}, a, ar$	r
4	$\frac{a}{r^3}, \frac{a}{r}, ar, ar^3$	r^2
5	$\frac{a}{r^2}, \frac{a}{r}, a, ar, ar^2$	r

Relation between AM and GM
Let A and G be the AM and GM of two positive real numbers a and b, respectively.

Then, $A = \frac{a+b}{2}$ and $G = \sqrt{ab}$

Now, $A - G = \frac{a+b}{2} - \sqrt{ab} = \frac{a + b - 2\sqrt{ab}}{2}$

$= \frac{(\sqrt{a} - \sqrt{b})}{2} \geq 0$

∴ $A \geq G$

Inserting *n* GM's between Two Positive Numbers
Let a and b be two positive numbers and $G_1, G_2, G_3, \ldots, G_n$ be the n GM's between a and b.
Then, $a, G_1, G_2, G_3, \ldots, G_n, b$ are in GP with common ratio r and number of terms is $n + 2$.

Topics to be Assessed Formatively

Sum of n Terms of Special Series

Sometimes, we can find the sum of given series directly by using some standard results. If we need to find the sum of special series without using the standard results, then it takes long time to solve it.

So, some standard result relates to sum of n terms of special series such as series of natural numbers, series of squares of natural numbers, series of cubes of natural numbers are given below

Sum of first n natural numbers

Sum of first n natural numbers 1, 2, 3, ..., n is denoted by Σn and defined as

$$1+2+3+\ldots+n=\sum n=\frac{n(n+1)}{2}$$

Sum of squares of first n natural numbers

We know that first n natural numbers are 1, 2, 3, ..., n. Then, sum of their squares is denoted by Σn^2 and defined as

$$1^2+2^2+3^2+\ldots+n^2=\Sigma n^2=\frac{n(n+1)(2n+1)}{6}$$

Sum of cubes of first n natural numbers

We know that first n natural numbers are 1, 2, 3, ..., n. Then, sum of their cubes is denoted by Σn^3 and defined as

$$1^3+2^3+3^3+\ldots+n^3=\Sigma n^3=\left[\frac{n(n+1)}{2}\right]^2$$

Example 1 **Find the sum of n terms of the series whose nth term is given by $(2n-1)^2$.**

Sol. Given, nth term $T_n=(2n-1)^2 \Rightarrow T_n=4n^2+1-4n$

Now, $S=\Sigma T_n=\Sigma(4n^2+1-4n)$

$$=4\Sigma n^2+\Sigma 1-4\Sigma n$$

$$=\frac{4n(n+1)(2n+1)}{6}+n-\frac{4n(n+1)}{2}$$

$$\left[\because \Sigma 1=n,\ \Sigma n=\frac{n(n+1)}{2} \text{ and } \Sigma n^2=\frac{n(n+1)(2n+1)}{6}\right]$$

$$=n\left[\frac{2(n+1)(2n+1)}{3}+\frac{1}{1}-\frac{2(n+1)}{1}\right]$$

$$=n\left[\frac{2(2n^2+n+2n+1)+3-6(n+1)}{3}\right]$$

$$=\frac{n[4n^2+6n+2+3-6n-6]}{3}$$

$$=\frac{n(4n^2-1)}{3}=\frac{n}{3}(2n+1)(2n-1)$$

$$[\because (a^2-b^2)=(a-b)(a+b)]$$

Example 2 **Find the sum of n terms of the series $1\cdot2+2\cdot3+3\cdot4+4\cdot5+\ldots$.**

Sol. Let $S=1\cdot2+2\cdot3+3\cdot4+4\cdot5+\ldots$

Then, nth term, $T_n=n(n+1)=n^2+n$

On taking summation from 1 to n both sides, we get

$$S_n=\sum_1^n T_n=\sum_1^n n^2+\sum_1^n n$$

$$=\frac{n(n+1)(2n+1)}{6}+\frac{n(n+1)}{2}$$

$$=\frac{n(n+1)}{6}[2n+1+3]$$

$$=\frac{n(n+1)}{6}[2n+4]$$

$$=\frac{n(n+1)(n+2)}{3}$$

Example 3 **Find the sum of the n terms of a series $3\times1^2+5\times2^2+7\times3^2+\ldots$**

Sol. Let given series be $S=3\times1^2+5\times2^2+7\times3^2+\ldots$

First, we will split the given series into two parts which are 3, 5, 7, ... and $1^2, 2^2, 3^2, \ldots$, then find the nth term of each part separately to find the nth term of the given series.

nth term of given series,

$$T_n=(n\text{th term of } 3, 5, 7, \ldots)\times(n\text{th term of } 1, 2, 3, \ldots)^2$$

$$=[3+(n-1)2][1+(n-1)1]^2$$

$$=(3+2n-2)(n)^2=(2n+1)n^2=2n^3+n^2$$

Now, taking summation both sides, we get

$$S=\Sigma T_n=2\Sigma n^3+\Sigma n^2$$

$$=2\left[\frac{n(n+1)}{2}\right]^2+\frac{n(n+1)(2n+1)}{6}$$

$$\left[\because \Sigma n^3 = \left\{\frac{n(n+1)}{2}\right\}^2 \text{ and } \Sigma n^2 = \frac{n(n+1)(2n+1)}{6}\right]$$

$$= \frac{n(n+1)}{2}\left[2\times\frac{n(n+1)}{2} + \frac{2n+1}{3}\right]$$

$$= \frac{n(n+1)}{2}\left[\frac{3n(n+1)+2n+1}{3}\right]$$

$$= \frac{n(n+1)}{6}\times(3n^2+3n+2n+1)$$

$$= \frac{n(n+1)(3n^2+5n+1)}{6}$$

Example 4 Find the sum of the following series upto n terms.

$$\frac{1^3}{1} + \frac{1^3+2^3}{1+3} + \frac{1^3+2^3+3^3}{1+3+5} + \dots$$

Sol. Here, nth term of numerator is $1^3+2^3+3^3+\dots+n^3$, so it can be written as Σn^3 and is nth term of denominator is $1+3+5+\dots+n$ which is an AP, so write its sum of n terms. Then, nth term

$$T_n = \frac{\Sigma n^3}{\frac{n}{2}[2\times 1+(n-1)2]} = \frac{\Sigma n^3}{\frac{n}{2}(2n)} = \frac{\Sigma n^3}{n^2}$$

$$T_n = \frac{\Sigma n^3}{n^2} = \frac{\left[\frac{n(n+1)}{2}\right]^2}{n^2} = \frac{(n+1)^2}{4} = \frac{1}{4}(n^2+2n+1)$$

On taking the summation from 1 to n both sides, we get

$$\Sigma T_n = \Sigma\left[\frac{1}{4}(n^2+2n+1)\right]$$

$$\Rightarrow \quad S_n = \Sigma\frac{(n^2+1+2n)}{4}$$

$$\therefore \quad S_n = \frac{1}{4}[\Sigma n^2 + \Sigma 1 + 2\,\Sigma n]$$

$$\Rightarrow \quad S_n = \frac{1}{4}\left[\frac{n(n+1)(2n+1)}{6} + n + \frac{2n(n+1)}{2}\right]$$

$$= \frac{n}{4}\left[\frac{2n^2+n+2n+1}{6} + 1 + n + 1\right]$$

$$= \frac{n}{4}\left[\frac{2n^2+3n+1+6n+12}{6}\right]$$

$$= \frac{n}{24}[2n^2+9n+13]$$

Example 5 If S_1, S_2, S_3 are the sum of first n natural numbers, their square and their cubes respectively, show that $9S_2^2 = S_3(1+8S_1)$.

Sol. Given, S_1 = Sum of first n natural numbers $= \Sigma n$

$$\Rightarrow \quad S_1 = \frac{n(n+1)}{2} \quad \dots(\text{i})$$

and S_2 = Sum of squares of first n natural numbers $= \Sigma n^2$

$$\Rightarrow \quad S_2 = \frac{n(n+1)(2n+1)}{6} \quad \dots(\text{ii})$$

and S_3 = Sum of cubes of first n natural numbers $= \Sigma n^3$

$$\Rightarrow \quad S_3 = \left[\frac{n(n+1)}{2}\right]^2 \quad \dots(\text{iii})$$

Now, we have to prove, $9S_2^2 = S_3(1+8S_1)$

Taking RHS $= S_3(1+8S_1)$

$$= \left[\frac{n(n+1)}{2}\right]^2\left[1+8\times\frac{n(n+1)}{2}\right]$$

[using Eqs. (i) and (iii)]

$$= \left[\frac{n(n+1)}{2}\right]^2[1+4n(n+1)]$$

$$= \left[\frac{n(n+1)}{2}\right]^2[4n^2+4n+1]$$

$$\left[\because \Sigma n^2 = \frac{n(n+1)(2n+1)}{6},\ \Sigma n = \frac{n(n+1)}{2},\ \Sigma 1 = n\right]$$

$$= \left[\frac{n(n+1)}{2}\right]^2(2n+1)^2$$

On multiplying and dividing by 9, we get

$$= 9\times\frac{n^2(n+1)^2}{4}\times\frac{(2n+1)^2}{9}$$

$$= 9\left[\frac{n(n+1)(2n+1)}{6}\right]^2$$

$$= 9\times S_2^2 = 9S_2^2 \quad \text{[using Eq. (ii)]}$$

= LHS **Hence proved.**

Difference Method of Finding nth Term

Sometimes the nth term of given series cannot be determined directly by observing. If the difference between two successive terms of given series i.e. $(T_2-T_1), (T_3-T_2), (T_4-T_3), \dots$ and so on are either in AP or in GP, then we determined its nth term by the method of difference and then find the sum of the series by using suitable formula.

This method can be understand with the help of following examples

Example 6 Find the sum to n terms of the series $3+15+35+63+\dots$

Sol. Given series is $3+15+35+63+\dots$

Here, the difference between two successive terms are $15-3=12$, $35-15=20$, $63-35=28, \dots$

Clearly, these difference are in AP.

So, we use the difference method to find its nth term.

Let T_n be the nth term and S_n be the sum to n terms of given series. Then,

$$S_n = 3+15+35+63+\ldots+T_{n-1}+T_n \quad \ldots\text{(i)}$$

Also, $S_n = 3+15+35+\ldots+T_{n-1}+T_n \quad \ldots$(ii)

On subtracting Eq. (ii) from Eq. (i), we get

$$0 = 3+12+20+28+\ldots+(T_n - T_{n-1}) - T_n$$

$$\Rightarrow T_n = 3+12+20+28+\ldots+(T_n - T_{n-1})$$

$$= 3+\left[\frac{(n-1)}{2}\{2\times 12+(n-1-1)\,8\}\right]$$

[$\because 12+20+28+\ldots+(T_n - T_{n-1})$ form an AP with first term 12, common difference 8 and number of terms $(n-1)$]

$$= 3+(n-1)(12+4n-8)$$

$$= 3+(n-1)(4n+4)$$

$$= 3+4n^2+4n-4n-4$$

$$\Rightarrow T_n = 4n^2-1$$

Now, on taking summation both sides, we get

$$S_n = \Sigma T_n = \Sigma(4n^2-1) = 4\Sigma n^2 - \Sigma 1$$

$$= 4\left\{\frac{n(n+1)(2n+1)}{6}\right\} - n$$

$$= \frac{n}{3}\{2(n+1)(2n+1)-3\}$$

$$= \frac{n}{3}\{4n^2+2n+4n+2-3\}$$

$$= \frac{n}{3}(4n^2+6n-1)$$

Summation of Some Special Series

Sometimes the given series is of the form

$$\frac{1}{a(a+d)}+\frac{1}{(a+d)(a+2d)}+\frac{1}{(a+2d)(a+3d)}+\ldots$$

For finding the sum of these types of series we firstly write its each term as the difference of two terms and then add them, to get the required sum.

Example 7 **Find the sum of n terms of series**

$$\frac{1}{1\times 2}+\frac{1}{2\times 3}+\frac{1}{3\times 4}+\ldots$$

Sol. Let the given series be

$$S = \frac{1}{1\times 2}+\frac{1}{2\times 3}+\frac{1}{3\times 4}+\ldots$$

Then, nth term $(T_n) = \dfrac{1}{n(n+1)}$

Now, we will split the denominator of the nth term into two parts or we will write T_n as the difference of two terms.

$$\therefore \quad T_n = \frac{1}{n(n+1)} = \frac{(n+1)-n}{n(n+1)}$$

$$= \frac{n+1}{n(n+1)} - \frac{n}{n(n+1)} = \frac{1}{n} - \frac{1}{n+1}$$

On putting $n = 1, 2, 3, 4, \ldots$ successively, we get

$$T_1 = \frac{1}{1}-\frac{1}{2},\quad T_2 = \frac{1}{2}-\frac{1}{3},\quad T_3 = \frac{1}{3}-\frac{1}{4}$$

....

....

$$T_n = \frac{1}{n} - \frac{1}{n+1}$$

On adding all these terms, we get

$$S = T_1 + T_2 + T_3 + \ldots + T_n$$

$$= \left(\frac{1}{1}-\frac{1}{2}\right)+\left(\frac{1}{2}-\frac{1}{3}\right)+\ldots+\left(\frac{1}{n}+\frac{1}{n+1}\right)$$

$$= 1 - \frac{1}{n+1}$$

$$= \frac{1}{1} - \frac{1}{n+1}$$

$$= \frac{n+1-1}{n+1}$$

$$\Rightarrow S = \frac{n}{n+1}$$

Exercise

Multiple Choice Questions

1. The sum of first n natural numbers, is

(a) $\dfrac{n(n+1)(2n+1)}{6}$ (b) $\dfrac{n(n+1)}{2}$

(c) $\left[\dfrac{n(n+1)}{2}\right]^2$ (d) $\dfrac{n(n+1)(n+2)(n+3)}{6}$

2. The sum of their squares, is

(a) $\dfrac{n(n+1)(2n+1)}{6}$ (b) $\dfrac{n(n+1)}{2}$

(c) $\left[\dfrac{n(n+1)}{2}\right]^2$ (d) $\dfrac{n(n+1)(n+2)(n+3)}{6}$

3. The sum of their cubes, is

(a) $\dfrac{n(n+1)(2n+1)}{6}$ (b) $\dfrac{n(n+1)}{2}$

(c) $\left[\dfrac{n(n+1)}{2}\right]^2$ (d) $\dfrac{n(n+1)(n+2)(n+3)}{6}$

4. If nth term is n^2+2^n, then the sum of first n terms of the series is

(a) $\frac{n(n+1)(2n+1)+2(2^n-1)}{6}$

(b) $\frac{n(n+1)(2n+1)}{6}+2(1-2^n)$

(c) $\frac{n(n+1)(2n+1)}{6}+2(2^n-1)$

(d) $\frac{n(n+1)(2n+1)+2(1-2^n)}{6}$

5. If the sum of first n natural numbers is $\frac{1}{5}$ times the sum of their squares, then the value of n is

(a) 5 (b) 6

(c) 7 (d) 8

Short Answer Type I Questions

6. Find the 20th term of the series;
$2\times4+4\times6+6\times8+\ldots+n$ terms.

7. Find the sum of n terms of the series whose nth term is $n(n+1)(n+4)$

8. Find the sum of the series $3.8+6.11+9.14+\ldots$ to n terms.

9. Find the sum upto n terms of the series
$3+7+13+21+31+\ldots$

10. Find the sum of the following series.
$$1\cdot n+2\cdot(n-1)+3\cdot(n-2)+\ldots+(n-1)\cdot2+n\cdot1$$

11. Find the sum upto n term of the series
$1^2+(1^2+2^2)+(1^2+2^2+3^2)+\ldots$

12. Find the sum to n terms of the following series.
$$1+5+12+22+35+\ldots$$

13. Find the nth term of the series $1+2+4+7+\ldots$

Short Answer Type II Questions

14. If T_n denotes the nth term of the series
$2+3+6+11+18+\ldots$, then find the value of T_{50}.

15. Find the sum to n terms of the following series.
$$\frac{3}{1^2\cdot2^2}+\frac{3}{2^2\cdot3^2}+\frac{3}{3^2\cdot4^2}+\ldots$$

16. Find the sum
$$\sum_{r=1}^{n}\frac{1}{(ar+b)(ar+a+b)}.$$

Long Answer Type Questions

17. Find the sum of the following series to n terms.
$$\frac{1}{1\cdot2\cdot3}+\frac{1}{2\cdot3\cdot4}+\frac{1}{3\cdot4\cdot5}+\ldots$$

18. Show that
$$\frac{1\times2^2+2\times3^2+\ldots+n\times(n+1)^2}{1^2\times2+2^2\times3+\ldots+n^2\times(n+1)}=\frac{3n+5}{3n+1}.$$

Answers

1. (b) 2. (a) 3. (c) 4. (c)

5. (c) 6. 1680 7. $\frac{n(n+1)(3n^3+23n+34)}{12}$

8. $3n(n+1)(n+3)$ 9. $\frac{n(n^2+3n+5)}{3}$

10. $\frac{n(n+1)(n+2)}{6}$ 11. $\frac{n(n+1)^2(n+2)}{12}$

12. $\frac{n^2(n+1)}{2}$ 13. $\frac{n^2-n+2}{2}$

14. 2403 15. $\frac{2n+n^2}{(n+1)^2}$

16. $\frac{n}{(a+b)\{(n+1)a+b\}}$ 17. $\frac{n(n+3)}{4(n+1)(n+2)}$

09

CHAPTER

A man is like a fraction whose numerator is what he is and denominator is what he thinks of himself. The larger the denominator, the smaller the fraction. — *Leo Tolstoy*

Straight Lines

YOUR STUDY GOALS

YOUR PRACTICE GOALS

- Objective Type Questions (28 Qs ; 15 MCQs, 10 A-R, 3 Case Based)
- Subjective Type Questions (35 Qs; 11 VSA, 12 SA, 12 LA)

9.1 A Recall of Coordinate Geometry

Distance Formula

The distance between two points $A(x_1, y_1)$ and $B(x_2, y_2)$ is given by

$$AB = \sqrt{(x_2 - x_1)^2 + (y_2 - y_1)^2}$$

or $$AB = \sqrt{(x_1 - x_2)^2 + (y_1 - y_2)^2}$$

Section Formulae (Internal Division)

If point $P(x, y)$ divides the line segment AB, obtained by joining $A(x_1, y_1)$ and $B(x_2, y_2)$ internally in the ratio $m_1 : m_2$, then coordinates of P are

$$x = \frac{m_1x_2 + m_2x_1}{m_1 + m_2},\ y = \frac{m_1y_2 + m_2y_1}{m_1 + m_2}$$

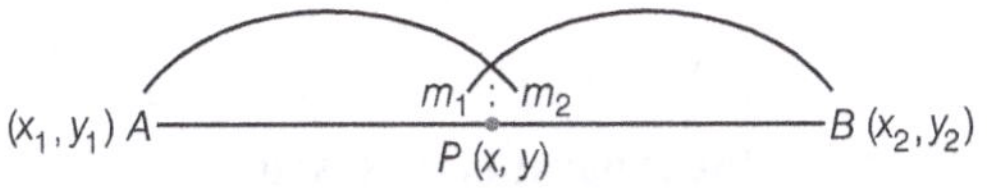

- If $m_1 = m_2$, then the coordinates of the mid-point of the line segment joining the points (x_1, y_1) and (x_2, y_2) are $\left(\frac{x_1 + x_2}{2}, \frac{y_1 + y_2}{2}\right)$.

Collinearity of Three Points

Three points A, B and C are collinear, if they lie on a same straight line,

i.e. if $$AB + BC = AC$$

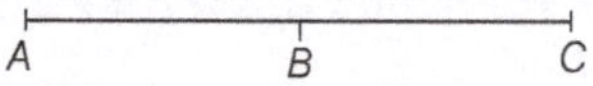

Area of a Triangle

If $A(x_1, y_1), B(x_2, y_2)$ and $C(x_3, y_3)$ are the vertices of a ΔABC, then area of ΔABC

$$= \frac{1}{2}|x_1(y_2 - y_3) + x_2(y_3 - y_1) + x_3(y_1 - y_2)|$$

$$\Delta = \frac{1}{2}|x_1y_2 + x_2y_3 + x_3y_1 - (x_1y_3 + x_2y_1 + x_3y_2)|$$

Note *Three points $A(x_1, y_1), B(x_2, y_2)$ and $C(x_3, y_3)$ are collinear, iff area of $\Delta ABC = 0$.*

9.2 Straight Line

A straight line is a curve such that all the points on the line segment joining any two points of it lie on it.

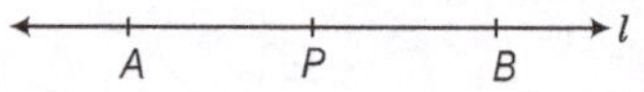

Angle of Inclination of a Line

An angle θ made by the line with positive X-axis in anti-clockwise direction is called angle of inclination of a line. A line in coordinate plane forms two angles with the X-axis, which are supplementary.

Thus, $$0° \le \theta \le 180°$$

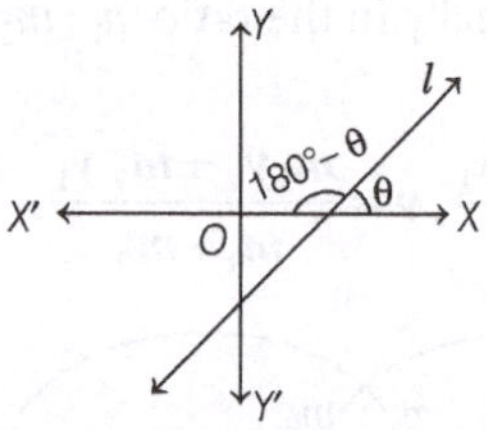

- When $\theta = 0°$, then **line is parallel to X-axis** (horizontal line).
- When $\theta = 90°$, then **line is perpendicular to X-axis** i.e. parallel to Y-axis (vertical line).

Slope or Gradient of a Line

If θ is the angle of inclination of a line l, then $\tan\theta$ is called the slope or gradient of the line l and it is denoted by m

i.e. $$m = \tan\theta,\ \theta \ne 90°$$

If $0° < \theta < 90°$, then $m > 0$ and if $90° < \theta < 180°$, then $m < 0$.

Slope of a Line Joining Two Points

If a line has two points (x_1, y_1) and (x_2, y_2), then the slope of line joining two points (x_1, y_1) and (x_2, y_2) is

$$m = \frac{y_2 - y_1}{x_2 - x_1},\ x_1 \ne x_2$$

Angle between Two Lines

Let l_1 and l_2 be two lines and their inclinations are α_1 and α_2 respectively and their slopes are $m_1 = \tan\alpha_1$ and $m_2 = \tan\alpha_2$.

Then, angle between them is

$$\tan\theta = \pm\frac{m_2 - m_1}{1 + m_1m_2}$$

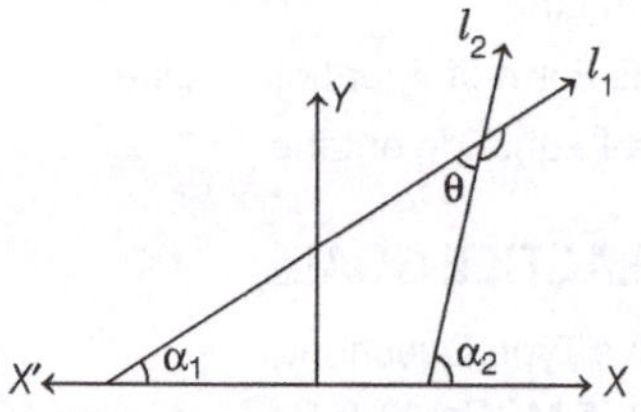

- If $\tan\theta$ is positive, then θ will be an acute angle.
- If $\tan\theta$ is negative, then θ will be an obtuse angle.

Condition of Parallelism of Two Lines

If two lines of slopes m_1 and m_2 are parallel, then the angle θ between them is 0°.

$\therefore$ $$\tan\theta = \tan 0° = 0 \quad [\because \tan 0° = 0]$$

$\Rightarrow$ $$\frac{m_2 - m_1}{1 + m_1m_2} = 0$$

$\Rightarrow$ $$m_2 - m_1 = 0$$

$\therefore$ $$m_1 = m_2$$

Condition of Perpendicularity of Two Lines

If two lines of slopes m_1 and m_2 are perpendicular, then the angle θ between them is $90°$.

$$\therefore \quad \tan\theta = \tan 90° = \frac{\sin 90°}{\cos 90°} = \frac{1}{0}$$

$$\Rightarrow \quad \frac{m_2 - m_1}{1 + m_1 m_2} = \frac{1}{0} \Rightarrow 1 + m_1 m_2 = 0$$

$$\therefore \quad \mathbf{m_1 m_2 = -1}$$

Thus, two lines are perpendicular, if and only if their slopes m_1 and m_2 satisfy the following condition

$$m_1 \cdot m_2 = -1 \quad \text{or} \quad m_1 = -\frac{1}{m_2}$$

TEACHER TIPS

- The slope of X-axis is $m = \tan 0° = 0$.
- The slope of a line, when $\theta = 90°$ is not defined i.e. the slope of Y-axis is not defined.
- Three points A, B and C lying in a plane are collinear, if slope of AB = slope of BC.

Example 1. Find the point on X-axis which is equidistant from the points (3, 2) and (−5, −2).

Sol. Let the point on X-axis be $P(x, 0)$, which is equidistant from (say) $A(3, 2)$ and (say) $B(-5, -2)$.

Since, P is equidistant from A and B.

$$\therefore \quad PA = PB$$

$$\Rightarrow \quad PA^2 = PB^2$$

$$\Rightarrow \quad (3 - x)^2 + (2 - 0)^2 = (-5 - x)^2 + (-2 - 0)^2 \quad \text{[by distance formula]}$$

$$\Rightarrow \quad 9 + x^2 - 6x + 4 = 25 + x^2 + 10x + 4$$

$$\Rightarrow \quad 16x + 16 = 0 \Rightarrow x = -1$$

Thus, point on X-axis is $(-1, 0)$.

Example 2. Show that four points (0, −1), (6, 7), (−2, 3) and (8, 3) are the vertices of a rectangle.

Sol. Let $A(0, -1)$, $B(6, 7)$, $C(-2, 3)$ and $D(8, 3)$ be the given points. Then,

$$AD = \sqrt{(8 - 0)^2 + (3 + 1)^2} \quad \text{[by distance formula]}$$

$$= \sqrt{64 + 16} = \sqrt{80} = 4\sqrt{5} \text{ units,}$$

$$CB = \sqrt{(6 + 2)^2 + (7 - 3)^2}$$

$$= \sqrt{64 + 16} = \sqrt{80} = 4\sqrt{5} \text{ units,}$$

$$AC = \sqrt{(-2 - 0)^2 + (3 + 1)^2}$$

$$= \sqrt{4 + 16} = \sqrt{20} = 2\sqrt{5} \text{ units}$$

and
$$BD = \sqrt{(8 - 6)^2 + (3 - 7)^2}$$

$$= \sqrt{4 + 16} = \sqrt{20} = 2\sqrt{5} \text{ units}$$

Thus, $AD = CB$ and $AC = BD$

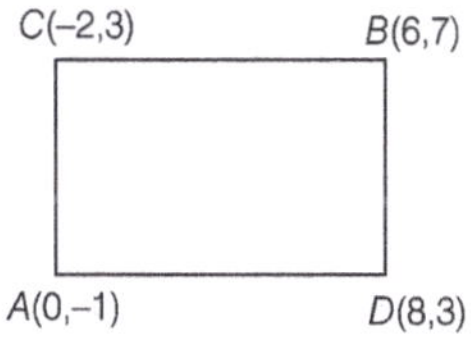

So, $ADBC$ is a parallelogram.

Now,
$$AB = \sqrt{(6 - 0)^2 + (7 + 1)^2}$$

$$= \sqrt{36 + 64} = \sqrt{100} = 10 \text{ units}$$

and
$$CD = \sqrt{(8 + 2)^2 + (3 - 3)^2} = \sqrt{100} = 10 \text{ units}$$

Clearly, $AB = CD$

Hence, $ADBC$ is a rectangle.

Example 3. Find the slope of the line

(a) passing through the points (4, −3) and (0, 5).

(b) making inclination of 30° with the positive direction of X-axis.

Sol. (a) The slope of the line passing through (4, −3) and (0, 5) is

$$m = \frac{5 - (-3)}{0 - 4} = \frac{8}{-4} = -2$$

(b) Here, $\theta = 30°$

$\therefore$ The slope of the line making inclination of 30° with the positive direction of X-axis,

$$m = \tan\theta$$

$$= \tan 30° = \frac{1}{\sqrt{3}}$$

Example 4. Find the slope of a line joining following two points

(i) $A(1, 2)$ and $B(3, -4)$.

(ii) $A(3, -2)$ and $B(7, -2)$.

Sol. We know that slope of a line joining two points (x_1, y_1) and (x_2, y_2) is $m = \dfrac{y_2 - y_1}{x_2 - x_1}$

(i) Given, $A(1, 2) \equiv (x_1, y_1)$ and $B(3, -4) \equiv (x_2, y_2)$

$\therefore$ Slope of line joining points A and B is

$$m = \frac{-4 - 2}{3 - 1} = \frac{-6}{2} = -3$$

(ii) Given, $A(3, -2) \equiv (x_1, y_1)$ and $B(7, -2) \equiv (x_2, y_2)$

$\therefore$ Slope of line joining the points A and B is

$$m = \frac{-2 - (-2)}{7 - 3} = \frac{0}{4} = 0$$

Example 5. Find the angle between the X-axis and the line joining the points (3, – 1) and (4, –2). NCERT

Sol. Given, $(x_1, y_1) \equiv (3, -1)$ and $(x_2, y_2) \equiv (4, -2)$

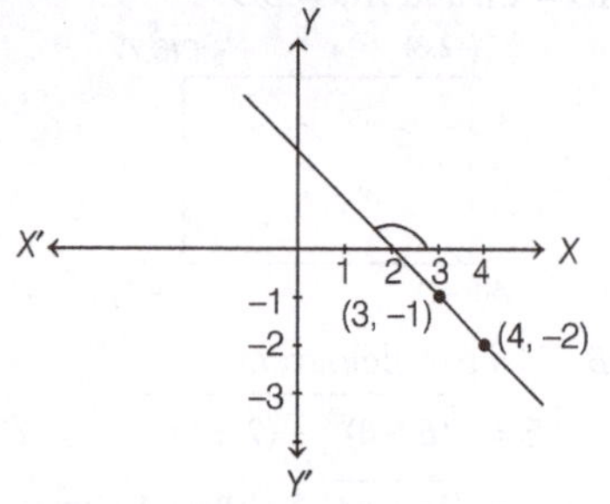

Slope of line, $m = \dfrac{y_2 - y_1}{x_2 - x_1} = \dfrac{-2+1}{4-3}$

$\Rightarrow \quad \tan\theta = \dfrac{-1}{1}$

$\Rightarrow \quad \tan\theta = -1$

$\Rightarrow \quad \tan\theta = -\tan 45^\circ \quad [\because 1 = \tan 45^\circ]$

$\Rightarrow \quad \tan\theta = \tan(180^\circ - 45^\circ)$

$[\because \tan(180^\circ - \theta) = -\tan\theta]$

$\Rightarrow \quad \tan\theta = \tan 135^\circ$

$\Rightarrow \quad \theta = 135^\circ$

Hence, the required angle between X-axis and the line joining the points $(3, -1)$ and $(4, -2)$ is 135°.

Example 6. A ray of light coming from the point (1, 2) is reflected at a point A on the X-axis and then passes through the point (5, 3). Find the coordinates of the point A. NCERT Exemplar

Sol. Let the coordinates of point A be $(x, 0)$. From the figure, the slope of the reflected ray is given by

$$\tan\theta = \frac{3}{5-x} \quad \text{...(i)}$$

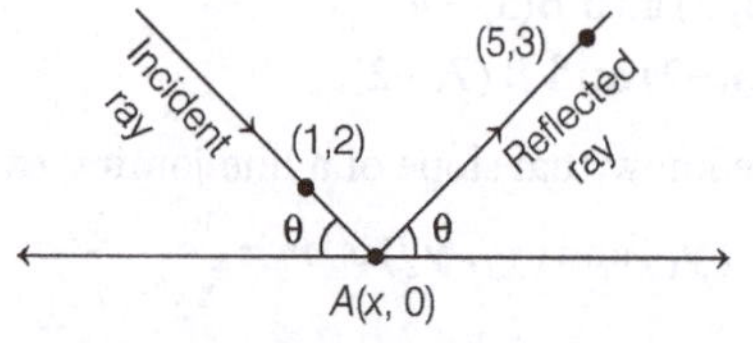

Again, the slope of the incident ray is given by

$$\tan(\pi - \theta) = \frac{-2}{x-1} \quad \left[\because m = \frac{y_2 - y_1}{x_2 - x_1}\right]$$

$\Rightarrow \quad -\tan\theta = \dfrac{-2}{x-1} \quad [\because \tan(\pi - \theta) = -\tan\theta]$

$\Rightarrow \quad \tan\theta = \dfrac{2}{x-1} \quad \text{...(ii)}$

From Eqs. (i) and (ii), we get

$$\frac{3}{5-x} = \frac{2}{x-1}$$

$\Rightarrow \quad 3x - 3 = 10 - 2x$

$\Rightarrow \quad x = \dfrac{13}{5}$

Therefore, the required coordinates of the point A are $\left(\dfrac{13}{5}, 0\right)$.

Example 7. Find the angle between the lines joining the points (0, 0), (2, 3) and the points (2, –2), (3, 5).

Sol. Let θ be the angle between the given lines.

We have,

m_1 = Slope of the line joining $(0, 0)$ and $(2, 3)$

$= \dfrac{3-0}{2-0} = \dfrac{3}{2}$

m_2 = Slope of the line joining $(2, -2)$ and $(3, 5)$

$= \dfrac{5+2}{3-2} = 7$

Now, $\tan\theta = \pm\left(\dfrac{m_2 - m_1}{1 + m_1 m_2}\right) = \pm\left(\dfrac{7 - 3/2}{1 + 7(3/2)}\right)$

$= \pm\left(\dfrac{11/2}{23/2}\right) = \pm\left(\dfrac{11}{23}\right)$

$\Rightarrow \quad \theta = \tan^{-1}\left(\dfrac{11}{23}\right)$ or $\pi - \tan^{-1}\left(\dfrac{11}{23}\right)$

Example 8. If the angle between two lines is $\dfrac{\pi}{4}$ and slope of one of the line is $\dfrac{1}{3}$, then find the slope of the other line.

Sol. Given, the angle between two lines is $\theta = \dfrac{\pi}{4}$.

Slope of a line, $m_1 = \dfrac{1}{3}$

Let slope of another line $= m_2$

We know that angle between two lines is

$$\tan\theta = \left|\frac{m_2 - m_1}{1 + m_1 m_2}\right|$$

$\Rightarrow \quad \tan\dfrac{\pi}{4} = \left|\dfrac{m_2 - \dfrac{1}{3}}{1 + \dfrac{m_2}{3}}\right|$

$\Rightarrow \quad \left|\dfrac{3m_2 - 1}{3 + m_2}\right| = 1$

$$\Rightarrow \quad \frac{3m_2 - 1}{3 + m_2} = \pm 1$$

On taking (+) sign, $\frac{3m_2 - 1}{3 + m_2} = 1$

$$\Rightarrow \quad 3m_2 - 1 = 3 + m_2$$
$$\Rightarrow \quad 2m_2 = 4$$
$$\Rightarrow \quad m_2 = 2$$

On taking (–) sign, $\frac{3m_2 - 1}{3 + m_2} = -1$

$$\Rightarrow \quad 3m_2 - 1 = -3 - m_2$$
$$\Rightarrow \quad 4m_2 = -2$$
$$\Rightarrow \quad m_2 = -\frac{1}{2}$$

$\therefore$ The slope of another line is 2 or $-\frac{1}{2}$.

Example 9. The slope of a line is double of the slope of another line. If tangent of the angle between them is $\frac{1}{3}$, then find the slopes of the lines. *NCERT*

Sol. If slope of one line is m, then the slope of the other line is $2m$. Let angle between these two lines be θ.

Then, $\tan\theta = \frac{1}{3}$ [given]

$$\Rightarrow \quad \left|\frac{2m - m}{1 + 2m \cdot m}\right| = \frac{1}{3} \quad \left[\because \tan\theta = \left|\frac{m_2 - m_1}{1 + m_1 \cdot m_2}\right|\right]$$

$$\Rightarrow \quad \frac{m}{1 + 2m^2} = \pm\frac{1}{3}$$

Here, two cases arise.

Case I When $\frac{m}{1 + 2m^2} = \frac{1}{3}$, then $3m = 1 + 2m^2$

$$\Rightarrow \quad 2m^2 - 3m + 1 = 0$$
$$\Rightarrow \quad 2m^2 - 2m - m + 1 = 0$$
$$\Rightarrow \quad 2m(m-1) - 1(m-1) = 0$$
$$\Rightarrow \quad (m-1)(2m-1) = 0$$
$$\Rightarrow \quad m - 1 = 0 \text{ or } 2m - 1 = 0$$
$$\Rightarrow \quad m = 1 \text{ or } m = \frac{1}{2}$$

As the slopes of the lines are m and $2m$, therefore the slopes of the lines are 1 and 2 or $\frac{1}{2}$ and 1.

Case II When $\frac{m}{1 + 2m^2} = \frac{-1}{3}$, then

$$3m = -1 - 2m^2$$
$$\Rightarrow \quad 2m^2 + 3m + 1 = 0$$
$$\Rightarrow \quad 2m^2 + 2m + m + 1 = 0$$
$$\Rightarrow \quad 2m(m+1) + 1(m+1) = 0$$
$$\Rightarrow \quad (2m+1)(m+1) = 0$$
$$\Rightarrow \quad 2m + 1 = 0 \text{ or } m + 1 = 0$$
$$\Rightarrow \quad m = \frac{-1}{2} \text{ or } m = -1$$

As the slopes of the lines are m and $2m$, therefore the slopes of the lines are $\frac{-1}{2}$ and -1 or -1 and -2.

Example 10. What is the value of y so that the line through $(3, y)$ and $(2, 7)$ is parallel to the line through $(-1, 4)$ and $(0, 6)$?

Sol. Let $A(3, y)$, $B(2, 7)$, $C(-1, 4)$ and $D(0, 6)$ be the given points.

Then, m_1 = Slope of the line $AB = \frac{7 - y}{2 - 3} = (y - 7)$

and m_2 = Slope of the line $CD = \frac{6 - 4}{0 - (-1)} = 2$

Since, AB and CD are parallel.

$$\therefore \quad m_1 = m_2$$
$$\Rightarrow \quad y - 7 = 2 \Rightarrow y = 9$$

Example 11. Line through the points $(-2, 6)$ and $(4, 8)$ is perpendicular to the line through the $(8, 12)$ and $(x, 24)$. Find the value of x. *NCERT*

Sol. Slope of the line through the points $(-2, 6)$ and $(4, 8)$ is

$$m_1 = \frac{8 - 6}{4 - (-2)} = \frac{2}{6} = \frac{1}{3}$$

Slope of the line through the points $(8, 12)$ and $(x, 24)$ is

$$m_2 = \frac{24 - 12}{x - 8} = \frac{12}{x - 8}$$

Since, two lines are perpendicular,

$$\therefore \quad m_1 m_2 = -1$$
$$\Rightarrow \quad \frac{1}{3} \times \frac{12}{x - 8} = -1$$
$$\Rightarrow \quad 4 = -(x - 8)$$
$$\therefore \quad x = 4$$

Example 12. Using slopes, prove that the points $A(-2, -1)$, $B(4, 0)$, $C(3, 3)$ and $D(-3, 2)$ are the vertices of a parallelogram.

Sol. Given points are $A(-2, -1)$, $B(4, 0)$, $C(3, 3)$ and $D(-3, 2)$.

Now, Slope of $AB = \frac{0 - (-1)}{4 - (-2)} = \frac{1}{6}$,

slope of $DC = \frac{3 - 2}{3 - (-3)} = \frac{1}{6}$,

$$\text{slope of } BC = \frac{3-0}{3-4} = \frac{3}{-1} = -3$$

and $$\text{slope of } AD = \frac{2-(-1)}{-3-(-2)} = \frac{3}{-1} = -3$$

$\because$ Slope of AB = Slope of DC, therefore $AB \parallel DC$
and slope of BC = Slope of AD, therefore $BC \parallel AD$.
Thus, $ABCD$ is a parallelogram. **Hence proved.**

Example 13. Without using Pythagoras theorem, show that $A(4, 4)$, $B(3, 5)$ and $C(-1, 1)$ are the vertices of a right angled triangle. NCERT

Sol. In ΔABC, we have, $m_1 = \text{Slope of } AB = \frac{4-5}{4-3} = -1$

and $m_2 = \text{Slope of } BC = \frac{5-1}{3-(-1)} = \frac{4}{4} = 1$

Clearly, $m_1 m_2 = -1$
This shows that AB is perpendicular to BC

i.e. $\angle ABC = \frac{\pi}{2}$

Hence, the given points are the vertices of a right angled triangle. **Hence proved.**

Example 14. Prove that $A(4, 3)$, $B(6, 4)$, $C(5, 6)$ and $D(3, 5)$ are angular points of a square.

Sol. Clearly, $AB = \sqrt{(6-4)^2 + (4-3)^2} = \sqrt{5}$,

$BC = \sqrt{(5-6)^2 + (6-4)^2} = \sqrt{5}$,

$CD = \sqrt{(5-3)^2 + (6-5)^2} = \sqrt{5}$

and $DA = \sqrt{(3-4)^2 + (5-3)^2} = \sqrt{5}$

$\therefore$ $AB = BC = CD = DA$

Now, $m_1 = \text{Slope of } AB = \frac{4-3}{6-4} = \frac{1}{2}$,

$m_2 = \text{Slope of } BC = \frac{6-4}{5-6} = -2$

and $m_3 = \text{Slope of } CD = \frac{5-6}{3-5} = \frac{1}{2}$

Clearly, $m_1 m_2 = \left(\frac{1}{2}\right)(-2) = -1$ and $m_1 = m_3$

Therefore, AB is perpendicular to BC and it is parallel to CD.

Thus, $AB = BC = CA = AD$
$AB \perp BC$ and AB is parallel to CD.
Hence, $ABCD$ is a square.

Quick Test 1

1. Find the point on the X-axis which is equidistant from the points (7, 6) and (3, 4). Competency Based Que
2. If A is a point on the X-axis with abscissa -5 and B is a point on the Y-axis with ordinate 8. Find the distance AB.
3. Using the distance formula, show that the points $A(3, -2)$, $B(5, 2)$ and $C(8, 8)$ are collinear.
4. Find the coordinate of the point which divides internally the join of $(-2, 14)$ and $(8, -6)$ in the ratio 2 : 3.
5. Find the area of ΔABC, the mid-points of whose sides AB, BC and CA are $D(3, -1)$, $E(5, 3)$ and $F(1, -3)$, respectively. Competency Based Que
6. Find the slope of a line whose inclination is
   (i) 30° (ii) 135°
7. Find the inclination of a line whose slope is
   (i) $\frac{1}{\sqrt{3}}$
   (ii) $-\sqrt{3}$
8. Find the slope of a line which passes through the points
   (i) $(0, -3)$ and $(2, 1)$
   (ii) $(-2, 3)$ and $(4, -6)$
   (iii) $(at_1^2, 2at_1)$ and $(at_2^2, 2at_2)$
9. Find the slope of a line perpendicular to the line, which passes through $(0, 8)$ and $(-5, 2)$.
10. What is the value of y so that the line through $(3, y)$ and $(2, 7)$ is parallel to the line through $(-1, 4)$ and $(0, 6)$? Competency Based Que

Answers

1. $\left(\frac{15}{2}, 0\right)$ 2. $\sqrt{89}$ 4. $(2, 6)$ 5. 8 sq. units 6. (i) $\frac{1}{\sqrt{3}}$ (ii) -1 7. (i) 30° (ii) 120° 8. (i) 2 (ii) $-\frac{3}{2}$ (iii) $\frac{2}{t_2 + t_1}$ 9. $\frac{-5}{6}$ 10. 9

For Solutions, See QR Code on Pg. 258

9.3 Various Form of Equation of Lines

Equation of Axes and Lines Parallel to Them

Equation of X-axis

Every point on X-axis having y-coordinate is zero.

$\therefore$ Equation of X-axis, $y = 0$

Equation of Line Parallel to X-axis (Or Equation of Horizontal Line)

Let L be a straight line parallel to X-axis at a distance a from it. Then, the ordinate of every point lying on the line is either a or $-a$.

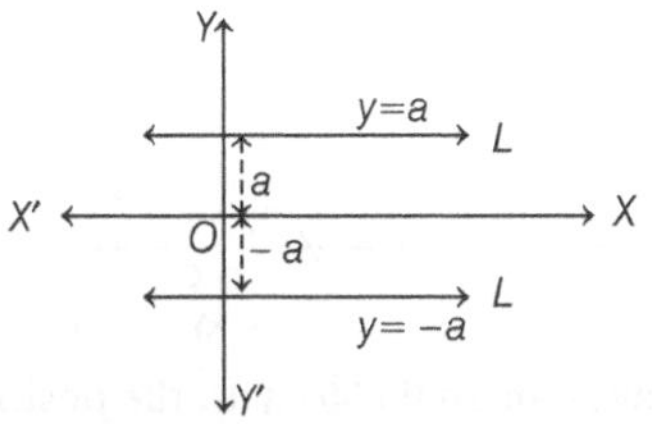

Equation of line parallel to X-axis (i.e. equation of horizontal line either above X-axis or below X-axis) is either

$$y = a \text{ or } y = -a$$

The choice of sign will depend upon the position of the line according as the line is on the positive or negative side of the Y-axis.

Equation of Y-axis

Every point on Y-axis having x-coordinate is zero.

$\therefore$ Equation of Y-axis, $x = 0$

Equation of Line Parallel to Y-axis (Or Equation of Vertical Line)

Let L be a straight line parallel to Y-axis at a distance b from it. Then, the abscissa of every point lying on the line is either b or $-b$.

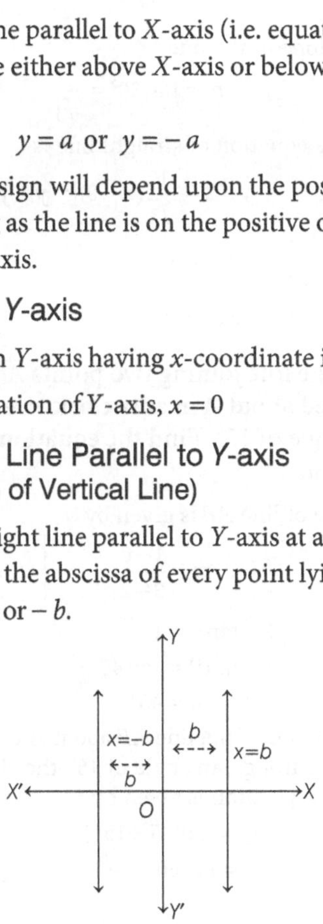

$\therefore$ Equation of line parallel to Y-axis (i.e. equation of vertical line either right of Y-axis or left of Y-axis) is either

$$x = b \text{ or } x = -b$$

Here, the choice of sign will depend upon the position of the line according as the line is on the positive or negative side of the X-axis.

Equation of a Line in Point-Slope Form

On the basis of number of point(s), the point slope form are of two types

(i) **Equation of a Line Passing Through a Point** The equation of the straight line having slope m and passes through given point $Q(x_0, y_0)$ is

$$y - y_0 = m(x - x_0)$$

(ii) **Equation of a Line Passing Through Two Points** The equation of a line passing through the points (x_1, y_1) and $(x_2, y_2,)$ is given by

$$y - y_1 = \frac{y_2 - y_1}{x_2 - x_1}(x - x_1)$$

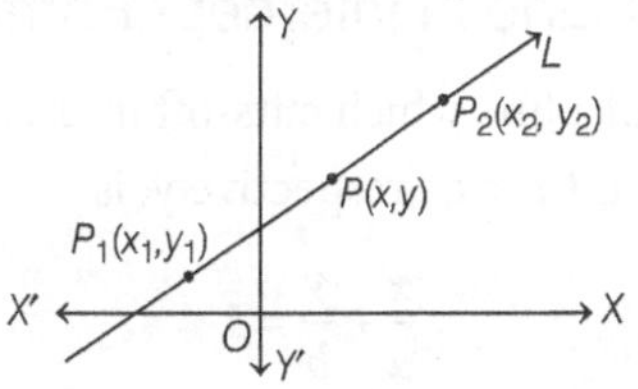

In the given points, if one point is (0, 0), then the equation of line is

$$y - y_1 = \frac{0 - y_1}{0 - x_1}(x - x_1) \text{ or } y - y_1 = \frac{y_1}{x_1}(x - x_1)$$

Intercept Cut by a Line on X and Y-axes

If any line intercept the X and Y-axes at points $A\ (a, 0)$ and $B\ (0, b)$, then $OA = a$ and $OB = b$ be the intercept cut by the X and Y-axes.

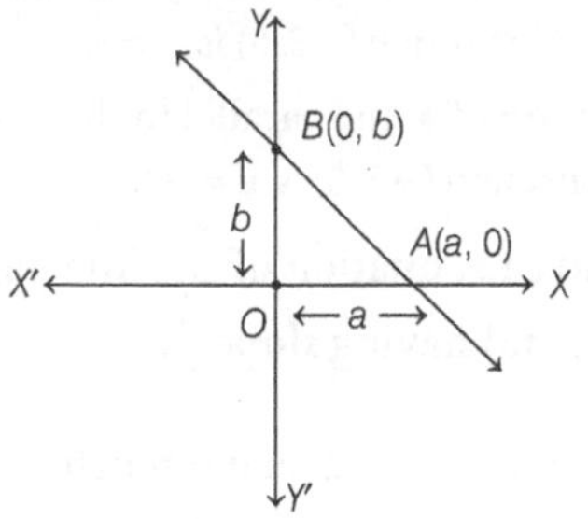

The line intercept X and Y-axes can be positive as well as negative.

Slope Intercept Form

Suppose a line L with slope m cuts the Y-axis at a distance c from the origin (distance c is called the y-intercept of the line).

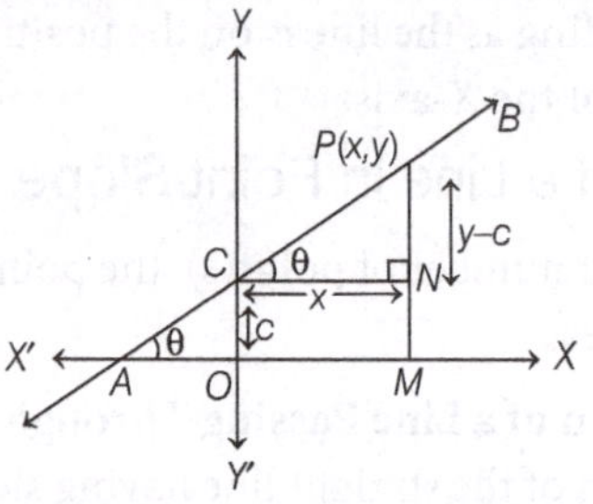

Then, equation of the line L is

$$y = mx + c$$

Suppose a line L with slope m cuts the X-axis at a distance d from the origin i.e. makes x-intercept, d. Then, equation of line L is

$$y = m(x - d)$$

Equation of Line in Intercept Form

The equation of a line which cuts-off intercepts a and b on the X-axis and Y-axis respectively, is

$$\frac{x}{a} + \frac{y}{b} = 1$$

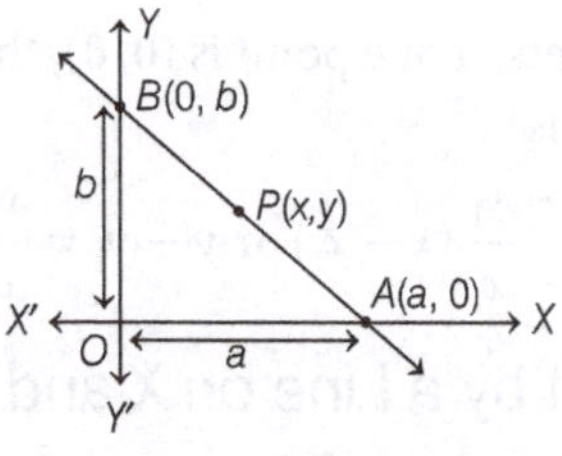

Example 1. Find the equation of the lines parallel to the axes and passing through the point (– 3, 5).

Sol. Clearly, the equation of a line parallel to the X-axis and passing through $(-3, 5)$ is $y = 5$.

The equation of a line parallel to the Y-axis and passing through $(-3, 5)$ is $x = -3$.

Example 2. Find the equation of the line passing through (–4, 3) and having slope $\frac{1}{2}$.

Sol. Equation of the line passing through the point (x_0, y_0) and having slope m is

$$y - y_0 = m(x - x_0) \quad \text{...(i)}$$

Given, m = slope of the line $= \frac{1}{2}$ and $x_0 = -4$, $y_0 = 3$

From Eq. (i), required equation of the line is

$$y - 3 = \frac{1}{2}(x + 4)$$

$$\Rightarrow \quad 2y - 6 = x + 4$$

$$\Rightarrow \quad x - 2y + 10 = 0$$

Example 3. Find the equation of line intersecting the Y-axis at a distance of 2 units above the origin and making an angle of 30° with the positive direction of the X-axis. NCERT

Sol. Given, the line intersects the Y-axis at distance of 2 units above the origin.

It shows that the line passes through (0, 2).

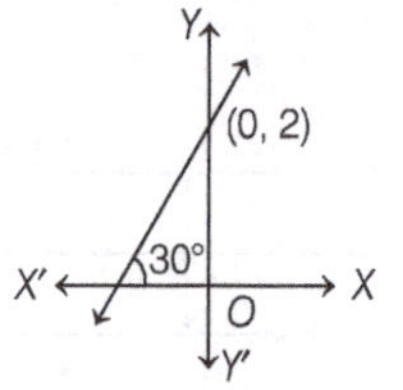

∵ It makes an angle 30° with the positive direction of X-axis.

So, the slope of the line,

$$m = \tan 30^\circ = \frac{1}{\sqrt{3}} \quad [\because m = \tan\theta]$$

Thus, the equation of straight line is

$$y - 2 = \frac{1}{\sqrt{3}}(x - 0) \quad [\because y - y_0 = m(x - x_0)]$$

$$\Rightarrow \quad \frac{x}{\sqrt{3}} - y + 2 = 0$$

$$\Rightarrow \quad x - \sqrt{3}\,y + 2\sqrt{3} = 0$$

Example 4. If the line joining two points $A(2, 0)$ and $B(3, 1)$ is rotated about A in anti-clockwise direction through an angle of 15°. Find the equation of the line in new position. NCERT Exemplar

Sol. The slope of line AB is given by

$$m = \frac{1-0}{3-2} \quad \left[\because m = \left(\frac{y_2 - y_1}{x_2 - x_1}\right)\right]$$

$$\Rightarrow \quad \tan\theta = 1 \quad [\because m = \tan\theta]$$

$$\Rightarrow \quad \tan\theta = \tan 45^\circ \quad [\because 1 = \tan 45^\circ]$$

$$\Rightarrow \quad \theta = 45^\circ$$

After rotation of the line AB about A is anti-clockwise direction through an angle of 15°, the slope of the line AC in new position is given by

$$m_1 = \tan(45^\circ + 15^\circ)$$

$$= \tan 60^\circ = \sqrt{3} \quad [\because \tan 60^\circ = \sqrt{3}]$$

∴ The equation of the new line AC is

$$(y - y_1) = m_1(x - x_1)$$

$$\Rightarrow \quad (y - 0) = \sqrt{3}(x - 2)$$

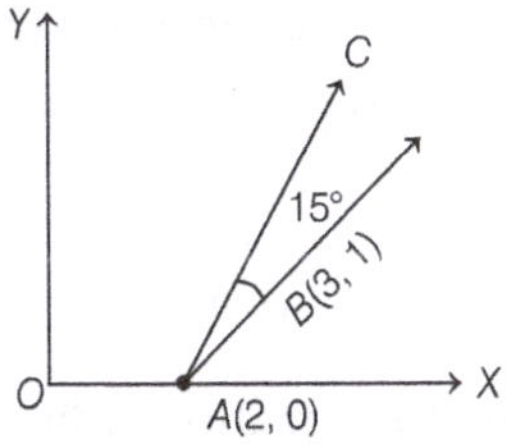

$$\Rightarrow \quad \sqrt{3}x - y - 2\sqrt{3} = 0$$

which is the required equation of the line.

Example 5. Find the equation of the perpendicular bisector of the line segment joining the points $A(2, 3)$ and $B(6, -5)$.

Sol. The slope of AB is given by

$$m = \frac{-5-3}{6-2} = -2 \quad \left[\because m = \frac{y_2 - y_1}{x_2 - x_1}\right]$$

∴ Slope of a line perpendicular to $AB = -\frac{1}{m} = \frac{1}{2}$

Let P be the mid-point of AB.

Then, the coordinates of P are $\left(\frac{2+6}{2}, \frac{3-5}{2}\right)$ i.e. $(4, -1)$.

Thus, the required line passes through $P(4, -1)$ and has slope $\frac{1}{2}$. So, its equation is

$$y + 1 = \frac{1}{2}(x - 4) \quad [\because y - y_0 = m(x - x_0)]$$

$$\therefore \quad x - 2y - 6 = 0$$

Example 6. A line perpendicular to the line segment joining the points (1, 0) and (2, 3) divides it in the ratio $1 : n$. Find the equation of the line. NCERT

Sol. Let the given points be $A(1, 0)$ and $B(2, 3)$.

Let the line PQ divide AB in the ratio $1 : n$ at R internally.

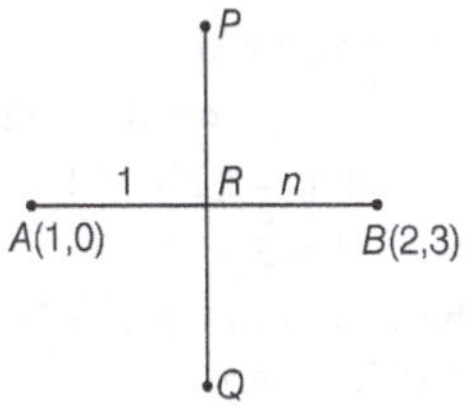

Then, coordinates of R

$$= \left(\frac{1 \times x_2 + n \times x_1}{1+n}, \frac{1 \times y_2 + n \times y_1}{1+n}\right)$$

[by internal division formula]

$$= \left(\frac{1 \times 2 + n \times 1}{1+n}, \frac{1 \times 3 + n \times 0}{1+n}\right)$$

$[\because x_1 = 1, y_1 = 0, x_2 = 2, y_2 = 3]$

$$= \left(\frac{n+2}{n+1}, \frac{3}{1+n}\right)$$

Let slope of line PQ be m.

$\because \quad PQ \perp AB$

∴ Slope of line $PQ \times$ Slope of line $AB = -1$

$[\because m_1 m_2 = -1]$

$$\Rightarrow \quad m \times \frac{y_2 - y_1}{x_2 - x_1} = -1$$

$$\Rightarrow \quad m \times \frac{3-0}{2-1} = -1$$

$[\because x_1 = 1, y_1 = 0, x_2 = 2, y_2 = 3]$

$$\Rightarrow \quad m \times 3 = -1$$

$$\Rightarrow \quad m = -\frac{1}{3}$$

Now, equation of line PQ using $y - y_0 = m(x - x_0)$ is

$$y - \frac{3}{1+n} = \frac{-1}{3}\left(x - \frac{n+2}{n+1}\right)$$

$$\left[\because R\left(\frac{n+2}{n+1}, \frac{3}{1+n}\right) = (x_0, y_0)\right]$$

$$\Rightarrow \quad \frac{3(n+1)y - 9}{1+n} = \frac{-x(n+1) + (n+2)}{n+1}$$

$$\Rightarrow \quad 3(n+1)y - 9 = -x(n+1) + (n+2)$$

$$\Rightarrow \quad x(n+1) + 3(n+1)y = n + 2 + 9$$

$$\Rightarrow \quad x(n+1) + 3(n+1)y = n + 11$$

which is the required equation of line.

Example 7. Find the equation of line passing through the points $(-1, 1)$ and $(2, -4)$. NCERT

Sol. Let the given points be $A(x_1, y_1) \equiv A(-1, 1)$ and $B(x_2, y_2) \equiv B(2, -4)$.

Then, equation of line AB is

$$y - 1 = \frac{-4-1}{2+1}(x + 1)$$

$$\left[\because y - y_1 = \frac{y_2 - y_1}{x_2 - x_1}(x - x_1)\right]$$

$$\Rightarrow \quad y - 1 = \frac{-5}{3}(x + 1)$$

$$\Rightarrow \quad 3y - 3 = -5x - 5$$

$$\therefore \quad 5x + 3y + 2 = 0$$

Example 8. **Find the equations of the sides of a triangle whose vertices are $A(-1, 8)$, $B(4, -2)$ and $C(-5, -3)$.**

Sol. Here, we use two points form to find the equations of sides.

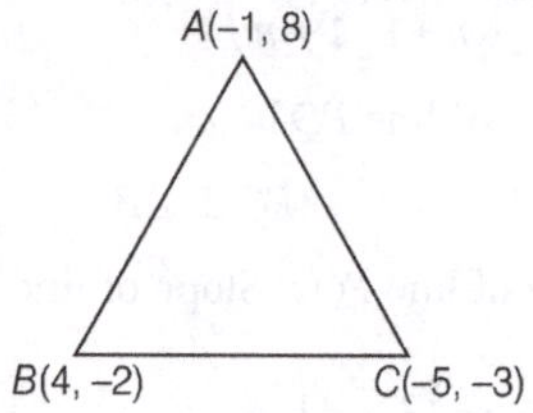

Equation of AB is $y-8=\frac{-2-8}{4+1}(x+1)$

$\Rightarrow \quad 5(y-8)+10(x+1)=0$

$\Rightarrow \quad 10x+5y-30=0$

$\Rightarrow \quad 2x+y-6=0$ [dividing both sides by 5]

Equation of BC is $y+2=\frac{-3+2}{-5-4}(x-4)$

$\Rightarrow \quad -9(y+2)+(x-4)=0$

$\Rightarrow \quad x-9y-22=0$

Equation of AC is $y-8=\frac{-3-8}{-5+1}(x+1)$

$\Rightarrow \quad -4(y-8)+11(x+1)=0$

$\therefore \quad 11x-4y+43=0$

Example 9. **Find the equations of the line which have slope 1/2 and cuts-off an intercept**
(i) –5 on Y-axis.
(ii) 4 on X-axis.

Sol. Given, $m=$ slope of the line $=\frac{1}{2}$

(i) Here, $c=$ intercept of the line on Y-axis $=-5$

Hence, required equation of the line is

$$y=\frac{1}{2}x-5$$

$\Rightarrow \quad x-2y-10=0$ $[\because y=mx+c]$

(ii) Here, $d=$ intercept of the line on X-axis $=4$.

Hence, required equation of the line is

$$y=\frac{1}{2}(x-4)$$

$\therefore \quad x-2y-4=0$ $[\because y=m(x-d)]$

Example 10. **Find the equation of the line intersecting the X-axis at a distance of 3 units to the left of origin with slope –2.**

Sol. Given, the line intersecting the X-axis to the left of origin. It means it cuts the negative X-axis at a distance of 3 units from the origin.

$\therefore$ x-intercept of line on X-axis $(d)=-3$

Hence, required equation of the line is given by

$y=-2(x+3)$ $[\because y=m(x-d)]$

$\Rightarrow \quad y=-2x-6$

$\therefore \quad 2x+y+6=0$

Example 11. **Find the equation of line which intercept the X and Y-axes from –3 units and 2 units, respectively.**

Sol. Given, intercept cut from X-axis, $a=-3$

and intercept cut from Y-axis, $b=2$

The equation of line in intercept form is

$$\frac{x}{a}+\frac{y}{b}=1.$$

$\therefore \quad \frac{x}{-3}+\frac{y}{2}=1$

$\Rightarrow \quad 2x-3y=-6$

$\Rightarrow \quad 2x-3y+6=0$

which is the required equation.

Example 12. **Find the equation of the lines which cuts-off intercepts on the axes whose sum and product are 1 and –6, respectively.**

Sol. Let a and b be the intercepts of the line on X and Y-axes, respectively.

Equation of the line will be $\frac{x}{a}+\frac{y}{b}=1$ ...(i)

Given, sum of intercepts, $a+b=1$ ...(ii)

and product of intercepts, $ab=-6$ ...(iii)

On putting the value of b from Eq. (ii) in Eq. (iii), we get

$a(1-a)=-6$

$\Rightarrow \quad a^2-a-6=0$

$\Rightarrow \quad a^2-3a+2a-6=0$

$\Rightarrow \quad (a-3)(a+2)=0$

$\Rightarrow \quad a=3$ or -2

From Eq. (ii), when $a=3$, $b=-2$

and when $a=-2$, $b=3$.

On putting the values of a and b in Eq. (i), required equations of the lines are

$$\frac{x}{3}+\frac{y}{-2}=1 \text{ and } \frac{x}{-2}+\frac{y}{3}=1$$

$\therefore \quad 2x-3y-6=0$

and $\quad 3x-2y+6=0$

Quick Test 2

1. Find the equation of the line through $(-2, 3)$ with slope -4.

2. Find the equation of line passing through $(0, 2)$ and $(3, -3)$.

3. Find the equation of a line, which is parallel to Y-axis and passes through $(-4, 3)$. *Competency Based Que*

4. If the line $\frac{x}{a}+\frac{y}{b}=1$ passes through the points $(2, -3)$ and $(4, -5)$, then find (a, b).

5. Find the equation of straight line, which passes through the point $(5, 6)$ and has intercepts on the axes equal in magnitude but opposite in sign.

6. Find the equation of the straight line, which passes through the point $(1, -2)$ and cuts-off equal intercepts from axes.

7. Find the equation of line passing through $(2, 2\sqrt{3})$ and inclined with X-axis at an angle of $75°$. *Competency Based Que*

8. The intercept cuts-off by a line from Y-axis is twice than that from X-axis and the line passes through the point $(1, 2)$. Find the equation of the line.

9. If the intercept of a line between the coordinate axes is divided by the point $(-5, 4)$ in the ratio $1:2$, then find the equation of the line. *Competency Based Que*

10. Find the equation of straight line which passes through $(3, 4)$ and the sum of whose intercepts on the coordinates axes is 14.

Answers

1. $4x + y + 5 = 0$ 2. $5x + 3y - 6 = 0$ 3. $x = -4$ 7. $(2+\sqrt{3})x - y - 4 = 0$ 8. $2x + y = 4$
4. $(-1, -1)$ 5. $x - y + 1 = 0$ 6. $x + y + 1 = 0$ 9. $8x - 5y + 60 = 0$ 10. $4x + 3y - 24 = 0$ and $x + y - 7 = 0$

For Solutions, See QR Code on Pg. 258

9.4 General Equation of Line

An equation of the form $Ax + By + C = 0$, where A, B and C are real constants and atleast one of A and B is non-zero, is called general linear equation or general equation of a line.

Different Forms of $Ax + By + C = 0$

The general equation of a line can be reduced into various forms of the equation of a line, which are given below.

Slope Intercept Form

If $B \neq 0$, then $Ax + By + C = 0$ can be written as

$$y = \frac{-A}{B}x - \frac{C}{B} \quad \text{or} \quad y = mx + c \qquad \text{...(i)}$$

where, $m = \frac{-A}{B}$ and $c = \frac{-C}{B}$

Intercept Form

If $C \neq 0$, then $Ax + By + C = 0$ can be written as

$$\frac{x}{\frac{-C}{A}} + \frac{y}{\frac{-C}{B}} = 1 \quad \text{or} \quad \frac{x}{a} + \frac{y}{b} = 1,$$

where $a = \frac{-C}{A}$

and $b = \frac{-C}{B}$

Angle Between Two Lines, having General Equations

If angle between two lines $A_1x + B_1y + C_1 = 0$ and $A_2x + B_2y + C_2 = 0$ is θ, then

$$\tan\theta = \left(\frac{-\frac{A_2}{B_2} + \frac{A_1}{B_1}}{1 + \frac{A_1}{B_1} \times \frac{A_2}{B_2}}\right)$$

- For acute angle, $\theta = \tan^{-1}\left|\frac{A_1B_2 - A_2B_1}{A_1A_2 + B_1B_2}\right|$
- For obtuse angle, $\theta = \pi - \tan^{-1}\left|\frac{A_1B_2 - A_2B_1}{A_1A_2 + B_1B_2}\right|$

Condition For Two Lines to be Parallel

If the lines $A_1x + B_1y + C_1 = 0$ and $A_2x + B_2y + C_2 = 0$ are parallel, then their slopes are equal

i.e. $m_1 = m_2$

$$\Rightarrow \quad -\frac{A_1}{B_1} = -\frac{A_2}{B_2}$$

$$\Rightarrow \quad \frac{A_1}{B_1} = \frac{A_2}{B_2}$$

Condition For Two lines to be Perpendicular

If the lines $A_1x + B_1y + C_1 = 0$ and $A_2x + B_2y + C_2 = 0$ are perpendicular, then the product of their slopes is equal to negative unity

i.e. $$m_1m_2 = -1$$

$$\Rightarrow \quad -\frac{A_1}{B_1} \times \left(-\frac{A_2}{B_2}\right) = -1$$

$$\Rightarrow \quad A_1A_2 + B_1B_2 = 0$$

Equation of a Line Parallel to a Given Line

Equation of a line parallel to a given line $ax + by + c = 0$ is $ax + by + \lambda = 0$, where λ is a constant.

Equation of a Line Perpendicular to a Given Line

Equation of a line perpendicular to a given line $ax + by + c = 0$ is $bx - ay + \lambda = 0$, where λ is a constant.

Distance of a Point from a Line

The distance of a point from a line is the length of perpendicular drawn from the point to the line.

Let $L: Ax + By + C = 0$ be a line, whose perpendicular distance from the point $P(x_1, y_1)$ is d. Then,

$$d = \frac{|Ax_1 + By_1 + C|}{\sqrt{A^2 + B^2}}$$

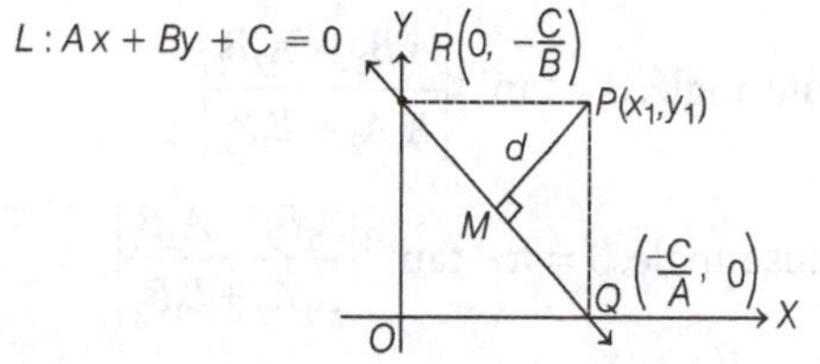

Distance between Two Parallel Lines

If two lines are parallel, then they have the same distance between them through out.

Therefore, the distance between two parallel lines can be find by the following two methods

- The distance between two parallel lines

$$y = mx + c_1 \quad \text{...(i)}$$
$$y = mx + c_2 \quad \text{...(ii)}$$

is given by $d = \dfrac{|c_1 - c_2|}{\sqrt{1 + m^2}}$

- If lines are given in general form i.e. given lines are $Ax + By + C_1 = 0$ and $Ax + By + C_2 = 0$, then distance between these lines

$$d = \frac{|C_1 - C_2|}{\sqrt{A^2 + B^2}}$$

Example 1. **Transform the equation of the line $3x + 2y - 7 = 0$ to**

(i) slope intercept form and also find the slope and y-intercept.

(ii) intercept form and also find the intercepts on the coordinate axes.

Sol. (i) Given equation is $3x + 2y - 7 = 0$.

It can be rewritten as

$$2y = -3x + 7$$

or $$y = -\frac{3}{2}x + \frac{7}{2} \quad \text{...(i)}$$

which is the required slope intercept form of the given line.

On comparing Eq. (i) with $y = mx + c$, we get

Slope, $m = -\dfrac{3}{2}$ and y-intercept, $c = \dfrac{7}{2}$

(ii) Given equation can be rewritten as

$$3x + 2y = 7$$

or $$\frac{x}{7/3} + \frac{y}{7/2} = 1 \quad \text{...(i)}$$

which is the required intercept form of the given line.

On comparing Eq. (i) with $\dfrac{x}{a} + \dfrac{y}{b} = 1$, we get

$$a = \frac{7}{3} \text{ and } b = \frac{7}{2}$$

Hence, x-intercept, $a = \dfrac{7}{3}$ and y-intercept, $b = \dfrac{7}{2}$

Example 2. **Find the angle between lines in which the lines intercept the coordinate axes at (3, 4) and (1, 8).**

Sol. Equation of line intercept the coordinate axes at (3, 4) is

$$\frac{x}{3} + \frac{y}{4} = 1 \quad \text{...(i)}$$

and equation of line intercept the coordinate axes at (1, 8) is

$$\frac{x}{1}+\frac{y}{8}=1 \quad \text{...(ii)}$$

On comparing Eqs. (i) and (ii) by $Ax + By + C = 0$, we get

$$A_1=\frac{1}{3}, B_1=\frac{1}{4}, A_2=1 \text{ and } B_2=\frac{1}{8}$$

Let angle between lines (i) and (ii) be θ. Then,

$$\tan\theta=\left|\frac{\frac{1}{3}\times\frac{1}{8}-1\times\frac{1}{4}}{1\times\frac{1}{3}+\frac{1}{4}\times\frac{1}{8}}\right| \quad \left[\because \tan\theta=\left|\frac{A_1B_2-A_2B_1}{A_1A_2+B_1B_2}\right|\right]$$

$$=\left|\frac{\frac{1}{24}-\frac{1}{4}}{\frac{1}{3}+\frac{1}{32}}\right|=\left|\frac{\frac{1-6}{24}}{\frac{32+3}{96}}\right|=\left|\frac{\frac{-5}{24}}{\frac{35}{96}}\right|=\frac{4}{7}$$

$$\therefore \quad \theta=\tan^{-1}\left(\frac{4}{7}\right)$$

Example 3. Find the angle between the lines $y=\sqrt{3}x+8$ and $y=\frac{1}{\sqrt{3}}x+7$

Sol. Given equation of line is $y=\sqrt{3}x+8$

$$\Rightarrow \quad \sqrt{3}x-y+8=0 \quad \text{...(i)}$$

and $$y=\frac{1}{\sqrt{3}}x+7$$

$$\Rightarrow \quad x-\sqrt{3}y+7\sqrt{3}=0 \quad \text{...(ii)}$$

Here, $A_1=\sqrt{3}, B_1=-1, C_1=8$

and $A_2=1, B_2=-\sqrt{3}, C_2=7\sqrt{3}$

Let angle between lines be θ.

Then, $$\tan\theta=\left|\frac{-3+1}{\sqrt{3}+\sqrt{3}}\right| \quad \left[\because \tan\theta=\left|\frac{A_1B_2-A_2B_1}{A_1A_2+B_1B_2}\right|\right]$$

$$\Rightarrow \quad \tan\theta=\left|\frac{-2}{2\sqrt{3}}\right| \Rightarrow \tan\theta=\frac{1}{\sqrt{3}}=\tan 30^\circ$$

$$\therefore \quad \theta=30^\circ$$

Example 4. Find the equation of the line, which is parallel to $3x-2y+5=0$ and passes through the point (5, −6).

Sol. The equation of any line parallel to the line $3x-2y+5=0$ is $3x-2y+\lambda=0$ ...(i)

This passes through (5, −6).

$$\therefore \quad 3\times5-2\times(-6)+\lambda=0 \Rightarrow \lambda=-27$$

On putting $\lambda=-27$ in Eq. (i), we get

$$3x-2y-27=0,$$

which is the required equation.

Example 5. Find the equation of perpendicular line to line $3x+5y+6=0$ and passing through a point (1, 2).

Sol. Given, equation of line is $3x+5y+6=0$.

$\therefore$ Slope of line is $m_1=-\frac{3}{5}$

$\Rightarrow$ Slope of perpendicular line is $m_2=-\frac{1}{m_1}=\frac{5}{3}$

Now, equation of line passing through a given point (1, 2) and perpendicular to a given line is

$$y-2=\frac{5}{3}(x-1) \quad \left[\because m_2=\frac{5}{3}\right]$$

$$\Rightarrow \quad 3y-6=5x-5$$

$$\Rightarrow \quad 3y-5x=1 \text{ or } 5x-3y+1=0$$

Example 6. Find the distance of the point (2, −3) from the line $2x-3y+6=0$.

Sol. Given equation of line is $2x-3y+6=0$

$\therefore$ Required distance of the point from the line

= The perpendicular distance from point to the line

$$=\frac{|2\times2-3(-3)+6|}{\sqrt{2^2+(-3)^2}}=\frac{|4+9+6|}{\sqrt{4+9}}=\frac{19}{\sqrt{13}}$$

Example 7. Find the points on X-axis whose perpendicular distance from the line $4x+3y=12$ is 4.
Competency Based Que

Sol. Given points are on X-axis, so let its coordinates be $(\alpha, 0)$.

Then, length of perpendicular from $(\alpha, 0)$ to the line $4x+3y-12=0$ is 4.

$$\therefore \quad \frac{|4\alpha+3(0)-12|}{\sqrt{(4)^2+(3)^2}}=4 \Rightarrow \frac{|4\alpha-12|}{\sqrt{16+9}}=4$$

$$\Rightarrow \quad |4\alpha-12|=4\times\sqrt{25}=4\times5$$

$$\Rightarrow \quad \alpha-3=\pm5 \Rightarrow \alpha=8 \text{ or } -2$$

Hence, the required points are (8, 0) and (−2, 0).

Example 8. Find the coordinates of a point on the line $x+y+3=0$, whose distance from the line $x+2y+2=0$ is $\sqrt{5}$.
Competency Based Que

Sol. Let the required point be $P(a, b)$.

Then, $$a+b+3=0 \quad \text{...(i)}$$

Also, $$\frac{|a+2b+2|}{\sqrt{1^2+2^2}}=\sqrt{5}$$

$$\Rightarrow \quad |a+2b+2|=5$$

$$\Rightarrow \quad a+2b+2=5 \text{ or } a+2b+2=-5$$

$$\Rightarrow \quad a+2b=3 \text{ or } a+2b=-7$$

Thus, $$a+2b=3 \quad \text{...(ii)}$$

or $\quad a+2b=-7$ ...(iii)

On solving Eqs. (i) and (ii), we get

$$a=-9 \text{ and } b=6$$

On solving Eqs. (i) and (iii), we get

$$a=1 \text{ and } b=-4$$

Hence, the required points are $(-9, 6)$ and $(1, -4)$.

Example 9. A variable line passes through a fixed point P. The algebraic sum of the perpendiculars drawn from the points (2, 0), (0, 2) and (1, 1) on the line is zero. Find the coordinates of the point P. NCERT Exemplar

Sol. Let slope of the line be m and the coordinates of fixed point P be (x_1, y_1).

Then, equation of line is $y - y_1 = m(x - x_1)$ ...(i)

Let the given points be $A(2, 0)$, $B(0, 2)$ and $C(1, 1)$.

Now, perpendicular distance from A

$$= \frac{0 - y_1 - m(2 - x_1)}{\sqrt{1+m^2}}$$

Perpendicular distance from $B = \dfrac{2 - y_1 - m(0 - x_1)}{\sqrt{1+m^2}}$

and perpendicular distance from C

$$= \frac{1 - y_1 - m(1 - x_1)}{\sqrt{1+m^2}}$$

Now, $\dfrac{\begin{bmatrix} -y_1 - 2m + mx_1 + 2 - y_1 \\ + mx_1 + 1 - y_1 - m + mx_1 \end{bmatrix}}{\sqrt{1+m^2}} = 0$

$\Rightarrow \quad -3y_1 - 3m + 3mx_1 + 3 = 0$

$\Rightarrow \quad -y_1 - m + mx_1 + 1 = 0$

$\Rightarrow \quad y_1 = -m + mx_1 + 1$

On substituting this value in Eq. (i), we get

$$y + m - mx_1 - 1 = mx - mx_1$$

$\Rightarrow \quad y - 1 = m(x - 1)$

Thus, $\quad (x_1, y_1) = (1, 1)$

Example 10. If p is the length of perpendicular from the origin on the line $\dfrac{x}{a} + \dfrac{y}{b} = 1$ and a^2, p^2 and b^2 are in AP, then show that $a^4 + b^4 = 0$. NCERT Exemplar, Competency Based Que

Sol. Given equation of line is $\dfrac{x}{a} + \dfrac{y}{b} = 1$ ...(i)

Perpendicular length from the origin to the line (i) is

$$p = \frac{1}{\sqrt{\dfrac{1}{a^2} + \dfrac{1}{b^2}}} = \frac{ab}{\sqrt{a^2+b^2}}$$

$\Rightarrow \quad p^2 = \dfrac{a^2b^2}{a^2+b^2}$ ...(ii)

Since, a^2, p^2 and b^2 are in AP.

$\therefore \quad 2p^2 = a^2 + b^2$

$\Rightarrow \quad \dfrac{2a^2b^2}{a^2+b^2} = a^2 + b^2$ [using Eq. (ii)]

$\Rightarrow \quad 2a^2b^2 = (a^2+b^2)^2$

$\Rightarrow \quad 2a^2b^2 = a^4 + b^4 + 2a^2b^2$

$\therefore \quad a^4 + b^4 = 0$ **Hence proved.**

Example 11. Find the distance between following parallel lines.

(i) $3x - 4y + 9 = 0$ and $6x - 8y - 15 = 0$

(ii) $l(x+y) + h = 0$ and $l(x+y) - r = 0$ NCERT

Sol. (i) Given parallel lines are

$$3x - 4y + 9 = 0 \quad \text{...(i)}$$

and $\quad 6x - 8y - 15 = 0$

$\Rightarrow \quad 3x - 4y - \dfrac{15}{2} = 0$ ...(ii)

On comparing Eqs. (i) and (ii) with $Ax + By + C_1 = 0$ and $Ax + By + C_2 = 0$, we get

$$A = 3, B = -4, C_1 = 9$$

and $\quad C_2 = \dfrac{-15}{2}$

$\therefore$ Required distance $= \dfrac{|C_1 - C_2|}{\sqrt{A^2+B^2}} = \dfrac{\left|9 + \dfrac{15}{2}\right|}{\sqrt{3^2 + (-4)^2}}$

$$= \frac{\left|\dfrac{18+15}{2}\right|}{\sqrt{9+16}} = \frac{33}{2\sqrt{25}}$$

$$= \frac{33}{2\times 5} = \frac{33}{10} \text{ units}$$

Hence, the required distance is $\dfrac{33}{10}$ units.

(ii) Given parallel lines are

$$lx + ly + h = 0 \text{ and } lx + ly - r = 0$$

On comparing with $Ax + By + C_1 = 0$ and $Ax + By + C_2 = 0$, we get

$$A = l, B = l, C_1 = h, C_2 = -r$$

$\therefore$ Required distance $= \dfrac{|C_1 - C_2|}{\sqrt{A^2 + B^2}}$

$$= \left|\frac{h+r}{\sqrt{l^2+l^2}}\right| = \left|\frac{h+r}{\sqrt{2}\,l}\right|$$

Quick Test 3

1. Transform the equation of the line $2x + 6y - 8 = 0$ to
   (i) slope intercept form and also find the slope and y-intercept.
   (ii) intercept form and also find the intercepts on the coordinate axes.
2. Find the angle between the straight lines, whose equations are $3x + y - 7 = 0$ and $x + 2y + 9 = 0$.
3. Find the equation of a line perpendicular to the line $x - 2y + 3 = 0$ and passing through the point $(1, -2)$.
4. Find the equation of the line passing through $(1, 2)$ and parallel to the line $y = 3x - 1$.
5. Find the coordinates of the foot of perpendicular from the point $(-1, 3)$ to the line $3x - 4y - 16 = 0$.
   Competency Based Que
6. Find the distance of the point $(3, -5)$ from the line $3x - 4y - 26 = 0$.
7. Find the distance between the parallel lines $3x - 4y + 7 = 0$ and $3x - 4y + 5 = 0$.
8. If p is the length of perpendicular from origin to the line whose intercept on the axes are a and b, then show that $\frac{1}{p^2} = \frac{1}{a^2} + \frac{1}{b^2}$.
   Competency Based Que
9. The perpendicular distance of a line from the origin is 5 units and its slope is -1. Find the equation of the line.
   Competency Based Que
10. Find the equation of the lines passing through the point $(1, 0)$ and at a distance $\frac{\sqrt{3}}{2}$ from the origin.
   NCERT Exemplar, Competency Based Que

Answers

1. (i) $y = -\frac{1}{3}x + \frac{4}{3}$, $m = \frac{-1}{3}$ and $c = \frac{4}{3}$
   (ii) $\frac{x}{4} + \frac{y}{\frac{4}{3}} = 1$, $a = 4$ and $b = \frac{4}{3}$
2. $\theta = 45^\circ$ or 135°
3. $y = -2x$
4. $3x - y - 1 = 0$
5. $\left(\frac{68}{25}, \frac{-49}{25}\right)$
6. $\frac{3}{5}$ unit
7. $\frac{2}{5}$ unit
9. $x + y + 5\sqrt{2} = 0$ or $x + y - 5\sqrt{2} = 0$
10. $\sqrt{3}x - y - \sqrt{3} = 0$ or $\sqrt{3}x + y - \sqrt{3} = 0$

For Solutions, See QR Code on Pg. 258

Exam Pattern Questions

Objective Type Questions

Multiple Choice Questions

1. The distance between the points (6, – 4) and (3, 0) is

(a) 7 units (b) 25 units (c) $\sqrt[3]{25}$ units (d) 5 units

2. The coordinates of a point which divides the line segment joining A (1, – 3) and B (– 3, 9) internally in the ratio 1 : 3, are given by

(a) (– 2, 6) (b) (0, 0) (c) $\left(\frac{-6}{4}, \frac{18}{4}\right)$ (d) $\left(\frac{10}{4}, \frac{-30}{4}\right)$

3. The line segment joining the points (–3, –5) and (1, – 3) is divided by Y-axis in the ratio

(a) 1 : 3 (b) 2 : 3 (c) 3 : 1 (d) 3 : 2

4. Area of triangle whose vertices are $(x_1, y_1), (x_2, y_2)$ and (x_3, y_3), is

(a) $|x_1(y_2 - y_3) + x_2(y_3 - y_1) + x_3(y_1 - y_2)|$

(b) $|x_1(y_2 - y_3) + x_2(y_1 - y_3) + x_3(y_1 - y_2)|$

(c) $\frac{1}{2}|x_1(y_2 - y_3) + x_2(y_3 - y_1) + x_3(y_1 - y_2)|$

(d) $\frac{1}{2}|x_1(y_2 - y_3) + x_2(y_1 - y_3) + x_3(y_1 - y_2)|$

5. The angle between the X-axis and the line joining the points (4, – 2) and (5, – 3) is **Competency Based Que**

(a) 45° (b) 135° (c) 90° (d) 180°

6. Two lines $ax + by = c$ and $a'x + b'y = c'$ are perpendicular, if **NCERT Exemplar**

(a) $aa' + bb' = 0$ (b) $ab' = ba'$
(c) $ab + a'b' = 0$ (d) $ab' + ba' = 0$

7. A line passes through (x_1, y_1) and (h, k). If slope of the line is m, then

(a) $y_1 - k = m(h - x_1)$ (b) $k - y_1 = m(h - x_1)$
(c) $h - x_1 = m(y_1 - k)$ (d) $h - x_1 = m(k - y_1)$

8. The inclination of the line $x - y + 3 = 0$ with the positive direction of X-axis is **NCERT Exemplar**

(a) 45° (b) 135° (c) –45° (d) –135°

9. If the line $2y = x + 5$ and $y = 3x + 7$ are equally inclined to the line $y = mx - 5$, then the value of m is

(a) $\frac{1+5\sqrt{2}}{7}$ (b) $\frac{5\sqrt{2}-1}{7}$ (c) $\frac{1-4\sqrt{2}}{7}$ (d) $\frac{-1-5\sqrt{2}}{7}$

10. The equation of line passes through the point (1, 2) and perpendicular to $x + y + 9 = 0$ is

(a) $y - x + 1 = 0$ (b) $y - x - 1 = 0$
(c) $y - x + 2 = 0$ (d) $y - x - 2 = 0$

11. Suppose a line L make x-intercept a and y-intercept b on the axes. Then, the equation of the line L is

(a) $\frac{x}{a} + \frac{y}{b} = 1$ (b) $\frac{x}{a} - \frac{y}{b} = 1$
(c) $\frac{y}{b} - \frac{x}{a} = 1$ (d) $ax + by = 1$

12. A line cutting off intercept –3 from Y-axis and the tangent at an angle to the X-axis is $\frac{3}{5}$, its equation is

(a) $5y - 3x + 15 = 0$ (b) $3y - 5x + 15 = 0$
(c) $5y - 3x - 15 = 0$ (d) $5y + 3x + 15 = 0$

13. The equation of the line which have slope –2 and cuts-off an intercept –6 on X-axis, is **Competency Based Que**

(a) $2x - y + 12 = 0$ (b) $2x + y + 12 = 0$
(c) $-2x + y + 12 = 0$ (d) $2x + y - 12 = 0$

14. The middle point of A (1, 2) and $B(x, y)$ is $C(2, 4)$. If BD is perpendicular to AB such that $CD = 3$ units, then what is the length of BD? **Competency Based Que**

(a) 22 units (b) 2 unit (c) 3 units (d) 32 units

15. The distance of the point $A(1, -3)$ from the line $2y - 3x = 4$ is

(a) 13 units (b) $\frac{7}{13}\sqrt{13}$ units
(c) $\sqrt{13}$ units (d) $\frac{\sqrt{13}}{13}$ unit

16. If the mid-point of the section of the straight line intercepted between the axes is (1, 1), then the equation of the line is **Competency Based Que**

(a) $2x + y = 3$ (b) $2x - y = 1$
(c) $x - y = 0$ (d) $x + y = 2$

17. In a ΔABC, if A is the point (1, 2) and equations of the median through B and C are respectively $x + y = 5$ and $x = 4$, then B is

(a) (1, 4) (b) (7, – 2)
(c) (4, 1) (d) None of these

18. **The reflection of the point (4, – 13) about the line $5x + y + 6 = 0$ is** **NCERT Exemplar**

(a) (–1, – 14) (b) (3, 4) (c) (0, 0) (d) (1, 2)

19. **What can be said regarding a line if its slope is negative?**

(a) θ is an acute angle
(b) θ is the obtuse angle
(c) Either the line is X-axis or it is parallel to the X-axis
(d) None of the above

20. **The equation of line which cuts-off equal and positive intercepts from the axes and passes through the point (α, β) is**

(a) $x + y = \alpha + \beta$ (b) $x + y = \beta$
(c) $x + y = \alpha$ (d) None of these

21. **Equation of a line perpendicular to line $2x + 3y + 4 = 0$ is**

(a) $3x - 2y + 6 = 0$ (b) $2x - 3y + 6 = 0$
(c) $3x + 2y - 6 = 0$ (d) $2x + 3y - 6 = 0$

22. **The distance of the point (2, – 3) from the line $3y = 2x + 4$ is**

(a) $\frac{13}{\sqrt{17}}$ units (b) $\frac{13}{\sqrt{11}}$ units
(c) $\frac{9}{\sqrt{13}}$ units (d) $\frac{17}{\sqrt{13}}$ units

23. **A point moves such that its distance from the point (4, 0) is half that of its distance from the line $x = 16$. The locus of the point is** **NCERT Exemplar**

(a) $3x^2 + 4y^2 = 192$ (b) $4x^2 + 3y^2 = 192$
(c) $x^2 + y^2 = 192$ (d) None of these

24. **The distance between the parallel lines $y = \sqrt{8}x + 6$ and $y = \sqrt{8}x + 4$ is** **Competency Based Que**

(a) $\frac{3}{2}$ units (b) $\frac{2}{3}$ unit
(c) $\frac{1}{3}$ unit (d) $\frac{4}{5}$ unit

25. **The distance between the lines $5x + 3y - 10 = 0$ and $15x + 9y + 5 = 0$ is**

(a) $\frac{35}{\sqrt{34}}$ units (b) $\frac{1}{3\sqrt{34}}$ unit
(c) $\frac{35}{3\sqrt{34}}$ units (d) $\frac{35}{2\sqrt{34}}$ units

Assertion-Reason Based Questions

Directions (Q. Nos. 26-40) *In the questions given below are two statements labelled as Assertion (A) and Reason (R). In the context of the two statements, which one of the following is correct?*

(a) Both A and R are correct; R is the correct explanation of A
(b) Both A and R are correct; R is not the correct explanation of A
(c) A is correct; R is incorrect
(d) R is correct; A is incorrect

26. **Assertion** (A) If θ is the inclination of a line l, then the slope or gradient of the line l is $\tan\theta$.

Reason (R) The slope of a line whose inclination is 90°, is not defined.

27. **Assertion** (A) The values of a for which the straight line $2x + 3y + 4 + a(6x - y + 12) = 0$ is perpendicular to the line $7x + 5y - 4 = 0$ is $\frac{-29}{37}$.

Reason (R) Two line having slopes m_1 and m_2 are perpendicular, if $m_1 = m_2$.

28. **Assertion** (A) The equation of a line which passes through (4, 5) and makes an angle of 30° with positive direction of X-axis is $x - \sqrt{3}y + (5\sqrt{3} - 4) = 0$.

Reason (R) Equation of line having slope m and passes through the point (x_1, y_1) is given by $y - y_1 = m(x - x_1)$.

29. **Assertion** (A) The inclination of the line l may be acute or obtuse.

Reason (R) When the slope of line is zero, then the line is parallel to X-axis. **Competency Based Que**

30. **Assertion** (A) Slope of the line passing through the points (5, – 3) and (5, 4) is 0.

Reason (R) If two lines having the same slope pass through a common point, then these lines will coincide.

31. **Assertion** (A) Equation of line for which $\tan\theta = \frac{1}{5}$, where θ is the inclination of the line and x - intercept is 4 is $x - 5y = 4$.

Reason (R) Equation of line in slope intercept form is $y = mx + c$, where c is the y - intercept.

32. **Assertion** (A) If $A(-2, -1)$, $B(4, 0)$, $C(3, 3)$ and $D(-3, 2)$ are vertices of a parallelogram, then mid-point of AC = mid-point of BD.

Reason (R) The points A, B and C are collinear. So, area of $\Delta ABC = 0$. **Competency Based Que**

33. **Assertion** (A) The equation of the line making intercepts a and b on X and Y-axes respectively is $\frac{x}{a} + \frac{y}{b} = 1$.

Reason (R) The slope of the line $ax + by + c = 0$ is $\frac{b}{a}$.

34. **Assertion** (A) The equation of a line parallel to the line $ax + by + c = 0$ is $ax - by - \lambda = 0$, where λ is a constant.

Reason (R) The equation of a line perpendicular to the line $ax + by + c = 0$ is $bx - ay + \lambda = 0$, where λ is a constant.

35. Assertion (A) The distance of the point (6, 4) from the straight line $2x - 5y + 14 = 0$ is $\frac{6}{\sqrt{34}}$.

Reason (R) The distance of the point (x_1, y_1) from the line $ax + by + c = 0$ is given by $\left|\frac{ax_1 + by_1 + c}{\sqrt{a^2 + b^2}}\right|$.

36. Assertion (A) The distance between parallel lines $9x - 12y = -54$ and $6x - 8y - 15 = 0$ is $\frac{51}{10}$.

Reason (R) Distance between the parallel lines $Ax + By + C_1 = 0$ and $Ax + By + C_2 = 0$, is given by $d = \frac{|C_1 - C_2|}{\sqrt{A^2 + B^2}}$.

Competency Based Que

37. Assertion (A) Equation of the horizontal line having distance p from the X-axis is either $y = p$ or $y = -p$.

Reason (R) Equation of the vertical line having distance q from the Y-axis is either $x = q$ or $x = -q$.

Competency Based Que

38. Assertion (A) Equation of line passing through the point $(-4, -3)$ and parallel to X-axis is $y + 3 = 0$.

Reason (R) Slope of X-axis is 0.

39. Assertion (A) Equation of the line which cuts-off an intercept 4 on the positive direction of X-axis and an intercept 3 on the negative direction of Y-axis is $3x - 4y = 12$.

Reason (R) Equation of a line which cuts-off intercept a and b respectively from X-axis and Y-axis is $\frac{x}{a} + \frac{y}{b} = 1$.

Competency Based Que

40. Assertion (A) The angle between the pair of lines $x - \sqrt{3}y - 5 = 0$ and $\sqrt{3}x + y - 7 = 0$ is 45°.

Reason (R) Angle between the line θ having slopes m_1 and m_2 is given by $\tan\theta = \left|\frac{m_1 - m_2}{1 + m_1 m_2}\right|$.

Case Study Based Questions

41. A person is standing at a point A of a triangular park ABC whose vertices are $A(2, 0)$, $B(3, 4)$ and $C(5, 6)$.

Competency Based Que

Based on the above information, answer the following questions.

(i) He wants to reach BC in least time. Find the equation of the path he should follow.

(a) $2x + y = 3$ (b) $2x + 3y = 4$

(c) $x + y = 2$ (d) $x + 4y = 7$

(ii) Find the shortest distance travelled by him to reach BC.

(a) $\frac{5}{2}\sqrt{2}$ units (b) $\frac{3}{2}\sqrt{2}$ units

(c) $\frac{4}{3}\sqrt{2}$ units (d) $\frac{7}{3}\sqrt{2}$ units

(iii) Suppose he meets BC at point D. Find the coordinates of the point D.

(a) $\left(\frac{5}{2}, \frac{7}{2}\right)$ (b) $\left(\frac{1}{2}, \frac{3}{2}\right)$ (c) $\left(\frac{3}{2}, \frac{1}{2}\right)$ (d) $\left(\frac{7}{2}, \frac{5}{2}\right)$

(iv) Find the area of the triangular park.

(a) 5 sq units (b) 10 sq units

(c) 3 sq units (d) None of these

42. If A and B are two points $(2, -3)$ and $(6, -5)$, respectively. If C is the point between A and B such that it divides the line AB in 1 : 3 ratio.

Based on the above information, answer the following questions.

(i) Find the distane between A and B.

(ii) Find equation of AB.

(iii) What are the coordinates of C?

(iv) Find the slope of line BC.

43. A parking lot in a school is triangular shaped with two of its vertices at $B(-2, 0)$ and $C(1, 12)$. The third vertex A is at the mid-point of the line joining the points (1, 1) and (3, 11).

Based on the above information, answer the following questions.

(i) The coordinate of A is

(a) $(-1, -5)$ (b) $(2, 6)$

(c) $(1, -1)$ (d) $(1, 5)$

(ii) The equation of line parallel to BC and passing through the vertex A is

(a) $4x - y = 2$ (b) $4y - x = 2$

(c) $4x - y = -2$ (d) $4y + x = 2$

(iii) Find the equation of line that passes through the points $B(-2, 0)$ and $C(1, 12)$.

(iv) Find the equation of line perpendicular to BC and passing through the vertex A.

Subjective Questions

Very Short Answer Type Questions

1. Three consecutive vertices of a parallelogram are $(-2, -1)$, $(1, 0)$ and $(4, 3)$, find the fourth vertex. **Competency Based Que**

2. For what value of k are the points $(8, 1)$, $(k, -4)$ and $(2, -5)$ collinear?

3. In what ratio Y-axis divides the line segment joining the points $(3, 4)$ and $(-2, 1)$? **Competency Based Que**

4. Show that the points $(a, 0)$, $(0, b)$ and $(3a, -2b)$ are collinear.

5. Find the equation of straight line cutting off an intercept -1 from Y-axis and makes an angle of 45° with positive direction of X-axis.

6. Find k, so that the line $2x + ky - 9 = 0$ may be perpendicular to $2x + 3y - 1 = 0$. **Competency Based Que**

7. Find the acute angle between lines $x + y = 0$ and $y = 0$.

8. Find the points on the X-axis whose distance from the line $\frac{x}{3} + \frac{y}{4} = 1$ are 4 units. **NCERT**

9. Find the distance between parallel lines $2x + 3y + 6 = 0$ and $4x + 6y = 5$.

10. Find the value of k for which the line $(k - 3)x - (4 - k^2)y + k^2 - 7k + 6 = 0$ is parallel to the X-axis. **Competency Based Que**

11. Find the value of k, if the straight line $2x + 3y + 4 + k(6x - y + 12) = 0$ is perpendicular to the line $7x + 5y - 4 = 0$.

Short Answer Type Questions

12. Without using Pythagoras theorem, show that the points $A(1,2)$, $B(1,5)$ and $C(5,2)$ are the vertices of a right angled triangle.

13. Prove that the line joining the mid-points of the two sides of a triangle is parallel to the third side.

14. If points $(a,0)$, $(0, b)$ and (x,y) are collinear using the concept of slope. Prove that $\frac{x}{a} + \frac{y}{b} = 1$

15. A quadrilateral has the vertices at the points $(4, 1)$, $(1, 7)$, $(-6, 0)$ and $(-1,-9)$. Show that the mid-points of the sides of this quadrilateral are the vertices of a parallelogram.

16. Prove that the points $A(2, 3)$, $B(6, 7)$, $C(6, 9)$ and $D(2, 5)$ are the vertices of a rectangle.

17. Prove that $A(4,3)$, $B(6, 4)$, $C(5, 6)$ and $D(3, 5)$ are the angular points of a square.

18. Find the equations of the bisectors of the angles between the coordinate axes.

19. If $A(2, 0)$, $B(0, 2)$ and $C(0, 7)$ are three vertices taken in order of an isosceles trapezium $ABCD$ in which $AB || DC$. Find the coordinates of D.

20. Find the equation of a line which is perpendicular to the line joining the points $(4, 2)$ and $(3, 5)$ and cuts-off an intercept of length 3 on Y-axis.

21. Find the equation of a line which divides the line segment joining of $(1, 0)$ and $(3, 0)$ in the ratio $2:1$ and perpendicular to it.

22. Find the equation of the right bisector of the line segment joining the points $A(1,0)$ and $B(2,3)$.

23. In what ratio is the line joining the points $(2, 3)$ and $(4, 1)$ divides the segment joining the points $(1, 2)$ and $(4, 3)$?

Long Answer Type Questions

24. Using the concept of slope, prove that medians of an equilateral triangle are perpendicular to the corresponding sides.

25. One side of a square makes an angle α with X-axis and one vertex of the square is at the origin. Prove that the equations of its diagonals are $x(\sin\alpha + \cos\alpha) = y(\cos\alpha - \sin\alpha)$ and $x(\cos\alpha - \sin\alpha) + y(\sin\alpha + \cos\alpha) = a$, where a is the length of the side of the square.

26. Find the equation of the line through the point $A(2,3)$ and making an angle of 45° with the X-axis. Also, determine the length of intercept on it between A and the line $x+y+1=0$.

27. Find the distance of the point $(2,3)$ from the line $2x-3y+9=0$ measured along a line $x-y+1=0$. **Competency Based Que**

28. Find the distance of the point $(2, 5)$ from the line $3x+y+4=0$ measured parallel to the line $3x-4y+8=0$. **Competency Based Que**

29. Find the equation of the straight line passing through the origin and bisecting the portion of the line $ax+by+c=0$ intercepted between the coordinate axes.

30. Find the value of m for which the lines $mx+(2m+3)y+m+6=0$ and $(2m+1)x+(m-1)y+m-9=0$ intersect at a point on Y-axis. **Competency Based Que**

31. Three sides AB, BC and CA of a ΔABC are $5x-3y+2=0$, $x-3y-2=0$ and $x+y-6=0$, respectively. Find the equation of the altitude through the vertex A.

32. A straight line L is perpendicular to the line $5x-y=1$. The area of the triangle formed by the line L and the coordinate axes is 5 sq units. Find the equation of the line L. **Competency Based Que**

33. The sides AB and AC of a ΔABC are $2x+3y=29$ and $x+2y=16$, respectively. If the mid-point of BC is (5, 6), then find the equation of BC.

34. If the slope of a line passing through the point A (3, 2) is $\frac{3}{4}$, then find the points on the line which are 5 units away from the point A. **NCERT Exemplar**

35. Find the perpendicular distance from the origin of the line joining the points $(\cos\theta, \sin\theta)$ and $(\cos\phi, \sin\phi)$. **NCERT**

Hints & Solutions

Objective Type Questions

1. (d) Using distance formula,

$$d=\sqrt{(x_2-x_1)^2+(y_2-y_1)^2}$$

Where, (x_1, y_1) and (x_2, y_2) are the coordinates of the two points.

Here, $(x_1, y_1)=(6,-4)$ and $(x_2, y_2)=(3,0)$

$\therefore$ $$d=\sqrt{(3-6)^2+(0-(-4))^2}$$
$$=\sqrt{9+16}=5 \text{ units}$$

2. (b) Using section formula,

$$x=\frac{mx_2+nx_1}{m+n}$$

and $$y=\frac{my_2+ny_1}{m+n}$$

Here, $(x_1, y_1)=(1,-3), (x_2, y_2)=(-3,9)$

and $m:n=1:3$

So, $$x=\frac{1\times(-3)+3\times 1}{1+3}=\frac{-3+3}{4}=0$$

and $$y=\frac{1\times 9+3\times(-3)}{1+3}=\frac{9-9}{4}=0$$

Hence, the required coordinates are (0, 0).

3. (c) Let Y-axis divides the line segment joining the points $(-3,-5)$ and $(1,-3)$ in the ratio $\lambda:1$.

$\therefore$ Coordinate of that point $=\left(\frac{\lambda-3}{\lambda+1}, \frac{-3\lambda-5}{\lambda+1}\right)$

This point lies on Y-axis. i.e coordinate of point $(0, b)$.

$$\therefore \frac{\lambda-3}{\lambda+1}=0 \Rightarrow \lambda-3=0 \Rightarrow \lambda=3$$

$\therefore$ Required ratio $=\lambda:1=3:1$

4. (c) If $A(x_1, y_1), B(x_2, y_2)$ and $C(x_3, y_3)$ are the vertices of a ΔABC, then area of ΔABC

$$=\frac{1}{2}|x_1(y_2-y_3)+x_2(y_3-y_1)+x_3(y_1-y_2)|$$

5. (b) Slope of the line passing through the points $(4,-2)$ and $(5,-3)$ is

$$m=\frac{y_2-y_1}{x_2-x_1}=\frac{-3-(-2)}{5-4}=-1$$

Slope $=m=\tan\theta$, where θ is the angle between line and positive X-axis.

So, $$m=\tan\theta \Rightarrow -1=\tan\theta$$

$\Rightarrow$ $$\tan\theta=-1=\tan(135°)$$

$\therefore$ $$\theta=135°$$

6. (a) Slops of the line $ax+by=c$ is $-\frac{a}{b}$ and the slope of the line $a'x+b'y=c'$ is $-\frac{a'}{b'}$. The lines are perpendicular if $\frac{-a}{b}\times\frac{-a'}{b'}=-1$

$\Rightarrow aa'=-bb' \Rightarrow aa'+bb'=0$

7. (b) The slope of line passing through (x_1, y_1) and (x_2, y_2) is given by

$$m=\frac{y_2-y_1}{x_2-x_1}$$

So, slope of line passing through (x_1, y_1) and (h, k) is

$$m=\frac{k-y_1}{h-x_1}$$

$\Rightarrow$ $$m(h-x_1)=k-y_1$$

$\therefore$ $$k-y_1=m(h-x_1)$$

8. (a) The given equation of line is $x-y+3=0$.

$\Rightarrow$ $y=x+3 \Rightarrow m=1$

$\therefore \tan\theta=1$

$\therefore$ $\theta=45°$.

9. (a) The given equation of lines are

$$2y=x+5 \Rightarrow y=\frac{1}{2}x+\frac{5}{2} \quad \text{...(i)}$$

and $$y=3x+7 \quad \text{... (ii)}$$

Slopes of lines (i) and (ii) are $m_1=\frac{1}{2}$ and $m_2=3$.

Since, given line $y = mx - 5$ is equally inclined with the lines (i) and (ii).

$$\therefore \quad \left|\frac{m-\frac{1}{2}}{1+m\times\frac{1}{2}}\right| = \left|\frac{3-m}{1+3m}\right|$$

$$\Rightarrow \quad \frac{2m-1}{2+m} = -\left(\frac{m-3}{1+3m}\right)$$

$$\Rightarrow \quad (2m-1)(1+3m) = -(2+m)(m-3)$$

$$\Rightarrow 6m^2 + 2m - 3m - 1 = -\{m^2 + 2m - 3m - 6\}$$

$$\Rightarrow \quad 7m^2 - 2m - 7 = 0$$

$$\Rightarrow \quad m = \frac{2 \pm \sqrt{4+196}}{14} = \frac{2 \pm \sqrt{200}}{14}$$

$$= \frac{2 \pm 10\sqrt{2}}{14} = \frac{1 \pm 5\sqrt{2}}{7}$$

$$\therefore \quad m = \frac{1+5\sqrt{2}}{7} \text{ and } m = \frac{1-5\sqrt{2}}{7}$$

10. (b) Let the slope of the line be m. Then, its equation passes through (1, 2) is given by

$$y - 2 = m(x-1) \quad \dots(i)$$

Again, given equation of line is perpendicular to the given line

$$x + y + a = 0 \Rightarrow y = -x - 9$$

Its slope is -1.

$\therefore$ We have, $m(-1) = -1 \Rightarrow m = 1$

$\therefore$ On putting $m = 1$ in Eq. (i), we get

$$y - 2 = 1(x-1) \Rightarrow y - x - 1 = 0$$

11. (a) The equation of a line which make x-intercept a and y-intercept b on axes is given by

$$\frac{x}{a} + \frac{y}{b} = 1$$

12. (a) Equation of line in intercept form is

$$\frac{x}{a} + \frac{y}{b} = 1 \quad \dots(i)$$

We have, $b = -3$ and $m = \tan\theta = \frac{3}{5}$

$$\Rightarrow \quad \frac{x}{a} - \frac{y}{3} = 1 \Rightarrow \frac{y}{3} = \frac{x}{a} - 1$$

$$\Rightarrow \quad y = \frac{3}{a}x - 3 \quad \dots(ii)$$

From Eq. (ii), slope is $\frac{3}{a} = \frac{3}{5} \Rightarrow a = 5$

On putting the value $a = 5$ in Eq (ii), we get

$$y = \frac{3}{5}x - 3$$

$$\Rightarrow \quad 5y - 3x + 15 = 0$$

13. (b) The equation of a line with slope m and x-intercept d is given by $\quad y = m(x-d)$.

So, $\quad y = -2(x - (-6))$

$\Rightarrow \quad y = -2x - 12$

$\therefore \quad 2x + y + 12 = 0$

14. (b) Given that mid-point of $A(1, 2)$ and $B(x, y)$ is $C(2, 4)$.

$$\frac{1+x}{2} = 2 \text{ and } \frac{2+y}{2} = 4$$

$\Rightarrow \quad x = 3$ and $y = 6$

So, the coordinates of B are (3, 6).

Given that $BD \perp AB$ and $CD = 3$ units

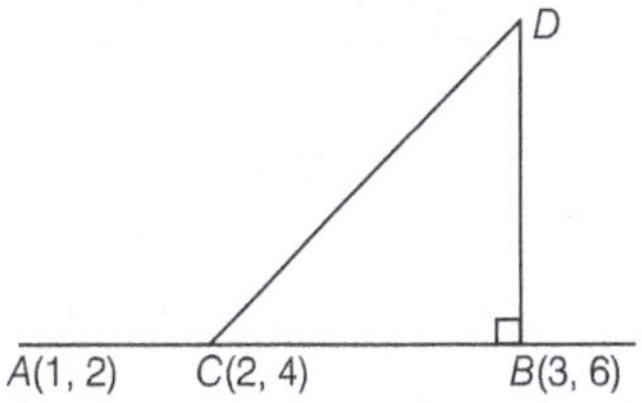

$$\therefore \quad BC = \sqrt{(2-3)^2 + (4-6)^2} = \sqrt{1+4} = \sqrt{5}$$

In right angled ΔCBD,

$$CD^2 = BC^2 + BD^2$$

$\Rightarrow \quad 9 = 5 + BD^2$

$\Rightarrow \quad BD^2 = 4$

$\therefore \quad BD = 2$ units

15. (c) The distance of the point $A(1, -3)$ from the line $2y - 3x - 4 = 0$ is length of perpendicular is

$$d = \left|\frac{2\times(-3) - 3\times 1 - 4}{\sqrt{2^2+3^2}}\right| = \left|\frac{-6-3-4}{\sqrt{13}}\right|$$

$$= \frac{13}{\sqrt{13}} = \sqrt{13} \text{ units}$$

16. (d) Given that (1, 1) is the mid-point of line AB.

Therefore, $\quad 1 = \frac{a+0}{2} \Rightarrow a = 2$

and $\quad 1 = \frac{0+b}{2} \Rightarrow b = 2$

$\therefore$ Required equation of line is $x + y = 2$.

17. (b) The equation of median through B is $x + y = 5$ and the point B lies on it.

Let the coordinates of B be $(x_1, 5 - x_1)$.

Now, CF is a median through C.

So, coordinates of F i.e. mid-point of AB is

$$\left(\frac{x_1+1}{2}, \frac{5-x_1+2}{2}\right).$$

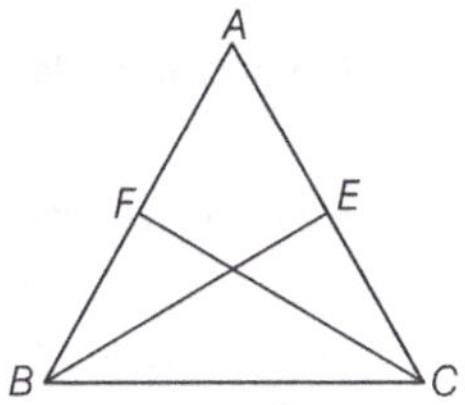

Since, point F lies on $x = 4$.

$\therefore \quad \dfrac{x_1 + 1}{2} = 4$

$\Rightarrow \quad x_1 + 1 = 8 \Rightarrow x_1 = 7$

Hence, the coordinates of B are $(7, -2)$.

18. (a) Let (h, k) be the point of reflection of the given point $(4, -13)$ about the line $5x + y + 6 = 0$. Then, mid point of the line segment joining $(4, -13)$ and (h, k) is given by $\left(\dfrac{4+h}{2}, \dfrac{-13+k}{2}\right)$.

And this point lies on the line $5x + y + 6 = 0$.

$$\Rightarrow \quad 5\left(\frac{4+h}{2}\right) + \left(\frac{-13+k}{2}\right) + 6 = 0$$

$$\Rightarrow \quad 20 + 5h + (-13 + k) + 12 = 0$$

$$\Rightarrow \quad 5h + k + 19 = 0 \quad \ldots\text{(i)}$$

Slope of the line joining the points (h, k) and $(4, -13)$ is given by $\dfrac{k+13}{h-4}$. This line is perpendicular to the given line and hence

$$(-5)\left(\frac{k+13}{h-4}\right) = -1 \Rightarrow 5k + 65 = h - 4$$

$$\Rightarrow h - 5k - 69 = 0 \quad \ldots\text{(ii)}$$

On solving Eqs. (i) and (ii), we get

$h = -1$ and $k = -14$.

Hence, $(-1, -14)$ is the reflection of the given point.

19. (b) When the slope is negative, then

$$\tan\theta = \text{Negative}$$

i.e. θ is obtuse.

Thus, the line makes an obtuse angle $\left(\theta > \dfrac{\pi}{2}\right)$ with the positive X-axis.

20. (a) Let the equation of the line be $\dfrac{x}{a} + \dfrac{y}{b} = 1$ which cuts-off intercepts a and b with the coordinate axes.

It is given that $a = b$, therefore the equation of the line is

$$\frac{x}{a} + \frac{y}{a} = 1 \Rightarrow x + y = a \quad \ldots\text{(i)}$$

But it passes through (α, β).

So, $\quad \alpha + \beta = a$

On puttinig this value in Eq. (i), we get

$$x + y = \alpha + \beta$$

21. (a) Given equation of line is

$$2x + 3y + 4 = 0 \quad \ldots\text{(i)}$$

On comparing Eq. (i) with $ax + by + c = 0$

$$a = 2 \text{ and } b = 3$$

We know that the equation of a line perpendicular to a given line $ax + by + c = 0$ is $bx - ay + \lambda = 0$, where λ is a constant.

$\therefore$ Required equation is $3x - 2y + 6 = 0$.

22. (d) Given equation of line is $2x - 3y + 4 = 0$

$\therefore$ Required distance of the point from the line = The perpendicular distance from point to the line.

$$= \frac{|2 \times 2 - 3 \times (-3) + 4|}{\sqrt{2^2 + (-3)^2}}$$

$$= \frac{|4 + 9 + 4|}{\sqrt{4+9}} = \frac{17}{\sqrt{13}} \text{ units}$$

23. (a) Let (h, k) be the coordinates of the moving point. Then,

$$\sqrt{(h-4)^2 + k^2} = \frac{1}{2}\frac{h-16}{\sqrt{1^2 + 0^2}} = \frac{1}{2}(h - 16)$$

$$\Rightarrow \quad (h-4)^2 + k^2 = \frac{1}{4}(h-16)^2$$

$$\Rightarrow 4(h^2 - 8h + 16 + k^2) = h^2 - 32h + 256$$

$$\Rightarrow \quad 3h^2 + 4k^2 = 192$$

Hence, the required locus of the point $3x^2 + 4y^2 = 192$.

24. (b) We know that the distance between two parallel line $y = mx + c_1$ and $y = mx + c_2$ is given by

$$d = \frac{|c_1 - c_2|}{\sqrt{1+m^2}}$$

$$\therefore \quad d = \frac{|6-4|}{\sqrt{1+(\sqrt{8})^2}} \Rightarrow d = \frac{2}{\sqrt{1+8}} = \frac{2}{3} \text{ unit}$$

25. (c) Distance between parallel lines is $\left|\dfrac{c_1 - c_2}{\sqrt{1+m^2}}\right|$, where m is the slope of the line.

We have,

$$5x + 3y - 10 = 0 \Rightarrow y = \frac{-5}{3}x + \frac{10}{3}$$

And $15x + 9y + 5 = 0 \Rightarrow y = -\dfrac{5}{3}x - \dfrac{5}{9}$

Here, $m = \dfrac{-5}{3}$ and $c_1 = \dfrac{10}{3}, c_2 = \dfrac{5}{9}$

$$\text{Required distance} = \left|\frac{c_1 - c_2}{\sqrt{1+m^2}}\right| = \left|\frac{\frac{10}{3} + \frac{5}{9}}{\sqrt{1+\left(\frac{-5}{3}\right)^2}}\right|$$

$$= \left|\frac{\frac{35}{9}}{\frac{\sqrt{9+25}}{3}}\right| = \frac{35}{3\sqrt{34}} \text{ units}$$

26. (b) We know that if θ is the inclination of a line, then slope of this line, $m = \tan\theta$.

But when θ is 90°, then $m = \tan 90° =$ not defined

Since, the slope of a line whose inclination is 90° is not defined.

Hence, Assertion and Reason both are correct but Reason is not the correct explanation of Assertion.

27. (c) We have,

$$2x + 3y + 4 + a(6k - y + 12) = 0$$

$$\Rightarrow (2 + 6a)x + (3 - a)y + (4 + 12a) = 0 \quad \ldots\text{(i)}$$

and $$7x + 5y - 4 = 0 \quad \ldots\text{(ii)}$$

Slope of line (i) is $m_1 = -\dfrac{2+6a}{3-a}$

Slope of line (ii) is $m_2 = -\dfrac{7}{5}$

Since, lines (i) and (ii) are perpendicular then

$m_1 \cdot m_2 = -1$

$$\Rightarrow \left(-\frac{2+6a}{3-a}\right)\cdot\left(-\frac{7}{5}\right) = -1$$

$$\Rightarrow \frac{14+42a}{15-5a} = -1 \Rightarrow 14 + 42a = -15 + 5a$$

$$\Rightarrow 37a = -29 \Rightarrow a = -\frac{29}{37}$$

Hence, Assertion is correct but Reason is incorrect.

28. (a) Given, slope $m = \tan 30° = \dfrac{1}{\sqrt{3}}$ and given point (4, 5)

$\therefore$ Equation of line is

$$y - 5 = \frac{1}{\sqrt{3}}(x - 4)$$

$$\Rightarrow \sqrt{3}(y-5) = x - 4 \Rightarrow \sqrt{3}y - 5\sqrt{3} = x - 4$$

$$\Rightarrow 0 = x - \sqrt{3}y + (5\sqrt{2} - 4$$

$$\Rightarrow x - \sqrt{3}y + (5\sqrt{3} - 4) = 0$$

Hence, Assertion and Reason both are correct and Reason is the correct explanation of Assertion.

29. (b) If slope $= \tan\theta = 0 \Rightarrow \theta = 0°$

$\therefore$ Slope of a line is zero, then line is parallel to X-axis.

If slope is positive, then $\tan\theta = +$ ve

$$\Rightarrow \theta = \text{acute}$$

i.e. line makes an acute angle $\left(0 < \theta < \dfrac{\pi}{2}\right)$ with the positive X-axis.

If slope is negative, then $\tan\theta = -$ ve

$$\Rightarrow \theta = \text{obtuse}$$

i.e. line makes an obtuse angle $\left(\theta > \dfrac{\pi}{2}\right)$ with the positive X-axis.

Hence, Assertion and Reason both are correct but Reason is not the correct explanation of Assertion.

30. (d) **Hint** Slope $= \dfrac{4-(-3)}{5-5}$ $\quad \left[\because m = \dfrac{y_2 - y_1}{x_2 - x_1}\right]$

$$= \frac{7}{0}$$

which is not defined.

Hence, Assertion is incorrect but Reason is correct.

31. (b) Given slope $= m = \tan\theta = \dfrac{1}{5}$, x - intercept = 4

So line passes through point (4, 0)

$\therefore$ Equation of required line is

$$y - 0 = \frac{1}{5}(x - 4) \Rightarrow 5y = x - 4$$

$$\Rightarrow x - 5y = 4$$

Hence, Assertion and Reason both are correct but Reason is not the correct explanation of Assertion.

32. (b) **Hint** Given, $A(-2,-1), B(4,0), C(3,3)$ and $D(-3,2)$.

Mid-point of $AC = \left(\dfrac{3-2}{2}, \dfrac{3-1}{2}\right)$ i.e. $\left(\dfrac{1}{2}, 1\right)$

Mid-point of $BD = \left(\dfrac{4+(-3)}{2}, \dfrac{0+2}{2}\right)$ i.e. $\left(\dfrac{1}{2}, 1\right)$

Hence, mid-point of $AC =$ mid-point of BD.

Reason is also correct.

33. (c) The equation of line making intercept a and b on X and Y-axes, respectively is given by $\dfrac{x}{a} + \dfrac{y}{b} = 1$

Assertion is correct.

For Reason,

Given line is

$$ax + by + c = 0$$

$$\Rightarrow y = -\frac{ax}{b} - \frac{c}{b}$$

On comparing above equation with $y = mx + c$, we get

$$\text{Slope, } m = -\frac{a}{b}$$

$\therefore$ Reason is incorrect.

34. (d) **Hint** The equation of a line parallel to the line $ax + by + c = 0$ is $ax + by + \lambda = 0$, where λ is constant.

$\therefore$ Assertion is incorrect.

Reason is correct.

35. (a) Required distance

$$= \left|\frac{12 - 20 + 14}{\sqrt{3^2 + (-5)^2}}\right| = \frac{6}{\sqrt{34}}$$

Hence, Assertion and Reason both are correct and Reason is the correct explanation of Assertion.

36. (a) Given parallel lines are $9x-12y=-54$

$\Rightarrow \quad 3x-4y+18=0 \quad \ldots(i)$

and $\quad 6x-8y-15=0$

$\Rightarrow \quad 3x-4y-\frac{15}{2}=0 \quad \ldots(ii)$

On comparing Eqs. (i) and (ii) with $Ax+By+C_1=0$ and $Ax+By+C_2=0$, we get

$$A=3, B=-4, C_1=18 \text{ and } C_2=\frac{-15}{2}$$

$$\text{Required distance}=\frac{|C_1-C_2|}{\sqrt{A^2+B^2}}=\frac{\left|18+\frac{15}{2}\right|}{\sqrt{3^2+(-4)^2}}$$

$$=\frac{51}{2\sqrt{25}}=\frac{51}{10} \text{ units}$$

Reason is also correct.

37. (b) Equation of the horizontal line having distance p from the X-axis either $y=p$ or $y=-p$.

$\therefore$ Assertion is correct.

Reason is also correct but Reason is not correct explanation of Assertion.

38. (b) Slope of a line parallel to X-axis = 0

Since, line passes through $(-4,-3)$ the equation of this line parallel to X-axis is

$$y-y_1=m(x-x_1)$$

$\Rightarrow \quad y+3=0(x-(-4)) \Rightarrow y+3=0$

Assertion is correct.

Also, we know slope of X-axis is 0.

Hence, Assertion and Reason both are correct but Reason is not the correct explanation of Assertion.

39. (a) We know that equation of a line which cuts-off intercept a and b respectively from X and Y-axes is

$$\frac{x}{a}+\frac{y}{b}=1 \quad \ldots(i)$$

Given, $a=4$ and $b=-3$

On putting $a=4$ and $b=-3$ in Eq. (i), we get

$$\frac{x}{4}+\frac{y}{-3}=1$$

$\Rightarrow \quad -3x+4y=-12 \Rightarrow 3x-4y=12$

Hence, Assertion and Reason both are correct and Reason is the correct explanation of Assertion.

40. (d) We are given equation of lines

$x-\sqrt{3}y-5=0 \quad \ldots(i)$

and $\quad \sqrt{3}x+y-7=0 \quad \ldots(ii)$

Slope of line (i)

$$m_1=\frac{1}{\sqrt{3}}$$

Slope of line (ii)

$$m_2=-\sqrt{3}$$

Here $m_1 \cdot m_2=\frac{1}{\sqrt{3}}\times(-\sqrt{3})=-1$

$\Rightarrow$ Given lines are perpendicular to each other.

Hence, Assertion is incorrect but Reason is correct.

41. (i) (c) Consider the figure given below.

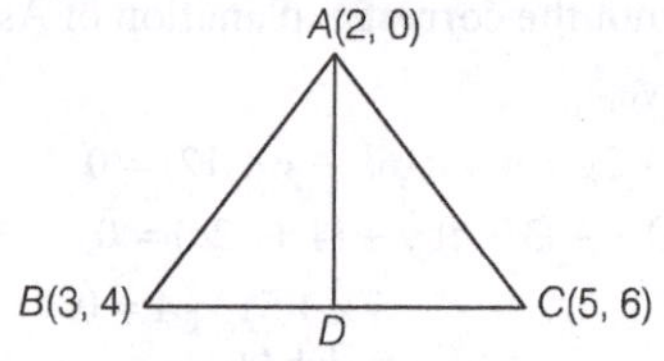

To reach BC in least time, he should follow the path AD.

Slope of $BC=\frac{6-4}{5-3}=1$

Slope of $AD=-1 \quad [\because AD \perp BC]$

Equation of AD is

$$y-0=-1(x-2)$$

$\Rightarrow \quad y=-x+2$

$\Rightarrow \quad x+y=2$

Hence, $x+y=2$ is the required equation of line which he should follow.

(ii) (b) Equation of BC is

$$\frac{y-4}{x-3}=1$$

$\Rightarrow \quad y-4=x-3$

$\Rightarrow \quad x-y+1=0$

Shortest distance travelled by him to reach BC, is

$$AD=\left|\frac{2-0+1}{\sqrt{1^2+1^2}}\right|$$

$$=\frac{3}{\sqrt{2}}=\frac{3\sqrt{2}}{2} \text{ units}$$

(iii) (b) Equation of AD is $x+y=2 \quad \ldots(i)$

Equation of BC is $x-y=-1 \quad \ldots(ii)$

On solving Eqs. (i) and (ii), we get

$$2x=1 \Rightarrow x=\frac{1}{2}$$

So, $\quad y=\frac{3}{2}$

Hence, the coordinates of point D are $\left(\frac{1}{2},\frac{3}{2}\right)$.

(iv) (c) Area of $\Delta ABC=\frac{1}{2}|(x_1(y_2-y_3)+x_2(y_3-y_1)+x_3(y_1-y_2)|$

$$=\frac{1}{2}|2(4-6)+3(6-0)+5(0-4)|$$

$$=\frac{1}{2}|-4+18-20|=\frac{6}{2}=3 \text{ sq units}$$

42. (i) Using distance formula,

$$AB = \sqrt{(6-2)^2 + (-5-(-3))^2} = \sqrt{(4)^2 + (-2)^2}$$
$$= \sqrt{16+4} = \sqrt{20} = 2\sqrt{5} \text{ units}$$

(ii) Slope of $AB = \dfrac{-5-(-3)}{6-2} = \dfrac{-2}{4} = -\dfrac{1}{2}$

Equation of AB is

$$y-(-3) = \frac{-1}{2}(x-2)$$
$$\Rightarrow \quad y+3 = \frac{-x}{2}+1 \Rightarrow \frac{x}{2}+y+2=0$$
$$\therefore \quad x+2y+4=0$$

(iii) Using section formula,

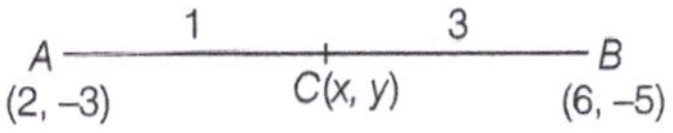

$$x = \frac{6\times1+3\times2}{1+3} = \frac{6+6}{4} = \frac{12}{4} = 3$$

and $y = \dfrac{1\times(-5)+3\times(-3)}{1+3} = \dfrac{-5-9}{4} = \dfrac{-14}{4} = \dfrac{-7}{2}$

Hence, the coordinates of C are $\left(3, \dfrac{-7}{2}\right)$.

(iv) **Hint** Slope of line $BC = \dfrac{\frac{-7}{2}-(-5)}{3-6} = \dfrac{\frac{-7}{2}+5}{-3}$

$$= \frac{\frac{3}{2}}{-3} = \frac{-1}{2}$$

43. (i) (b) A is the mid-point of the points (1, 1) and (3, 11).

$\therefore \quad A = \left(\dfrac{1+3}{2}, \dfrac{1+11}{2}\right)$ [by mid-point formula]

Therefore, the coordinates of A are (2, 6).

(ii) (a) $\because$ Equation of line BC is

$$y-y_1 = \frac{y_2-y_1}{x_2-x_1}(x-x_1)$$
$$\therefore \quad y-0 = \frac{12-0}{1+2}(x-(-2))$$
$$\Rightarrow \quad y = 4x+8$$
$$\Rightarrow \quad 4x-y+8=0$$

Equation of line parallel to BC and passes through (2, 6) is

$$y-6 = 4(x-2) \Rightarrow 4x-y=2$$

(iii) Equation of line passes through two points $B(-2, 0)$ and $C(1, 12)$ is

$$y-y_1 = \frac{y_2-y_1}{x_2-x_1}(x-x_1)$$
$$\Rightarrow \quad y-0 = \frac{12-0}{1-(-2)}(x-(-2))$$
$$\Rightarrow \quad y = 4(x+2)$$
$$\therefore \quad 4x-y+8=0$$

(iv) **Hint** Equation of line perpendicular to BC and passes through (2, 6) is

$$y-6 = -\frac{1}{4}(x-2)$$
$$\therefore \quad x+4y=26$$

Subjective Questions

1. Given, vertices of a parallelogram are $(-2, -1)$, (1, 0) and (4, 3).

Let the fourth vertex be (x, y).

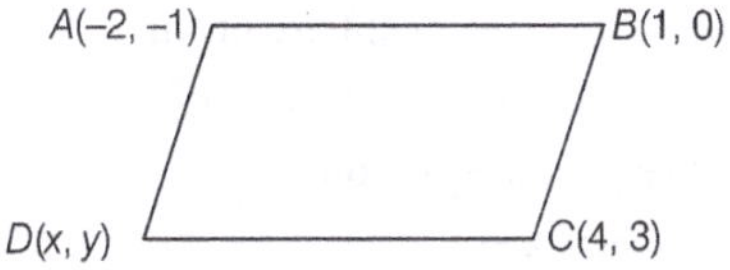

We know that in a parallelogram, diagonals bisect each other

i.e. Mid-point of AC = Mid-point BD

$$\Rightarrow \quad \left(\frac{-2+4}{2}, \frac{-1+3}{2}\right) = \left(\frac{x+1}{2}, \frac{y+0}{2}\right)$$
$$\Rightarrow \quad (1, 1) = \left(\frac{x+1}{2}, \frac{y+0}{2}\right)$$
$$\Rightarrow \quad x+1=2, y+0=2$$
$$\therefore \quad x=1,\ y=2$$

2. The points $A(8, 1)$, $B(k, -4)$ and $C(2, -5)$ are collinear, if ar $(\Delta ABC) = 0$.

$$\therefore \quad \frac{1}{2}|8(-4+5)+k(-5-1)+2(1-(-4))| = 0$$
$$\Rightarrow \quad \frac{1}{2}|8-6k+10| = 0$$
$$\Rightarrow \quad 6k = 18$$
$$\therefore \quad k = 3$$

3. Let the given points be $A(x_1, y_1) = (3, 4)$ and $B(x_2, y_2) = (-2, 1)$. Let $m:n$ be the ratio in which the line segment is divided by Y-axis i.e. the coordinates of the point are $(0, y)$.

$$\therefore \quad (0, y) = \left(\frac{mx_2+nx_1}{m+n}, \frac{my_2+ny_1}{m+n}\right)$$
$$\Rightarrow \quad (0, y) = \left(\frac{-2m+3n}{m+n}, \frac{m+4n}{m+n}\right)$$
$$\Rightarrow \quad 0 = \frac{-2m+3n}{m+n}, y = \frac{m+4n}{m+n}$$
$$\Rightarrow \quad -2m+3n=0 \Rightarrow 2m=3n$$
$$\therefore \quad m:n = 3:2$$

4. Let the given points be $A(x_1, y_1) = (a, 0), B(x_2, y_2) = (0, b)$ and $C(x_3, y_3) = (3a, -2b)$

$\because$ The points A, B and C are collinear.

$\therefore$ Area of $\Delta ABC = 0$

$$\Rightarrow \frac{1}{2}|x_1(y_2 - y_3) + x_2(y_3 - y_1) + x_3(y_1 - y_2)|$$

$$= \frac{1}{2}|a(b - (-2b)) + 0(-2b - 0) + 3a(0 - b)|$$

$$= \frac{1}{2}|3ab + 0 - 3ab| = 0$$

Hence, the given points are collinear.

5. As slope and y-intercept are given

$$c = -1 \text{ and } m = \tan 45^\circ = 1$$

So, the required equation is

$$y = x - 1$$

6. Hint Slope of line $2x + ky - 9 = 0$ is $-\frac{2}{k}$.

Slope of line $2x + 3y - 1 = 0$ is $-\frac{2}{3}$.

Since, both the lines are perpendicular.

$$\therefore \quad \frac{-2}{k} \times \frac{-2}{3} = -1 \Rightarrow k = \frac{-4}{3}$$

7. Hint Given lines are $x + y = 0$ and $y = 0$.

$$\theta = \tan^{-1}\left|\frac{1-0}{0+1}\right| = \tan^{-1}(1) = 45^\circ$$

8. Let the required point be $(\alpha, 0)$, then length of the perpendicular from $(\alpha, 0)$ on $\frac{x}{3} + \frac{y}{4} = 1$ or $4x + 3y - 12 = 0$ is 4.

$$\therefore \quad \left|\frac{4\alpha + 3 \times 0 - 12}{\sqrt{4^2 + 3^2}}\right| = 4$$

$$\Rightarrow \quad |4\alpha - 12| = 20$$

$$\Rightarrow \quad \alpha - 3 = \pm 5$$

$$\Rightarrow \quad \alpha = 8, -2$$

Hence, the required points are $(8, 0)$ and $(-2, 0)$.

9. Lines are $2x + 3y + 6 = 0$ and $4x + 6y = 5$

Here, coefficients of x and y are proportional, so we make them same as

$$2x + 3y + 6 = 0 \quad \text{...(i)}$$

and

$$2x + 3y - \frac{5}{2} = 0 \quad \text{...(ii)}$$

$\therefore$ Distance between lines (i) and (ii) is

$$d = \left|\frac{6 - \left(-\frac{5}{2}\right)}{\sqrt{4+9}}\right| = \left|\frac{12+5}{2\sqrt{13}}\right| = \frac{17}{2\sqrt{13}} \text{ units}$$

10. Hint Write the given equation in the form $y = mx + c$ and compare with it.

$$\text{Slope} = \frac{(k-3)}{4-k^2}$$

If parallel to the X-axis, then

$$\text{slope} = 0$$

$$\Rightarrow \quad k - 3 = 0$$

$$\Rightarrow \quad k = 3$$

11. Slope of the line $2x + 3y + 4 + k(6x - y + 12) = 0$ is $\left\{-\frac{(2+6k)}{3-k}\right\}$ and slope of the line $7x + 5y - 4 = 0$ is $\left(\frac{-7}{5}\right)$.

If lines are perpendicular, then

$$\left\{-\frac{(2+6k)}{3-k}\right\} \times \frac{(-7)}{5} = -1 \quad [\because m_1 \times m_2 = -1]$$

$$\Rightarrow \quad 14 + 42k = -15 + 5k$$

$$\therefore \quad k = -\frac{29}{37}$$

12. Hint Slope of AB = Not defined

Slope of $BC = \frac{-3}{4}$ and slope of $AC = 0$

$\because AB$ is parallel to Y-axis and AC is parallel to X-axis.

Thus, $\angle A = 90^\circ$

[$\because$ angle between X-axis and Y-axis is 90°]

13. Hint Let $A(x_1, y_1)$, $B(x_2, y_2)$ and $C(x_3, y_3)$ are the vertices of ΔABC. Also, let D, E and F are the mid-points of sides AB, BC and CA, respectively.

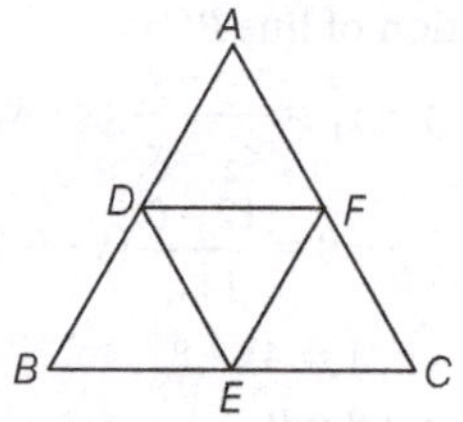

Then, coordinates of $D = \left(\frac{x_1 + x_2}{2}, \frac{y_1 + y_2}{2}\right)$,

coordinates of $E = \left(\frac{x_2 + x_3}{2}, \frac{y_2 + y_3}{2}\right)$

and coordinate of $F = \left(\frac{x_1 + x_3}{2}, \frac{y_1 + y_3}{2}\right)$

Now, show that slope of DF = slope of BC,

slope of DE = slope of AC

and slope of EF = slope of AB

14. Let the given points be $A(a, 0)$, $B(0, b)$ and $C(x, y)$.
Now,

Slope of AB = Slope of BC [$\because A, B$ and C are collinear]

$$\Rightarrow \quad \frac{b}{-a} = \frac{y-b}{x} \Rightarrow xb = -ay + ab$$

$$\Rightarrow \quad ay + bx = ab$$

$$\therefore \quad \frac{x}{a} + \frac{y}{b} = 1$$

Hence proved.

15. Hint Let $ABCD$ be the given quadrilateral.

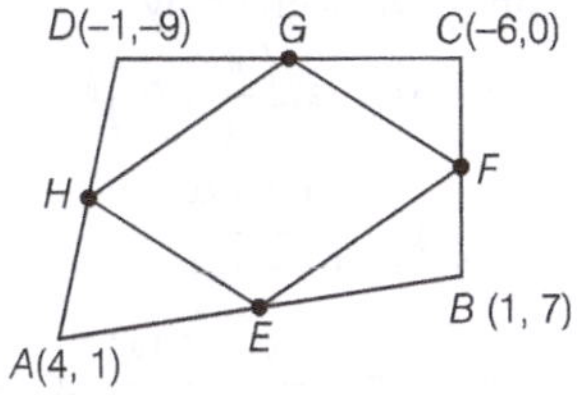

Also, let E, F, G and H be the mid-points of sides AB, BC, CD and DA, respectively.

Now, find the coordinates of E, F, G and H and then show that slope of EF = slope of GH and slope of FG = slope of EH.

16. Hint If A, B, C and D are the vertices of a rectangle, then following condition will be satisfied.

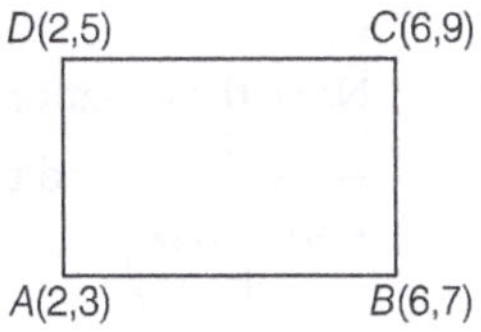

(i) Slope of AB = Slope of DC
(ii) Slope of AD = Slope of BC
(iii) Slope of $AB \times$ Slope of $BC = -1$

17. Hint If given points A, B, C and D are the angular points of a square, then following condition will be satisfied.

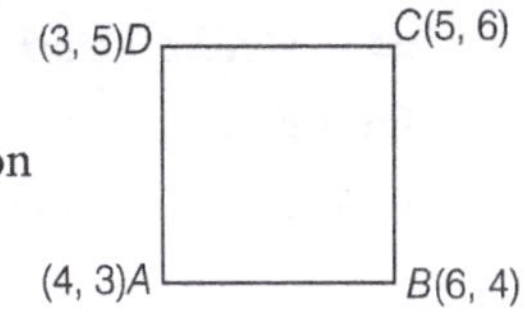

(i) Slope of AB = Slope of DC
(ii) Slope of AD = Slope of BC
(iii) Slope of $AC \times$ Slope of $BD = -1$
(iv) Slope of $AB \times$ Slope of $BC = -1$

18. Hint Bisectors of the angles between the coordinate axes, pass through origin and make an angle of 45° or 135° with positive direction of X-axis.

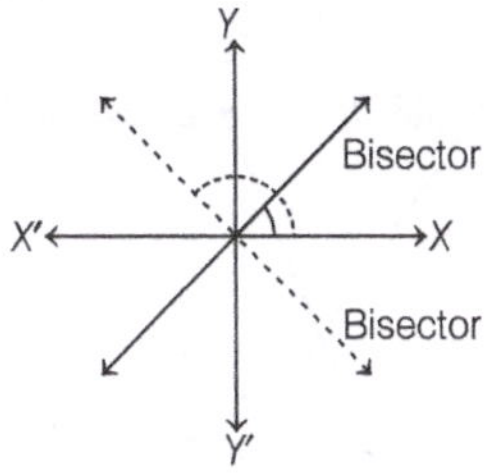

Ans. $y = \pm x$

19. Hint Given, $ABCD$ is an isosceles trapezium and $AB \parallel DC$.

We have, $AD = BC$ and Slope of AB = Slope of DC

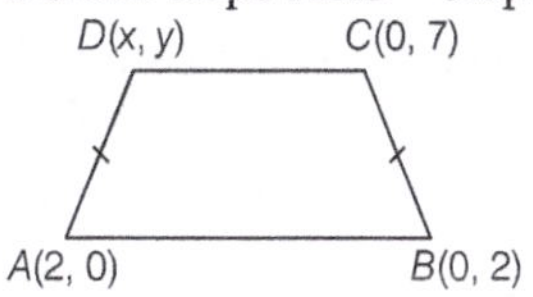

Let the coordinates of D be (x, y).

Then, we have $x^2 - 4x + y^2 = 21$ and $x + y = 7$

On solving these two equations, we get

$$(x, y) = (2, 5) \text{ or } (7, 0)$$

20. Hint Slope of line joining the points (4, 2) and (3, 5) is -3.

$\therefore$ Slope of line perpendicular to it is $\frac{1}{3}$.

$\because$ The perpendicular line cuts-off an intercept of length 3 units on Y-axis.

$\therefore$ Equation of the line is $y = \frac{1}{3}x + 3$

$$\Rightarrow \quad x - 3y + 9 = 0.$$

21. Slope of perpendicular line $= \dfrac{-1}{\text{Slope of } AB}$

$$= \frac{-1}{0} = \text{Not defined}$$

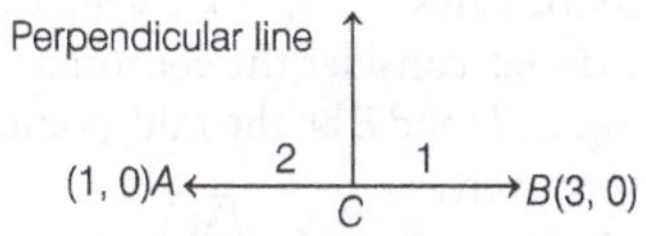

$\Rightarrow$ Perpendicular line is parallel to Y-axis.

$\because$ The coordinates of C are $\left(\frac{7}{3}, 0\right)$.

$\therefore$ Equation of the perpendicular line is $x = \frac{7}{3}$

i.e. $3x = 7$.

22. Hint Mid-point of $AB = \left(\frac{3}{2}, \frac{3}{2}\right)$

Slope of right bisector $= \dfrac{-1}{\text{Slope of } AB} = \dfrac{-1}{3}$

Equation of right bisector is given by

$$y - y_1 = m(x - x_1)$$

$$\therefore \quad y - \frac{3}{2} = \frac{-1}{3}\left(x - \frac{3}{2}\right) \Rightarrow x + 3y - 6 = 0$$

23. The equation of the line joining the points (2, 3) and (4, 1) is

$$y - 3 = \frac{1-3}{4-2}(x-2)$$

$$\Rightarrow \quad y - 3 = -x + 2 \Rightarrow x + y - 5 = 0 \quad \ldots\text{(i)}$$

Suppose, the line joining (2, 3) and (4, 1) divides the segment joining (1, 2) and (4, 3) at point P in the ratio $\lambda : 1$.

Then, the coordinates of P are $\left(\frac{4\lambda+1}{\lambda+1}, \frac{3\lambda+2}{\lambda+1}\right)$.

Clearly, P lies on line (i).

$\therefore \quad \frac{4\lambda+1}{\lambda+1} + \frac{3\lambda+2}{\lambda+1} - 5 = 0 \Rightarrow \lambda = 1$

Hence, the required ratio is $\lambda : 1$ i.e. $1 : 1$.

24. Hint Let ABC be an equilateral triangle of side $2a$.

For sake of convenience, consider BC along X-axis such that mid-point of BC be the origin. Then, the coordinates of B and C are $(-a, 0)$ and $(a, 0)$.

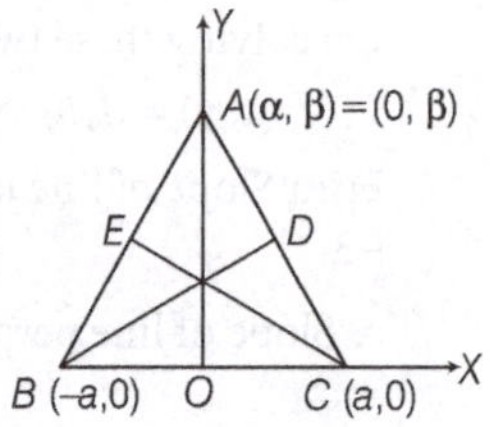

Let the coordinates of A be (α, β).

Since, ABC is an equilateral triangle.

$\therefore \quad AB = AC = BC \Rightarrow AB = AC$

$\Rightarrow \quad AB^2 = AC^2 \Rightarrow \alpha = 0$

Also, $\quad AB^2 = BC^2$

$\Rightarrow \quad a^2 + \beta^2 = (2a)^2 \Rightarrow \beta^2 = 3a^2$

$\Rightarrow \quad \beta = \pm\sqrt{3}a$

Thus, the coordinates of A are $(0, \pm\sqrt{3}a)$. Consider A on positive Y-axis, i.e. consider the coordinates of A as $(0, \sqrt{3}a)$. Now, let D and E be the mid-points of sides AC and AB, respectively.

Then, coordinates of $D = \left(\frac{a}{2}, \frac{\sqrt{3}a}{2}\right)$

and coordinates of $E = \left(\frac{-a}{2}, \frac{\sqrt{3}a}{2}\right)$.

Now, prove that

Slope of $BD \times$ Slope of $AC = -1$ and

Slope of $CE \times$ Slope of $AB = -1$

25. Let $OABC$ be the given square such that $OA = a$ and $\angle AOX = \alpha$.

Now, draw $AD \perp OX$.

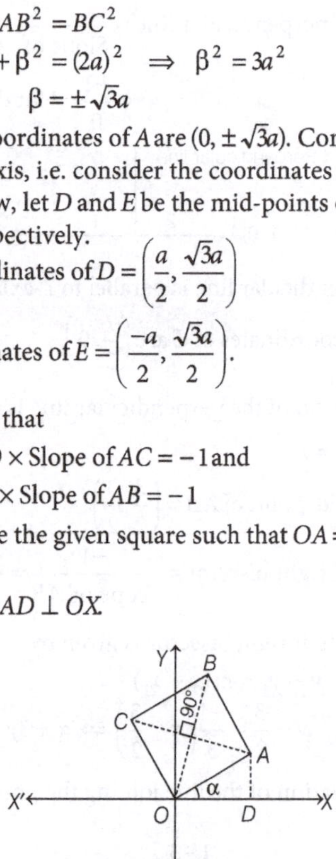

In ΔODA, $\cos\alpha = \frac{OD}{OA} = \frac{OD}{a}$

$\Rightarrow \quad OD = a\cos\alpha$

and $\quad \sin\alpha = \frac{AD}{OA} = \frac{AD}{a}$

$\Rightarrow \quad AD = a\sin\alpha$

Thus, the coordinates of A are $(a\cos\alpha, a\sin\alpha)$.

Clearly, the diagonal OB makes an angle $\frac{\pi}{4} + \alpha$ with positive X-axis and passes through (0, 0).

So, the equation of OB is given by

$$y - 0 = \tan\left(\frac{\pi}{4} + \alpha\right)(x - 0)$$

$$\Rightarrow \quad y = \frac{1 + \tan\alpha}{1 - \tan\alpha} \cdot x$$

$$\Rightarrow \quad y = \frac{\cos\alpha + \sin\alpha}{\cos\alpha - \sin\alpha} \cdot x$$

$\Rightarrow \quad y(\cos\alpha - \sin\alpha) = (\cos\alpha + \sin\alpha)x$

Now, as $OB \perp AC$.

$\therefore$ Slope of $AC = \frac{-1}{\text{Slope of } OB} = \frac{-1}{\tan\left(\frac{\pi}{4} + \alpha\right)}$

Now, the equation of diagonal AC, having slope $\frac{-1}{\tan\left(\frac{\pi}{4} + \alpha\right)}$ and passes through $(a\cos\alpha, a\sin\alpha)$ is

given by $(y - a\sin\alpha) = \frac{-1}{\tan\left(\frac{\pi}{4} + \alpha\right)}(x - a\cos\alpha)$

$\Rightarrow \quad x(\cos\alpha - \sin\alpha) + y(\cos\alpha + \sin\alpha) = a$

26. Hint Equation of line, passing through $A(2, 3)$ and making an angle of 45° with X-axis, is given by

$$y - 3 = \tan 45°(x - 2)$$

$\Rightarrow \quad x - y + 1 = 0$

Let it intersect the line $x + y + 1$ at point D.

Then, on solving these equations, we get $D(-1, 0)$.

Now, length of intercept between A and the line $x + y + 1 = 0$ = length of $AD = 3\sqrt{2}$.

27. Hint Let the given point be $A(2, 3)$, which lie on the line $x - y + 1 = 0$.

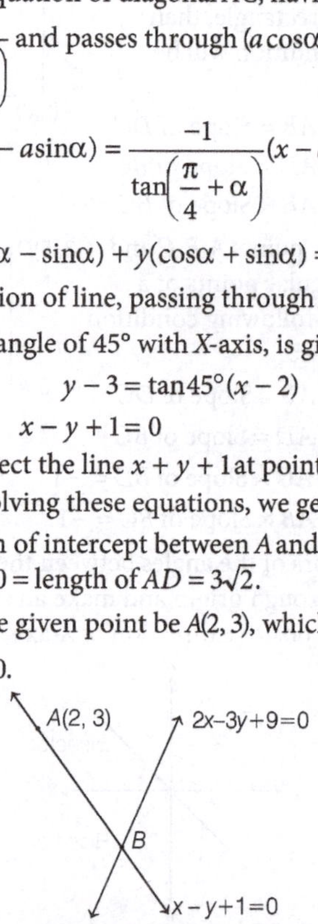

Let $x - y + 1 = 0$ intersect $2x - 3y + 9 = 0$ at point B.

Then, required distance = length of $AB = 4\sqrt{3}$ units.

28. Equation of the line parallel to $3x - 4y + 8 = 0$ is given by $3x - 4y + k = 0$.

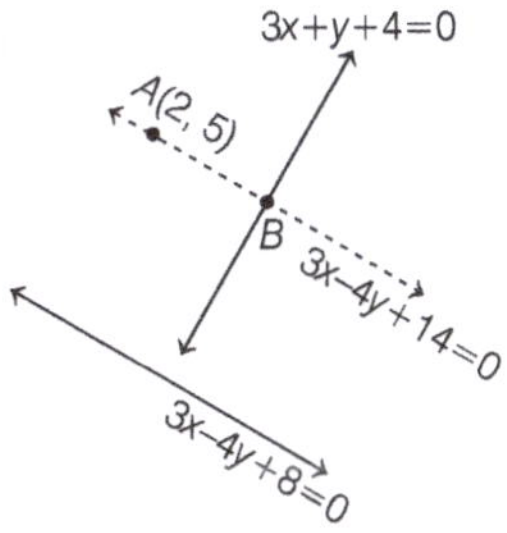

Now, required parallel line should pass through (2, 5).

We have, $3(2) - 4(5) + k = 0$

$\Rightarrow \quad k = 20 - 6 = 14$

Thus, distance of the point (2, 5) from the line $3x + y + 4 = 0$ will be measured along the line $3x - 4y + 14 = 0$.

Now, on solving $3x + y + 4 = 0$ and $3x - 4y + 14 = 0$, we get $x = -2$ and $y = 2$

Thus, the coordinates of B are $(-2, 2)$.

Hence, required distance = length of AB = 5 units.

29. Clearly, the line $ax+by+c = 0$ intersects X-axis and Y-axis at

$A\left(\frac{-c}{a}, 0\right)$ and $B\left(0, \frac{-c}{b}\right)$.

Mid-point of AB is $C\left(\frac{-c}{2a}, \frac{-c}{2b}\right)$.

Now, the equation of line OC is given by

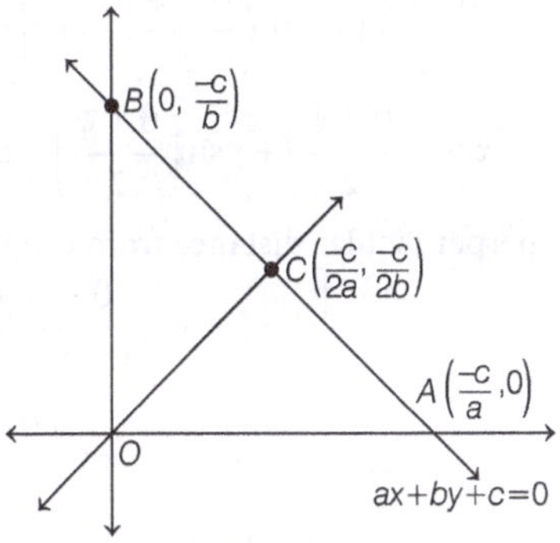

$$y - 0 = \frac{-c/2b}{-c/2a}(x - 0) \Rightarrow ax - by = 0$$

30. Given, equation of lines are

$$mx + (2m+3)y + m + 6 = 0 \quad \ldots(i)$$

and $(2m+1)x + (m-1)y + m - 9 = 0 \quad \ldots(ii)$

Line (i) intersects Y-axis at $\left(0, \frac{-m+9}{2m+3}\right)$ and line (ii) intersects Y-axis at $\left(0, -\frac{(m-9)}{m-1}\right)$.

Now, for lines to be intersect at a point on Y-axis.

$$-\left(\frac{m+6}{2m+3}\right) = -\left(\frac{m-9}{m-1}\right)$$

$$\Rightarrow \quad \frac{m+6}{2m+3} = \frac{m-9}{m-1}$$

$$\Rightarrow \quad m^2 - m + 6m - 6 = 2m^2 + 3m - 18m - 27$$

$$\Rightarrow \quad m^2 + 5m - 6 = 2m^2 - 15m - 27$$

$$\Rightarrow \quad m^2 - 20m - 21 = 0 \Rightarrow m = -1, 21$$

31. **Hint** Let $AD \perp BC$. Then,

$$\text{slope of } AD = \frac{-1}{\text{Slope of } BC} = \frac{-1}{\left(\frac{1}{3}\right)} = -3$$

Now, solve the equation of AB and AC, to get the coordinates of vertex A. This will give $A(2,4)$.

Now, the equation of AD is given by

$$(y-4) = -3(x-2)$$

$$\therefore \quad 3x + y - 10 = 0$$

32. Slope of line $L = \frac{-1}{\text{Slope of } 5x - y = 1} = \frac{-1}{5}$

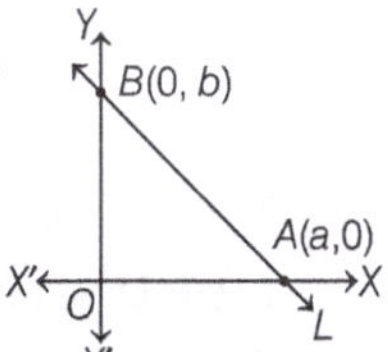

Let line L intersect X-axis at $A(a,0)$ and Y-axis at $B(0,b)$.

Then, area of $\Delta OAB = \frac{1}{2}ab$

$$\Rightarrow \quad \frac{1}{2}ab = 5 \Rightarrow ab = 10 \quad \ldots(i)$$

Also, slope of line $L = \frac{-b}{a} \Rightarrow \frac{-b}{a} = -\frac{1}{5} \Rightarrow b = \frac{a}{5} \quad \ldots(ii)$

On solving Eqs. (i) and (ii) we get

$$a = 5\sqrt{2}, b = \sqrt{2} \text{ or } a = -5\sqrt{2}, b = -\sqrt{2}$$

Now, the equation of line L is

$$\frac{x}{5\sqrt{2}} + \frac{y}{\sqrt{2}} = 1$$

or $\quad \frac{x}{-5\sqrt{2}} + \frac{y}{-\sqrt{2}} = 1$

$$\therefore \quad x + 5y - 5\sqrt{2} = 0 \text{ or } x + 5y + 5\sqrt{2} = 0$$

33. Let the coordinates of B and C be (x_1, y_1) and (x_2, y_2), respectively. Then, we have

$$2x_1 + 3y_1 = 29 \quad \ldots(i)$$

and $\quad x_2 + 2y_2 = 16 \quad \ldots(ii)$

Given that, the mid-point of BC is (5, 6). Therefore,

$$\frac{x_1 + x_2}{2} = 5 \Rightarrow x_1 + x_2 = 10 \quad \ldots(iii)$$

and $\quad \frac{y_1 + y_2}{2} = 6 \Rightarrow y_1 + y_2 = 12 \quad \ldots(iv)$

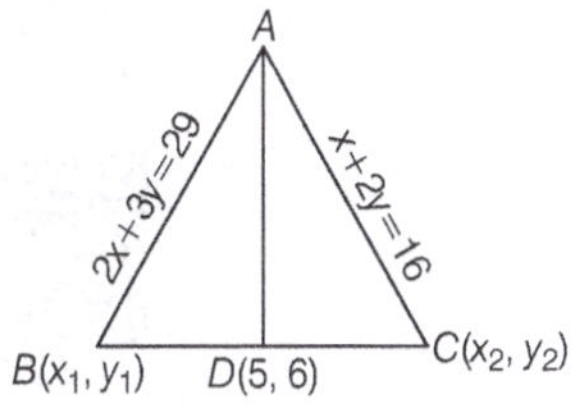

On substituting the values of x_2 and y_2 from Eqs. (iii) and (iv) in Eq. (ii), we get

$$(10-x_1)+2(12-y_1)=16$$
$$\Rightarrow \quad 10-x_1+24-2y_1=16$$
$$\Rightarrow \quad x_1+2y_1=18 \qquad \ldots(v)$$

Now, on solving Eqs. (i) and (v), we get

$$x_1=4 \text{ and } y_1=7$$
$$\Rightarrow \quad B(x_1, y_1)=B(4, 7)$$

Now, the equation of BC is given by

$$y-7=\frac{6-7}{5-4}(x-4) \Rightarrow x+y-11=0$$

34. Equation of the line passing through (3, 2) having slope $\frac{3}{4}$ is given by

$$y-2=\frac{3}{4}(x-3) \qquad [\because y-y_0=m(x-x_0)]$$
$$\Rightarrow \quad 4y-3x+1=0 \qquad \ldots(i)$$

Let (h, k) be the required point on the line such that distance between (h, k) and $(3, 2)$ is 5.

$$\Rightarrow \quad (h-3)^2+(k-2)^2=25 \qquad \ldots(ii)$$

[by distance formula]

Since, point (h, k) lies on the line (i).

$$4k-3h+1=0 \qquad \ldots(iii)$$
$$\Rightarrow \quad k=\frac{3h-1}{4} \qquad \ldots(iv)$$

On putting the value of k in Eq. (ii) and simplifying, we get

$$25h^2-150h-175=0$$
$$\Rightarrow \quad h^2-6h-7=0 \Rightarrow (h+1)(h-7)=0$$
$$\Rightarrow \quad h=-1 \text{ or } h=7$$

On putting these values of h in Eq. (iv), we get

$$k=-1 \text{ or } k=5$$

Therefore, the coordinates of the required points are either $(-1, -1)$ or $(7, 5)$.

35. Let the given points be $A(\cos\theta, \sin\theta)$ and $B(\cos\phi, \sin\phi)$.

Then, equation of line AB is

$$y-\sin\theta=\frac{\sin\phi-\sin\theta}{\cos\phi-\cos\theta}(x-\cos\theta)$$

$$\left[\because y-y_1=\frac{y_2-y_1}{x_2-x_1}(x-x_1)\right]$$

$$\Rightarrow \quad y-\sin\theta=\frac{2\cos\left(\frac{\phi+\theta}{2}\right)\sin\left(\frac{\phi-\theta}{2}\right)}{-2\sin\left(\frac{\phi+\theta}{2}\right)\sin\left(\frac{\phi-\theta}{2}\right)}(x-\cos\theta)$$

$$\left[\because \sin C-\sin D=2\cos\left(\frac{C+D}{2}\right)\sin\left(\frac{C-D}{2}\right) \text{ and } \cos C-\cos D=-2\sin\left(\frac{C+D}{2}\right)\sin\left(\frac{C-D}{2}\right)\right]$$

$$\Rightarrow \quad y\sin\left(\frac{\phi+\theta}{2}\right)-\sin\theta\sin\left(\frac{\phi+\theta}{2}\right)=-x\cos\left(\frac{\phi+\theta}{2}\right)+\cos\theta\cos\left(\frac{\theta+\phi}{2}\right)$$

$$\Rightarrow \quad x\cos\left(\frac{\theta+\phi}{2}\right)+y\sin\left(\frac{\theta+\phi}{2}\right)-\left\{\cos\theta\cos\left(\frac{\theta+\phi}{2}\right)+\sin\theta\sin\left(\frac{\theta+\phi}{2}\right)\right\}=0$$

$$\Rightarrow \quad x\cos\left(\frac{\theta+\phi}{2}\right)+y\sin\left(\frac{\theta+\phi}{2}\right)-\cos\left(\theta-\frac{\theta+\phi}{2}\right)=0$$

$$\Rightarrow \quad x\cos\left(\frac{\theta+\phi}{2}\right)+y\sin\left(\frac{\theta+\phi}{2}\right)-\cos\left(\frac{\theta-\phi}{2}\right)=0$$

Now, perpendicular distance from origin

$$d=\frac{\left|0+0-\cos\left(\frac{\theta-\phi}{2}\right)\right|}{\sqrt{\cos^2\left(\frac{\theta+\phi}{2}\right)+\sin^2\left(\frac{\theta+\phi}{2}\right)}}$$

$$\left[\because \text{distance from } (x_1, y_1) \text{ to the line } ax+by+c=0 \text{ is } \frac{|ax_1+by_1+c|}{\sqrt{a^2+b^2}}\right]$$

$$=\left|\cos\left(\frac{\theta-\phi}{2}\right)\right|$$

QR Code for Quick Test 1

QR Code for Quick Test 2

QR Code for Quick Test 3

Mind Map

Straight Line

A straight line is a curve such that all the points on the line segment joining any two points of it lie on it.

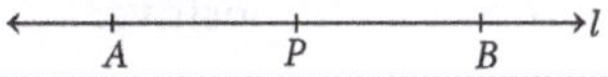

Angle of Inclination of a Line

An angle θ made by the line with positive X-axis in anti-clockwise direction is called angle of inclination of *a*.

(i) When $\theta = 0°$, then **line is parallel to X-axis** (horizontal line).

(ii) When $\theta = 90°$, then **line is perpendicular to X-axis** i.e. parallel to Y-axis (vertical line).

Slope of a Line Joining Two Points

If a line has two points (x_1, y_1) and (x_2, y_2), then the slope of line joining two points (x_1, y_1) and (x_2, y_2) is

$$m = \frac{y_2 - y_1}{x_2 - x_1}, x_1 \neq x_2$$

Slope or Gradient of a Line

If θ is the angle of inclination of a line *l*, then tan θ is called the slope or gradient of the line *l* and it is denoted by *m*

i.e. $m = \tan\theta, \theta \neq 90°$

If $0° < \theta < 90°$, then $m > 0$ and if $90° < \theta < 180°$, then $m < 0$.

Equation of a Line in Point Slope Form

On the basis of number of point(s), the point slope form are of two types

(i) **Equation of a Line Passing Through a Point** The equation of the straight line having slope *m* and passes through given point $Q(x_0, y_0)$ is

$y - y_0 = m(x - x_0)$

(ii) **Equation of a Line Passing Through Two Points** The equation of a line passing through the points (x_1, y_1) and (x_2, y_2) is given by

$$y - y_1 = \frac{y_2 - y_1}{x_2 - x_1}(x - x_1)$$

General Equation of Line

An equation of the form $Ax + By + C = 0$, where A, B and C are variable.

Distance of a Point from a Line

The distance of a point from a line is the length of perpendicular drawn from the point to the line.

Let $L : Ax + By + C = 0$ be a line, whose perpendicular distance from the point $P(x_1, y_1)$ is d. Then,

$$d = \frac{|Ax_1 + By_1 + C|}{\sqrt{A^2 + B^2}}$$

Distance Between Two Parallel Lines

- The distance between two parallel lines $y = mx + c_1$ and $y = mx + c_2$ is given by $d = \dfrac{|c_1 - c_2|}{\sqrt{1 + m^2}}$
- If lines are given in general form i.e. given lines are $Ax + By + C_1 = 0$ and $Ax + By + C_2 = 0$, then distance between these lines

$$d = \frac{|C_1 - C_2|}{\sqrt{A^2 + B^2}}$$

Equation of a Line Parallel to a Given Line

Equation of a line parallel to a given line $ax + by + c = 0$ is $ax + by + \lambda = 0$, where λ is a constant.

Equation of a Line Perpendicular to a Given Line

Equation of a line perpendicular to a given line $ax + by + c = 0$ is $bx - ay + \lambda = 0$, where λ is a constant.

Angle Between Two Lines, having General Equations

If angle between two lines $A_1x + B_1y + C_1 = 0$ and $A_2x + B_2y + C_2 = 0$ is θ, then

$$\tan\theta = \left(\frac{-\frac{A_2}{B_2} + \frac{A_1}{B_1}}{1 + \frac{A_1}{B_1} \times \frac{A_2}{B_2}}\right)$$

For acute angle, $\theta = \tan^{-1}\left|\dfrac{A_1B_2 - A_2B_1}{A_1A_2 + B_1B_2}\right|$

For obtuse angle,

$$\theta = \pi - \tan^{-1}\left|\frac{A_1B_2 - A_2B_1}{A_1A_2 + B_1B_2}\right|$$

Various Form of Equation of Lines

Slope Intercept Form

Suppose a line L with slope m cuts the Y-axis at a distance c from the origin (distance c is called the y-intercept of the line).

$$y = mx + c$$

Suppose a line L with slope m cuts the X-axis at a distance d from the origin i.e. makes x-intercept, d.

Then, equation of line L is

$$y = m(x - d)$$

Equation of X-axis

Every point on X-axis having y-coordinate is zero.

∴ Equation of X-axis, $y = 0$

Equation of Line Parallel to X-axis (Or Equation of Horizontal Line)

Let L be a straight line parallel to X-axis at a distance a from it. Then, the ordinate of every point lying on the line is either a or $-a$.

Equation of Y-axis

Every point on Y-axis having x-coordinate is zero.

∴ Equation of Y-axis, $x = 0$

Equation of Line Parallel to Y-axis (Or Equation of Vertical Line)

Let L be a straight line parallel to Y-axis at a distance b from it. Then, the abscissa of every point lying on the line is either b or $-b$.

Equation of Line in Intercept Form

The equation of a line which cuts-off intercepts a and b on the X-axis and Y-axis respectively, is

$$\frac{x}{a} + \frac{y}{b} = 1$$

Topics to be Assessed Formatively

General Equation of Line and Equation of Family of Lines

An equation of the form $Ax+By+C=0$, where A, B and C are real constants and atleast one of A and B is non-zero, is called **general linear equation** or **general equation** of a line.

Different Forms of $Ax + By + C = 0$

The general equation of a line can be reduced into various forms of the equation of a line, which are given below

Slope Intercept Form

If $B \neq 0$, then $Ax+By+C=0$ can be written as

$$\boxed{y=\frac{-A}{B}x-\frac{C}{B}} \quad \text{or} \quad y=mx+c \qquad \text{...(i)}$$

where, $m=\frac{-A}{B}$ and $c=\frac{-C}{B}$

Intercept Form

If $C \neq 0$, then $Ax+By+C=0$ can be written as

$$\boxed{\frac{x}{\frac{-C}{A}}+\frac{y}{\frac{-C}{B}}=1} \quad \text{or} \quad \frac{x}{a}+\frac{y}{b}=1$$

where, $a=\frac{-C}{A}$ and $b=\frac{-C}{B}$

Normal Form

Let $x\cos\omega+y\sin\omega=p$ be the normal form of the line represented by the equation $Ax+By+C=0$ or $Ax+By=-C$.

Thus, both the equations are same and therefore

$$\frac{A}{\cos\omega}=\frac{B}{\sin\omega}=\frac{-C}{p}$$

which gives $\cos\omega=\frac{-Ap}{C}$ and $\sin\omega=\frac{-Bp}{C}$

Now, $\sin^2\omega+\cos^2\omega=\left(-\frac{Ap}{C}\right)^2+\left(\frac{-Bp}{C}\right)^2$

$\Rightarrow \quad 1=\frac{A^2p^2}{C^2}+\frac{B^2p^2}{C^2}$

$\Rightarrow \quad p^2=\frac{C^2}{A^2+B^2}$

$\Rightarrow \quad p=\pm\frac{C}{\sqrt{A^2+B^2}}$

Therefore, $\cos\omega=\pm\frac{A}{\sqrt{A^2+B^2}}$

and $\sin\omega=\pm\frac{B}{\sqrt{A^2+B^2}}$

Thus, the normal form of the equation $Ax+By+C=0$ is

$$\boxed{x\cos\omega+y\sin\omega=p}$$

where, $\cos\omega=\pm\frac{A}{\sqrt{A^2+B^2}}$, $\sin\omega=\pm\frac{B}{\sqrt{A^2+B^2}}$

and $p=\pm\frac{C}{\sqrt{A^2+B^2}}$

Proper choice of signs is mode so that p should be positive.

Method to Transform General Equation to Normal Form

In order to transform the general equation of a line to the normal form, we use the following steps

(i) Shift the constant term on the RHS and make it positive.

(ii) Divide both sides by $\sqrt{(\text{coeff. of } x)^2+(\text{coeff. of } y)^2}$.

The equation so obtained is in the normal form.

Example 1 **Transform the equation of the line $3x+2y-7=0$ to**

(i) slope intercept form and also find the slope and y-intercept.

(ii) intercept form and also find the intercepts on the coordinate axes.

(iii) normal form and also find the inclination of the perpendicular segment from the origin on the line with the axis and its length.

Sol. (i) Given equation is $3x + 2y - 7 = 0$.

It can be rewritten as $2y = -3x + 7$

or $$y = -\frac{3}{2}x + \frac{7}{2} \quad \text{...(i)}$$

which is the required slope intercept form of the given line.

On comparing Eq. (i) with $y = mx + c$, we get

Slope, $m = -\frac{3}{2}$ and y-intercept, $c = \frac{7}{2}$

(ii) Given equation can be rewritten as

$$3x + 2y = 7$$

or $$\frac{x}{7/3} + \frac{y}{7/2} = 1 \quad \text{...(i)}$$

which is the required intercept form of the given line.

On comparing Eq. (i) with $\frac{x}{a} + \frac{y}{b} = 1$, we get

$$a = \frac{7}{3} \text{ and } b = \frac{7}{2}$$

Hence, x-intercept, $a = \frac{7}{3}$ and y-intercept, $b = \frac{7}{2}$

(iii) Given equation is $3x + 2y - 7 = 0$. ...(i)

On comparing Eq. (i) with $Ax + By + C = 0$, we get

$$A = 3, B = 2 \text{ and } C = -7$$

$$\therefore \sqrt{A^2 + B^2} = \sqrt{3^2 + 2^2} = \sqrt{9 + 4} = \sqrt{13}$$

On dividing both sides of Eq. (i) by $\sqrt{13}$, we get

$$\frac{3}{\sqrt{13}}x + \frac{2}{\sqrt{13}}y - \frac{7}{\sqrt{13}} = 0$$

$$\Rightarrow \frac{3}{\sqrt{13}}x + \frac{2}{\sqrt{13}}y = \frac{7}{\sqrt{13}} \quad \text{...(ii)}$$

which is the required normal form of the given line.

On comparing Eq. (ii) with $x \cos\omega + y \sin\omega = p$, we get

$$\cos\omega = \frac{3}{\sqrt{13}}, \sin\omega = \frac{2}{\sqrt{13}} \text{ and } p = \frac{7}{\sqrt{13}}$$

Since, $\cos\omega$ and $\sin\omega$ both are positive, therefore ω is in the first quadrant and is equal to $\tan^{-1}\frac{2}{3}$

Hence, for the given line, we have

$$\omega = \tan^{-1}\frac{2}{3} \text{ and } p = \frac{7}{\sqrt{13}}.$$

Example 2 Reduce the lines $3x - 4y + 4 = 0$ and $4x - 3y + 12 = 0$ to the normal form and hence determine which line is nearer to the origin.

Sol. Given equation to the first line is

$$3x - 4y + 4 = 0 \Rightarrow -3x + 4y = 4$$

Dividing throughout by $\sqrt{(-3)^2 + 4^2}$, we obtain

$$-\frac{3x}{\sqrt{(-3)^2 + 4^2}} + \frac{4y}{\sqrt{(-3)^2 + 4^2}} = \frac{4}{\sqrt{(-3)^2 + 4^2}}$$

or $$-\frac{3}{5}x + \frac{4}{5}y = \frac{4}{5}$$

This is the normal form of $3x - 4y + 4 = 0$ from which we find that the length of the perpendicular from the origin to it is given by $p_1 = \frac{4}{5}$.

The equation of the second line is

$4x - 3y + 12 = 0$ or $-4x + 3y = 12$

Dividing throughout by $\sqrt{(-4)^2 + 3^2}$, we obtain

$$-\frac{4x}{\sqrt{(-4)^2 + 3^2}} + \frac{3y}{\sqrt{(-4)^2 + 3^2}} = \frac{12}{\sqrt{(-4)^2 + 3^2}}$$

or $$-\frac{4}{5}x + \frac{3}{5}y = \frac{12}{5}$$

This is the normal form of $4x - 3y + 12 = 0$ from which we find that the length of the perpendicular from the origin to it is given, by $p_2 = \frac{12}{5}$.

Since, $p_2 > p_1$, therefore the line $3x - 4y + 4 = 0$ is nearer to the origin.

Angle Between Two lines, having General Equations

Let general equations of lines be $A_1x + B_1y + C_1 = 0$ and $A_2x + B_2y + C_2 = 0$.

Then, slope of given lines are $m_1 = -\frac{A_1}{B_1}$ and $m_2 = -\frac{A_2}{B_2}$

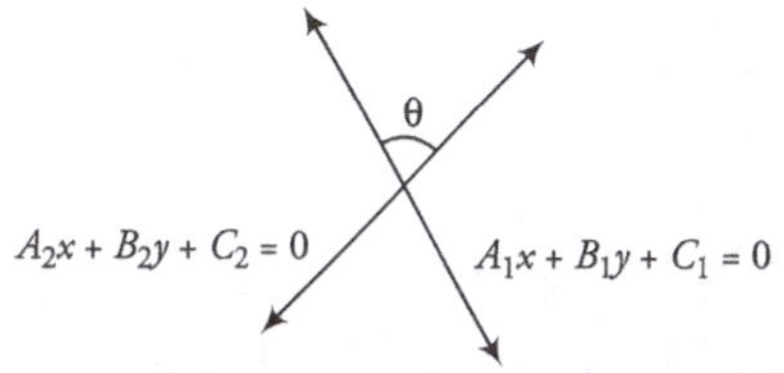

Let θ be the angle between two lines, then

$$\tan\theta = \pm\left(\frac{m_2 - m_1}{1 + m_1m_2}\right) = \pm\left(\frac{-\frac{A_2}{B_2} + \frac{A_1}{B_1}}{1 + \frac{A_1}{B_1} + \frac{A_2}{B_2}}\right)$$

Note

(i) For acute angle, we take $\theta = \tan^{-1}\left|\frac{A_1B_2 - A_2B_1}{A_1A_2 + B_1B_2}\right|$

(ii) For obtuse angle, we take $\theta = \pi - \tan^{-1}\left|\frac{A_1B_2 - A_2B_1}{A_1A_2 + B_1B_2}\right|$

Example 3 **Find the angle between the lines $x-2y+3=0$ and $3x+y-1=0$.**

Sol. Let m_1 and m_2 be the slopes of the straight lines $x-2y+3=0$ and $3x+y-1=0$.

Then, $m_1=-\frac{1}{-2}=\frac{1}{2}$ and $m_2=-\frac{3}{1}=-3$

Let θ be the angle between the given lines. Then,

$$\tan\theta=\pm\left(\frac{m_2-m_1}{1+m_1m_2}\right)=\pm\left(\frac{-3-\frac{1}{2}}{1-\frac{3}{2}}\right)=\pm 7$$

$\Rightarrow \quad \theta=\tan^{-1}(7)$ or $\pi-\tan^{-1}(7)$

Example 4 **Find the equation of one of the sides of an isosceles right angled triangle whose hypotenuse is given by $3x+4y=4$ and the opposite vertex of the hypotenuse is (2, 2).**

Sol. Let slope of line AC be m and the angle between line AC and BC be θ.

Then, $\tan\theta=\left|\frac{m+\frac{3}{4}}{1-\frac{3m}{4}}\right| \Rightarrow \tan 45^\circ=\left|\frac{m+\frac{3}{4}}{1-\frac{3m}{4}}\right| \quad [\because \theta=45^\circ]$

$\Rightarrow \quad \pm 1=\left(\frac{m+\frac{3}{4}}{1-\frac{3m}{4}}\right)$

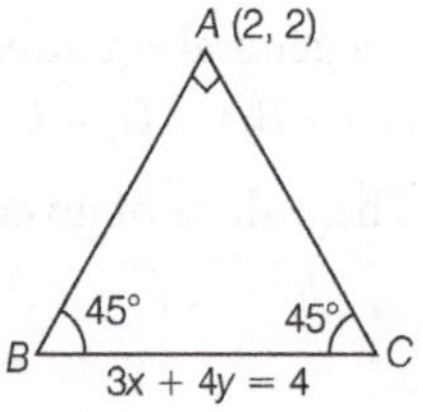

Taking positive sign,

$$1=\frac{m+\frac{3}{4}}{1-\frac{3m}{4}}$$

$\Rightarrow m+\frac{3}{4}=1-\frac{3m}{4}$

$\Rightarrow m+\frac{3m}{4}=1-\frac{3}{4} \Rightarrow \frac{7m}{4}=\frac{1}{4}\Rightarrow m=\frac{1}{7}$

Taking negative sign, $-1=\left(\frac{m+\frac{3}{4}}{1-\frac{3m}{4}}\right)$

$\Rightarrow \quad -1+\frac{3m}{4}=m+\frac{3}{4}$

$\Rightarrow m-\frac{3m}{4}=-1-\frac{3}{4}\Rightarrow \frac{m}{4}=\frac{-7}{4}\Rightarrow m=-7$

Now, equation of side AC is $y-2=\frac{1}{7}(x-2)$

$\Rightarrow \quad 7y-14=x-2 \Rightarrow x-7y+12=0$

or $\quad y-2=-7(x-2)$

$\Rightarrow \quad y-2=-7x+14 \Rightarrow 7x+y-16=0$

Condition For Two lines to be Parallel

If the lines $A_1x+B_1y+C_1=0$ and $A_2x+B_2y+C_2=0$ are parallel, then their slopes are equal

i.e. $\quad m_1=m_2\Rightarrow -\frac{A_1}{B_1}=-\frac{A_2}{B_2}\Rightarrow \frac{A_1}{B_1}=\frac{A_2}{B_2}$

Condition For Two lines to be Perpendicular

If the lines $A_1x+B_1y+C_1=0$ and $A_2x+B_2y+C_2=0$ are perpendicular, then the product of their slopes is equal to negative unity,

i.e. $\quad m_1m_2=-1$

$\Rightarrow \quad -\frac{A_1}{B_1}\times\left(-\frac{A_2}{B_2}\right)=-1 \Rightarrow A_1A_2+B_1B_2=0$

Example 5 **A line passing through the points $(a, 2a)$ and $(-2, 3)$ is perpendicular to the line $4x+3y+5=0$, find the value of a.**

Sol. Let m_1 be the slope of the line joining $A(a, 2a)$ and $B(-2, 3)$.

Then, $\quad m_1=\frac{2a-3}{a+2}$

Let m_2 be the slope of the line $4x+3y+5=0$.

Then, $\quad m_2=-\frac{4}{3}$.

Since, given lines are perpendicular.

Therefore, $m_1m_2=-1$

$\Rightarrow \quad \frac{2a-3}{a+2}\times-\frac{4}{3}=-1$

$\Rightarrow \quad 8a-12=3a+6 \Rightarrow a=\frac{18}{5}$

Equation of Perpendicular and Parallel lines Passing Through a Given point

Let the general equation of a line be $ax+by+c=0$.

Then, the slope of line is $m_1=-\frac{a}{b}$

Let m_2 be the slope of perpendicular line, then by using the condition of perpendicularity,

$$m_1m_2=-1$$

$\Rightarrow \quad -\frac{a}{b}\times m_2=-1$

$\Rightarrow \quad m_2=\frac{b}{a}$

Now, the equation of a line passing through (x_1, y_1) and perpendicular to the given line is $y - y_1 = \frac{b}{a}(x - x_1)$

$\Rightarrow \quad bx - ay - (bx_1 - ay_1) = 0$

Again let m_2', be the slope of parallel line, then by using the condition of parallelism.

$$m_1 = m_2' \Rightarrow m_2' = -\frac{a}{b}$$

Now, the equation of a line passing through (x_1, y_1) and parallel to the given line is

$$y - y_1 = -\frac{a}{b}(x - x_1) \Rightarrow \quad ax + by - (ax_1 + by_1) = 0$$

Example 6 Find the equation of perpendicular line to line $3x + 5y + 6 = 0$ and passing through a point (1, 2).

Sol. Given equation of line is $3x + 5y + 6 = 0$.

$\therefore$ Slope of line is $m_1 = -\frac{3}{5}$

$\Rightarrow$ Slope of perpendicular line is $m_2 = -\frac{1}{m_1} = \frac{5}{3}$

Now, equation of line passing through a given point (1, 2) and perpendicular to a given line is

$$y - 2 = \frac{5}{3}(x - 1) \qquad \left[\because m_2 = \frac{5}{3}\right]$$

$\Rightarrow \quad 3y - 6 = 5x - 5 \Rightarrow 3y - 5x = 1$

Example 7 Find the equation of line passing through the point $(a\cos^3\theta, a\sin^3\theta)$ and perpendicular to the line $x\sec\theta + y \operatorname{cosec}\theta = a$.

Sol. Given point is $(a\cos^3\theta, a\sin^3\theta)$ and the line is $x\sec\theta + y\operatorname{cosec}\theta = a$

$\because \quad$ Slope of given line $= -\frac{\sec\theta}{\operatorname{cosec}\theta} = -\tan\theta$

$\therefore$ Slope of required line $= \frac{1}{\tan\theta} = \cot\theta$

Now, equation of the required line having slope $\cot\theta$ and passing through $(a\cos^3\theta, a\sin^3\theta)$, is

$$y - a\sin^3\theta = \cot\theta\,(x - a\cos^3\theta)$$

$\Rightarrow \quad y - a\sin^3\theta = \frac{\cos\theta}{\sin\theta}(x - a\cos^3\theta)$

$\Rightarrow \quad y\sin\theta - a\sin^4\theta = x\cos\theta - a\cos^4\theta$

$\Rightarrow \quad x\cos\theta - y\sin\theta = a\cos^4\theta - a\sin^4\theta$

$\Rightarrow \quad x\cos\theta - y\sin\theta = a[(\cos^2\theta + \sin^2\theta)(\cos^2\theta - \sin^2\theta)]$

$\Rightarrow \quad x\cos\theta - y\sin\theta = a\cos 2\theta$

Exercise

Objective Type Questions

1 If $Ax + By + C = 0, B \neq 0$, then slope intercept form of this equation is

(a) $y = \frac{A}{B}x + \frac{C}{B}$ (b) $y = \frac{A}{B}x - \frac{C}{B}$

(c) $y = -\frac{A}{B}x + \frac{C}{B}$ (d) $y = -\frac{A}{B}x - \frac{C}{B}$

2 The equation of the line, which makes intercepts a and b on the X and Y-axes respectively, is $2x - 3y + 6 = 0$, where a and b respectively are

(a) 3, 2 (b) – 3, –2 (c) 3, – 2 (d) – 3, 2

3 If the normal form of the equation $Ax+By+C=0$ is $x\cos\omega + y\sin\omega = p$, then p is equal to

(a) $\pm\frac{C}{\sqrt{A^2+B^2}}$ (b) $\pm\frac{C}{\sqrt{A^2-B^2}}$ (c) $\pm\frac{C}{\sqrt{C^2+B^2}}$

(d) $\pm\frac{B}{\sqrt{A^2+C^2}}$

4 If the normal form of the equation $\sqrt{3}x + y - 8 = 0$ is $x\cos\omega + y\sin\omega = p$, then p and ω respectively are

(a) 4, 45° (b) 4, 30° (c) 3, 45° (d) 3, 30°

5 The equation of a line perpendicular to the line $x - 2y + 3 = 0$ and passing through the point (1, – 2) is

(a) $y = 2x$ (b) $x = 2y$ (c) $x = -2y$ (d) $y = -2x$

short answer Type I Questions

6 Reduce the following equations into slope intercept form and find their slopes and the y-intercepts.

(i) $x + 7y = 0$ (ii) $y = 0$ *[NCERT]*

7 Reduce the following equations into intercept form and find their intercepts on the axes.

(i) $6x + y - 6 = 0$ (ii) $2y + 5 = 0$

8 Reduce the equation $\sqrt{3}x + y = 4$ into normal form and hence find the values of p and α.

9 Find the condition, if the two lines $ax + by = c$ and $a'x + b'y = c'$ are perpendicular.

10 Find the equation of the line perpendicular to the line $x - 7y + 5 = 0$ and having x-intercept 3.

11 Find the inclination of the line $x - y + 3 = 0$ with the positive direction of X-axis.

12 Find the equation of the line passing through (1, 2) and parallel to the line $y = 3x - 1$.

13 Find the equation of the line passing through the point of intersection of $2x + y = 5$ and $x + 3y + 8 = 0$ and parallel to the line $3x + 4y = 7$.

14 Show that the straight lines given by $x(a + 2b) + y(a + 3b) = a + b$ for different values of a and b pass through a fixed point.

15 Find the values of θ and p, if the equation $x\cos\theta + y\sin\theta = p$ is the normal form of the line $\sqrt{3}x + y + 2 = 0$.

16 In what ratio, the line joining $(-1, 1)$ and $(5, 7)$ is divided by the line $x + y = 4$?

17 Find the equation of the line passing through $(1, 2)$ and perpendicular to $x + y + 7 = 0$.

Short answer Type II Questions

18 Find the coordinates of the foot of perpendicular from the point $(-1, 3)$ to the line $3x - 4y - 16 = 0$.

19 Show that the tangent of an angle between the lines $\frac{x}{a} + \frac{y}{b} = 1$ and $\frac{x}{a} - \frac{y}{b} = 1$ is $\frac{2ab}{a^2 - b^2}$.

20 Find the angle between the lines $\sqrt{3}x + y = 1$ and $x + \sqrt{3}y = 1$.

21 For what values of a and b the intercepts cuts-off on the coordinates axes by the line $ax + by + 8 = 0$ are equal in length but opposite in signs to those cuts-off by the line $2x - 3y + 6 = 0$ on the axes?

Long Answer Type Questions

22 Find the values of k for which the line $(k - 3)x - (4 - k^2)y + k^2 - 7k + 6 = 0$ is

(i) parallel to the X-axis.

(ii) parallel to the Y-axis.

(iii) passing through origin.

23 Find the angle between the lines $y = (2 - \sqrt{3})(x + 5)$ and $y = (2 + \sqrt{3})(x - 7)$.

24 Show that the equation of the line passing through the origin and making an angle θ with the line $y = mx + c$ is $\frac{y}{x} = \pm\frac{m + \tan\theta}{1 - m\tan\theta}$.

25 Find the equation of the lines which pass through the point $(3, -2)$ and are inclined at 60° to the line $\sqrt{3}x + y = 1$.

Answers

1. (d) 2. (d) 3. (a) 4. (b)

5. (d) 6. (i) $m = -\frac{1}{7}, c = 0$ (ii) $m = 0, c = 0$

7. (i) x-intercept $= 1$ and y-intercept $= 6$
(ii) x-intercept $= 0$ and y-intercept $= -\frac{5}{2}$

8. $\alpha = 30^\circ$ and $p = 2$ units 9. $aa' + bb' = 0$

10. $7x + y = 21$ 11. $\theta = 45^\circ$ 12. $3x - y - 1 = 0$

13. $3x + 4y + 3 = 0$ 15. $\theta = 210^\circ$ and $p = 1$

16. $k : 1 = 1 : 2$ (internally)

17. $y - x - 1 = 0$

18. $\left(\frac{68}{25}, \frac{-49}{25}\right)$

20. $\theta = 30^\circ$ or 150° 21. $a = -\frac{8}{3}, b = 4$

22. (i) $k = 3$ (ii) $k = \pm 2$
(iii) $k = 1, 6$

23. $\theta = \frac{\pi}{3}$ or $\frac{2\pi}{3}$ 24. $\frac{y}{x} = \frac{m \pm \tan\theta}{1 \mp m\tan\theta} = \pm\frac{m + \tan\theta}{1 - m\tan\theta}$

25. $y + 2 = 0$ and $\sqrt{3}x - y - 2 - 3\sqrt{3} = 0$

10 (a)

CHAPTER

"Mathematics is the most beautiful and most powerful creation of the human spirit."
—*Stefan Banach*

Conic Sections : Circle

YOUR STUDY GOALS

10.1.1 Right Circular Cone
10.1.2 Conic Sections
10.1.3 Circle

YOUR PRACTICE GOALS

- Objective Type Questions (16 Qs; 10 MCQs, 5 A-R, 1 Case Based)
- Subjective Type Questions
- (13 Qs; 5 VSA, 4 SA, 4 LA)

10.1.1 Right Circular Cone

The locus of a line passing through a fixed point say A and making a constant angle θ with a fixed line AB passing through the fixed point A is called right circular cone.

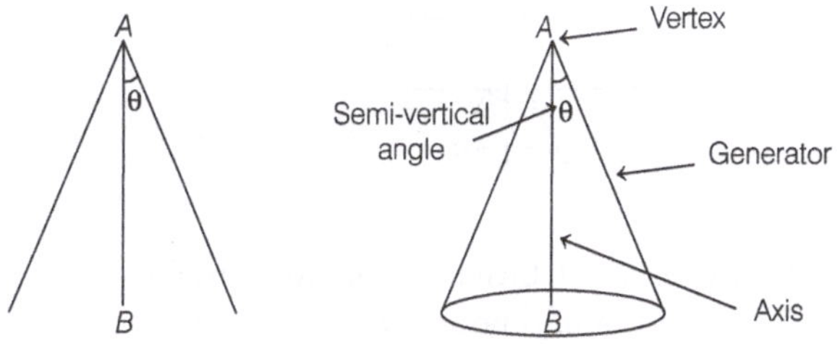

Double Napped Right Circular Cone

The locus of the line l is the surface generated by l, when it rotates around the line AB so that it always makes an angle θ with AB, is called a double napped right circular cone or cone extending indefinitely far in both directions.

The two parts of the generated surface separated by vertex V are called **nappes.**

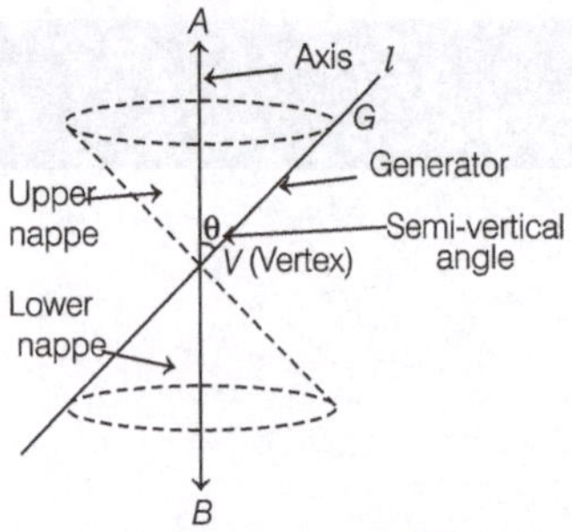

10.1.2 Conic Sections

On intersecting a right circular cone by a plane in different positions, different sections, so obtained are called conic sections or conics.

Section of a Doubled Napped Right Circular Cone by a Plane

Consider a double napped right circular cone, having semi-vertical angle θ.

Let α be the angle between the plane and the axis of cone.

Case I When the plane passes through vertex of the cone

(i) When $\theta < \alpha \le 90°$, then the section is a **point**.

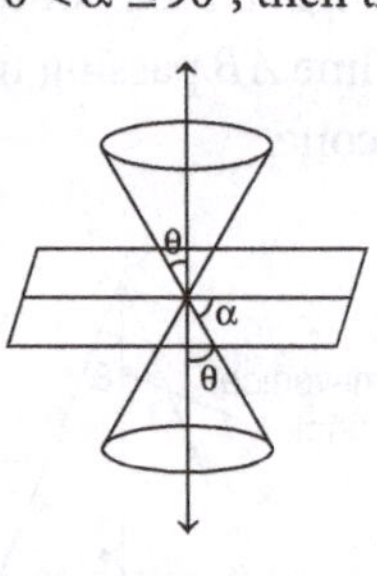

(ii) When $\theta = \alpha$, then the plane contains the generator of cone and the section is a **straight line**. It is a **degenerated case of a parabola.**

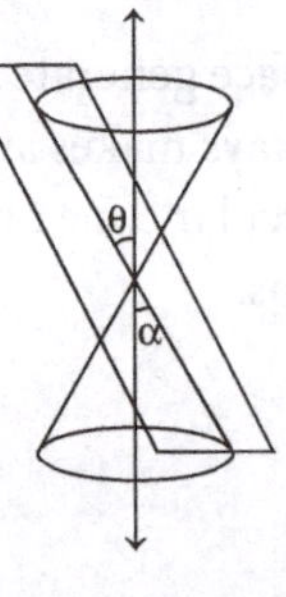

(iii) When $0 \le \alpha < \theta$, then section is a pair of straight lines. It is a **degenerated case of a hyperbola.**

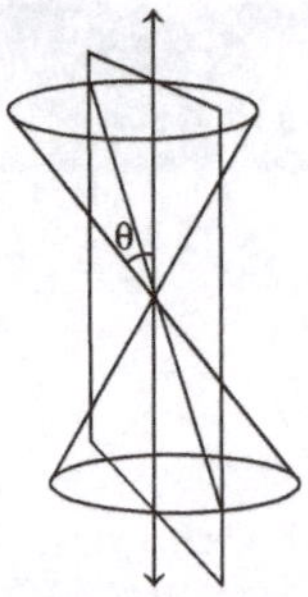

Case II When the plane does not pass through vertex of the cone

(i) When $\alpha = 90°$, then cutting plane section is a **circle.**

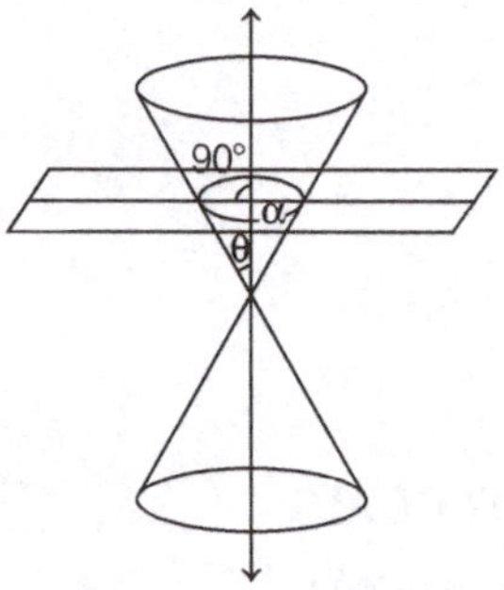

(ii) When $\theta < \alpha < 90°$, then cutting plane section is an **ellipse.**

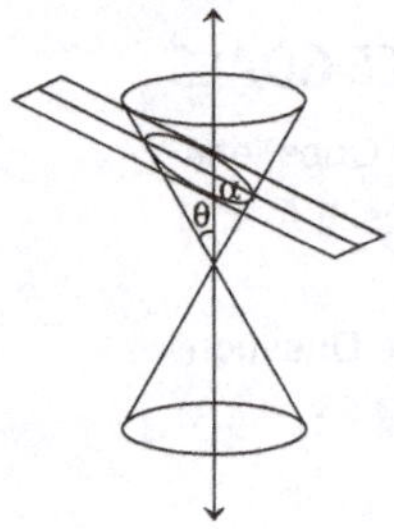

(iii) When $\alpha = \theta$ i.e. plane is parallel to a generator, then cutting plane is **parabola.**

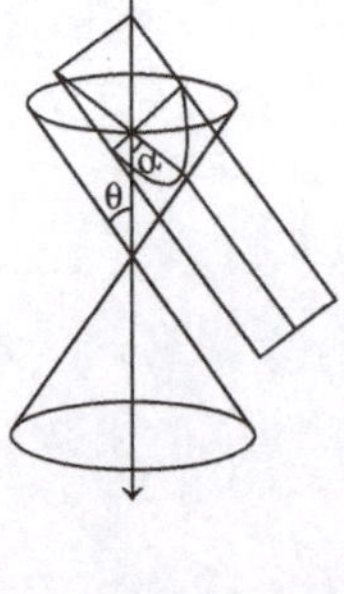

(iv) When $0 \leq \alpha < \theta$, then the plane intersects both the nappies and section is a **hyperbola**.

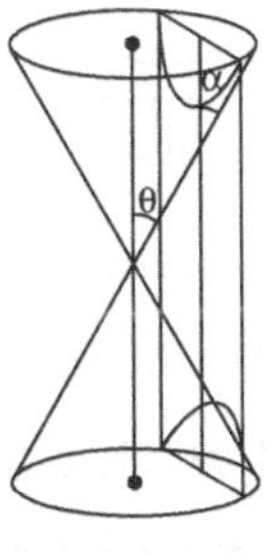

Geometrical Definition of Conic Section

A conic is the locus of a point which moves in a plane in such a way that the ratio of its distance from a fixed point and a fixed line is a constant. Then,

(i) the fixed point is called **focus** and is denoted by F.

(ii) the fixed straight line ZM is called **directrix**.

(iii) the constant ratio is called **eccentricity** and is denoted by e.

In the given figure, $\dfrac{PF}{PM} = e = \text{constant}$

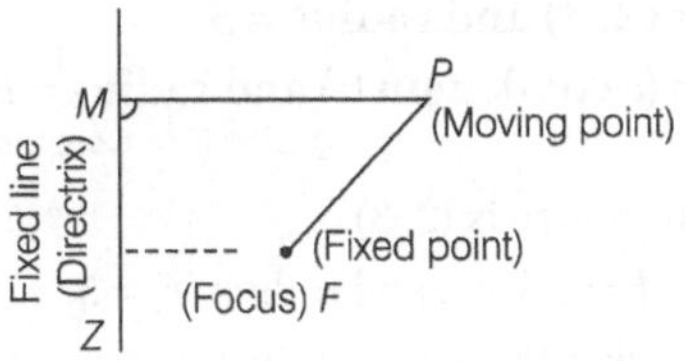

On the basis of the value of e, we get the following cases

(a) **Circle** If eccentricity, $e = 0$, then the conic is called circle.

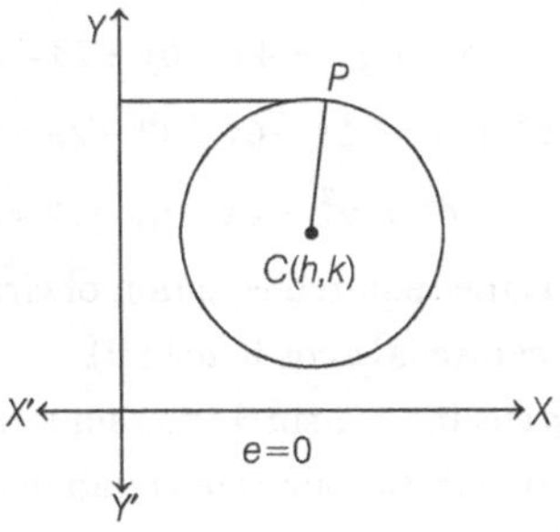

(b) **Parabola** If eccentricity, $e = 1$, then the conic is called parabola.

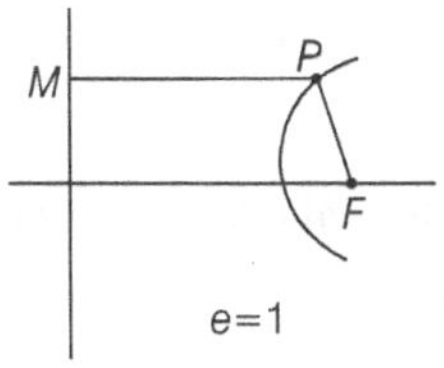

(c) **Ellipse** If eccentricity, $e < 1$, then the conic is called ellipse.

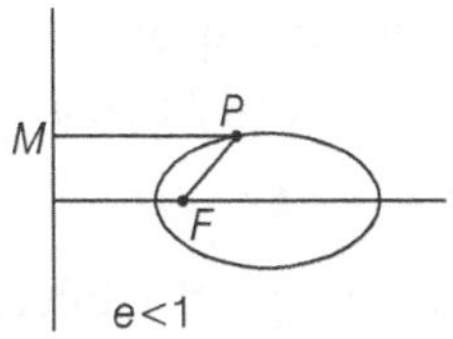

(d) **Hyperbola** If eccentricity, $e > 1$, then the conic is called hyperbola.

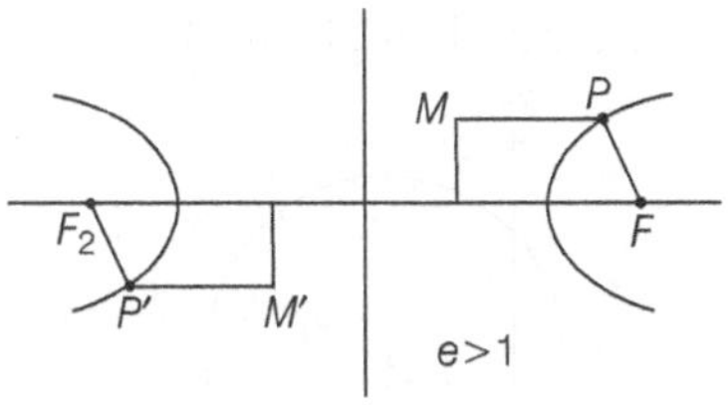

10.1.3 Circle

A circle is the set of all points in a plane that are equidistant from a fixed point in that plane.

- **Centre** The fixed point C is called the centre of the circle.
- **Radius** The constant distance CP from the centre C to a point on the circle, is called radius r.

Standard Equation of a Circle

Let $C(h, k)$ be the **centre** of the circle, $P(x, y)$ be any point on the circumference of the circle and r be the **radius** of the circle.

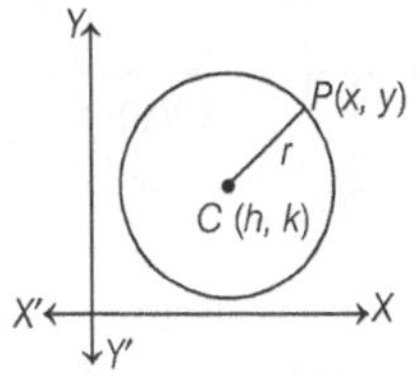

Then, equation of circle in standard form is

$$(x - h)^2 + (y - k)^2 = r^2$$

This equation is also known as **central form** of the equation of a circle.

If the centre of circle is at origin and radius r, then equation of circle is

$$x^2 + y^2 = r^2$$

Equation of Circle in Special Cases

We know that the standard equation of the circle with centre at (h, k) and radius equal to r, is

$$(x-h)^2 + (y-k)^2 = r^2$$

It can be change according to the different conditions, which are given below

Case I When the Circle Passes Through the Origin

The circle passes through origin $(0, 0)$.

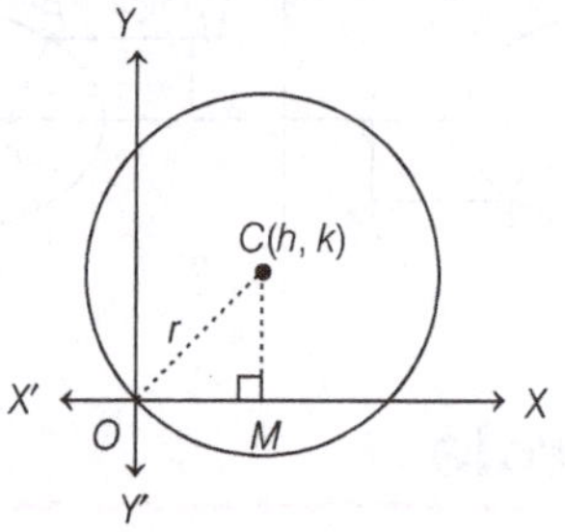

$\therefore$ Radius $= r =$ Distance between points O and C

$$= \sqrt{h^2 + k^2}$$

Hence, the equation of a circle passing through the origin is

$$(x-h)^2 + (y-k)^2 = h^2 + k^2$$

i.e. $$x^2 + y^2 - 2hx - 2ky = 0$$

Case II When the Centre Lies on X-axis or Y-axis

If centre lies on X-axis, then $k = 0$.

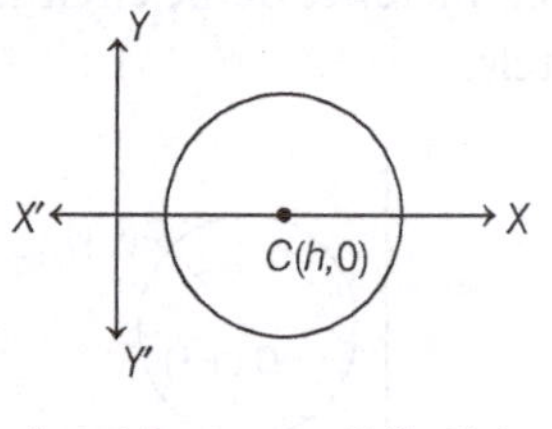

$\therefore$ Equation of circle with centre $C(h, 0)$ is

$$(x-h)^2 + y^2 = r^2$$

If centre lies on Y-axis, then $h = 0$.

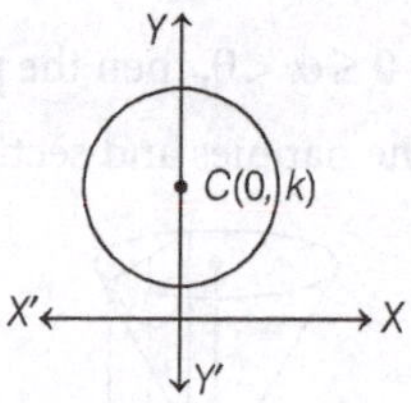

$\therefore$ Equation of circle with centre $C(0, k)$ is

$$x^2 + (y-k)^2 = r^2$$

Position of a Point w.r.t. Circle

Let $\quad S \equiv x^2 + y^2 + 2gx + 2fy + c = 0$

or $\quad S \equiv (x-h)^2 + (y-k)^2 = r^2$

be the equation of the circle and $P(x_1, y_1)$ be any point in the plane of the circle.

Then, $\quad S_1 \equiv x_1^2 + y_1^2 + 2gx_1 + 2fy_1 + c = 0$

or $\quad S_1 \equiv (x_1 - h)^2 + (y_1 - k)^2 = r^2$

Case I If $S_1 > 0$, then the point lies *outside* the circle.

Case II If $S_1 = 0$, then the point lies *on* the circle.

Case III If $S_1 < 0$, then the point lies *inside* the circle.

Example 1. Find the equation of the circle with
(i) centre = (2, 3) and radius = 5
(ii) centre = $(a\cos\theta, a\sin\theta)$ and radius = a

Competency Based Que

Sol. (i) Given, centre is (2, 3).

$\therefore \quad h = 2, k = 3$ and radius $(r) = 5$

On putting these values in equation of circle $(x-h)^2 + (y-k)^2 = r^2$, we get

$$(x-2)^2 + (y-3)^2 = 5^2$$

$\Rightarrow \quad x^2 + 4 - 4x + y^2 + 9 - 6y = 25$

$[\because (A-B)^2 = A^2 + B^2 - 2AB]$

$\Rightarrow \quad x^2 + y^2 - 4x - 6y + 13 = 25$

$\Rightarrow \quad x^2 + y^2 - 4x - 6y + 13 - 25 = 0$

$\therefore \quad x^2 + y^2 - 4x - 6y - 12 = 0$

which is the required equation of circle.

(ii) Given, centre is $(a\cos\theta, a\sin\theta)$.

$\therefore h = a\cos\theta, k = a\sin\theta$ and radius $(r) = a$

On putting these values in equation of circle $(x-h)^2 + (y-k)^2 = r^2$, we get

$$(x - a\cos\theta)^2 + (y - a\sin\theta)^2 = a^2$$

$$\Rightarrow \quad x^2 + a^2\cos^2\theta - 2ax\cos\theta + y^2$$
$$+ a^2\sin^2\theta - 2ay\sin\theta = a^2$$
$$\Rightarrow \quad x^2 + y^2 - 2ax\cos\theta - 2ay\sin\theta$$
$$+ a^2(\sin^2\theta + \cos^2\theta) = a^2$$
$$\Rightarrow x^2 + y^2 - 2a(x\cos\theta + y\sin\theta) + a^2 = a^2$$
$$[\because \sin^2\theta + \cos^2\theta = 1]$$
$$\therefore \quad x^2 + y^2 - 2a(x\cos\theta + y\sin\theta) = 0$$

which is the required equation of circle.

Example 2. Find the equation of circle whose centre is (1, 2) and which passes through the point (4, 6).

Sol. Coordinates of centre of given circle are (1, 2) and it passes through the point (4, 6).

Then, radius of the circle is equal to the distance from the centre to a point on a circle.

$\therefore$ Radius of circle $= \sqrt{(1-4)^2 + (2-6)^2} = \sqrt{9+16} = 5$

By distance formula,

$$\text{Distance} = \sqrt{(x_2 - x_1)^2 + (y_2 - y_1)^2}$$

Hence, the required equation of the circle is

$$(x-1)^2 + (y-2)^2 = (5)^2$$
$$\Rightarrow \quad x^2 + y^2 - 2x - 4y - 20 = 0$$

Example 3. Find the centre and radius of each of the following circle.

(i) $x^2 + (y+2)^2 = 9$ (ii) $x^2 + y^2 + 6x - 4y + 4 = 0$

Sol. (i) Given equation of circle is

$$x^2 + (y+2)^2 = 9$$
$$\Rightarrow (x-0)^2 + \{y-(-2)\}^2 = 3^2$$

On comparing with $(x-h)^2 + (y-k)^2 = r^2$, we get

$$h = 0, k = -2 \text{ and } r = 3$$

Hence, centre of circle = (0, – 2) and radius = 3 units.

(ii) Given equation of circle is

$$x^2 + y^2 + 6x - 4y + 4 = 0$$
$$\Rightarrow \quad (x^2 + 6x) + (y^2 - 4y) = -4$$
$$\Rightarrow \quad (x^2 + 6x + 9 - 9) + (y^2 - 4y + 4 - 4) = -4$$
$$\Rightarrow \quad (x^2 + 6x + 9) + (y^2 - 4y + 4) = -4 + 4 + 9$$
$$\Rightarrow \quad (x+3)^2 + (y-2)^2 = 9$$
$$\Rightarrow \quad \{x-(-3)\}^2 + (y-2)^2 = 3^2$$

On comparing with $(x-h)^2 + (y-k)^2 = r^2$, we get

$$h = -3, k = 2 \text{ and } r = 3$$

Hence, centre of circle = (–3, 2) and radius = 3.

Example 4. Find the equation of the circle whose centre is (a, b) and passes through the origin.

Sol. We know that circle passes through the origin, so radius of circle will be equal to the distance between point (a, b) and origin.

$\therefore$ Radius of circle

$$= \text{Distance between points } (0, 0) \text{ and } (a, b)$$
$$= \sqrt{(0-a)^2 + (0-b)^2} = \sqrt{a^2 + b^2}$$

[by distance formula]

$\because$ Centre $= (h, k) = (a, b)$

On putting these values in equation of circle $(x-h)^2 + (y-k)^2 = r^2$, we get

$$(x-a)^2 + (y-b)^2 = (\sqrt{a^2 + b^2})^2$$
$$\Rightarrow x^2 + a^2 - 2ax + y^2 + b^2 - 2by = a^2 + b^2$$
$$[\because (A-B)^2 = A^2 - 2AB + B^2]$$
$$\Rightarrow \quad x^2 + y^2 - 2ax - 2by = 0$$

which is the required equation of circle.

Example 5. If the circle passes through the point (2,3), then find the equation of the circle whose radius is 5 units and centre lies on X-axis. Competency Based Que

Sol. Given, radius, $r = 5$ units, point lies on circle = (2, 3) and centre of circle is on X-axis.

$\therefore$ Its y-coordinate will be zero.

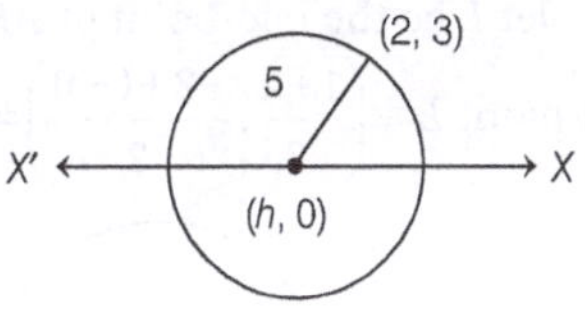

Let centre of circle $= (h, 0)$.

Now, distance from the centre to a point on the circle = Radius of circle

$$\therefore \quad \sqrt{(h-2)^2 + (0-3)^2} = 5 \quad \text{[by distance formula]}$$
$$\Rightarrow \quad \sqrt{h^2 + 4 - 4h + 9} = 5$$
$$[\because (A-B)^2 = A^2 + B^2 - 2AB]$$
$$\Rightarrow \quad h^2 + 4 - 4h + 9 = 25 \quad \text{[squaring on both sides]}$$
$$\Rightarrow \quad h^2 - 4h + 13 - 25 = 0$$
$$\Rightarrow \quad h^2 - 4h - 12 = 0$$
$$\Rightarrow \quad h^2 - 6h + 2h - 12 = 0$$
$$\Rightarrow \quad h(h-6) + 2(h-6) = 0$$
$$\Rightarrow \quad (h-6)(h+2) = 0$$
$$\Rightarrow \quad h - 6 = 0 \text{ or } h + 2 = 0$$
$$\Rightarrow \quad h = 6 \text{ or } h = -2$$

So, the centre of circle is (6, 0) or (– 2, 0).

When centre $(h, k) = (6, 0)$ and radius $(r) = 5$, then the equation of circle is

$$(x-6)^2 + (y-0)^2 = (5)^2 \quad [\because (x-h)^2 + (y-k)^2 = r^2]$$

$$\Rightarrow x^2 + 36 - 12x + y^2 = 25 \quad [\because (a-b)^2 = a^2 - 2ab + b^2]$$

$$\Rightarrow x^2 + y^2 - 12x + 11 = 0$$

When centre $(h, k) = (-2, 0)$ and radius $(r) = 5$, then the equation of circle is

$$(x+2)^2 + (y-0)^2 = (5)^2 \quad [\because (x-h)^2 + (y-k)^2 = r^2]$$

$$\Rightarrow x^2 + 4 + 4x + y^2 = 25$$

$$\Rightarrow x^2 + y^2 + 4x + 4 - 25 = 0$$

$$\Rightarrow x^2 + y^2 + 4x - 21 = 0$$

Hence the equations of circle are

$$x^2 + y^2 - 12x + 11 = 0$$

and

$$x^2 + y^2 + 4x - 21 = 0$$

Example 6. **Find the equation of the circle passing through the points (1, – 2) and (4, – 3) and whose centre lies on the line $3x + 4y = 7$.**

Sol. Let points $A \equiv (1, -2)$ and $B \equiv (4, -3)$

Given line is $3x + 4y = 7$ ...(i)

Again, let L be the mid-point of AB.

$\therefore$ Mid-piont $L \equiv \left(\frac{1+4}{2}, \frac{-2+(-3)}{2}\right) = \left(\frac{5}{2}, -\frac{5}{2}\right)$

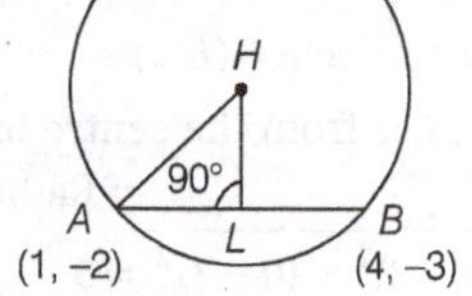

Slope of $AB = \frac{y_1 - y_2}{x_1 - x_2} = \frac{-2+3}{1-4} = -\frac{1}{3}$

$\therefore$ Slope of $HL = 3$ $[\because HL \perp AB]$

Then, equation of HL will be $y + \frac{5}{2} = 3\left(x - \frac{5}{2}\right)$

$$\Rightarrow 2y + 5 = 6x - 15$$

$$\Rightarrow 6x - 2y = 20$$

$$\therefore 3x - y = 10 \quad \text{...(ii)}$$

Centre H of the circle lies on lines (i) and (ii), therefore it is the point of intersection of lines (i) and (ii).

On multiplying Eq. (ii) by 4 and adding it to Eq. (i), we get

$$15x = 47$$

$$\Rightarrow x = \frac{47}{15}$$

On subtracting Eq. (ii) from Eq. (i), we get

$$5y = -3$$

$$\Rightarrow y = -\frac{3}{5}$$

$$\therefore \text{Centre } H \equiv \left(\frac{47}{15}, -\frac{3}{5}\right)$$

Now, radius of the circle

$$= HA = \sqrt{\left(\frac{47}{15} - 1\right)^2 + \left(-\frac{3}{5} + 2\right)^2}$$

$$= \sqrt{\left(\frac{32}{15}\right)^2 + \left(\frac{7}{5}\right)^2}$$

$$= \sqrt{\frac{32^2 + 21^2}{15^2}} = \frac{\sqrt{1465}}{15}$$

Hence, the equation of required circle is

$$\left(x - \frac{47}{15}\right)^2 + \left(y + \frac{3}{5}\right)^2 = \frac{1465}{15^2}$$

$$\Rightarrow (15x - 47)^2 + 9(5y + 3)^2 = 1465$$

$$\Rightarrow 225x^2 + 2209 - 1410x + 225y^2 + 81 + 270y = 1465$$

$$\Rightarrow 225x^2 + 225y^2 - 1410x + 270y + 825 = 0$$

$$\Rightarrow 15x^2 + 15y^2 - 94x + 18y + 55 = 0$$

Example 7. **Does the point (–2.5, 3.5) lie inside, outside or on the circle $x^2 + y^2 = 25$?** Competency Based Que

Sol. Given, equation of circle is

$$S_1 \equiv x^2 + y^2 = 25$$

or

$$S_1 \equiv x^2 + y^2 - 25 = 0 \quad \text{...(i)}$$

Now, put $x = -2.5$ and $y = 3.5$ in Eq. (i), we get

$$S_1 = (-2.5)^2 + (3.5)^2 - 25$$

$$= 6.25 + 12.25 - 25 = 18.5 - 25 = -6.5 < 0$$

Since, $S_1 < 0$, so the given point (–2.5, 3.5) lies inside the circle.

Exam Pattern Questions

Objective Type Questions

Multiple Choice Questions

1. The vertex separates the double napped right circular cone into ...M... parts called ...N... . Here, M and N respectively stand for **Competency Based Que**

(a) three, nappes (b) two, shapes
(c) two, nappes (d) three, cones

2. Different kinds of conic sections are obtained depending on **Competency Based Que**

(a) the position of the intersecting plane with respect to the cone
(b) angle made by intersecting plane with the vertical axis of the cone
(c) Both (a) and (b)
(d) Neither (a) nor (b)

3. A circle is the set of all points in a plane that are equidistant from a ...P... point in the ...Q... . Here, P and Q respectively are **Competency Based Que**

(a) any, space (b) fixed, space
(c) any, plane (d) fixed, plane

4. If the equation of the circle with centre at (h, k) and radius r is $x^2 + y^2 = r^2$. Then, h and k respectively are

(a) (1, 1) (b) (−1, −1)
(c) (0, 0) (d) None of these

5. The centre and radius of the circle $3x^2 + 3y^2 = 7$ is

(a) $(0, 0); \sqrt{7}$ (b) $(0, 0); \sqrt{3}$
(c) $(0, 0); \sqrt{\frac{7}{3}}$ (d) $(0, 0); \frac{7}{3}$

6. Centre of the circle $(x+1)^2 + (y-3)^2 = 5$ is

(a) $(\sqrt{5}, 0)$ (b) (5, 0) (c) (1, −3) (d) (−1, 3)

7. If the line $y = \sqrt{3}x + k$ touches the circle $x^2 + y^2 = 16$, then the value of k is **NCERT Exemplar**

(a) 4 (b) ± 4 (c) 8 (d) ± 8

8. Find the equation of the circle whose centre is (−1, 3) and radius 5 units. **Competency Based Que**

(a) $(x-1)^2 + (y-3)^2 = 25$ (b) $(x+1)^2 + (y-3)^2 = 25$
(c) $(x-1)^2 + (y+3)^2 = 5$ (d) $(x+1)^2 + (y-3)^2 = 5$

9. The equation of the circle in the first quadrant touching each coordinate axis at a distane of one unit from the origin is **NCERT Exemplar**

(a) $x^2 + y^2 - 2x - 2y + 1 = 0$ (b) $x^2 + y^2 - 2x - 2y - 1 = 0$
(c) $x^2 + y^2 - 2x - 2y = 0$ (d) None of these

10. The equation of the circle having centre $(1, -2)$ and passing through the point of intersection of the lines $3x + y = 14$ and $2x + 5y = 18$ is **NCERT Exemplar**

(a) $x^2 + y^2 - 2x + 4y - 20 = 0$
(b) $x^2 - y^2 - 2x - 4y - 20 = 0$
(c) $x^2 ; y^2 + 2x - 4y - 20 = 0$
(d) $x^2 + y^2 + 2x + 4y - 20 = 0$

Assertion-Reason Based Questions

Directions (Q. Nos. 11-15) *In the questions given below are two statements labelled as Assertion (A) and Reason (R). In the context of the two statements, which one of the following is correct?*

(a) Both A and R are correct; R is the correct explanation of A
(b) Both A and R are correct; R is not the correct explanation of A
(c) A is correct; R is incorrect
(d) R is correct; A is incorrect

11. **Assertion** (A) The equation of circle whose centre is (−2, −3) and radius is 10, is $(x+2)^2 + (y+3)^2 = 100$.

Reason (R) The equation of circle with centre (h, k) having radius r given by $(x-h)^2 + (y-k)^2 = r^2$. **Competency Based Que**

12. **Assertion** (A) The centre and radius of the circle $x^2 + (y-2)^2 = 16$ are (0, 2) and 4, respectively.

Reason (R) The equation $x^2 + y^2 = 1$ represents a unit circle with centre (0, 0).

13. **Assertion** (A) The equation of circle whose centre is (3, 1) and passes through the point (6, 5) is $(x-3)^2 + (y-1)^2 = 25$.

Reason (R) The equation $(x-h)^2 + y^2 = r^2$ represents a circle with centre $(h, 0)$ and radius r.

14. **Assertion** (A) $2x^2 + 2y^2 + 3y + 10 = 0$ represents a circle.

Reason (R) A equation of the form $ax^2 + ay^2 + 2gx + 2fy + c = 0$ represents a circle. **Competency Based Que**

15. **Assertion** (A) The equation of the circle with centre $(-a, -b)$ and radius $\sqrt{a^2 - b^2}$ is $x^2 + y^2 + 2ax + 2by - 2b^2 = 0$.

Reason (R) The equation of a circle whose centre is (h,k) and radius r units is $(x-h)^2 + (y-k)^2 = r^2$. **Competency Based Que**

Case Study Based Question

16. Jaya made a circle having equation $x^2 + y^2 + 8x - 16y + 64 = 0$ to design on the wall of her house. There is a mirror facing the wall.

Competency Based Que

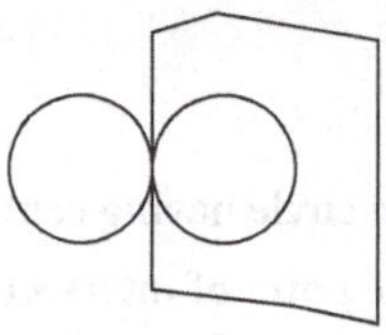

On the basis of above information, answer the following questions.

(i) What is the radius of the circle drawn by Jaya?

(ii) Find the centre of the mirror image of the circle in the line mirror $x = 0$.

(iii) Find the centre of the circle drawn by Jaya on her wall.

(iv) Find the equation of the mirror image of the circle drawn by Jaya in the line mirror $x = 0$.

Subjective Questions

Very Short Answer Type Questions

1. Find the equation of circle with centre $= (-3, 2)$ and radius $= 5$

2. Find the equation of circle with centre $= (a, a)$ and radius $= a\sqrt{2}$

3. Find the equation of the circle whose centre is $(2, -5)$ and which passes through the point $(3, 2)$.

Competency Based Que

4. Find the centre and radius of the circle $(x+5)^2 + (y-3)^2 = 36$.

5. Find the centre and radius of the circle $(x-2)^2 + (y-4)^2 = 65$.

Short Answer Type Questions

6. Find the centre and radius of each of the following circles.

(i) $\left(x - \frac{1}{2}\right)^2 + \left(y + \frac{1}{3}\right)^2 = \frac{1}{4}$ (ii) $x^2 + y^2 - 4x + 6y = 5$

7. Find the equation of the circle which passes through the points $(2, -2)$ and $(3, 4)$ and whose centre lies on the line $x + y = 2$.

Competency Based Que

8. Find the equation of circle whose centre is $(1, 2)$ and touches X-axis.

9. Does the point $(-1.5, 2.5)$ lies inside or outside or on the circle $x^2 + y^2 = 25$?

NCERT

Long Answer Type Questions

10. A man running a rececourse notes that the sum of the distances from the two flag posts from him is always 10 m and the distance between the flag posts is 8 m. Find the equation of the path traced by the man.

NCERT, Competency Based Que

11. Find the equation of the circle passing through $(0, 0)$ and making intercepts a and b on the coordinate axes.

NCERT, NCT 2012, 2013

12. A circle of radius 5 units touches the coordinate axes in the first quadrant. If the circle makes one complete roll on X-axis along the positive direction of X-axis, then find its equation in the new position.

Competency Based Que

13. Show that the points (x, y) given by $x = \frac{2at}{1+t^2}$ and $y = \frac{a(1-t^2)}{1+t^2}$ lies on a circle for all real values of t such that $-1 \le t \le 1$, where a is any given real numbers.

Competency Based Que

Hints & Solutions

Objective Type Questions

1. (c) We know that the vertex separates the double napped right circular cone into two parts called nappers.

2. (c) Different kinds of conic sections are obtained depending on the position of the intersecting plane with respect to the cone and angle made by intersecting plane with the verticle axis of the cone.

3. (d) A circle is the set of all points in a plane that are equidistant from a fixed point in the plane.

4. (c) Given, $x^2+y^2=r^2$

$\Rightarrow \quad (x-0)^2+(y-0)^2=r^2$

Therefore, centre $(h,k)=(0,0)$

5. (c) Circle is $3x^2+3y^2=7$

$\Rightarrow \quad x^2+y^2=\frac{7}{3}$

$\Rightarrow \quad (x-0)^2+(y-0)^2=\left(\sqrt{\frac{7}{3}}\right)^2$

$\therefore$ Centre $\equiv(0,0)$ and radius $=\sqrt{\frac{7}{3}}$.

6. (d) As circle is $[x-(-1)]^2+[y-3]^2=5$. So, centre is $(-1,3)$.

7. (d) If the line touches the circle, then distance from centre is equal to radius. Centre and radius of circle $x^2+y^2=16$ are (0,0) and 4, respectively.

8. (b) Given, centre is $(-1,3)$ and radius is 5.

$\therefore$ Circle is $(x+1)^2+(y-3)^2=(5)^2$

i.e. $\quad (x+1)^2+(y-3)^2=25$

9. (a) From option (a),

$$x^2+y^2-2x-2y+1=0$$

$\Rightarrow \quad x^2-2x+1+y^2-2y+1=1$

$\Rightarrow \quad (x-1)^2+(y-1)^2=1$

Which is a equation of circle touching both the axes with its centre (1, 1) and radius 1 unit.

10. (a) We have,

centre (1, – 2)

Given equations of lines are

$$3x+y-14=0 \quad \ldots\text{(i)}$$

and $\quad 2x+5y-18=0 \quad \ldots\text{(ii)}$

From Eqs. (i) and (ii), we get

$$x=4 \text{ and } y=2$$

$\therefore$ Point of intersection is (4, 2)

and radius of circle $=\sqrt{(1-4)^2+(-2-2)^2}$

$=\sqrt{9+16}=5$

Therefore, required equation of circle is

$$(x-1)^2+(y+2)^2=25$$

$\Rightarrow \quad x^2+y^2-2x+4y-20=0$

11. (a) Equation of circle with centre $(-2,-3)$ and radius 10 is given by $(x-(-2))^2+(y-(-3))^2=10^2$

$\Rightarrow \quad (x+2)^2+(y+3)^2=100$

12. (b) Given equation of circle is $x^2+(y-2)^2=16$

$\Rightarrow \quad (x-0)^2+(y-(-2))^2=4^2$

This represents equation of circle with centre (0, 2) and radius 4.

13. (b) Equation of circle with centre (3, 1) and radius a is

$(x-3)^2+(y-1)^2=a^2$

$\because$ The circle passes through the point (6, 5).

$\therefore \quad (6-3)^2+(5-1)^2=a^2$

$\Rightarrow \quad 9+16=a^2$

$\Rightarrow \quad a^2=25$

Thus, the equation of circle is

$$(x-3)^2+(y-1)^2=25$$

14. (a) Given equation

$2x^2+2y^2+3y+10=0$ is the form of

$ax^2+ay^2+2gx+2fy+c=0$.

Therefore, it represents a circle.

15. (d) Given, centre $(-a,-b)$ and radius $r=\sqrt{a^2-b^2}$

Equation of circle is

$$(x+a)^2+(y+b)^2=a^2-b^2$$

$\Rightarrow \quad x^2+y^2+2ax+2by+2b^2=0$

16. (i) We have, the equation of circle drawn by Jaya is

$$x^2+y^2+8x-16y+64=0$$

$\Rightarrow \quad x^2+8x+16+(y^2-16y+64)=16$

$\Rightarrow \quad (x+4)^2+(y-8)^2=4^2$

$\Rightarrow \quad \{x-(-4)\}^2+(y-8)^2=4^2 \quad \ldots\text{(i)}$

From Eq. (i), we get $h=-4$, $k=8$ and $r=4$

Hence, radius $r=4$ units.

(ii)

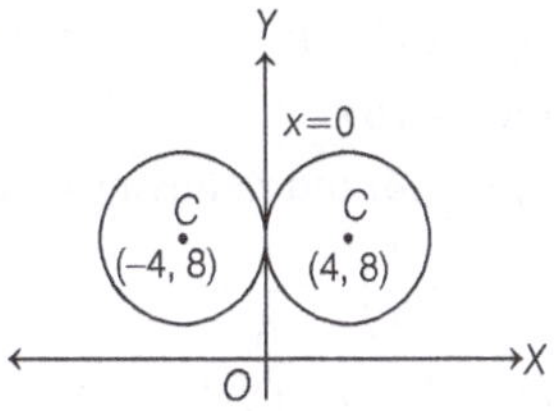

It is clear from the above figure, the circle in the line mirror has centre (4,8) and radius 4 units.

(iii) We have, from Eq. (i)

$$\{x-(-4)\}^2+(y-8)^2=4^2$$

So, $h=-4$ and $k=8$

Therefore, the centre of the circle drawn by Jaya is (–4,8).

(iv) From part (ii), we have centre of mirror image of circle is (4,8) and radius is 4 units.

So, its equation is

$$(x-4)^2+(y-8)^2=4^2$$

$$\Rightarrow \quad x^2+y^2-8x-16y+64=0$$

Subjective Questions

1. Given, centre is (–3,2) and radius = 5 units

$\therefore \quad (h, k) \equiv (-3, 2)$ and $r=5$

On putting these values in equation of circle $(x-h)^2+(y-k)^2=r^2$, we get

$$(x-(-3))^2+(y-2)^2=5^2$$

$$\Rightarrow \quad (x+3)^2+(y-2)^2=25$$

2. Given, centre $(h,k)=(a,a)$ and radius $(r)=a\sqrt{2}$

On putting these values in equation of circle $(x-h)^2+(y-k)^2=r^2$, we get

$$(x-a)^2+(y-a)^2=2a^2$$

3. Given, centre of circle is (2,–5) and circle passes through (3, 2).

Since, radius of the circle is equal to the distance from the centre to a point on a circle.

$\therefore$ Radius of circle

$$=\sqrt{(2-3)^2+(-5-2)^2}=\sqrt{1+49}=\sqrt{50}$$

Therefore, equation of the circle is

$$(x-2)^2+(y+5)^2=50$$

4. Given, equation of the circle is $(x+5)^2+(y-3)^2=36$

i.e. $(x-(-5))^2+(y-3)^2=6^2$

On comparing with $(x-h)^2+(y-k)^2=r^2$, we get $h=-5, k=3$ and $r=6$

$\therefore$ Centre $(h, k) \equiv (-5,3)$ and radius = 6 units

5. Given, equation of circle is $(x-2)^2+(y-4)^2=65$

On comparing with $(x-h)^2+(y-k)^2=r^2$, we get $h=2, k=4$ and $r=\sqrt{65}$

$\therefore$ Centre $(h, k) \equiv (2, 4)$ and radius $(r)=\sqrt{65}$

6. (i) $\left(x-\frac{1}{2}\right)^2+\left(y+\frac{1}{3}\right)^2=\frac{1}{4}$

$$\Rightarrow \left(x-\frac{1}{2}\right)^2+\left(y-\left(-\frac{1}{3}\right)\right)^2=\left(\frac{1}{2}\right)^2$$

$\therefore$ Centre $\equiv\left(\frac{1}{2},-\frac{1}{3}\right)$ and radius $=\frac{1}{2}$ unit

(ii) $x^2+y^2-4x+6y=5$

$$\Rightarrow x^2-4x+4+y^2+6y+9-4-9=5$$

$$\Rightarrow \quad (x-2)^2+(y+3)^2=18$$

$$\Rightarrow \quad (x-2)^2+(y-(-3))^2=(3\sqrt{2})^2$$

$\therefore$ Centre $\equiv (2,-3)$ and radius $=3\sqrt{2}$ units

7. Let points $A \equiv (2,-2)$ and $B \equiv (3, 4)$

Given line is $x+y=2$ ...(i)

Let L be the mid-point of AB.

$\therefore$ Mid-point, $L \equiv\left(\frac{2+3}{2},\frac{-2+4}{2}\right)=\left(\frac{5}{2},1\right)$

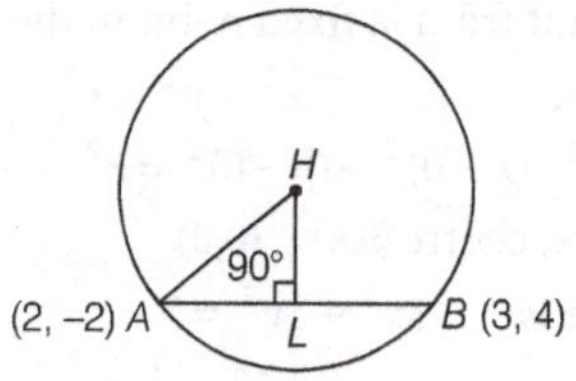

Slop of $AB=\frac{y_2-y_1}{x_2-x_1}=\frac{4-(-2)}{3-2}=6$

$\therefore \quad$ Slop of $HL=-\frac{1}{6}$ $\quad [\because HL \perp AB]$

Then, equation of HL will be $y-1=-\frac{1}{6}\left(x-\frac{5}{2}\right)$

$$\Rightarrow \quad -6y+6=x-\frac{5}{2} \Rightarrow -12y+12=2x-5$$

$$\Rightarrow \quad 2x+12y=17 \quad \text{...(ii)}$$

Centre H of the circle lies on lines (i) and (ii), therefore it is the point of intersection of lines (i) and (ii),

$$x=\frac{7}{10} \text{ and } y=\frac{13}{10}$$

$\therefore \quad$ Centre, $H \equiv\left(\frac{7}{10},\frac{13}{10}\right)$

Now, radius of the circle

$$=HA=\sqrt{\left(\frac{7}{10}-2\right)^2+\left(\frac{13}{10}+2\right)^2}$$

$$=\sqrt{\left(\frac{13}{10}\right)^2+\left(\frac{33}{10}\right)^2}=\sqrt{12.58}$$

$\therefore$ The equation of required circle is

$$(x-0.7)^2+(y-1.3)^2=12.58$$

8. Given, the circle touches the X-axis and centre is (1, 2).

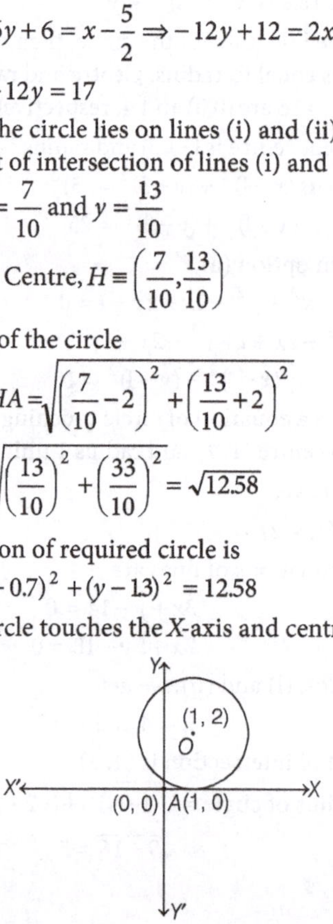

$\because$ Radius of circle is equal to distance between O and A.

$\therefore$ Radius of circle $= \sqrt{(1-1)^2 + (2-0)^2} = 2$

Now, equation of circle is

$$(x-1)^2 + (y-2)^2 = 2^2 = 4$$

9. Given, equation of circle is

$$S_1 \equiv x^2 + y^2 = 25$$

or $\quad S_1 \equiv x^2 + y^2 - 25 = 0 \quad \ldots(i)$

Now, put $x = -1.5$ and $y = 2.5$ in Eq (i), we get

$$S_1 = (-1.5)^2 + (2.5)^2 - 25$$
$$= 2.25 + 6.25 - 25 = -16.5 < 0$$

$\because \quad S_1 < 0$

So, the given point $(-1.5, 2.5)$ lies inside the circle.

10. Let the mid-point of the distance between the flag posts represent the centre with coordinates (0, 0). Then, coordinates of two flag posts are $(-4, 0)$ and $(4, 0)$.

Let (x, y) lie on the path traced by the man.

$$\therefore \sqrt{(x+4)^2 + (y-0)^2} + \sqrt{(x-4)^2 + (y-0)^2} = 10$$

$$\Rightarrow \sqrt{x^2 + 8x + 16 + y^2} = 10 - \sqrt{x^2 - 8x + 16 + y^2}$$

On squaring both sides, we get

$$x^2 + 8x + 16 + y^2$$
$$= 100 + x^2 - 8x + 16 + y^2 - 20\sqrt{x^2 - 8x + 16 + y^2}$$

$$\Rightarrow \quad 16x - 100 = -20\sqrt{x^2 - 8x + 16 + y^2}$$

On dividing both sides by 4, we get

$$4x - 25 = -5\sqrt{x^2 - 8x + 16 + y^2}$$

Again, on squaring, we get

$$16x^2 + 625 - 200x = 25(x^2 - 8x + 16 + y^2)$$

$$\Rightarrow \quad 16x^2 + 625 - 200x = 25x^2 - 200x + 400 + 25y^2$$

Hence, $9x^2 + 25y^2 = 225$ is the required equation of path.

11. Circle is passing through origin. Let it meet axes at the point A and B. Then, AB is diameter and coordinates of ends are $A(a, 0)$ and $B(0, b)$.

$\therefore$ Coordinates of centre are $P\left(\frac{a}{2}, \frac{b}{2}\right)$ and radius is $\frac{1}{2}\sqrt{a^2 + b^2}$.

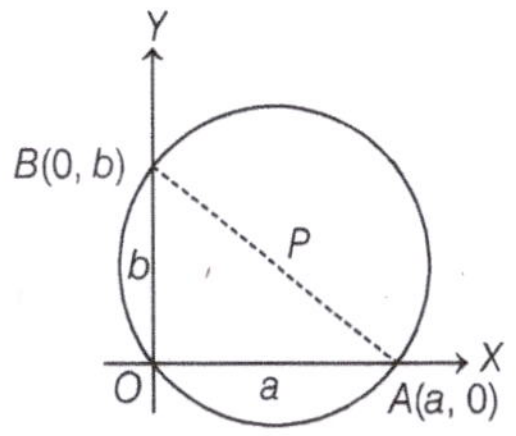

$\therefore$ Equation of circle is

$$\left(x - \frac{a}{2}\right)^2 + \left(y - \frac{b}{2}\right)^2 = \left(\frac{1}{2}\sqrt{a^2 + b^2}\right)^2$$

$$\Rightarrow x^2 + \frac{a^2}{4} - ax + y^2 + \frac{b^2}{4} - by = \frac{1}{4}(a^2 + b^2)$$

$\Rightarrow x^2 + y^2 - ax - by = 0$ is the required equation.

12. Hint Let C be the centre of the circle in its initial position and D be its centre in the new position.

Then, $C \equiv (5, 5)$ and $D = (5 + 10\pi, 5)$

Now, centre of the circle in the new position is $(5 + 10\pi, 5)$ and its radius is 5, therefore its equation will be

$$(x - 5 - 10\pi)^2 + (y-5)^2 = 5^2$$

$$\Rightarrow \quad x^2 + 25 + 100\pi^2 - 10x - 20\pi x + 100\pi + y^2 + 25 - 10y = 25$$

Ans. $x^2 + y^2 - 10(2\pi + 1)x - 10y + 100\pi^2 + 100\pi + 25 = 0$

13. Hint Given, $x = \frac{2at}{1+t^2}$ and $y = \frac{a(1-t^2)}{1+t^2}$

On squaring and adding, we get

$$x^2 + y^2 = \left(\frac{2at}{1+t^2}\right)^2 + \left(\frac{a(1-t^2)}{1+t^2}\right)^2$$

$$= \frac{4a^2t^2}{(1+t^2)^2} + \frac{a^2(1-t^2)^2}{(1+t^2)^2}$$

$$= \frac{a^2(4t^2 + 1 + t^4 - 2t^2)}{(1+t^2)^2}$$

$$= \frac{a^2[t^4 + 1 + 2t^2]}{(1+t^2)^2} = \frac{a^2(t^2+1)^2}{(t^2+1)^2} = a^2$$

$$\Rightarrow \quad x^2 + y^2 = a^2$$

Mind Map

Conic Sections

On intersecting a right circular cone by a plane in different positions, different sections, so obtained are called conic sections or

Geometrical Definition of Conic Section

A conic is the locus of a point which moves in a plane in such a way that the ratio of its distance from a fixed point and a fixed line is a constant. Then,

(i) the fixed point is called **focus** and is denoted by F.
(ii) the fixed straight line ZM is called **directrix**.
(iii) the constant ratio is called **eccentricity** and is denoted by e.

In the given figure, $\frac{PF}{PM} = e =$ constant

On the basis of the value of e, we get the following cases.

If $e = 0$, then conic is circle.
If $e = 1$, then conic is parabolic.
If $e < 1$, then conic is ellipse.
If $e > 1$, then conic is hyperbola.

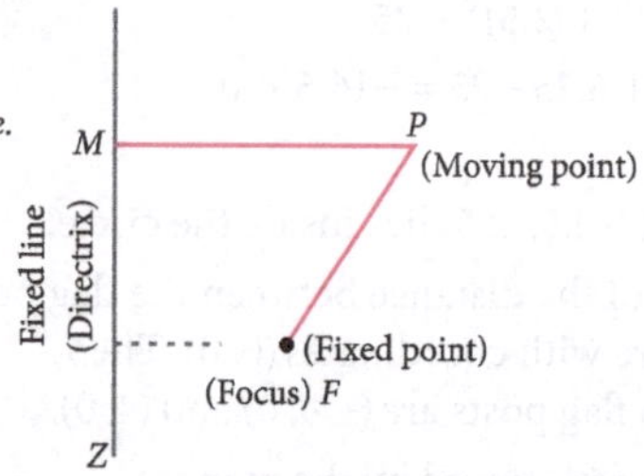

Circle

A circle is the set of all points in a plane that are equidistant from a fixed point in that plane.

Centre The fixed point C is called the centre of the circle.

Radius The constant distance CP from the centre C to a point on the circle, is called radius r.

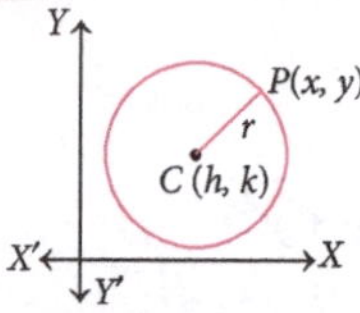

Standard Equation of a Circle

Let $C(h, k)$ be the **centre** of the circle, $P(x, y)$ be any point on the circumference of the circle and r be the **radius** of the circle.

Then, equation of circle in standard form is

$$(x-h)^2 + (y-k)^2 = r^2$$

This equation is also known as **central form** of the equation of a circle.

If the centre of circle is at origin and radius r, then equation of circle is

$$x^2 + y^2 = r^2$$

Equation of Circle in Special Cases

We know that the standard equation of the circle with centre at (h, k) and radius equal to r, is

$$(x-h)^2 + (y-k)^2 = r^2$$

Position of a Point w.r.t. Circle

Let $S \equiv x^2 + y^2 + 2gx + 2fy + c = 0$

or $S \equiv (x-h)^2 + (y-k)^2 = r^2$

be the equation of the circle and $P(x_1, y_1)$ be any point in the plane of the circle.

Then, $S_1 \equiv x_1^2 + y_1^2 + 2gx_1 + 2fy_1 + c = 0$

or $S_1 \equiv (x_1-h)^2 + (y_1-k)^2 = r^2$

Case I If $S_1 > 0$, then the point lies **outside** the circle.

Case II If $S_1 = 0$, then the point lies **on** the circle.

Case III If $S_1 < 0$, then the point lies **inside** the circle.

Case II When the Centre Lies on X-axis or Y-axis

If centre lies on X-axis, then $k = 0$.

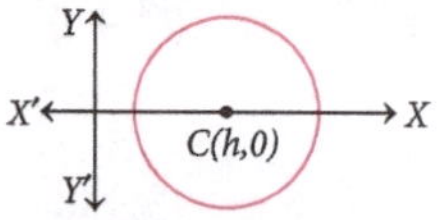

$\therefore$ Equation of circle with centre $C(h, 0)$ is

$$(x-h)^2 + y^2 = r^2$$

If centre lies on Y-axis, then $h = 0$.

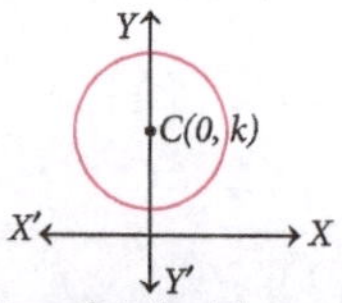

$\therefore$ Equation of circle with centre $C(0, k)$ is

$$x^2 + (y-k)^2 = r^2$$

Case I When the Circle Passes Through the Origin

The circle passes through origin (0, 0).

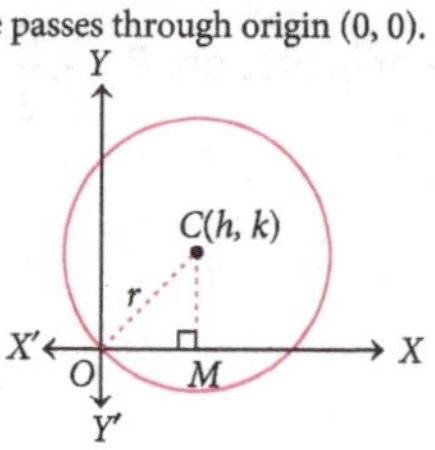

$\therefore$ Radius $= r =$ Distance between points O and C

$= \sqrt{h^2 + k^2}$

Hence, the equation of a circle passing through the origin is

$$(x-h)^2 + (y-k)^2 = h^2 + k^2$$

i.e. $\quad x^2 + y^2 - 2hx - 2ky = 0$

10 (b)

CHAPTER

Conic Section : Parabola

If people do not believe that Mathematics is simple, it is only because they do not realise how complicated life is.

—*John Von Neumann*

YOUR STUDY GOALS

YOUR PRACTICE GOALS

- Objective Type Questions (16 Qs; 10 MCQs, 5 A-R, 1 Case Based)
- Subjective Type Questions
- (18 Qs; 6 VSA, 8 SA, 4 LA)

10.2.1 Parabola

A parabola is the locus of a point which moves in a plane such that its distance from a fixed point is always equal to its distance from a fixed straight line in the same plane.

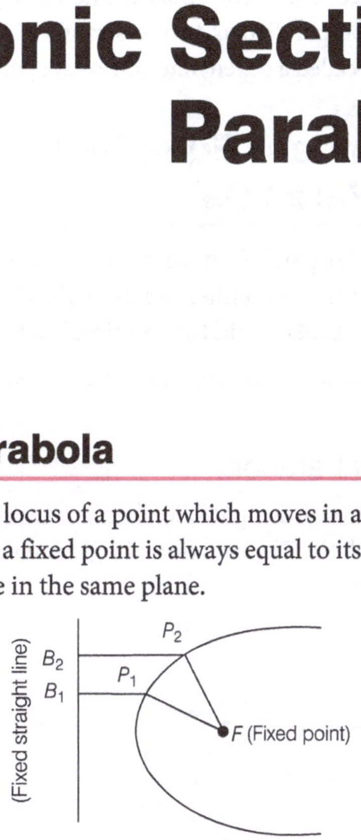

From the figure, $P_1F = P_1B_1$ and $P_2F = P_2B_2$

Here,

- The fixed line is called the **directrix** and the fixed point is called **focus** of the parabola.
- A line through the focus and perpendicular to the directix is called the **axis** of the parabola.

- Point of intersection of parabola with the axis is called the **vertex** of the parabola.

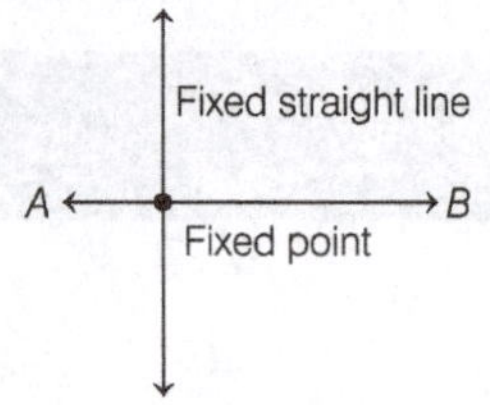

If the fixed point lies on the fixed line, then the set of points in the plane, which are equidistant from the fixed point and the fixed line is the straight line through the fixed point and perpendicular to the straight line. The straight line AB is called a degenerate case of the parabola.

10.2.2 Standard Equations of Parabola

If the equation of the parabola having vertex at origin and the axis of symmetry is either X-axis or Y-axis, then the equation of parabola is said to be in simplest form.

There are four standard equations of parabola which are given below.

Right Handed Parabola

If parabola opens towards right side, then it is called right handed parabola. The equation of right handed parabola is of the form $y^2 = 4ax, a > 0$

Some important terms related to this parabola are

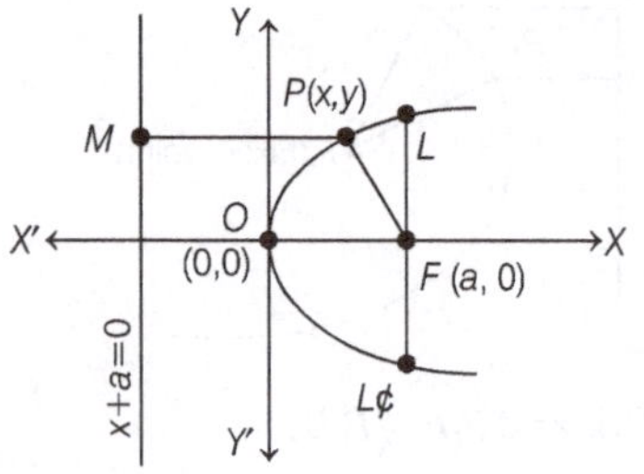

(i) Vertex is $O\,(0, 0)$.
(ii) Axis is the line $y = 0$.
(iii) Focus is $F\,(a, 0)$.
(iv) Directrix is the line $x + a = 0$.
(v) Latusrectum of a parabola is a line segment through focus and perpendicular to the axis of the parabola, whose end points lie on the parabola i.e. $LL' = 4a$.
(vi) Coordinates of latusrectum $= (a, \pm 2a)$.
(vii) Equation of latusrectum is $x = a$ or $x - a = 0$.
(viii) It is symmetrical about X-axis.
(ix) Focal distance of any point $x + a$.

Left Handed Parabola

If parabola opens towards left side, then it is called left handed parabola. The equation of the left handed parabola is of the form

$$y^2 = -4ax, a > 0$$

Some important terms related to this parabola are

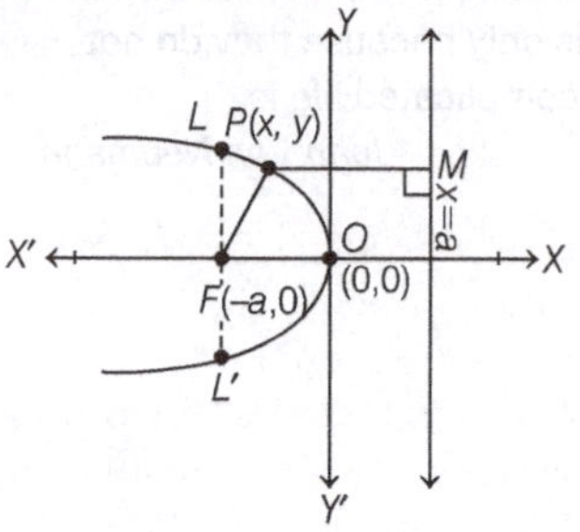

(i) Vertex is $O\,(0, 0)$.
(ii) Axis is the line $y = 0$.
(iii) Focus is $F(-a, 0)$.
(iv) Directrix is the line $x - a = 0$.
(v) Length of the latusrectum is $4a$.
(vi) Coordinates of latusrectum $= (-a, \pm 2a)$.
(vii) Equation of latusrectum is $x = -a$ or $x + a = 0$.
(viii) It is symmetrical about X-axis.
(ix) Focal distance of a point $P(x, y)$ is $a - x$.

Upward Parabola

If parabola opens upward, then it is called upward parabola. The equation of the upward parabola is of the form

$$x^2 = 4ay, a > 0$$

Some important terms related to this parabola are

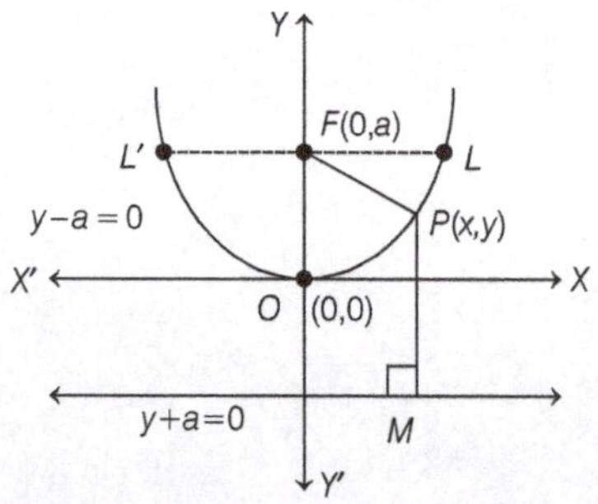

(i) Focus is $F(0, a)$.
(ii) Vertex is $O(0, 0)$.
(iii) Directrix is the line $y + a = 0$.
(iv) Axis is the line $x = 0$.
(v) Length of latusrectum is $4a$.
(vi) Coordinates of latusrectum $= (\pm 2a, a)$.
(vii) Equation of latusrectum is $y = a$ or $y - a = 0$.
(viii) It is symmetrical about Y-axis.
(ix) Focal distance of a point $P(x, y)$ is $a + y$.

Downward Parabola

If parabola opens downwards, then it is called downward parabola. The equation of downward parabola is of the form

$$x^2 = -4ay, a > 0$$

Some important terms related to this parabola are

(i) Focus is $F(0, -a)$.
(ii) Vertex is $O(0, 0)$.
(iii) Directrix is the line $y - a = 0$.

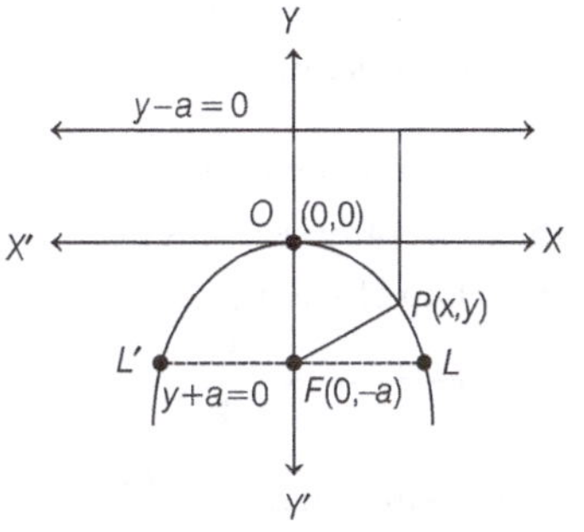

(iv) Axis is the line $x = 0$.
(v) Length of latusrectum is $4a$.
(vi) Coordinates of latusrectum $= (\pm 2a, (-a))$.
(vii) Equation of latusrectum is $y = -a$ or $y + a = 0$.
(viii) It is symmetrical about Y-axis.
(ix) Focal distance of a point $P(x, y)$ is $a - y$.

Table for Different Types of Parabola

Parabola	Vertex	Focus	Latusrectum	Coordinates of latusrectum	Axis	Directrix	Symmetry	Focal Distance of a Point $P(x, y)$
$y^2 = 4ax$	(0, 0)	$(a, 0)$	$4a$	$(a, \pm 2a)$	$y = 0$	$x = -a$	X-axis	$a + x$
$y^2 = -4ax$	(0, 0)	$(-a, 0)$	$4a$	$(-a, \pm 2a)$	$y = 0$	$x = a$	X-axis	$a - x$
$x^2 = 4ay$	(0, 0)	$(0, a)$	$4a$	$(\pm 2a, a)$	$x = 0$	$y = -a$	Y-axis	$a + y$
$x^2 = -4ay$	(0, 0)	$(0, -a)$	$4a$	$(\pm 2a, -a)$	$x = 0$	$y = a$	Y-axis	$a - y$

Example 1. **In each of the following questions, find the coordinates of the focus, axis of the parabola, the equation of the directrix and the length of the latusrectum.**

(i) $y^2 = 12x$ **(ii) $y^2 = -8x$**
(iii) $y^2 = 10x$ **(iv) $x^2 = -9y$**

Sol. (i) Given, equation of parabola is $y^2 = 12x$, which is of the form $y^2 = 4ax$ i.e focus lies on the positive direction of X-axis.

Here, $4a = 12 \Rightarrow a = 3$

$\therefore$ Focus $= (a, 0) = (3, 0)$

Axis $= X$-axis

Directrix, $x = -a \Rightarrow x = -3$

and length of latusrectum $= 4a = 4 \times 3 = 12$

(ii) Given, equation of parabola is $y^2 = -8x$, which is of the form $y^2 = -4ax$ i.e. focus lies on the negative direction of X-axis.

Here, $4a = 8 \Rightarrow a = 2$

$\therefore$ Focus $= (-a, 0) = (-2, 0)$

Axis $= X$-axis

Directrix, $x = a \Rightarrow x = 2$

and length of latusrectum $= 4a = 4 \times 2 = 8$

(iii) Given, equation of parabola is $y^2 = 10x$, which is of the form $y^2 = 4ax$ i.e. focus lies on the positive direction of X-axis.

Here, $4a = 10$

$\Rightarrow a = \frac{5}{2}$

$\therefore$ Focus $= (a, 0) = \left(\frac{5}{2}, 0\right)$

Axis = X-axis

Directrix, $x = -a \Rightarrow x = -\frac{5}{2}$

and length of latusrectum $= 4a = 4 \times \frac{5}{2} = 10$

(iv) Given, equation of parabola is $x^2 = -9y$, which is of the form $x^2 = -4ay$ i.e. focus lies on the negative direction of Y-axis.

Here, $4a = 9 \Rightarrow a = \frac{9}{4}$

$\therefore$ Focus $= (0, -a) = \left(0, -\frac{9}{4}\right)$

Axis = Y-axis

Directrix, $y = a \Rightarrow y = \frac{9}{4}$

and length of latusrectum $= 4a = 9$

Example 2. Find the equation of the parabola, whose focus is the point (4, 0) and whose directrix is $x = -4$. Also, find the length of latusrectum. Competency Based Que

Sol. Here, focus is (4, 0) and directrix is $x + 4 = 0$.

Let $P(x, y)$ be any moving point. Then, draw $PM \perp ZZ'$ from P to the directrix.

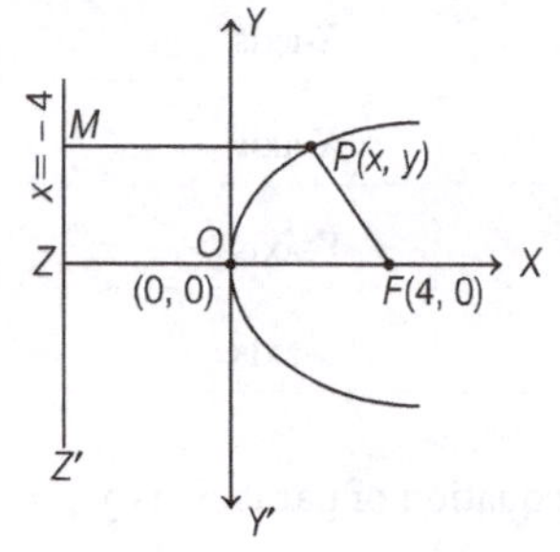

Now, $FP = PM$ [by definition of parabola]

$\Rightarrow FP^2 = PM^2$

$\Rightarrow (x-4)^2 + (y-0)^2 = \frac{(x+4)^2}{(\sqrt{1})^2}$

$\Rightarrow x^2 - 8x + 16 + y^2 = x^2 + 8x + 16$

$\Rightarrow y^2 = 16x$

which is the required equation of parabola.

$\therefore$ Length of latusrectum = 16

Example 3. Find the equation of the parabola with vertex at the orign, the axis along the X-axis and passing through the point (2, 3).

Sol. Equation of parabola, whose axis is X-axis and passing through the point (2, 3), will be of the form

$$y^2 = 4ax \quad \text{...(i)}$$

On putting $x = 2$ and $y = 3$ in Eq. (i), we get $9 = 4a(2)$

$\Rightarrow a = \frac{9}{8}$

Now, on substituting $a = \frac{9}{8}$ in Eq. (i), we get

$$y^2 = 4 \times \frac{9}{8} \times x$$

$\Rightarrow 8y^2 = 36x$

$\Rightarrow 2y^2 = 9x$, which is the required equation of parabola.

Example 4. Find the equation of parabola when the vertex is at (0, 0) and focus is at (0, – 4).

Sol. Here, the vertex is at (0, 0) and focus is at (0, –4), which lies on Y-axis. So, Y-axis is the axis of the parabola.

$\therefore$ Equation of parabola is of the form

$x^2 = -4ay \Rightarrow x^2 = -4(4)y$ $[\because a = 4]$

$\therefore x^2 = -16y$

Example 5. Find the equation of the parabola that satisfies the condition, vertex (0, 0) and focus at (–4, 0).

Sol. Here, the vertex is at (0, 0) and focus is at (–4, 0), which lies on X-axis. So, X-axis is the axis of parabola.

$\therefore$ Equation of parabla is of the form

$y^2 = 4ax$

$\Rightarrow y^2 = 4(-4)x$ $[\because a = -4]$

$\therefore y^2 = -16x$

Example 6. If a parabolic reflector is 20 cm in diameter and 5 cm deep, then find the focus. Competency Based Que

Sol. Let POQ be the parabolic reflector which is 20 cm in diameter and 5 cm deep.

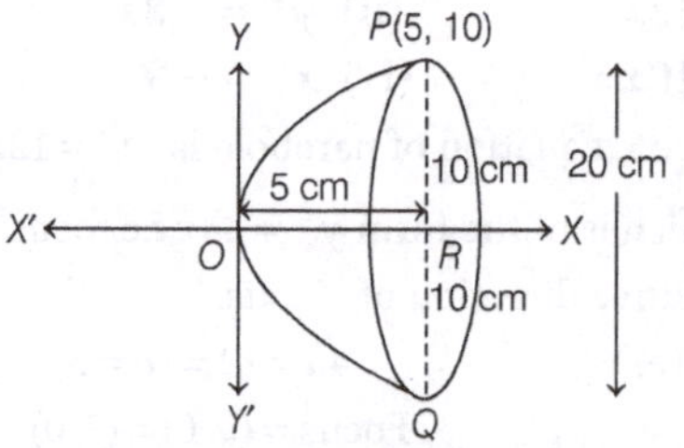

Then, $PQ = 20$ cm and $OR = 5$ cm, where R is the mid-point of PQ. We take OX as X-axis and OY as Y-axis. The equation of parabola may be taken as $y^2 = 4ax$.

Since, the point $P(5, 10)$ lies on the parabola.

$\therefore \quad 10^2 = 4a(5) \Rightarrow a = 5$

Therefore, the coordinates of the focus are $(a, 0)$ i.e. $(5, 0)$. Hence, the focus is the mid-point of the given diameter.

Example 7. An arch is in the form of a parabola with its axis vertical. The arch is 10 m high and 5 m wide at the base. How wide it is 2 m from the vertex of the parabola? *Competency Based Que*

Sol. Here, axis is vertical, so let arch of parabola be in the form

$$x^2 = 4ay \quad \text{...(i)}$$

Given, $OB = 10$ m

and $AC = 5 \text{ m} \Rightarrow AB = \frac{5}{2} \text{ m} \quad \left[\because AB = BC = \frac{AC}{2}\right]$

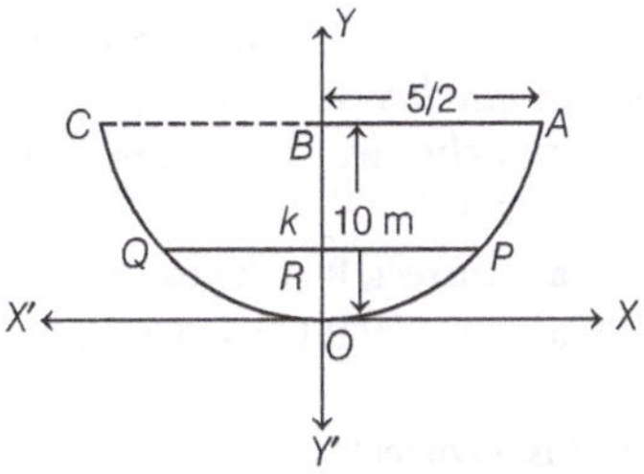

Hence, coordinates of $A = \left(\frac{5}{2}, 10\right)$ will satisfy Eq. (i)

i.e. $\left(\frac{5}{2}\right)^2 = 4a \times 10$

$\Rightarrow \frac{25}{4} = 40a \Rightarrow a = \frac{5}{32}$

From Eq. (i), $x^2 = 4 \times \frac{5}{32} y \Rightarrow x^2 = \frac{5}{8} y$

Now, let $OR = 2$ and $PQ = k \Rightarrow RP = \frac{k}{2}$

Therefore, $P = \left(\frac{k}{2}, 2\right)$ will lie on parabola.

$\therefore \left(\frac{k}{2}\right)^2 = \frac{5}{8} \times 2 \Rightarrow \frac{k^2}{4} = \frac{5}{4}$

$\Rightarrow k = \sqrt{5} = 2.23$ m (approx.)

Example 8. Find the area of the triangle formed by the lines joining the vertex of the parabola $x^2 = 12y$ to the ends of its latusrectum. *NCERT*

Sol. We know that latusrectum is a line perpendicular to the axis and passing through focus whose length is $4a$.

Given, $x^2 = 12y$, which is of the form $x^2 = 4ay$.

$\therefore \quad 4a = 12 =$ Length of ACB

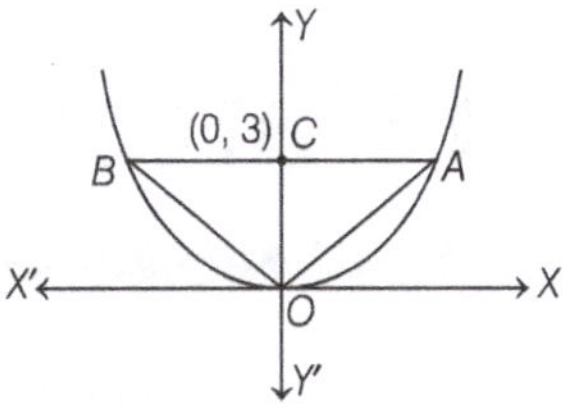

Focus $C = (0, 3)$

$\therefore$ Area of $\Delta OAB = \frac{1}{2} \times AB \times OC = \frac{1}{2} \times 12 \times 3$

$= 6 \times 3 = 18$ sq units

Example 9. The cable of a uniformly loaded suspension bridge hangs in the form of a parabola. The roadway which is horizontal and 100 m long is supported by vertical wires attached to the cable, the longest wire being 30 m and the shortest being 6 m. Find the length of a supporting wire attached to the roadway 18 m from the middle. *Competency Based Que*

Sol. Here, wires are vertical.

Let equation of the parabola be in the form

$$x^2 = 4ay \quad \text{...(i)}$$

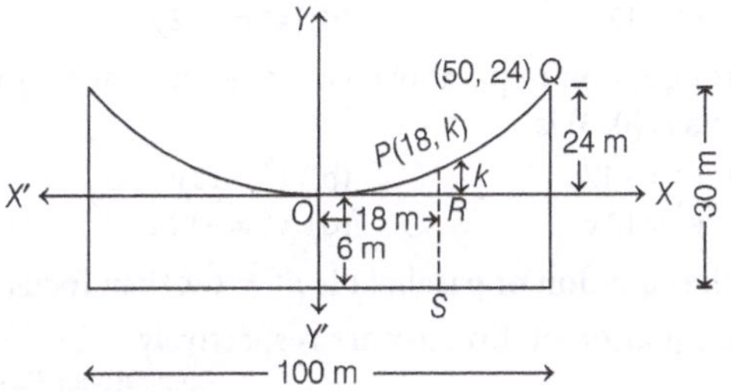

Focus is at the middle of the cable and shortest and longest vertical supports are 6 m and 30 m and roadway in 100 m long.

Clearly, the coordinates of $Q(50, 24)$ will satisfy Eq. (i).

$\therefore \quad (50)^2 = 4a \times 24$

$\Rightarrow \quad 2500 = 96a \Rightarrow a = \frac{2500}{96}$

Hence, from Eq. (i), $x^2 = 4 \times \frac{2500}{96} y \Rightarrow x^2 = \frac{2500}{24} y$

Let $\quad PR = k$ m

Then, point $P(18, k)$ will satisfy the equation of parabola.

$\therefore$ From Eq. (i), $(18)^2 = \frac{2500}{24} \times k$

$\Rightarrow \quad 324 = \frac{2500}{24} k$

$\Rightarrow \quad k = \frac{324 \times 24}{2500} = \frac{324 \times 6}{625} = \frac{1944}{625}$

$\Rightarrow \quad k = 3.11$

$\therefore$ Required length $= 6 + k = 6 + 3.11 = 9.11$ m (approx.)

Exam Pattern Questions

Objective Type Questions

Multiple Choice Questions

1. Equation of parabola is $y^2 = -4ax$, then the coordinates of latusrectum are

(a) $(-a, \pm 2a)$ (b) $(a, \pm 2a)$
(c) $(\pm 2a, a)$ (d) $(-2a, \pm a)$

2. Equation of parabola is $x^2 = -18y$, then the length of latusrectum is

(a) $\frac{9}{2}$ (b) 18 (c) 9 (d) $\frac{9}{4}$

3. The equation of parabola whose focus is (3, 0) and directrix $x = -3$ is

(a) $y^2 = 12x$ (b) $y^2 = -12x$
(c) $x^2 = 12y$ (d) $x^2 = -12y$

4. The equation of parabola whose vertex is at origin and focus is (0, 3) is

(a) $x^2 = -12y$ (b) $x^2 = 12y$
(c) $y^2 = 12x$ (d) $y^2 = -12x$

5. If the equation of parabola is $x^2 = 6y$, then focus and the equation of directrix are respectively.

Competency Based Que

(a) $\left(0, \frac{-3}{2}\right)$ and $y = \frac{3}{2}$ (b) $\left(0, \frac{3}{2}\right)$ and $y = \frac{3}{2}$
(c) $\left(0, \frac{3}{2}\right)$ and $y = \frac{-3}{2}$ (d) None of these

6. The equation of parabola when the vertex is at (0, 0) and focus is at (0, –4) is

(a) $y^2 = 4x$ (b) $y^2 = 16x$
(c) $x^2 = 4y$ (d) $x^2 - 16y = 0$

7. The equation of parabola is $x^2 = 12y$. If its coordinates of latusrectum is $(\pm 2a, a)$, then a is equal to

Competency Based Que

(a) 3 (b) 2 (c) 1 (d) –2

8. The coordinates of focus of parabola $y^2 = 3x$ are

(a) (0.75, 0) (b) (0, 0.75)
(c) (0, – 3) (d) (–3, 0)

9. The equation of parabola whose vertex is at origin and focus is (0, 3), is

Competency Based Que

(a) $x^2 = 12y$ (b) $y^2 = 12x$
(c) $x^2 = -12y$ (d) $y^2 = -12x$

10. The equation of parabola whose axis is along Y-axis vertex at origin and passing through (–2, 5), is

Competency Based Que

(a) $x^2 = -5y$
(b) $y^2 = -5x$
(c) $5y^2 = 4x$
(d) $5x^2 = 4y$

Assertion-Reason Based Questions

Directions (Q. Nos. 11-15) *In the questions given below are two statements labelled as Assertion (A) and Reason (R). In the context of the two statements, which one of the following is correct?*

(a) Both A and R are correct; R is the correct explanation of A
(b) Both A and R are correct; R is not the correct explanation of A
(c) A is correct; R is incorrect
(d) R is correct; A is incorrect

11. **Assertion** (A) For the parabola $y^2 = 16x$, the coordinates of focus are (4, 0).

Reason (R) The equation $y^2 = 4ax$ represents the parabola with vertex (0, 0).

12. **Assertion** (A) For the parabola $x^2 = -16x$, length of latusrectum is 16.

Reason (R) The equation $y^2 = -4ax$ represents a parabola with focus $(a, 0)$.

Competency Based Que

13. **Assertion** (A) For the parabola $x^2 = -6y$, equation of directrix is $y = \frac{3}{2}$.

Reason (R) For the parabola $y^2 = 4ax$, equation of axis is $x = 0$.

14. **Assertion** (A) The directrix of the parabola $y^2 = \frac{8}{3}x$ is $3x = -2$.

Reason (R) The equation of directrix of the form of parabola $y^2 = 4ax$ is $x = -a$.

Competency Based Que

15. **Assertion** (A) The equation of the parabola whose vertex is at origin and focus (–2,0) is $y^2 = -8x$.

Reason (R) The equation of parabola with vertex (0,0) and focus $(0, a)$ is given by $x^2 = 4ay$.

Case Study Based Questions

16. Water flowing in a fountain follows trajectories as shown below.

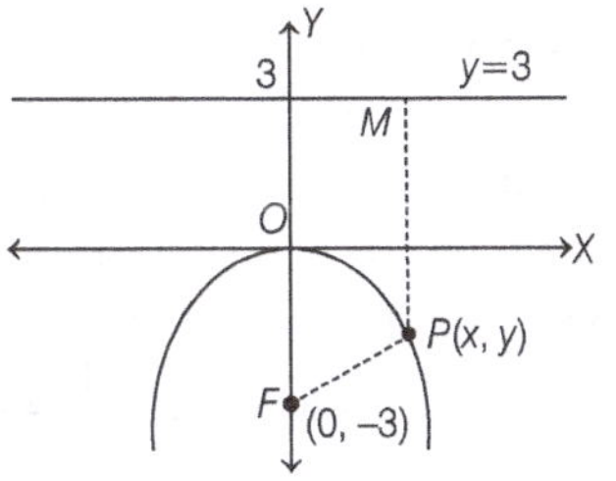

A student observes this and draws its parabolic slope on a plane paper by taking axes as shown in the given figure that $PF = PM$.

Based on the given information, answer the following questions. **Competency Based Que**

(i) What is the equation of parabola?

(ii) Find the length of latusrectum.

(iii) What is the directrix of parabola $x^2 = -12y$?

(iv) Write the axis of symmetry of parabola $x^2 = -12y$.

Subjective Questions

Very Short Answer Type Questions

1. Find the equation of the parabola whose focus is (2, 0) and directrix is $x = -2$.

2. Find the equation of the parabola whose focus is $(0, -3)$ and directrix is $y = 3$.

3. Find the equation of the parabola with vertex at the origin and focus at $(-2, 0)$.

4. If the equation of the parabola is $x^2 = -8y$, then find the equation of directrix. **Competency Based Que**

5. Find the equation of the parabola which is symmetric about the Y-axis and passes through the point $(2, -3)$.

6. Find the equation of parabola whose centre is at origin, axis along the X-axis and passes through the piont (2, 3). **Competency Based Que**

Short Answer Type Questions

7. Find the coordinates of the focus and the vertex, the equations of directrix and the axis of the following

(i) $x^2 = 10y$ (ii) $3x^2 = 8y$

8. Find the coordinates of the focus and the vertex, equation of directrix and length of latusrectum of the following

(i) $y^2 = 10x$ (ii) $5y^2 = -16x$

9. Find the equation of the parabola with vertex at the origin, the axis along the X-axis and passing through the point (2, 3). **Competency Based Que**

10. Find the focus, directrix and eccentricity of the parabola $3y^2 = 8x$.

11. Find the coordinates of the focus and the vertex, the equations of the directrix and the axis and length of latusrectum of the parabola $y^2 = -16x$.

12. For the following curve, find the coordinates of the focus, axis, equation of the directrix and length of the latusrectum : $x^2 = -16y$. **Competency Based Que**

13. Write the equation of the parabola with vertex at origin and directrix $y = 2$.

14. Find the equation of the parabola that satisfies the given condition vertex (0, 0); focus $(-2, 0)$. **NCERT**

Long Answer Type Questions

15. Find the area of the triangle formed by the lines joining the vertex of the parabola $x^2 = -36y$ to the ends of the latusrectum. **Competency Based Que**

16. An equilateral triangle is inscribed in the parabola $y^2 = 4ax$, where one vertex is at the vertex of the parabola. Find the length of the side of the triangle.

17. A beam is supported at its ends by supports which are 12 m apart. Since, the load is connected at this centre, there is a deflection of 3 cm at the centre and the deflected beam is in the shape of a parabola. How far from the centre is the reflection 1 cm? **Competency Based Que**

18. The focus of a parabolic mirror as shown in figure is at a distance 6 cm from its vertex. If the mirror is 20 cm deep, then find the distance LM. **Competency Based Que**

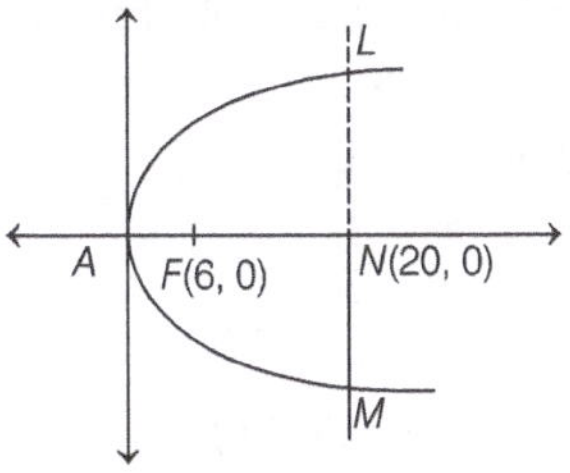

Hints & Solutions

Objective Type Questions

1. (a) We have,

equation of parabola $y^2 = -4ax$

$\therefore$ Coordinates of latusrectum $= (-a, \pm 2a)$

2. (b) Given parabola is

$x^2 = -18y$, which is the form of $x^2 = -4ay$ i.e. focus lies on the negative direction of Y-axis.

Here, $4a = 18$

$\Rightarrow a = \frac{18}{4} = \frac{9}{2}$

$\therefore$ Length of latusrectum $= 4a$

$= 4 \times \frac{9}{2} = 18$

3. (a) As directrix is of the form $x + a = 0$ and focus $(a, 0)$.

Here, $a = 3$

Equation of parabola is $y^2 = 4ax$.

$\therefore y^2 = 4 \times 3x \Rightarrow y^2 = 12x$

4. (b) As vertex is origin and focus is on the Y-axis.

$\therefore$ Parabola is $x^2 = 4ay$.

$\because a = 3$

$\therefore$ Parabola is $x^2 = 12y$.

5. (c) Given equation of the parabola is $x^2 = 4ay \Rightarrow a = \frac{3}{2}$

Focus of the parabola is $\left(0, \frac{3}{2}\right)$.

Its directrix is $y = -a \Rightarrow y = \frac{-3}{2}$.

6. (d) Since, the vertex of parabola is at (0, 0) and focus is at (0,–4) which lies on Y-axis. So, the axis of the parabola.

Here, $a = -4$

$\therefore$ Required equation of parabola is $x^2 = -4ay$

$\Rightarrow x^2 = -4(-4)y$ $[\because a = -4]$

$\Rightarrow x^2 = 16y$

$\Rightarrow x^2 - 16y = 0$

7. (a) Given equation of parabola is $x^2 = 12y$. It is the form of $x^2 = 4ay$.

The coordinates of latusrectum will be $(\pm 2a, a)$.

i.e. $(\pm 2a, a) = (\pm 2 \times 3, 3) = (\pm 6, 3)$

Hence, $a = 3$

8. (a) As parabola $y^2 = 3x$

Here, $4a = 3$

$\Rightarrow a = \frac{3}{4} = 0.75$

Now, focus is $(a, 0)$ i.e. $(0.75, 0)$.

9. (a) As vertex is at origin (0,0) and focus is on the Y-axis.

$\therefore$ Parabola is $x^2 = 4ay$

Here, $a = 3$ $[\because (0, a) \equiv (0, 3)]$

Therefore, the required equation of parabola is $x^2 = 12y$.

10. (d) As parabola $x^2 = 4ay$ passes through $(-2, 5)$.

Therefore, $4 = 4 \times a \times 5$

$\Rightarrow 4 = 20a \Rightarrow a = \frac{4}{20} = \frac{1}{5}$

$\therefore$ Equation of parabola is $5x^2 = 4y$

11. (b) $\because y^2 = 16x \Rightarrow 4a = 16 \Rightarrow a = 4$

$\therefore$ Coordinates of focus are $(a, 0)$ i.e. $(4, 0)$.

12. (c) We have, $x^2 = -16x$

$\Rightarrow 4a = 16 \Rightarrow a = 4$

$\therefore$ Length of latusrectum $= 4a = 16$.

13. (c) We have, $x^2 = -6y$

$\Rightarrow 4a = 6 \Rightarrow a = \frac{3}{2}$

Equation of directrix is $y = a$.

$\therefore y = \frac{3}{2}$

14. (a) Given parabola is $y^2 = \frac{8}{3}x$

Here, $4a = \frac{8}{3} \Rightarrow a = \frac{2}{3}$

$\therefore$ Directrix is $x = -a$

$\Rightarrow x = -\frac{2}{3}$

$\Rightarrow 3x = -2 \Rightarrow 3x + 2 = 0$

15. (b) Vertex (0,0) and focus (–2,0)

$\therefore a = -2$

$\because$ Equation of parabola is $y^2 = 4ax$.

$\Rightarrow y^2 = 4(-2)x$

$y^2 = -8x$

Reason is also correct but Reason is not the correct explanation of Assertion.

16. (i) Given,

$PF = PM \Rightarrow \sqrt{(x-0)^2 + (y+3)^2} = (y-3)$

$\Rightarrow x^2 + (y+3)^2 = (y-3)^2$

$\Rightarrow x^2 = (y-3)^2 - (y+3)^2 = -12y$

$\therefore x^2 = -12y$ is the required equation of parabola.

(ii) We have, equation of parabola $x^2 = -12y$

$\Rightarrow x^2 = 4 \times (-3y)$

$\therefore$ Length of the latusrectum $= 4a$

$= 4 \times 3 = 12$

(iii) Given parabola, $x^2 = -12y$

$\Rightarrow \quad x^2 = 4\times(-3y)$

but $\quad x^2 = 4ay$

$\therefore \quad a = -3$

$\therefore$ Its directrix $y + a = 0$

$\Rightarrow \quad y - 3 = 0$

(iv) Equation of given parabola is $x^2 = -12y$.

Replacing x by $-x$, we get

$(-x)^2 = -12y$

$\Rightarrow \quad x^2 = -12y$

So, there is no change in the equation of parabola.

$\therefore$ It is symmetric about Y-axis.

Subjective Questions

1. **Hint** $\because$ Directrix is $x = -2$ and the focus is (2, 0).

$\therefore$ Required equation of parabola is $y^2 = 4ax$

$\Rightarrow \quad y^2 = 4\times 2x \quad [\because a = 2]$

Ans. $y^2 = 8x$

2. Given, focus of parabola is $(0,-3)$ and its directrix $y = 3$.
Let $P(x, y)$ be any moving point. Then, draw $PM \perp ZZ'$ from to the directrix.

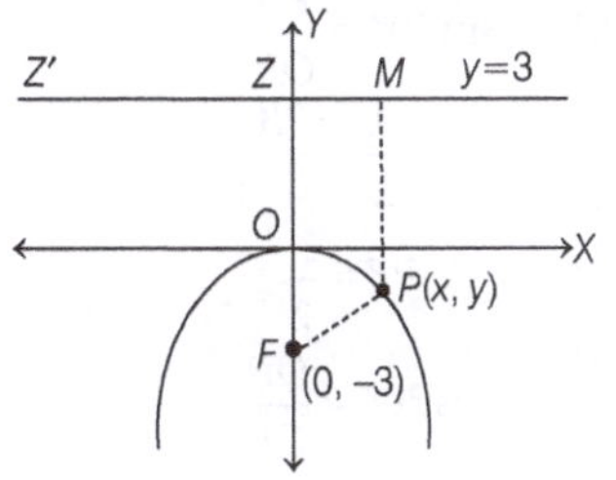

Now $\quad FP = PM$ [by definition of parabola]

$\Rightarrow \quad FP^2 = PM^2$

$\Rightarrow \quad (x-0)^2 + (y+3)^2 = (3-y)^2$

$\Rightarrow \quad x^2 + y^2 + 6y + 9 = y^2 - 6y + 9$

$\therefore \quad x^2 = -12y$

3. Equation of parabola, whose vertex is at origin and focus at $(-a, 0)$, is $y^2 = -4ax$

Given, focus $= (-2, 0)$

$\therefore \quad a = 2$

Hence, equation of parabola is

$y^2 = -4(2)x \Rightarrow y^2 = -8x$

4. Given equation of parabola is $x^2 = -8y$.

Which is of the form $x^2 = -4ay$.

Here, $\quad 4a = 8 \Rightarrow a = 2$

$\therefore$ Equation of directrix is $y = a$ i.e. $y = 2$

5. Since, the parabola is symmetric about Y-axis and its vertex at the origin, then the equation of parabola is of the form $x^2 = 4ay$.

Now, parabola passes through $(2, -3)$.

On putting $x = 2$ and $y = -3$ in equation of parabola, we get

$\Rightarrow \quad 2^2 = 4a(-3) \Rightarrow a = -\frac{1}{3}$

$\therefore$ Equation of parabola is $x^2 = -\frac{4}{3}y$

6. Parabola is $y^2 = 4ax \quad$ ...(i)

It passes through the point (2, 3).

$\therefore \quad 9 = 4a\times 2 \Rightarrow a = \frac{9}{8}$

On substituting $a = \frac{9}{8}$ in Eq. (i), we get

$y^2 = 4\times\frac{9}{8}\times x \Rightarrow 2y^2 = 9x,$

which is the required equation of parabola.

7. (i) Given parabola is $x^2 = 10y$ i.e. $x^2 = 4\left(\frac{5}{2}\right)y$

Here, $a = \frac{5}{2}$

$\therefore$ Focus, $(0, a) \equiv \left(0, \frac{5}{2}\right)$, vertex $= (0, 0)$

Equation of directrix, $y = -a$ i.e. $y = \frac{-5}{2}$ or $2y + 5 = 0$

and axis is $x = 0$

(ii) $3x^2 = 8y$ i.e. $x^2 = 4\left(\frac{2}{3}\right)y \Rightarrow a = \frac{2}{3}$

$\therefore$ Focus, $(0, a) \equiv \left(0, \frac{2}{3}\right)$, vertex $= (0, 0)$

Equation of directrix, $y = \frac{-2}{3}$ i.e. $3y + 2 = 0$ and axis is $x = 0$

8. (i) Given, equation of parabola is $y^2 = 10x$.

$\Rightarrow \quad y^2 = 4\left(\frac{5}{2}\right)x \Rightarrow a = \frac{5}{2}$

$\therefore$ Focus, $(a, 0) \equiv \left(\frac{5}{2}, 0\right)$, vertex $= (0, 0)$

Equation of directrix is $x = -a$ i.e. $x = \frac{-5}{2}$

or $2x + 5 = 0$ and length of latusrectum $= 4a = 10$

(ii) Given, equation of parabola is $5y^2 = -16x$

i.e. $\quad y^2 = 4\left(\frac{-4}{5}\right)x \Rightarrow a = \frac{-4}{5}$

$\therefore$ Focus, $(a, 0) \equiv \left(\frac{-4}{5}, 0\right)$, vertex $= (0, 0)$

Equation of directrix is $x = -a$

i.e. $x = -\left(\frac{-4}{5}\right)$ or $5x - 4 = 0$

and length of latusrectum $= |4(a)| = \frac{16}{5}$ units

9. Given, vertex of parabola at origin and axis of parabola is X-axis.

So, equation of parabola is $y^2 = 4ax$.

$\because$ Equation of parabola passing through the point (2, 3).

$\therefore \quad (3)^2 = 4a(2)$

$\Rightarrow \quad 9 = 8a$

$\Rightarrow \quad a = \frac{9}{8}$

$\therefore$ Equation of parabola is $y^2 = 4\left(\frac{9}{8}\right)x$ i.e. $y^2 = \frac{9}{2}x$

10. Given parabola is $3y^2 = 8x \Rightarrow y^2 = \frac{8}{3}x$

$\therefore \quad 4a = \frac{8}{3} \Rightarrow a = \frac{2}{3}$

$\therefore \quad$ Focus, $(a, 0) \equiv \left(\frac{2}{3}, 0\right)$

As it is parabola, so eccentricity = 1.

Directrix is $x = -a \Rightarrow x = -\frac{2}{3} \Rightarrow 3x + 2 = 0$

11. Given parabola is $y^2 = -16x$

Here, $\quad 4a = -16 \Rightarrow a = -4$

$\therefore$ Focus $(a, 0)$ i.e. $(-4, 0)$

Vertex (0, 0)

Equation of axis is $y = 0$

Equation of directrix is $x + a = 0$ i.e. $x = 4$

Length of latusrectum $= 4a = |-16| = 16$.

12. Consider the equation $x^2 = -16y$

On comparing with $x^2 = 4ay$, we get

$4a = -16 \Rightarrow a = -4$

$\therefore$ Coordinates of focus are $(0, a)$ i.e. $(0, -4)$

Axis is Y-axis.

Equation of directrix is $y + a = 0$

$\Rightarrow \quad y - 4 = 0$

Length of latusrectum $= |4a| = 16$.

13. Vertex (0, 0);

Directrix $y = 2 \Rightarrow a = -2$

Parabola is $x^2 = 4(-2)y$

$\Rightarrow \quad x^2 = -8y$

14. $\because$ Vertex is (0, 0) and focus is (–2, 0).

$\therefore \quad a = -2$

Equation is $y^2 = 4(-2)x$

$\Rightarrow \quad y^2 = -8x$

15. The given parabola is $x^2 = -36y$

On comparing this equation with $x^2 = -4ay$, we get

$4a = 36 \Rightarrow a = 9$

$\therefore$ Focus, $S(0, -a) \equiv (0, -9)$

Let AB be the latusrectum of the given parabola.

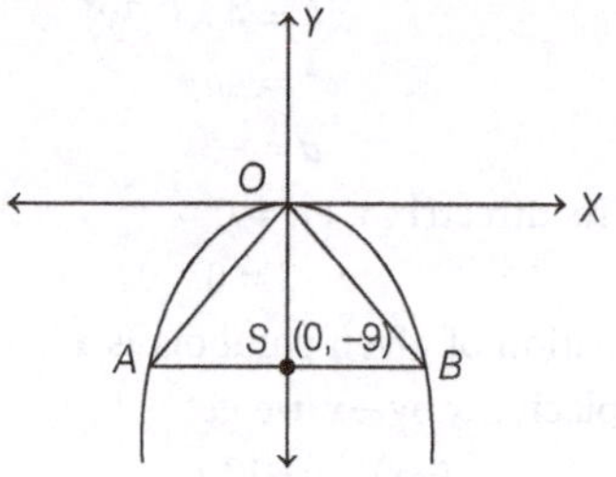

At $y = -9$, $x^2 = -36(-9) \Rightarrow x = \pm 18$

$\therefore$ The coordinates of A and B are $(-18, -9)$ and $(18, -9)$, respectively.

Now, area of

$\Delta OAB = \frac{1}{2}|x_1(y_2 - y_3) + x_2(y_3 - y_1) + x_3(y_1 - y_2)|$

$= \frac{1}{2}|0(-9+9) + (-18)(-9-0) + 18(0+9)|$

$= \frac{1}{2}|162 + 162| = 162$ sq units

16. First, we draw the parabola in the positive side of X-axis and inside that, draw an equilateral ΔOAB.

Let $OB = \quad = OA = AB \Rightarrow \angle BOA = 60^\circ \Rightarrow \angle BOP = 30^\circ$

In ΔBOP, $\quad \sin 30^\circ = \frac{PB}{OB}$

$\Rightarrow \quad \frac{1}{2} = \frac{PB}{}$

$\Rightarrow \quad PB = \frac{}{2}$

and $\quad \cos 30^\circ = \frac{OP}{OB}$

$\Rightarrow \quad \frac{\sqrt{3}}{2} = \frac{OP}{} \Rightarrow OP = \frac{\sqrt{3}}{2}$

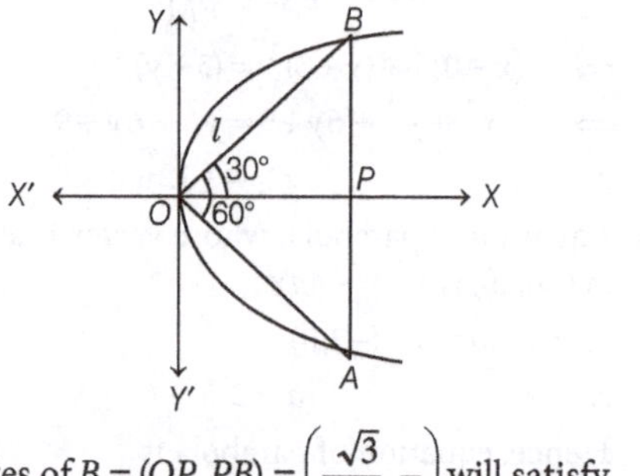

$\therefore$ Coordinates of $B = (OP, PB) = \left(\frac{\sqrt{3}}{2}, \frac{}{2}\right)$ will satisfy

$y^2 = 4ax$

i.e. $\quad \left(\frac{}{2}\right)^2 = \frac{4a \times \sqrt{3}}{2}$

$\Rightarrow \quad \frac{^2}{4} = \frac{4a\sqrt{3}}{2} \Rightarrow \quad = 8\sqrt{3}a$

Hence, the length of side of the triangle is $8\sqrt{3}a$ units.

17.

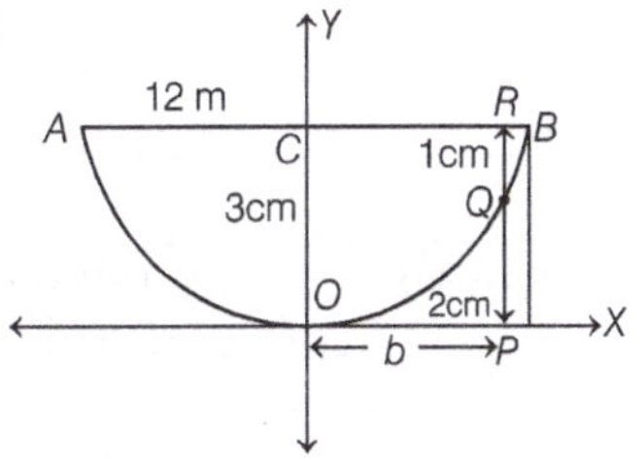

The beam takes the shape of a parabola, whose equation is of the form

$$x^2 = 4ay \quad \ldots\text{(i)}$$

Since, the point $B\left(6, \frac{3}{100}\right)$ lies on it, we get

$$(6)^2 = 4a\left(\frac{3}{100}\right)$$

$$\Rightarrow \quad a = 300 \text{ m}$$

Let RQ be the deflection.

Then, $RQ = \frac{1}{100}$ m and Q is $\left(b, \frac{2}{100}\right)$

$$\therefore \quad b^2 = 4\,(300) \times \frac{2}{100}$$

$$\Rightarrow \quad b = 2\sqrt{6} \text{ m}$$

which is required distance.

18. **Hint** Vertex $\equiv (0, 0)$ and focus $(a, 0) \equiv (6, 0)$

$\therefore$ Equation of parabola is $y^2 = 4\,(6)x$

At $x = 20, y^2 = 24\,(20) \Rightarrow y = \pm\, 4\sqrt{30}$

$\Rightarrow L \equiv (20, 4\sqrt{30})$ and $M(20, -4\sqrt{30})$

Ans. $LM = 8\sqrt{30}$ cm

Mind Map

Table for Different Types of Parabola

Parabola	Vertex	Focus	Latusercutim	Coordinates of latusrectum	Axis	Directrix	Symmetry	Focal Distance of a Point $P(x, y)$
$y^2 = 4ax$	(0, 0)	$(a, 0)$	$4a$	$(a, \pm 2a)$	$y = 0$	$x = -a$	X-axis	$a + x$
$y^2 = -4ax$,	(0, 0)	$(-a, 0)$	$4a$	$(-a, \pm 2a)$	$y = 0$	$x = a$	X-axis	$a - x$
$x^2 = 4ay$,	(0, 0)	$(0, a)$	$4a$	$(\pm 2a, a)$	$x = 0$	$y = -a$	Y-axis	$a + y$
$x^2 = -4ay$,	(0, 0)	$(0, -a)$	$4a$	$(\pm 2a, -a)$	$x = 0$	$y = a$	Y-axis	$a - y$

Parabola

Parabola

A parabola is the locus of a point which moves in a plane such that its distance from a fixed point is always equal to its distance from a fixed straight line in the same plane.

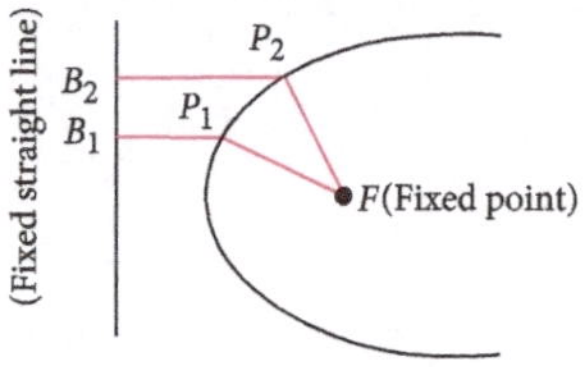

From the figure, $P_1F = P_1B_1$ and $P_2F = P_2B_2$

Here,

- The fixed line is called the **directrix** and the fixed point is called **focus** of the parabola.
- A line through the focus and perpendicular to the directix is called the **axis** of the parabola.

Standard Equations of Parabola

Right Handed Parabola

If parabola opens towards right side, then it is called right handed parabola. The equation of right handed parabola is of the form

$$y^2 = 4ax, a > 0$$

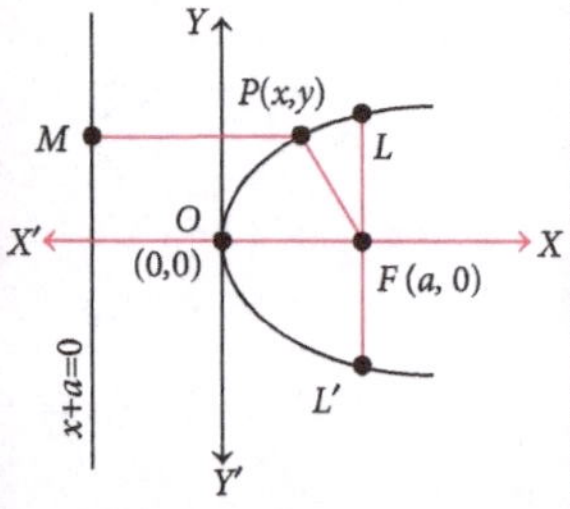

Left Handed Parabola

If parabola opens towards left side, then it is called left handed parabola. The equation of the left handed parabola is of the form

$$y^2 = -4ax, a > 0$$

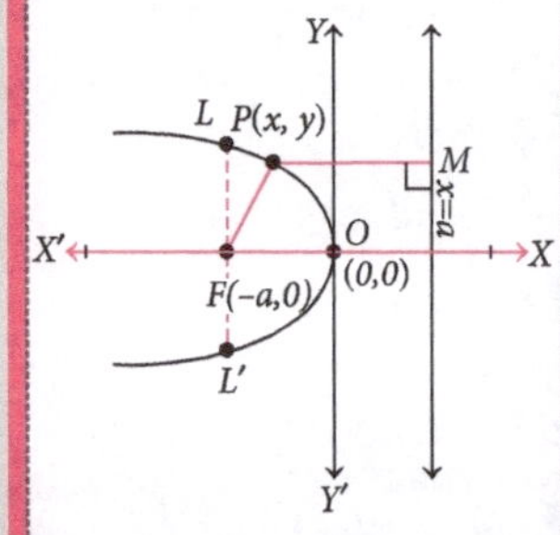

Upward Parabola

If parabola opens upward, then it is called upward parabola.
The equation of the upward parabola is of the form

$$x^2 = -4ay, a > 0$$

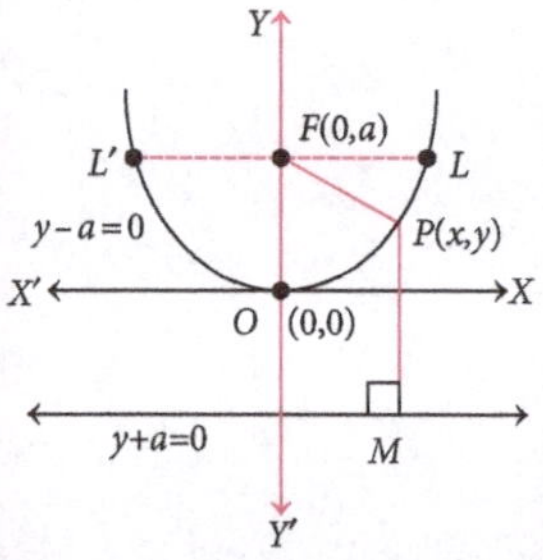

Downward Parabola

If parabola opens downwards, then it is called downward parabola.
The equation of downward parabola is of the form

$$x^2 = -4ay, a > 0$$

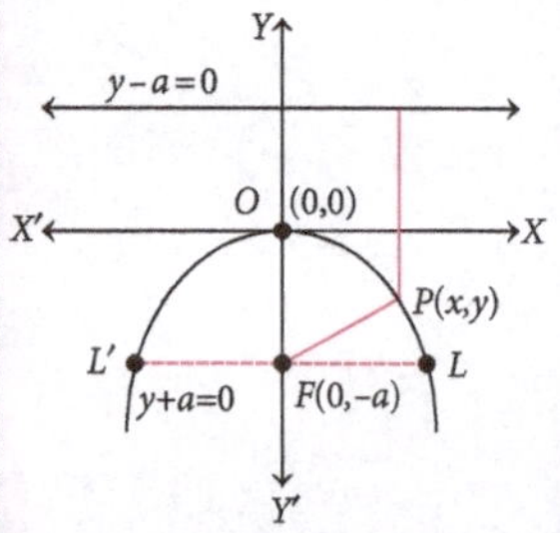

10(c)

CHAPTER

Maths is only place where truth and beauty mean the same thing. —*Danica Mckellar*

Conic Section : Ellipse

▼ **YOUR STUDY GOALS**

▼ **YOUR PRACTICE GOALS**

- Objective Type Questions (13 Qs; 8 MCQs, 4 A-R, 1 Case Based)
- Subjective Type Questions
- (18 Qs; 8 VSA, 5 SA, 5 LA)

10.3.1 Ellipse

An ellipse is the locus of a point in a plane which moves in the plane in such a way that the ratio of its distance from a fixed point (called **focus**) in the same plane to its distance from a fixed straight line (called **directrix**) is always constant, which is always less than unity.

The constant ratio is denoted by e and is called the **eccentricity** of the ellipse.

In the given figure, F is the **focus**, AA' is the **directrix** and P is any point on ellipse.

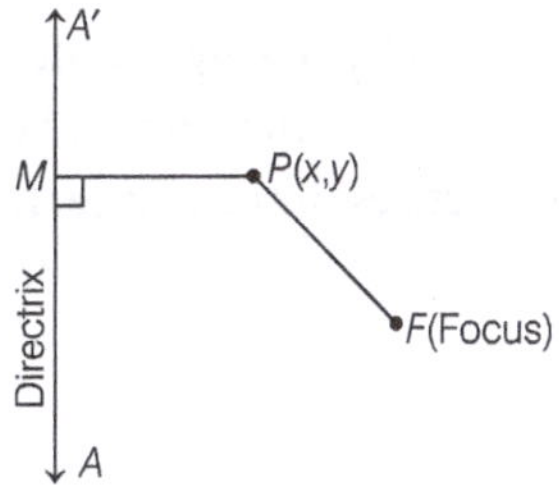

Then, by definition,

$$\frac{PF}{PM} = e, \ \ 0 < e < 1 \ \ \Rightarrow \ \ PF = e \cdot PM$$

In other words, we can say that an ellipse is the set of all points in a plane, the sum of whose distance from two fixed points in the plane is a constant.

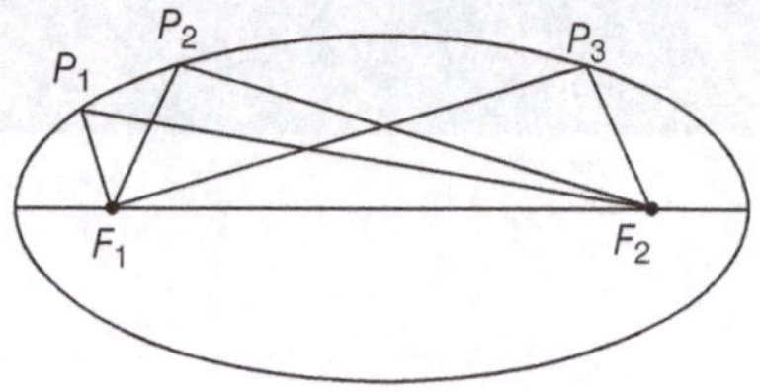

$\therefore\ P_1F_1+P_1F_2=P_2F_1+P_2F_2=P_3F_1+P_3F_2=\text{Constant}$

Terms Related to an Ellipse

- **Focus** Two fixed points are called the foci (plural of focus) of the ellipse and denoted by F_1 and F_2. The distance between two foci F_1 and F_2 is $2c$.
- **Centre** The mid-point of the line segment joining the foci, is called centre of ellipse.
- **Major Axis** The line segment through the foci of the ellipse, is called the major axis. The length of major axis is denoted by $2a$.
- **Minor Axis** The line segment through the centre and perpendicular to the major axis is called minor axis. The length of minor axis is denoted by $2b$.
- **Vertices** The end points of the major axis, are called vertices of the ellipse.

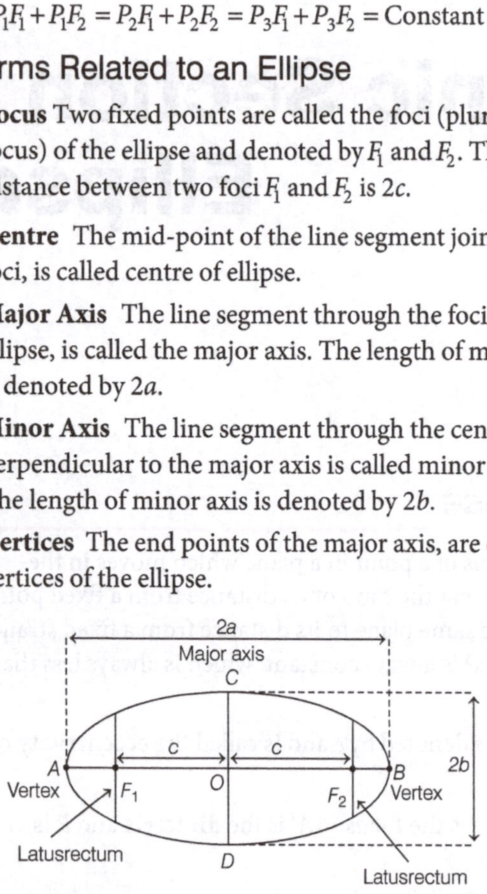

- **Eccentricity** The eccentricity of ellipse is the ratio of the distance from the centre of the ellipse to one of the foci and to one of the vertices of the ellipse. It is denoted by e.

Thus, $e=\dfrac{c}{a}$

$\Rightarrow\ c=ae$

Since, $c<a$

$\Rightarrow\ \dfrac{c}{a}<1$

$\Rightarrow\ e<1$

- **Latusrectum** Latusrectum of an ellipse is a line segment perpendicular to the major axis through any of the foci and whose end points lie on the ellipse.

Thus, the length of latusrectum $=2l=\dfrac{2b^2}{a}$

Relation among Semi-major, Semi-minor Axes and Distance of Focus from the Centre

Let a and b be the semi-major and semi-minor axes, respectively. Again, let c be the distance of the focus from the centre.

$\therefore$ Relation among a, b and c is given $c=\sqrt{a^2-b^2}$.

10.3.2 Standard Equations of an Ellipse

The standard equation of ellipse is $\dfrac{x^2}{a^2}+\dfrac{y^2}{b^2}=1, a>b,$ having centre and major axis lies on X-axis and minor axis lies on Y-axis. It is also called **horizontal ellipse**.

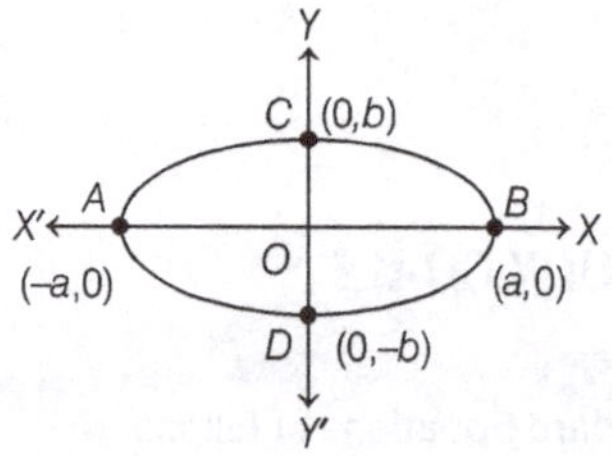

Some important terms related to horizontal ellipse are

(i) Centre is $O\,(0, 0)$.

(ii) Vertices are $(\pm a, 0)$.

(iii) Foci are $F_1(-c, 0)$ and $F_2(c, 0)$, where $c=\sqrt{a^2-b^2}$.

(iv) Length of the major axis, $AB=2a$

(v) Length of the minor axis, $CD=2b$

(vi) Eccentricity, $e=\dfrac{c}{a}=\dfrac{\sqrt{a^2-b^2}}{a}$

(vii) Equation of major axis, $y=0$

(viii) Equation of minor axis, $x=0$

(ix) Length of latusrectum $=\dfrac{2b^2}{a}$

(x) Vertices of latusrectum $=\left(\pm c, \pm\dfrac{b^2}{a}\right)$

(xi) Focal length $=2c$

Other form of an Ellipse

The another form of the equation of an ellipse is $\frac{x^2}{b^2}+\frac{y^2}{a^2}=1, a>b$, having centre at origin and major axis lies on Y-axis and minor axis lies on X-axis. It is also called **vertical ellipse**.

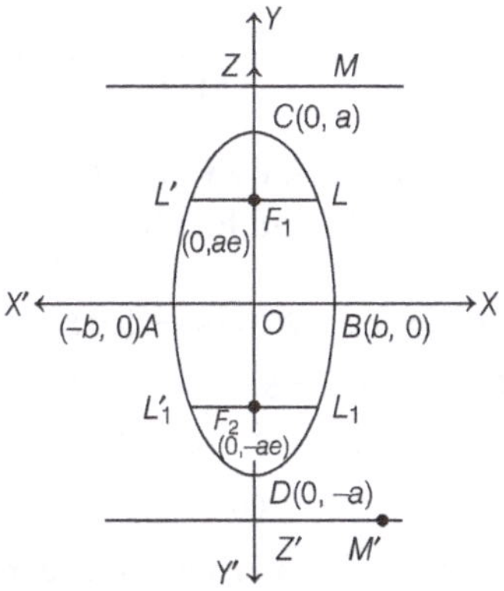

Some important terms related to vertical ellipse are

(i) Centre is $O(0, 0)$.

(ii) Vertices are $C\ (0, a)$ and $D\ (0, -a)$.

(iii) Foci are $F_1(0, c)$ and $F_2(0, -c)$.

(iv) Length of the major axis, $CD = 2a$

(v) Length of the minor axis, $AB = 2b$

(vi) Eccentricity, $e=\frac{c}{a}=\frac{\sqrt{a^2-b^2}}{a}$

(vii) Equation of major axis, $x = 0$

(viii) Equation of minor axis, $y = 0$

(ix) Length of latusrectum $=\frac{2b^2}{a}$

(x) Coordinates of latusrectum $=\left(\pm\frac{b^2}{a}, \pm c\right)$

Comparison between Two Standard Ellipse

S.No.	Terms	Horizontal ellipse $\frac{x^2}{a^2}+\frac{y^2}{b^2}=1, a>b>0$	Vertical ellipse $\frac{x^2}{b^2}+\frac{y^2}{a^2}=1, a>b>0$
1.	Shape	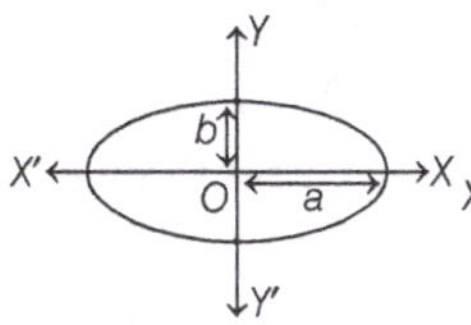	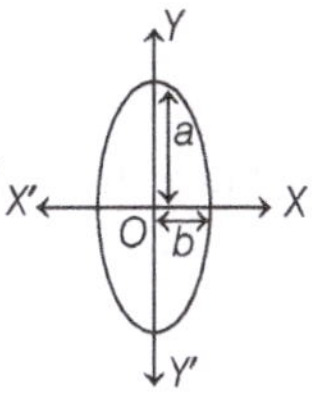
2.	Centre	$O(0, 0)$	$O(0, 0)$
3.	Vertices	$(\pm a, 0)$	$(0, \pm a)$
4.	Major axis	$2a$	$2a$
5.	Minor axis	$2b$	$2b$
6.	Value of c	$c=\sqrt{a^2-b^2}$	$c=\sqrt{a^2-b^2}$
7.	Equation of major axis	$y = 0$	$x = 0$
8.	Equation of minor axis	$x = 0$	$y = 0$
9.	Directrices	$x=\pm\frac{a^2}{c}$	$y=\pm\frac{a^2}{c}$
10.	Foci	$(\pm c, 0)$	$(0, \pm c)$
11.	Eccentricity	$e=\frac{c}{a}=\sqrt{1-\frac{b^2}{a^2}}$	$e=\frac{c}{a}=\sqrt{1-\frac{b^2}{a^2}}$
12.	Length of latusrectum	$\frac{2b^2}{a}$	$\frac{2b^2}{a}$

Example 1. **Find the coordinates of the foci, the vertices, the length of major axis, the minor axis, the eccentricity and the length of the latusrectum of the ellipse $\frac{x^2}{4}+\frac{y^2}{25}=1$.** NCERT

Sol. Given equation of ellipse is $\frac{x^2}{4}+\frac{y^2}{25}=1$.

Since, the denominator of $\frac{x^2}{4}$ is smaller than the denominator of $\frac{y^2}{25}$.

So, the major axis is along Y-axis.

On comparing the given equation with $\frac{x^2}{b^2}+\frac{y^2}{a^2}=1$, we get

$$b^2 = 4 \text{ and } a^2 = 25$$

$$\Rightarrow \quad b = 2 \text{ and } a = 5$$

Here, a and b are lengths, so we take only positive sign.

$$\because \quad c^2 = a^2 - b^2 = 25 - 4 = 21$$

$$\Rightarrow \quad c = \sqrt{21}$$

Here, the major axis is along Y-axis.

$\therefore$ Foci $= (0, \pm c) = (0, \pm\sqrt{21})$

Vertices $= (0, \pm a) = (0, \pm 5)$

Length of major axis $= 2a = 2\times 5 = 10$

Length of minor axis $= 2b = 2\times 2 = 4$

Eccentricity, $e = \frac{c}{a} = \frac{\sqrt{21}}{5}$

Length of latusrectum $= \frac{2b^2}{a} = \frac{2\times 4}{5} = \frac{8}{5}$

Example 3. **Find the equation of the ellipse whose axes are along the coordinate axes, vertices are (± 5, 0) and foci at (± 4, 0).** *NCERT, Competency Based Que*

Sol. Given, vertices $\equiv(\pm 5, 0)$ and foci $\equiv(\pm 4, 0)$.

$\because$ *y*-coordinates of vertices are zero. So, let equation of ellipse be $\frac{x^2}{a^2}+\frac{y^2}{b^2}=1, a>b$

Now, vertices $\equiv(\pm a, 0) = (\pm 5, 0)$

$\therefore$ $a=5$ and foci $\equiv(\pm ae, 0) = (\pm 4, 0)$

$\therefore$ $ae = 4 \Rightarrow e = \frac{4}{a} = \frac{4}{5}$ $[\because a=5]$

Now, $b^2 = a^2(1-e^2) = 25\left(1-\frac{16}{25}\right)$

$= 25\frac{(25-16)}{25} = 9$

Hence, the required equation of ellipse is $\frac{x^2}{25}+\frac{y^2}{9}=1.$

Example 4. **Find the equation of the ellipse, if foci are (± 3, 0) and $a = 4$.** *NCERT*

Sol. Given, foci are on *X*-axis. So, the major axis will be along the *X*-axis.

So, the equation of ellipse is of the form

$$\frac{x^2}{a^2}+\frac{y^2}{b^2}=1,\ a>b$$

Also, we have $a=4$ and $c=3$

$\because$ $b^2 = a^2 - c^2$

$\Rightarrow$ $b=\sqrt{a^2-c^2}=\sqrt{16-9}=\sqrt{7}$

Hence, the equation of an ellipse is $\frac{x^2}{16}+\frac{y^2}{7}=1.$

Example 5. **Find the equation of the ellipse, if length of major axis is 26 and foci (± 5, 0).** *NCERT, Competency Based Que*

Sol. We have, foci (±5, 0) which are on *X*-axis, so the equation of ellipse is of the form

$$\frac{x^2}{a^2}+\frac{y^2}{b^2}=1, a>b$$

Since, foci $\equiv (\pm c, 0) = (\pm 5, 0)$ $\Rightarrow c=5$

Length of major axis $= 2a = 26 \Rightarrow a=13$

We know that $c^2 = a^2 - b^2$

$\Rightarrow$ $b=\sqrt{a^2-c^2}$

$=\sqrt{169-25}=\sqrt{144}$

$\therefore$ $b=12$

Hence, the equation of an ellipse is

$$\frac{x^2}{a^2}+\frac{y^2}{b^2}=1 \text{ i.e. } \frac{x^2}{169}+\frac{y^2}{144}=1$$

Example 6. **Find the equation of the ellipse having length of minor axis = 16 and foci (0, ± 6).** *NCERT, Competency Based Que*

Sol. Length of minor axis, $2b=16 \Rightarrow b=8$

Foci are $(0, \pm 6) \Rightarrow c=6$

$\therefore$ $a^2 = b^2 + c^2 = (8)^2 + (6)^2 = 64+36$

$\Rightarrow$ $a^2 = 100$

Here, coefficient of *x* in foci is zero. So, major axis is on *Y*-axis.

Then, equation of ellipse is

$$\frac{x^2}{64}+\frac{y^2}{100}=1$$

Example 7. **Find the equation of the ellipse, if the ends of major axis are (± 3, 0) and ends of minor axis are (0, ± 2).** *NCERT*

Sol. Since, the vertices are on *X*-axis. Therefore, the equation is of the form $\frac{x^2}{a^2}+\frac{y^2}{b^2}=1, a>b.$

Since, major and minor axes are (± 3, 0) and (0, ± 2), respectively.

$\therefore$ $a=3$

and $b=2$

Hence, the equation of an ellipse is $\frac{x^2}{9}+\frac{y^2}{4}=1.$

Example 8. **Find the equation of ellipse with centre at the origin, major axis on the *Y*-axis and passing through the points (3, 2) and (1, 6).** *NCERT, Competency Based Que*

Sol. We have, major axis of the ellipse lies on *Y*-axis.

So, the equation of ellipse is

$$\frac{x^2}{b^2}+\frac{y^2}{a^2}=1,\ a>b$$

Since, this ellipse passes through the points (3, 2) and (1, 6).

$\therefore$ $\frac{9}{b^2}+\frac{4}{a^2}=1$ ...(i)

and $$\frac{1}{b^2}+\frac{36}{a^2}=1 \quad \text{...(ii)}$$

On solving Eqs. (i) and (ii), we get

$$a^2=40, b^2=10$$

Hence, the equation of ellipse is $\frac{x^2}{10}+\frac{y^2}{40}=1.$

Example 9. Find the equation of ellipse, if it satisfies the condition $b=3, c=4$, centre at origin, foci on the X-axis.

Sol. Given, foci lies on X-axis. So, the equation of ellipse will be of the form

$$\frac{x^2}{a^2}+\frac{y^2}{b^2}=1, a>b \quad \text{...(i)}$$

Also, given that $b=3$ and $c=4$

$\therefore$ $c^2=a^2-3^2$

$\Rightarrow$ $(4)^2=a^2-9$

$\Rightarrow$ $16=a^2-9$

$\Rightarrow$ $a^2=16+9 \Rightarrow a^2=25$

On putting the values of $a^2=25$ and $b^2=9$ in Eq. (i), we get

$$\frac{x^2}{25}+\frac{y^2}{9}=1$$

Example 10. Find the equation of an ellipse whose foci are (± 4, 0) and the eccentricity is $\frac{1}{3}$.

Sol. We have foci $(\pm 4, 0)$ which lie on X-axis, so the equation of ellipse is

$$\frac{x^2}{a^2}+\frac{y^2}{b^2}=1, a>b$$

Since, foci $\equiv(\pm c, 0)=(\pm 4, 0)$

$\Rightarrow$ $c=4$

We know that $e=\frac{c}{a}$

$\Rightarrow$ $a=\frac{c}{e}=\frac{4}{1/3}=12$ $\left[\because e=\frac{1}{3}\right]$

$\therefore$ $a=12$

Now, $c^2=a^2-b^2$

$\Rightarrow$ $(4)^2=(12)^2-(b)^2$

$\Rightarrow$ $b^2=144-16=128$

$\therefore$ $a^2=(12)^2=144$ and $b^2=128$

Hence, the equation of the ellipse is $\frac{x^2}{144}+\frac{y^2}{128}=1.$

Example 11. An arch is the form of a semi-ellipse. It is 8 m wide and 2 m high of the centre. Find the height of the arch at a point 1.5 m from one end.

NCERT, Competency Based Que

Sol. Clearly, equation of ellipse takes the form

$$\frac{x^2}{a^2}+\frac{y^2}{b^2}=1 \quad \text{...(i)}$$

Here, it is given that $2a=8$ and $b=2$

$\Rightarrow$ $a=4$ and $b=2$

Put the values of a and b in Eq. (i), we get

$$\frac{x^2}{16}+\frac{y^2}{4}=1$$

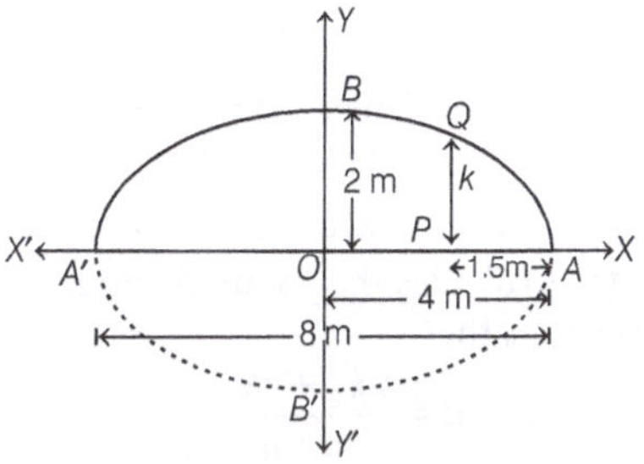

Given, $AP=1.5\text{ m}$

$\Rightarrow$ $OP=OA-AP=4-1.5$

$\Rightarrow$ $OP=2.5\text{ m}$

Let $PQ=k$

Then, the coordinates $Q(2.5, k)$ will satisfy the equation of ellipse.

i.e. $$\frac{(2.5)^2}{16}+\frac{k^2}{4}=1$$

$\Rightarrow$ $$\frac{6.25}{16}+\frac{k^2}{4}=1$$

$\Rightarrow$ $$\frac{k^2}{4}=\frac{1}{1}-\frac{6.25}{16}=\frac{16-6.25}{16}$$

$\Rightarrow$ $$\frac{k^2}{4}=\frac{9.75}{16}$$

$\Rightarrow$ $$k^2=\frac{9.75}{4}$$

$\Rightarrow$ $k^2=2.4375$

$\therefore$ $k=1.56\text{ m}$ (approx.)

Example 12. A rod of length 12 cm moves with its ends always touching the coordinate axes. Determine the equation of the locus of a point P on the rod, which is 3 cm from the end in contact with the X-axis.

NCERT, Competency Based Que

Sol. Let AB be the rod of length 12 cm.

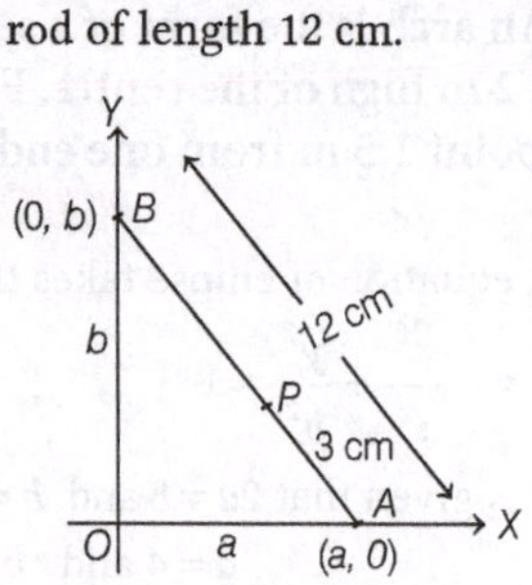

Let $OA = a$ and $OB = b$

Using Pythagoras theorem in ΔOAB,

$$a^2 + b^2 = (12)^2$$

$$\Rightarrow \quad a^2 + b^2 = 144 \quad \text{...(i)}$$

Let $AP = 3$ cm. Then, $PB = AB - AP = 12 - 3 = 9$ cm

$$\therefore \quad AP : PB = 3 : 9 = 1 : 3$$

Let $P = (h, k)$ for which locus to be found.

If $P(h, k)$ divides the points A and B internally in the ratio $m : n$, then

$$h = \frac{mx_2 + nx_1}{m + n}$$

and $$k = \frac{my_2 + ny_1}{m + n}$$

$$\therefore \quad h = \frac{1 \times 0 + 3 \times a}{1 + 3}$$

$$\Rightarrow \quad h = \frac{3a}{4} \Rightarrow a = \frac{4h}{3}$$

and $$k = \frac{1 \times b + 3 \times 0}{1 + 3}$$

$$\Rightarrow \quad k = \frac{b}{4} \Rightarrow b = 4k$$

Put the values of $a = \dfrac{4h}{3}$ and $b = 4k$ in Eq. (i), we get

$$\left(\frac{4h}{3}\right)^2 + (4k)^2 = 144$$

$$\Rightarrow \quad \frac{16h^2}{9} + 16k^2 = 144$$

$$\Rightarrow \quad \frac{h^2}{9} + k^2 = 9$$

$$\Rightarrow \quad \frac{h^2}{81} + \frac{k^2}{9} = 1$$

Hence, the locus of a point $P(h, k)$ is $\dfrac{x^2}{81} + \dfrac{y^2}{9} = 1$, which is an ellipse.

Exam Pattern Questions

Objective Type Questions

Multiple Choice Questions

1. Equation of ellipse is $\frac{x^2}{a^2}+\frac{y^2}{b^2}=1, a>b>0$. Its length of latusrectum is **Competency Based Que**

(a) $2a$ (b) $2b$ (c) $\frac{2b^2}{a}$ (d) $\frac{2a^2}{b}$

2. For ellipse $\frac{x^2}{36}+\frac{y^2}{9}=1$, eccentricity is

(a) $\frac{\sqrt{3}}{2}$ (b) $\frac{1}{2}$
(c) $\frac{\sqrt{3}}{4}$ (d) $\frac{1}{\sqrt{3}}$

3. The distance between foci for the ellipse $x^2+4y^2=1$ is

(a) 1 (b) $\frac{1}{\sqrt{3}}$ (c) $\frac{\sqrt{3}}{2}$ (d) $\sqrt{3}$

4. The equation of ellipse whose vertices are $(0, \pm 5)$ and eccentricity is $\frac{1}{5}$ is given by **Competency Based Que**

(a) $\frac{x^2}{16}+\frac{y^2}{9}=1$ (b) $\frac{x^2}{25}+\frac{y^2}{9}=1$
(c) $\frac{x^2}{25}+\frac{y^2}{24}=1$ (d) $\frac{x^2}{24}+\frac{y^2}{25}=1$

5. The eccentricity of the ellipse with foci on the X-axis. If its latusrectum be equal to half of major axis is

(a) $\frac{1}{\sqrt{3}}$ (b) $\frac{1}{\sqrt{2}}$ (c) $\frac{2}{\sqrt{5}}$ (d) $\frac{1}{\sqrt{2}}$

6. The eccentricity of the ellipse with foci at $(\pm 3, 0)$ and $x=\frac{16}{3}$ as one of the directrices is

(a) $\frac{1}{3}$ (b) $\frac{3}{4}$ (c) $\frac{1}{\sqrt{3}}$ (d) $\frac{2}{\sqrt{5}}$

7. The vertices of a ellipse are $(\pm 5, 0)$ and its foci are $(\pm 3, 0)$. The equation of the ellipse is **Competency Based Que**

(a) $\frac{x^2}{25}+\frac{y^2}{9}=1$ (b) $\frac{x^2}{9}+\frac{y^2}{25}=1$
(c) $\frac{x^2}{25}+\frac{y^2}{16}=1$ (d) $\frac{x^2}{16}+\frac{y^2}{25}=1$

8. The length of the latusrectum of the ellipse $16x^2+y^2=16$ is

(a) 2 (b) $\frac{1}{2}$ (c) 1 (d) 4

Assertion-Reason Based Questions

Directions (Q. Nos. 9-12) *In the questions given below are two statements labelled as Assertion (A) and Reason (R). In the context of the two statements, which one of the following is correct?*

(a) Both A and R are correct; R is the correct explanation of A
(b) Both A and R are correct; R is not the correct explanation of A
(c) A is correct; R is incorrect
(d) R is correct; A is incorrect

9. **Assertion** (A) For the ellipse $16x^2+25y^2=400$, length of minor axis is 20.

Reason (R) For the ellipse $\frac{x^2}{a^2}+\frac{y^2}{b^2}=1, a>b$, equation of major axis is $y=0$. **Competency Based Que**

10. **Assertion** (A) For the ellipse $\frac{x^2}{25}+\frac{y^2}{16}=1$, length of major axis is 40.

Reason (R) For the ellipse $\frac{x^2}{a^2}+\frac{y^2}{b^2}=1, a>b$, equation of major axis is $y=0$.

11. **Assertion** (A) For the ellipse $9x^2+25y^2=225$, eccentricity is $\frac{4}{5}$.

Reason (R) For the ellipse $\frac{x^2}{a^2}+\frac{y^2}{b^2}=1, a<b$, equation of minor axis is $y=0$.

12. **Assertion** (A) For the horizontal ellipse $3x^2+5y^2=32$ the vertices is $\left(\pm\sqrt{\frac{32}{3}}, 0\right)$.

Reason (R) For the horizontal ellipse $\frac{x^2}{a^2}+\frac{y^2}{b^2}=1, a>b$, the vertices is $(\pm a, 0)$. **Competency Based Que**

Case Study Based Questions

13. Mrs. Narang is a sincere citizen of 'Gaur Society' in New Delhi. She makes a green garden in her plot which is in the shape of ellipse as shown in the figure below.

Competency Based Que

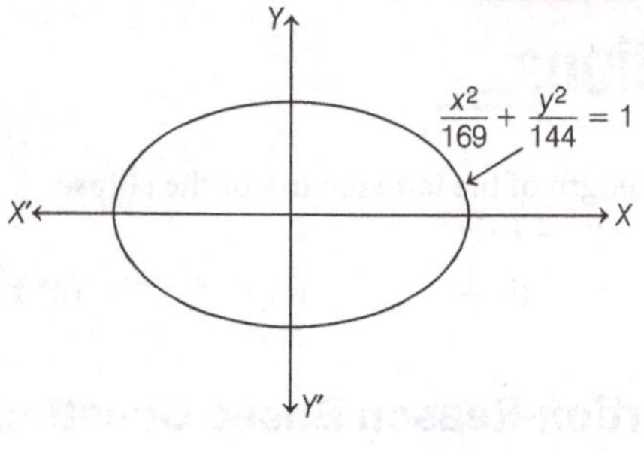

(i) Value of eccentricity (e) is

(a) $\frac{12}{13}$ (b) $\frac{5}{12}$ (c) $\frac{5}{13}$ (d) $\frac{15}{13}$

(ii) Length of latusrectum is

(a) $\frac{288}{12}$ (b) $\frac{144}{13}$

(c) $\frac{144}{12}$ (d) $\frac{288}{13}$

(iii) Length of major axis is

(a) 24 (b) 26 (c) 22 (d) 28

(iv) Foci of the ellipse are

(a) $(\pm 5, 0)$ (b) $(\pm 12, 0)$

(c) $(0, \pm 5)$ (d) $(\pm 13, 0)$

Subjective Questions

Very Short Answer Type Questions

1. Find the equation of ellipse, having ends of major and minor axes are $(0, \pm\sqrt{5})$ and $(\pm 1, 0)$. NCERT

2. Find the equation of the ellipse, having eccentricity $\frac{1}{2}$ and semi-major axis is 4. Competency Based Que

3. Eccentricity of the ellipse is $\frac{5}{3}$. Is the statement true? Give reason.

4. Find the eccentricity of the ellipse $\frac{x^2}{4}+\frac{y^2}{9}=1$.

5. Find the length of the latusrectum of the ellipse $3x^2+y^2=12$.

6. Find the equation of the ellipse, if foci are $(\pm 5, 0)$ and $a=6$.

7. Find the equation of the ellipse, if length of major axis is 22 and foci are $(\pm 3, 0)$. Competency Based Que

8. Find the equation of the ellipse, whose foci are $(0, \pm 5)$ and vertices are $(0, \pm 13)$. NCERT, Competency Based Que

Short Answer Type Questions

9. Find the vertices, length of major axis, length of minor axis, eccentricity, foci and length of latusrectum of the ellipse $\frac{x^2}{64}+\frac{y^2}{25}=1$.

10. Find the vertices, length major axis, length of minor axis, eccentricity, foci and length of latusrectum of the ellipse $\frac{x^2}{100}+\frac{y^2}{400}=1$. Competency Based Que

11. Find the equation of the ellipse, whose axes along coordinate axes, passing through (4, 3) and (– 1, 4). NCERT

12. Find the equation of the ellipse, if the centre is at (0, 0), major axis on the X-axis and passing through the points (3, 2) and (1, 6).

13. Find the equation of the ellipse whose axes are along the coordinate axes, vertices are $(0, \pm 10)$ and eccentricity $e=\frac{4}{5}$. Competency Based Que

Long Answer Type Questions

14. Find the lengths of the major and the minor axes, the coordinates of the foci, the vertices, the eccentricity and the equations of the directrices of the ellipse $\frac{x^2}{100}+\frac{y^2}{25}=1$

15. Find the equation of the ellipse whose vertices are $(0, \pm 6)$ and eccentricity is 1/3.

16. Find the coordinates of the foci, the vertices, the eccentricity and the length of the latusrectum of the ellipse $\frac{x^2}{49}+\frac{y^2}{25}=1$ Competency Based Que

17. Find the equation of the ellipse, whose major axis is on X-axis and passes through (4, 3) and (6, 2). NCERT, Competency Based Que

Hints & Solutions

Objective Type Questions

1. (c) Length of latusrectum of horizontal ellipse $\frac{x^2}{a^2}+\frac{y^2}{b^2}=1$, $a>b>0$ is given by $\frac{2b^2}{a}$.

2. (a) Equation of ellipse is
$$\frac{x^2}{a^2}+\frac{y^2}{b^2}=1, a>b$$
Then, eccentricity e is given by
$$e=\sqrt{1-\frac{b^2}{a^2}}=\sqrt{1-\frac{9}{36}}=\sqrt{\frac{3}{4}}=\frac{\sqrt{3}}{2}$$

3. (d) Consider equation of ellipse is
$x^2+4y^2=1$
$\Rightarrow \frac{x^2}{1}+\frac{y^2}{\frac{1}{4}}=1$, here $a^2=1, b^2=\frac{1}{4}$
Therefore, $c=\sqrt{a^2-b^2}=\sqrt{1-\frac{1}{4}}=\frac{\sqrt{3}}{2}$
$\therefore$ Required distance between foci $=2c=\sqrt{3}$.

4. (d) Given that vertices of ellipse are $(0, \pm 5)$.
So, $a=5$
Also given, $e=\frac{1}{5}$
Using $b^2=a^2(1-e^2)$, we get
$$b^2=25\left(1-\frac{1}{25}\right)\Rightarrow b^2=24$$
Equation of ellipse is $\frac{x^2}{b^2}+\frac{y^2}{a^2}=1$
$$\Rightarrow \quad \frac{x^2}{24}+\frac{y^2}{25}=1$$

5. (d) According to given condition,
$$\frac{2b^2}{a}=\frac{1}{2}(2a)$$
$$\Rightarrow \quad 2b^2=a^2$$
$$\therefore \quad e=\frac{\sqrt{a^2-b^2}}{a}$$
$$=\sqrt{\frac{2b^2-b^2}{2b^2}}=\frac{1}{\sqrt{2}}$$

6. (b) We have, $ae=3$, $\frac{a}{e}=\frac{16}{3}$
$\therefore a^2=16 \Rightarrow a=4$
Therefore, $e=\frac{3}{4}$
Hence, required eccentricity is $\frac{3}{4}$.

7. (c) As the vertices of the given ellipse are on the X-axis.
Let equation of the ellipse be $\frac{x^2}{a^2}+\frac{y^2}{x^2}=1$, where $a>b$.
Vertices are $(\pm 5, 0) \Rightarrow a=5$
Foci are $(\pm 3, 0) \Rightarrow c=3$
$$c^2=a^2-b^2$$
$$\Rightarrow \quad b^2=a^2-c^2=5^2-3^2=16$$
$$\Rightarrow \quad b=4$$
So, equation of the ellipse is $\frac{x^2}{25}+\frac{y^2}{16}=1$.

8. (b) Consider ellipse, $16x^2+y^2=16$
$\Rightarrow \frac{x^2}{1}+\frac{y^2}{16}=1$
$\Rightarrow a^2=16$ and $b^2=1$
Length of latusrectum $=\frac{2b^2}{a}=\frac{2\times 1}{4}=\frac{1}{2}$

9. (d) We have, $16x^2+25y^2=400$
$$\Rightarrow \quad \frac{16x^2+25y^2}{400}=1$$
$$\Rightarrow \quad \frac{16x^2}{100}+\frac{25y^2}{400}=1$$
$$\Rightarrow \quad \frac{x^2}{25}+\frac{y^2}{16}=1 \Rightarrow \frac{x^2}{5^2}+\frac{y^2}{4^2}=1$$
$\Rightarrow a=5, b=4, a>b$
$\therefore$ Length of minor axis $=2b=8$
Hence, Assertion is incorrect and Reason is correct.

10. (d) We have,
$$\frac{x^2}{25}+\frac{y^2}{16}=1$$
$$\Rightarrow \quad \frac{x^2}{5^2}+\frac{y^2}{4^2}=1$$
$$\therefore \quad a=5, b=4, a>b$$
So, length of the major axis $=2a=2\times 5=10$
Hence, Assertion is incorrect and Reason is correct.

11. (c) We have, ellipse $9x^2+25y^2=225$
$$\Rightarrow \quad \frac{9x^2}{225}+\frac{25y^2}{225}=1$$
$$\Rightarrow \quad \frac{x^2}{5^2}+\frac{y^2}{3^2}=1$$
So, $a=5$ and $b=3$
Now, $c=\sqrt{a^2-b^2}=\sqrt{25-9}=4$
Thus, eccentricity, $e=\frac{c}{a}=\frac{4}{5}$

12. (a) We have, $3x^2 + 5y^2 = 32$

$\Rightarrow \frac{x^2}{\frac{32}{3}} + \frac{y^2}{\frac{32}{5}} = 1$

Here, $a^2 = \frac{32}{3}$ and $b^2 = \frac{32}{5}$

$\therefore a = \sqrt{\frac{32}{3}}$

$\therefore$ Vertices $= (\pm a, 0) = \left(\pm\sqrt{\frac{32}{3}}, 0\right)$

13. (i) (c) Given equation of ellipse is $\frac{x^2}{169} + \frac{y^2}{144} = 1$.

Here, $a^2 = 169$ and $b^2 = 144$

$\Rightarrow a = 13$ and $b = 12$

Now, $c = \sqrt{a^2 - b^2} = \sqrt{169 - 144} = 5$

$\therefore$ Eccentricity, $e = \frac{c}{a} = \frac{5}{13}$

(ii) (d) Length of latusrectum $= \frac{2b^2}{a} = \frac{2 \times 144}{13} = \frac{288}{13}$

(iii) (b) Length of major axis, $2a = 2 \times 13 = 26$

(iv) (a) Foci of the ellipse, $(\pm c, 0) = (\pm 5, 0)$

Subjective Questions

1. Given, end points of major axis $= (0, \pm\sqrt{5})$

i.e. $(0, \pm a) = (0, \pm\sqrt{5})$

and end points of minor axis $= (\pm 1, 0)$

i.e. $(\pm b, 0) = (\pm 1, 0)$

$\therefore$ Equation of the ellipse is $\frac{x^2}{b^2} + \frac{y^2}{a^2} = 1$

i.e. $\frac{x^2}{1} + \frac{y^2}{5} = 1$

2. **Hint** Given, $e = \frac{1}{2}$ and $a = 4$

$\therefore e^2 = \frac{a^2 - b^2}{a^2}$

$\Rightarrow \frac{1}{4} = \frac{16 - b^2}{16}$

$\Rightarrow b^2 = 12$

$\therefore$ Equation of an ellipse is $\frac{x^2}{a^2} + \frac{y^2}{b^2} = 1$.

$\Rightarrow \frac{x^2}{16} + \frac{y^2}{12} = 1$

Ans. $3x^2 + 4y^2 = 48$

3. Statement is false as eccentricity of an ellipse is less than 1.

4. Given, equation of ellipse is

$\frac{x^2}{4} + \frac{y^2}{9} = 1$

which is of the form

$\frac{x^2}{b^2} + \frac{y^2}{a^2} = 1$, where $a > b$

On comparing both equations, we get

$b^2 = 4$ and $a^2 = 9$

$\Rightarrow b = 2$ and $a = 3$

Now, $c^2 = a^2 - b^2 = 9 - 4 = 5$

$\Rightarrow c = \sqrt{5}$

Since, eccentricity, $e = \frac{c}{a}$

$\Rightarrow e = \frac{\sqrt{5}}{3}$

5. Given, equation of ellipse is $3x^2 + y^2 = 12$

$\Rightarrow \frac{3x^2}{12} + \frac{y^2}{12} = 1 \Rightarrow \frac{x^2}{4} + \frac{y^2}{12} = 1$

which is of the form $\frac{x^2}{b^2} + \frac{y^2}{a^2} = 1$ where $a > b$

$\therefore a^2 = 12$ and $b^2 = 4$

$\Rightarrow a = 2\sqrt{3}$ and $b = 2$

Now, length of latusrectum $= \frac{2b^2}{a} = \frac{8}{2\sqrt{3}} = \frac{4\sqrt{3}}{3}$

6. Given, foci $(\pm c, 0) = (\pm 5, 0)$ and $a = 6$

So, major axis of ellipse is along X-axis.

$\therefore$ Equation of ellipse will be form of

$\frac{x^2}{a^2} + \frac{y^2}{b^2} = 1, a > b$

Now, $c^2 = a^2 - b^2$

$\Rightarrow b^2 = 36 - 25 = 11$

$\therefore$ Equation of the ellipse is $\frac{x^2}{36} + \frac{y^2}{11} = 1$

7. Given, foci $(\pm c, 0) = (\pm 3, 0)$

So, major axis is along the X-axis.

$\therefore$ Equation of ellipse will be form of

$\frac{x^2}{a^2} + \frac{y^2}{b^2} = 1, a > b$

Since, $2a = 22 \Rightarrow a = 11$

Now, $c^2 = a^2 - b^2 \Rightarrow b^2 = 121 - 9 = 112$

$\therefore$ Equation of ellipse is $\frac{x^2}{121} + \frac{y^2}{112} = 1$

8. Given, vertices are $(0, \pm 13)$ and foci are $(0, \pm 5)$.

Here, the vertices are on the Y-axis

Therefore, the equation of the ellipse will be of the form

$\frac{x^2}{b^2} + \frac{y^2}{a^2} = 1, a > b$

$\therefore a = 13$ and $c = 5$

Now, $b^2 = a^2 - c^2 = 169 - 25 = 144$

Thus, the equation of the ellipse is

$$\frac{x^2}{144} + \frac{y^2}{169} = 1$$

9. Given, equation of ellipse is $\frac{x^2}{64} + \frac{y^2}{25} = 1$

Since, the denominator of $\frac{x^2}{64}$ is greater than the denominator of $\frac{y^2}{25}$.

So, the major axis is along X-axis.

On comparing the given equation with

$$\frac{x^2}{a^2} + \frac{y^2}{b^2} = 1,$$

we get $a^2 = 64$ and $b^2 = 25$

$\Rightarrow \quad a = 8$ and $b = 5$

Here, a and b are lengths. So, we take only positive sign.

$\therefore \quad c^2 = a^2 - b^2 = 64 - 25 = 39$

$\Rightarrow \quad c = \sqrt{39}$

Here, the major axis is along X-axis.

$\therefore$ vertices $(\pm a, 0) = (\pm 8, 0)$

Length of major axis $= 2a = 16$

Length of minor axis $= 2b = 10$

Eccentricity $(e) = \frac{c}{a} = \frac{\sqrt{39}}{8}$

Foci $(\pm c, 0) = (\pm \sqrt{39}, 0)$

and length of latusrectum $= \frac{2b^2}{a} = \frac{25}{4}$

10. Given, equation of ellipse is $\frac{x^2}{100} + \frac{y^2}{400} = 1$

Since, the denominator of $\frac{x^2}{100}$ is smaller than the denominator of $\frac{y^2}{400}$.

So, the major axis is along Y-axis.

On comparing the given equation with $\frac{x^2}{b^2} + \frac{y^2}{a^2} = 1$,

we get

$b^2 = 100$ and $a^2 = 400$

$\Rightarrow \quad b = 10$ and $a = 20$

$\therefore \quad c^2 = a^2 - b^2 = 400 - 100 = 300$

$\Rightarrow \quad c = 10\sqrt{3}$

Here, the major axis is along Y-axis.

$\therefore$ Vertices $(0, \pm a) = (0, \pm 20)$

Length of major axis $= 2a = 40$

Length of minor axis $= 2b = 20$

Eccentricity, $e = \frac{c}{a} = \frac{10\sqrt{3}}{20} = \frac{\sqrt{3}}{2}$

Foci, $(0, \pm c) = (0, \pm 10\sqrt{3})$

and length of latusrectum $= \frac{2b^2}{a} = 10$

11. Let the equation of the ellipse be

$$\frac{x^2}{a^2} + \frac{y^2}{b^2} = 1$$

Since, the point (4, 3) and (−1, 4) lie on the ellipse, we get

$$\frac{16}{a^2} + \frac{9}{b^2} = 1 \quad \text{...(i)}$$

and $$\frac{1}{a^2} + \frac{16}{b^2} = 1 \quad \text{...(ii)}$$

On solving Eqs. (i) and (ii), we get

$$a^2 = \frac{247}{7}, b^2 = \frac{247}{15}$$

Hence, the required equation is

$$\frac{x^2}{\left(\frac{247}{7}\right)} + \frac{y^2}{\left(\frac{247}{15}\right)} = 1$$

$\Rightarrow \quad 7x^2 + 15y^2 = 247$

12. Let equation of ellipse is $\frac{x^2}{a^2} + \frac{y^2}{b^2} = 1$

Since, the ellipse passes through the point (3, 2).

$$\therefore \quad \frac{9}{a^2} + \frac{4}{b^2} = 1 \quad \text{...(i)}$$

Also, the ellipse passes through the point (1, 6).

$$\therefore \quad \frac{1}{a^2} + \frac{36}{b^2} = 1 \quad \text{...(ii)}$$

On solving Eqs. (i) and (ii), we get

$a^2 = 10$ and $b^2 = 40$

Thus, the equation of required ellipse is

$$\frac{x^2}{10} + \frac{y^2}{40} = 1$$

13. Let the equation of the ellipse be

$$\frac{x^2}{a^2} + \frac{y^2}{b^2} = 1 \quad \text{...(i)}$$

The coordinates of vertices are $(0, \pm b)$

i.e. $(0, \pm 10)$

$\therefore \quad b = 10$

Now, $\quad a^2 = b^2 (1 - e^2)$

$\Rightarrow \quad a^2 = 100\left(1 - \frac{16}{25}\right)$

$\Rightarrow \quad a^2 = 36$

On putting $a^2 = 36$ and $b^2 = 100$ in Eq. (i), we get

$$\frac{x^2}{36} + \frac{y^2}{100} = 1$$

14. Given ellipse is $\frac{x^2}{100}+\frac{y^2}{25}=1$.

Here, $a^2=100$ and $b^2=25$

$\Rightarrow \quad a=10$ and $b=5$

Vertices $(\pm a, 0)=(\pm 10, 0)$,

$$c=\sqrt{a^2-b^2}=\sqrt{100-25}$$

$$=\sqrt{75}=5\sqrt{3}$$

Eccentricity $(e)=\frac{c}{a}=\frac{5\sqrt{3}}{10}=\frac{\sqrt{3}}{2}$

Foci $(\pm c, 0)$ i.e. $(\pm 5\sqrt{3}, 0)$

Length of major axis $=2a=20$

Length of minor axis $=2b=10$

Equations of directrices, $x=\pm\frac{a}{e}$

$$\Rightarrow \quad x=\pm\frac{10\times 2}{\sqrt{3}} \Rightarrow x=\pm\frac{20}{\sqrt{3}}$$

15. It is given that $a=6$ and $e=\frac{1}{3}$

Using $b^2=a^2(1-e^2)$, we get

$b^2=32$

Equation of ellipse is $\frac{x^2}{b^2}+\frac{y^2}{a^2}=1$

$$\Rightarrow \quad \frac{x^2}{32}+\frac{y^2}{36}=1$$

16. Hint Given ellipse is $\frac{x^2}{49}+\frac{y^2}{36}=1$.

Here, $a^2=49$ and $b^2=36$

$$\Rightarrow \quad c=\sqrt{49-36}=\sqrt{13}$$

Vertices $=(\pm 7, 0)$, Foci $=(\pm\sqrt{13}, 0)$

17. Let the major axis be along X-axis. Then, the equation of ellipse is

$$\frac{x^2}{a^2}+\frac{y^2}{b^2}=1, a>b \quad \text{...(i)}$$

Since, the equation of ellipse is passing through the points (4, 3) and (6, 2). So, the point (4, 3) lies on it.

$$\therefore \quad \frac{16}{a^2}+\frac{9}{b^2}=1 \quad \text{...(ii)}$$

Also, the point (6, 2) lies on it.

$$\therefore \quad \frac{36}{a^2}+\frac{4}{b^2}=1 \quad \text{...(iii)}$$

On multiplying Eq. (ii) by 4 and Eq. (iii) by 9 and then subtracting Eq. (ii) from Eq. (iii), we get

$$\frac{324}{a^2}-\frac{64}{a^2}=9-4$$

$$\Rightarrow \quad \frac{260}{a^2}=5$$

$$\Rightarrow \quad a^2=\frac{260}{5}=52$$

$$\Rightarrow \quad a^2=52$$

On putting the value of a^2 in Eq. (ii), we get

$$\frac{16}{52}+\frac{9}{b^2}=1$$

$$\Rightarrow \quad \frac{9}{b^2}=1-\frac{16}{52}=\frac{36}{52}$$

$$\Rightarrow \quad b^2=13$$

Hence, the equation of ellipse is

$$\frac{x^2}{52}+\frac{y^2}{13}=1$$

[put $a^2=52$ and $b^2=13$ in Eq. (i)]

Mind Map

Ellipse

An ellipse is the locus of a point in a plane which moves in the plane in such a way that the ratio of its distance from a fixed point (called **focus**) in the same plane to its distance from a fixed straight line (called **directrix**) is always constant, which is always less than unity.

The constant ratio is denoted by e and is called the **eccentricity** of the ellipse. Also, $e < 1$

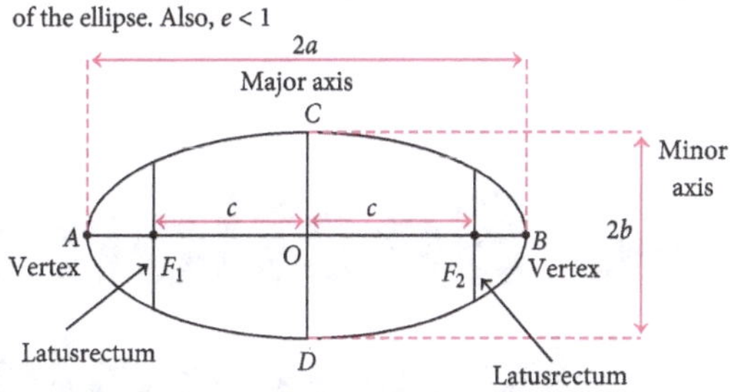

Terms Related to an Ellipse

- **Focus** Two fixed points are called the foci (plural of focus) of the ellipse and denoted by F_1 and F_2. The distance between two foci F_1 and F_2 is $2c$.
- **Centre** The mid-point of the line segment joining the foci, is called centre of ellipse.
- **Major Axis** The line segment through the foci of the ellipse, is called the major axis. The length of major axis is denoted by $2a$.
- **Minor Axis** The line segment through the centre and perpendicular to the major axis is called minor axis. The length of minor axis is denoted by $2b$.
- **Vertices** The end points of the major axis, are called vertices of the ellipse.
- **Eccentricity** The eccentricity of ellipse is the ratio of the distance from the centre of the ellipse to one of the foci and to one of the vertices of the ellipse. It is denoted by e. Always $e < 1$.
- **Latusrectum** Latusrectum of an ellipse is a line segment perpendicular to the major axis through any of the foci and whose end points lie on the ellipse. Thus, the length of latusrectum $= 2l = \frac{2b^2}{a}$

Ellipse

Standard Equations of an Ellipse

Horizontal Ellipse

The standard equation of horizontal ellipse is

$\frac{x^2}{a^2} + \frac{y^2}{b^2} = 1, a > b$

having centre and major axis lies on X-axis and minor axis lies on Y-axis. It is also called horizontal ellipse.

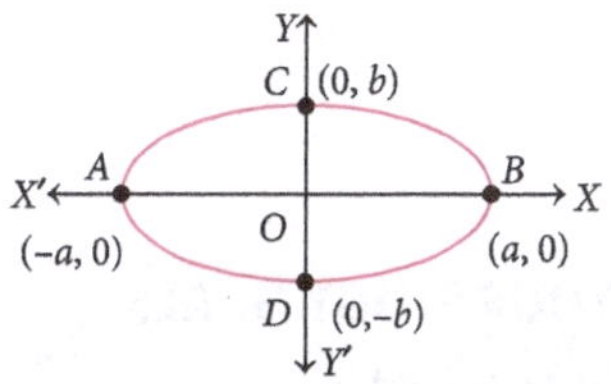

Vertical Ellipse

The another form of the equation of a vertical ellipse is $\frac{x^2}{b^2} + \frac{y^2}{a^2} = 1$, $a > b$ having centre at origin and major axis lies on Y-axis and minor axis lies on X-axis. It is also called vertical ellipse.

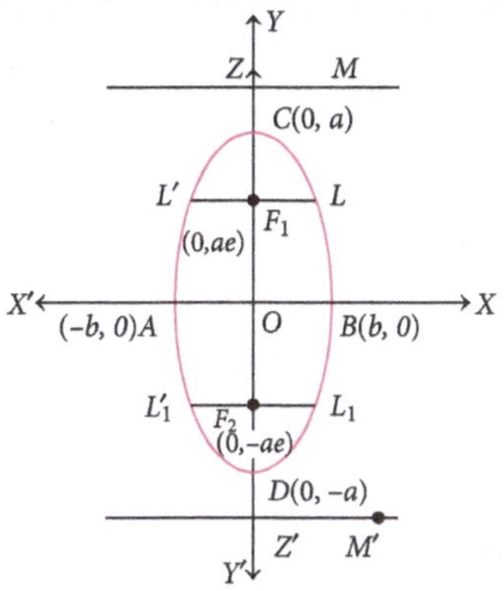

S.No.	Terms	Horizontal ellipse $\frac{x^2}{a^2} + \frac{y^2}{b^2} = 1, a > b > 0$	Vertical ellipse $\frac{x^2}{b^2} + \frac{y^2}{a^2} = 1, a > b > 0$
1.	Centre	$O(0, 0)$	$O(0, 0)$
2.	Vertices	$(\pm a, 0)$	$(0, \pm a)$
3.	Major axis	$2a$	$2a$
4.	Minor axis	$2b$	$2b$
5.	Value of c	$c = \sqrt{a^2 - b^2}$	$c = \sqrt{a^2 - b^2}$
6.	Equation of major axis	$y = 0$	$x = 0$
7.	Equation of minor axis	$x = 0$	$y = 0$
8.	Directrices	$x = \pm \frac{a^2}{c}$	$y = \pm \frac{a^2}{c}$
9.	Foci	$(\pm c, 0)$	$(0, \pm c)$
10.	Eccentricity	$e = \frac{c}{a} = \sqrt{1 - \frac{b^2}{a^2}}$	$e = \frac{c}{a} = \sqrt{1 - \frac{b^2}{a^2}}$
11.	Length of latusrectum	$\frac{2b^2}{a}$	$\frac{2b^2}{a}$

10(d)

CHAPTER

Mathematics is a game played according to certain simple rules with meaningless marks on paper.
— *David Hilbert*

Conic Section : Hyperbola

▼ **YOUR STUDY GOALS**

▼ **YOUR PRACTICE GOALS**

- Objective Type Questions (16 Qs; 10 MCQs, 5 A-R, 1 Case Based)
- Subjective Type Questions
- (16 Qs; 7 VSA, 6 SA, 3 LA)

10.4.1 Hyperbola

A hyperbola is the locus of a point in a plane which moves in the plane in such a way that the ratio of its distance from a fixed point (called **focus**) in the same plane to its distance from a fixed straight line
(called **directrix**) is always constant, which is always greater than unity.

The constant ratio is denoted by e and is called the eccentricity of the hyperbola.

In the given figure, F is the focus, AA' is the directrix and P is any point on hyperbola.

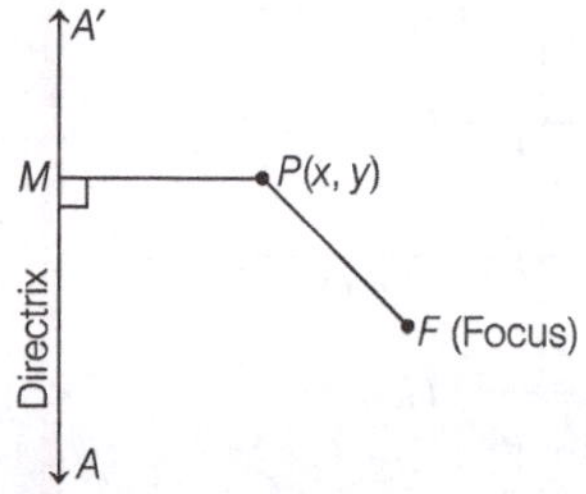

Then, by definition, $\frac{PF}{PM} = e$, where $e > 1$

$\Rightarrow \qquad PF = e \cdot PM$

In other words, we can say that a hyperbola is the set of all points in a plane, the difference of whose distances from two fixed points in the plane is a constant.

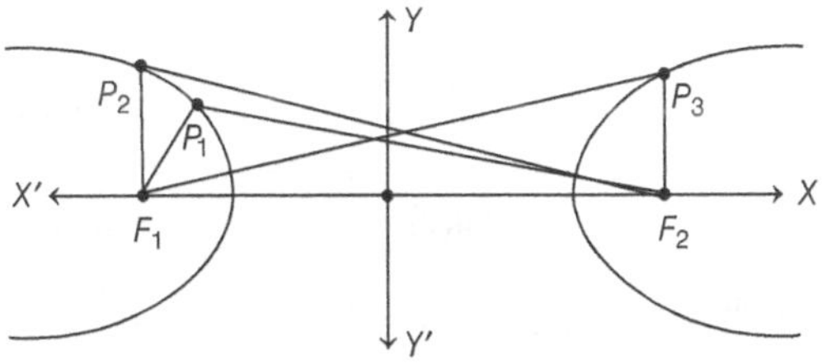

$\therefore \quad P_1F_2 - P_1F_1 = P_2F_2 - P_2F_1 = P_3F_1 - P_3F_2$

The term difference used in the definition means the distance to the farther point minus the distance to the closer point.

Terms Related to Hyperbola

- **Focus** The two fixed points are called the foci of the hyperbola and denoted by F_1 and F_2. The distance between two foci F_1 and F_2 is $2c$.
- **Centre** The mid-point of the line segment joining the foci, is called centre of hyberbola.
- **Transverse axis** The line through the foci, is called the transverse axis.
- **Conjugate axis** The line through the centre and perpendicular to the transverse axis, is called the conjugate axis.
- **Vertices** The points at which the hyperbola intersects the transverse axis are called the vertices of the hyperbola. The distance between two vertices is $2a$.

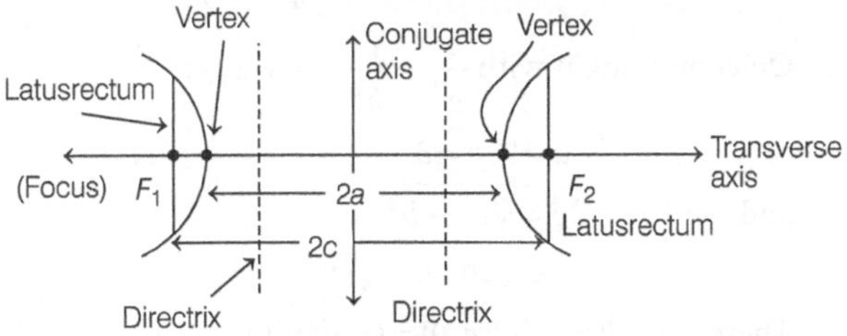

- **Eccentricity** Eccentricity of the hyperbola is the ratio of the distance of anyone focus from the centre and the distance of anyone vertex from the centre and it is denoted by e.

$\therefore \quad e = \dfrac{c}{a}$ and $c > a$

$\Rightarrow \quad \dfrac{c}{a} > 1$

$\Rightarrow \quad e > 1$

- **Directrix** Directrix is a line perpendicular to the transverse axis and cuts it at a distance of $\dfrac{a^2}{c}$ from the centre i.e. $x = \pm\dfrac{a^2}{c}$ or $y = \pm\dfrac{a^2}{c}$
- **Latusrectum** Latusrectum of a hyperbola is a line segment perpendicular to the transverse axis through any of the foci and whose end points lie on the hyperbola.

Thus, length of latusrectum $= 2l = \dfrac{2b^2}{a}$.

10.4.2 Standard Equations of Hyperbola

We have two standard equations of hyperbola

(i) $\dfrac{x^2}{a^2} - \dfrac{y^2}{b^2} = 1$

Whose X-axis is transverse axis and Y-axis as conjugate axis.

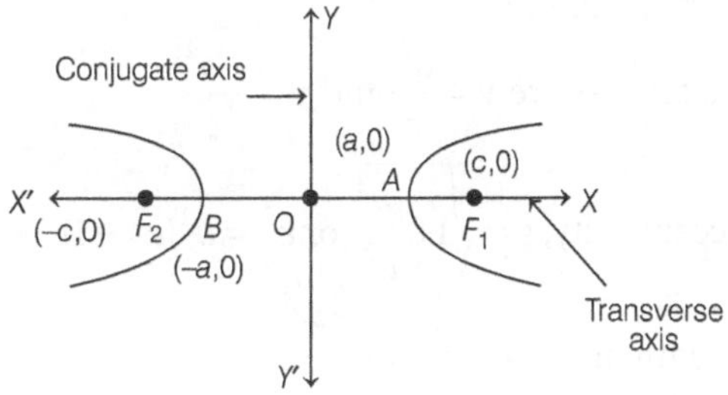

Some important terms related to standard hyperbola are

(i) Vertices are $A(a, 0)$ and $A'(-a, 0)$.

(ii) Centre is $O(0, 0)$.

(iii) Length of transverse axis is $2a$.

(iv) Length of conjugate axis is $2b$.

(v) Foci are $F_1(c, 0)$ and $F_2(-c, 0)$.

(vi) Directrices are $x = \dfrac{a^2}{c}$ and $x = -\dfrac{a^2}{c}$.

(vii) Eccentricity, $e = \sqrt{\dfrac{a^2 + b^2}{a^2}}$ or $b^2 = a^2(e^2 - 1)$

(viii) Length of latusrectum $= \dfrac{2b^2}{a}$

(ix) Coordinates of latusrectum $= \left(\pm ae, \pm\dfrac{b^2}{a}\right)$

(x) Focal length $= 2ae$

(ii) $\dfrac{y^2}{a^2}-\dfrac{x^2}{b^2}=1$

Whose X-axis is conjugate axis and Y-axis is transverse axis.

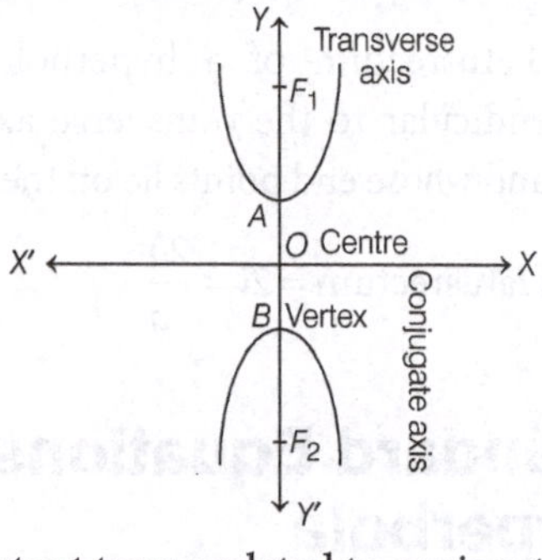

Some important terms related to conjugate hyperbola are

(i) Vertices are $A(0, a)$ and $B(0, -a)$.

(ii) Centre is O (0, 0).

(iii) Transverse axis is $2a$.

(iv) Conjugate axis is $2b$.

(v) Foci are $(0, \pm c)$.

(vi) Directrices are $y=\dfrac{a^2}{c}$ and $y=\dfrac{-a^2}{c}$.

(vii) Eccentricity, $e=\sqrt{1+\dfrac{b^2}{a^2}}$ or $b^2=a^2(e^2-1)$

(viii) Length of latusrectum $=\dfrac{2b^2}{a}$

(ix) Coordinates of latusrectum $=\left(\pm\dfrac{b^2}{a}, \pm ae\right)$

(x) Focal length $= 2ae$

Comparison between Two Standard Hyperbola

S.No.	Terms	Hyperbola	Conjugate hyperbola
	Equation of hyperbola	$\dfrac{x^2}{a^2}-\dfrac{y^2}{b^2}=1$	$\dfrac{y^2}{a^2}-\dfrac{x^2}{b^2}=1$
1.	Shape	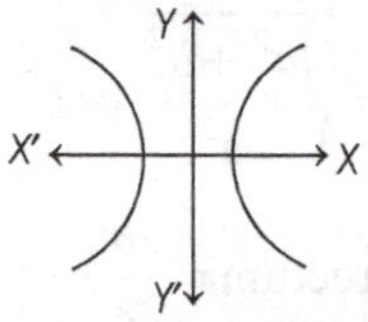	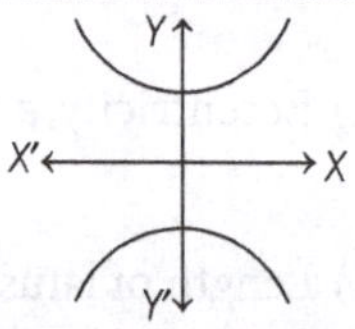
2.	Transverse axis	$2a$	$2a$
3.	Conjugate axis	$2b$	$2b$
4.	Value of c	$c=\sqrt{a^2+b^2}$	$c=\sqrt{a^2+b^2}$
5.	Vertices	$(\pm a, 0)$	$(0, \pm a)$
6.	Directrices	$x=\pm\dfrac{a^2}{c}$ or $\pm\dfrac{a}{e}$	$y=\pm\dfrac{a^2}{c}$ or $\pm\dfrac{a}{e}$
7.	Foci	$(\pm ae, 0)$ or $(\pm c, 0)$	$(0, \pm ae)$ or $(0, \pm c)$
8.	Eccentricity	$e=\sqrt{1+\dfrac{b^2}{a^2}}$ or $\dfrac{c}{a}$	$e=\sqrt{1+\dfrac{b^2}{a^2}}$ or $\dfrac{c}{a}$
9.	Length of latusrectum	$\dfrac{2b^2}{a}$	$\dfrac{2b^2}{a}$
10.	Focal distances	$ex \pm a$	$ey \pm b$

TEACHER TIPS

Equilateral Hyperbola A hyperbola in which $a=b$ is called an equilateral hyperbola.

Example 1. Find the coordinates of the foci and the vertices, the eccentricity, the length of the latusrectum of the given hyperbolas

(i) $\dfrac{x^2}{49}-\dfrac{y^2}{9}=1$

(ii) $5y^2-9x^2=36$

Sol. (i) We have, equation of hyperbola $\dfrac{x^2}{49}-\dfrac{y^2}{9}=1$

On comparing it with $\dfrac{x^2}{a^2}-\dfrac{y^2}{b^2}=1$, we get

$$a=7, b=3$$

and
$$c=\sqrt{a^2+b^2}$$
$$=\sqrt{49+9}=\sqrt{58}$$

Therefore, Foci $=(\pm c, 0)=(\pm\sqrt{58}, 0)$

Vertices $=(\pm a, 0)=(\pm 7, 0)$

Eccentricity, $e=\dfrac{c}{a}=\dfrac{\sqrt{58}}{7}$

Length of latusrectum $=\dfrac{2b^2}{a}$

$$=\frac{2\times 3^2}{7}=\frac{18}{7}$$

(ii) We have, equation of hyperbola is $5y^2 - 9x^2 = 36$.

It can be written as $\dfrac{y^2}{\left(\dfrac{6}{\sqrt{5}}\right)^2} - \dfrac{x^2}{(2)^2} = 1$

On comparing with $\dfrac{y^2}{a^2} - \dfrac{x^2}{b^2} = 1$, we get

$$a^2 = \left(\frac{6}{\sqrt{5}}\right)^2 \text{ and } b^2 = 2^2$$

$$\Rightarrow \quad a = \frac{6}{\sqrt{5}} \text{ and } b = 2$$

Value of $c = \sqrt{a^2 + b^2} = \sqrt{\dfrac{36}{5} + 4} = \sqrt{\dfrac{56}{5}} = \dfrac{2\sqrt{14}}{\sqrt{5}}$

Therefore, Foci $= (0, \pm c) = \left(0, \dfrac{\pm 2\sqrt{14}}{\sqrt{5}}\right)$

Vertices $= (0, \pm a) = \left(0, \pm \dfrac{6}{\sqrt{5}}\right)$

Eccentricity, $e = \dfrac{c}{a} = \dfrac{2\sqrt{\dfrac{14}{5}}}{\dfrac{6}{\sqrt{5}}} = \dfrac{\sqrt{14}}{3}$

Length of latusrecturm,

$$\frac{2b^2}{a} = \frac{2 \times 4}{\dfrac{6}{\sqrt{5}}} = \frac{4\sqrt{5}}{3}$$

Example 2. Find the equation of the hyperbola, whose vertices are $(0, \pm 5)$ and foci $(0, \pm 8)$. NCERT

Sol. We have, vertices $(0, \pm a) = (0, \pm 5) \Rightarrow a = 5$

and foci $(0, \pm c) = (0, \pm 8) \Rightarrow c = 8$

We know that $c^2 = a^2 + b^2 \Rightarrow 64 = 25 + b^2$

$\Rightarrow \quad b^2 = 64 - 25 \Rightarrow b^2 = 39$

Here, the foci and vertices lie on Y-axis, therefore the equation of hyperbola is of the form $\dfrac{y^2}{a^2} - \dfrac{x^2}{b^2} = 1$

i.e. $\dfrac{y^2}{25} - \dfrac{x^2}{39} = 1$

Example 3. Find the equation of the hyperbola having foci $(0, \pm 4)$ and transverse axis of length 6.

Sol. We have, the foci of hyperbola lies on Y-axis.

So, the equation of hyperbola is

$$\frac{y^2}{a^2} - \frac{x^2}{b^2} = 1$$

Length of its transverse axis $= 2a$

$\therefore \quad 2a = 6 \Rightarrow a = 3$

Foci $\equiv (0, \pm c) = (0, \pm 4) \Rightarrow c = 4$

We know that $c^2 = a^2 + b^2$

$\Rightarrow \quad 16 = 9 + b^2$

$\Rightarrow \quad b^2 = 7$

Thus, $\quad a^2 = 9$ and $b^2 = 7$

Hence, the equation of hyperbola is $\dfrac{y^2}{9} - \dfrac{x^2}{7} = 1$.

Example 4. Find the equation of the hyperbola whose foci are at $(0, \pm 6)$ and length of whose conjugate axis is $2\sqrt{11}$. Competency Based Que

Sol. We have, foci of the hyperbola lie on Y-axis.

So, the equation of hyperbola is

$$\frac{y^2}{a^2} - \frac{x^2}{b^2} = 1$$

Foci $\equiv (0, \pm c) = (0, \pm 6) \Rightarrow c = 6$

Length of its conjugate axis $= 2b$

$\therefore \quad 2b = 2\sqrt{11}$

$\Rightarrow \quad b = \sqrt{11}$

Now, $\quad c^2 = a^2 + b^2 \quad [\because c = 6 \text{ and } b = \sqrt{11}]$

$\Rightarrow \quad 6^2 = a^2 + (\sqrt{11})^2$

$\Rightarrow \quad a^2 = 36 - 11 = 25$

Thus, $\quad a^2 = 25, b^2 = 11$

Hence, the equation of hyperbola is

$$\frac{y^2}{25} - \frac{x^2}{11} = 1$$

Example 5. Find the equation of hyperbola, if vertices are at $(\pm 7, 0)$ and $e = \dfrac{4}{3}$. NCERT, Competency Based Que

Sol. We have, vertices of hyperbola lie on X-axis.

So, the equation of hyperbola is

$$\frac{x^2}{a^2} - \frac{y^2}{b^2} = 1 \quad \text{...(i)}$$

Vertices $\equiv (\pm a, 0) = (\pm 7, 0) \Rightarrow a = 7$

Now, $\quad e = \sqrt{1 + \dfrac{b^2}{a^2}}$

$\Rightarrow \quad e^2 = 1 + \dfrac{b^2}{a^2}$

$\Rightarrow \quad \left(\dfrac{4}{3}\right)^2 = 1 + \dfrac{b^2}{7^2} \quad [\because e = 4/3 \text{ and } a = 7]$

$\Rightarrow \quad \frac{b^2}{49} = \frac{16}{9} - 1$

$\Rightarrow \quad b^2 = \frac{343}{9}$

Thus, $\quad a^2 = 49, b^2 = \frac{343}{9}$

Hence, the equation of hyperbola is

$$\frac{x^2}{49} - \frac{9y^2}{343} = 1$$

$\Rightarrow \quad 7x^2 - 9y^2 = 343$ [from Eq. (i)]

Example 6. **Find the equation of hyperbola whose foci are $(0, \pm 12)$ and the length of latusrectum is 36.**

Sol. We have, foci of the hyperbola lie on Y-axis.

So, the equation of hyperbola is

$$\frac{y^2}{a^2} - \frac{x^2}{b^2} = 1 \quad \text{...(i)}$$

Foci $\equiv (0, \pm c) = (0, \pm 12) \Rightarrow c = 12$

Length of its latusrectum $= \frac{2b^2}{a}$

$\Rightarrow \quad \frac{2b^2}{a} = 36 \Rightarrow b^2 = 18a$

Now, $\quad c^2 = a^2 + b^2$

$\Rightarrow \quad (12)^2 = a^2 + 18a$

$\Rightarrow \quad a^2 + 18a - 144 = 0$

$\Rightarrow \quad (a + 24)(a - 6) = 0$

$\Rightarrow \quad a = 6$ [$\because a = -24$ is neglected]

Thus, $a^2 = 36$ and $b^2 = 18a = 18 \times 6 = 108$

Hence, the equation of hyperbola is

$$\frac{y^2}{36} - \frac{x^2}{108} = 1$$

$\Rightarrow \quad 3y^2 - x^2 = 108$ [from Eq. (i)]

Example 7. **Referred to the principal axes as the axes of coordinates, find the equation of hyperbola whose foci are at $(0, \pm\sqrt{10})$ and which passes through the point (2, 3).**

Competency Based Que

Sol. Since, the vertices are on Y-axis.

So, let the equation of the required hyperbola be

$$\frac{x^2}{a^2} - \frac{y^2}{b^2} = -1 \quad \text{...(i)}$$

It passes through (2, 3).

$\therefore \quad \frac{4}{a^2} - \frac{9}{b^2} = -1$

$\Rightarrow \quad \frac{4}{b^2(e^2 - 1)} - \frac{9}{b^2} = -1$ $\quad [\because a^2 = b^2(e^2 - 1)]$

$\Rightarrow \quad \frac{4}{b^2e^2 - b^2} - \frac{9}{b^2} = -1$ ...(ii)

The coordinates of foci are given to be $(0, \pm\sqrt{10})$.

$\therefore \quad be = \sqrt{10} \Rightarrow b^2e^2 = 10$ ...(iii)

From Eqs. (ii) and (iii), we get

$$\frac{4}{10 - b^2} - \frac{9}{b^2} = -1$$

$\Rightarrow \quad 4b^2 - 9(10 - b^2) = -b^2(10 - b^2)$

$\Rightarrow \quad 13b^2 - 90 = -10b^2 + b^4$

$\Rightarrow \quad b^4 - 23b^2 + 90 = 0$

$\Rightarrow \quad (b^2 - 18)(b^2 - 5) = 0$

$\Rightarrow \quad b^2 = 18$ or $b^2 = 5$

Now, $\quad a^2 = b^2(e^2 - 1)$

$\Rightarrow \quad a^2 = (be)^2 - b^2 \Rightarrow a^2 = 10 - b^2$

If $b^2 = 18$, then $a^2 = 10 - b^2$

$\Rightarrow \quad a^2 = 10 - 18 = -8$,

which is not possible.

$\therefore \quad b^2 = 5$

and hence $a^2 = 10 - b^2 \Rightarrow 10 - 5 = 5$

On substituting the values of a^2 and b^2 in Eq. (i), we get

$$\frac{x^2}{5} - \frac{y^2}{5} = -1$$

$\Rightarrow \quad x^2 - y^2 = -5$

as the required equation of hyperbola.

Exam Pattern Questions

Objective Type Questions

Multiple Choice Questions

1. The length of the axis of the hyperbola $\frac{y^2}{16} - \frac{x^2}{49} = 1$ are, respectively

(a) 8, 14 (b) 6, 8
(c) 6, 10 (d) 5, 8

2. The vertices and the coordinates of foci of hyperbola $\frac{x^2}{25} - \frac{y^2}{4} = 1$ are, respectively

(a) $(\pm 5, 0), (\pm \sqrt{29}, 0)$ (b) $(\pm \sqrt{29}, 0), (\pm \sqrt{5}, 0)$
(c) $(\pm \sqrt{2}, 0), (\pm \sqrt{17}, 0)$ (d) $(\pm \sqrt{17}, 0), (\pm \sqrt{2}, 0)$

3. Equation of hyperbola is $\frac{x^2}{9} - \frac{y^2}{16} = 1$. The length of its latusrectum will be

(a) 18 (b) $\frac{31}{2}$
(c) $\frac{32}{3}$ (d) $\frac{15}{4}$

4. The eccentricity of the hyperbola $16x^2 - 9y^2 = 144$ is **Competency Based Que**

(a) $\frac{1}{3}$ (b) $\frac{5}{3}$ (c) $\frac{5}{\sqrt{3}}$ (d) $\frac{5}{2}$

5. For hyperbola $x^2 - y^2 = 1$, the coordinates of foci and the eccentricity, respectively are **Competency Based Que**

(a) $(\pm \sqrt{5}, 0), \sqrt{3}$ (b) $(0, \pm \sqrt{5}), \sqrt{2}$
(c) $(\pm \sqrt{3}, 0), \sqrt{5}$ (d) $(\pm \sqrt{2}, 0), \sqrt{2}$

6. For equation of hyperbola $\frac{x^2}{a^2} - \frac{y^2}{b^2} = 1$, which option is incorrect.

(a) Length of transverse axis is $2a$
(b) Length of conjugate axis is $2b$
(c) Length of latusrectum $= \frac{2b^2}{a}$
(d) Eccentricity, $e = \sqrt{\frac{a^2 - b^2}{a^2}}$

7. Equation of transverse axis of the hyperbola $\frac{y^2}{9} - \frac{x^2}{4} = 1$

(a) $x = 0$ (b) $y = 0$
(c) $x = 2$ (d) $y = 2$

8. For the given hyperbola $\frac{x^2}{3} - y^2 = 1$, length of conjugate axis is

(a) 1 (b) 2 (c) 4 (d) 3

9. Eccentricity of hyperbola $\frac{x^2}{9} - \frac{y^2}{16} = 1$ is **Competency Based Que**

(a) $\frac{5}{9}$ (b) $\frac{5}{3}$ (c) $\frac{5}{16}$ (d) $\frac{5}{4}$

10. A hyperbola in which length of transverse axis is equal to the length of conjugate axis, is called **Competency Based Que**

(a) equilateral hyperbola (b) obtuse hyperbola
(c) acute hyperbola (d) None of these

Assertion-Reason Based Questions

Directions (Q. Nos. 11-15) *In the questions given below are two statements labelled as Assertion (A) and Reason (R). In the context of the two statements, which one of the following is correct?*

(a) Both A and R are correct; R is the correct explanation of A
(b) Both A and R are correct; R is not the correct explanation of A
(c) A is correct; R is incorrect
(d) R is correct; A is incorrect

11. Assertion (A) For the hyperbola $\frac{x^2}{25} - \frac{y^2}{20} = 1$, length of latusrectum is 8.

Reason (R) For the hyperbola $\frac{x^2}{a^2} - \frac{y^2}{b^2} = 1$, eccentricity is given by $e = \sqrt{1 + \frac{a^2}{b^2}}$. **Competency Based Que**

12. Assertion (A) For the hyperbola $9x^2 - 16y^2 = 144$, eccentricity is $\frac{5}{4}$.

Reason (R) For the hyperbola $\frac{x^2}{a^2} - \frac{y^2}{b^2} = 1$, length of transverse axis is $2a$.

13. Assertion (A) The vertices of the hyperbola $\frac{y^2}{16} - \frac{x^2}{49} = 1$ are $(0, \pm 4)$.

Reason (R) The vertices of the hyperbola $\frac{y^2}{a^2} - \frac{x^2}{b^2} = 1$ are $(0, \pm a)$. **Competency Based Que**

14. Assertion (A) For hyperbola $\frac{y^2}{9} - \frac{x^2}{16} = 1$, the length of latusrectum is $\frac{18}{3}$.

Reason (R) For hyperbola $\frac{y^2}{a^2} - \frac{x^2}{b^2} = 1$, the value of c is given by $\sqrt{a^2 + b^2}$.

15. Assertion (A) $\frac{x^2}{36} - \frac{y^2}{36} = 1$ is a equilateral hyperbola.

Reason (R) A hyperbola in which $a = b$ is called an equilateral hyperbola.

Case Based Questions

16. Mr. Sharma draw the shape of the hyperbola whose equation is $16x^2 - 9y^2 = 576$. Shape is shown as in the below figure.

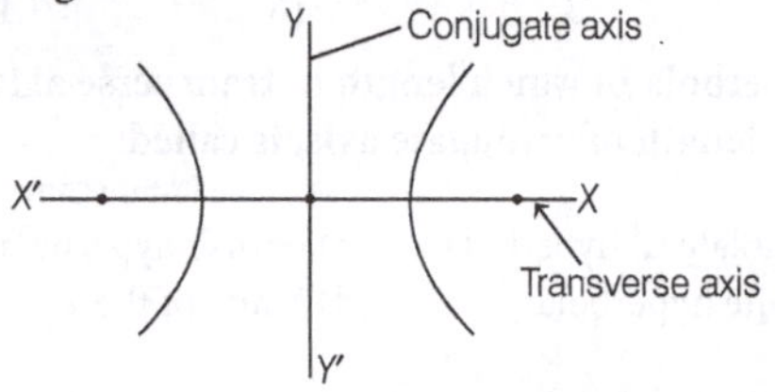

On the basis of above information, answer the following questions.

(i) The coordinates of the foci of the hyperbola is
(a) $(\pm 10, 0)$
(b) $(\pm 6, 0)$
(c) $(\pm 3, 0)$
(d) $(\pm 8, 0)$

(ii) The vertices of the hyperbola are
(a) $(\pm 3, 0)$ (b) $(\pm 6, 0)$
(c) $(\pm 4, 0)$ (d) $(\pm 5, 0)$

(iii) The eccentricity of the hyperbola $16x^2 - 9y^2 = 576$ is given by
(a) $\frac{7}{3}$
(b) $\frac{8}{3}$
(c) $\frac{5}{3}$
(d) $\frac{9}{4}$

(iv) Find the length of latusrectum of the hyperbola $16x^2 - 9y^2 = 576$.

Subjective Questions

Very Short Answer Type Questions

1. Find the foci of the hyperbola $9x^2 - 16y^2 = 144$.

2. Find the latusrectum of the hyperbola $16x^2 - 9y^2 = 144$.

3. Find the equation of hyperbola, having vertices $(\pm 5, 0)$ and foci at $(\pm 7, 0)$. **Competency Based Que**

4. Find the equation of hyperbola, if length of transverse axis is 10 and conjugate axis is 8.

5. Represent the equation of hyperbola $9x^2 - 4y^2 = -36$ in standard form and find whether the foci lie on the X-axis or the Y-axis.

6. Find the equation of the hyperbola with vertices at $(0, \pm 6)$ and eccentricity $\frac{5}{3}$. Also, find its foci. **Competency Based Que**

7. Represent the hyperbola $3x^2 - 7y^2 = 21$ in standard form and hence find whether foci lie on the X-axis or the Y-axis. **Competency Based Que**

Short Answer Type Questions

8. Find the lengths of the axes, coordinates of the vertices, coordinates of the foci, eccentricity and length of the latusrectum of each of the following hyperbola.
(i) $3x^2 - 2y^2 = 6$ (ii) $25x^2 - 9y^2 = 225$

9. Find the coordinates of the foci, vertices, eccentricity and length of latusrectum of the hyperbola $49y^2 - 16x^2 = 784$.

10. Find the equation of hyperbola, having foci $(\pm 5, 0)$ and length of transverse axis is 8. **NCERT**

11. Find the equation of the hyperbola with vertices at $(0, \pm 6)$ and $e = \frac{5}{3}$. Find its foci. **NCERT Exemplar**

12. Find the equation of the hyperbola whose vertices are $(\pm 2, 0)$ and the eccentricity is 2.

13. Find the equation of hyperbola, if vertices are at $(0, \pm 5)$ and $e = \frac{4}{3}$.

Long Answer Type Questions

14. Find the length of the axes, vertices, foci, eccentricity and length of the latusrectum of the following hyperbolas
(i) $9x^2 - 16y^2 = 144$ (ii) $9y^2 - 4x^2 = 36$

15. Find the equation of hyperbola, having foci $(\pm 4, 0)$ and length of latusrectum is 12. **NCERT**

16. Find the equation of the hyperbola, having foci $(\pm 3\sqrt{5}, 0)$ and length of latusrectum is 8.

Hints & Solutions

Objective Type Questions

1. (a) We have, equation of hyperbola is $\frac{y^2}{16}-\frac{x^2}{49}=1$

On comparing with $\frac{y^2}{a^2}-\frac{x^2}{b^2}=1$, we get

$$a^2=16 \text{ and } b^2=49$$

$\Rightarrow$ $a=4$ and $b=7$

$\therefore$ Length of axes are $2a$ and $2b$ i.e. 8, 14

2. (a) Given, $\frac{x^2}{25}-\frac{y^2}{4}=1$

On comparing with $\frac{x^2}{a^2}-\frac{y^2}{b^2}=1$, we get

$$a^2=25 \text{ and } b^2=4$$

$\Rightarrow$ $a=5$ and $b=2$

Now, value of $c=\sqrt{a^2+b^2}=\sqrt{25+4}=\sqrt{29}$

$\therefore$ Length of axes are $2a$ and $2b$ i.e. 10 and 4

Vertices $(\pm a, 0)=(\pm 5, 0)$

Foci $(\pm c, 0)=(\pm\sqrt{29}, 0)$

3. (c) Given, $\frac{x^2}{9}-\frac{y^2}{16}=1$

On comparing $\frac{x^2}{a^2}-\frac{y^2}{b^2}=1$, we get

$$a^2=9 \text{ and } b^2=16$$

$\Rightarrow$ $a=3$ and $b=4$

$\therefore$ The length of latusrectum $=\frac{2b^2}{a}=\frac{32}{3}$

4. (b) Given, $16x^2-9y^2=144$

$\Rightarrow$ $\frac{x^2}{9}-\frac{y^2}{16}=1$

On comparing with $\frac{x^2}{a^2}-\frac{y^2}{b^2}=1$, we get

$$a^2=9 \text{ and } b^2=16$$

$\Rightarrow$ $a=3$ and $b=4$

Now, value of $c=\sqrt{a^2+b^2}$

$=\sqrt{9+16}=5$

$\therefore$ The eccentricity, $e=\frac{c}{a}=\frac{5}{3}$

5. (d) Given, $x^2-y^2=1$

On comparing $\frac{x^2}{a^2}-\frac{y^2}{b^2}=1$, we get

$a^2=1$ and $b^2=1$

$\Rightarrow a=1$ and $b=1$

Now, value of $c=\sqrt{a^2+b^2}=\sqrt{2}$

$\therefore$ Foci $(\pm c, 0)=(\pm\sqrt{2}, 0)$

and eccentricity, $e=\frac{c}{a}=\sqrt{2}$

6. (d) For hyperbola $\frac{x^2}{a^2}-\frac{y^2}{b^2}=1$

Lenght of transverse axis is $2a$.

Length of conjugate axis is $2b$.

Eccentricity, $e=\sqrt{\frac{a^2+b^2}{a^2}}$

Length of latusrectum $=\frac{2b^2}{a}$

7. (a) Given hyperbola is $\frac{y^2}{9}-\frac{x^2}{4}=1$, foci lie on the Y-axis.

$\therefore$ Transverse axis is the Y-axis and equation is $x=0$.

8. (b) As $a^2=3$ and $b^2=1$.

So, length of conjugate axis $=2b=2$.

9. (b) $c=\sqrt{9+16}=5$

Eccentricity $=\frac{c}{a}=\frac{5}{3}$

10. (a) A hyperbola in which length of transverse axis is equal to the length of conjugate axis, is called equilateral hyperbola.

11. (c) We have, $\frac{x^2}{25}-\frac{y^2}{20}=1$

Here, $a=5$ and $b=\sqrt{20}$

Length of latusrectum $=\frac{2b^2}{a}=\frac{2\times 20}{5}=8$

12. (b) Given hyperbola is $9x^2-16y^2=144$

$\Rightarrow$ $\frac{x^2}{16}-\frac{y^2}{9}=1$

Here, $a^2=16$ and $b^2=9$

$\Rightarrow$ $c=\sqrt{16+9}=5$

Eccentricity, $e=\frac{c}{a}=\frac{5}{4}$

13. (a) Given hyperbola is

$$\frac{y^2}{16}-\frac{x^2}{49}=1$$

It is the form of conjugate hyperbola

$$\frac{y^2}{a^2}-\frac{x^2}{b^2}=1$$

Here, $a=4$ and $b=7$.

$\therefore$ Vertices $=(0, \pm a)=(0, \pm 4)$

14. (d) Given hyperbola is

$$\frac{y^2}{9}-\frac{x^2}{16}=1$$

On comparing with conjugate hyperbola

$$\frac{y^2}{a^2}-\frac{x^2}{b^2}=1, \text{ we get}$$

$a=3$ and $b=4$

$\therefore$ Length of latusrectum $=\frac{2b^2}{a}=\frac{32}{3}$

$\therefore$ Assertion is incorrect and Reason is correct.

15. (a) We have, $\frac{x^2}{36}-\frac{y^2}{36}=1$

On comparing with $\frac{x^2}{a^2}-\frac{y^2}{b^2}=1$, we get

$$a^2=36 \text{ and } b^2=36$$

$\Rightarrow$ $a=6$ and $b=6$

Here $a=b$, hence given hyperbola is a equilateral hyperbola.

16. Given, $16x^2-9y^2=576$

$\Rightarrow$ $\frac{x^2}{36}-\frac{y^2}{64}=1$

On comparing it with $\frac{x^2}{a^2}-\frac{y^2}{b^2}=1$, we get

$a=6$ and $b=8$

Now, transverse axis, $2a=12$

and conjugate axis, $2b=16$

Value of $c=\sqrt{a^2+b^2}=\sqrt{36+64}=10$

(i) (a) Foci of hyperbola $=(\pm c, 0)=(\pm 10, 0)$

(ii) (b) Vertices of the hyperbola $=(\pm a, 0)=(\pm 6, 0)$

(iii) (c) Eccentricity $e=\frac{c}{a}=\frac{10}{6}=\frac{5}{3}$

(iv) Length of latusrectum $=\frac{2b^2}{a}=\frac{2\times 64}{6}=\frac{64}{3}$

Subjective Questions

1. Given, $9x^2-16y^2=144$

i.e. $\frac{x^2}{16}-\frac{y^2}{9}=1$

On comparing $\frac{x^2}{a^2}-\frac{y^2}{b^2}=1$, we get

$$a^2=16 \text{ and } b^2=9 \Rightarrow a=4 \text{ and } b=3$$

Since, $c=\sqrt{a^2+b^2}=\sqrt{16+9}=5$

$\therefore$ Foci $(\pm c, 0)=(\pm 5, 0)$

2. Given, $16x^2-9y^2=144$

i.e. $\frac{x^2}{9}-\frac{y^2}{16}=1$

On comparing $\frac{x^2}{a^2}-\frac{y^2}{b^2}=1$, we get

$$a^2=9 \text{ and } b^2=16$$

$\because$ Length of latusrectum $=\frac{2b^2}{a}$

$\therefore$ Length of latusrectum $=\frac{2(16)}{3}=\frac{32}{3}$

3. We have, vertices $(\pm a, 0)=(\pm 5, 0)$

and foci $(\pm c, 0)=(\pm 7, 0)$

We know that $c^2=a^2+b^2$

$\Rightarrow$ $49=25+b^2 \Rightarrow b^2=24$

Here, the foci and vertices lie on X-axis, therefore the equation of hyperbola is of the form

$$\frac{x^2}{a^2}-\frac{y^2}{b^2}=1$$

i.e. $\frac{x^2}{25}-\frac{y^2}{24}=1$

4. Given, $2a=10$ and $2b=8$

$\Rightarrow a=5$ and $b=4$

$\therefore$ Equation of hyperbola is $\frac{x^2}{5^2}-\frac{y^2}{4^2}=1$

i.e. $\frac{x^2}{25}-\frac{y^2}{16}=1$

5. Given equation is $9x^2-4y^2=-36$

$\Rightarrow$ $-\frac{x^2}{4}+\frac{y^2}{9}=1$ is standard form

Here, foci lie on the Y-axis.

6. Since, the vertices are at $(0, \pm 6)$

i.e. $(0, \pm 6)=(0, \pm a)$

$\Rightarrow$ $a=6$

$$e=\frac{5}{3} \Rightarrow e=\frac{\sqrt{a^2+b^2}}{a}$$

$$\frac{25}{9}=1+\frac{b^2}{36}$$

$\Rightarrow$ $\frac{b^2}{36}=\frac{16}{9} \Rightarrow b^2=64$

So, equation of the hyperbola is $\frac{y^2}{36}-\frac{x^2}{64}=1$

$$c=\sqrt{a^2+b^2}=\sqrt{36+64}=10$$

Foci $=(0, \pm c)=(0, \pm 10)$

7. Given hyperbola is $3x^2-7y^2=21$.

$\Rightarrow$ $\frac{x^2}{7}-\frac{y^2}{3}=1$

As it is of the form $\frac{x^2}{a^2}-\frac{y^2}{b^2}=1$, foci lie on the X-axis.

8. (i) Given, $3x^2 - 2y^2 = 6$ i.e. $\frac{x^2}{2} - \frac{y^2}{3} = 1$

On comparing $\frac{x^2}{a^2} - \frac{y^2}{b^2} = 1$, we get

$a^2 = 2$ and $b^2 = 3 \Rightarrow a = \sqrt{2}$ and $b = \sqrt{3}$

Now, value of $c = \sqrt{a^2 + b^2} = \sqrt{5}$

$\therefore$ Length of axes are $2a$ and $2b$ i.e. $2\sqrt{2}$ and $2\sqrt{3}$.

Vertices $(\pm a, 0) = (\pm \sqrt{2}, 0)$

Foci $(\pm c, 0) = (\pm \sqrt{5}, 0)$,

eccentricity, $e = \frac{c}{a} = \frac{\sqrt{5}}{\sqrt{2}}$

and length of latusrectum $= \frac{2b^2}{a} = \frac{6}{\sqrt{2}} = 3\sqrt{2}$

(ii) Given, $25x^2 - 9y^2 = 225$

i.e. $\frac{x^2}{9} - \frac{y^2}{25} = 1$

On comparing $\frac{x^2}{a^2} - \frac{y^2}{b^2} = 1$, we get

$a^2 = 9$ and $b^2 = 25 \Rightarrow a = 3$ and $b = 5$

Now, value of $c = \sqrt{a^2 + b^2}$

$= \sqrt{9 + 25} = \sqrt{34}$

$\therefore$ Length of axes are $2a$ and $2b$ i.e. 6 and 10.

Vertices $(\pm a, 0) = (\pm 3, 0)$,

foci $(\pm c, 0) = (\pm \sqrt{34}, 0)$

and eccentricity, $e = \frac{c}{a} = \frac{\sqrt{34}}{3}$

and length of latusrectum $= \frac{2b^2}{a} = \frac{50}{3}$ = units

9. Given, $49y^2 - 16x^2 = 784$

$\Rightarrow \frac{y^2}{16} - \frac{x^2}{49} = 1$ [dividing both sides by 784]

On comparing it with $\frac{y^2}{a^2} - \frac{x^2}{b^2} = 1$, we get

$a = 4$ and $b = 7$

Now, transverse axis, $2a = 8$

Conjugate axis, $2b = 14$

Value of $c = \sqrt{a^2 + b^2} = \sqrt{16 + 49} = \sqrt{65}$

Vertices $\equiv (0, \pm a) = (0, \pm 4)$,

foci $\equiv (0, \pm c) = (0, + \sqrt{65})$,

eccentricity, $e = \frac{c}{a} = \frac{\sqrt{65}}{4}$

and length of latusrectum $= \frac{2b^2}{a} = \frac{2 \times 49}{4} = \frac{49}{2}$

10. Since, the foci of the given hyperbola are of the form $(\pm c, 0)$, it is a horizontal hyperbola.

Let the required equation be $\frac{x^2}{a^2} - \frac{y^2}{b^2} = 1$

$\because$ Length of its transverse axis $= 2a$

$\therefore \quad 2a = 8 \Rightarrow a = 4 \Rightarrow a^2 = 16$

Let its foci be $(\pm c, 0)$.

Then, $c = 5$ [$\because$ foci are $(\pm 5, 0)$]

$\therefore \quad b^2 = c^2 - a^2 = 5^2 - 4^2 = 9$

Hence, the required equation is

$$\frac{x^2}{16} - \frac{y^2}{9} = 1$$

11. Hint Let vertices $\equiv (0, \pm b) = (0, \pm 6)$

$b = 6$ and $e = \frac{5}{3}$

$\therefore \quad e^2 = \frac{a^2 + b^2}{b^2}$

$\Rightarrow \quad \frac{25}{9} = \frac{36 + a^2}{36}$

$\Rightarrow \quad a^2 = 64$

Hence, the required equation of hyperbola is

$$-\frac{x^2}{64} + \frac{y^2}{36} = 1 \Rightarrow \frac{y^2}{36} - \frac{x^2}{64} = 1$$

Ans. foci $= (0, \pm be) = (0, \pm 10)$

12. Hint $(\pm a, 0) = (\pm 2, 0)$ and $e = 2 = \sqrt{1 + \frac{b^2}{a^2}}$

Ans. $\frac{x^2}{4} - \frac{y^2}{12} = 0$

13. Since, vertices are on Y-axis (with origin as the centre).

Therefore, equation of hyperbola will be $\frac{y^2}{a^2} - \frac{x^2}{b^2} = 1$.

Distance between centre and vertices $= a = 5$

$\because \quad e^2 = 1 + \frac{b^2}{a^2}$

$\left(\frac{4}{3}\right)^2 = 1 + \frac{b^2}{5^2}$

$\therefore \quad \frac{16}{9} - 1 = \frac{b^2}{5^2}$

$\Rightarrow \quad b^2 = \frac{175}{9}$

$\therefore$ Required equation is $\frac{y^2}{25} - \frac{9}{175} = 1$

14. (i) Given, $9x^2 - 16y^2 = 144$

$\Rightarrow \quad \frac{x^2}{16} - \frac{y^2}{9} = 1$ [dividing both sides by 144]

On comparing it with $\frac{x^2}{a^2} - \frac{y^2}{b^2} = 1$, we get

$a = 4$ and $b = 3$

Now, transverse axis, $2a = 8$

Conjugate axis, $2b = 6$

Value of $c = \sqrt{a^2 + b^2} = \sqrt{16 + 9} = 5$

Vertices $\equiv (\pm a, 0) = (\pm 4, 0)$, Foci $\equiv (\pm c, 0) = (\pm 5, 0)$

Eccentricity $= e = \frac{c}{a} = \frac{5}{4}$

and length of latusrectum $= \frac{2b^2}{a} = \frac{2 \times 3^2}{4} = \frac{9}{2}$

(ii) **Hint** Solve as part (i).

Ans. Transverse axis = 4, Conjugate axis = 6,

Vertices $\equiv (0, \pm 2)$, Foci $\equiv (0, \pm \sqrt{13})$

Eccentricity $= \frac{\sqrt{13}}{2}$, Length of latusrectum = 9

15. The given foci are $(\pm 4, 0)$ and length of latusrectum is 12.

Since, the foci are on the X-axis, the equation of the hyperbola is represented as

$$\frac{x^2}{a^2} - \frac{y^2}{b^2} = 1$$

Since, X-axis is the transverse axis, coordinates of foci $= (\pm c, 0)$

$\therefore \quad c = 4$

$\because$ Length of latusrectum $= \frac{2b^2}{a}$

$\therefore \quad 12 = \frac{2b^2}{a}$

$\Rightarrow \quad b^2 = 6a$

Now, $\quad c^2 = a^2 + b^2$

$\Rightarrow \quad 16 = a^2 + 6a$

$\Rightarrow \quad a^2 + 6a - 16 = 0$

$\Rightarrow \quad a^2 + 8a - 2a - 16 = 0$

$\Rightarrow \quad a(a + 8) - 2(a + 8) = 0$

$\Rightarrow \quad (a + 8)(a - 2) = 0$

$\Rightarrow \quad a = -8$ or $a = 2$

but a is non-negative, $a = 2$

$\therefore \quad b^2 = 6a$

$\Rightarrow \quad b^2 = 12$

Thus, the required equation is

$$\frac{x^2}{4} - \frac{y^2}{12} = 1$$

16. **Hint** Let foci $(\pm c, 0) = (\pm 3\sqrt{5}, 0)$

$\Rightarrow \quad c = 3\sqrt{5}$

and length of latusrectum $= \frac{2b^2}{a} = 8$

$\Rightarrow \quad b^2 = 4a$

On putting $c = 3\sqrt{5}$ and $b^2 = 4a$ in $c^2 = a^2 + b^2$, we get

$45 = a^2 + 4a$

$\Rightarrow \quad a^2 + 4a - 45 = 0$

$\Rightarrow \quad a = -9$ or $a = 5$

But $\quad a \neq -9 \Rightarrow a = 5$

On putting $a = 5$ in $b^2 = 4a$, we get $b^2 = 20$

Ans. The required equation of hyperbola is $\frac{x^2}{25} - \frac{y^2}{20} = 1.$

Mind Map

Hyperbola

A hyperbola is the locus of a point in a plane which moves in the plane in such a way that the ratio of its distance from a fixed point (called **focus**) in the same plane to its distance from a fixed straight line (called **directrix**) is always constant, which is always greater than unity.

i.e. eccentricity $e > 1$

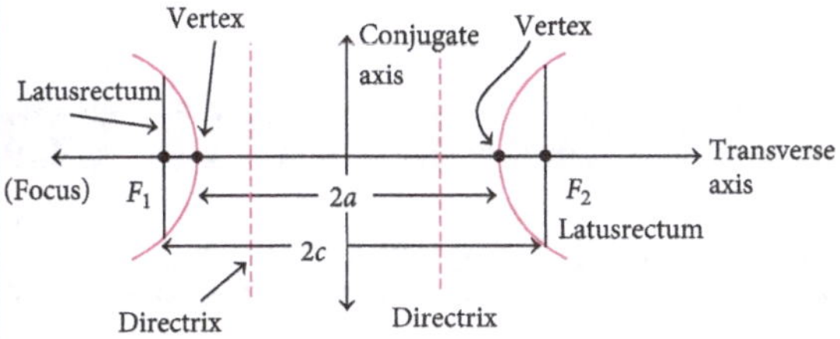

Terms Related to Hyperbola

- Focus The two fixed points are called the foci of the hyperbola and denoted by F_1 and F_2. The distance between two foci and F_1 and F_2 is $2c$.
- **Centre** The mid-point of the line segment joining the foci, is called centre of hyberbola.
- **Transverse axis** The line through the foci, is called the transverse axis.
- **Conjugate axis** The line through the centre and perpendicular to the transverse axis, is called the conjugate axis.
- **Vertices** The points at which the hyperbola intersects the transverse axis are called the vertices of the hyperbola. The distance between two vertices is $2a$.
- **Eccentricity** Eccentricity of the hyperbola is the ratio of the distance of anyone focus from the centre and the distance of anyone vertex from the centre and it is denoted by e. Also $e > 1$
- **Directrix** Directrix is a line perpendicular to the transverse axis and cuts it at a distance of $\frac{a^2}{c}$ from the centre i.e. $x = \pm \frac{a^2}{c}$ or $y = \pm \frac{a^2}{c}$
- **Latusrectum** Latusrectum of a hyperbola is a line segment perpendicular to the transverse axis through any of the foci and whose end points lie on the hyperbola. Thus, length of latusrectum $= 2l = \frac{2b^2}{a}$

Hyperbola

S.No.	Terms	Hyperbola	Conjugate hyperbola
1.	Transverse axis	$2a$	$2a$
2.	Conjugate axis	$2b$	$2b$
3.	Value of c	$c = \sqrt{a^2 + b^2}$	$c = \sqrt{a^2 + b^2}$
4.	Vertices	$(\pm a, 0)$	$(0, \pm a)$
5.	Directrices	$x = \pm \frac{a^2}{c}$ or $\pm \frac{a}{e}$	$y = \pm \frac{a^2}{c}$ or $\pm \frac{a}{e}$
6.	Foci	$(\pm ae, 0)$ or $(\pm c, 0)$	$(0, \pm ae)$ or $(0, \pm c)$
7.	Eccentricity	$e = \sqrt{1 + \frac{b^2}{a^2}}$ or $\frac{c}{a}$	$e = \sqrt{1 + \frac{b^2}{a^2}}$ or $\frac{c}{a}$
8.	Length of latusrectum	$\frac{2b^2}{a}$	$\frac{2b^2}{a}$
9.	Focal distances	$ex \pm a$	$ey \pm b$

Standard Equations of Hyperbola

Standard Hyperbola

$$\frac{x^2}{a^2} - \frac{y^2}{b^2} = 1$$

Whose X-axis is transverse axis and Y-axis as conjugate axis.

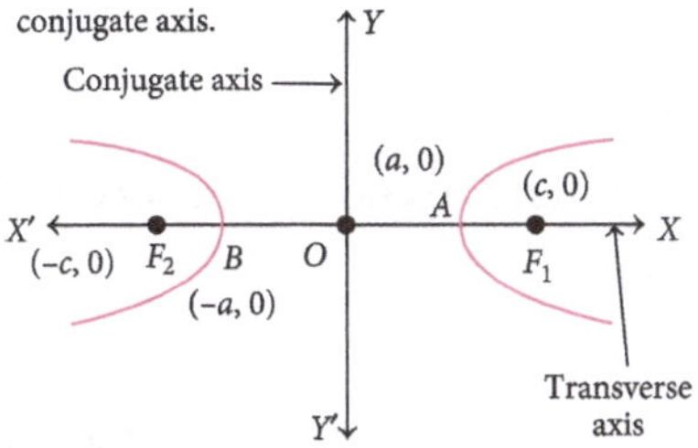

Conjugate Hyperbola

$$\frac{y^2}{a^2} - \frac{x^2}{b^2} = 1$$

Whose X-axis is conjugate axis and Y-axis is transverse axis.

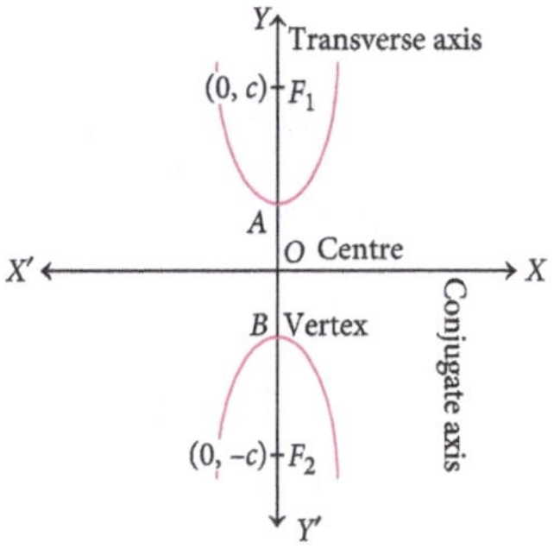

11
CHAPTER

Beauty is the first test, there is no permanent place in the world for ugly mathematics.
—*Godfrey Harold hardy*

Introduction to Three Dimensional Geometry

YOUR STUDY GOALS

11.1 Coordinate Axes and Coordinate Planes in Three Dimensional Spaces

11.2 Distance between Two Points

YOUR PRACTICE GOALS

- Objective Type Questions (23 Qs; 11 MCQs, 10 A-R, 2 Case Based)
- Subjective Type Questions (13 Qs; 4 VSA, 3 SA, 6 LA)

11.1 Coordinate Axes and Coordinate Planes in Three Dimensional Spaces

Coordinate Axes

Let XOX', YOY' and ZOZ' be the three mutually perpendicular lines, intersecting at O.

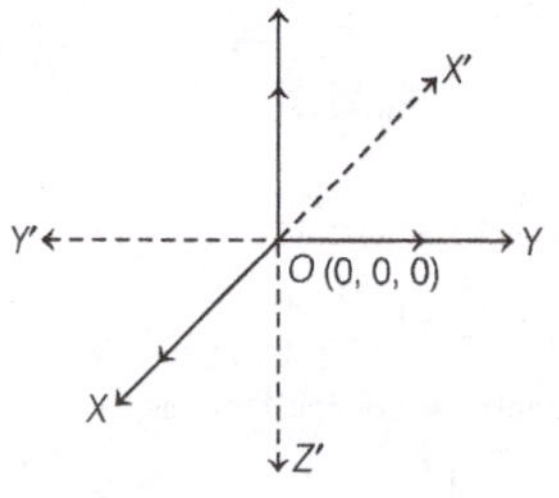

The point $O(0, 0, 0)$ is called the **origin** and the lines (XOX', YOY' and ZOZ') are called **rectangular coordinate axes** say X, Y and Z, respectively.

Thus, in the given figure

- $X'OX$ is called the X-axis.
- $Y'OY$ is called the Y-axis.
- $Z'OZ$ is called the Z-axis.

Coordinate Planes

The three coordinate axes define three mutually perpendicular planes XOY, YOZ and ZOX (or XY, YZ and ZX) called **coordinate planes** which divide the space into eight parts called **octants.**

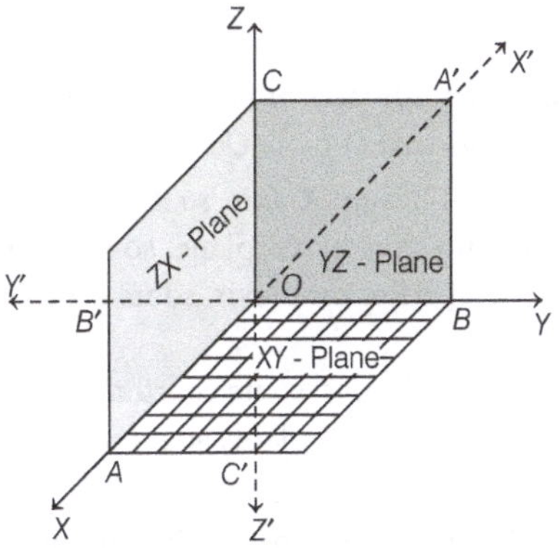

Thus, in the given figure,

- XOY is called the XY-plane.
- YOZ is called the YZ-plane.
- ZOX is called the ZX-plane.
- $XOYZ, X'OYZ, XOY'Z, X'OY'Z, XOYZ', X'OYZ'$, $XOY'Z'$ and $X'OY'Z'$ are called **octants.**

Coordinates of a Point in Space

Let P be a point in space. Through P, draw three planes parallel to the coordinate axes to meet the axes at A, B and C, respectively.

Let $OA = x$, $OB = y$ and $OC = z$. These three numbers taken in order are called **coordinates of a point *P*** and written as $P(x, y, z)$.

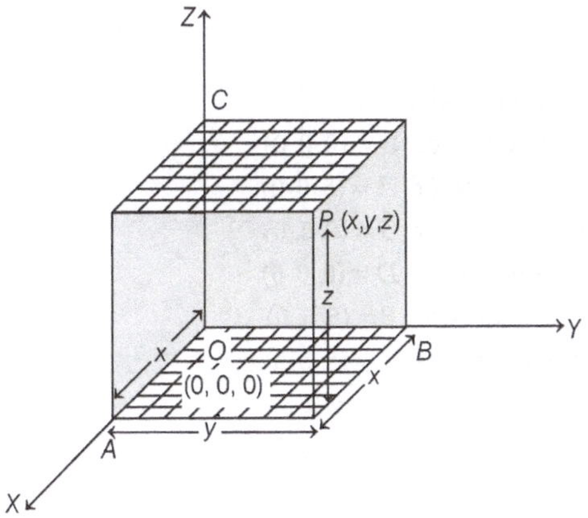

Thus, ***x*-coordinate of a point *P*** = Length of perpendicular from P to YZ-plane with proper sign.

y*-coordinate of a point *P = Length of perpendicular from P to ZX-plane with proper sign.

z*-coordinate of a point *P = Length of perpendicular from P to XY-plane with proper sign.

Sign of Coordinates

The sign of coordinates of the points in the octants in which the space is divided are given in the following table.

Octants / Coordinates	I	II	III	IV	V	VI	VII	VIII
	$XOYZ$	$X'OYZ$	$X'OY'Z$	$XOY'Z$	$XOYZ'$	$X'OYZ'$	$X'OY'Z'$	$XOY'Z'$
x	+	−	−	+	+	−	−	+
y	+	+	−	−	+	+	−	−
z	+	+	+	+	−	−	−	−

Coordinates of a Point on Coordinate Axes

Let $P(x, y, z)$ be any point in space and we draw three perpendicular lines from P to the coordinate axes meet at points A, B and C. The coordinates of a point P on X-axis will be of the form $(x, 0, 0)$.

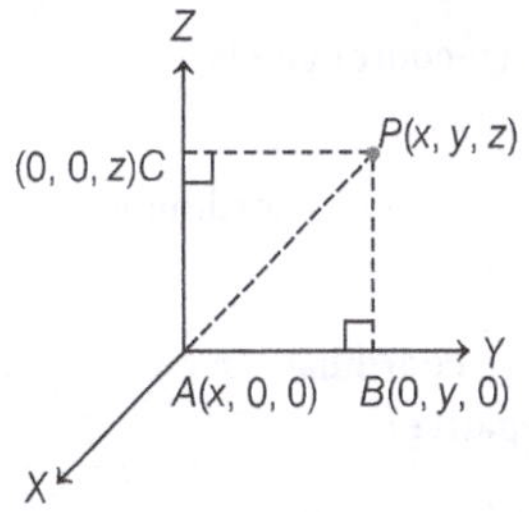

Here, x is the distance of foot of perpendicular from P to X-axis from the origin with suitable sign.

Thus, coordinates of a point on

(i) X axis will be of the form $(x, 0, 0)$.

(ii) Y-axis will be of the form $(0, y, 0)$.

(iii) Z-axis will be of the form $(0, 0, z)$.

Coordinates of a Point in Coordinate Planes

Let $P(x, y, z)$ be any point in space and we draw three perpendicular lines from P to the coordinate planes XY, YZ and ZX, respectively.

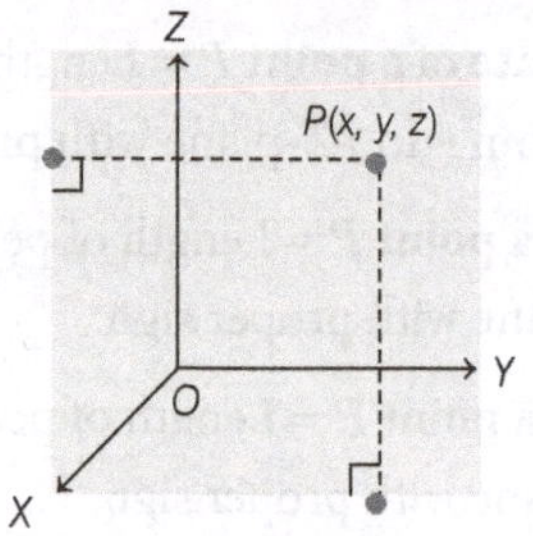

Then, the coordinates of a point in

(i) XY-plane will be of the form $(x, y, 0)$.
[$\because$ In XY-plane, the z-coordinate of a point is zero]

(ii) YZ-plane will be of the form $(0, y, z)$ and

(iii) ZX-plane will be of the form $(x, 0, z)$.

Example 1. Name the octants in which the following points lie.

(1, 2, 3), (4, −2, 3), (4, −2, −5), (4, 2, −5), (−4, 2, −5), (−4, 2, 5), (−3, −1, 6), (2, −4, −7) NCERT

Sol. The name of octants in which given points lie, are given in the following table.

Point	Octant	Name
(1, 2, 3)	I (all the coordinates are positive)	$XOYZ$
(4, −2, 3)	IV (y-coordinate is negative)	$XOY'Z$
(4, −2, −5)	VIII (y and z-coordinates are negative)	$XOY'Z'$
(4, 2, −5)	V (z-coordinate is negative)	$XOYZ'$
(−4, 2, −5)	VI (x and z-coordinates are negative)	$X'OYZ'$
(−4, 2, 5)	II (x-coordinate is negative)	$X'OYZ$
(−3, −1, 6)	III (x and y-coordinates are negative)	$X'OY'Z$
(2, −4, −7)	VIII (y and z-coordinates are negative)	$XOY'Z'$

Example 2. In the given figure, if the coordinates of point P are (a, b, c), then write the coordinates of A, D, B, C and E.

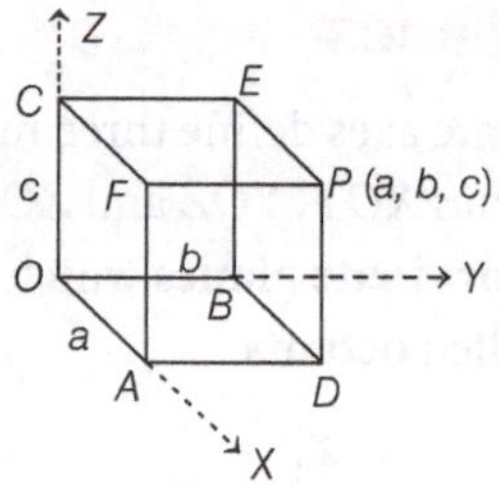

Sol. Given, the coordinates of point P are (a, b, c), which shows that $OA = a$, $OB = b$ and $OC = c$.

Now, point A lies on X-axis, so its coordinates are $(a, 0, 0)$. Point D lies in XY-plane, so its coordinates are $(a, b, 0)$. Point B lies on Y-axis, so its coordinates are $(0, b, 0)$.

Point C lies on Z-axis, so its coordinates are $(0, 0, c)$ and point E lies in YZ-plane, so its coordinates are $(0, b, c)$.

Hence, the coordinates of required points are $A(a, 0, 0)$, $D(a, b, 0)$, $B(0, b, 0)$, $C(0, 0, c)$ and $E(0, b, c)$.

Example 3. What are the coordinates of the vertices of a cube whose edge is 5 units, one of whose vertices coincides with the origin and three edges passing through the origin coincides with the positive direction of the axes through the origin? Competency Based Que

Sol. Given, edge of a cube is 5 units.

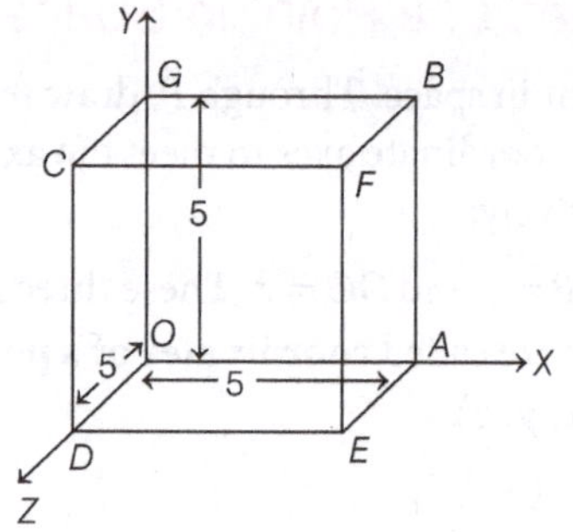

It is clear that

Coordinates of $O = (0, 0, 0)$

Coordinates of $A = (5, 0, 0)$

Coordinates of $G = (0, 5, 0)$

Coordinates of $D = (0, 0, 5)$

Coordinates of $B = (5, 5, 0)$

Coordinates of $F = (5, 5, 5)$

Coordinates of $E = (5, 0, 5)$

Coordinates of $C = (0, 5, 5)$

Example 4. **Let L, M and N be the feet of the perpendiculars drawn from the point P (3, 4, 5) on the XY, YZ and ZX-planes, respectively. Find the distance of these points L, M and N from the point P.** *NCERT Exemplar*

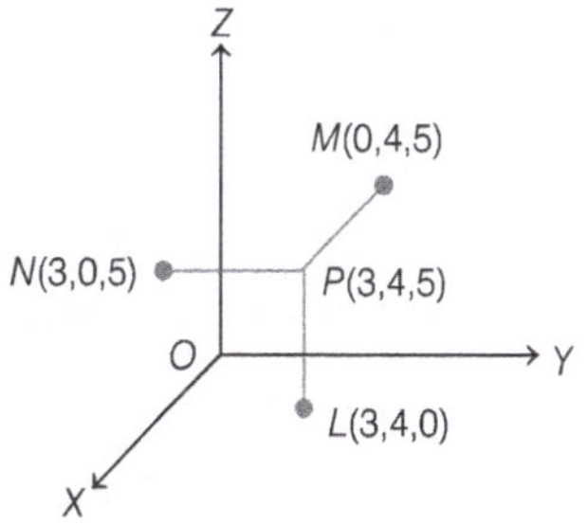

Sol. L is the foot of perpendicular drawn from the point $P(3, 4, 5)$ to the XY-plane.

Therefore, the coordinates of the point L are (3, 4, 0). The distance between the points (3, 4, 5) and (3, 4, 0) is 5 units.

Similarly, the lengths of the foot of perpendiculars on YZ and ZX-planes are 3 and 4 units, respectively.

Quick Test 1

1. Name the octant in which each of the following points lies. *NCERT Exemplar*

(i) (4, −2, 3) (ii) (4, −2, −5) (iii) (4, 2, −5)
(iv) (−3, −1, 6) (v) (−3, 1, −2) (vi) (2, 3, −4)
(vii) (4, 6, 8) (viii) (−4, −1, −7)

2. In the given figure, if the coordinates of the point P are (3, 4, 6), then write the coordinates of A, D, B, C and E. *ompetency Based Que*

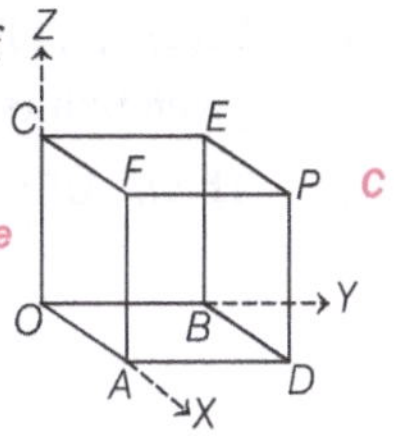

3. If a point is on the X-axis, then find its y-coordinate and z-coordinate.

4. If a point is on the ZX-plane, then find its coordinates.

5. L is the foot of the perpendicular drawn from a point $P(6, 7, 8)$ on the XY-plane. What are the coordinates of point L? *NCERT Exemplar, Competency Based Que*

6. Let L, M, N be the feet of the perpendiculars drawn from a point $P(3, 4, 5)$ on the X, Y and Z-axes, respectively. Find the coordinates of L, M and N. *NCERT Exemplar*

7. If A, B and C are the feet of perpendiculars from a point P on the XY, YZ and ZX-planes respectively, then find the distance of A, B and C from P, where the point P is
(i) (3, 4, 2)
(ii) (−5, 3, 7)

8. If A, B and C are the feet of perpendiculars from a point P on XY, YZ and ZX-planes respectively, then find the coordinates of A, B and C in each of the following, where the point P is *Competency Based Que*
(i) (−5, 3, 7)
(ii) (4, −3, −5)

Answers

1. (i) Fourth octant (ii) Eighth octant
(iii) Fifth octant (iv) Third octant
(v) Sixth octant (vi) Fifth octant
(vii) First octant (viii) Seventh octant
2. (3, 0, 0), (3, 4, 0), (0, 4, 0), (0, 0, 6), (0, 4, 6)
3. Coordinates of any point on the X-axis is $(x, 0, 0)$. Therefore, both y and z-coordinates are zero.
4. $(x, 0, z)$
5. (6, 7, 0)
6. L (3, 0, 0), M (0, 4, 0) and N (0, 0, 5)
7. (i) 2 units, 3units, 4 units (ii) 7 units, 5 units, 3 units
8. (i) $A(-5, 3, 0)$, $B(0, 3, 7)$, $C(-5, 0, 7)$
(ii) $A(4, -3, 0)$, $B(0, -3, -5)$, $C(4, 0, -5)$

For Solutions, See QR Code on Pg. 324

11.2 Distance between Two Points

The distance between two points $P(x_1, y_1, z_1)$ and $Q(x_2, y_2, z_2)$ is given by

$$PQ = \sqrt{(x_2 - x_1)^2 + (y_2 - y_1)^2 + (z_2 - z_1)^2}$$

or $$PQ = \sqrt{(x_1 - x_2)^2 + (y_1 - y_2)^2 + (z_1 - z_2)^2}$$

If P is the origin and $Q(x, y, z)$ is a point in space, then

$$PQ = \sqrt{(x-0)^2 + (y-0)^2 + (z-0)^2}$$

or $$PQ = \sqrt{x^2 + y^2 + z^2}$$

Condition for Collinearity

Three points A, B and C are said to be collinear, if

$$AB + BC = AC$$

Example 1. **Find the distance between the points $A(2, 3, 1)$ and $B(1, -2, 0)$.**

Sol. Given points are $A(2, 3, 1)$ and $B(1, -2, 0)$.

Here, $x_1 = 2, y_1 = 3, z_1 = 1$ and $x_2 = 1, y_2 = -2, z_2 = 0$

$\therefore$ The distance between the points A and B

$$= \sqrt{(x_2 - x_1)^2 + (y_2 - y_1)^2 + (z_2 - z_1)^2}$$
$$= \sqrt{(1-2)^2 + (-2-3)^2 + (0-1)^2}$$
$$= \sqrt{(-1)^2 + (-5)^2 + (-1)^2}$$
$$= \sqrt{1+25+1} = \sqrt{27} = 3\sqrt{3} \text{ units}$$

Example 2. **Find the equation of set of point P such that $PA^2 + PB^2 = 2k^2$, where A and B are the points $(3, 4, 5)$ and $(-1, 3, -7)$, respectively.** NCERT

Sol. Given points are $A(3, 4, 5)$ and $B(-1, 3, -7)$.

Let the coordinates of point P be (x, y, z).

Then, $PA^2 = (x-3)^2 + (y-4)^2 + (z-5)^2$

$[\because \text{distance} = \sqrt{(x_2 - x_1)^2 + (y_2 - y_1)^2 + (z_2 - z_1)^2}]$

and $PB^2 = (x+1)^2 + (y-3)^2 + (z+7)^2$

By the given condition $PA^2 + PB^2 = 2k^2$,

$$(x-3)^2 + (y-4)^2 + (z-5)^2 + (x+1)^2 + (y-3)^2 + (z+7)^2 = 2k^2$$
$$\Rightarrow x^2 + 9 - 6x + y^2 + 16 - 8y + z^2 + 25 - 10z + x^2 + 2x + 1 + y^2 + 9 - 6y + z^2 + 49 + 14z = 2k^2$$
$$\Rightarrow 2x^2 + 2y^2 + 2z^2 - 4x - 14y + 4z = 2k^2 - 109$$

which is the required equation.

Example 3. **Find the equation of the curve formed by the set of all points whose distances from the points $(3, 4, -5)$ and $(-2, 1, 4)$ are equal.** Competency Based Que

Sol. Let $P(x, y, z)$ be any point on the given curve and let $A(3, 4, -5)$ and $B(-2, 1, 4)$ be the given points.

Then, $PA = PB \Rightarrow PA^2 = PB^2$

$$\Rightarrow (x-3)^2 + (y-4)^2 + (z+5)^2 = (x+2)^2 + (y-1)^2 + (z-4)^2$$
$$\Rightarrow x^2 + 9 - 6x + y^2 + 16 - 8y + z^2 + 25 + 10z = x^2 + 4 + 4x + y^2 + 1 - 2y + z^2 + 16 - 8z$$
$$\Rightarrow 10x + 6y - 18z - 29 = 0$$

Hence, the required curve is

$$10x + 6y - 18z - 29 = 0$$

Example 4. **Find the coordinates of a point equidistant from the four points $O(0, 0, 0)$, $A(l, 0, 0)$, $B(0, m, 0)$ and $C(0, 0, n)$.** NCERT Exemplar

Sol. Let $P(x, y, z)$ be required point.

Then, $OP = PA = PB = PC$

Now, $OP = PA \Rightarrow OP^2 = PA^2$

$$\Rightarrow (0-x)^2 + (0-y)^2 + (0-z)^2 = (x-l)^2 + (y-0)^2 + (z-0)^2$$
$$\Rightarrow x^2 + y^2 + z^2 = x^2 - 2lx + l^2 + y^2 + z^2$$
$$\Rightarrow 2lx = l^2$$
$$\Rightarrow x = \frac{l}{2}$$

Similarly, $OP = PB \Rightarrow y = \frac{m}{2}$ and $OP = PC \Rightarrow z = \frac{n}{2}$

Hence, the coordinates of the required point are $\left(\frac{l}{2}, \frac{m}{2}, \frac{n}{2}\right)$.

Example 5. **Show that the points $(0, 7, 10)$, $(-1, 6, 6)$ and $(-4, 9, 6)$ are the vertices of a right angled isosceles triangle.** Competency Based Que

Sol. Let $A(0, 7, 10)$, $B(-1, 6, 6)$ and $C(-4, 9, 6)$ be the given points.

Then, $AB = \sqrt{(-1-0)^2 + (6-7)^2 + (6-10)^2}$ [using the distance formula]

$$= \sqrt{1+1+16} = \sqrt{18} = 3\sqrt{2} \text{ units}$$

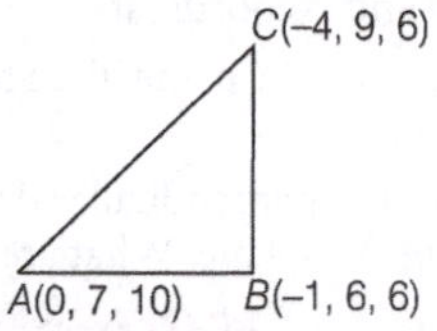

$$BC = \sqrt{(-4+1)^2 + (9-6)^2 + (6-6)^2}$$
$$= \sqrt{9+9+0}$$
$$= \sqrt{18} = 3\sqrt{2} \text{ units}$$

and $AC = \sqrt{(-4-0)^2 + (9-7)^2 + (6-10)^2}$

$$= \sqrt{16+4+16}$$

$\Rightarrow AC = \sqrt{36} = 6$ units ...(i)

Now, $AB^2 + BC^2 = (3\sqrt{2})^2 + (3\sqrt{2})^2 = 18 + 18 = 36$ units

$\therefore AB^2 + BC^2 = AC^2$ [from Eq. (i)]

Also, $AB = BC = 3\sqrt{2}$ units

Hence, ΔABC is a right angled isosceles triangle.

Example 6. **Show that the points $(-2, 6, -2)$, $(0, 4, -1)$, $(-2, 3, 1)$ and $(-4, 5, 0)$ are the vertices of a square.** Competency Based Que

Sol. Let $A(-2, 6, -2)$, $B(0, 4, -1)$, $C(-2, 3, 1)$ and $D(-4, 5, 0)$ be the given points.

$\therefore AB = \sqrt{(0+2)^2 + (4-6)^2 + (-1+2)^2}$

[using the distance formula]

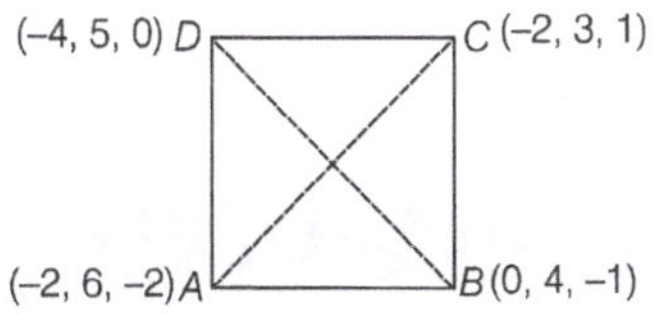

$= \sqrt{4+4+1} = \sqrt{9} = 3$ units

$BC = \sqrt{(-2-0)^2 + (3-4)^2 + (1+1)^2}$

$= \sqrt{4+1+4}$

$= \sqrt{9} = 3$ units

$CD = \sqrt{(-4+2)^2 + (5-3)^2 + (0-1)^2}$

$= \sqrt{4+4+1}$ units

$= \sqrt{9} = 3$

$AD = \sqrt{(-4+2)^2 + (5-6)^2 + (0+2)^2}$

$= \sqrt{4+1+4}$

$= \sqrt{9} = 3$ units

Here, $AB = BC = CD = DA$.

So, *ABCD* is a square or a rhombus.

Now, $AC = \sqrt{(-2+2)^2 + (3-6)^2 + (1+2)^2}$

$[\because \text{distance} = \sqrt{(x_2-x_1)^2 + (y_2-y_1)^2 + (z_2-z_1)^2}]$

$= \sqrt{0+9+9}$

$= \sqrt{18}$ units

and $BD = \sqrt{(-4-0)^2 + (5-4)^2 + (0+1)^2}$

$= \sqrt{16+1+1} = \sqrt{18}$ units

Since, diagonals, $AC = BD$

Hence, *ABCD* is a square.

Example 7. Verify that the points (3, – 2, 4), (1, 0, – 2) and (–1, 2, – 8) are collinear. Competency Based Que

Sol. Let $A(3, -2, 4)$, $B(1, 0, -2)$ and $C(-1, 2, -8)$ be the given points.

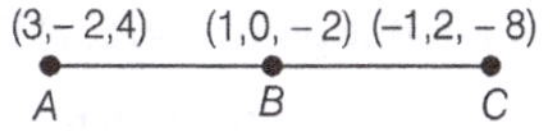

Then, $AB = \sqrt{(1-3)^2 + (0+2)^2 + (-2-4)^2}$ [using distance formula]

$= \sqrt{4+4+36} = \sqrt{44} = 2\sqrt{11}$ units

$BC = \sqrt{(-1-1)^2 + (2-0)^2 + (-8+2)^2}$ [using distance formula]

$= \sqrt{4+4+36} = \sqrt{44} = 2\sqrt{11}$ units

and $CA = \sqrt{(3+1)^2 + (-2-2)^2 + (4+8)^2}$ [using distance formula]

$= \sqrt{16+16+144} = \sqrt{176} = 4\sqrt{11}$ units

We observe that

$AB + BC = 2\sqrt{11} + 2\sqrt{11} = 4\sqrt{11} = CA$

Hence, *A*, *B* and *C* are collinear.

Quick Test 2

1. Find the distance between the following pairs of points.
(i) $(-1, 3, -4)$ and $(1, -3, 4)$
(ii) $(2, -1, 3)$ and $(-2, 1, 3)$

2. Find the distance from the origin to (6, 6, 7). NCERT Exemplar

3. How far apart are the points (2, 0, 0) and (-3, 0, 0)?

4. Find the equation of the set of points which are equidistant from the points (1, 2, 3) and (3, 2, –1). NCERT

5. Prove that the triangle formed by joining the three points whose coordinates are $A(1, 2, 3)$, $B(2, 3, 1)$ and $C(3, 1, 2)$ is an equilateral triangle. Competency Based Que

6. Show that $D(-1, 4, -3)$ is the circumcentre of ΔABC with vertices $A(3, 2, -5)$, $B(-3, 8, -5)$ and $C(-3, 2, 1)$. Competency Based Que

7. Are the points $A(3, 6, 9)$, $B(10, 20, 30)$ and $C(25, -41, 5)$, the vertices of a right angled triangle?

8. Show that ΔABC with vertices $A(0, 4, 1)$, $B(2, 3, -1)$ and $C(4, 5, 0)$ is right angled triangle. NCERT Exemplar

9. Show that the points $P(0, 7, 10)$, $Q(-1, 6, 6)$ and $R(-4, 9, 6)$ form a right angled isosceles triangle. NCERT Exemplar, Competency Based Que

10. Using distance formula, show that the points $A(1, -1, 3)$, $B(2, -4, 5)$ and $C(5, -13, 11)$ are collinear.

Answers

1. (i) $2\sqrt{26}$ units (ii) $2\sqrt{5}$ units 2. 11 units 3. 5 units 4. $x - 2z = 0$ 7. No

For Solutions, See QR Code on Pg. 324

Exam Pattern Questions

Objective Type Questions

Multiple Choice Questions

1. **In which octant do the point (−4, 5, 2) lie?**

(a) Octant I (b) Octant II
(c) Octant III (d) Octant IV

2. **A point lies on the X-axis. Which of the following represent the point?**

(a) $(0, 0, x)$ (b) $(0, x, 0)$
(c) $(x, 0, 0)$ (d) None of these

3. **Into how many octants does the coordinate planes divide the space?**

(a) 10 (b) 8 (c) 6 (d) 4

4. **Which of the following statement(s) is/are correct?** Competency Based Que

(a) The coordinates of the origin O are (0, 0, 0).
(b) The coordinates of any point on the X-axis will be as $(0, y, z)$.
(c) The coordinates of any point in the YZ-plane will be $(x, 0, 0)$.
(d) All of the above are correct.

5. **Which of the following statement(s) is/are correct?** Competency Based Que

(a) The X-axis and Y-axis taken together determine a plane known as XY-plane.
(b) The coordinates of a point in the XY-plane are of the form $(0, 0, z)$.
(c) Coordinate planes divide the space into six octants.
(d) All of the above are correct.

6. **In which plane the point $(0, y, z)$ will lie?** Competency Based Que

(a) XY-plane (b) YZ-plane
(c) ZX-plane (d) None of these

7. **The point, which lies in VI octant, is**

(a) (5, 6, 4) (b) (−5, 6, 4)
(c) (5, − 6, − 4) (d) (−5, 6, − 4)

8. **The Y-axis and Z-axis take together determine a plane known as**

(a) XY-plane (b) XZ-plane
(c) YZ-plane (d) None of these

9. **Distance of point (5, 4, 3) from the origin (0, 0, 0) is**

(a) $\sqrt{50}$ units (b) $\sqrt{5}$ units
(c) 5 units (d) 4 units

10. **The distance between the points (5, 4, 1) and (3, 2, 2) is**

(a) 2 units (b) 3 units
(c) 4 units (d) 5 units

11. **The perpendicular distance of the point (8, 6, 10) from Y-axis is** Competency Based Que

(a) 10 units (b) $\sqrt{41}$ units
(c) $10\sqrt{2}$ units (d) $2\sqrt{41}$ units

Assertion-Reason Based Questions

Directions (Q. Nos. 12-21) *In the questions given below are two statements labelled as Assertion (A) and Reason (R). In the context of the two statements, which one of the following is correct?*

(a) Both A and R are correct; R is the correct explanation of A.
(b) Both A and R are correct; R is not the correct explanation of A.
(c) A is correct; R is incorrect.
(d) R is correct; A is incorrect.

12. **Assertion** (A) Coordinates of any point on X-axis are $(x, 0, 0)$.

Reason (R) Point on X-axis has y and z coordinates are 0.

13. **Assertion** (A) The point (3, 2, 5) lies in $XOYZ$ octant.

Reason (R) Signs of coordinates of point (x, y, z) in $XOYZ$ is (+, +, +). Competency Based Que

14. **Assertion** (A) The point $(x, 0, z)$ lies in ZX-plane.

Reason (R) The point $(x, y, 0)$ lies in XY-plane.

15. **Assertion** (A) The coordinate plane divides the space into six parts called octants.

Reason (R) The point (4, − 1, 2) lies in the IV octant.

16. **Assertion** (A) The point (4, 0, 0) lies on the X-axis.

Reason (R) The point (0, 4, 0) lies on Z-axis.

17. **Assertion** (A) The distance between the points (−3, 3, 0) and (0, 1, − 4) is $\sqrt{29}$ units.

Reason (R) The distance between the points $P(x_1, y_1, z_1)$ and $Q(x_2, y_2, z_2)$ is given by

$$PQ = \sqrt{(x_2 - x_1)^2 + (y_2 - y_1)^2 + (z_2 - z_1)^2}.$$

Competency Based Que

18. **Assertion** (A) The distance between the points (1, 2, 4) and (3, 8, 8) is $2\sqrt{14}$.

Reason (R) The distance between the points (3, 8, 8) and (7, 20, 16) is $2\sqrt{14}$.

19. **Assertion** (A) Three points $A(-2, 4, 0)$, $B(1, 2, -1)$ and $C(-1, 1, 2)$ form the vertices of an equilateral triangle.

Reason (R) In an equilateral triangle, all three sides of a triangle are equal.

20. **Assertion** (A) Points $A(1, 2, 3)$, $B(-1, -2, -1)$ and $C(2, 3, 2)$ and $D(4, 7, 6)$ are the vertices of a parallelogram $ABCD$.

Reason (R) In a parallelogram, opposite sides are equal and diagonals are unequal. **Competency Based Que**

21. **Assertion** (A) Three points $P(1, 2, 3)$, $Q(-1, -1, -1)$ and $R(3, 5, 7)$ are collinear.

Reason (R) Three points A, B and C are said to be collinear, if $AB + BC = AC$. **Competency Based Que**

Case Study Based Questions

22. Aman and his brother were walking in a large park. They saw a kite flying in the sky. The position of kite, Aman and his brother are (20, 30, 10), (4, 3, 7) and (5, 3, 7), respectively. **Competency Based Que**

On the basis of above information, answer the following questions.

(i) The distance between Aman and kite is

(a) 41.32 units (b) 31.52 units
(c) 38.32 units (d) 40.39 units

(ii) The distance between Aman's brother and kite is

(a) 31.30 units (b) 29.00 units
(c) 38.43 units (d) 31.03 units

(iii) The coordinates of Aman lie in

(a) II octant
(b) I octant
(c) III octant
(d) IV octant

(iv) The coordinates of points in the XY-plane are of the form.

(a) $(0, 0, z)$ (b) $(x, y, 0)$
(c) $(x, 0, y)$ (d) $(0, x, y)$

23. Utkarsh and his friends went camping for 2 nights and 3 days. There they setup a tent, which is triangular in shape. The vertices of the tent are $A(4, 5, 9)$, $B(3, 2, 5)$, $C(5, 2, 5)$, $D(-3, 2, -5)$ and $E(-4, 5, -9)$, respectively.

The vertex A is tied up by the rope at the ends F and G and the vertex E is tied up at the ends I and H. **Competency Based Que**

On the basis of above information, answer the following questions.

(i) If M denotes the position of their bags inside the tent and it is just in middle of the vertices B and D, then find the coordinates of M and the length AE.

(ii) Find the octants in which points A, B, C, D and E will lie.

(iii) Find the equation of the set of points which are equidistant from the points $(-1, 2, 3)$ and $(3, 2, -1)$.

Subjective Questions

Very Short Answer Type Questions

1. Name the octant in which the point $(-6, -5, 8)$ lies.

2. A point is in the ZX-plane. What can you say about its y-coordinate? **Competency Based Que**

3. Find the distance of point (3, 2, 5) from the YZ-plane.

4. Find the distance between the points $(-2, 5, 0)$ and $(4, 7, 2)$.

Short Answer Type Questions

5. Name the octants in which the following points lie $(3, 2, 1)$, $(3, -2, 4)$, $(5, -2, -5)$, $(-5, 2, -4)$, $(-5, 2, 4)$, $(-6, -1, 3)$, $(7, -4, -2)$.

6. Find the equation of the set of the points P such that its distances from the points $A(3, 4, -5)$ and $B(-2, 1, 4)$ are equal.

7. The centroid of a triangle ABC is at the point $(1, 1, 1)$. If the coordinates of A and B are $(3, -5, 7)$ and $(-1, 7, -6)$, respectively, then find the coordinates of the point C. **Competency Based Que**

Long Answer Type Questions

8. Show that the points $(0, 4, 1)$, $(2, 3, -1)$, $(4, 5, 0)$ and $(2, 6, 2)$ are the vertices of a square.

9. Prove that the points $(5, 3, 2)$, $(3, 2, 5)$ and $(2, 5, 3)$ are the vertices of an equilateral triangle. **Competency Based Que**

10. Show that the points $(a, b, c), (b, c, a)$ and (c, a, b) are the vertices of an equilateral triangle. **Competency Based Que**

11. Show that the points $(-1, -6, 10), (1, -3, 4), (-5, -1, 1)$ and $(-7, -4, 7)$ are the vertices of a rhombus.

12. If the origin is the centroid of the ΔPQR with vertices $P(2a, -2, 6), Q(-4, 3b, -10)$ and $R(8, 14, 2c)$, then find the values of a, b and c. Also find the length of PQ. **Competency Based Que**

13. Using distance formula, check whether the points $(2, -1, 0), (-2, 3, -2)$ and $(0, 3, 0)$ are collinear or not.

Hints & Solutions

Objective Type Questions

1. (b) Given point is $(-4, 5, 2)$.
Here, the x-coordinate is negative but y and z-coordinates are positive. Therefore, $(-4, 5, 2)$ lies in octant II.

2. (c) At X-axis, y and z-coordinates are zero.

3. (b) The coordinate planes divide the three dimensional spaces into eight octants.

4. (a) The coordinates of the origin O are $(0, 0, 0)$.
The coordinates of any point on the X-axis will be as $(x, 0, 0)$.
The coordinates of any point in the YZ-plane will be $(0, y, z)$.

5. (a) The X-axis and Y-axis taken together determine a plane known as XY-plane.
The coordinates of a point in the XY-plane are of the form $(x, y, 0)$.
The coordinate planes divide the space into eight octants.

6. (b) The point $(0, y, z)$ will lie in YZ-plane.

7. (d) The point which lies in VI octant is $(-5, 6, -4)$ because the x and z-coordinates are negative.

8. (c) The Y-axis and Z-axis represent the YZ-plane.

9. (a) Distance of point $(5, 4, 3)$ from origin $(0, 0, 0)$ is
$$\sqrt{5^2+4^2+3^2} = \sqrt{25+16+9} = \sqrt{50} \text{ units}$$

10. (b) Distance between points $(5, 4, 1)$ and $(3, 2, 2)$ is given by
$$\sqrt{(5-3)^2+(4-2)^2+(1-2)^2} = \sqrt{4+4+1}$$
$$= \sqrt{9} = 3 \text{ units}$$

11. (d) Let L be the foot of perpendicular from point $P(8, 6, 10)$ on Y-axis.
Then, the coordinates of L are $(0, 6, 0)$.
Now, the distance between the points $L(0, 6, 0)$ and $P(8, 6, 10)$ is
$$\sqrt{(8-0)^2+(6-6)^2+(10-0)^2} \text{ [using distance formula]}$$
$$= \sqrt{64+0+100} = \sqrt{164} = 2\sqrt{41} \text{ units}$$

12. (a) The coordinates of any point on X-axis is given by $(x, 0, 0)$.

13. (a) Since, all the coordinates of the point $(3, 2, 5)$ are positive, therefore it lies in $XOYZ$ octant.

14. (b) The coordinates of point in ZX-plane are $(x, 0, z)$.
The coordinates of point in XY-plane are $(x, y, 0)$.

15. (d) The coordinate plane divides the space into eight parts called octants.
Assertion is incorrect.
Since, y-coordinate is negative in point $(4, -1, 2)$, therefore it lies in IV octant.
Reason is correct.

16. (c) The point $(4, 0, 0)$ lies on the X-axis.
Assertion is correct.
The point $(0, 4, 0)$ lies on the Y-axis.
Reason is incorrect.

17. (a) Let the points be $P(-3, 3, 0)$ and $Q(0, 1, -4)$.
$$PQ = \sqrt{(0+3)^2+(1-3)^2+(-4-0)^2}$$
$$= \sqrt{9+4+16} = \sqrt{29}$$

18. (c) Distance between the points $(1, 2, 4)$ and $(3, 8, 8)$ is
$$\sqrt{(3-1)^2+(8-2)^2+(8-4)^2} = \sqrt{4+36+16} = 2\sqrt{14}$$
Assertion is correct.
Distance between the points $(3, 8, 8)$ and $(7, 20, 16)$ is
$$\sqrt{(7-3)^2+(20-8)^2+(16-8)^2} = \sqrt{16+144+64}$$
$$= 4\sqrt{14}$$
Reason is incorrect.

19. (a) Given points are $A(-2, 4, 0), B(1, 2, -1)$ and $C(-1, 1, 2)$.
$$AB = \sqrt{(1+2)^2+(2-4)^2+(-1-0)^2}$$
$$= \sqrt{9+4+1} = \sqrt{14}$$
$$BC = \sqrt{(-1-1)^2+(1-2)^2+(2+1)^2}$$
$$= \sqrt{4+1+9} = \sqrt{14}$$
and $$AC = \sqrt{(-1+2)^2+(1-4)^2+(2-0)^2}$$
$$= \sqrt{1+9+4} = \sqrt{14}$$

Here, $AB = BC = AC$

Hence, the given points are the vertices of an equilateral triangle.

20. (a) Given points are $A(1, 2, 3)$, $B(-1, -2, -1)$, $C(2, 3, 2)$ and $D(4, 7, 6)$.

$$AB = \sqrt{(-1-1)^2 + (-2-2)^2 + (-1-3)^2}$$
$$= \sqrt{4+16+16} = 6$$
$$BC = \sqrt{(2+1)^2 + (3+2)^2 + (2+1)^2}$$
$$= \sqrt{9+25+9} = \sqrt{43}$$
$$CD = \sqrt{(4-2)^2 + (7-3)^2 + (6-2)^2}$$
$$= \sqrt{4+16+16} = 6$$
$$AD = \sqrt{(4-1)^2 + (7-2)^2 + (6-3)^2}$$
$$= \sqrt{9+25+9} = \sqrt{43}$$
$$AC = \sqrt{(2-1)^2 + (3-2)^2 + (2-3)^2}$$
$$= \sqrt{1+1+1} = \sqrt{3}$$

and $$BD = \sqrt{(4+1)^2 + (7+2)^2 + (6+1)^2}$$
$$= \sqrt{25+81+49} = \sqrt{155}$$

Here, $AB = CD, BC = AD$ and $AC \neq BD$.

Here, the given points are the vertices of a parallelogram $ABCD$.

21. (a) Given points are $P(1, 2, 3)$, $Q(-1, -1, -1)$ and $R(3, 5, 7)$.

$$PQ = \sqrt{(-1-1)^2 + (-1-2)^2 + (-1-3)^2}$$
$$= \sqrt{(-2)^2 + (-3)^2 + (-4)^2} = \sqrt{29},$$
$$QR = \sqrt{(3+1)^2 + (5+1)^2 + (7+1)^2}$$
$$= \sqrt{4^2 + 6^2 + 8^2} = \sqrt{116} = 2\sqrt{29}$$

and $$PR = \sqrt{(3-1)^2 + (5-2)^2 + (7-3)^2}$$
$$= \sqrt{2^2 + 3^2 + 4^2} = \sqrt{29}$$

Here, $PQ + QR = PR$

Hence, the three points are collinear.

22. (i) (b) Required distance

$$= \sqrt{(20-4)^2 + (30-3)^2 + (10-7)^2}$$
$$= \sqrt{16^2 + 27^2 + 3^2} = \sqrt{994} = 31.52 \text{ units}$$

(ii) (d) Required distance

$$= \sqrt{(20-5)^2 + (30-3)^2 + (10-7)^2}$$
$$= \sqrt{15^2 + 27^2 + 3^2}$$
$$= \sqrt{255+729+9} = \sqrt{963} = 31.03 \text{ units}$$

(iiii) (b) Since, in the coordinate of Aman (4, 3, 7), all are positive, then the coordinates lie in the I quadrant.

(iv) (b) For XY-plane, $z = 0$

$\Rightarrow$ The coordinates are of the form $(x, y, 0)$.

23. (i) As M is the middle point of $B(3, 2, 5)$ and $D(-3, 2, -5)$.

$\therefore$ The coordinates of M are

$$\left(\frac{3-3}{2}, \frac{2+2}{2}, \frac{5-5}{2}\right) = (0, 2, 0)$$

The length AE is

$$\sqrt{(-4-4)^2 + (5-5)^2 + (-9-9)^2} = \sqrt{64+0+324}$$
$$= \sqrt{388} = 2\sqrt{97} \text{ units}$$

(ii)

Point	Octant
$A(4,5,9)$	I (all coordinates are positive)
$B(3,2,5)$	I (all coordinates are positive)
$C(5,2,5)$	I (all coordinates are positive)
$D(-3,2,-5)$	VI (x ane z-coordinates are negative)
$E(-4,5,-9)$	VI (x and z-coordinates are negative)

(iii) Let $P(x, y, z)$ be the point equidistant from two points $A(-1, 2, 3)$ and $B(3, 2, -1)$.

Thus, $PA = PB$

On squaring both sides, we get

$$PA^2 = PB^2$$
$$\Rightarrow \quad (x+1)^2 + (y-2)^2 + (z-3)^2 = (x-3)^2 + (y-2)^2 + (z+1)^2$$
$$\Rightarrow \quad x^2 + 2x + 1 + y^2 - 4y + 4 + z^2 + 9 - 6z = x^2 + 9 - 6x + y^2 + 4 - 4y + z^2 + 1 + 2z$$
$$\Rightarrow \quad 2x - 4y - 6z + 14 = -6x - 4y + 2z + 14$$
$$\Rightarrow \quad 2x - 6z + 6x - 2z = 0$$
$$\Rightarrow \quad 8x - 8z = 0 \Rightarrow x - z = 0$$

Subjective Questions

1. Point $(-6, -5, 8)$ has x and y-coordinates negative, so it lies in III octant.

2. In the ZX-plane, y-coordinate will be zero.

3. Distance of point (3, 2, 5) from the YZ-plane is the value of the x-coordinate i.e. 3 units.

4. Let points be $A(-2, 5, 0)$ and $B(4, 7, 2)$.

Then, $$AB = \sqrt{(4+2)^2 + (7-5)^2 + (2-0)^2}$$
$$= \sqrt{36+4+4} = \sqrt{44}$$

5.

Point	Octant
(3, 2, 1)	I (all the coordinates are positive)
(3, −2, 4)	IV (y-coordinate is negagive)
(5, −2, −5)	VIII (y and z-coordinates are negative)
(−5, 2, −4)	VI (x and z-coordinates are negative)
(−5, 2, 4)	II (x-coordinate is negative)
(−6, −1, 3)	III (x and y-coordinates are negative)
(7, −4, −2)	VIII (y and z-coordinates are negative)

6. Let $P(x, y, z)$ be any point such that $PA = PB$

Now, $\sqrt{(x-3)^2+(y-4)^2+(z+5)^2}$

$$= \sqrt{(x+2)^2+(y-1)^2+(z-4)^2}$$

$$\Rightarrow \quad (x-3)^2+(y-4)^2+(z+5)^2 = (x+2)^2+(y-1)^2+(z-4)^2$$

$$\therefore \quad 10x+6y-18z-29=0$$

7. Let the coordinates of C be (x, y, z) and the coordinates of the centroid G be $(1, 1, 1)$.

Then, $\dfrac{x+3-1}{3} = 1$ i.e. $x = 1$

$\dfrac{y-5+7}{3} = 1$ i.e. $y = 1$

$\dfrac{z+7-6}{3} = 1$ i.e. $z = 2$

Hence, coordinates of C are $(1, 1, 2)$.

8. Let $A(0, 4, 1)$, $B(2, 3, -1)$, $C(4, 5, 0)$ and $D(2, 6, 2)$ be the given points.

$$AB = \sqrt{(2-0)^2+(3-4)^2+(-1-1)^2} = \sqrt{9} = 3,$$

$$BC = \sqrt{(4-2)^2+(5-3)^2+(0+1)^2} = \sqrt{9} = 3,$$

$$CD = \sqrt{(2-4)^2+(6-5)^2+(2-0)^2} = \sqrt{9} = 3$$

and $DA = \sqrt{(0-2)^2+(4-6)^2+(1-2)^2} = \sqrt{9} = 3$

Since, $AB = BC = CD = DA$. So, $ABCD$ is a square or a rhombus.

Now, $AC = \sqrt{(4-0)^2+(5-4)^2+(0-1)^2} = \sqrt{18}$

and $BD = \sqrt{(2-2)^2+(6-3)^2+(2+1)^2} = \sqrt{18}$

Since, $AC = BD$

Hence, $ABCD$ is a square.

9. Let $A(5, 3, 2)$, $B(3, 2, 5)$ and $C(2, 5, 3)$ be the given points.

$$AB = \sqrt{(3-5)^2+(2-3)^2+(5-2)^2} = \sqrt{14},$$

$$BC = \sqrt{(2-3)^2+(5-2)^2+(3-5)^2} = \sqrt{14}$$

and $CA = \sqrt{(5-2)^2+(3-5)^2+(2-3)^2} = \sqrt{14}$

Since, $AB = BC = CA$

Hence, ABC is an equilateral triangle.

10. Let $A = (a, b, c)$, $B = (b, c, a)$ and $C = (c, a, b)$

$$AB = \sqrt{(b-a)^2+(c-b)^2+(a-c)^2},$$

$$BC = \sqrt{(c-b)^2+(a-c)^2+(b-a)^2}$$

and $CA = \sqrt{(a-c)^2+(b-a)^2+(c-b)^2}$

Here, $AB = BC = CA$

Hence, the given points are the vertices of an equilateral triangle.

11. Let $A(-1, -6, 10)$, $B(1, -3, 4)$, $C(-5, -1, 1)$ and $D(-7, -4, 7)$ be the given points.

$$AB = \sqrt{(1+1)^2+(-3+6)^2+(4-10)^2} = 7,$$

$$BC = \sqrt{(-5-1)^2+(-1+3)^2+(1-4)^2} = 7,$$

$$CD = \sqrt{(-7+5)^2+(-4+1)^2+(7-1)^2} = 7,$$

$$AD = \sqrt{(-7+1)^2+(-4+6)^2+(7-10)^2} = 7,$$

$$AC = \sqrt{(-5+1)^2+(-1+6)^2+(1-10)^2} = \sqrt{122}$$

and $BD = \sqrt{(-7-1)^2+(-4+3)^2+(7-4)^2}$

$$= \sqrt{74}$$

Here, $AB = BC = CD = AD$ and $AC \neq BD$.

Hence, the given points are the vertices of a rhombus.

12. Given centroid of the triangle PQR is origin i.e. $(0, 0, 0)$.

Then, $\dfrac{2a-4+8}{3} = 0 \Rightarrow 2a = -4 \Rightarrow a = -2,$

$\dfrac{-2+3b+14}{3} = 0 \Rightarrow 3b = -12 \Rightarrow b = -4$

and $\dfrac{6-10+2c}{3} = 0 \Rightarrow 2c = 4 \Rightarrow c = 2$

Thus, the values of a, b and c are -2, -4 and 2, respectively.

Therefore, the coordinates of points P and Q are $P(2a, -2, 6) \equiv (-4, -2, 6)$ and $Q(-4, 3b, -10) \equiv (-4, -12, -10)$

$$\therefore PQ = \sqrt{(-4+4)^2+(-12+2)^2+(-10-6)^2}$$

[using distance formula]

$$PQ = \sqrt{0+100+256} = \sqrt{356} = 2\sqrt{89} \text{ units.}$$

Hence, the length of PQ is $2\sqrt{89}$ units.

13. Let $A = (2, -1, 0)$, $B = (-2, 3, -2)$ and $C = (0, 3, 0)$

$$AB = \sqrt{(-2-2)^2+(3+1)^2+(-2-0)^2} = 6,$$

$$BC = \sqrt{(0+2)^2+(3-3)^2+(0+2)^2} = 2\sqrt{2}$$

and $AC = \sqrt{(0-2)^2+(3+1)^2+(0-0)^2} = 2\sqrt{5}$

Here, $AB + BC \neq AC$

Hence, the given points are not collinear.

QR Code for Quick Test 1

QR Code for Quick Test 2

Mind Map

Introduction to Three Dimensional Geometry

Coordinates of a Point in Space

Let P be a point in space and $OA = x$, $OB = y$ and $OC = z$. These three numbers taken in order are called coordinates of a point P and written as $P(x, y, z)$.

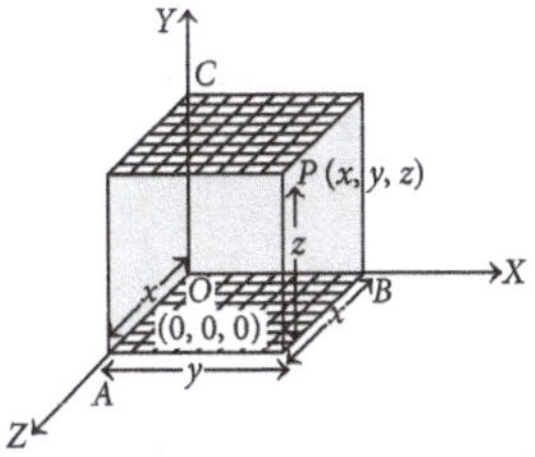

Thus, ***x*-coordinate of a point *P*** = Length of perpendicular from P to YZ-plane with proper sign.

y*-coordinate of a point *P = Length of perpendicular from P to ZX-plane with proper sign.

z*-coordinate of a point *P = Length of perpendicular from P to XY-plane with proper sign.

Sign of Coordinates

The sign of coordinates of the points in the octants in which the space is divided are given in the following table.

Octants / Coordinates	I	II	III	IV	V	VI	VI	VIII
	$XOYZ$	$X'OYZ$	$X'OY'Z$	$XOY'Z$	$XOYZ'$	$X'OYZ'$	$X'OY'Z'$	$XOY'Z'$
x	+	−	−	+	+	−	−	+
y	+	+	−	−	+	+	−	−
z	+	+	+	+	−	−	−	−

Coordinates of a Point on Coordinate Axes

Let $P(x, y, z)$ be any point in space and we draw three perpendicular lines from P to the coordinate axes meet at points A, B and C.

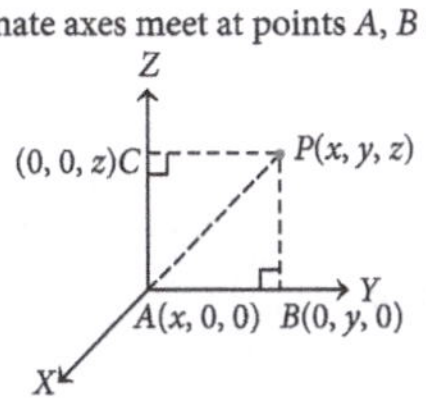

Thus, coordinates of a point on
(i) X–axis will be of the from $(x, 0, 0)$.
(ii) Y–axis will be of the from $(0, y, 0)$.
(iii) Z–axis will be of the from $(0, 0, z)$.

Coordinates of a Point in Coordinate Planes

Let $P(x, y, z)$ be any point in space and we draw three perpendicular lines from P to the coordinate planes XY, YZ and ZX, respectively.

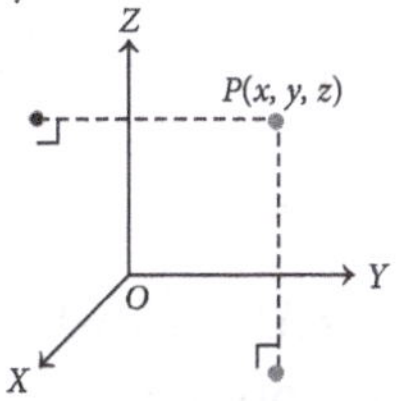

Then, the coordinates of a point in
(i) XY-plane will be of the form $(x, y, 0)$.
[$\therefore$ in XY-plane, the z-coordinate of a point is zero]
(ii) YZ-plane will be of the form $(0, y, z)$ and
(iii) ZX-plane will be of the form $(x, 0, z)$

Distance between Two Points

The distance between two points $P(x_1, y_1, z_1)$ and $Q(x_2, y_2, z_2)$ is given by

$$PQ = \sqrt{(x_2 - x_1)^2 + (y_2 - y_1)^2 + (z_2 - z_1)^2}$$

or $PQ = \sqrt{(x_1 - x_2)^2 + (y_1 - y_2)^2 + (z_1 - z_2)^2}$

If P is the origin and $Q(x, y, z)$ is a point in space, then

$$PQ = \sqrt{(x - 0)^2 + (y - 0)^2 + (z - 0)^2}$$

or $PQ = \sqrt{x^2 + y^2 + z^2}$

Condition for Collinearity

Three points A, B and C are said to be collinear, if $AB + BC = AC$

Coordinate Axes and Coordinate Plane in Three Dimensional Space

Coordinate Axes

Let XOX', YOY' and ZOZ' be the three mutually perpendicular lines, intersecting at O.

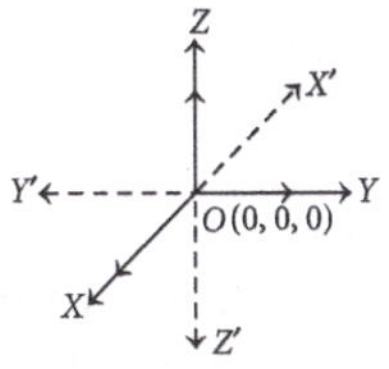

The point $O(0, 0, 0)$ is called the **origin** and the lines (XOX', YOY' and ZOZ') called **rectangular coordinate axes** say X, Y and Z, respectively.

Coordinate Planes

The three coordinate axes define three mutually perpendicular planes XOY, YOZ and ZOX (or XY, YZ and ZX) called **coordinate planes** which divide the space into eight parts called **octants**.

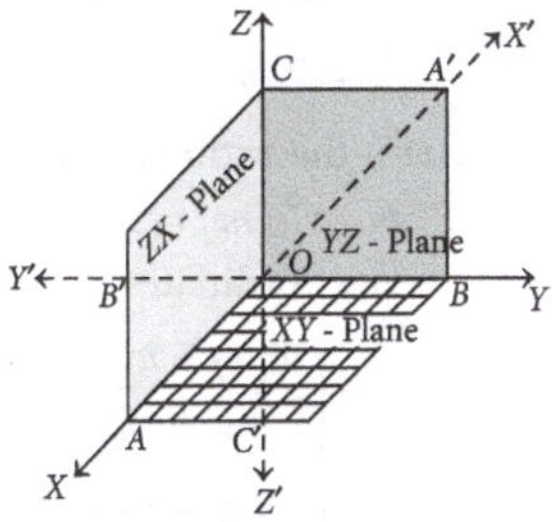

- XOY is called the XY-plane.
- YOZ is called the YZ-plane.
- ZOX is called the ZX-plane.

Topics to be Assessed Formatively

Section Formulae

Let $A(x_1, y_1, z_1)$ and $B(x_2, y_2, z_2)$ be the end points of a line segment AB and C be any point on AB which divide AB in the ratio $m_1 : m_2$.

I. For Internal Ratio

If C divides AB internally in the ratio $m_1 : m_2$, then the coordinates of C are

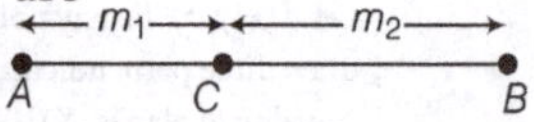

$$\left(\frac{m_1x_2 + m_2x_1}{m_1 + m_2}, \frac{m_1y_2 + m_2y_1}{m_1 + m_2}, \frac{m_1z_2 + m_2z_1}{m_1 + m_2}\right)$$

II. For External Ratio

If C divides AB externally in the ratio $m_1 : m_2$, then the coordinates of C are

$$\left(\frac{m_1x_2 + (-m_2)x_1}{m_1 + (-m_2)}, \frac{m_1y_2 + (-m_2)y_1}{m_1 + (-m_2)}, \frac{m_1z_2 + (-m_2)z_1}{m_1 + (-m_2)}\right)$$

$$= \left(\frac{m_1x_2 - m_2x_1}{m_1 - m_2}, \frac{m_1y_2 - m_2y_1}{m_1 - m_2}, \frac{m_1z_2 - m_2z_1}{m_1 - m_2}\right)$$

EXAMPLE 1 Find the coordinates of the point which divides the line segment joining the points (–2, 3, 5) and (1, – 4, 6) in the ratio

(i) 2 : 3 internally (ii) 2 : 3 externally

Sol. (i) Let $P(x, y, z)$ be any point which divides the line segment joining points $A(-2, 3, 5)$ and $B(1, -4, 6)$ in the ratio 2 : 3 internally.

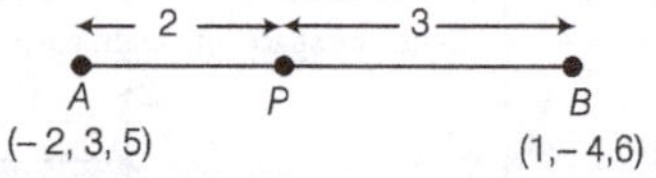

Here, the ratio is 2:3.

$\therefore$ $m_1 = 2, m_2 = 3$

Then, coordinates of point P

$$= \left[\frac{m_1x_2 + m_2x_1}{m_1 + m_2}, \frac{m_1y_2 + m_2y_1}{m_1 + m_2}, \frac{m_1z_2 + m_2z_1}{m_1 + m_2}\right]$$

$$= \left[\frac{2\times1 + 3\times(-2)}{2+3}, \frac{2(-4) + 3(3)}{2+3}, \frac{2(6) + 3(5)}{2+3}\right]$$

$$= \left(\frac{2-6}{5}, \frac{-8+9}{5}, \frac{12+15}{5}\right) = \left(\frac{-4}{5}, \frac{1}{5}, \frac{27}{5}\right)$$

(ii) Let $P(x, y, z)$ be any point which divides the line segment joining points $A(-2, 3, 5)$ and $B(1, -4, 6)$ in the ratio 2 : 3 externally.

Here, the ratio is 2:3.

$\therefore$ $m_1 = 2, m_2 = 3$

Then, coordinates of point P

$$= \left[\left(\frac{m_1x_2 - m_2x_1}{m_1 - m_2}, \frac{m_1y_2 - m_2y_1}{m_1 - m_2}, \frac{m_1z_2 - m_2z_1}{m_1 - m_2}\right)\right]$$

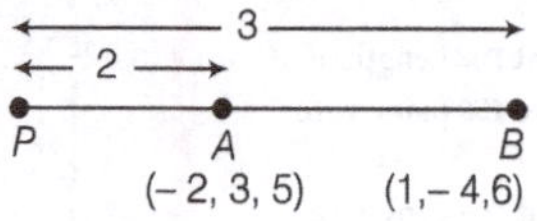

$$= \left[\frac{2(1) - (3)(-2)}{2-(3)}, \frac{2(-4) - (3)\times3}{2-(3)}, \frac{2(6) - (3)(5)}{2-(3)}\right]$$

$$= \left(\frac{2+6}{2-3}, \frac{-8-9}{2-3}, \frac{12-15}{2-3}\right)$$

$$= \left(\frac{8}{-1}, \frac{-17}{-1}, \frac{-3}{-1}\right) = (-8, 17, 3)$$

Applications of section formulae

Coordinates of the Mid-Point

If C is the mid-point of AB, then $m_1 : m_2 = 1:1$ so that

$$x = \frac{x_1 + x_2}{2}, y = \frac{y_1 + y_2}{2} \text{ and } z = \frac{z_1 + z_2}{2}$$

These are the coordinates of the mid-point of the line segment joining $A(x_1, y_1, z_1)$ and $B(x_2, y_2, z_2)$.

TEACHER TIPS

The coordinates of the point C which divides AB in the ratio $k:1$, i.e. $\frac{m_1}{m_2} = \frac{k}{1}$, are given as

$$\left(\frac{kx_2 + x_1}{k+1}, \frac{ky_2 + y_1}{k+1}, \frac{kz_2 + z_1}{k+1}\right)$$

EXAMPLE 2 Three consecutive vertices of a parallelogram $ABCD$ are $A(6, -2, 4)$, $B(2, 4, -8)$ and $C(-2, 2, 4)$. Find the coordinates of the fourth vertex.

Sol. Let the coordinates of the fourth vertex of a parallelogram be $D(x, y, z)$.

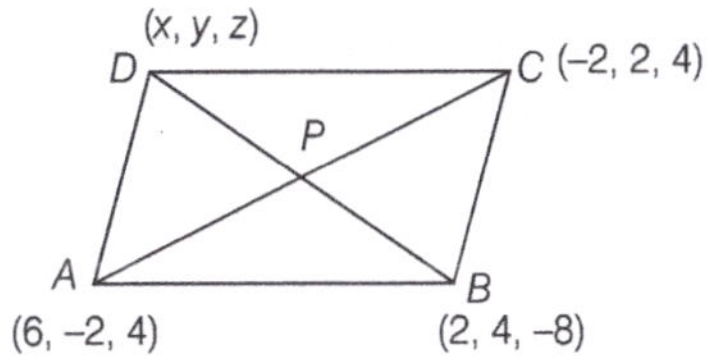

We know that the diagonals of a parallelogram bisect each other.

$\therefore$ Mid-point of AC = Mid-point of BD

$$\Rightarrow \left(\frac{6-2}{2}, \frac{-2+2}{2}, \frac{4+4}{2}\right) = \left(\frac{x+2}{2}, \frac{y+4}{2}, \frac{z-8}{2}\right)$$

$$\Rightarrow \quad (2, 0, 4) = \left(\frac{x+2}{2}, \frac{y+4}{2}, \frac{z-8}{2}\right)$$

On comparing the coordinates both sides, we get

$$\frac{x+2}{2} = 2, \quad \frac{y+4}{2} = 0 \text{ and } \frac{z-8}{2} = 4$$

$\Rightarrow$ $x + 2 = 4$, $y + 4 = 0$ and $z - 8 = 8$

$\Rightarrow$ $x = 2$, $y = -4$ and $z = 16$

Hence, the coordinates of fourth vertex D of a parallelogram are $(2, -4, 16)$.

EXAMPLE 3 Verify that (–1, 2, 1), (1, –2, 5), (4, –7, 8) and (2, – 3, 4) are the vertices of a parallelogram.

Sol. Let $A(-1, 2, 1)$, $B(1, -2, 5)$, $C(4, -7, 8)$ and $D(2, -3, 4)$ be the vertices of a quadrilateral $ABCD$.

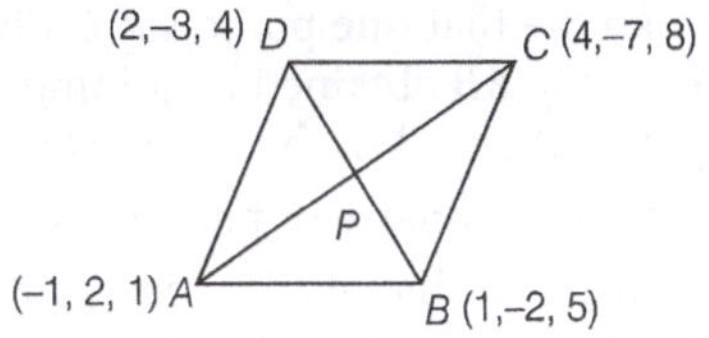

Then, mid-point of AC

$$= \left(\frac{-1+4}{2}, \frac{2-7}{2}, \frac{1+8}{2}\right) = \left(\frac{3}{2}, \frac{-5}{2}, \frac{9}{2}\right)$$

$$\left[\because \text{coordinates of mid-point } = \left(\frac{x_1+x_2}{2}, \frac{y_1+y_2}{2}, \frac{z_1+z_2}{2}\right)\right]$$

and mid- point of BD $= \left(\frac{1+2}{2}, \frac{-2-3}{2}, \frac{5+4}{2}\right)$

$$= \left(\frac{3}{2}, \frac{-5}{2}, \frac{9}{2}\right)$$

Here, mid-points of both the diagonals are same i.e. they bisect each other. Hence, $ABCD$ is a parallelogram.

EXAMPLE 4 Find the ratio in which the line segment joining the points (4, 8, 10) and (6, 10, –8) is divided by the *YZ*-plane.

Sol. Let YZ-plane divides the line segment joining the points $A(4, 8, 10)$ and $B(6, 10, -8)$ at $P(x, y, z)$ in the ratio $k:1$.

Then, the coordinates of $P = \left(\frac{4+6k}{k+1}, \frac{8+10k}{k+1}, \frac{10-8k}{k+1}\right)$

$$\left[\because \text{coordinates of internal ratio, } \left(\frac{m_2x_1+m_1x_2}{m_1+m_2}, \frac{m_2y_1+m_1y_2}{m_1+m_2}, \frac{m_2z_1+m_1z_2}{m_1+m_2}\right)\right]$$

Since, P lies on the YZ-plane, so its x-coordinate is zero,

$$\therefore \quad \frac{4+6k}{k+1} = 0 \Rightarrow k = -\frac{2}{3}$$

Therefore, YZ-plane divides AB externally in the ratio $2:3$.

EXAMPLE 5 A point *R* with *x*-coordinate 4 lies on the line segment joining the points *P*(2, – 3, 4) and *Q*(8, 0, 10). Find the coordinates of the point *R*.

Sol. Let the point $R(x, y, z)$ divides PQ in the ratio $k:1$.

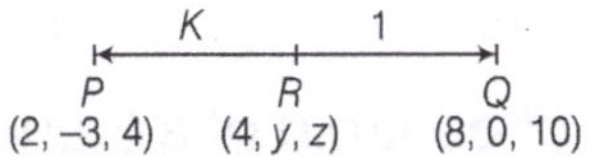

Given, x-coordinate of $R = 4$.

$$\therefore \quad \frac{k \times 8 + 1 \times 2}{k+1} = 4 \Rightarrow \frac{8k+2}{k+1} = 4$$

$$\Rightarrow \quad 8k + 2 = 4k + 4 \Rightarrow 8k - 4k = 4 - 2$$

$$\Rightarrow \quad 4k = 2 \Rightarrow k = \frac{2}{4} = \frac{1}{2}$$

So, $\quad k:1 = \frac{1}{2}:1$ or $1:2$.

Hence, the point R divides PQ internally in the ratio $1:2$.

Therefore, y-coordinate of R

$$= \left(\frac{1 \times 0 + 2 \times (-3)}{1+2}\right) = \left(\frac{-6}{3}\right) = -2$$

and z-coordinate of R

$$= \left(\frac{1 \times 10 + 2 \times 4}{1+2}\right) = \left(\frac{10+8}{3}\right) = \frac{18}{3} = 6$$

Hence, coordiantes of point R are $(4, -2, 6)$.

EXAMPLE 6 Find the ratio in which the join of the points *P*(2, – 1, 3) and *Q*(4, 3, 1) is divided by the point $R\left(\frac{20}{7}, \frac{5}{7}, \frac{15}{7}\right)$.

Sol. Given points are $P(2, -1, 3)$ and $Q(4, 3, 1)$.

Let the required ratio be $\lambda:1$.

Then, the coordinates of R are $\left(\frac{4\lambda+2}{\lambda+1}, \frac{3\lambda-1}{\lambda+1}, \frac{\lambda+3}{\lambda+1}\right)$

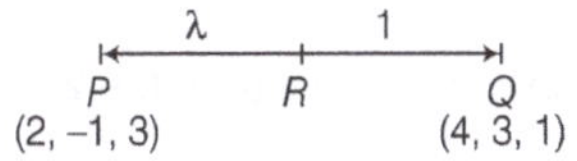

But, the coordinates of R are $\left(\frac{20}{7}, \frac{5}{7}, \frac{15}{7}\right)$. [given]

$\therefore \quad \frac{4\lambda + 2}{\lambda + 1} = \frac{20}{7}$ [$\because$ x-coordinate of R]

$\Rightarrow \quad 28\lambda + 14 = 20\lambda + 20 \Rightarrow 8\lambda = 6 \Rightarrow \lambda = \frac{3}{4}$

So, the required ratio is $\frac{3}{4}:1$, i.e. $3:4$.

Coordinates of the Point which Trisect the Line Segment

If R_1 and R_2 trisect the line segment AB, then

$$AR_1 = R_1R_2 = R_2B$$

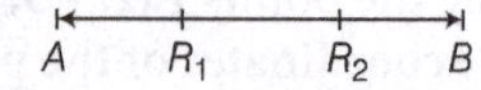

Thus, for finding the coordinates of R_1, take ratio $1:2$ and for finding the coordinates of R_2 take ratio $2:1$.

Coordinates of centroid of a triangle

If the vertices of a ΔABC are $A(x_1, y_1, z_1), B(x_2, y_2, z_2)$ and $C(x_3, y_3, z_3)$, then the centroid of a ΔABC is

$$G\left(\frac{x_1 + x_2 + x_3}{3}, \frac{y_1 + y_2 + y_3}{3}, \frac{z_1 + z_2 + z_3}{3}\right)$$

EXAMPLE 7 **Find the coordinates of the points which trisect the line segment joining the points $P(4, 2, -6)$ and $Q(10, -16, 6)$.**

Sol. Let R_1 and R_2 be two points which trisect the joining of P and Q. Then, $PR_1 = R_1R_2 = R_2Q$

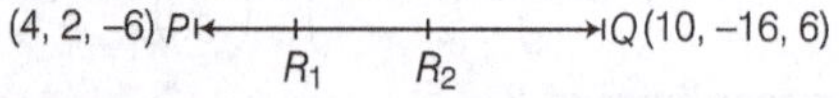

Point R_1 divides the join of PQ in the ratio $1:2$.

$\therefore$ Coordinates of R_1

$$= \left(\frac{1 \times 10 + 2 \times 4}{1 + 2}, \frac{1 \times (-16) + 2 \times 2}{1 + 2}, \frac{1 \times 6 + 2 \times (-6)}{1 + 2}\right)$$

$$= \left(\frac{10 + 8}{3}, \frac{-16 + 4}{3}, \frac{6 - 12}{3}\right) = \left(\frac{18}{3}, \frac{-12}{3}, \frac{-6}{3}\right) = (6, -4, -2)$$

Also, point R_2 divides the join of PQ in the ratio $2:1$.

$\therefore$ Coordinates of R_2

$$= \left(\frac{2 \times 10 + 1 \times 4}{2 + 1}, \frac{2 \times (-16) + 1 \times 2}{2 + 1}, \frac{2 \times 6 + 1 \times (-6)}{1 + 2}\right)$$

$$= \left(\frac{20 + 4}{3}, \frac{-32 + 2}{3}, \frac{12 - 6}{3}\right)$$

$$= \left(\frac{24}{3}, \frac{-30}{3}, \frac{6}{3}\right) = (8, -10, 2)$$

Hence, required points are $(6, -4, -2)$ and $(8, -10, 2)$.

EXAMPLE 8 **If the origin is the centroid of the ΔPQR with vertices $P(2a, 2, 6)$, $Q(-4, 3b, -10)$ and $R(8, 14, 2c)$, then find the values of a, b and c.**

Sol. Given vertices of ΔPQR are $P(2a, 2, 6)$, $Q(-4, 3b, -10)$ and $R(8, 14, 2c)$. Centroid of ΔPQR is

$$\left(\frac{x_1 + x_2 + x_3}{3}, \frac{y_1 + y_2 + y_3}{3}, \frac{z_1 + z_2 + z_3}{3}\right)$$

$$= \left(\frac{2a - 4 + 8}{3}, \frac{2 + 3b + 14}{3}, \frac{6 - 10 + 2c}{3}\right) = \left(\frac{2a + 4}{3}, \frac{3b + 16}{3}, \frac{2c - 4}{3}\right)$$

Also, given that centroid of triangle is origin i.e. (0, 0, 0).

$\therefore \quad \frac{2a + 4}{3} = 0 \Rightarrow 2a + 4 = 0 \Rightarrow a = -2$

$\frac{3b + 16}{3} = 0 \Rightarrow 3b + 16 = 0$

$\Rightarrow \quad b = \frac{-16}{3}$

and $\quad \frac{2c - 4}{3} = 0 \Rightarrow 2c - 4 = 0 \Rightarrow c = 2$

Hence, $\quad a = -2, b = \frac{-16}{3}$ and $c = 2$.

Method to Show Collinearity by Section Formula

To show that three points say A, B and C are collinear by using section formula, *we use the following steps*

***Step* I** First, assume that one point, say C divide the line segment, say AB obtained by joining the other two points A and B in the ratio $k:1$ internally.

***Step* II** Write the coordinates of point C in terms of k.

***Step* III** Now, find the value of k by comparing the x-coordinate, y-coordinate and z-coordinate separately of given point C with respective coordinates obtained in step II of C.

***Step* IV** If value of k obtained by comparing each coordinate separately is same, then given points will be collinear otherwise not.

EXAMPLE 9 **Using section formula, show that the points $(2, -3, 4)$, $(-1, 2, 1)$ and $\left(0, \frac{1}{3}, 2\right)$ are collinear.**

Sol. Let $C\left(0, \frac{1}{3}, 2\right)$ divides the join of $A(2, -3, 4)$ and $B(-1, 2, 1)$ in the ratio $k:1$.

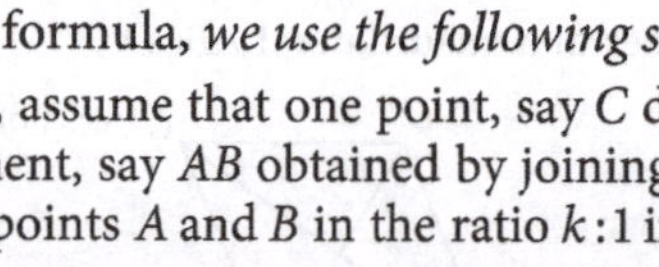

Then, coordinates of C are

$$\left(\frac{-k+2}{k+1}, \frac{2k-3}{k+1}, \frac{k+4}{k+1}\right) \quad \ldots\text{(i)}$$

[using internal ratio formula]

But coordinates of C are $\left(0, \frac{1}{3}, 2\right)$. [given] ...(ii)

On comparing Eqs. (i) and (ii), we get

$$\frac{-k+2}{k+1} = 0 \Rightarrow -k+2 = 0 \Rightarrow k = 2$$

$$\frac{2k-3}{k+1} = \frac{1}{3} \Rightarrow 6k - 9 = k + 1 \Rightarrow 5k = 10 \Rightarrow k = 2$$

and $\frac{k+4}{k+1} = 2 \Rightarrow k + 4 = 2k + 2 \Rightarrow k = 2$

From each of these equations, we get $k = 2$

Since, from each equations, we get same value of k.

Therefore, the given points are collinear and C divides AB internally in the ratio 2 : 1. **Hence proved.**

Problem Based on Distance and Section Formula

EXAMPLE 10 Three points A (3, 2, 0), B (5, 3, 2) and C(−9, 6, −3) are forming a triangle. The bisector AD of $\angle BAC$ meets BC at D. Find the coordinates of D.

Sol. Since, AD is the bisector of $\angle BAC$.

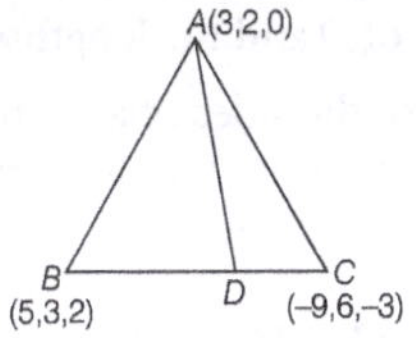

$$\Rightarrow \frac{BD}{DC} = \frac{AB}{AC} \quad \text{[by angle bisector theorem] ...(i)}$$

Now, $AB = \sqrt{(5-3)^2 + (3-2)^2 + (2-0)^2}$

$$[\because \text{distance} = \sqrt{(x_2 - x_1)^2 + (y_2 - y_1)^2 + (z_2 - z_1)^2}]$$

$$= \sqrt{2^2 + 1^2 + 2^2} = \sqrt{4+1+4} = \sqrt{9} = 3 \text{ units}$$

and $AC = \sqrt{(-9-3)^2 + (6-2)^2 + (-3-0)^2}$

$$= \sqrt{(-12)^2 + (4)^2 + (-3)^2}$$

$$= \sqrt{144 + 16 + 9} = \sqrt{169} = 13 \text{ units}$$

Then, from Eq. (i), $\frac{BD}{DC} = \frac{3}{13}$

Hence, D divides BC in the ratio 3 : 13. Then, coordinates of D are

$$\left[\frac{3(-9)+13(5)}{3+13}, \frac{3(6)+13(3)}{3+13}, \frac{3(-3)+13(2)}{3+13}\right]$$

[using internal ratio formula]

$$= \left(\frac{-27+65}{16}, \frac{18+39}{16}, \frac{-9+26}{16}\right)$$

$$= \left(\frac{38}{16}, \frac{57}{16}, \frac{17}{16}\right)$$

$$= \left(\frac{19}{8}, \frac{57}{16}, \frac{17}{16}\right)$$

EXAMPLE 11 Find the length of the medians of the triangle with vertices A (0, 0, 6), B (0, 4, 0) and C (6, 0, 0).

Sol. ABC is a triangle with vertices A (0, 0, 6), B (0, 4, 0) and C (6, 0, 0). Let points D, E and F be the mid-points of BC, AC and AB, respectively. So, AD, BE and CF will be the medians of the triangle.

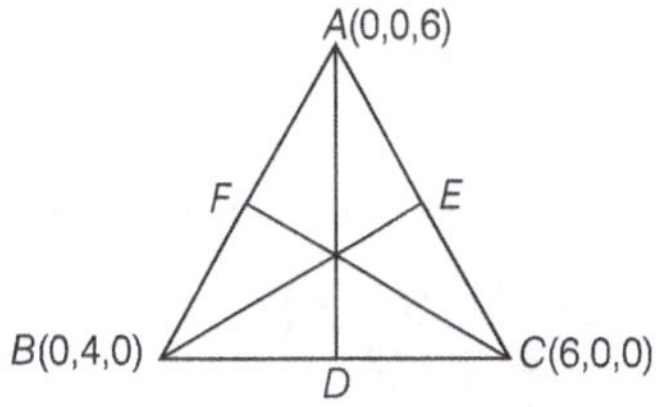

Coordinates of point

$$D = \left(\frac{0+6}{2}, \frac{4+0}{2}, \frac{0+0}{2}\right) = (3, 2, 0)$$

$$\left[\because \text{coordinates of mid-point } \left(\frac{x_1+x_2}{2}, \frac{y_1+y_2}{2}, \frac{z_1+z_2}{2}\right)\right]$$

Coordinates of point $E = \left(\frac{0+6}{2}, \frac{0+0}{2}, \frac{6+0}{2}\right) = (3, 0, 3)$

and coordinates of point F

$$= \left(\frac{0+0}{2}, \frac{0+4}{2}, \frac{6+0}{2}\right) = (0, 2, 3)$$

Now, length of median AD

= Distance between points A and D

$$AD = \sqrt{(0-3)^2 + (0-2)^2 + (6-0)^2}$$

$$[\because \text{distance} = \sqrt{(x_1 - x_2)^2 + (y_1 - y_2)^2 + (z_1 - z_2)^2}]$$

$$= \sqrt{9 + 4 + 36}$$

$$= \sqrt{49} = 7$$

Similarly, $BE = \sqrt{(0-3)^2 + (4-0)^2 + (0-3)^2}$

$$= \sqrt{9 + 16 + 9} = \sqrt{34}$$

and $CF = \sqrt{(6-0)^2 + (0-2)^2 + (0-3)^2}$

$$= \sqrt{36 + 4 + 9}$$

$$= \sqrt{49} = 7$$

Hence, length of the medians are 7, $\sqrt{34}$ and 7.

Exercise

Objective Type qUESTIONS

1 The coordinates of the point R, which divides the segment joining $P(x_1, y_1, z_1)$ and $Q(x_2, y_2, z_2)$ internally in the ratio $k:1$, are

(a) $\left(\frac{kx_2 - x_1}{1-k}, \frac{ky_2 - y_1}{1-k}, \frac{kz_2 - z_1}{1-k}\right)$

(b) $\left(\frac{kx_2 + x_1}{1+k}, \frac{ky_2 + y_1}{1+k}, \frac{kz_2 + z_1}{1+k}\right)$

(c) $\left(\frac{kx_2 + x_1}{1-k}, \frac{ky_2 + y_1}{1-k}, \frac{kz_2 + z_1}{1-k}\right)$

(d) None of the above

2 If R is the mid-point of the segment joining $P(x_1, y_1, z_1)$ and $Q(x_2, y_2, z_2)$, then coordinates of R are

(a) $\left(\frac{x_1 - x_2}{2}, \frac{y_1 - y_2}{2}, \frac{z_1 - z_2}{2}\right)$

(b) $\left(\frac{x_1 + x_2}{2}, \frac{y_1 + y_2}{2}, \frac{z_1 + z_2}{2}\right)$

(c) $\left(\frac{x_2 - x_1}{2}, \frac{y_2 - y_1}{2}, \frac{z_2 - z_1}{2}\right)$

(d) None of the above

3 The coordinates of the point P which is five-sixth of the way from $A(-2, 0, 6)$ to $B(10, -6, -12)$, is

(a) (8, 5, 9) (b) (−8, 5, 9)
(c) (8, −5, −9) (d) None of these

4 The ratio in which the line joining (2, 4, 5) and (3, 5, −4) is divided by the YZ-plane, is

(a) 2 : 3 (b) 3 : 2 (c) −2 : 3 (d) 4 : −3

5 A point on XOZ-plane divides the join of (5, −3, −2) and (1, 2, −2) at

(a) $\left(\frac{13}{5}, 0, -2\right)$ (b) $\left(\frac{13}{5}, 0, 2\right)$
(c) (5, 0, 2) (d) (5, 0, −2)

Very Short Answer Type Questions

6 Find the coordinates of the point which divides the join of $P(2, -1, 4)$ and $Q(4, 3, 2)$ in the ratio 2 : 3 (i) internally (ii) externally.

7 Find the ratio in which the line segment joining the points (2, 4, 5) and (3, 5, −4) is divided by the XZ-plane.

8 Find the ratio in which the line segment joining the points (4, 4, −10) and (−2, 2, 4) is divided by the YZ-plane.

9 Find the ratio in which the line segment joining the points (2, 4, −3) and (−3, 5, 4) divided the XY-plane.

Short Answer Type I Questions

10 Find the centroid of a triangle, the mid-point of whose sides are $D(1, 2, -3)$, $E(3, 0, 1)$ and $F(-1, 1, -4)$.

11 Find the third vertex of triangle whose centroid is origin and two vertices are (2, 4, 6) and (0, −2, 5).

12 Using section formula, show that the three points $A(-2, 3, 5)$, $B(1, 2, 3)$ and $C(7, 0, -1)$ are collinear.

13 Show that the three points $A(2, 3, 4)$, $B(-1, 2, -3)$ and $C(-4, 1, -10)$ are collinear and find the ratio in which C divides AB.

14 Verify that $A(-1, 2, 1)$, $B(1, -2, 5)$, $C(4, -7, 8)$ and $D(2, -3, 4)$ are the vertices of a parallelogram.

Short Answer Type II Questions

15 Prove that the points (0, −1, −7), (2, 1, −9) and (6, 5, −13) are collinear. Find the ratio in which the first point divides the join of the other two.

16 Find the coordinate of the points which trisect the line segment joining the points $A(2, 1, -3)$ and $B(5, -8, 3)$.

17 The vertices of the triangles are $A(5, 4, 6)$, $B(1, -1, 3)$ and $C(4, 3, 2)$. The internal bisector of $\angle A$ meets BC at D. Find the coordinates of D and the length AD.

18 The mid-point of the sides of a triangle are (1, 5, −1), (0, 4, −2) and (2, 3, 4). Find its vertices and also find the centroid of the triangle.

Answers

1. (b) 2. (b) 3. (c) 4. (c)

5. (a) 6. (i) $\left(\frac{14}{5}, \frac{3}{5}, \frac{16}{5}\right)$ (ii) (−2, −9, 8)

7. 4 : 5 externally 8. 2 : 1 internally

9. 3 : 4 internally

10. (1, 1, −2)

11. (−2, −2, −11)

13. 2 : 1 externally

15. 1 : 3 externally

16. (3, −2, −1) and (4, −5, 1)

17. $\left(\frac{23}{8}, \frac{3}{2}, \frac{19}{8}\right)$, $\frac{\sqrt{1530}}{8}$

18. The vertices of the triangle are $A(1, 2, 3)$, $B(3, 4, 5)$ and $C(-1, 6, -7)$. Also, centroid of the triangle is $G(1, 4, 1/3)$.

12 CHAPTER

Mathematics knows no races or geographic boundries; for Mathematics, the cultural world is one country.

—David Hilberry

YOUR STUDY GOALS

YOUR PRACTICE GOALS

- Objective Type Questions (26 Qs ; 13 MCQs, 11 A-R, 2 Case Based)
- Subjective Type Questions (18 Qs; 6 VSA, 5 SA, 7 LA)

Limits and Derivatives

12.1 Definition of Limit

If $f(x)$ approaches to a real number l, when x approaches to a (through lesser or greater values to a) i.e. if $f(x) \to l$ when $x \to a$, then l is called limit of the function $f(x)$.

In symbolic form, it can be written as $\lim_{x \to a} f(x) = l$.

Left Hand Limit and Right Hand Limit

There are two ways through which x can approach to a, either from left side (from values lower than a) or from right side (from values higher than a).

Therefore, two types of limit exists; **left hand limit** and **right hand limit**.

- Left hand limit can be written as $\lim_{x \to a^-} f(x)$ is the expected value of f at $x = a$, when we have the values of f near x to the left of a.
- Similarly, right hand limit can be written as $\lim_{x \to a^+} f(x)$ is the expected value of f at $x = a$, when we have the values of f near x to the right of a.

Existence of Limit

If the left hand limit and right hand limit both exist (i.e. finite and unique) and coincide (i.e. same), then we say that limit exists and their common value is called the limit of $f(x)$ at $x = a$.

i.e. if $\lim_{x\to a^-} f(x) = \lim_{x\to a^+} f(x) = l$,

then $\lim_{x\to a} f(x) = l$

TEACHER TIPS

- $x \to a - 0$ or $x \to a^-$ is read as x tends to a from left and it means that x is very close to a but it is always less than a.
- $x \to a + 0$ or $x \to a^+$ is read as x tends to a from right and it means that x is very close to a but it is always greater than a.
- $x \to a$ is read as x tends to a and it means that x is very close to a but it is never equal to a.

Algebra of Limits

Let f and g be two real functions with common domain D such that $\lim_{x\to a} f(x)$ and $\lim_{x\to a} g(x)$ exists. Then,

- Limit of sum of two functions is sum of the limits of the functions.

  i.e. $\lim_{x\to a} (f+g)(x) = \lim_{x\to a} f(x) + \lim_{x\to a} g(x)$

- Limit of difference of two functions is difference of the limits of the function.

  i.e. $\lim_{x\to a} (f-g)(x) = \lim_{x\to a} f(x) - \lim_{x\to a} g(x)$

- Limit of product of a constant and one function is the product of that constant and limit of a function.

  i.e. $\lim_{x\to a} [c \cdot f(x)] = c \lim_{x\to a} f(x)$,

  where c is a constant.

- Limit of product of two functions is product of the limits of the functions.

  i.e. $\lim_{x\to a} [f(x)\cdot g(x)] = \lim_{x\to a} f(x) \cdot \lim_{x\to a} g(x)$

- Limit of quotient of two functions is quotient of the limits of the functions.

  i.e. $\lim_{x\to a} \dfrac{f(x)}{g(x)} = \dfrac{\lim_{x\to a} f(x)}{\lim_{x\to a} g(x)}$,

  where $\lim_{x\to a} g(x) \neq 0$

Limits of Polynomial Function

A function f is said to be a **polynomial function** of degree n, if $f(x) = a_0 + a_1x + a_2x^2 + \ldots + a_nx^n$, where a_i's are real numbers and $a_n \neq 0$.

Let $f(x) = a_0 + a_1x + a_2x^2 + \ldots + a_nx^n$ be a polynomial function.

Then, limit of a polynomial function $f(x)$

$$= \lim_{x\to a} f(x) = \lim_{x\to a} [a_0 + a_1x + a_2x^2 + \ldots + a_nx^n]$$

$$= \lim_{x\to a} a_0 + \lim_{x\to a} a_1 x + \lim_{x\to a} a_2 x^2 + \ldots + \lim_{x\to a} a_n x^n$$

$$= a_0 + a_1 \lim_{x\to a} x + a_2 \lim_{x\to a} x^2 + \ldots + a_n \lim_{x\to a} x^n$$

$$= a_0 + a_1a + a_2a^2 + \ldots + a_na^n$$

$$= f(a)$$

Example 1. Suppose the function is defined by

$$f(x) = \begin{cases} x^2, & x \neq 2 \\ 0, & x = 2 \end{cases}$$

(i) Find the left hand limit of $f(x)$ at $x = 2$.

(ii) Find the right hand limit of $f(x)$ at $x = 2$.

Sol. (i) Given, $f(x) = \begin{cases} x^2, & x \neq 2 \\ 0, & x = 2 \end{cases}$

$\therefore$ Left hand limit at $x = 2$ is

$$\lim_{x\to 2^-} f(x) = \lim_{x\to 2^-} x^2 \quad \ldots(i)$$

On putting $x = 2 - h$ and changing the limit $x \to 2^-$ by $h \to 0$ in Eq. (i), we get

$$\lim_{x\to 2^-} f(x) = \lim_{h\to 0} (2-h)^2 = (2-0)^2 = 4$$

(ii) Right hand limit at $x = 2$ is

$$\lim_{x\to 2^+} f(x) = \lim_{x\to 2^+} x^2 \quad \ldots(ii)$$

On putting $x = 2 + h$ and changing the limit $x \to 2^+$ by $h \to 0$ in Eq. (ii), we get

$$\lim_{x\to 2^+} f(x) = \lim_{h\to 0} (2+h)^2 = (2+0)^2 = 4$$

Example 2. Let us consider two functions $f(x) = x^2 + 4$ and $g(x) = x - 3$ such that $f(x)$ and $g(x)$ exist at $x = 5$. Find the limit of the following functions at $x = 5$.

(i) $f(x) + g(x)$ (ii) $f(x) - g(x)$

(iii) $f(x) \times g(x)$ (iv) $\dfrac{f(x)}{g(x)}$

Sol. Given functions are $f(x) = x^2 + 4$

and $g(x) = x - 3$

Clearly, $\lim_{x\to 5} f(x) = \lim_{x\to 5} x^2 + 4$

$= (5)^2 + 4 = 25 + 4$

$= 29$ ...(i)

and $\lim_{x \to 5} g(x) = \lim_{x \to 5} (x - 3)$

$= (5 - 3) = 2$ ...(ii)

(i) $\lim_{x \to 5} [f(x) + g(x)] = \lim_{x \to 5} f(x) + \lim_{x \to 5} g(x)$

$= 29 + 2$ [from Eqs. (i) and (ii)]

$= 31$

(ii) $\lim_{x \to 5} [f(x) - g(x)] = \lim_{x \to 5} f(x) - \lim_{x \to 5} g(x)$

$= 29 - 2$

$= 27$ [from Eqs. (i) and (ii)]

(iii) $\lim_{x \to 5} [f(x) \times g(x)] = \lim_{x \to 5} f(x) \times \lim_{x \to 5} g(x)$

$= 29 \times 2$

$= 58$ [from Eqs. (i) and (ii)]

(iv) $\lim_{x \to 5} \frac{f(x)}{g(x)} = \frac{\lim_{x \to 5} f(x)}{\lim_{x \to 5} g(x)}$

$= \frac{29}{2} = 14.5$ [from Eqs. (i) and (ii)]

Example 3. **Evaluate the limit $\lim_{x \to 3} (4x^3 - 2x^2 - x + 1)$.**

Sol. $\lim_{x \to 3} (4x^3 - 2x^2 - x + 1)$

$= 4 \lim_{x \to 3} x^3 - 2 \lim_{x \to 3} x^2 - \lim_{x \to 3} x + \lim_{x \to 3} 1$

$= 4(3)^3 - 2(3)^2 - 3 + 1 = 108 - 18 - 2 = 88$

Example 4. **Evaluate the left hand and right hand limits of the following function at $x = 2$.**

$$f(x) = \begin{cases} 2x+3, \text{ if } x \le 2 \\ x+5, \text{ if } x > 2 \end{cases}$$

Does $\lim_{x \to 2} f(x)$ exist?

Sol. Given, $f(x) = \begin{cases} 2x + 3, \text{ if } x \le 2 \\ x + 5, \text{ if } x > 2 \end{cases}$

$\text{LHL} = \lim_{x \to 2^-} f(x) = \lim_{x \to 2^-} 2x + 3$

$[\because f(x) = 2x + 3, \text{ if } x \le 2]$

$= \lim_{h \to 0} [2(2 - h) + 3]$ [put $x = 2 - h$ and when $x \to 2^-$, then $h \to 0$]

$= 2(2 - 0) + 3 = 4 + 3 = 7$

and $\text{RHL} = \lim_{x \to 2^+} f(x) = \lim_{x \to 2^+} (x + 5)$

$[\because f(x) = x + 5, \text{ if } x > 2]$

$= \lim_{h \to 0} (2 + h + 5)$ [put $x = 2 + h$ and when $x \to 2^+$, then $h \to 0$]

$= 2 + 0 + 5 = 7$

$\because$ LHL of f(at $x = 2$) = RHL of f (at $x = 2$)

$\therefore \lim_{x \to 2} f(x)$ exists and it is equal to 7.

Example 5. **Evaluate the left hand and right hand limits of the function defined by**

$$f(x) = \begin{cases} 1 + x^2, \text{if } 0 \le x \le 1 \\ 2 - x^2, \text{if } x > 1 \end{cases} \text{ at } x = 1.$$

Also, show that $\lim_{x \to 1} f(x)$ does not exist.

Sol. We have, $f(x) = \begin{cases} 1 + x^2, \text{ if } 0 \le x \le 1 \\ 2 - x^2, \text{ if } x > 1 \end{cases}$

At $x = 1$, $\text{LHL} = \lim_{x \to 1^-} f(x) = \lim_{x \to 1^-} (1 + x^2)$

$[\because f(x) = 1 + x^2, \text{ if } 0 \le x \le 1]$

$= \lim_{h \to 0} [1 + (1 - h)^2]$ [put $x = 1 - h$ and when $x \to 1^-$, then $h \to 0$]

$= 1 + (1 - 0)^2 = 2$

$\text{RHL} = \lim_{x \to 1^+} f(x) = \lim_{x \to 1^+} (2 - x^2)$

$[\because f(x) = 2 - x^2, \text{ if } x > 1]$

$= \lim_{h \to 0} [2 - (1 + h)^2]$ [put $x = 1 + h$ and when $x \to 1^+$, then $h \to 0$]

$= 2 - 1 = 1$

Since, LHL $\ne$ RHL

Therefore, $\lim_{x \to 1} f(x)$ does not exist.

Example 6. **Suppose the function is defined by**

$$f(x) = \begin{cases} \dfrac{|x - 3|}{x - 3}, & x \ne 3 \\ 0, & x = 3 \end{cases}$$

(i) Find the left hand limit of $f(x)$ at $x = 3$.

(ii) Find the right hand limit of $f(x)$ at $x = 3$.

Sol. (i) Given, $f(x) = \begin{cases} \dfrac{|x - 3|}{x - 3}, & x \ne 3 \\ 0, & x = 3 \end{cases}$

$\therefore$ Left hand limit at $x = 3$ is

$\lim_{x \to 3^-} f(x) = \lim_{x \to 3^-} \frac{|x - 3|}{x - 3}$ ...(i)

On putting $x = 3 - h$ and changing the limit $x \to 3^-$ by $h \to 0$ in Eq. (i), we get

$\lim_{x \to 3^-} f(x) = \lim_{x \to 3^-} \frac{|x - 3|}{x - 3} = \lim_{h \to 0} \frac{|-h|}{-h}$

$\Rightarrow \lim_{x \to 3^-} f(x) = \lim_{h \to 0} \frac{h}{(-h)}$ $[\because |-x| = x, \text{ for } x \ge 0]$

$= -1$

(ii) Right hand limit at $x = 3$ is

$\lim_{x \to 3^+} f(x) = \lim_{x \to 3^+} \frac{|x - 3|}{x - 3}$ ...(ii)

On putting $x=3+h$ and changing the limit $x \to 3^+$ by $h \to 0$ in Eq. (ii), we get

$$\lim_{x \to 3^+} f(x) = \lim_{x \to 3^+} \frac{|x-3|}{x-3} = \lim_{h \to 0} \frac{|h|}{h}$$

$$\Rightarrow \lim_{x \to 3^+} f(x) = \lim_{h \to 0} \frac{h}{h} = 1 \quad [\because |x| = x, \text{ for } x \geq 0]$$

Alternate Method

Given, $f(x) = \begin{cases} \dfrac{|x-3|}{x-3}, & x \neq 3 \\ 0, & x = 3 \end{cases}$

$$= \begin{cases} \dfrac{-(x-3)}{(x-3)}, & \text{if } x<3 \\ \dfrac{x-3}{x-3}, & \text{if } x>3 \\ 0, & \text{if } x=3 \end{cases}$$

$$= \begin{cases} -1, & \text{if } x<3 \\ 1, & \text{if } x>3 \\ 0, & \text{if } x=3 \end{cases}$$

(i) LHL $= \lim_{x \to 3^-} f(x) = \lim_{x \to 3^-} (-1)$ $[\because f(x) = -1, \text{ for } x<3]$

$= -1$

(ii) RHL $= \lim_{x \to 3^+} f(x) = \lim_{x \to 3^+} (1)$ $[\because f(x) = 1, \text{ for } x>3]$

$= 1$

Example 7. **For what integers m and n does $\lim_{x \to 0} f(x)$ and $\lim_{x \to 1} f(x)$ exist, if**

$$f(x) = \begin{cases} mx^2+n, & x<0 \\ nx+m, & 0 \leq x \leq 1 \\ nx^2+m, & x>1 \end{cases}$$

NCERT

Sol. Given, $f(x) = \begin{cases} mx^2+n, & x<0 \\ nx+m, & 0 \leq x \leq 1 \\ nx^2+m, & x>1 \end{cases}$

Limit at $x=0$

LHL $= \lim_{x \to 0^-} f(x) = \lim_{x \to 0^-} mx^2+n$

$= \lim_{h \to 0} m(0-h)^2+n$

[put $x=0-h$ and as $x \to 0^-$, then $h \to 0$]

$= n$

RHL $= \lim_{x \to 0^+} f(x) = \lim_{x \to 0^+} nx+m$

$= \lim_{h \to 0} n(0+h)+m = m$

[put $x=0+h$ and as $x \to 0^+$, then $h \to 0$]

Now, for $\lim_{x \to 0} f(x)$ to be exists.

LHL = RHL $\Rightarrow n=m$

Hence, $\lim_{x \to 0} f(x)$ exist when $m=n$.

Limit at $x=1$

Here, LHL $= \lim_{x \to 1^-} f(x) = \lim_{x \to 1^-} (nx+m)$

$= \lim_{h \to 0} [n(1-h)+m]$

[put $x=1-h$ and as $x \to 1^-$, then $h \to 0$]

$= n+m$

RHL $= \lim_{x \to 1^+} f(x) = \lim_{x \to 1^+} (nx^2+m)$

$= \lim_{h \to 0} n(1+h)^2+m$

[put $x=1+h$ and as $x \to 1^+$, then $h \to 0$]

$= n+m$

Now, for $\lim_{x \to 1} f(x)$ to be exists.

$\lim_{x \to 1^+} f(x) = \lim_{x \to 1^-} f(x)$

$\Rightarrow n+m = n+m$, which is true for all $n, m \in Z$.

Hence, $\lim_{x \to 1} f(x)$ exists for all $n, m \in Z$.

Example 8. **If $f(x) = \begin{cases} |x|+1, & x<0 \\ 0, & x=0 \\ |x|-1, & x>0 \end{cases}$, for what value(s) of a does $\lim_{x \to a} f(x)$ exist?** NCERT

Sol. We have, $f(x) = \begin{cases} -x+1, & x<0 \\ 0, & x=0 \\ x-1, & x>0 \end{cases}$

$$\left[\because |x| = \begin{cases} x, & \text{if } x \geq 0 \\ -x, & \text{if } x<0 \end{cases}\right]$$

Case I When $a=0$

In this case, LHL $= \lim_{x \to 0^-} f(x) = \lim_{h \to 0} f(0-h)$

$= \lim_{h \to 0} -(0-h)+1$

$[\because \text{for } x<0, f(x) = -x+1]$

$= \lim_{h \to 0}(h+1) = \lim_{h \to 0} h + \lim_{h \to 0} 1 = 0+1 = 1$

and RHL $= \lim_{x \to 0^+} f(x) = \lim_{h \to 0} f(0+h)$

$= \lim_{h \to 0}(0+h)-1 = -1$ $[\because \text{for } x>0, f(x) = x-1]$

$\because$ LHL $\neq$ RHL

$\therefore$ For $a=0$, $\lim_{x \to a} f(x)$ does not exists.

Case II When $a<0$

In this case, $\lim_{x \to a} f(x) = \lim_{x \to a}(-x+1)$

$[\because \text{for } x<0, f(x) = -x+1]$

$= -a+1$, which is a fixed real number.

$\therefore$ For $a<0$, $\lim_{x \to a} f(x)$ exists.

Case III When $a>0$

In this case, $\lim_{x \to a} f(x) = \lim_{x \to a}(x-1)$

$[\because \text{for } x>0, f(x) = x-1]$

$= a-1$, which is a fixed real number.

$\therefore$ For $a>0$, $\lim_{x \to a} f(x)$ exists.

Hence, from cases I, II and III, we conclude that $\lim_{x \to a} f(x)$ exists for all $a \neq 0$.

Quick Test 1

1. Find

(i) $\lim_{x \to 2^-} (x^2 - x + 1)$ (ii) $\lim_{x \to 2^+} \frac{x-3}{x^2+4}$

(iii) $\lim_{x \to 0^+} \frac{2}{x^{1/5}}$ (iv) $\lim_{x \to -8^+} \frac{2x}{x-8}$

(v) $\lim_{x \to 1^+} \frac{x^3+x-1}{x+1}$

2. Find the limits.

(i) $\lim_{x \to 1} [x^2 - x - 1]$ (ii) $\lim_{x \to 3} [x^2(x+1)]$

(iii) $\lim_{x \to -1} (1 + x + x^2 + \ldots + x^9)$

3. Evaluate the left hand and right hand limits of the following functions at $x=1$.

$f(x) = \begin{cases} 5x-4, & \text{if } 0 < x \le 1 \\ 4x^2 - 3x, & \text{if } 1 < x < 2 \end{cases}$ Does $\lim_{x \to 1} f(x)$ exist?

4. Let $f(x) = \begin{cases} 2x+3, & \text{if } x \le 0 \\ 3(x+1), & \text{if } x > 0 \end{cases}$, then evaluate

(i) $\lim_{x \to 0} f(x)$ (ii) $\lim_{x \to 1} f(x)$

5. Evaluate $\lim_{x \to 0} f(x)$, where $f(x) = \begin{cases} \frac{x}{|x|}, & x \ne 0 \\ 0, & x = 0 \end{cases}$.

NCERT, Competency Based Que

6. If $f(x) = \begin{cases} \frac{x-|x|}{x}, & \text{if } x \ne 0 \\ 2, & \text{if } x = 0 \end{cases}$, then show that $\lim_{x \to 0} f(x)$ does not exist.

7. If $f(x) = \begin{cases} x+2, & x \le -1 \\ cx^2, & x > -1 \end{cases}$, then find c when $\lim_{x \to -1} f(x)$ exists.

NCERT Exemplar, Competency Based Que

8. Suppose $f(x) = \begin{cases} a+bx, & x<1 \\ 4, & x=1, \\ b-ax, & x>1 \end{cases}$ and if $\lim_{x \to 1} f(x) = f(1)$, then what are the possible values of a and b?

NCERT, Competency Based Que

Answers

1. (i) 3 (ii) $-\frac{1}{8}$ (iii) ∞ (iv) 1 (v) $\frac{1}{2}$

2. (i) -1 (ii) 36 (iii) 0

3. $\lim_{x \to 1} f(x)$ exists and it is equal to 1.

4. (i) 3 (ii) 6

5. At $x = 0$, limit does not exist.

6. $\lim_{x \to 0} f(x)$ does not exist. **7.** $c = 1$ **8.** $a = 0, b = 4$

For Solutions, See QR Code on Pg. 357

12.2 Limits of Rational Functions

A function f is said to be a rational function, if

$f(x) = \frac{g(x)}{h(x)}$, where $g(x)$ and $h(x)$ are polynomial functions such that $h(x) \ne 0$.

Then, $\lim_{x \to a} f(x) = \lim_{x \to a} \frac{g(x)}{h(x)} = \frac{\lim_{x \to a} g(x)}{\lim_{x \to a} h(x)} = \frac{g(a)}{h(a)}$

However, if $h(a) = 0$, then two cases arise

(i) $g(a) \ne 0$ (ii) $g(a) = 0$

In the first case, we say that the limit does not exist.

In the second case,

we have, $h(a) = 0$ and $g(a) = 0$

So, we can write as $g(x) = (x-a)^n g_1(x)$

and $h(x) = (x-a)^m h_1(x)$

$$\therefore \quad \lim_{x \to a} f(x) = \frac{\lim_{x \to a} g(x)}{\lim_{x \to a} h(x)} = \frac{\lim_{x \to a} (x-a)^n g_1(x)}{\lim_{x \to a} (x-a)^m h_1(x)}$$

$$= \frac{\lim_{x \to a} (x-a)^{n-m} g_1(x)}{\lim_{x \to a} h_1(x)}$$

Now, three cases arise

Case I When $n > m$, then $\lim_{x \to a} f(x) = \frac{0 \cdot g_1(a)}{h_1(a)} = 0$

Case II When $n = m$, then $\lim_{x \to a} f(x) = \frac{g_1(a)}{h_1(a)}$

Case III When $n < m$, then

$\lim_{x \to a} f(x) =$ Limit does not exist.

Limit of a rational function can be find with the help of following methods

Direct Substitution Method

In this method, we substitute the point, to which the variable tends to in the given limit. If it give us a real number, then the number so obtained is the limit of the function and if it does not give us a real number, then use other methods.

Factorisation Method

Let $\lim_{x \to a} \frac{f(x)}{g(x)}$ reduces to the form $\frac{0}{0}$. When we substitute $x = a$, then we factorise $f(x)$ and $g(x)$ and then cancel out the common factor to evaluate the limit.

Rationalisation Method

If we get $\frac{0}{0}$ form and numerator or denominator or both have radical sign, then we rationalise the numerator or denominator or both by multiplying their conjugate to remove $\frac{0}{0}$ form and then find limit by direct substitution method.

Using Some Standard Limits

Theorem Let n be any positive integer.

Then, $$\lim_{x \to a} \frac{x^n - a^n}{x - a} = n a^{n-1}$$

The expression in the above theorem, for the limit is true, if n is any rational number and a is positive.

Example 1. **Find the limits of the following.**

(i) $\lim_{x \to 2} \frac{x^2 - 4}{x + 3}$ (ii) $\lim_{x \to -1} \frac{(x-1)^2 + 3x^2}{(x^4 + 1)^2}$

(iii) $\lim_{x \to 0} \frac{\sqrt{2+x} + \sqrt{2-x}}{2+x}$ NCERT

Sol. (i) $\lim_{x \to 2} \frac{x^2 - 4}{x + 3} = \frac{4-4}{2+3} = \frac{0}{5} = 0$

(ii) $\lim_{x \to -1} \frac{(x-1)^2 + 3x^2}{(x^4 + 1)^2}$

$$= \frac{(-1-1)^2 + 3(-1)^2}{((-1)^4 + 1)^2} = \frac{(-2)^2 + 3(1)}{(1+1)^2} = \frac{4+3}{2^2} = \frac{7}{4}$$

(iii) $\lim_{x \to 0} \frac{\sqrt{2+x} + \sqrt{2-x}}{2+x} = \frac{\sqrt{2+0} + \sqrt{2-0}}{2+0}$

$$= \frac{\sqrt{2} + \sqrt{2}}{2} = \frac{2\sqrt{2}}{2} = \sqrt{2}$$

Example 2. **Evaluate** $L = \lim_{x \to 2} \frac{x^3 - 8}{(x-2)}$.

Sol. We have, $L = \lim_{x \to 2} \frac{x^3 - 8}{x - 2}$

Let $f(x) = x^3 - 8$ and $g(x) = x - 2$

Here, $\lim_{x \to 2} f(x) = \lim_{x \to 2} x^3 - 8 = 2^3 - 8 = 0$

and $\lim_{x \to 2} g(x) = \lim_{x \to 2} x - 2 = 2 - 2 = 0$

Thus, we get $\frac{0}{0}$ form.

Now, factorise $f(x)$ and $g(x)$ such that $(x - 2)$ is a common factor.

Here, $f(x) = x^3 - 8 = (x^3 - 2^3)$

$= (x-2)(x^2 + 4 + 2x)$ and $g(x) = x - 2$

$\therefore \quad L = \lim_{x \to 2} \frac{(x-2)(x^2 + 4 + 2x)}{(x-2)}$

On cancelling the common factor $(x - 2)$, we get

$$L = \lim_{x \to 2} (x^2 + 4 + 2x)$$
$$= (2)^2 + 4 + 2(2)$$
$$= 4 + 4 + 4 = 12$$

Hence, $\lim_{x \to 2} \frac{x^3 - 8}{x - 2} = 12$

Example 3. **Evaluate** $\lim_{x \to 1} \left(\frac{2}{1 - x^2} + \frac{1}{x - 1} \right)$.

Sol. We have, $\lim_{x \to 1} \left(\frac{2}{1 - x^2} + \frac{1}{x - 1} \right) = \lim_{x \to 1} \left(\frac{2}{1 - x^2} - \frac{1}{1 - x} \right)$ $[\infty - \infty \text{ form}]$

$= \lim_{x \to 1} \frac{2 - (1+x)}{1 - x^2}$ $\left[\frac{0}{0} \text{ form}\right]$

$= \lim_{x \to 1} \frac{1 - x}{1 - x^2} = \lim_{x \to 1} \frac{1}{1 + x} = \frac{1}{2}$

Example 4. **Evaluate** $\lim_{x \to 2} \left[\frac{x^2 - 4}{x^3 - 4x^2 + 4x} \right]$.

Sol. On putting $x = 2$, we get the form $\frac{0}{0}$. So, let us first factorise it.

Consider $\lim_{x \to 2} \frac{x^2 - 4}{x^3 - 4x^2 + 4x} = \lim_{x \to 2} \frac{(x+2)(x-2)}{x(x-2)^2}$

$= \lim_{x \to 2} \frac{(x+2)}{x(x-2)} = \frac{2+2}{2(2-2)} = \frac{4}{0}$

which is not defined.

$\therefore \lim_{x \to 2} \left[\frac{x^2 - 4}{x^3 - 4x^2 + 4x} \right]$ does not exist.

Example 5. Find the limit

$$\lim_{x\to1}\left[\frac{x-2}{x^2-x}-\frac{1}{x^3-3x^2+2x}\right].$$ *NCERT*

Sol. Here, $\left[\frac{x-2}{x^2-x}-\frac{1}{x^3-3x^2+2x}\right]$

$$=\left[\frac{x-2}{x(x-1)}-\frac{1}{x(x^2-3x+2)}\right]$$

$$=\left[\frac{x-2}{x(x-1)}-\frac{1}{x(x-1)(x-2)}\right]$$

$$=\frac{(x-2)^2-1}{x(x-1)(x-2)}=\left[\frac{x^2-4x+4-1}{x(x-1)(x-2)}\right]$$

$$=\left[\frac{x^2-4x+3}{x(x-1)(x-2)}\right]=\frac{(x-3)(x-1)}{x(x-1)(x-2)}$$

$$\therefore \lim_{x\to1}\left[\frac{x-2}{x^2-x}-\frac{1}{x^3-3x^2+2x}\right]$$

$$=\lim_{x\to1}\frac{(x-3)(x-1)}{x(x-1)(x-2)}$$

$$=\lim_{x\to1}\frac{x-3}{x(x-2)}=\frac{1-3}{1(1-2)}=\frac{-2}{-1}=2$$

Example 6. Evaluate $\lim_{x\to a}\frac{\sqrt{a+2x}-\sqrt{3x}}{\sqrt{3a+x}-2\sqrt{x}}$. *NCERT Exemplar*

Sol. We have, $\lim_{x\to a}\frac{\sqrt{a+2x}-\sqrt{3x}}{\sqrt{3a+x}-2\sqrt{x}}$

$$=\lim_{x\to a}\frac{\sqrt{a+2x}-\sqrt{3x}}{\sqrt{3a+x}-2\sqrt{x}}\times\frac{\sqrt{a+2x}+\sqrt{3x}}{\sqrt{a+2x}+\sqrt{3x}}$$

[multiply numerator and denominator by $\sqrt{a+2x}+\sqrt{3x}$]

$$=\lim_{x\to a}\frac{a+2x-3x}{(\sqrt{3a+x}-2\sqrt{x})(\sqrt{a+2x}+\sqrt{3x})}$$

$[\because (A-B)(A+B)=A^2-B^2]$

$$=\lim_{x\to a}\frac{(a-x)(\sqrt{3a+x}+2\sqrt{x})}{(\sqrt{a+2x}+\sqrt{3x})(\sqrt{3a+x}-2\sqrt{x})(\sqrt{3a+x}+2\sqrt{x})}$$

[multiply numerator and denominator by $\sqrt{3a+x}+2\sqrt{x}$]

$$=\lim_{x\to a}\frac{(a-x)[\sqrt{3a+x}+2\sqrt{x}]}{(\sqrt{a+2x}+\sqrt{3x})(3a+x-4x)}$$

$$=\lim_{x\to a}\frac{(a-x)(\sqrt{3a+x}+2\sqrt{x})}{(\sqrt{a+2x}+\sqrt{3x})(3a-3x)}$$

$$=\lim_{x\to a}\frac{\sqrt{3a+x}+2\sqrt{x}}{3(\sqrt{a+2x}+\sqrt{3x})}=\frac{\sqrt{3a+a}+2\sqrt{a}}{3(\sqrt{a+2a}+\sqrt{3a})}$$

$$=\frac{4\sqrt{a}}{3\times2\sqrt{3}\sqrt{a}}=\frac{2}{3\sqrt{3}}=\frac{2\sqrt{3}}{9}$$

Example 7. Evaluate $\lim_{x\to1}\frac{(2x-3)(\sqrt{x}-1)}{2x^2+x-3}$.

Sol. We have, $\lim_{x\to1}=\frac{(2x-3)(\sqrt{x}-1)}{2x^2+x-3}$

When $x=1$, the expression $\frac{(2x-3)(\sqrt{x}-1)}{2x^2+x-3}$ becomes of the form $\frac{0}{0}$.

So, rationalising $(\sqrt{x}-1)$ in the numerator by multiplying and dividing its conjugate i.e. $\sqrt{x}+1$.

We have, $\lim_{x\to1}\frac{(2x-3)(\sqrt{x}-1)}{2x^2+x-3}$

$$=\lim_{x\to1}\frac{(2x-3)(\sqrt{x}-1)(\sqrt{x}+1)}{(\sqrt{x}+1)(2x^2+x-3)}$$

$$=\lim_{x\to1}\frac{(2x-3)(x-1)}{(\sqrt{x}+1)(2x^2+x-3)} \quad \left[\frac{0}{0}\text{ form}\right]$$

$$=\lim_{x\to1}\frac{(2x-3)(x-1)}{(\sqrt{x}+1)(2x+3)(x-1)}$$

$$=\lim_{x\to1}\frac{2x-3}{(\sqrt{x}+1)(2x+3)}=-\frac{1}{10}$$

Example 8. Evaluate $\lim_{x\to2}\frac{x^{10}-1024}{x-2}$.

Sol. When $x=2$, the expression $\frac{x^{10}-1024}{x-2}$ becomes of the form $\frac{0}{0}$.

Now, $\lim_{x\to2}\frac{x^{10}-1024}{x-2}=\lim_{x\to2}\frac{x^{10}-2^{10}}{x-2}=10(2^{10-1})=5120$

Example 9. Find the limit $\lim_{x\to0}\frac{\sqrt{1+x}-1}{x}$. *NCERT*

Sol. We have, $\lim_{x\to0}\frac{\sqrt{1+x}-1}{x}$

On putting $y=1+x$, then $y\to1$ as $x\to0$.

$$\therefore \lim_{x\to0}\frac{\sqrt{1+x}-1}{x}=\lim_{y\to1}\frac{\sqrt{y}-1}{y-1} \quad [\because x+1=y\Rightarrow x=y-1]$$

$$=\lim_{y\to1}\frac{y^{\frac{1}{2}}-1^{\frac{1}{2}}}{y-1}=\frac{1}{2}(1)^{\frac{1}{2}-1} \quad \left[\because \lim_{x\to a}\frac{x^n-a^n}{x-a}=na^{n-1}\right]$$

$$=\frac{1}{2}$$

Example 10. If $\lim_{x\to 1}\frac{x^4-1}{x-1}=\lim_{x\to k}\frac{x^3-k^3}{x^2-k^2}$, then find the value of k. *NCERT Exemplar*

Sol. Given, $\lim_{x\to 1}\frac{x^4-1}{x-1}=\lim_{x\to k}\frac{x^3-k^3}{x^2-k^2}$

$$\Rightarrow 4(1)^{4-1}=\lim_{x\to k}\frac{\frac{x^3-k^3}{x-k}}{\frac{x^2-k^2}{x-k}}\quad\left[\because \lim_{x\to a}\frac{x^n-a^n}{x-a}=na^{n-1}\right]$$

$$\Rightarrow 4=\frac{\lim_{x\to k}\frac{x^3-k^3}{x-k}}{\lim_{x\to k}\frac{x^2-k^2}{x-k}}\quad\left[\because \lim_{x\to a}\frac{f(x)}{g(x)}=\frac{\lim_{x\to a}f(x)}{\lim_{x\to a}g(x)}\right]$$

$$\Rightarrow 4=\frac{3k^2}{2k}\Rightarrow 4=\frac{3}{2}k$$

$$\Rightarrow k=\frac{8}{3}$$

Example 11. Evaluate $\lim_{x\to 1}\frac{x^7-2x^5+1}{x^3-3x^2+2}$. *NCERT Exemplar*

Sol. We have, $\lim_{x\to 1}\frac{x^7-2x^5+1}{x^3-3x^2+2}$ $\left[\text{which is }\frac{0}{0}\text{ from}\right]$

$$=\lim_{x\to 1}\frac{x^7-x^5-x^5+1}{x^3-x^2-2x^2+2}=\lim_{x\to 1}\frac{x^5(x^2-1)-1(x^5-1)}{x^2(x-1)-2(x^2-1)}$$

$$=\lim_{x\to 1}\frac{\frac{x^5(x^2-1)}{(x-1)}-\frac{1(x^5-1)}{(x-1)}}{\frac{x^2(x-1)}{(x-1)}-\frac{2(x^2-1)}{(x-1)}}$$

[dividing numerator and denominator by $(x-1)$]

$$=\frac{\lim_{x\to 1}x^5(x+1)-\lim_{x\to 1}\left(\frac{x^5-1}{x-1}\right)}{\lim_{x\to 1}x^2-\lim_{x\to 1}2(x+1)}$$

$$=\frac{1\times 2-5\times(1)^4}{1-2\times 2}=\frac{2-5}{1-4}=\frac{-3}{-3}=1$$

Example 12. Evaluate $\lim_{x\to 0}\frac{(1+x)^6-1}{(1+x)^2-1}$.

Sol. We have, $\lim_{x\to 0}\frac{(1+x)^6-1}{(1+x)^2-1}$

Put $1+x=y$, then $y\to 1$ as $x\to 0$

$$\therefore \lim_{x\to 0}\frac{(1+x)^6-1}{(1+x)^2-1}=\lim_{y\to 1}\frac{y^6-1}{y^2-1}=\lim_{y\to 1}\left(\frac{\frac{y^6-1}{y-1}}{\frac{y^2-1}{y-1}}\right)$$

[dividing numerator and denominator by $y-1$]

$$=\lim_{y\to 1}\frac{y^6-1}{y-1}\div\lim_{y\to 1}\frac{y^2-1}{y-1}=\frac{6(1)^{6-1}}{2(1)^{2-1}}=\frac{6}{2}=3$$

Example 13. Evaluate $\lim_{x\to 2}\frac{(2x+4)^{1/3}-2}{x-2}$.

Sol. We have, $\lim_{x\to 2}\frac{(2x+4)^{1/3}-2}{x-2}$

On putting $2x+4=y$, then $y\to 8$ as $x\to 2$

$$\therefore \lim_{x\to 2}\frac{(2x+4)^{1/3}-2}{x-2}=\lim_{y\to 8}\frac{y^{1/3}-2}{\frac{y-4}{2}-2}\quad\left[\because 2x+4=y\Rightarrow x=\frac{y-4}{2}\right]$$

$$=2\lim_{y\to 8}\frac{y^{1/3}-(8)^{1/3}}{y-4-4}=2\lim_{y\to 8}\frac{y^{1/3}-8^{1/3}}{y-8}=2\cdot\frac{1}{3}(8)^{\frac{1}{3}-1}$$

$$=2\cdot\frac{1}{3}(2^3)^{-\frac{2}{3}}=\frac{2}{3}\cdot(2)^{-2}=\frac{2}{3}\times\frac{1}{4}=\frac{1}{6}$$

Quick Test 2

Evaluate the following limits.

1. $\lim_{x\to 0}\frac{ax+b}{cx+1}$
2. $\lim_{x\to 3}\frac{\sqrt{2x+3}}{x+3}$
3. $\lim_{x\to -1}\frac{x^{10}+x^5+1}{x-1}$
4. $\lim_{x\to 1}\frac{ax^2+bx+c}{cx^2+bx+a}$
5. $\lim_{x\to 2}\frac{x^2-4}{x-2}$
6. $\lim_{x\to -2}\frac{\frac{1}{x}+\frac{1}{2}}{x+2}$
7. $\lim_{x\to 2}\frac{x^2-3x+2}{x^2+x-6}$
8. $\lim_{x\to 3}\frac{x^4-81}{2x^2-5x-3}$
9. $\lim_{x\to 1}\frac{(\sqrt{3+x}-\sqrt{5-x})}{x^2-1}$ *Comptency Based Que*
10. If $\lim_{x\to b}\frac{x^3-b^3}{x-b}=\lim_{x\to 1}\frac{x^4-1}{x-1}$, then find all possible values of b. *Comptency Based Que*

Answers

1. b 2. $\frac{1}{2}$ 3. $-\frac{1}{2}$ 4. 1 5. 4 6. $-\frac{1}{4}$ 7. $\frac{1}{5}$ 8. $\frac{108}{7}$ 9. $\frac{1}{4}$ 10. $\pm\frac{2}{\sqrt{3}}$

For Solutions, See QR Code on Pg. 357

12.3 Limits of Trigonometric Functions

Evaluation of Trigonometric Limits when the Variable Tends to Zero

To evaluate the trigonometric limits, when the variable x is measured in radians and tends to zero, then

- $\lim_{x\to 0}\frac{\sin x}{x}=1$
- $\lim_{x\to 0}\frac{\tan x}{x}=1$
- $\lim_{x\to 0}\frac{1-\cos x}{x}=0$

Evaluation of Trigonometric Limits when Variable Tends to a Non-zero Quantity

Suppose, we have limit of the form, $\lim_{x\to a} f(x)$.

In such cases, first we convert the limit tends to zero.

For this, replace $x\to a$ by $h\to 0$ and x by $(a+h)$ and then solve the limit.

Example 1. Evaluate $\lim_{x\to 0}\frac{\sin 3x}{5x}$.

Sol. $\lim_{x\to 0}\frac{\sin 3x}{5x}=\lim_{x\to 0}\frac{\sin 3x}{5x\times\frac{3}{3}}$

$=\lim_{x\to 0}\frac{3}{5}\cdot\frac{\sin 3x}{3x}$

$=\frac{3}{5}\cdot\lim_{3x\to 0}\frac{\sin 3x}{3x}$ [as $x\to 0$, therefore $3x\to 0$]

$=\frac{3}{5}\times 1$ $\left[\because \lim_{\theta\to 0}\frac{\sin\theta}{\theta}=1\right]$

$=\frac{3}{5}$

Example 2. Evaluate $\lim_{\theta\to 0}\frac{\sin a\theta}{\sin b\theta}$.

Sol. $\lim_{\theta\to 0}\frac{\sin a\theta}{\sin b\theta}=\lim_{\theta\to 0}\frac{\sin a\theta}{\theta}\div\lim_{\theta\to 0}\frac{\sin b\theta}{\theta}$

$=a\lim_{a\theta\to 0}\frac{\sin a\theta}{a\theta}\div b\lim_{b\theta\to 0}\frac{\sin b\theta}{b\theta}$

[as $\theta\to 0$, then $a\theta\to 0$ and as $\theta\to 0$, then $b\theta\to 0$]

$=a(1)\div b(1)=\frac{a}{b}$

Example 3. Evaluate $\lim_{x\to 0}\frac{1-\cos 4x}{x}$.

Sol. $\lim_{x\to 0}\frac{1-\cos 4x}{x}=\lim_{x\to 0}\frac{2\sin^2 2x}{x}\times\frac{4x}{4x}$

[$\because 1-\cos 2\theta=2\sin^2\theta$]

$=\lim_{x\to 0}2\left(\frac{\sin 2x}{2x}\right)^2\times 4x$

$=2\times 1\times 0$ $\left[\because \lim_{x\to 0}\frac{\sin x}{x}=1\right]$

$=0$

Example 4. Evaluate $\lim_{x\to 0}\frac{\sin ax+bx}{ax+\sin bx}$; $a, b, a+b\neq 0$. NCERT

Sol. $\lim_{x\to 0}\frac{\sin ax+bx}{ax+\sin bx}=\lim_{x\to 0}\frac{\frac{\sin ax}{x}+\frac{bx}{x}}{\frac{ax}{x}+\frac{\sin bx}{x}}$

[dividing both numerator and denominator by x]

$=\frac{\lim_{x\to 0}\frac{\sin(ax)}{ax}\times a+\lim_{x\to 0}b}{\lim_{x\to 0}a+\lim_{x\to 0}\frac{\sin bx}{bx}\times b}$

$\left[\because \lim_{x\to a}\frac{f(x)}{g(x)}=\frac{\lim_{x\to a}f(x)}{\lim_{x\to a}g(x)}\text{ and } \lim_{x\to a}(f(x)+g(x))=\lim_{x\to a}f(x)+\lim_{x\to a}g(x)\right]$

$=\frac{a\lim_{x\to 0}\frac{\sin ax}{ax}+\lim_{x\to 0}b}{\lim_{x\to 0}a+b\lim_{x\to 0}\frac{\sin bx}{bx}}$

$=\frac{a(1)+b}{a+b(1)}$ $\left[\because \lim_{x\to 0}\frac{\sin x}{x}=1\right]$

$=\frac{a+b}{a+b}=1$

Example 5. Evaluate $\lim_{x\to 0}\frac{\tan 2x-\sin 2x}{x^3}$.

Sol. $\lim_{x\to 0}\frac{\tan 2x-\sin 2x}{x^3}$

$=\lim_{x\to 0}\frac{\frac{\sin 2x}{\cos 2x}-\sin 2x}{x^3}$

$=\lim_{x\to 0}\frac{\sin 2x-\sin 2x\cos 2x}{x^3\cos 2x}$

$=\lim_{x\to 0}\frac{\sin 2x(1-\cos 2x)}{x^3\cos 2x}$

$$= \lim_{x\to 0} \frac{\tan 2x}{x} \cdot \lim_{x\to 0} \frac{2\sin^2 x}{x^2}$$

[using product of limits and $\cos 2\theta = 1 - 2\sin^2\theta$]

$$= 2\lim_{x\to 0} \frac{\tan 2x}{2x} \times 2\lim_{x\to 0}\left(\frac{\sin x}{x}\right)^2$$

$$= 2(1) \times 2(1)^2 \left[\because \lim_{x\to 0} \frac{\sin x}{x} = \lim_{x\to 0} \frac{\tan x}{x} = 1\right]$$

$$= 4$$

Example 6. Evaluate $\lim_{x\to 0} \frac{\cos ax - \cos bx}{\cos cx - 1}$.

NCERT Exemplar, Comptency Based Que

Sol. $\lim_{x\to 0} \frac{\cos ax - \cos bx}{\cos cx - 1}$

$$= \lim_{x\to 0} \frac{-2\sin\left[\left(\frac{a+b}{2}\right)x\right]\sin\left[\left(\frac{a-b}{2}\right)x\right]}{-2\sin^2\frac{cx}{2}}$$

$$\left[\because \cos C - \cos D = -2\sin\left(\frac{C+D}{2}\right)\sin\left(\frac{C-D}{2}\right) \text{ and } \cos 2\theta = 1 - 2\sin^2\theta\right]$$

$$= \lim_{x\to 0} \frac{\sin\left(\frac{a+b}{2}\right)x \cdot \sin\left(\frac{a-b}{2}\right)x}{x^2} \cdot \frac{x^2}{\sin^2\frac{cx}{2}}$$

$$= \lim_{x\to 0} \frac{\sin\left(\frac{a+b}{2}\right)x}{\left(\frac{a+b}{2}\right)x \cdot \left(\frac{2}{a+b}\right)} \cdot \frac{\sin\left[\left(\frac{a-b}{2}\right)x\right]}{\left(\frac{a-b}{2}\right)x \cdot \left(\frac{2}{a-b}\right)} \cdot \frac{\left(\frac{cx}{2}\right)^2 \times \frac{4}{c^2}}{\sin^2\frac{cx}{2}}$$

$$= \left(\frac{a+b}{2} \times \frac{a-b}{2} \times \frac{4}{c^2}\right) \quad \left[\because \lim_{x\to 0} \frac{\sin x}{x} = 1\right]$$

$$= \frac{a^2 - b^2}{c^2}$$

Example 7. Evaluate $\lim_{x\to \pi} \frac{\sin x}{\pi - x}$.

Comptency Based Que

Sol. $\lim_{x\to \pi} \frac{\sin x}{\pi - x} = \lim_{h\to 0} \frac{\sin(\pi + h)}{\pi - (\pi + h)}$

[put $x = \pi + h$ as $x \to \pi$, then $h \to 0$]

$$= \lim_{h\to 0} \frac{-\sin h}{-h}$$

$$= \lim_{h\to 0} \frac{\sin h}{h} = 1$$

Example 8. Evaluate $\lim_{x\to \pi/6} \frac{\sqrt{3}\sin x - \cos x}{x - \frac{\pi}{6}}$.

NCERT Exemplar, Comptency Based Que

Sol. $\lim_{x\to \pi/6} \frac{\sqrt{3}\sin x - \cos x}{x - \frac{\pi}{6}}$

$$= \lim_{x\to \pi/6} \frac{2\left[\sin x \cos\frac{\pi}{6} - \cos x \sin\frac{\pi}{6}\right]}{\left(x - \frac{\pi}{6}\right)}$$

$$= 2\lim_{x\to \pi/6} \frac{\sin\left(x - \frac{\pi}{6}\right)}{\left(x - \frac{\pi}{6}\right)}$$

$[\because \sin A \cos B - \cos A \sin B = \sin(A - B)]$

$$= 2 \quad \left[\because \lim_{\theta\to 0} \frac{\sin\theta}{\theta} = 1 \text{ and } x \to \frac{\pi}{6} \Rightarrow \left(x - \frac{\pi}{6}\right) \to 0\right]$$

Example 9. Evaluate $\lim_{x\to \pi/6} \frac{2\sin^2 x + \sin x - 1}{2\sin^2 x - 3\sin x + 1}$ by using factorisation method.

NCERT Exemplar

Sol. $\lim_{x\to \pi/6} \frac{2\sin^2 x + \sin x - 1}{2\sin^2 x - 3\sin x + 1}$

$$= \lim_{x\to \pi/6} \frac{(2\sin x - 1)(\sin x + 1)}{(2\sin x - 1)(\sin x - 1)}$$

$$= \lim_{x\to \pi/6} \frac{\sin x + 1}{\sin x - 1}$$

$$= \frac{1 + \sin\frac{\pi}{6}}{\sin\frac{\pi}{6} - 1}$$

$$= \frac{1 + \frac{1}{2}}{\frac{1}{2} - 1} = -3$$

Quick Test 3

Direction (Q. Nos. 1 to 9) Evaluate the following limits.

1. $\lim_{x\to 0} x \operatorname{cosec} 3x$

2. $\lim_{x\to 0} x \cot 4x$

3. $\lim_{x\to 0} \frac{\sin 3x}{\sin 7x}$

4. $\lim_{x\to 0} \frac{ax + x\cos x}{b \sin x}$ Competency Based Que

5. $\lim_{x\to 0} \frac{1-\cos 5x}{x}$

6. $\lim_{x\to 0} (\operatorname{cosec} x - \cot x)$ Competency Based Que

7. $\lim_{x\to 0} \frac{\sin x}{x(1+\cos x)}$

8. $\lim_{x\to \pi} \frac{\sin(\pi - x)}{\pi(\pi - x)}$ Competency Based Que

9. $\lim_{x\to \pi/2} \frac{\tan 2x}{x - \frac{\pi}{2}}$

10. If $f(x) = \begin{cases} \frac{k\cos x}{\pi - 2x}, & \text{when } x \neq \frac{\pi}{2} \\ 3, & \text{when } x = \frac{\pi}{2} \end{cases}$ and $\lim_{x\to \pi/2} f(x) = f\left(\frac{\pi}{2}\right)$, then find the value of k. Competency Based Que

Answers

1. $\frac{1}{3}$ **2.** $\frac{1}{4}$ **3.** $\frac{3}{7}$ **4.** $\frac{a+1}{b}$ **5.** 0 **6.** 0 **7.** $\frac{1}{2}$ **8.** $\frac{1}{\pi}$ **9.** 2 **10.** 6

For Solutions, See QR Code on Pg. 357

12.4 Derivative of a Function

Suppose, f is a real valued function and a is a point in its domain. Then, derivative of f at a is defined by

$$\lim_{h\to 0} \frac{f(a+h) - f(a)}{h}$$

provided this limit exists.

The derivative of $f(x)$ at a is denoted by

$$f'(a) \text{ or } \left[\frac{df(x)}{dx}\right]_{x=a}$$

12.5 First Principle of Derivative

Suppose f is a real valued function, the function defined by $\lim_{h\to 0} \frac{f(x+h) - f(x)}{h}$, wherever the limit exists, is defined to be the derivative of f at x and is denoted by $f'(x)$. This definition of derivative is called the first principle of derivative.

Thus, $$f'(x) = \lim_{h\to 0} \frac{f(x+h) - f(x)}{h}$$

Sometimes, $f'(x)$ is denoted by $\frac{d}{dx}[f(x)]$ or if $y = f(x)$, then it is denoted by $\frac{dy}{dx}$ and referred to as derivative of $f(x)$ or y with respect to x.

It is also denoted by $D[f(x)]$. Derivative of f at $x = a$ is also given by substituting $x = a$ in $f'(x)$ and it is denoted by

$$\left.\frac{d}{dx} f(x)\right|_a \text{ or } \left.\frac{df}{dx}\right|_a \text{ or } \left(\frac{df}{dx}\right)_{x=a}.$$

Example 1. Find the derivative of $f(x) = 4x + 5$ at $x = 3$.

Sol. Given, $f(x) = 4x + 5$

We know that at $x = a$, $f'(a) = \lim_{h\to 0} \frac{f(a+h) - f(a)}{h}$

$$\therefore f'(3) = \lim_{h\to 0} \frac{f(3+h) - f(3)}{h}$$

$$= \lim_{h\to 0} \frac{4(3+h) + 5 - (4\times 3 + 5)}{h}$$

$$= \lim_{h\to 0} \frac{12 + 4h + 5 - 17}{h} = \lim_{h\to 0} \frac{4h}{h} = 4$$

Example 2. Find the derivative of the function $f(x) = 2x^2 + 3x - 5$ at $x = -1$. Also, prove that $f'(0) + 3f'(-1) = 0$. NCERT

Sol. Clearly, $f'(-1) = \lim_{h\to 0} \frac{f(-1+h) - f(-1)}{h}$

$$\left[\because f'(a) = \lim_{h\to 0} \frac{f(a+h) - f(a)}{h}\right]$$

$$= \lim_{h\to 0} \frac{[2(-1+h)^2 + 3(-1+h) - 5] - [2(-1)^2 + 3(-1) - 5]}{h}$$

$$= \lim_{h\to 0} \frac{[2(1 + h^2 - 2h) - 3 + 3h - 5] - [2 - 3 - 5\}]}{h}$$

$$= \lim_{h\to 0} \frac{2h^2 - h}{h} = \lim_{h\to 0} (2h-1) = 2(0) - 1 = -1$$

and $f'(0) = \lim_{h\to 0} \frac{f(0+h) - f(0)}{h}$

$$\left[\because f'(a) = \lim_{h\to 0} \frac{f(a+h)-f(a)}{h}\right]$$

$$= \lim_{h\to 0} \frac{[2(0+h)^2 + 3(0+h) - 5] - [2(0)^2 + 3(0) - 5]}{h}$$

$$= \lim_{h\to 0} \frac{2h^2 + 3h}{h} = \lim_{h\to 0}(2h+3) = 2(0) + 3 = 3$$

Now, $f'(0) + 3f'(-1) = 3 - 3 = 0$ **Hence proved.**

Example 3. Find the derivative of $f(x) = \frac{1}{x}$ from first principle.

Sol. We have, $f(x) = \frac{1}{x}$

By using first principle, $f'(x) = \lim_{h\to 0} \frac{f(x+h) - f(x)}{h}$

$$\therefore f'(x) = \lim_{h\to 0} \frac{\frac{1}{x+h} - \frac{1}{x}}{h} \qquad \left[\because f(x) = \frac{1}{x} \quad \therefore f(x+h) = \frac{1}{x+h}\right]$$

$$= \lim_{h\to 0} \frac{1}{h}\left[\frac{x-(x+h)}{x(x+h)}\right] = \lim_{h\to 0} \frac{1}{h}\left[\frac{-h}{x(x+h)}\right]$$

$$= \lim_{h\to 0}\left[\frac{-1}{x(x+h)}\right] = \frac{-1}{x^2}$$

Example 4. Find the derivative of $f(x) = x^n$, where n is positive integer, by first principle. NCERT

Sol. By definition of first principle,

$$f'(x) = \lim_{h\to 0} \frac{f(x+h) - f(x)}{h} = \lim_{h\to 0} \frac{(x+h)^n - x^n}{h} \qquad \text{...(i)}$$

By binomial theorem, we get

$$(x+h)^n = {}^nC_0 x^n + {}^nC_1 x^{n-1} h + {}^nC_2 x^{n-2} h^2 + \dots + {}^nC_n h^n$$

$$\Rightarrow (x+h)^n = x^n + nhx^{n-1} + \frac{n(n-1)}{2} h^2 x^{n-2} + \dots + h^n$$

$$\Rightarrow (x+h)^n - x^n = nhx^{n-1} + \frac{n(n-1)}{2} h^2 x^{n-2} + \dots + h^n$$

On putting this value in Eq. (i), we get

$$f'(x) = \lim_{h\to 0} \frac{nhx^{n-1} + \frac{n(n-1)}{2} h^2 x^{n-2} + \dots + h^n}{h}$$

$$= \lim_{h\to 0} \frac{h\left[nx^{n-1} + \frac{n(n-1)}{2} hx^{n-2} + \dots + h^{n-1}\right]}{h}$$

$$= \lim_{h\to 0}\left[nx^{n-1} + \frac{n(n-1)}{2} hx^{n-2} + \dots + h^{n-1}\right] = nx^{n-1}$$

Hence, $f'(x)$ or $\frac{d}{dx} f(x) = nx^{n-1}$

Example 5. Find the derivative of $(x-1)(x-2)$ from first principle. NCERT

Sol. Let $f(x) = (x-1)(x-2) = x^2 - 3x + 2$

By first principle of derivative, we have

$$f'(x) = \lim_{h\to 0} \frac{f(x+h) - f(x)}{h}$$

$$= \lim_{h\to 0} \frac{[(x+h)^2 - 3(x+h) + 2] - [x^2 - 3x + 2]}{h}$$

$$= \lim_{h\to 0} \frac{[x^2 + h^2 + 2xh - 3x - 3h + 2] - [x^2 - 3x + 2]}{h}$$

$$= \lim_{h\to 0} \frac{2hx + h^2 - 3h}{h} = \lim_{h\to 0} \frac{h(2x + h - 3)}{h} = 2x - 3$$

Example 6. Find the derivative of $\left(\frac{x+1}{x-1}\right)$ from the first principle. NCERT

Sol. Let $f(x) = \frac{x+1}{x-1}$

By first principle of derivative, we have

$$f'(x) = \lim_{h\to 0} \frac{f(x+h) - f(x)}{h} = \lim_{h\to 0} \frac{\left[\frac{(x+h)+1}{(x+h)-1} - \frac{x+1}{x-1}\right]}{h}$$

$$= \lim_{h\to 0} \frac{(x+h+1)(x-1) - (x+1)(x+h-1)}{h(x+h-1)(x-1)}$$

$$= \lim_{h\to 0} \frac{(x^2 - x + xh - h + x - 1) - (x^2 + xh - x + x + h - 1)}{h(x+h-1)(x-1)}$$

$$= \lim_{h\to 0} \frac{-2h}{h(x+h-1)(x-1)} = \lim_{h\to 0} \frac{-2}{(x+h-1)(x-1)}$$

$$= \frac{-2}{(x-1)^2}$$

Example 7. Find the derivative of the following function by using first principle. NCERT Exemplar

(i) sin x

(ii) sec x

Sol. (i) Let $f(x) = \sin x$

By first principle of derivative, we get

$$f'(x) = \lim_{h\to 0} \frac{f(x+h) - f(x)}{h}$$

$$= \lim_{h\to 0} \frac{\sin(x+h) - \sin x}{h}$$

$$= \lim_{h\to 0} \frac{2\cos\left(\frac{x+h+x}{2}\right)\cdot \sin\left(\frac{x+h-x}{2}\right)}{h}$$

$$\left[\because \sin C - \sin D = 2\cos\left(\frac{C+D}{2}\right)\sin\left(\frac{C-D}{2}\right)\right]$$

$$= \lim_{h\to 0} \frac{2\cos\left(x+\frac{h}{2}\right)\cdot \sin\frac{h}{2}}{h}$$

$$= \lim_{h\to 0} \cos\left(x+\frac{h}{2}\right)\cdot \lim_{\frac{h}{2}\to 0} \frac{\sin\frac{h}{2}}{\frac{h}{2}} \qquad \left[\therefore\ h\to 0 \Rightarrow \frac{h}{2}\to 0\right]$$

$$= \lim_{h\to 0} \cos\left(x+\frac{h}{2}\right)\times 1 \qquad \left[\therefore\ \lim_{\theta\to 0}\frac{\sin\theta}{\theta}=1\right]$$

$$= \cos(x+0) = \cos x \qquad [\text{put } h=0]$$

$$\therefore\ \frac{d}{dx}(\sin x) = \cos x$$

(ii) Let $f(x) = \sec x$

By first principle of derivative, we have

$$f'(x) = \lim_{h\to 0} \frac{f(x+h)-f(x)}{h}$$

$$\therefore\quad f'(x) = \lim_{h\to 0} \frac{\sec(x+h)-\sec x}{h}$$

$$= \lim_{h\to 0} \frac{\frac{1}{\cos(x+h)} - \frac{1}{\cos x}}{h}$$

$$= \lim_{h\to 0} \frac{\cos x - \cos(x+h)}{h\times \cos x\cdot \cos(x+h)}$$

$$= \lim_{h\to 0} \left[\frac{-2\sin\left(\frac{x+x+h}{2}\right)\cdot \sin\frac{(x-x-h)}{2}}{h\cdot \cos x\cdot \cos(x+h)}\right]$$

$$\left[\because \cos C - \cos D = -2\sin\left(\frac{C+D}{2}\right)\sin\left(\frac{C-D}{2}\right)\right]$$

$$= \lim_{h\to 0}\left[\frac{-2\sin\left(x+\frac{h}{2}\right)\cdot\left(-\sin\frac{h}{2}\right)}{h\cdot \cos x\cos(x+h)}\right]$$

$$= \lim_{h\to 0} \frac{\sin\left(x+\frac{h}{2}\right)}{\cos(x+h)\cdot \cos x}\cdot \lim_{h\to 0} \frac{\sin\frac{h}{2}}{\frac{h}{2}}$$

$$= \frac{\sin x}{\cos^2 x}\times 1 \qquad \left[\because \lim_{\theta\to 0}\frac{\sin\theta}{\theta}=1\right]$$

$$= \frac{\sin x}{\cos x}\cdot\frac{1}{\cos x}$$

$$= \tan x\cdot \sec x$$

Example 8. **Differentiate the function $\cos(x^2+1)$ by the first principle.** *[NCERT Exemplar]*

Sol. Let $f(x) = \cos(x^2+1)$

We know by first principle,

$$f'(x) = \lim_{h\to 0}\frac{f(x+h)-f(x)}{h}$$

$$\therefore\ f'(x) = \lim_{h\to 0}\frac{\cos[(x+h)^2+1] - \cos(x^2+1)}{h}$$

$$[\because f(x) = \cos(x^2+1)]$$

$$= \lim_{h\to 0}\frac{-2\sin\frac{(x+h)^2+1+x^2+1}{2}\sin\frac{(x+h)^2+1-(x^2+1)}{2}}{h}$$

$$\left[\because \cos C - \cos D = -2\sin\left(\frac{C+D}{2}\right)\sin\left(\frac{C-D}{2}\right)\right]$$

$$= \lim_{h\to 0}\frac{-2\sin\frac{(x+h)^2+x^2+2}{2}\sin\frac{(x+h)^2-x^2}{2}}{h}$$

$$= \lim_{h\to 0}\frac{\left\{-2\sin\frac{(x+h)^2+x^2+2}{2}\left[\sin\frac{(x+h)^2-x^2}{2}\right]\cdot\left[\frac{(x+h)^2-x^2}{2}\right]\right\}}{h\times\frac{(x+h)^2-x^2}{2}}$$

$$= \lim_{h\to 0}\frac{-2\sin\frac{(x+h)^2+x^2+2}{2}\times\frac{x^2+h^2+2xh-x^2}{2}}{h}$$

$$\left[\because \lim_{h\to 0}\frac{\sin\left(\frac{(x+h)^2-x^2}{2}\right)}{\frac{(x+h)^2-x^2}{2}}=1\right]$$

$$= \lim_{h\to 0}\frac{-2\sin\left[\frac{(x+h)^2+x^2+2}{2}\right](2x+h)\frac{h}{2}}{h}$$

$$= -2\sin\left(\frac{2x^2+2}{2}\right)\times 2x\times\frac{1}{2}$$

$$= -2x\sin(x^2+1)$$

Example 9. **Find the derivative of $x \sin x$ from first principle.** NCERT

Sol. We have, $f(x) = x \sin x$

By first principle of derivative,

$$f'(x) = \lim_{h \to 0} \frac{f(x+h) - f(x)}{h}$$

$$\therefore \quad f'(x) = \lim_{h \to 0} \frac{(x+h)\sin(x+h) - x\sin x}{h}$$

$$= \lim_{h\to 0} \frac{(x+h)[\sin x \cdot \cos h + \cos x \cdot \sin h] - x \sin x}{h}$$

$[\because \sin(A+B) = \sin A \cos B + \cos A \sin B]$

$$= \lim_{h\to 0} \frac{\left[\begin{array}{l} x\sin x \cdot \cos h + x \cdot \cos x \cdot \sin h \\ \quad + h\sin x \cdot \cos h + h\cos x \cdot \sin h - x \sin x \end{array}\right]}{h}$$

$$= \lim_{h\to 0} \frac{\left[\begin{array}{l} x\sin x(\cos h - 1) + x \cdot \cos x \cdot \sin h \\ \quad + h(\sin x \cdot \cos h + \cos x \cdot \sin h) \end{array}\right]}{h}$$

$$= \lim_{h\to 0} \frac{x\sin x(\cos h - 1)}{h} + \lim_{h \to 0} x \cdot \cos x \cdot \frac{\sin h}{h} + \lim_{h\to 0} \frac{h(\sin x \cdot \cos h + \cos x \cdot \sin h)}{h}$$

$$= x \sin x \lim_{h\to 0}\left[\frac{-(1-\cos h)}{h}\right] + x \cdot \cos x(1) + \sin x$$

$$= - x\sin x \cdot \lim_{\frac{h}{2} \to 0} \frac{2\sin^2 \frac{h}{2}}{h \times \frac{h}{4}} \times \frac{h}{4} + x \cos x + \sin x$$

$$= - x \cdot \sin x \cdot \frac{2}{4} \lim_{\frac{h}{2} \to 0} \left(\frac{\sin \frac{h}{2}}{\frac{h}{2}}\right)^2 \times h + x \cos x + \sin x$$

$$= - x \sin x \cdot \frac{1}{2}(1) \times 0 + x \cdot \cos x + \sin x \quad \left[\because \lim_{\theta\to 0} \frac{\sin\theta}{\theta} = 1\right]$$

$$= x \cos x + \sin x$$

Example 10. **Find the derivative of $f(x)$ from the first principle, where $f(x)$ is $\sin x + \cos x$.**

Sol. We have, $f(x) = \sin x + \cos x$

By first principle of derivative, $f'(x) = \dfrac{f(x+h) - f(x)}{h}$

$$= \lim_{h\to 0} \frac{\sin(x+h) + \cos(x+h) - \sin x - \cos x}{h}$$

$$= \lim_{h\to 0} \frac{\sin x \cos h + \cos x \sin h + \cos x \cos h - \sin x \sin h - \sin x - \cos x}{h}$$

$\left[\begin{array}{l} \because \quad \sin(A+B) = \sin A \cos B + \cos A \sin B \\ \text{and } \cos(A+B) = \cos A \cos B - \sin A \sin B \end{array}\right]$

$$= \lim_{h\to 0} \frac{\sin h(\cos x - \sin x) + \sin x(\cos h - 1) + \cos x(\cos h - 1)}{h}$$

$$= \lim_{h\to 0} \frac{\sin h}{h}(\cos x - \sin x) + \lim_{h\to 0} \sin x \frac{(\cos h - 1)}{h} + \lim_{h\to 0} \cos x \frac{(\cos h - 1)}{h}$$

$$= \cos x - \sin x + \sin x \times 0 + \cos x \times 0$$

$$\left[\because \lim_{h\to 0} \frac{\sin h}{h} = 1 \text{ and } \lim_{h\to 0}\left(\frac{\cos h - 1}{h}\right) = 0\right]$$

$$= \cos x - \sin x$$

Quick Test 4

1 Find the derivative of $f(x) = 3$ at $x = 0$ and at $x = 3$.

2 Using first principle, find the derivative of $99x$ at $x = 100$. NCERT

Directions (Q. Nos. 3-10) *Find the derivative of the following functions from first principle.*

3 $x^3 - 27$ **4** $\frac{1}{x^3}$ **5** $x^{2/3}$

6 $(x^2 + 1)(x - 5)$ **7** $\operatorname{cosec} x$ **8** $\sin(x + 1)$

9 $x \cos x$ **10** $(\sin x - \cos x)$

Answers

1. $f'(0) = 0, f'(3) = 0$ 2. 99 3. $3x^2$ 4. $\frac{-3}{x^4}$ 5. $\frac{2}{3}x^{-1/3}$ 6. $3x^2 - 10x + 1$ 7. $-\operatorname{cosec} x \cot x$ 8. $\cos(x + 1)$ 9. $\cos x - x \sin x$ 10. $\cos x + \sin x$

For Solutions, See QR Code on Pg. 357

12.6 Algebra of Derivative of Functions

Let f and g be two functions such that their derivatives are defined in a common domain. Then,

- Derivative of sum of two functions is sum of the derivatives of the functions.
$$\frac{d}{dx}[f(x)+g(x)]=\frac{d}{dx}f(x)+\frac{d}{dx}g(x)$$
- Derivative of difference of two functions is difference of the derivatives of the functions.
$$\frac{d}{dx}[f(x)-g(x)]=\frac{d}{dx}f(x)-\frac{d}{dx}g(x)$$
- Derivative of product of two functions is given by the following product rule.
$$\frac{d}{dx}[f(x)\cdot g(x)]=f(x)\frac{d}{dx}g(x)+g(x)\frac{d}{dx}f(x)$$
This is also known as Leibnitz product rule of derivative.
- Derivative of quotient of two functions is given by the following quotient rule.
$$\frac{d}{dx}\left[\frac{f(x)}{g(x)}\right]=\frac{g(x)\frac{d}{dx}f(x)-f(x)\frac{d}{dx}g(x)}{[g(x)]^2}, g(x)\neq 0$$

TEACHER TIPS

Theorem 1 Derivative of $f(x)=x^n$ is nx^{n-1} for any positive integer n.

The above theorem is true for all powers of x i.e. n can be any real number.

Theorem 2 Derivative of Polynomial Functions

Let $f(x)=a_nx^n+a_{n-1}x^{n-1}+...+a_1x+a_0$

be a polynomial function, where a_i's are all real numbers and $a_n\neq 0$. Then, the derivative function is given by

$$\frac{d}{dx}[f(x)]=na_nx^{n-1}+(n-1)a_{n-1}x^{n-2}+...+2a_2x+a_1$$

12.7 Derivative of Trigonometric Functions

To find the derivative of trigonometric functions, we use the algebra of derivative and the following formulae

- $\frac{d}{dx}(\sin x)=\cos x$
- $\frac{d}{dx}(\cos x)=-\sin x$
- $\frac{d}{dx}(\tan x)=\sec^2 x$
- $\frac{d}{dx}(\sec x)=\sec x\cdot\tan x$
- $\frac{d}{dx}(\cot x)=-\text{cosec}^2 x$
- $\frac{d}{dx}(\text{cosec } x)=-\text{cosec } x\cdot\cot x$

Example 1. **Differentiate $2x^3-4x^2+6x+8$ w.r.t. x.**

Sol. Let $y=2x^3-4x^2+6x+8$

On differentiating both sides w.r.t. x, we get

$$\frac{dy}{dx}=\frac{d}{dx}(2x^3-4x^2+6x+8)$$
$$=2\frac{d}{dx}(x^3)-4\frac{d}{dx}(x^2)+6\frac{d}{dx}(x)+\frac{d}{dx}(8)$$
$$=2(3x^2)-4(2x)+6(1)+0 \quad \left[\because \frac{d}{dx}(x^n)=nx^{n-1}\right]$$
$$=6x^2-8x+6$$

Example 2. **If $u=7t^4-2t^3-8t-5$, then find $\frac{du}{dt}$ at $t=2$.**

Sol. We have, $u=7t^4-2t^3-8t-5$

On differentiating both sides w.r.t. t, we get

$$\frac{du}{dt}=\frac{d}{dt}[7t^4-2t^3-8t-5]$$
$$=7(4t^3)-2(3t^2)-8(1)-0 \quad \left[\because \frac{d}{dx}(x^n)=nx^{n-1}\right]$$
$$=28t^3-6t^2-8$$

Now, $\left(\frac{du}{dt}\right)_{t=2}=28(2)^3-6(2)^2-8$
$$=224-24-8=192$$

Example 3. **Differentiate the following functions w.r.t. x.**

(i) $\left(x+\frac{1}{x}\right)^3$ (ii) $(ax+b)(cx+d)^2$

(iii) $\frac{px^2+qx+r}{(ax+b)}$ NCERT

Sol. (i) Let $y=\left(x+\frac{1}{x}\right)^3=x^3+3x\times\frac{1}{x^2}+3x^2\times\frac{1}{x}+\frac{1}{x^3}$

$[\because (A+B)^3=A^3+3A^2B+3AB^2+B^3]$

$$=x^3+\frac{3}{x}+3x+\frac{1}{x^3}$$

On differentiating both sides w.r.t. x, we get

$$\frac{dy}{dx}=\frac{d}{dx}(x^3)+\frac{d}{dx}\left(\frac{3}{x}\right)+\frac{d}{dx}(3x)+\frac{d}{dx}\left(\frac{1}{x^3}\right)$$

$$=3x^2+3\frac{d}{dx}(x^{-1})+3\times1+\frac{d}{dx}(x^{-3})$$

$$=3x^2+\frac{-3}{x^2}+3-\frac{3}{x^4}$$

(ii) Let $y=(ax+b)(cx+d)^2$

On differentiating both sides w.r.t. x, we get

$$\frac{dy}{dx}=(ax+b)\frac{d}{dx}(cx+d)^2+(cx+d)^2\frac{d}{dx}(ax+b)$$

[using product rule of derivative]

$$=(ax+b)\frac{d}{dx}(c^2x^2+d^2+2cxd)+(cx+d)^2(a\times1+0)$$

$[\because (A+B)^2=A^2+2AB+B^2]$

$$=(ax+b)[c^2(2x)+0+2c\times1\times d]+(cx+d)^2\times a$$

$$=(ax+b)(2c^2x+2cd)+a(cx+d)^2$$

$$=(ax+b)\,2c\,(cx+d)+a(cx+d)^2$$

$$=(cx+d)[2c(ax+b)+a(cx+d)]$$

$$=(cx+d)(2acx+2bc+acx+ad)$$

$$=(cx+d)(3acx+2bc+ad)$$

(iii) Let $y=\dfrac{px^2+qx+r}{(ax+b)}$

On differentiating both sides w.r.t. x, we get

$$\frac{dy}{dx}=\frac{\left[(ax+b)\frac{d}{dx}(px^2+qx+r)-(px^2+qx+r)\frac{d}{dx}(ax+b)\right]}{(ax+b)^2}$$

[by quotient rule]

$$=\frac{(ax+b)(2px+q+0)-(px^2+qx+r)(a\times1+0)}{(ax+b)^2}$$

$$=\frac{(ax+b)(2px+q)-(px^2+qx+r)\,a}{(ax+b)^2}$$

$$=\frac{(2apx^2+aqx+2bpx+bq)-(apx^2+aqx+ra)}{(ax+b)^2}$$

$$=\frac{2apx^2+aqx+2bpx+bq-apx^2-aqx-ra}{(ax+b)^2}$$

$$\frac{dy}{dx}=\frac{apx^2+2bpx+bq-ra}{(ax+b)^2}$$

Example 4. **Differentiate the following functions w.r.t. x.**

(i) $\cos(x+a)$ **(ii) $\dfrac{\sin(x+a)}{\cos x}$**

(iii) $f(x)=\sin 2x$ **(iv) $f(x)=(\cos^2 x-\sin^2 x)$**

(v) $x^2\cos x$

Sol. (i) Let $y=\cos(x+a)$

$$y=\cos x\cos a-\sin x\sin a$$

$[\because \cos(A+B)=\cos A\cos B-\sin A\sin B]$

On differentiating both sides w.r.t. x, we get

$$\frac{dy}{dx}=\cos a\frac{d}{dx}(\cos x)-\sin a\frac{d}{dx}(\sin x)$$

$$=\cos a(-\sin x)-\sin a(\cos x)$$

$$=-\cos a\sin x-\sin a\cos x=-\sin(x+a)$$

$[\because \sin A\cos B+\cos A\sin B=\sin(A+B)]$

(ii) Let $y=\dfrac{\sin(x+a)}{\cos x}=\dfrac{\sin x\cos a+\cos x\sin a}{\cos x}$

$[\because \sin(A+B)=\sin A\cos B+\cos A\sin B]$

$$=\frac{\sin x\cos a}{\cos x}+\frac{\cos x\sin a}{\cos x}=\cos a\tan x+\sin a$$

On differentiating both sides w.r.t. x, we get

$$\frac{dy}{dx}=\cos a\frac{d}{dx}(\tan x)+\frac{d}{dx}(\sin a)$$

$$=\cos a\sec^2x+0 \quad \left[\because \frac{d}{dx}(\text{constant})=0\right]$$

$$=\frac{\cos a}{\cos^2 x} \quad \left[\because \sec x=\frac{1}{\cos x}\right]$$

(iii) We have, $f(x)=\sin 2x=2\sin x\cos x$

On differentiating both sides w.r.t. x, we get

$$f'(x)=2\left[\sin x\frac{d}{dx}(\cos x)+\cos x\frac{d}{dx}(\sin x)\right]$$

[using product rule of derivative]

$$=2[\sin x\cdot(-\sin x)+\cos x\cdot\cos x]$$

$$=2(-\sin^2x+\cos^2x)$$

$$=2(\cos^2x-\sin^2x)=2\cos 2x$$

(iv) We have, $f(x)=(\cos^2x-\sin^2x)$

$$f(x)=(\cos x-\sin x)(\cos x+\sin x)$$

On differentiating both sides w.r.t. x, we get

$$f'(x)=(\cos x-\sin x)\frac{d}{dx}(\cos x+\sin x)+(\cos x+\sin x)\frac{d}{dx}(\cos x-\sin x)$$

[using product rule of derivative]

$$=(\cos x-\sin x)(-\sin x+\cos x)+(\cos x+\sin x)(-\sin x-\cos x)$$

$$=(\cos x-\sin x)(\cos x-\sin x)-(\cos x+\sin x)(\cos x+\sin x)$$

$$=(\cos x-\sin x)^2-(\cos x+\sin x)^2$$

$$=(\cos^2x+\sin^2x-2\sin x\cos x)-(\cos^2x+\sin^2x+2\sin x\cos x)$$

$= -2\sin x\cos x - 2\sin x\cos x$

$= -\sin 2x - \sin 2x = -2\sin 2x$

(v) Let $y = x^2 \cos x$

On differentiating both sides w.r.t. x, we get

$$\frac{dy}{dx} = \frac{d}{dx}(x^2\cos x) = x^2\frac{d}{dx}(\cos x) + \cos x\frac{d}{dx}(x^2)$$

[using product rule of derivative]

$= x^2(-\sin x) + \cos x(2x)$

$= 2x\cos x - x^2\sin x$

Example 5. **Find the derivative of $\frac{4x+5\sin x}{3x+7\cos x}$.** *NCERT*

Sol. Let $f(x) = \frac{4x+5\sin x}{3x+7\cos x}$

On differentiating both sides w.r.t. x, we get

$$f'(x) = \frac{\left[(3x+7\cos x)\frac{d}{dx}(4x+5\sin x) - (4x+5\sin x) \times \frac{d}{dx}(3x+7\cos x)\right]}{(3x+7\cos x)^2}$$

$$\left[\because \frac{d}{dx}\left(\frac{f(x)}{g(x)}\right) = \frac{g(x)\frac{d}{dx}f(x) - f(x)\frac{d}{dx}g(x)}{[g(x)]^2}\right]$$

$$= \frac{(3x+7\cos x)(4+5\cos x) - (4x+5\sin x)[3+7(-\sin x)]}{(3x+7\cos x)^2}$$

$$= \frac{[(12x+15x\cos x + 28\cos x + 35\cos^2 x) - (12x - 28x\sin x + 15\sin x - 35\sin^2 x)]}{(3x+7\cos x)^2}$$

$$= \frac{[35(\sin^2 x + \cos^2 x) + 28x\sin x + 15x\cos x + 28\cos x - 15\sin x]}{(3x+7\cos x)^2}$$

$$= \frac{35 + 28x\sin x + 15x\cos x - 15\sin x + 28\cos x}{(3x+7\cos x)^2}$$

Example 6. **Differentiate $(x+\sec x)\cdot(x-\tan x)$ w.r.t. x.** *NCERT*

Sol. We have, $y = (x+\sec x)(x-\tan x)$

On differentiating both sides w.r.t. x, we get

$$\frac{dy}{dx} = \frac{d}{dx}[(x+\sec x)(x-\tan x)]$$

$$= (x+\sec x)\frac{d}{dx}(x-\tan x) + (x-\tan x)\frac{d}{dx}(x+\sec x)$$

[using product rule of derivative]

$$= (x+\sec x)(1-\sec^2 x) + (x-\tan x)(1+\sec x\cdot\tan x)$$

Example 7. **Differentiate the following functions w.r.t. x.**

(i) $1-2\sin^2\frac{x}{2}$ **(ii) $2\cos^2\frac{x}{2}-1$**

(iii) $3\sin\frac{x}{3} - 4\sin^3\frac{x}{3}$

Sol. (i) Let $y = 1 - 2\sin^2\frac{x}{2}$

$\Rightarrow y = \cos x$ $[\because \cos 2\theta = 1 - 2\sin^2\theta]$

On differentiating both sides w.r.t. x, we get

$$\frac{dy}{dx} = \frac{d}{dx}(\cos x) = -\sin x$$

(ii) Let $y = 2\cos^2\frac{x}{2} - 1$

$\Rightarrow y = \cos x$ $[\because \cos 2\theta = 2\cos^2\theta - 1]$

On differentiating both sides w.r.t. x, we get

$$\frac{dy}{dx} = \frac{d}{dx}(\cos x) = -\sin x$$

(iii) Let $y = 3\sin\frac{x}{3} - 4\sin^3\frac{x}{3}$

$= \sin 3\left(\frac{x}{3}\right)$ $[\because \sin 3\theta = 3\sin\theta - 4\sin^3\theta]$

$\Rightarrow y = \sin x$

On differentiating both sides w.r.t. x, we get

$$\frac{dy}{dx} = \frac{d}{dx}(\sin x) = \cos x$$

Example 8. **Find the derivative of $\frac{\sec x + \tan x}{\sec x - \tan x}$.** *NCERT, NCERT Exemplar*

Sol. Let $y = \frac{\sec x + \tan x}{\sec x - \tan x} = \frac{\frac{1}{\cos x} + \frac{\sin x}{\cos x}}{\frac{1}{\cos x} - \frac{\sin x}{\cos x}} = \frac{1+\sin x}{1-\sin x}$

On differentiating both sides w.r.t. x, we get

$$\frac{dy}{dx} = \frac{d}{dx}\left(\frac{\sec x + \tan x}{\sec x - \tan x}\right) = \frac{d}{dx}\left(\frac{1+\sin x}{1-\sin x}\right)$$

$$= \frac{(1-\sin x)\frac{d}{dx}(1+\sin x) - (1+\sin x)\frac{d}{dx}(1-\sin x)}{(1-\sin x)^2}$$

[using quotient rule of derivative]

$$= \frac{(1-\sin x)(0+\cos x) - (1+\sin x)(0-\cos x)}{(1-\sin x)^2}$$

$$= \frac{2\cos x}{(1-\sin x)^2}$$

Quick Test 5

1 Find the derivative of $2x^4+x$. *NCERT Exemplar*

2 If $f(x)=x^{100}+x^{99}+...+x+1$, then find $f'(1)$. *NCERT Exemplar, Competency Based Que*

3 For some constants a and b, find the derivative of the following functions.

(i) $(x-a)(x-b)$ (ii) $(ax^2+b)^2$

(iii) $\frac{x-a}{x-b}$ (iv) $\frac{1}{ax^2+bx+c}$

(v) $\frac{x^4+x^3+x^2+1}{x}$ (vi) $\frac{x^2+x+1}{\sqrt{x}}$

(vii) $(5x^3+3x-1)(x-1)$ (viii) $x^{-4}(3-4x^{-5})$

(ix) $\frac{2}{x+1}-\frac{x^2}{3x-1}$ (x) $\frac{3x+4}{5x^2-7x+9}$ *NCERT Exemplar*

4 Find the derivative of $\frac{(x-1)(x-2)}{(x-3)(x-4)}$. *Competency Based Que*

5 If $y=\frac{1+\frac{1}{x^2}}{1-\frac{1}{x^2}}$, then find $\frac{dy}{dx}$. *NCERT Exempler*

6 Find the derivative of the following functions.

(i) $\sin x\cos x$ (ii) $5\sec x+4\cos x$

(iii) $3\cot x+5\operatorname{cosec} x$ (iv) $5\sin x+6\cos x+7$

(v) $2\tan x-7\sec x$ *NCERT*

(vi) $\operatorname{cosec} x\cot x$

(vii) $(\sec x-1)(\sec x+1)$

(viii) $(ax^2+\cot x)(p+q\cos x)$ *NCERT Exemplar*

(ix) $(x^2+1)\cos x$ (x) $\sin(x+a)$

7 Find the derivative of the following functions.

(i) $(x+\cos x)(x-\tan x)$ (ii) $x^4(5\sin x-3\cos x)$

(iii) $\frac{x}{1+\tan x}$ (iv) $(3x+5)(1+\tan x)$

(v) $\frac{x^5-\cos x}{\sin x}$ (vi) $\frac{a+b\sin x}{c+d\cos x}$

(vii) $\frac{\cos x}{1+\sin x}$ (viii) $\frac{\cos(x-a)}{\cos x}$

(ix) $\frac{1-\cos x}{1+\cos x}$ (x) $x^2\sin x+\cos 2x$

8 Find the derivative of the following functions.

(i) $\frac{x\tan x}{(\sec x+\tan x)}$ (ii) $\frac{\sin x-x\cos x}{x\sin x+\cos x}$

9 If $y=\frac{\sin(x+9)}{\cos x}$, then find $\frac{dy}{dx}$ at $x=0$. *Competency Based Que*

10 If $y=\frac{\sin x+\cos x}{\sin x-\cos x}$, then find $\frac{dy}{dx}$ at $x=0$. *NCERT Exemplar*

Answers

1. $8x^3+1$ **2.** 5050

3. (i) $2x-a-b$ (ii) $4ax(ax^2+b)$

(iii) $\frac{a-b}{(x-b)^2}$ (iv) $\frac{-(2ax+b)}{(ax^2+bx+c)^2}$

(v) $\frac{3x^4+2x^3+x^2-1}{x^2}$ (vi) $\frac{3}{2}x^{\frac{1}{2}}+\frac{1}{2}x^{-\frac{1}{2}}-\frac{1}{2}x^{-\frac{3}{2}}$

(vii) $20x^3-15x^2+6x-4$ (viii) $-12x^{-5}+36x^{-10}$

(ix) $\frac{-2}{(x+1)^2}-\frac{x(3x-2)}{(3x-1)^2}$ (x) $\frac{55-15x^2-40x}{(5x^2-7x+9)^2}$

4. $\frac{-4x^2+20x-22}{(x-3)^2(x-4)^2}$ **5.** $-\frac{4x}{(x^2-1)^2}$

6. (i) $\cos 2x$ (ii) $5\sec x\tan x-4\sin x$

(iii) $-3\operatorname{cosec}^2 x-5\operatorname{cosec} x\cot x$

(iv) $5\cos x-6\sin x$ (v) $2\sec^2 x-7\sec x\tan x$

(vi) $-\operatorname{cosec}^3 x-\cot^2 x\cdot\operatorname{cosec} x$ (vii) $2\tan x\sec^2 x$

(viii) $-q\sin x(ax^2+\cot x)+(p+q\cos x)(2ax-\operatorname{cosec}^2 x)$

(ix) $-x^2\sin x-\sin x+2x\cos x$

(x) $\cos(x+a)$

7. (i) $(x+\cos x)(-\tan^2 x)+(x-\tan x)(1-\sin x)$

(ii) $x^3[5x\cos x+3x\sin x+20\sin x-12\cos x]$

(iii) $\frac{1+\tan x-x\sec^2 x}{(1+\tan x)^2}$

(iv) $3x\sec^2 x+5\sec^2 x+3\tan x+3$

(v) $\frac{5x^4\sin x-x^5\cos x+1}{\sin^2 x}$

(vi) $\frac{bc\cos x+ad\sin x+bd}{(c+d\cos x)^2}$

(vii) $\frac{-1}{1+\sin x}$ (viii) $\sec^2 x\cdot\sin a$

(ix) $\frac{2\sin x}{(1+\cos x)^2}$ (x) $x^2\cos x+2x\sin x-2\sin 2x$

8. (i) $\frac{x\sec x(\sec x-\tan x)+\tan x}{(\sec x+\tan x)}$

(ii) $\frac{x^2}{(x\sin x+\cos x)^2}$

9. $\cos 9$ **10.** -2

For Solutions, See QR Code on Pg. 357

Exam Pattern Questions

Objective Type Questions

Multiple Choice Questions

1. If $f(x)=\begin{cases} x^2-1, & x\le 1 \\ -x^2-1, & x>1 \end{cases}$, then

(a) LHL $\neq$ RHL , at $x=1$
(b) LHL = RHL $=-\ 2$, at $x=1$
(c) LHL = RHL $=-1$, at $x=1$
(d) None of the above

2. The value of $\lim\limits_{x\to 0}\dfrac{1-\cos 4x}{1-\cos 6x}$ is equal to **Competency Based Que**

(a) $\frac{4}{9}$ (b) $\frac{1}{2}$
(c) $\frac{-1}{2}$ (d) -1

3. The value of $\lim\limits_{r\to 2} 2\pi r$ is

(a) 2π (b) π
(c) 4π (d) 6π

4. The value of $\lim\limits_{x\to 2}\dfrac{x^2+3x-10}{x^2-9x+14}$ is

(a) $\frac{7}{5}$ (b) $-\frac{7}{5}$
(c) $\frac{6}{5}$ (d) $-\frac{4}{5}$

5. If $f(x)=\begin{cases} x & , \ x<1 \\ 4 & , \ x=1 \\ x+4 & , \ x>1 \end{cases}$, then $\lim\limits_{x\to 1} f(x)$ is equal to **Competency Based Que**

(a) 1 (b) 5
(c) 4 (d) Does not exist

6. $\lim\limits_{x\to 0}\dfrac{(1+x)^n-1}{x}$ is equal to **Competency Based Que**

(a) 0 (b) 1
(c) n (d) $-n$

7. The derivative of the function $f(x)=3x$ at $x=2$ is

(a) 0 (b) 1
(c) 2 (d) 3

8. If $\dfrac{d}{dx}\left\{\dfrac{1}{f(x)}\right\}=\dfrac{p}{q}$, where p and q respectively are

(a) $f'(x), \{f(x)\}^2$ (b) $-f'(x), \{f(x)\}^2$
(c) $-f'(x), f(x)$ (d) $f(x), f'(x)$

9. Which of the following is/are true?

Statement I The derivative of $f(x)=\sin 2x$ is $2(\cos^2 x-\sin^2 x)$.

Statement II The derivative of $g(x)=\cot x$ is $-\operatorname{cosec}^2 x$.

(a) Both Statement I and Statement II are true.
(b) Only Statement I is true.
(c) Only Statement II is true.
(d) Both Statement I and Statement II are false.

10. If $y=\left\{\dfrac{1-\tan x}{1+\tan x}\right\}$ and $\dfrac{dy}{dx}=\dfrac{k}{1+\sin 2x}$, then the value of k is **Competency Based Que**

(a) 1 (b) 2 (c) -1 (d) -2

11. If $f(x)=p$, where p is a constant, then $f'(4)$ is **Competency Based Que**

(a) 4 (b) 4C
(c) -4 (d) 0

12. Derivative of $6x^5-\tan^2 x$ with respect to x is

(a) $30x^4-2\tan x\sec^2 x$ (b) $30x^5-2\tan x$
(c) $30x^4-\tan x\sec^2 x$ (d) $30x^4-2\sec^2 x$

13. If $y=\sqrt{2x}+\dfrac{1}{\sqrt{2x}}$, then $\dfrac{dy}{dx}$ at $x=2$ is

(a) $\frac{1}{2}$ (b) $\frac{1}{8}$ (c) $\frac{5}{8}$ (d) $\frac{3}{8}$

Assertion-Reason Based Questions

Directions (Q. Nos. 14-24) *In the questions given below are two statements labelled as Assertion (A) and Reason (R). In the context of the two statements, which one of the following is correct?*

(a) Both A and R are correct; R is the correct explanation of A.
(b) Both A and R are correct; R is not the correct explanation of A.
(c) A is correct; R is incorrect.
(d) R is correct; A is incorrect.

14. **Assertion** (A) The value of $\lim\limits_{x\to 2^+}\dfrac{x^2-1}{2x+4}$ is $\dfrac{3}{8}$.

Reason (R) $\lim\limits_{x\to a^+} f(x)=\lim\limits_{h\to 0} f(a+h)$.

15. Assertion (A) For a function $f(x) = \begin{cases} \dfrac{\sin x}{x}, & x \neq 0 \\ 2, & x = 0 \end{cases}$, $\lim_{x \to 0}$ does not exist.

Reason (R) For a function $f(x)$, at $x = a$, limit will exists if LHL (at $x = a$) = RHL (at $x = a$).

16. Assertion (A) $\lim_{x \to 0} \dfrac{\sin ax}{\sin bx} = \dfrac{a}{b}$

Reason (R) $\lim_{x \to 0} \dfrac{\sin mx}{mx} = 1$

Competency Based Que

17. Assertion (A) $\lim_{x \to 0} \dfrac{\tan x}{x} = 1$

Reason (R) If $\lim_{x \to 0} f(x) = l_1$, $\lim_{x \to 0} g(x) = l_2$, then $\lim_{x \to 0} (f(x) + g(x)) = l_1 + l_2$

18. Assertion (A) $\lim_{x \to 5} \dfrac{x^3 - 125}{x - 5} = 75$

Reason (R) $\lim_{x \to a} \dfrac{x^n - a^n}{x - a} = a^{n-1}$

Competency Based Que

19. Assertion (A) The derivative of $f(x) = \cos x$ at $x = 0$ is 0.

Reason (R) For a function $f(x)$,

$$f'(a) = \lim_{h \to 0} \frac{f(a+h) - f(a)}{h}$$

20. Assertion (A) $f(x) = x^2 + 5 \Rightarrow f'(2) = 4$

Reason (R) $g(x) = 2x^2 + 3x + 5$, then $g'(x) = 4x + 3$

21. Assertion (A) If $y = 1 + \dfrac{x}{1!} + \dfrac{x^2}{2!} + \dfrac{x^3}{3!} + \ldots$, then $\dfrac{dy}{dx} = y$.

Reason (R) If $y = p(x)$, then $\dfrac{dy}{dx} = p'(x)$.

Competency Based Que

22. Assertion (A) $\dfrac{d}{dx}\left(\dfrac{1}{x}\right) = -\dfrac{1}{x^2}$

Reason (R) $\dfrac{d}{dx}(x^n) = nx^{n-1}$

23. Assertion (A) $\dfrac{d}{dx}\left(\dfrac{x+3}{x^2+1}\right) = \dfrac{1 - 6x - x^2}{(x^2+1)^2}$

Reason (R) $\dfrac{d}{dx}\left(\dfrac{u}{v}\right) = \dfrac{vu' - uv'}{v^2}$

24. Assertion (A) $\dfrac{d}{dx}(\sin x^2) = 2x \cos(x^2)$

Reason (R) $\dfrac{d}{dx}(\sin x) = \cos x$

Case Study Based Questions

25. Mr. Abhishek has a rectangular plot, which is used for growing vegetables. Perimeter of plot is 100 m. Length and width of plot are x m and y m respectively.

Based on the above information, answer the following questions.

(i) Relation between x and y is
(a) $x + y = 100$
(b) $x + y = 50$
(c) $x + y = 25$
(d) $x = y$

(ii) Area function, $A(x)$ is
(a) $x^2 - 5$ (b) $50x - x^2$
(c) $x^2 - 50x$ (d) $50 - x$

(iii) Derivative of $A(x)$ w.r.t. x is
(a) $2x$ (b) $-2x$
(c) $50 - 2x$ (d) $2x - 50$

(iv) The value of x for which $A'(x) = 0$ is
(a) 25 (b) 50
(c) 100 (d) 0

(v) The value of $A(x)$ at $x = 25$ is
(a) 250 (b) 1250
(c) 525 (d) 625

26. Consider the following functions.

$u(x) = \sqrt{x}$, $v(x) = \cot x$, $f(x) = u(x) \times v(x)$.

$g(x) = \dfrac{u(x)}{v(x)}$ and $h(x) = \dfrac{v(x)}{u(x)}$

Based on the above information, answer the following questions.

(i) Derivative of $u(x)$ is ......
(ii) Derivative of $v(x)$ is .....
(iii) Derivative of $f(x)$ is .....
(iv) Derivative of $g(x)$ is .....
(v) Derivative of $h(x)$ is .....

Subjective Questions

Very Short Answer Type Questions

1. Evaluate the following.

(i) $\lim_{x \to 1} \frac{x^2 + 3x + 2}{x^2 + 1}$

(ii) $\lim_{x \to 4} \frac{x^3 - 2x^2 - 9x + 4}{x^2 - 2x - 8}$

(iii) $\lim_{x \to 1} \frac{x^2 - \sqrt{x}}{\sqrt{x} - 1}$

(iv) $\lim_{x \to 0} \frac{(1 + x)^5 - 1}{3x + 5x^2}$

(v) $\lim_{x \to 2} \frac{x - 2}{\sqrt{x^2 - 4} + \sqrt{x - 2}}$

2. Find the positive integer n, so that $\lim_{x \to 3} \frac{x^n - 3^n}{x - 3} = 108.$

Competency Based Que

3. Find k, so that $\lim_{x \to -1} f(x)$ may exist where

$$f(x) = \begin{cases} x^3 - 3x + 7, & x \le -1 \\ 3x + k \quad, & x > -1 \end{cases}$$

Competency Based Que

4. Evaluate $\lim_{x \to 2} \frac{x\sqrt{x} - \sqrt{8}}{x - 2}$.

5. Differentiate $y = (x^2 - 4x + 5)(x^3 - 2)$ w.r.t. x.

6. If $f(x) = 1 + x + \frac{x^2}{2} + \ldots + \frac{x^{100}}{100}$, then find $f'(1)$.

Competency Based Que

Short Answer Type Questions

7. Evaluate the following.

(i) $\lim_{x \to \alpha} \frac{x \sin\alpha - \alpha \sin x}{x - \alpha}$

Competency Based Que

(ii) $\lim_{x \to 0} \frac{1 - \cos x \cos 2x \cos 3x}{\sin^2 2x}$

(iii) $\lim_{x \to 0} \frac{8}{x^8}\left(1 - \cos\frac{x^2}{2} - \cos\frac{x^2}{4} + \cos\frac{x^2}{2}\cos\frac{x^2}{4}\right)$

(iv) $\lim_{x \to 0} \frac{3\sin x - \sin 3x}{x(\cos 2x - \cos 4x)}$

Competency Based Que

8. For the function f, given by $f(x) = x^2 - 6x + 8$ prove that $f'(5) - 3f'(2) = f'(8)$.

9. Differentiate $\sin^3 x$ w.r.t. x from first principle.

10. If $y = \frac{2\sin^2 x + 3\cos x - 1}{\sin x}$, then find $\frac{dy}{dx}$.

11. If $y = a\sin x + b\cos x$, then show that

$$y^2 + \left(\frac{dy}{dx}\right)^2 = a^2 + b^2.$$

Long Answer Type Questions

12. Evaluate the left hand and right hand limits of the following function at $x = 1$

$$f(x) = \begin{cases} 1 + x^3, & 0 \le x \le 1 \\ 2 - x^2, & x > 1 \end{cases}$$

Does $\lim_{x \to 1} f(x)$ exist?

13. Let $f(x) = \begin{cases} \frac{5x}{|x| + 4x}, & x \ne 0 \\ 0 \quad, & x = 0 \end{cases}$

Show that $\lim_{x \to 0} f(x)$ does not exist.

14. Evaluate $\lim_{x \to 1} \frac{x^4 - 3x^3 + 2}{x^3 - 5x^2 + 3x + 1}$.

15. Evaluate $\lim_{x \to \sqrt{2}} \frac{x^4 - 4}{x^2 + 3x\sqrt{2} - 8}$.

16. Evaluate $\lim_{x \to 0} \frac{\tan x - \sin x}{x^3}$.

Competency Based Que

17. Prove that the derivative of $\sin x$ with respect to x is $\cos x$ i.e. $\frac{d}{dx}(\sin x) = \cos x$.

18. Find $\frac{dy}{dx}$, when $y = \frac{x^2 \sin x}{1 - x}$.

Hints & Solutions

Objective Type Questions

1. (a) $\text{LHL} = \lim_{x \to 1^-} f(x) = \lim_{x \to 1^-} (x^2 - 1)$

$= (1^2 - 1) = 0$

$\text{RHL} = \lim_{x \to 1^+} f(x) = \lim_{x \to 1^+} (-x^2 - 1)$

$= -1^2 - 1 = -2$

$\therefore$ LHL $\neq$ RHL at $x = 1$

2. (a) $\lim_{x \to 0} \frac{1 - \cos 4x}{1 - \cos 6x} = \lim_{x \to 0} \frac{2 \sin^2 2x}{2 \sin^2 3x}$

$$= \lim_{x \to 0} \frac{\left(\frac{\sin 2x}{2x}\right)^2 \times 4x^2}{\left(\frac{\sin 3x}{3x}\right)^2 \times 9x^2}$$

$$= \frac{4}{9} \frac{\lim_{x \to 0} \left(\frac{\sin 2x}{2x}\right)^2}{\lim_{x \to 0} \left(\frac{\sin 3x}{3x}\right)^2}$$

$$= \frac{4}{9} \times \frac{1}{1} = \frac{4}{9}$$

3. (c) $\lim_{r \to 2} 2\pi r = 2\pi(2) = 4\pi$

4. (b) $\lim_{x \to 2} \frac{x^2 + 3x - 10}{x^2 - 9x + 14} = \lim_{x \to 2} \frac{(x+5)(x-2)}{(x-7)(x-2)}$

$$= \lim_{x \to 2} \frac{x+5}{x-7} = \frac{2+5}{2-7}$$

$$= \frac{-7}{5}$$

5. (d) $\text{LHL} = \lim_{x \to 1^-} f(x) = \lim_{x \to 1^-} x = 1$

$\text{RHL} = \lim_{x \to 1^+} f(x) = \lim_{x \to 1^+} x + 4 = 1 + 4 = 5$

$\because$ LHL $\neq$ RHL $\neq f(1)$

$\therefore$ Limit does not exist.

6. (c) $\lim_{x \to 0} \frac{(1+x)^n - 1}{(1+x) - 1} = \lim_{1+x \to 1} \frac{(1+x)^n - 1^n}{(1+x) - 1} = n(1)^{n-1} = n$

7. (d) We have, $f(x) = 3x \Rightarrow f'(x) = 3$

At $x = 2, f'(2) = 3$

8. (b) Given, $\frac{d}{dx}\left\{\frac{1}{f(x)}\right\} = \frac{p}{q}$

$$\Rightarrow \frac{f(x)\frac{d}{dx}(1) - 1\frac{d}{dx}(f(x))}{(f(x))^2} = \frac{p}{q}$$

$$\Rightarrow \frac{0 - f'(x)}{(f(x))^2} = \frac{p}{q}$$

$$\Rightarrow \frac{-f'(x)}{(f(x))^2} = \frac{p}{q}$$

Hence, $p = -f'(x)$ and $q = (f(x))^2$

9. (a) **Statement I** $f(x) = \sin 2x \Rightarrow f'(x) = 2\cos 2x$

$\therefore \quad f'(x) = 2(\cos^2 x - \sin^2 x)$

Statement II $g(x) = \cot x \Rightarrow g'(x) = -\operatorname{cosec}^2 x$

Both Statement I and Statement II are true.

10. (d) We have, $y = \frac{1 - \tan x}{1 + \tan x}$

$$\frac{dy}{dx} = \frac{(1 + \tan x)(-\sec^2 x) - (1 - \tan x)(\sec^2 x)}{(1 + \tan x)^2}$$

$$= \frac{-\sec^2 x - \tan x \sec^2 x - \sec^2 x + \tan x \sec^2 x}{(1 + \tan x)^2}$$

$$= \frac{-2\sec^2 x}{(1 + \tan x)^2} = \frac{-2(1 + \tan^2 x)}{\left(1 + \frac{\sin x}{\cos x}\right)^2}$$

$$= \frac{-2\left(1 + \frac{\sin^2 x}{\cos^2 x}\right)}{\left(\frac{\cos x + \sin x}{\cos x}\right)^2}$$

$$= \frac{-2(\cos^2 x + \sin^2 x)}{\cos^2 x + \sin^2 x + 2\sin x \cos x} = \frac{-2}{1 + \sin 2x}$$

Given, $\frac{dy}{dx} = \frac{k}{1 + \sin 2x} = \frac{-2}{1 + \sin 2x}$

$\therefore \quad k = -2$

11. (d) Given, $f(x) = p \Rightarrow f'(x) = 0$

At $x = 4$, we get $f'(4) = 0$

12. (a) We have, $6x^5 - \tan^2 x$

$$\frac{d}{dx}(6x^5 - \tan^2 x) = \frac{d}{dx}(6x^5) - \frac{d}{dx}(\tan^2 x)$$

$$= 30x^4 - 2\tan x \sec^2 x$$

13. (d) Given, $y = \sqrt{2x} + \frac{1}{\sqrt{2x}}$

$$\Rightarrow \frac{dy}{dx} = \frac{1}{2\sqrt{2x}}(2) + \left(-\frac{1}{2(2x)^{3/2}}(2)\right)$$

$$= \frac{1}{\sqrt{2x}} - \frac{1}{2\sqrt{2}x^{3/2}}$$

At $x = 2$,

$$\frac{dy}{dx} = \frac{1}{\sqrt{4}} - \frac{1}{2\sqrt{2}(2\sqrt{2})}$$

$$= \frac{1}{2} - \frac{1}{8} = \frac{3}{8}$$

14. (a) $\lim_{x\to 2^+} \frac{x^2 - 1}{2x + 4} = \lim_{x\to 0} \frac{(2+h)^2 - 1}{2(2+h) + 4}$

$$= \lim_{x\to 0} \frac{4 + h^2 + 4h - 1}{2h + 8}$$

$$= \lim_{x\to 0} \frac{h^2 + 4h + 3}{2h + 8}$$

$$= \frac{0 + 0 + 3}{0 + 8} = \frac{3}{8}$$

15. (d) $\text{LHL} = \lim_{x\to 0^-} f(x)$

$$= \lim_{x\to 0^-} \frac{\sin x}{x} = 1$$

$\text{RHL} = \lim_{x\to 0^+} f(x)$

$$= \lim_{x\to 0^+} \frac{\sin x}{x} = 1$$

$\Rightarrow$ LHL = RHL

Hence, limit exist.

Assertion is false.

16. (a) $\lim_{x\to 0} \frac{\sin ax}{\sin bx} = \lim_{x\to 0} \frac{\frac{\sin ax}{ax} \times ax}{\frac{\sin bx}{bx} \times bx}$

$$= \frac{\lim_{x\to 0}\left(\frac{\sin ax}{ax}\right)\cdot ax}{\lim_{x\to 0}\left(\frac{\sin bx}{bx}\right)\cdot bx} = \frac{a}{b}$$

17. (b) $\lim_{x\to 0} \frac{\tan x}{x} = \lim_{x\to 0} \frac{\sin x}{\cos x \times x}$

$$= \lim_{x\to 0}\left(\frac{\sin x}{x}\right)\left(\frac{1}{\cos x}\right)$$

$$= 1 \times 1 \qquad \left[\because \lim_{x\to 0} \frac{\sin x}{x} = 1\right]$$

So, Assertion and Reason both are correct but Reason is not the correct explanation of Assertion.

18. (c) We know that $\lim_{x\to a} \frac{x^3 - a^n}{x - a} = na^{n-1}$

$$\therefore \lim_{x\to 5} \frac{x^3 - 125}{x - 5} = \lim_{x\to 5} \frac{x^3 - 5^3}{x - 5}$$

$$= 3(5)^{3-1} = 3 \times 5^2 = 75$$

19. (a) Given, $f(x) = \cos x$

$$f'(a) = \lim_{h\to 0} \frac{f(a+h) - f(a)}{h}$$

$$f'(0) = \lim_{h\to 0} \frac{\cos(0+h) - \cos(0)}{h}$$

$$= \lim_{h\to 0} \frac{\cos h - 1}{h}$$

$$= \lim_{h\to 0} \frac{-2\sin^2 \frac{h}{2}}{h}$$

$$= \lim_{h\to 0} \frac{-2\sin^2 \frac{h}{2}}{\frac{h^2}{4}} \times \frac{h}{4}$$

$$= \lim_{h\to 0} -1 \times \frac{h}{2} = 0$$

20. (b) Given, $f(x) = x^2 + 5 \Rightarrow f'(x) = 2x$

$\therefore f'(2) = 4$

So, Assertion is correct.

$g(x) = 2x^2 + 3x + 5 \Rightarrow g'(x) = 4x + 3$

So, Assertion and Reason both are correct but Reason is not the correct explanation of Assertion.

21. (b) given, $y = 1 + \frac{x}{1!} + \frac{x^2}{2!} + \frac{x^3}{3!} + \ldots$

$$\frac{dy}{dx} = \frac{d}{dx}\left(1 + \frac{x}{1!} + \frac{x^2}{2!} + \frac{x^3}{3!} + \ldots\right)$$

$$\Rightarrow \frac{dy}{dx} = 1 + \frac{2x}{2} + \frac{3x^2}{3\times 2!} + \ldots$$

$$= 1 + \frac{x}{1!} + \frac{x^2}{2!} + \ldots = y$$

$$\Rightarrow \frac{dy}{dx} = y$$

So, Assertion and Reason both are correct but Reason is not the correct explanation of Assertion.

22. (a) $\frac{d}{dx}\left(\frac{1}{x}\right) = \frac{d}{dx}(x^{-1}) = (-1)x^{-1-1}$

$$= (-1)x^{-2} = \frac{-1}{x^2}$$

So, Assertion and Reason both are correct and Reason is the correct explanation of Assertion.

23. (a) $\frac{d}{dx}\left(\frac{x+3}{x^2+1}\right) = \frac{(x^2+1)\frac{d}{dx}(x+3) - (x+3)\frac{d}{dx}(x^2+1)}{(x^2+1)^2}$

$$= \frac{(x^2+1) - (x+3)(2x)}{(x^2+1)^2}$$

$$= \frac{x^2 + 1 - 2x^2 - 6x}{(x^2+1)^2}$$

$$= \frac{1-6x-x^2}{(1+x^2)^2}$$

So, Assertion and Reason both are correct and Reason is the correct explanation of Assertion.

24. (b) $\frac{d}{dx}(\sin x^2) = \frac{d}{dx^2}(\sin x^2) \times \frac{dx^2}{dx}$

$$= \cos(x^2) \times 2x = 2x \cos(x^2)$$

So, Assertion and Reason both are correct but Reason is not the correct explanation of Assertion.

25. (i) (b) Given, perimeter of plot is 100 m.

$$2(x+y) = 100$$

$$\Rightarrow \quad x+y = 50$$

(ii) (b) Area, $\quad A = xy$

We have, $\quad x+y = 50$

$$\Rightarrow \quad y = 50 - x$$

$$\therefore \quad A(x) = x(50-x) = 50x - x^2$$

(iii) (c) We have, $A(x) = 50x - x^2$

$$\therefore \quad A'(x) = 50 - 2x$$

(iv) (a) We have,

$$A'(x) = 50 - 2x$$

$$A'(x) = 0$$

$$\Rightarrow \quad 50 - 2x = 0 \Rightarrow 2x = 50 \Rightarrow x = 25$$

(v) (d) We have, $A(x) = 50x - x^2$

At $x = 25$, $A(25) = 50 \times 25 - (25)^2$

$$= 1250 - 625 = 625$$

26. (i) Given, $\quad u(x) = \sqrt{x}$

$$\Rightarrow \quad \frac{d}{dx}(u(x)) = \frac{1}{2\sqrt{x}}$$

(ii) Given, $\quad v(x) = \cot x$

$$\Rightarrow \quad \frac{d}{dx}(v(x)) = -\operatorname{cosec}^2 x$$

(iii) Given, $f(x) = u(x) \times v(x)$

$$f(x) = \sqrt{x} \cot x$$

$$\Rightarrow \quad \frac{d}{dx}(f(x)) = \sqrt{x}\frac{d}{dx}(\cot x) + \cot x \frac{d}{dx}(\sqrt{x})$$

$$= \sqrt{x}(-\operatorname{cosec}^2 x) + \cot x \cdot \frac{1}{2\sqrt{x}}$$

$$= \frac{\cot x}{2\sqrt{x}} - \sqrt{x}\operatorname{cosec}^2 x$$

(iv) Given, $g(x) = \frac{u(x)}{v(x)} \Rightarrow g(x) = \frac{\sqrt{x}}{\cot x}$

$$\Rightarrow \quad \frac{d}{dx}(g(x)) = \frac{\cot x \frac{d}{dx}(\sqrt{x}) - \sqrt{x}\frac{d}{dx}(\cot x)}{(\cot x)^2}$$

$$= \frac{\frac{\cot x}{2\sqrt{x}} - \sqrt{x} - \operatorname{cosec}^2 x}{(\cot x)^2}$$

$$= \frac{(2x \operatorname{cosec}^2 x + \cot x)}{2\sqrt{x}(\cot x)^2}$$

(v) Given, $h(x) = \frac{v(x)}{u(x)}$

So, $\quad h(x) = \frac{\cot x}{\sqrt{x}}$

$$\Rightarrow \quad \frac{d}{dx}(h(x)) = \frac{d}{dx}\left(\frac{\cot x}{\sqrt{x}}\right)$$

$$= \frac{\sqrt{x}\frac{d}{dx}(\cot x) - \cot x \frac{d}{dx}(\sqrt{x})}{(\sqrt{x})^2}$$

$$= \frac{\sqrt{x}(-\operatorname{cosec}^2 x) - \cot x \frac{1}{2\sqrt{x}}}{x}$$

$$= -\frac{(2x \operatorname{cosec}^2 x + \cot x)}{2x\sqrt{x}}$$

Subjective Questions

1. (i) $\lim_{x \to 1} \frac{x^2+3x+2}{x^2+1} = \frac{(1)^2 + 3(1) + 2}{(1)^2 + 1}$

$$= \frac{1+3+2}{1+1} = \frac{6}{2} = 3$$

(ii) $\lim_{x \to 4} \frac{x^3 - 2x^2 - 9x + 4}{x^2 - 2x - 8} = \lim_{x \to 4} \frac{(x-4)(x^2+2x-1)}{(x-4)(x+2)}$

$$= \lim_{x \to 4} \frac{x^2 + 2x - 1}{x+2}$$

$$= \frac{(4)^2 + 2(4) - 1}{4+2}$$

$$= \frac{16+8-1}{6} = \frac{23}{6}$$

(iii) $\lim_{x \to 1} \frac{x^2 - \sqrt{x}}{\sqrt{x} - 1} = \lim_{x \to 1} \frac{(\sqrt{x})^4 - \sqrt{x}}{(\sqrt{x} - 1)}$

$$= \lim_{x \to 1} \frac{\sqrt{x}[(\sqrt{x})^3 - 1]}{(\sqrt{x}-1)}$$

$$= \lim_{x \to 1} \frac{\sqrt{x}(\sqrt{x}-1)(x+1+\sqrt{x})}{(\sqrt{x}-1)}$$

$$= \lim_{\sqrt{x} \to 1} \sqrt{x}(x+1+\sqrt{x})$$

[as $x \to 1$, so $\sqrt{x} \to \sqrt{1}$ i.e. $\sqrt{x} \to 1$]

$$= 3$$

(iv) $\lim_{x\to 0}\dfrac{(1+x)^5-1}{3x+5x^2}$ $\left[\dfrac{0}{0}\text{ form}\right]$

$$=\frac{\lim_{x\to 0}\dfrac{(1+x)^5-1}{x}}{\lim_{x\to 0}\dfrac{3x+5x^2}{x}}=\frac{\lim_{(1+x)\to 1}\dfrac{(1+x)^5-1}{(1+x)-1}}{\lim_{x\to 0}(3+5x)}=\frac{5}{3}$$

(v) $\lim_{x\to 2}\dfrac{x-2}{\sqrt{x^2-4}+\sqrt{x-2}}$

$$=\lim_{x\to 2}\frac{x-2}{\sqrt{x-2}\,[\sqrt{x+2}+1]}$$

$$=\lim_{x\to 2}\frac{\sqrt{x-2}}{(\sqrt{x+2}+1)}$$

$$=0$$

2. We have, $\lim_{x\to 3}\dfrac{x^n-3^n}{x-3}=108$

We know that $\lim_{x\to a}\dfrac{x^n-a^n}{x-a}=na^{n-1}$

$\therefore$ $\lim_{x\to 3}\dfrac{x^n-3^n}{x-3}=n(3)^{n-1}$

$\Rightarrow$ $n(3)^{n-1}=108$

$\Rightarrow$ $n(3)^{n-1}=2^2\times 3^3$

$\Rightarrow$ $n(3)^{n-1}=4(3)^3$

Hence, $n=4$

3. $\lim_{x\to -1^-}f(x)=\lim_{x\to -1^-}(x^3-3x+7)$ $[\because f(x)=x^3-3x+7 \text{ for } x\le -1]$

$=(-1)^3-3(-1)+7=-1+3+7=9$

and $\lim_{x\to -1^+}f(x)=\lim_{x\to -1^+}(3x+k)$ $[\because f(x)=3x+k \text{ for } x>-1]$

$=3\times(-1)+k=-3+k$

Since, $\lim_{x\to -1}f(x)$ exists.

$\lim_{x\to -1^-}f(x)=\lim_{x\to -1^+}f(x)$

$\Rightarrow$ $9=-3+k \Rightarrow k=12$

4. $\lim_{x\to 2}\dfrac{x\sqrt{x}-\sqrt{8}}{x-2}=\lim_{x\to 2}\dfrac{x^{3/2}-2^{3/2}}{x-2}$

$$=\frac{3}{2}(2)^{\frac{1}{2}}=\frac{3}{2}\sqrt{2}=\frac{3\sqrt{2}}{2}$$

5. $\dfrac{dy}{dx}=(x^2-4x+5)\dfrac{d}{dx}(x^3-2)+(x^3-2)\dfrac{d}{dx}(x^2-4x+5)$

$=(x^2-4x+5)\times 3x^2+(x^3-2)(2x-4)$

$=5x^4-16x^3+15x^2-4x+8$

6. As $f(x)=1+x+\dfrac{x^2}{2}+\ldots+\dfrac{x^{100}}{100}$

$f'(x)=0+1+x+x^2+\ldots+\dfrac{100x^{99}}{100}$

$=1+x+x^2+..+x^{99}$

$f'(1)=1+1+1+\ldots+1(100\text{ times})=100$

7. (i) **Hint** Let $x=\alpha+h$, then as $x\to\alpha$, $h\to 0$

Now, $\lim_{x\to\alpha}\dfrac{x\sin\alpha-\alpha\sin x}{x-\alpha}$

$$=\lim_{h\to 0}\frac{(\alpha+h)\sin\alpha-\alpha\sin(\alpha+h)}{\alpha+h-\alpha}$$

$$=\lim_{h\to 0}\frac{\alpha\sin\alpha+h\sin\alpha-\alpha\sin(\alpha+h)}{h}$$

$$=\lim_{h\to 0}\frac{\alpha\cdot 2\cos\left(\dfrac{2\alpha+h}{2}\right)\sin\left(-\dfrac{h}{2}\right)}{h}+\frac{h\sin\alpha}{h}$$

$$=\lim_{h\to 0}\frac{\alpha\cdot 2\cos\left(\dfrac{2\alpha+h}{2}\right)\cdot\dfrac{\sin\left(\dfrac{-h}{2}\right)}{\left(\dfrac{-h}{2}\right)}\cdot\left(\dfrac{-h}{2}\right)}{h}+\sin\alpha$$

$=\sin\alpha-\alpha\cos\alpha$

(ii) **Hint** Let $\cos x\cos 2x\cos 3x=\dfrac{1}{2}(2\cos x\cos 3x\cos 2x)$

$$=\frac{1}{2}[(\cos 2x+\cos 4x)\cos 2x]$$

$$=\frac{1}{4}(2\cos^2 2x+2\cos 2x\cos 4x)$$

$$=\frac{1}{4}(1+\cos 4x+\cos 2x+\cos 6x)$$

Now, $\lim_{x\to 0}\dfrac{1-\cos x\cos 2x\cos 3x}{\sin^2 2x}$

$$=\lim_{x\to 0}\frac{1-\dfrac{1}{4}(1+\cos 4x+\cos 2x+\cos 6x)}{\sin^2 2x}$$

$$=\frac{7}{4}$$

(iii) $\lim_{x\to 0}\dfrac{8}{x^8}\left(1-\cos\dfrac{x^2}{2}-\cos\dfrac{x^2}{4}+\cos\dfrac{x^2}{2}\cos\dfrac{x^2}{4}\right)$

$$=\lim_{x\to 0}\frac{8}{x^8}\left[\left(1-\cos\frac{x^2}{2}\right)-\cos\frac{x^2}{4}\left(1-\cos\frac{x^2}{2}\right)\right]$$

$$=\lim_{x\to 0}\frac{8}{x^8}\left(1-\cos\frac{x^2}{2}\right)\left(1-\cos\frac{x^2}{4}\right)$$

$$=\lim_{x\to 0}\frac{8}{x^8}2\sin^2\frac{x^2}{4}\cdot 2\sin^2\frac{x^2}{8}=\frac{1}{32}$$

(iv) $\lim_{x \to 0} \dfrac{3\sin x - \sin 3x}{x(\cos 2x - \cos 4x)}$ $\left[\dfrac{0}{0}\text{ form}\right]$

$$= \lim_{x \to 0} \frac{3\sin x - 3\sin x + 4\sin^3 x}{x\left[2\sin\left(\dfrac{4x+2x}{2}\right)\sin\left(\dfrac{4x-2x}{2}\right)\right]}$$

$$= \lim_{x \to 0} \frac{2\sin^3 x}{x \sin 3x \sin x} = \frac{2}{3}\lim_{x \to 0} \frac{\left(\dfrac{\sin x}{x}\right)^2}{\dfrac{\sin 3x}{3x}} = \frac{2}{3}$$

8. Given, $f(x) = x^2 - 6x + 8$

On differentiating both sides w.r.t. x, we get

$$f'(x) = 2x - 6 \quad \ldots\text{(i)}$$

Now, $f'(5) - 3f'(2) = 2 \times 5 - 6 - 3(2 \times 2 - 6)$ [from Eq. (i)]

$$= 10 - 6 - 12 + 18 = 10 \quad \ldots\text{(ii)}$$

and $f'(8) = 2 \times 8 - 6$ [from Eq. (ii)]

$$= 10 \quad \ldots\text{(iii)}$$

From Eqs. (ii) and (iii), we get

$$f'(5) - 3f'(2) = f'(8)$$

9. Let $f(x) = \sin^3 x$ ...(i)

By definition of first principle of derivative, we get

$$f'(x) = \lim_{h \to 0} \frac{f(x+h) - f(x)}{h}$$

$$= \lim_{h \to 0} \frac{\sin^3(x+h) - \sin^3 x}{h} \quad \text{[from Eq. (i)]}$$

$$= \lim_{h \to 0} \frac{\left[\begin{array}{c}\{\sin(x+h) - \sin x\}\{\sin^2(x+h) \\ + \sin^2 x + \sin(x+h)\sin x\}\end{array}\right]}{h}$$

$$= \lim_{h \to 0} \frac{\left[\begin{array}{c}2\sin\dfrac{h}{2}\cos\left(x+\dfrac{h}{2}\right)\{\sin^2(x+h) \\ + \sin^2 x + \sin(x+h)\sin x\}\end{array}\right]}{h}$$

$$= \lim_{h \to 0} \frac{\sin h/2}{h/2} \cdot \lim_{h \to 0} \cos\left(x + \frac{h}{2}\right) \cdot \lim_{h \to 0}[\sin^2(x+h) + \sin^2 x + \sin(x+h)\sin x]$$

$$= 1 \cdot \cos x \cdot (\sin^2 x + \sin^2 x + \sin^2 x)$$

$$= 3\sin^2 x \cos x$$

10. Given, $y = \dfrac{2\sin^2 x + 3\cos x - 1}{\sin x}$

$$= \frac{2\sin^2 x}{\sin x} + 3\frac{\cos x}{\sin x} - \frac{1}{\sin x}$$

$$\Rightarrow \quad y = 2\sin x + 3\cot x - \operatorname{cosec} x$$

On differentiating both sides w.r.t. x, we get

$$\frac{dy}{dx} = 2\cos x + 3(-\operatorname{cosec}^2 x) - (-\operatorname{cosec} x \cot x)$$

$$= 2\cos x - 3\operatorname{cosec}^2 x + \operatorname{cosec} x \cdot \cot x$$

11. We have, $y = a\sin x + b\cos x$

$$\Rightarrow \quad \frac{dy}{dx} = a\cos x - b\sin x$$

$$\Rightarrow \quad \left(\frac{dy}{dx}\right)^2 = (a\cos x - b\sin x)^2$$

$$\Rightarrow \quad y^2 + \left(\frac{dy}{dx}\right)^2 = (a\sin x + b\cos x)^2 + (a\cos x - b\sin x)^2$$

$$= a^2 + b^2$$

12. Hint We have, $f(x) = \begin{cases} 1 + x^3, & 0 \le x \le 1 \\ 2 - x^2, & x > 1 \end{cases}$

$$\text{LHL} = \lim_{x \to 1^-} f(x)$$

$$= \lim_{x \to 1^-} (1 + x^3) = 1 + 1^3 = 2$$

$$\text{RHL} = \lim_{x \to 1^+} (2 - x^2) = 2 - 1^2 = 1$$

$\because$ LHL $\ne$ RHL $\ne f(1)$

$\therefore \lim_{x \to 1} f(x)$ does not exist.

13. We have, $f(x) = \begin{cases} \dfrac{5x}{|x| + 4x}, & x \ne 0 \\ 0, & x = 0 \end{cases}$

$$= \begin{cases} \dfrac{5x}{-x + 4x}, & x < 0 \\ \dfrac{5x}{x + 4x}, & x > 0 \\ 0, & x = 0 \end{cases}$$

$$= \begin{cases} \dfrac{5}{3}, & x < 0 \\ 1, & x > 0 \\ 0, & x = 0 \end{cases}$$

$$\text{LHL} = \lim_{x \to 0^-} f(x) = \frac{5}{3}$$

$$\text{RHL} = \lim_{x \to 0^+} f(x) = 1$$

$\because$ LHL $\ne$ RHL $\ne f(0)$

$\therefore = \lim_{x \to 0} f(x)$ does not exist.

14. $\lim_{x \to 1} \dfrac{x^4 - 3x^3 + 2}{x^3 - 5x^2 + 3x + 1}$ $\left[\dfrac{0}{0}\text{ form}\right]$

$$= \lim_{x \to 1} \frac{x^4 - x^3 - 2x^3 + 2x^2 - 2x^2 + 2x - 2x + 2}{x^3 - x^2 - 4x^2 + 4x - x + 1}$$

$$= \lim_{x \to 1} \frac{x^3(x-1) - 2x^2(x-1) - 2x(x-1) - 2(x-1)}{x^2(x-1) - 4x(x-1) - 1(x-1)}$$

$$= \lim_{x \to 1} \frac{(x-1)(x^3 - 2x^2 - 2x - 2)}{(x-1)(x^2 - 4x - 1)}$$

$$= \lim_{x \to 1} \frac{x^3 - 2x^2 - 2x - 2}{x^2 - 4x - 1} = \frac{1^3 - 2 \times 1^2 - 2 \times 1 - 2}{1^2 - 4 \times 1 - 1}$$

$$= \frac{1-2-2-2}{1-4-1} = \frac{-5}{-4} = \frac{5}{4}$$

15. $\lim_{x \to \sqrt{2}} \frac{x^4 - 4}{x^2 + 3x\sqrt{2} - 8}$ $\left[\frac{0}{0} \text{ form}\right]$

$$= \lim_{x \to \sqrt{2}} \frac{(x^2-2)(x^2+2)}{x^2 + 4x\sqrt{2} - \sqrt{2}x - 8}$$

$$= \lim_{x \to \sqrt{2}} \frac{(x-\sqrt{2})(x+\sqrt{2})(x^2+2)}{x(x+4\sqrt{2}) - \sqrt{2}(x+4\sqrt{2})}$$

$$= \lim_{x \to \sqrt{2}} \frac{(x-\sqrt{2})(x+\sqrt{2})(x^2+2)}{(x+4\sqrt{2})(x-\sqrt{2})}$$

$$= \lim_{x \to \sqrt{2}} \frac{(x+\sqrt{2})(x^2+2)}{(x+4\sqrt{2})}$$

$$= \frac{(\sqrt{2}+\sqrt{2})((\sqrt{2})^2+2)}{\sqrt{2}+4\sqrt{2}} = \frac{(2\sqrt{2})4}{5\sqrt{2}} = \frac{8}{5}$$

16. $\lim_{x \to 0} \frac{\frac{\sin x}{\cos x} - \frac{\sin x}{1}}{x^3}$

$$= \lim_{x \to 0} \frac{\sin x}{\cos x}\left(\frac{1 - \cos x}{x^3}\right) = \lim_{x \to 0} \frac{\tan x}{x} \times \frac{2\sin^2 \frac{x}{2}}{x^2}$$

$$= \lim_{x \to 0} \frac{\tan x}{x} \times \lim_{x \to 0} \frac{2\sin^2 \frac{x}{2}}{4 \times \left(\frac{x}{2}\right)^2}$$

$$= 1 \times \frac{2}{4} \lim_{x \to 0} \left(\frac{\sin \frac{x}{2}}{\frac{x}{2}}\right)^2 = 1 \times \frac{1}{2} \times 1 = \frac{1}{2}$$

17. Let $f(x) = \sin x$. Then, $f(x+h) = \sin(x+h)$

$$\therefore \quad \frac{d}{dx}(f(x)) = \lim_{h \to 0} \frac{f(x+h) - f(x)}{h}$$

$$\Rightarrow \frac{d}{dx}(f(x)) = \lim_{h \to 0} \frac{\sin(x+h) - \sin x}{h}$$

$$\Rightarrow \frac{d}{dx}(f(x)) = \lim_{h \to 0} \frac{2\sin\left(\frac{h}{2}\right)\cos\left(\frac{2x+h}{2}\right)}{h}$$

$$\left[\because \sin C - \sin D = 2\sin\frac{C-D}{2}\cos\frac{C+D}{2}\right]$$

$$\Rightarrow \frac{d}{dx}(f(x)) = \lim_{h \to 0} \frac{(2\sin h/2)\cos(x+h/2)}{2(h/2)}$$

$$\Rightarrow \frac{d}{dx}(f(x)) = \lim_{h \to 0} \cos\left(x + \frac{h}{2}\right) \lim_{h \to 0} \frac{\sin(h/2)}{(h/2)}$$

$$\Rightarrow \frac{d}{dx}(f(x)) = (\cos x) \times 1 = \cos x \quad \left[\because \lim_{h \to 0} \frac{\sin(h/2)}{(h/2)} = 1\right]$$

18. Given, $y = \frac{x^2 \sin x}{1-x}$

On differentiating w.r.t. x, we get

$$\frac{dy}{dx} = \frac{d}{dx}\left(\frac{x^2 \sin x}{1-x}\right)$$

$$= \frac{(1-x)\frac{d}{dx}(x^2 \sin x) - (x^2 \sin x)\frac{d}{dx}(1-x)}{(1-x)^2}$$

[Quotient rule]

$$= \frac{(1-x)\left\{x^2 \frac{d}{dx}(\sin x) + \sin x \frac{d}{dx}(x^2)\right\} - x^2 \sin x(0-1)}{(1-x)^2}$$

$$= \frac{(1-x)\{x^2 \cos x + \sin x(2x)\} + x^2 \sin x}{(1-x)^2}$$

$$= \frac{(1-x)x^2 \cos x + 2x(1-x)\sin x + x^2 \sin x}{(1-x)^2}$$

$$= \frac{(1-x)x^2 \cos x + x(2-x)\sin x}{(1-x)^2}, x \neq 1$$

QR Code for Quick Test 1	QR Code for Quick Test 2	QR Code for Quick Test 3	QR Code for Quick Test 4	QR Code for Quick Test 5

Mind Map

Limits and Derivatives

Derivative of Trigonometric Functions

To find the derivative of trigonometric functions, we use the algebra of derivative and the following formulae

$\frac{d}{dx}(\sin x) = \cos x$

$\frac{d}{dx}(\cos x) = -\sin x$

$\frac{d}{dx}(\tan x) = \sec^2 x$

$\frac{d}{dx}(\sec x) = \sec x \times \tan x$

$\frac{d}{dx}(\cot x) = -\operatorname{cosec}^2 x$

$\frac{d}{dx}(\operatorname{cosec} x) = -\operatorname{cosec} x \times \cot x$

First Principle of Derivative

Suppose f is a real valued function, the function defined by

$$\lim_{h\to 0}\frac{f(x+h)-f(x)}{h}$$

wherever the limit exists, is defined to be the derivative of f at x and is denoted by $f'(x)$.
This definition of derivative is called the first principle of derivative.

Algebra of Derivative of Functions

Let f and g be two functions such that their derivatives are defined in a common domain. Then,

- Derivative of sum of two functions is sum of the derivatives of the functions.
$$\frac{d}{dx}[f(x)+g(x)] = \frac{d}{dx}f(x) + \frac{d}{dx}g(x)$$
- Derivative of difference of two functions is difference of the derivatives of the functions.
$$\frac{d}{dx}[f(x)-g(x)] = \frac{d}{dx}f(x) - \frac{d}{dx}g(x)$$
- Derivative of product of two functions is given by the following product rule.
$$\frac{d}{dx}[f(x)\times g(x)] = f(x)\frac{d}{dx}g(x) + g(x)\frac{d}{dx}f(x)$$
This is also known as Leibnitz product rule of derivative.
- Derivative of quotient of two functions is given by the following quotient rule.
$$\frac{d}{dx}\left[\frac{f(x)}{g(x)}\right] = \frac{g(x)\frac{d}{dx}f(x) - f(x)\frac{d}{dx}g(x)}{[g(x)]^2}, g(x)\neq 0$$

Derivative of a function

Suppose, f is a real valued function and a is a point in its domain. Then, derivative of f at a is defined by

$$\lim_{h\to 0}\frac{f(a+h)-f(a)}{h}$$

provided this limit exists.
The derivative of $f(x)$ at a is denoted by

$$f'(a) \text{ or } \left[\frac{df(x)}{dx}\right]_{x=a}$$

Existence of Limit

If the left hand limit and right hand limit both exist (i.e. finite and unique) and coincide (i.e. same), then we say that limit exists and their common value is called the limit of $f(x)$ at $x = a$.
i.e. if $\lim_{x\to a^-} f(x) = \lim_{x\to a^+} f(x) = l$, then $\lim_{x\to a} f(x) = l$

Limits of Rational Functions

A function f is said to be a rational function, if $f(x) = \frac{g(x)}{h(x)}$, where $g(x)$ and $h(x)$ are polynomial functions such that $h(x) \neq 0$.
Then, $\lim_{x\to a} f(x) = \lim_{x\to a}\frac{g(x)}{h(x)} = \frac{\lim_{x\to a} g(x)}{\lim_{x\to a} h(x)} = \frac{g(a)}{h(a)}$

Algebra of Limits

Let f and g be two real functions with common domain D, such that $\lim_{x\to a} f(x)$ and $\lim_{x\to a} g(x)$ exists. Then,

- Limit of sum of two functions is sum of the limits of the functions.
i.e. $\lim_{x\to a}(f+g)(x) = \lim_{x\to a} f(x) + \lim_{x\to a} g(x)$
- Limit of difference of two functions is difference of the limits of the function.
i.e. $\lim_{x\to a}(f-g)(x) = \lim_{x\to a} f(x) - \lim_{x\to a} g(x)$
- Limit of product of a constant and one function is the product of that constant and limit of a function.
i.e. $\lim_{x\to a}[c\cdot f(x)] = c\lim_{x\to a} f(x)$,
where c is a constant.
- Limit of product of two functions is product of the limits of the function.
i.e. $\lim_{x\to a}[f(x).g(x)] = \lim_{x\to a} f(x).\lim_{x\to a} g(x)$
- Limit of quotient of two functions is quotient of the limits of the functions.
i.e. $\lim_{x\to a}\frac{f(x)}{g(x)} = \frac{\lim_{x\to a} f(x)}{\lim_{x\to a} g(x)}$,

where $\lim_{x\to a} g(x) \neq 0$

Limit

If $f(x)$ approaches to a real number l, when x approaches to a (through lesser or greater values to a) i.e. if $f(x) \to l$, when $x \to a$, then l is called limit of the function $f(x)$.
In symbolic form, it can be written as $\lim_{x\to a} f(x) = l$.

Limits of Trigonometric Functions

Evaluation of Trigonometric Limits when the Variable Tends to Zero
To evaluate the trigonometric limits, when the variable x is measured in radians and tends to zero, then

$$\lim_{x\to 0}\frac{\sin x}{x} = 1, \lim_{x\to 0}\frac{\tan x}{x} = 1 \text{ and } \lim_{x\to 0}\frac{1-\cos x}{x} = 1$$

Evaluation of Trigonometric Limits when Variable Tends to a Non-zero Quantity

Suppose, we have limit of the form, $\lim_{x\to 0} f(x)$
In such cases, first we convert the limit tends to zero. For this, replace $x \to a$ by $h \to 0$ and x by $(a+h)$ and then solve the limit.

Topics to be Assessed Formatively

Derivatives of Composite Functions

Derivative of a composite function can be found out by chain rule.

Chain Rule

Chain rule is applied when the given function is the function of function, i.e. a function is in the form of $fog(x)$ or $f[g(x)]$.

Let y be a real valued function which is a composite of two functions, say $y = f(u)$ and $u = g(x)$.

Then, $\dfrac{dy}{dx} = \dfrac{dy}{du} \cdot \dfrac{du}{dx} = f'(u) \cdot g'(x)$

i.e. $\dfrac{d}{dx}[f\{g(x)\}] = f'\{g(x)\} \cdot g'(x)$

Extension of Chain Rule

Extension of chain rule is applied when the given function (say y) is a composite of three functions.

e.g. If $y = f(u)$, $u = g(v)$ and $v = h(x)$, then

$$\frac{dy}{dx} = \frac{dy}{du} \cdot \frac{du}{dv} \cdot \frac{dv}{dx} = f'(u) \cdot g'(v) \cdot h'(x)$$

i.e. $\dfrac{d}{dx}[f(g(h(x)))] = f'[g(h(x)]. g'(h(x)) \cdot h'(x)$.

Example 1 **Differentiate $\sin(x^2 + 5)$ w.r.t. x.**

Sol. Let $y = \sin u$, where $u = x^2 + 5$. Using chain rule,

$$\frac{dy}{dx} = \frac{dy}{du} \cdot \frac{du}{dx} = \cos u \cdot \frac{du}{dx} = \cos(x^2+5)\frac{d}{dx}(x^2+5)$$

$$= \cos(x^2 + 5) \cdot (2x)$$

$$= 2x \cdot \cos(x^2 + 5)$$

Example 2 **Differentiate $\sin^2(3x + 1)$ w.r.t. x.**

Sol. Let $y = \sin^2(3x + 1)$

Here, y is a composite of three functions.

Let $y = t^2$, where $t = \sin u$ and $u = 3x + 1$.

Then, $\dfrac{dy}{dx} = \dfrac{dy}{dt} \cdot \dfrac{dt}{du} \cdot \dfrac{du}{dx}$ [by chain rule]

$$= \frac{d}{dt}[t^2] \cdot \frac{d}{du}(\sin u) \cdot \frac{d}{dx}(3x+1)$$

$$= 2t(\cos u) \cdot 3 = 6t \cos u$$

$$= 6 \sin u \cos u \qquad [\because t = \sin u]$$

$$= 3 \sin 2u \qquad [\because 2\sin\theta\cos\theta = \sin 2\theta]$$

$$= 3 \sin 2(3x + 1)$$

$$= 3\sin(6x + 2) \qquad [\text{put } u = 3x + 1]$$

Example 3 **Differentiate $\sin 5x \cdot \cos 7x$ w.r.t. x.**

Sol. Let $y = \sin 5x \cdot \cos 7x = \dfrac{1}{2}[2\sin 5x \cdot \cos 7x]$

$$= \frac{1}{2}[\sin(5x + 7x) + \sin(5x - 7x)]$$

$$[\because 2\sin A\cos B = \sin(A+B) + \sin(A-B)]$$

$$= \frac{1}{2}[\sin 12x - \sin 2x]$$

$$\Rightarrow y = \frac{1}{2}\sin 12x - \frac{1}{2}\sin 2x$$

On differentiating both sides w.r.t. x, we get

$$\frac{dy}{dx} = \frac{1}{2} \cdot \frac{d}{dx}(\sin 12x) - \frac{1}{2} \cdot \frac{d}{dx}(\sin 2x)$$

[by difference rule]

$$= \frac{1}{2}\cos 12x \frac{d}{dx}(12x) - \frac{1}{2}\cos 2x \frac{d}{dx}(2x)$$

$$= \frac{1}{2}\cos 12x \cdot 12 - \frac{1}{2}\cos 2x \cdot 2$$

$$= 6\cos 12x - \cos 2x$$

Example 4 **Differentiate $2\sqrt{\cot(x^2)}$ w.r.t. x.**

Sol. Let $y = 2\sqrt{\cot(x^2)}$.

On differentiating both sides w.r.t. x, we get

$$\frac{dy}{dx} = \frac{d}{dx} 2(\cot x^2)^{1/2} = 2 \cdot \frac{1}{2} \{\cot(x^2)\}^{\frac{1}{2}-1} \cdot \frac{d}{dx}\cot(x^2)$$

$$\left[\text{using chain rule } \frac{d}{dx}[fg(x)] = f'\{g(x)\}\frac{d}{dx}g(x)\right]$$

$$= \frac{1}{\sqrt{\cot(x^2)}}[-\operatorname{cosec}^2 x^2]\frac{d}{dx}(x^2) \quad [\text{using chain rule}]$$

$$= -\frac{\operatorname{cosec}^2(x^2)\, 2x}{\sqrt{\cot(x^2)}} = \frac{-2x \operatorname{cosec}^2(x^2)}{\sqrt{\cot(x^2)}}$$

Exercise

Multiple Choice Questions

1. The differential coefficient of $\sin(\cos(x^2))$ with respect to x is

(a) $-2x\sin x^2 \cos(\cos x^2)$
(b) $2x\sin(x^2)\cos(x^2)$
(c) $2x\sin(x^2)\cos(x^2)\cos x$
(d) None of the above

2. If $y = \sqrt{3x+2} + \dfrac{1}{\sqrt{2x^2+4}}$, then $\dfrac{dy}{dx}$ is equal to

(a) $\dfrac{3}{2\sqrt{3x+2}} - \dfrac{2x}{(2x^2+4)^{3/2}}$

(b) $\dfrac{3}{2\sqrt{3x+2}} + \dfrac{2x}{(2x^2+4)^{3/2}}$

(c) $\dfrac{3}{2\sqrt{3x+2}} + \dfrac{2}{(2x^2+4)^{3/2}}$

(d) None of the above

Very short answer Type Questions

3. Differentiate the following functions.

(i) $\tan(x^2+5)$ (ii) $\tan(\sin x)$

(iii) $\operatorname{cosec}(\sqrt{x^2+2})$ (iv) $\sec\left(\dfrac{5}{x}\right)$

Short Answer Type I Questions

4. If $f(x) = \sin 2x - \cos 2x$, then find $f'\left(\dfrac{\pi}{6}\right)$.

5. If $f(x) = |\cos x|$, then find $f'\left(\dfrac{3\pi}{4}\right)$.

6. If $f(x) = |\cos x - \sin x|$, then find $f'\left(\dfrac{\pi}{6}\right)$.

7. Differentiate $\sqrt{\tan\sqrt{x}}$ w.r.t. x.

8. Differentiate $\cos x^3 \cdot \sin^2(x^5)$ w.r.t. x.

9. If $y = \sin[\sqrt{\sin\sqrt{x}}]$, then find $\dfrac{dy}{dx}$.

10. If $y = x\tan x + \sec x$, then find the value of $\dfrac{dy}{dx}$ at $x = \dfrac{\pi}{4}$.

11. If $y = \sin^3\sqrt{ax^2+bx+c}$, then find $\dfrac{dy}{dx}$.

Answers

1. (a) **2.** (a)

3. (i) $2x\sec^2(x^2+5)$

(ii) $\sec^2(\sin x)\cdot\cos x$

(iii) $-\operatorname{cosec}\sqrt{x^2+2}\cot\sqrt{x^2+2}\times\dfrac{x}{\sqrt{x^2+2}}$

(iv) $\sec\left(\dfrac{5}{x}\right)\tan\left(\dfrac{5}{x}\right)\times\dfrac{-5}{x^2}$

4. $1+\sqrt{3}$

5. $\dfrac{1}{\sqrt{2}}$

6. $\dfrac{-1}{2}(1+\sqrt{3})$

7. $\dfrac{\sec^2(\sqrt{x})}{4\sqrt{x}\sqrt{\tan\sqrt{x}}}$

8. $10x^4\sin x^5\cos x^5\cos x^3 - 3x^2\sin x^3\sin^2 x^5$

9. $\dfrac{\cos(\sqrt{\sin\sqrt{x}})}{4\sqrt{x}\sqrt{\sin\sqrt{x}}}\cdot\cos\sqrt{x}$

10. $\dfrac{\pi}{2}+1+\sqrt{2}$

11. $\dfrac{3}{2}\sin^2\sqrt{ax^2+bx+c}\cdot\cos\sqrt{ax^2+bx+c}\cdot\dfrac{(2ax+b)}{\sqrt{ax^2+bx+c}}$

13

CHAPTER

Maths is like going to the gym for your brain. It sharpens your mind.

—Danica Mckellar

▼ **YOUR STUDY GOALS**

13.1 Measure of Dispersion
13.2 Range
13.3 Mean Deviation
13.4 Variance and Standard Deviation
13.5 Shortcut Method Or Step-Deviation Method

▼ **YOUR PRACTICE GOALS**

- Objective Type Questions (27 Qs ; 15 MCQs, 10 A-R, 2 Case Based)
- Subjective Type Questions (16 Qs; 5 VSA, 6SA, 5LA)

Statistics

13.1 Measure of Dispersion

The dispersion or scatter in a data is measured on the basis of the observations and the types of measure of central tendency used.

There are following measures of dispersion.

1. Range
2. Quartile deviation
3. Mean deviation
4. Standard deviation.

13.2 Range

Range is the difference of maximum and minimum values of data.

Range = Maximum value – Minimum value

13.3 Mean Deviation

The arithmetic mean of the absolute deviations of all the values taken from a central value or a fixed number a. The mean devation from a is denoted as MD (a).

Thus, mean deviation from a,

$$\text{MD}(a) = \frac{\textbf{[Sum of absolute values of deviations from a]}}{\textbf{Number of observations}}$$

Mean Deviation for Ungrouped Data

Let $x_1, x_2, \ldots, x_n$ be n observations. Then, mean deviation about mean or median can be determined by following steps:

I. Find the mean or median of given observations using the suitable formula.

II. Find the deviation of each observation x_i from $\bar{x}$ (mean) or M (median) and then take their absolute values i.e. $|x_i - \bar{x}|$ or $|x_i - M|$.

III. Find the sum of absolute values of deviations obtained in Step II

i.e. $\sum_{i=1}^{n} |x_i - \bar{x}|$ or $\sum_{i=1}^{n} |x_i - M|$

IV. Now, find mean deviation about mean or median using the formula $\dfrac{\sum_{i=1}^{n} |x_i - \bar{x}|}{n}$ or $\dfrac{\sum_{i=1}^{n} |x_i - M|}{n}$,

where n is the number of observations.

Mean Deviation for Grouped Data

Data can be grouped into two ways.

(a) For Discrete Frequency Distribution

Let the given data have n distinct values $x_1, x_2, \ldots, x_n$ and their corresponding frequencies are $f_1, f_2, \ldots, f_n$, respectively.

Then, this data can be represented in the tabular form, as

x_i	x_1	x_2	x_3	...	x_n
f_i	f_1	f_2	f_3	...	f_n

and is called discrete frequency distribution.

Here, mean deviation about mean or median is given by

$$\frac{\sum_{i=1}^{n} f_i |x_i - A|}{N}, \text{ where } N = \sum_{i=1}^{n} f_i = \text{total frequency}$$

and A = mean or median.

Method to find the Mean Deviation about Mean

For finding the mean deviation about mean, we use the following steps

I. Find the mean of given observations using the formula $\bar{x} = \dfrac{\Sigma f_i x_i}{\Sigma f_i}$.

II. Find the deviation of each observation x_i from $\bar{x}$ and take their absolute values i.e. $|x_i - \bar{x}|$ and then find $f_i|x_i - \bar{x}|$.

III. Find the sum of absolute values of deviations obtained in Step II i.e. $\Sigma f_i |x_i - \bar{x}|$.

IV. Now, find the mean deviation about mean by using the formula, $\dfrac{\Sigma f_i |x_i - \bar{x}|}{\Sigma f_i}$.

Method to find the Mean Deviation about Median

For finding the mean deviation about median, we use the following steps

I. Arrange the given data either in ascending order or descending order.

II. Make a cumulative frequency table.

III. Find the median by using the formula,

- **If $\Sigma f_i = N$ is even,** then

$$\text{Median} = \frac{\text{Value of}\left(\frac{N}{2}\right)\text{th observation} + \text{Value of}\left(\frac{N}{2}+1\right)\text{th observation}}{2}$$

- **If $\Sigma f_i = N$ is odd,** then

$$\text{Median} = \text{Value of}\left(\frac{N+1}{2}\right)\text{th observation}$$

IV. Find the absolute values of deviations of observations x_i from M.

V. Find the product of frequency with absolute deviations i.e. $f_i|x_i - M|$.

VI. Find the mean deviation from median by using the formula, $\text{MD} = \dfrac{\Sigma f_i|x_i - M|}{\Sigma f_i}$.

(b) For Continuous Frequency Distribution

A continuous frequency distribution is a series in which the data is classified into different class intervals without gaps alongwith their respective frequencies.

Mean Deviation about Mean

For calculating mean deviation from mean of a continuous frequency distribution, the procedure is same as for a discrete frequency distribution. The only difference is that here, we have to obtain the mid-points of the various classes and take the deviations of these mid-points from the mean.

Mean Deviation about Median

For calculating mean deviation from median of a continuous frequency distribution, the procedure is same as about mean. The only difference is that here, we replace mean by median and median is calculated by the following formula,

$$M = l + \frac{\frac{N}{2} - cf}{f} \times h$$

l, f, h and cf are respectively the lower limit, the frequency, the width of median class and cumulative frequency of class just preceding the median class.

Shortcut Method for Calculating the Mean Deviation about Mean

Sometimes, the data is too large and then the calculation by the previous method is tedious. So, we apply the step deviation method.

In this method, we take an assumed mean, which is in the middle or just close to it, in the data. The process of taking step deviation is the change of scale on the number line as shown in the figure given below

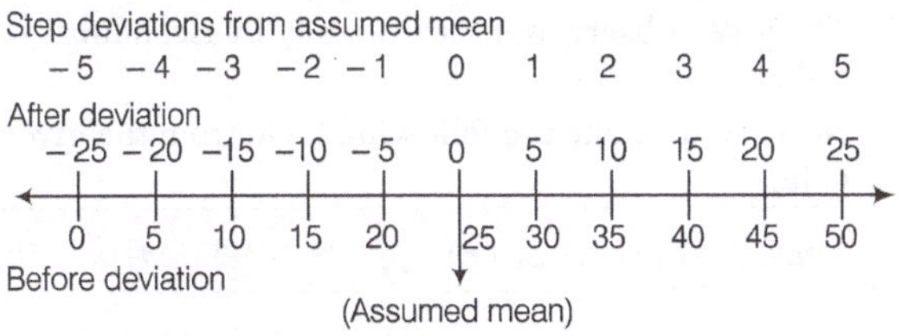

For step deviation method, we denote the new variable by u_i and it is defined as $u_i = \frac{x_i - a}{h}$.

Where, a is the assumed mean and h is the common factor or length of the class interval.

The mean $\bar{x}$ by step deviation method is given by

$$\bar{x} = a + \frac{\sum_{i=1}^{n} f_i u_i}{N} \times h$$

Example 1. **Find the mean deviation about the mean for the following data.** *NCERT*

38, 70, 48, 40, 42, 55, 63, 46, 54, 44

Sol. Given observations are

38, 70, 48, 40, 42, 55, 63, 46, 54 and 44

Here, number of observations, $n = 10$

$\therefore$ Mean,

$$\bar{x} = \frac{(38 + 70 + 48 + 40 + 42 + 55 + 63 + 46 + 54 + 44)}{10}$$

$$= \frac{500}{10} = 50$$

Let us make the table for deviation and absolute deviation

x_i	$x_i - \bar{x}$	$\lvert x_i - \bar{x} \rvert$
38	$38 - 50 = -12$	12
70	$70 - 50 = 20$	20
48	$48 - 50 = -2$	2
40	$40 - 50 = -10$	10
42	$42 - 50 = -8$	8
55	$55 - 50 = 5$	5
63	$63 - 50 = 13$	13
46	$46 - 50 = -4$	4
54	$54 - 50 = 4$	4
44	$44 - 50 = -6$	6
Total		84

Now, $$\text{MD} = \frac{\sum_{i=1}^{10} \lvert x_i - \bar{x} \rvert}{10} = \frac{84}{10} = 8.4$$

Example 2. **Find the mean deviation about the median for the data 34, 66, 30, 38, 44, 50, 40, 60, 42, 51.**

Sol. The given data can be arranged in ascending order as 30, 34, 38, 40, 42, 44, 50, 51, 60, 66. Here, total number of observations are 10. i.e. $n = 10$, which is even.

$\therefore$ Median,

$$M = \frac{\left(\frac{n}{2}\right)\text{th observation} + \left(\frac{n}{2} + 1\right)\text{th observation}}{2}$$

$$= \frac{\left(\frac{10}{2}\right)\text{th observation} + \left(\frac{10}{2} + 1\right)\text{th observation}}{2}$$

$= \frac{(\text{5th observation} + \text{6th observation})}{2} = \frac{42+44}{2} = \frac{86}{2}$

$= 43$

Let us make the table for absolute deviation

x_i	$\|x_i - M\|$
30	$\|30-43\|=13$
34	$\|34-43\|=9$
38	$\|38-43\|=5$
40	$\|40-43\|=3$
42	$\|42-43\|=1$
44	$\|44-43\|=1$
50	$\|50-43\|=7$
51	$\|51-43\|=8$
60	$\|60-43\|=17$
66	$\|66-43\|=23$
Total	87

Now, mean deviation about median,

$$\text{MD} = \frac{\sum_{i=1}^{10} |x_i - M|}{10} = \frac{87}{10} = 8.7$$

Example 3. Find the mean deviation about the mean for the following data.

x_i	2	5	6	8	10	12
f_i	2	8	10	7	8	5

Competency Based Que

Sol. Let us make the following table from the given data.

x_i	f_i	$f_i x_i$	$\|x_i - \bar{x}\|$	$f_i\|x_i - \bar{x}\|$
2	2	4	5.5	11
5	8	40	2.5	20
6	10	60	1.5	15
8	7	56	0.5	3.5
10	8	80	2.5	20
12	5	60	4.5	22.5
Total	40	300		92

Here, $N = \Sigma f_i = 40, \Sigma f_i x_i = 300$

Now, mean $(\bar{x}) = \frac{1}{N} \Sigma f_i x_i = \frac{1}{40} \times 300 = 7.5$

$\therefore$ Mean deviation about the mean,

$$\text{MD}(\bar{x}) = \frac{1}{N} \Sigma f_i |x_i - \bar{x}| = \frac{1}{40} \times 92 = 2.3$$

Hence, the mean deviation about mean is 2.3.

Example 4. Find the mean deviation from the median for the following data. *NCERT*

x_i	15	21	27	30	35
f_i	3	5	6	7	8

Sol. The given observations are already in ascending order. Now, let us make the following table from the given data.

x_i	f_i	cf	$\|x_i - M\|$	$f_i\|x_i - M\|$
15	3	3	15	45
21	5	8	9	45
27	6	14	3	18
30	7	21	0	0
35	8	29	5	40
Total	29		32	148

Here, $N = \Sigma f_i = 29$, which is odd.

So, the median $= \left(\frac{N+1}{2}\right)$th observation

i.e. $\frac{29+1}{2} = 15$th observation, which is equal to 30.

[$\because$ 15th observation lie in the cumulative frequency 21 and its corresponding observation is 30]

Thus, the median $(M) = 30$.

Hence, the mean deviation from the median

$$= \frac{\Sigma f_i |x_i - 30|}{\Sigma f_i} = \frac{148}{29} = 5.1$$

Example 5. Find the mean deviation from the median of the following frequency distribution.

Age (in years)	10	12	15	18	21	23
Frequency	3	5	4	10	8	4

Sol. The given observations are already in ascending order.

Now, let us make the following table from the given data.

Age (x_i)	Frequency (f_i)	cf	$\|x_i - 18\|$	$f_i\|x_i - 18\|$
10	3	3	8	24
12	5	8	6	30
15	4	12	3	12
18	10	22	0	0
21	8	30	3	24
23	4	34	5	20
Total	$N = 34$			110

Here, $\Sigma f_i = N = 34$, which is even.

∴ Median

$$= \frac{\text{Value of}\left(\frac{34}{2}\right)\text{th observation} + \text{Value of}\left(\frac{34}{2}+1\right)\text{th observation}}{2}$$

$$= \frac{\text{Value of 17th observation} + \text{Value of 18th observation}}{2}$$

$$= \frac{18+18}{2} = 18$$

[∵ both of these observations lie in the cumulative frequency 22 and its corresponding observation is 18]

Hence, the mean deviation from the median

$$= \frac{\Sigma f_i |x_i - M|}{\Sigma f_i} = \frac{110}{34} = 3.24 \text{ yr}$$

Example 6. Find the mean deviation about the mean for the following data.

Marks obtained	Number of students
10-20	2
20-30	3
30-40	8
40-50	14
50-60	8
60-70	3
70-80	2

Competency Based Que

Sol. Let us make the following table from the given data.

Marks obtained	Number of students (f_i)	Mid-points (x_i)	$f_i x_i$	$\lvert x_i - \bar{x}\rvert$	$f_i\lvert x_i - \bar{x}\rvert$
10-20	2	15	30	30	60
20-30	3	25	75	20	60
30-40	8	35	280	10	80
40-50	14	45	630	0	0
50-60	8	55	440	10	80
60-70	3	65	195	20	60
70-80	2	75	150	30	60
Total	40		1800		400

Here, $N = \sum_{i=1}^{7} f_i = 40$ and $\sum_{i=1}^{7} f_i x_i = 1800$,

Therefore, mean $(\bar{x}) = \frac{1}{N}\sum_{i=1}^{7} f_i x_i = \frac{1800}{40} = 45$

Now, mean deviation,

$$\text{MD}(\bar{x}) = \frac{1}{N}\sum_{i=1}^{7} f_i |x_i - \bar{x}|$$

$$= \frac{1}{40} \times 400 = 10 \qquad \left[\because \sum_{i=1}^{7} f_i |x_i - \bar{x}| = 400\right]$$

Hence, the mean deviation about mean is 10.

Example 7. Calculate the mean deviation about median for the following data. *NCERT*

Class	0-10	10-20	20-30	30-40	40-50	50-60
Frequency	6	7	15	16	4	2

Sol. Let us make the following table from the given data.

Class	f_i	cf	Mid-point (x_i)	$\lvert x_i - M\rvert$	$f_i\lvert x_i - M\rvert$
0-10	6	6	5	23	138
10-20	7	13	15	13	91
20-30	15	28	25	3	45
30-40	16	44	35	7	112
40-50	4	48	45	17	68
50-60	2	50	55	27	54
Total	50				508

Here, $\frac{N}{2} = \frac{50}{2} = 25$

which lies in the cumulative frequency 28 . Therefore, 20-30 is the median class.

So, we have, $l = 20$, $cf = 13$, $f = 15$, $h = 10$ and $N = 50$

Now, Median, $M = l + \frac{\frac{N}{2} - cf}{f} \times h$

$$= 20 + \frac{25-13}{15} \times 10$$

$$= 20 + 8 = 28$$

Hence, the mean deviation about median is given by

$$\text{MD}(M) = \frac{1}{N}\sum_{i=1}^{6} f_i |x_i - M|$$

$$= \frac{1}{50} \times 508 = 10.16$$

Hence, the mean deviation about median is 10.16.

Example 8. Calculate the mean deviation about median for the age distribution of 100 persons given below. NCERT

Age	Number
16-20	5
21-25	6
26-30	12
31-35	14
36-40	26
41-45	12
46-50	16
51-55	9

Competency Based Que

Sol. Let us make the following table from the given data.

Class	f_i	cf	Mid-value (x_i)	$\lvert x_i - M\rvert$	$f_i\lvert x_i - M\rvert$
15.5-20.5	5	5	18	$\lvert 18-38\rvert = 20$	100
20.5-25.5	6	11	23	$\lvert 23-38\rvert = 15$	90
25.5-30.5	12	23	28	$\lvert 28-38\rvert = 10$	120
30.5-35.5	14	37	33	$\lvert 33-38\rvert = 5$	70
35.5-40.5	26	63	38	$\lvert 38-38\rvert = 0$	0
40.5-45.5	12	75	43	$\lvert 43-38\rvert = 5$	60
45.5-50.5	16	91	48	$\lvert 48-38\rvert = 10$	160
50.5-55.5	9	100	53	$\lvert 53-38\rvert = 15$	135
Total	$N = 100$				735

Here, $\frac{N}{2} = 50$

So, the median class is 35.5-40.5.

$\therefore \quad l = 35.5,\ cf = 37,\ f = 26$ and $h = 5$

Now, median $(M) = l + \dfrac{\frac{N}{2} - cf}{f} \times h$

$$= 35.5 + \frac{50-37}{26} \times 5$$

$$= 35.5 + \frac{13}{26} \times 5$$

$$= 35.5 + 2.5 = 38$$

and mean deviation about the median

$$= \frac{\Sigma f_i\lvert x_i - M\rvert}{\Sigma f_i} = \frac{735}{100} = 7.35$$

Hence, the mean deviation about median is 7.35.

Example 9. Find the mean deviation from the mean of the following data by shortcut or step deviation method.

Class	Frequency
0-100	4
100-200	8
200-300	9
300-400	10
400-500	7
500-600	5
600-700	4
700-800	3

Competency Based Que

Sol. Let us make the following table for step deviation and product of frequency with absolute deviation.

Class	(f_i)	Mid-points (x_i)	$u_i = \frac{x_i - 450}{100}$ $(a = 450, h = 100)$	$f_i u_i$	$\lvert x_i - \bar{x}\rvert = \lvert x_i - 358\rvert$	$f_i\lvert x_i - \bar{x}\rvert$
0-100	4	50	−4	−16	308	1232
100-200	8	150	−3	−24	208	1664
200-300	9	250	−2	−18	108	972
300-400	10	350	−1	−10	8	80
400-500	7	450	0	0	92	644
500-600	5	550	1	5	192	960
600-700	4	650	2	8	292	1168
700-800	3	750	3	9	392	1176
Total	50			−46		7896

Here, $N = 50$,

assumed mean, $a = 450$,

$\Sigma f_i u_i = -46$ and $h = 100$

$\therefore \quad$ Mean $(\bar{x}) = a + h\left(\frac{1}{N}\Sigma f_i u_i\right)$

$$= 450 + 100 \times \left(-\frac{46}{50}\right) = 358$$

Now, mean deviation $= \frac{1}{N}\Sigma f_i\lvert x_i - \bar{x}\rvert$

$$= \frac{7896}{50} = 157.92$$

Quick Test 1

1. Find the mean deviation of the data 3, 10, 10, 4, 7, 10, 5 from the mean.
2. Find the mean deviation from the median for the given data 3, 9, 5, 3, 12, 10, 18, 4, 7, 9, 21 NCERT
3. Find the mean deviation from the mean for the following data. Competency Based Que

x_i	10	30	50	70	90
f_i	4	24	28	16	8

4. Find the mean deviation from the median for the following data. NCERT

x_i	15	21	27	30	35
f_i	2	6	6	7	8

5. Calculate the mean deviation from the mean of the following distribution.

Marks	0-10	10-20	20-30	30-40	40-50
Number of students	5	8	15	16	6

6. Calculate the mean deviation from the median for the following data. Competency Based Que

Wages per day	20-30	30-40	40-50	50-60	60-70
Number of workers	3	8	12	9	8

7. Calculate the mean deviation from the median for the given data.

Class	0-10	10-20	20-30	30-40	40-50
Frequency	5	10	20	5	10

8. Find the mean deviation from the mean for the following data by shortcut method. Competency Based Que

Class	Frequency
10-20	8
20-30	10
30-40	15
40-50	25
50-60	20
60-70	18
70-80	9
80-90	5

Answers

1. 2.57 2. 5.27 3. 16 4. 4.9 5. 9.44 6. 10.125 7. 9 8. 14.95

For Solutions, See QR Code on Pg. 384

13.4 Variance and Standard Deviation

Mean of the squares of the deviations from the mean is called the **variance**, to be denoted by σ^2.

The positive square root of variance is called **standard deviation** and it is denoted by σ.

Variance and Standard Deviation of Ungrouped Data

Variance of n observations $x_1, x_2, x_3, \dots x_n$ is given by

$$\sigma^2 = \frac{\sum_{i=1}^{n}(x_i-\bar{x})^2}{n} \text{ or } \frac{\sum_{i=1}^{n} x_i^2}{n} - \left(\frac{\sum_{i=1}^{n} x_i}{n}\right)^2$$

We know that standard deviation $= \sqrt{\text{Variance}}$

$$\therefore \quad \sigma = \sqrt{\frac{\sum_{i=1}^{n}(x_i-\bar{x})^2}{n}} \text{ or } \sqrt{\frac{\sum_{i=1}^{n} x_i^2}{n} - \left(\frac{\sum_{i=1}^{n} x_i}{n}\right)^2}$$

$$\text{or} \quad \frac{1}{n}\sqrt{n\sum_{i=1}^{n} x_i^2 - \left(\sum_{i=1}^{n} x_i\right)^2}$$

Method to Find Variance and Standard Deviation

To find the standard deviation or variance when deviations are taken from actual mean, we use the following steps.

I. Calculate the mean $\bar{x}$ of the given observation $x_1, x_2, \dots, x_n$.

II. Take the deviations of the observations from their mean i.e. find $x_i - \bar{x}; i = 1, 2, \dots, n$.

III. Square the deviations obtained in Step II and then find the sum i.e. $\sum_{i=1}^{n}(x_i - \bar{x})^2$.

IV. Find the variance and standard deviation by using the formula, variance, $\sigma^2 = \frac{\sum_{i=1}^{n}(x_i - \bar{x})^2}{n}$

and standard deviation $= \sqrt{\sigma^2} = \sigma$

Variance and Standard Deviation of a Discrete Frequency Distribution

Let the discrete frequency distribution be
$x: x_1, x_2, x_3, \ldots, x_n$ and $f: f_1, f_2, f_3, \ldots, f_n$.
Then, by

Direct Method

$$\text{Variance } (\sigma^2) = \frac{1}{N}\sum_{i=1}^{n} f_i(x_i - \bar{x})^2$$

or $$\sigma^2 = \frac{1}{N}\Sigma f_i x_i^2 - \left(\frac{\Sigma f_i x_i}{N}\right)^2$$

and standard deviation,

$$\sigma = \sqrt{\frac{1}{N}\sum_{i=1}^{n} f_i(x_i - \bar{x})^2} \quad \text{or } \sigma = \frac{1}{N}\sqrt{N\Sigma f_i x_i^2 - (\Sigma f_i x_i)^2}$$

where, $N = \sum_{i=1}^{n} f_i$

Shortcut Method

$$\text{Variance, } \sigma^2 = \frac{1}{N}\sum_{i=1}^{n} f_i d_i^2 - \left(\frac{\Sigma f_i d_i}{N}\right)^2$$

and standard deviation,

$$\sigma = \sqrt{\frac{1}{N}\sum_{i=1}^{n} f_i d_i^2 - \left(\frac{\Sigma f_i d_i}{N}\right)^2}$$

where, $d_i = x_i - a$ is deviation from assumed mean and a = assumed mean

Variance and Standard Deviation of a Continuous Frequency Distribution

Direct Method

In this method, we first replace each class by its mid-point, then this method becomes similar to the discrete frequency distribution. If there is a frequency distribution of n classes and each class defined by its mid-point x_i with corresponding frequency f_i, then variance and standard deviation are respectively,

$$\sigma^2 = \frac{1}{N}\sum_{i=1}^{n} f_i(x_i - \bar{x})^2 \text{ and } \sigma = \sqrt{\frac{1}{N}\sum_{i=1}^{n} f_i(x_i - \bar{x})^2}$$

or $$\sigma^2 = \frac{1}{N^2}[N\Sigma f_i x_i^2 - (\Sigma f_i x_i)^2]$$

and $$\sigma = \frac{1}{N}\sqrt{N\Sigma f_i x_i^2 - (\Sigma f_i x_i)^2}$$

13.4 Shortcut Method Or Step-Deviation Method

Sometimes the values of mid-points x_i of different classes in a continuous distribution are large and so the calculation of mean and variance becomes tedious and time consuming. For this, we use the **step deviation method** to find mean and variance.

$$\text{Variance, } \sigma^2 = h^2\left[\frac{1}{N}\sum_{i=1}^{n} f_i u_i^2 - \left(\frac{\sum_{i=1}^{n} f_i u_i}{N}\right)^2\right]$$

and standard deviation,

$$\sigma = h\sqrt{\frac{1}{N}\sum_{i=1}^{n} f_i u_i^2 - \left(\frac{\sum_{i=1}^{n} f_i u_i}{N}\right)^2}$$

where, $u_i = \frac{x_i - a}{h}$, a = assumed mean

and h = width of class interval

Method to find Variance and Standard Deviation by Shortcut Method

To find variance and standard deviation, use the following steps

I. Select an assumed mean, say a and then calculate $u_i = \frac{x_i - a}{h}$, where h = width of class interval or common factor.

II. Multiply the frequency of each class with the corresponding u_i and obtain $\Sigma f_i u_i$.

III. Square the values of u_i and multiply them with the corresponding frequencies and obtain $\Sigma f_i u_i^2$.

IV. Substitute the values of $\Sigma f_i u_i$, $\Sigma f_i u_i^2$ and $\Sigma f_i = N$ in the formula,

$$\text{Variance}(\sigma^2) = h^2\left[\frac{1}{N}\Sigma f_i u_i^2 - \left(\frac{1}{N}\Sigma f_i u_i\right)^2\right]$$

and standard deviation $= \sqrt{\sigma^2}$ to get the required values.

TEACHER TIPS

- If each observation is multiplied by a constant k, the variance of the resulting observations becomes k^2 times the original variance.
- Adding or subtracting a positive number to or from each observation of a group does not affect the variance.

Example 1. **Find the mean, variance and standard deviation for the following data, 6, 7, 10, 12, 13, 4, 8, 12.**

Sol. Given observations are 6, 7, 10, 12, 13, 4, 8, 12

Number of observations = 8

$\therefore$ Mean $(\bar{x}) = \frac{6+7+10+12+13+4+8+12}{8} = \frac{72}{8} = 9$

Now, let us make the following table for deviation.

x_i	$x_i - \bar{x}$	$(x_i - \bar{x})^2$
6	−3	9
7	−2	4
10	1	1
12	3	9
13	4	16
4	−5	25
8	−1	1
12	3	9
Total		74

$\therefore$ Sum of squares of deviations, $\sum_{i=1}^{8}(x_i - \bar{x})^2 = 74$

Hence, variance, $\sigma^2 = \frac{\sum_{i=1}^{8}(x_i - \bar{x})^2}{n} = \frac{74}{8} = 9.25$

and standard deviation, $\sqrt{\sigma^2} = \sqrt{9.25} = 3.04$

Example 2. **Find the variance and standard deviation of the following data.** *NCERT, Competency Based Que*

x_i	4	8	11	17	20	24	32
f_i	3	5	9	5	4	3	1

Sol. Let us make the following table from the given data.

x_i	f_i	$f_i x_i$	$x_i - \bar{x} = x_i - 14$	$(x_i - \bar{x})^2$	$f_i(x_i - \bar{x})^2$
4	3	12	−10	100	300
8	5	40	−6	36	180
11	9	99	−3	9	81
17	5	85	3	9	45
20	4	80	6	36	144
24	3	72	10	100	300
32	1	32	18	324	324
Total	30	420			1374

Here, we have, $N = \Sigma f_i = 30$, $\Sigma f_i x_i = 420$

$\therefore \quad \bar{x} = \frac{\sum_{i=1}^{7} f_i x_i}{N} = \frac{420}{30} = 14$

Hence, variance $(\sigma^2) = \frac{1}{N}\sum_{i=1}^{7} f_i (x_i - \bar{x})^2$

$$\left[\because \sigma^2 = \frac{1}{N}\left\{\sum_{i=1}^{n} f_i (x_i - \bar{x})^2\right\}\right]$$

$= \frac{1}{30} \times 1374 = 45.8$

and standard deviation, $\sigma = \sqrt{\sigma^2} = \sqrt{45.8} = 6.77$

Alternate Method

Let us make the following table from the given data.

x_i	f_i	$f_i x_i$	x_i^2	$f_i x_i^2$
4	3	12	16	48
8	5	40	64	320
11	9	99	121	1089
17	5	85	289	1445
20	4	80	400	1600
24	3	72	576	1728
32	1	32	1024	1024
Total	30	420		7254

Here, $N = \sum_{i=1}^{7} f_i = 30$, $\sum_{i=1}^{7} f_i x_i = 420$

and $\sum_{i=1}^{7} f_i x_i^2 = 7254$

$$\therefore \text{ Variance, } \sigma^2 = \frac{1}{N}\sum_{i=1}^{7} f_i x_i^2 - \left(\frac{\sum_{i=1}^{7} f_i x_i}{N}\right)^2$$

$$= \frac{1}{30} \times 7254 - \left(\frac{420}{30}\right)^2$$

$$= \frac{7254}{30} - 196$$

$$= 241.8 - 196 = 45.8$$

and standard deviation, $\sigma = \sqrt{45.8} = 6.77$

Example 3. Find the variance and standard deviation for the following distribution using shortcut method. *NCERT*

x_i	60	61	62	63	64	65	66	67	68
f_i	2	1	12	29	25	12	10	4	5

Sol. Let the assumed mean be $a = 64$

Now, let as make a table from the given data.

x_i	f_i	$d_i = x_i - 64$	$f_i d_i$	d_i^2	$f_i d_i^2$
60	2	$60-64=-4$	−8	16	32
61	1	$61-64=-3$	−3	9	9
62	12	$62-64=-2$	−24	4	48
63	29	$63-64=-1$	−29	1	29
64	25	$64-64=0$	0	0	0
65	12	$65-64=1$	12	1	12
66	10	$66-64=2$	20	4	40
67	4	$67-64=3$	12	9	36
68	5	$68-64=4$	20	16	80
Total	100		0	60	286

Here, $N = \Sigma f_i = 100$; $\Sigma f_i d_i = 0$ and $\Sigma f_i d_i^2 = 286$

$$\text{Now variance, } \sigma^2 = \left[\frac{1}{N}\Sigma f_i d_i^2 - \left(\frac{1}{N}\Sigma f_i d_i\right)^2\right]$$

$$= \frac{1}{100} \times 286 - \left(\frac{1}{100} \times 0\right)^2 = 2.86$$

and standard deviation, $\sigma = \sqrt{2.86} = 1.69$

Example 4. Calculate the variance and standard deviation for the following distribution.

Class	30-40	40-50	50-60	60-70	70-80	80-90	90-100
Frequency	3	7	12	15	8	3	2

Sol. Let us construct the following table.

Class	Frequency (f_i)	Mid-point (x_i)	$f_i x_i$	$(x_i - \bar{x})^2$	$f_i(x_i - \bar{x})^2$
30-40	3	35	105	729	2187
40-50	7	45	315	289	2023
50-60	12	55	660	49	588
60-70	15	65	975	9	135
70-80	8	75	600	169	1352
80-90	3	85	255	529	1587
90-100	2	95	190	1089	2178
Total	50		3100		10050

Here, $N = 50$ and $\Sigma f_i x_i = 3100$

$$\therefore \quad \text{Mean, } \bar{x} = \frac{1}{N}\Sigma f_i x_i = \frac{3100}{50} = 62$$

$$\text{Now, variance, } \sigma^2 = \frac{1}{N}\Sigma f_i (x_i - \bar{x})^2$$

$$= \frac{1}{50} \times 10050 = 201$$

and standard deviation, $\sigma = \sqrt{201} = 14.18$

Example 5. The diameter of circles (in mm) drawn in a design are given below.

Diameter (in mm)	33-36	37-40	41-44	45-48	49-52
Number of circles	15	17	21	22	25

Calculate the standard deviation and mean diameter of circles. *NCERT, Competency Based Que*

Sol. Let us first make the data continuous by making classes as 32.5-36.5, 36.5-40.5, 40.5-44.5, 44.5-48.5, 48.5-52.5.

Let assumed mean is 42.5 and $h = 4$

$$\text{Then,} \quad u_i = \frac{x_i - a}{h} = \frac{x_i - 42.5}{4}$$

Now, let us make the following table from the given data.

Class interval	Mid-value (x_i)	f_i	$u_i = \frac{x_i - 42.5}{4}$	$f_i u_i$	u_i^2	$f_i u_i^2$
32.5-36.5	34.5	15	−2	−30	4	60
36.5-40.5	38.5	17	−1	−17	1	17
40.5-44.5	42.5	21	0	0	0	0
44.5-48.5	46.5	22	1	22	1	22
48.5-52.5	50.5	25	2	50	4	100
Total		100		25		199

Now, mean $(\bar{x}) = a + \frac{\Sigma f_i u_i}{\Sigma f_i} \times h = 42.5 + 4 \times \frac{25}{100}$

[using step deviation method]

$= 42.5 + 1 = 43.5$

Standard deviation, σ

$$= \sqrt{h^2 \left[\frac{1}{N} \Sigma f_i u_i^2 - \left(\frac{1}{N} \Sigma f_i u_i\right)^2\right]}$$

[by shortcut method]

$$= \sqrt{(4)^2 \left[\frac{1}{100} \times 199 - \left(\frac{1}{100} \times 25\right)^2\right]}$$

$$= \sqrt{16\left(1.99 - \frac{1}{16}\right)} = \sqrt{16\left(\frac{31.84 - 1}{16}\right)}$$

$= \sqrt{30.84} = 5.55$

Hence, the standard deviation and mean diameter of circles are 5.55 and 43.5 respectively.

Example 6. Calculate the mean and standard deviation of the following cumulative data. *Competency Based Que*

Wages (in ₹)	0-15	15-30	30-45	45-60	60-75	75-90	90-105	105-120
Number of workers	12	30	65	107	157	202	222	230

Sol. We are given the cumulative frequency distribution. So, first we will prepare the frequency distribution as given below

Class interval	cf	Mid-value (x_i)	f_i	$u_i = \frac{x_i - 67.5}{15}$	$f_i u_i$	$f_i u_i^2$
0-15	12	7.5	12	−4	−48	192
15-30	30	22.5	18	−3	−54	162
30-45	65	37.5	35	−2	−70	140
45-60	107	52.5	42	−1	−42	42
60-75	157	67.5	50	0	0	0
75-90	202	82.5	45	1	45	45
90-105	222	97.5	20	2	40	80
105-120	230	112.5	8	3	24	72
Total			230		−105	733

Here, $a = 67.5, h = 15, N = \Sigma f_i = 230,$

$\Sigma f_i u_i = -105$ and $\Sigma f_i u_i^2 = 733$

Now, mean $= a + h\left(\frac{1}{N} \Sigma f_i u_i\right) = 67.5 + 15\left(\frac{-105}{230}\right)$

$= 67.5 - 6.85 = 60.65$

Standard deviation,

$$\sigma = \sqrt{h^2 \left[\frac{1}{N} \sum f_i u_i^2 - \left(\frac{1}{N} \sum f_i u_i\right)^2\right]}$$

$$= \sqrt{225\left[\frac{733}{230} - \left(-\frac{105}{230}\right)^2\right]}$$

$= \sqrt{225\,[3.187 - (0.46)^2]}$

$= \sqrt{225\,(3.1870 - 0.2116)}$

$= \sqrt{669.465} = 25.87$

Example 7. If each of the observation $x_1, x_2, \ldots, x_n$ is increased by a, where a is a negative or positive number, then show that the variance remains unchanged. *NCERT*

Sol. Let $\bar{x}$ be the mean of $x_1, x_2, \ldots, x_n$.

Then, the variance is given by

$$\sigma_1^2 = \frac{1}{n} \sum_{i=1}^{n} (x_i - \bar{x})^2 \quad \ldots\text{(i)}$$

If a is added to each observation, then the new observation will be $y_i = x_i + a$ ...(ii)

Let the mean of the new observation be $\bar{y}$. Then,

$$\bar{y} = \frac{1}{n} \sum_{i=1}^{n} y_i = \frac{1}{n} \sum_{i=1}^{n} (x_i + a) = \frac{1}{n}\left[\sum_{i=1}^{n} x_i + \sum_{i=1}^{n} a\right]$$

$$= \frac{1}{n}\left[\sum_{i=1}^{n} x_i + na\right] = \bar{x} + a \text{ i.e. } \bar{y} = \bar{x} + a \quad \ldots\text{(iii)}$$

Now, new variance, σ_2^2 is given by

$$\sigma_2^2 = \frac{1}{n} \sum_{i=1}^{n} (y_i - \bar{y})^2 = \frac{1}{n} \sum_{i=1}^{n} (x_i + a - \bar{x} - a)^2$$

[using Eqs. (ii) and (iii)]

$$= \frac{1}{n} \sum_{i=1}^{n} (x_i - \bar{x})^2 = \sigma_1^2 \quad \text{[using Eq. (i)]}$$

Hence, the variance remains unchanged.

Example 8. Given $\bar{x}$ is the mean and σ^2 is variance of n observations $x_1, x_2, \ldots, x_n$. Prove that the mean and variance of the observations $ax_1, ax_2, ax_3, \ldots, ax_n$ are $a\bar{x}$ and $a^2\sigma^2$, respectively, $(a \neq 0)$. *NCERT, Competency Based Que*

Sol. Mean of $ax_1, ax_2, \ldots, ax_n = \frac{ax_1 + ax_2 + \ldots + ax_n}{n}$

$$= a\left(\frac{x_1 + x_2 + x_3 + \ldots + x_n}{n}\right)$$

$$= a\bar{x} \quad \left[\because \bar{x} = \frac{x_1 + x_2 + \ldots + x_n}{n}\right]$$

Variance of $ax_1, ax_2, ..., ax_n$

$$= \frac{\Sigma(ax_i - a\bar{x})^2}{n}$$

$$= \frac{(ax_1 - a\bar{x})^2 + (ax_2 - a\bar{x})^2 + ... + (ax_n - a\bar{x})^2}{n}$$

$$= \frac{a^2[(x_1 - \bar{x})^2 + (x_2 - \bar{x})^2 + ... + (x_n - \bar{x})^2]}{n}$$

$$= a^2 \frac{\Sigma(x_i - \bar{x})^2}{n} = a^2\sigma^2 \quad \left[\because \sigma^2 = \frac{\Sigma(x_i - \bar{x})^2}{n}\right]$$

Hence proved.

Example 9. **The mean and variance of 7 observations are 8 and 16, respectively. If five of them are 2, 4, 10, 12, 14. Find the remaining two observations.** NCERT

Sol. Let the remaining two observations be a and b.

Given, $\bar{x} = 8$

and $\sigma^2 = 16$

$\because \bar{x} = 8$

$$\therefore \frac{\text{Sum of all observations}}{\text{Number of observations}} = 8$$

$$\Rightarrow \frac{2 + 4 + 10 + 12 + 14 + a + b}{7} = 8$$

$$\Rightarrow 42 + a + b = 56$$

$$\Rightarrow a + b = 14 \quad ...(i)$$

and $$\sigma^2 = \frac{\Sigma x_i^2}{n} - \left(\frac{\Sigma x_i}{n}\right)^2$$

$$\Rightarrow \sigma^2 = \frac{\Sigma x_i^2}{n} - (\bar{x})^2$$

$$\Rightarrow 16 = \frac{4 + 16 + 100 + 144 + 196 + a^2 + b^2}{n} - (8)^2$$

$$\Rightarrow 16 + 64 = \frac{460 + a^2 + b^2}{7}$$

$$\Rightarrow 7 \times (80) = 460 + a^2 + b^2$$

$$\Rightarrow a^2 + b^2 = 560 - 460$$

$$\Rightarrow a^2 + b^2 = 100 \quad ...(ii)$$

Now, from Eq. (i), putting $b = 14 - a$ in Eq. (ii), we get

$$a^2 + (14 - a)^2 = 100$$

$$\Rightarrow a^2 + 196 + a^2 - 28a = 100$$

$$\Rightarrow 2a^2 - 28a + 196 - 100 = 0$$

$$\Rightarrow 2a^2 - 28a + 96 = 0$$

$$\Rightarrow a^2 - 14a + 48 = 0$$

$$\Rightarrow a^2 - 8a - 6a + 48 = 0$$

$$\Rightarrow a(a - 8) - 6(a - 8) = 0$$

$$\Rightarrow (a - 6)(a - 8) = 0$$

$$\Rightarrow a = 6 \text{ or } 8$$

From Eq. (i), $b = 8$ or 6

Hence, the remaining two observations are 6, 8.

Example 10. **For a group of 200 candidates, the mean and standard deviation were found to be 40 and 15, respectively. Later on it was found that the score 43 was misread as 34. Find the correct mean and correct standard deviation.** Competency Based Que

Sol. We have, $n = 200$, $\bar{x} = 40$, $\sigma = 15$

$$\therefore \bar{x} = \frac{1}{n}\Sigma x_i \Rightarrow \Sigma x_i = n\bar{x} = 200 \times 40 = 8000$$

Now, corrected Σx_i = Incorrect Σx_i

– (incorrect value) + (correct value)

$$= 8000 - 34 + 43 = 8009$$

$$\therefore \text{Corrected Mean} = \frac{\text{Corrected } \Sigma x_i}{n} = \frac{8009}{200} = 40.045$$

Again, $\sigma = 15 \Rightarrow \sigma^2 = 15^2$

$$\Rightarrow \frac{1}{200}\Sigma x_i^2 - \left(\frac{\Sigma x_i}{200}\right)^2 = 225$$

$$\Rightarrow \frac{\Sigma x_i^2}{200} - \left(\frac{8000}{200}\right)^2 = 225$$

$$\Rightarrow \frac{\Sigma x_i^2}{200} - 1600 = 225$$

$$\Rightarrow \frac{\Sigma x_i^2}{200} = 1825$$

$$\Rightarrow \Sigma x_i^2 = 1825 \times 200 = 365000$$

$$\Rightarrow \text{Incorrect } \Sigma x_i^2 = 365000$$

$\therefore$ Corrected Σx_i^2 = (Incorrect Σx_i^2)

– (square of incorrect value)

+ (square of correct value)

$$\Rightarrow \text{Corrected } \Sigma x_i^2 = 365000 - (34)^2 + (43)^2 = 365693$$

So, corrected

$$\sigma = \sqrt{\frac{1}{n}\text{correct } \Sigma x_i^2 - \left(\frac{\text{Corrected } \Sigma x_i}{n}\right)^2}$$

$$= \sqrt{\frac{365693}{200} - \left(\frac{8009}{200}\right)^2}$$

$$= \sqrt{1828.465 - 1603.602} = 14.995$$

Quick Test 2

1. The marks obtained by 7 students are 8, 9, 11, 13, 14, 15, 21. Find the mean, variance and standard deviation of these marks. *Competency Based Que*

2. The following relates to a sample of size 60, $\Sigma x^2 = 18000$ and $\Sigma x = 960$, find the variance.

3. Calculate the mean and variance of the given data.

x_i	2	4	6	8	10	12	14	16
f_i	4	4	5	15	8	5	4	5

4. Calculate the mean and variance for the following data.

Class	Frequency
0-30	2
30-60	3
60-90	5
90-120	10
120-150	3
150-180	5
180-210	2

5. The following table shows the marks obtained by 100 candidates in an examination. Calculate the mean and standard deviation.

Marks	1-10	11-20	21-30	31-40	41-50	51-60
Number of candidates	3	16	26	31	16	8

Competency Based Que

6. The mean and variance of 15 observations are 3.8 and 4, respectively. If each observation is increased by 11, then find the mean and variance of the resulting observation.

7. The mean and standard deviation of 6 observations are 8 and 4 respectively. If each observation is multiplied by 3, then find the new mean and new standard deviation of the resulting observations. *Competency Based Que*

8. The mean and standard deviation of 20 observations are found to be 10 and 2, respectively. On rechecking, it was found that an observation 8 was incorrect. Calculate the correct mean and standard deviation in each of the following cases: *Competency Based Que*

(i) If wrong item is omitted.

(ii) If it is replaced by 12.

Answers

1. 13, 16.29, 4.04
2. 44
3. 9, 15.08
4. 107, 2276.1
5. 32, 12.36
6. 14.8, 4
7. 24, 12
8. (i) 10.1, 1.997 (ii) 10.2, 1.99

For Solutions, See QR Code on Pg. 384

Exam Pattern Questions

Objective Type Questions

Multiple Choice Questions

1. Which of the following is/are used as the measures of dispersion?

(a) Range (b) Quartile deviation
(c) Standard deviation (d) All of these

2. Mean deviation about median for continuous frequency distribution is calculated by using the formula **Competency Based Que**

(a) $\text{MD}(M) = \frac{1}{N}\sum_{i=1}^{n} f_i |x_i + M|$

(b) $\text{MD}(M) = \frac{1}{N}\sum_{i=1}^{n} f_i |x_i - M|$

(c) $\text{MD}(M) = \frac{1}{N}\sum_{i=1}^{n} \frac{f_i}{|x_i - M|}$

(d) $\text{MD}(M) = \frac{1}{N}\sum_{i=1}^{n} \frac{|x_i - M|}{f_i}$

3. Mean deviation of x observation $x_1, x_2, ..., x_n$ from their mean $\bar{x}$ is given by **NCERT Exemplar**

(a) $\sum_{i=1}^{n} (x_i - \bar{x})$ (b) $\frac{1}{N}\sum_{i=1}^{n} |x_i - \bar{x}|$

(c) $\sum_{i=1}^{n} (x_i - \bar{x})^2$ (d) $\frac{1}{N}\sum_{i=1}^{n} (x_i - \bar{x})^2$

4. The mean deviation of the data 3, 10, 12, 4, 5, 10, 5 from the mean is

(a) 3.14 (b) 3
(c) 2.57 (d) 2

5. The marks obtained by 9 students in Mathematics test are 55, 64, 25, 33, 50, 39, 40, 63, 59. Then, the mean deviation from the median is

(a) 11.52 (b) 12.67
(c) 11 (d) 11.56

6. The standard deviation of the data 8, 6, 10, 14, 13, 9, 10 is **Competency Based Que**

(a) $\sqrt{\frac{46}{7}}$ (b) 7

(c) $\sqrt{\frac{64}{7}}$ (d) $\sqrt{7}$

7. If $x_1 + x_2 + x_3 + ... + x_n$ be n observations and $\bar{x}$ be their arithmetic mean. Then, formula for the standard deviation is given by **NCERT Exemplar**

(a) $\sum(x_i - \bar{x})^2$ (b) $\frac{\sum(x_i - \bar{x})^2}{x}$

(c) $\sqrt{\frac{\sum(x_i - \bar{x})^2}{n}}$ (d) $\frac{\sum x_i^2}{n} - \bar{x}^2$

8. If x_1, x_2, x_3, x_4 and x_5 be the observations with mean m and standard deviation s, then the standard deviation of the observations kx_1, kx_2, kx_3, kx_4 and kx_5 is **Competency Based Que**

(a) $k + s$ (b) $\frac{s}{k}$ (c) ks (d) s

9. The standard deviations of first 10 natural numbers is

(a) 5.5 (b) 3.87 (c) 2.97 (d) 2.87

10. The variance of first 20 natural numbers is

(a) $\frac{133}{4}$ (b) $\frac{379}{12}$ (c) $\frac{133}{2}$ (d) $\frac{399}{4}$

11. If a, b, c, d and e be the observations with mean m and standard deviations S, then the standard deviation of the observation $a + K, b + K, c + K, d + K$ and $e + K$ is **NCERT Exemplar, Competency Based Que**

(a) S (b) KS (c) $S + K$ (d) $\frac{S}{K}$

12. Consider the following data 1, 2, 3, 4, 5, 6, 7, 8, 9, 10 If 1 is added to each number, then variance of the numbers, so obtained is **Competency Based Que**

(a) 6.5 (b) 2.87 (c) 3.87 (d) 8.25

13. If the variance is 625, what is the standard deviation?

(a) 25 (b) 15
(c) 5 (d) None of these

14. If each observation of a data whose standard deviation is σ is multiplied by a, then new standard deviation is

(a) σ (b) σ^2 (c) $a\sigma$ (d) $\frac{\sigma}{a}$

15. The mean deviation of the series $a, a + d, a + 2d, ..., a + nd$ from the mean is **Competency Based Que**

(a) $\frac{n+1}{2n+1}|d|$ (b) $\frac{n(n+1)}{2n+1}|d|$

(c) $\frac{n(n-1)}{2n+1}|d|$ (d) None of these

Assertion-Reason Based Questions

Directions (Q. Nos. 16-25) *In the questions given below are two statements labelled as Assertion (A) and Reason (R). In the context of the two statements, which one of the following is correct?*

(a) Both A and R are correct; R is the correct explanation of A.
(b) Both A and R are correct; R is not the correct explanation of A.
(c) A is correct; R is incorrect.
(d) R is correct; A is incorrect.

16. **Assertion** (A) The range of the observations 5, 6, 2, 1, 13, 5, 7, 8, 11, 15 is 14.

Reason (R) The range is the difference between two extreme observations of the distribution.

17. **Assertion** (A) Mean of the observations

$$= \frac{\text{Product of deviations}}{\text{Number of Observations}}$$

Reason (R) To find the dispersion of the values of x from mean $\bar{x}$, we take absolute measure of dispersion.

18. **Assertion** (A) Mean deviation about the median for the data 5, 2, 4, 9, 7, 6, 8 is 1.85.

Reason (R) Median of a data is the value among the observations which appears maximum number of times.

19. **Assertion** (A) Mean deviation about the mean of the first ten natural numbers is 49.5.

Reason (R) Mean deviation about any central tendency m is given by MD $(m) = \frac{\Sigma|x_i - m|}{n}$. **Competency Based Que**

20. **Assertion** (A) To find standard deviation for grouped data we use formula $\sigma = \frac{\sum f_i(x_i - \bar{x})^2}{\sum f_i}$.

Reason (R) For finding standard deviation, deviation are taken from arithmetic mean of the data.
Competency Based Que

21. **Assertion** (A) Mean deviation about the median of the observations 38, 70, 48, 34, 63, 42, 55, 44, 53, 47 is 8.4.

Reason (R) MD (Median) $= \frac{\sum | x_i - \text{Median} |}{n}$

22. **Assertion** (A) A set of n values $x_1, x_2, .., x_n$ has standard deviation σ. The standard deviation of n values $x_1 + k, x_2 + k, ..., x_n + k$ is σ^2.

Reason (R) The formula for standard deviation is

$$\sigma = \sqrt{\frac{1}{n}\sum_{i=1}^{n}(x_i - \bar{x})^2}$$

Competency Based Que

23. **Assertion** (A) The variance of data is 144, then the standard deviation of data is 12.

Reason (R) Standard deviation, $SD(\sigma) = \sqrt{\text{Variance}(\sigma^2)}$

24. **Assertion** (A) Variance of the data 2, 4, 5, 6, 7, 8, 17 is 23.33. Then, variance of 4, 8, 10, 12, 16, 34 is 93.32.

Reason (R) Mean of the squares of the deviations from mean is called the variance.

25. **Assertion** (A) Let $x_1, x_2, x_3, .., x_n$ be n observations and let $\bar{x}$ be their arithmetic mean and σ^2 be the variance. Arithmetic mean of $2x_1, 2x_2, .., 2x_n$ is $4\bar{x}$.

Reason (R) Variance of $2x_1, 2x_2, .., 2x_n$ is $4\sigma^2$.
Competency Based Que

Case Study Based Questions

26. The given table represents the number of persons in a society affected by covid-19 of different age.

Age of Person (in years)	Number of Persons
10-20	6
20-30	10
30-40	10
40-50	15
50-60	9

Based on the above information, answer the following questions. **Competency Based Que**

(i) Find the mean age of a person affected by covid-19.
(ii) Find the number of persons above 40 years age of person affected by covid-19.
(iii) Find the variance of the age.
(iv) Find the standard deviation of the age.

27. Mahendra is doing one of his project. For this he asked shoe size of 10 of his class-mates which are as follows.

6, 5, 5, 6, 8, 6, 7, 7, 8, 8

Based on the above information, answer the following questions. **Competency Based Que**

(i) What would be the mean shoe size for the data?
(a) 6 (b) 6.25 (c) 6.6 (d) 7

(ii) What would be the median for the data.
(a) 5 (b) 6 (c) 6.5 (d) 7

(iii) What would be the mean deviation about mean for the data?
(a) 1 (b) 0.6 (c) 0.5 (d) 1.5

(iv) What would be the mean deviation about median for the data?
(a) 0.75 (b) 1 (c) 1.25 (d) 1.5

(v) Different measure of dispersion are
(a) Range (b) Mean Deviation
(c) Standard Deviation (d) All of these

Subjective Questions

Very Short Answer Type Questions

1. Write the formula for mean deviation for continuous frequency distribution.

2. Find the mean deviation from mean for the following data : 11, 13, 4, 7, 8, 6, 15, 14, 3, 19.

3. Find the mean deviation from the median for the data : 3, 3, 5, 9, 10, 12, 12, 12, 18, 22, 22

4. If $\sum f_i y_i = 12, \sum f_i y_i^2 = 84, h = 10$ and $N = 40$, then find variance. **Competency Based Que**

5. Find the standard deviation from the given information.

$N = 10, \sum_{i=1}^{10} x_i = 80, \sum_{i=1}^{10} x_i^2 = 1000.$

Competency Based Que

Short Answer Type Questions

6. The frequency distribution

x_i	A	$2A$	$3A$	$4A$	$5A$	$6A$
f_i	2	1	1	1	1	1

Where, A is a positive integer, has a variance of 160. Determine the value of A. **Competency Based Que**

7. Find the mean and standard deviation for the following data.

Class interval	Frequency
0-10	3
10-20	2
20-30	4
30-40	6
40-50	5
50-60	5
60-70	5
70-80	2
80-90	8
90-100	5

8. Calculate mean deviation (from median) for the following data. **NCERT**

Marks	0-10	10-20	20-30	30-40	40-50
Number of students	5	8	15	16	6

9. Find the mean and standard deviation of the following frequency distribution. **NCERT**

x_i	6	10	14	18	24	28	30
f_i	2	4	7	12	8	4	3

10. Find the mean and standard deviation for the following data. **Competency Based Que**

Age (in years)	Number of Teachers
25-30	30
30-35	23
35-40	20
40-45	14
45-50	10
50-55	3

11. Find the variance and standard deviation for the following distribution.

x_i	4.5	14.5	24.5	34.5	44.5	54.5	64.5
f_i	1	5	12	22	17	9	4

Long Answer Type Questions

12. An original frequency table with mean 11 and variance 9.9 was lost but the following table derived from it was found. Construct the original table. **Competency Based Que**

Value of deviation (u)	−2	−1	0	1	2
Frequency (f)	1	6	7	4	2

13. The runs of two players for 10 innings each are as follows.

A	58	59	60	54	65	66	52	75	69	52
B	94	26	92	65	96	78	14	34	98	13

Who may be regarded as more consistent player?

14. The mean and variance of eight observations are 9 and 9.25, respectively. If six of the observations are 6, 7, 10, 12, 12 and 13, then find the remaining two observations. **NCERT, Competency Based Que**

15. From a frequency distribution consisting of 18 observations, the mean and the standard deviation were found to be 7 and 4, respectively. But on comparison of the original data, it was found that a figure 12 was misread as 21 in calculations. Calculate the correct mean and standard deviation. **Competency Based Que**

16. The mean of 5 observations is 4.4 and their variance is 8.24. If three of the observations are 1, 2 and 6, find the other two obsevations. **NCERT**

Hints & Solutions

Objective Type Questions

1. (d) The measures of dispersion are Range, Quartile deviation and standard deviation.

2. (b) The formula for mean deviation about median for continuous frequency distribution is MD (M)

$$= \frac{1}{N}\sum_{i=1}^{n} f_i |x_i - M|$$

3. (b) $MD = \frac{1}{N}\sum_{i=1}^{n} |x_i - \bar{x}|$

4. (a) The given observations are 3, 10, 12, 4, 5, 10, 5.

$$\therefore \quad \bar{x} = \frac{3+10+12+4+5+10+5}{7} = \frac{49}{7} = 7$$

x_i	$\lvert x_i - \bar{x}\rvert$
3	4
10	3
12	5
4	3
5	2
10	3
5	2
.Total	22

$$MD(\bar{x}) = \frac{\sum |x_i - \bar{x}|}{n} = \frac{22}{7} = 3.14$$

5. (d) Marks obtained are 55, 64, 25, 33, 50, 39, 40, 63, 59.

Write it in ascending order 25, 33, 39, 40, 50, 55, 59, 63, 64.

Here, $n = 9$, 80 median is $\left(\frac{n+1}{2}\right)$th term i.e. 5th term.

So Median, $M = 50$.

Now,

x_i	$\lvert x_i - \bar{x}\rvert$
25	25
33	17
39	11
40	10
50	0
55	5
59	9
63	13
64	14
Total	104

$$MD(\bar{x}) = \frac{\sum |x_i - \bar{x}|}{n} = \frac{104}{9} = 11.555 = 11.56$$

6. (a) We have data 8, 6, 10, 14, 13, 9, 10

$$\therefore \quad n = 7$$

x_i	x_i^2
8	64
6	36
10	100
14	196
13	169
9	81
10	100
$\sum x_i = 70$	$\sum x_i^2 = 746$

$$\therefore \quad SD = \sqrt{\frac{\sum x_i^2}{n} - \left(\frac{\sum x_i}{n}\right)^2}$$

$$= \sqrt{\frac{746}{7} - \left(\frac{70}{7}\right)^2}$$

$$= \sqrt{\frac{746}{7} - 100} = \sqrt{\frac{746 - 700}{7}} = \sqrt{\frac{46}{7}}$$

7. (c) $SD = \sigma = \sqrt{\frac{\sum (x_i - \bar{x})^2}{n}}$

8. (c) Let x_1, x_2, x_3, x_4, x_5 be the observations with mean m and standard deviation s.

Here, $$m = \frac{\sum x_i}{n}$$

and $$s = \sqrt{\frac{\sum x_i^2}{5} - \left(\frac{\sum x_i}{5}\right)^2}$$

$$\therefore \quad SD = \sqrt{\frac{k^2 \sum x_i^2}{5} - \left(\frac{k\sum x_i}{5}\right)^2}$$

$$= \sqrt{\frac{k^2 \sum x_i^2}{5} - k^2\left(\frac{\sum x_i}{5}\right)^2}$$

$$= k\sqrt{\frac{\sum x_i^2}{5} - \left(\frac{\sum x_i}{5}\right)^2}$$

$$= ks$$

9. (d) We know that SD of first n natural numbers is $\sqrt{\dfrac{n^2-1}{12}}$.

Here, $n = 10$

$$\therefore \quad SD = \sqrt{\frac{100-1}{12}} = \sqrt{\frac{99}{12}}$$

$$= \sqrt{8.25} = 2.87$$

10. (a) $\sigma^2 = \dfrac{\sum x_i^2}{n} - \left(\dfrac{\sum x_i}{n}\right)^2$

$$= \frac{1}{20}[1^2 + 2^2 + \ldots + 20^2] - \left[\frac{1}{20}(1+2+\ldots+20)\right]^2$$

$$= \frac{1}{20} \times \frac{20 \times 21 \times (40+1)}{6} - \left[\frac{1}{20} \times \frac{20 \times 21}{2}\right]^2$$

$$= \frac{21 \times 41}{6} - \frac{441}{4} = \frac{287}{2} - \frac{441}{4} = \frac{574 - 441}{4} = \frac{133}{4}$$

11. (a) Given observations are a, b, c, d and e.

$$\therefore \text{Mean} = \frac{a+b+c+d+e}{5}$$

$$\Rightarrow \quad m = \frac{a+b+c+d+e}{5}$$

$$\Rightarrow a+b+c+d+e = 5m$$

Now, mean of $a+K, b+K, c+K, d+K, e+K$

$$= \frac{a+K+b+K+c+K+d+K+e+K}{5}$$

$$= \frac{a+b+c+d+e+5K}{5}$$

$$= \frac{5m+5K}{5} = m+K$$

$$\therefore \quad SD = \sqrt{\frac{\left(\sum x_i + K\right)^2}{N} - \left(\frac{\sum (x_i+K)}{N}\right)^2}$$

$$= \sqrt{\frac{\sum (x_i^2 + 2x_i K + K^2)}{N} - (m+K)^2}$$

$$= \sqrt{\frac{\sum (x_i^2 + 2x_i K + K^2)}{5} - (m+K)^2}$$

$$= \sqrt{\frac{\sum x_i^2}{5} + 2K\frac{\sum x_i}{5} + K^2 \times \frac{5}{5} - (m+K)^2}$$

$$= \sqrt{\frac{\sum x_i^2}{5} + 2Km + K^2 - (m+K)^2}$$

$$= \sqrt{\frac{\sum x_i^2}{5} + 2Km + K^2 - m^2 - 2mK - K^2}$$

$$= \sqrt{\frac{\sum x_i^2}{5} - m^2}$$

$$= S$$

12. (d) The given numbers are 1, 2, 3, 4, 5, 6, 7, 8, 9, 10. If 1 is added to each numbers, then the new numbers attained are 2, 3, 4, 5, 6, 7, 8, 9, 10, 11

Now,

$$\sum x_i = 2+3+4+5+6+7+8+9+10+11 = 65$$

and $\sum x_i^2 = 4+9+16+25+36+49+64+81 + 100+121$

$$= 505$$

$\therefore$ Variance of the numbers, so obtained

$$= \frac{\sum x_i^2}{10} - \left(\frac{\sum x_i}{10}\right)^2$$

$$= \frac{505}{10} - \left(\frac{65}{10}\right)^2$$

$$= 50.5 - 42.25$$

$$= 8.25$$

13. (a) Variance $= 625$

$$\therefore \quad SD, \sigma = \sqrt{\text{Variance}}$$

$$= \sqrt{625} = 25$$

14. (c) Given, $SD = \sigma = \sqrt{\dfrac{\sum (x_i - \bar{x})}{n}}$

Here, $\bar{x}$ represents the arithmetic mean.

On multiplying each x_i by a, we get

$$\text{New mean} = \frac{1}{n}\sum a x_i = a \times \frac{1}{n}\sum x_i = a\bar{x}$$

$$\therefore \quad \text{New SD} = \sqrt{\frac{\sum (a x_i - a\bar{x})^2}{n}}$$

$$= \sqrt{\frac{\sum a^2 (x_i - \bar{x})}{n}}$$

$$= |a| \sqrt{\frac{\sum (x_i - \bar{x})^2}{n}} = |a| \sigma$$

$$= a\sigma$$

15. (b) Number of terms in the series $= 2n + 1$

$$\therefore \text{ Mean, } \bar{x} = \frac{a + (a+d) + (a+2d) + \ldots + (a+2nd)}{(2n+1)}$$

$$= \frac{1}{2n+1}\left[\frac{2n+1}{2}(a + a + 2nd)\right]$$

$$= \frac{(2a+2nd)}{2} = a + nd$$

Also,

$$\sum |x_i - \bar{x}| = |-nd| + |(1-n)d| + \ldots + 0 + |d| + \ldots + |nd|$$

$$= 2|d|\,[n + (n-1) + \ldots + 1]$$

$$= 2|d|\frac{n(n+1)}{2} = n(n+1)\,|d|$$

$\therefore$ Mean deviation from the mean

$$= \frac{\sum |x_i - \bar{x}|}{N} = \frac{n(n+1)}{2n+1} |d|$$

16. (a) We have given data in ascending order 1, 2, 5, 5, 6, 7, 8, 11, 13, 15.

$\therefore$ Range $= 15 - 1 = 14$

17. (d) Clearly, Assertion is incorrect but Reason is correct.

18. (c) Rearranging the given data in ascending order, 2, 4, 5, 6, 7, 8, 9

Since, the number of observations is odd.

$\therefore$ Median $= \left(\frac{n+1}{2}\right)$th term

$= 6$

Now, $\sum |x_i - M| = 4 + 2 + 1 + 0 + 1 + 2 + 3 = 13$

Mean deviation about median $= \frac{\sum |x_i - M|}{n} = \frac{13}{7} = 1.85$

Assertion is correct and Reason is incorrect.

19. (d) First ten natural numbers are 1, 2, 3, 4, 5, 6, 7, 8, 9, 10.

Mean, $\bar{x} = \frac{1+2+3+4+5+6+7+8+9+10}{10}$

$= \frac{55}{10} = 5.5$

Mean deviation $= \frac{\sum |x - \bar{x}|}{n}$

$= \frac{4.5 + 3.5 + 2.5 + 1.5 + 0.5 + 0.5 + 1.5 + 2.5 + 3.5 + 4.5}{10}$

$= \frac{25}{10} = 2.5$

20. (a) The formula for standard deviation for grouped data is $\sigma = \frac{\sum f_i (x_i - \bar{x})^2}{\sum f_i}$.

21. (a) We have data 34, 38, 42, 44, 47, 48, 53, 55, 63, 70.

Here, $n = 10$

Median = AM of 5th and 6th observations.

$= \frac{47 + 48}{2} = 47.5$

MD (47.5) $= \frac{\sum |x_i - 47.5|}{10}$

$$= \frac{|34 - 47.5| + |38 - 47.5| + |42 - 47.5| + |44 - 47.5| + |47 - 47.5| + |48 - 47.5| + |53 - 47.5| + |55 - 47.5| + |63 - 47.5| + |70 - 47.5|}{10}$$

$$= \frac{13.5 + 9.5 + 5.5 + 3.5 + 0.5 + 0.5 + 5.5 + 7.5 + 15.5 + 22.5}{10}$$

$= \frac{84}{10} = 8.4$

22. (d) **Hint** If each observation is increased by a constant k, then the standard deviation is unchanged.

23. (a) We know that SD, $\sigma = \sqrt{\text{Variance}}$

$\therefore$ $\sigma = \sqrt{144} = 12$

24. (b) As each observation is multiplied by 2, then variance is multiplied by $2^2 = 4$

i.e. $23.33 \times 4 = 93.32$

25. (d) Given, $\frac{x_1 + x_2 + \ldots + x_n}{n} = \bar{x}$

$\therefore$ Mean of $2x_1, 2x_2, \ldots, 2x_n$

$= \frac{2x_1 + 2x_2 + \ldots + 2x_n}{n} = \frac{2(x_1 + x_2 + \ldots + x_n)}{n}$

$= 2\bar{x}$

So, Assertion is incorrect.

We know that if each observation is multiplied by k, then the variance gets multiplied by k^2.

So, Assertion is incorrect but Reason is correct.

26. (i) Let us make the following table.

Class	f_i	x_i	$y_i = \frac{x_i - 35}{10}$	y_i^2	$f_i y_i$	$f_i y_i^2$
10-20	6	15	−2	4	−12	24
20-30	10	25	−1	1	−10	10
30-40	10	35	0	0	0	0
40-50	15	45	1	1	15	15
50-60	9	55	2	4	18	36
Total	$N = 50$				11	85

Here, $h = 10$ and $A = 35$

Therefore, $\bar{x} = A + \frac{\sum f_i y_i}{N} \times h = 35 + \frac{11}{50} \times 10 = 37.2$

(ii) Number of persons above 40 years $= 15 + 9 = 24$

(iii) Variance, $\sigma^2 = \frac{h^2}{N^2}[N \sum f_i y_i^2 - (\sum f_i y_i)^2]$

$= \frac{10^2}{50^2} [50 \times 85 - (11)^2]$

$= \frac{100}{2500} [4250 - 121]$

$= 165.16$

(iv) Standard Deviation, $\sigma = \sqrt{\sigma^2} = \sqrt{165.16} = 12.85$

27. (i) (c) Mean $= \frac{6+5+5+6+8+6+7+7+8+8}{10}$

$= \frac{66}{10} = 6.6$

(ii) (c) Rearrange the given data in ascending order.

5, 5, 6, 6, 6, 7, 7, 8, 8, 8

Since, the number of observations are even,

$$\text{Median} = \frac{\left(\frac{n}{2}\right)\text{th} + \left(\frac{n}{2}+1\right)\text{th}}{2} = \frac{6+7}{2} = 6.5$$

(iii) (a) Table for mean deviation.

x_i	$x_i - \bar{x} = x_i - 6.6$	$\lvert x_i - \bar{x}\rvert = \lvert x_i - 6.6\rvert$
6	– 0.6	0.6
5	– 1.6	1.6
5	– 1.6	1.6
6	– 0.6	0.6
8	1.4	1.4
6	– 0.6	0.6
7	0.4	0.4
7	0.4	0.4
8	1.4	1.4
8	1.4	1.4
Total		10

$$\text{Mean Deviation} = \frac{\sum\lvert x_i - \bar{x}\rvert}{n} = \frac{10}{10} = 1$$

(iv) (b) Table for mean deviation about median.

x_i	$x_i - M = x_i - 6.5$	$\lvert x_i - M\rvert = \lvert x_i - 6.5\rvert$
5	– 1.5	1.5
5	– 1.5	1.5
6	– 0.5	0.5
6	– 0.5	0.5
6	– 0.5	0.5
7	0.5	0.5
7	0.5	0.5
8	1.5	1.5
8	1.5	1.5
8	1.5	1.5
Total		10

$$\text{Mean deviation} = \frac{\sum\lvert x_i - M\rvert}{n} = \frac{10}{10} = 1$$

(v) (d) Measures of dispersion are range, mean deviation and standard deviation.

Subjective Questions

1. $\text{MD}(M) = \dfrac{\sum f_i\lvert x_i - M\rvert}{\sum f_i}$

2. Mean,

$$\bar{x} = \frac{11+13+4+7+8+6+15+14+3+19}{10} = \frac{100}{10} = 10$$

Table for mean deviation

x_i	$x_i - \bar{x} = x_i - 10$	$\lvert x_i - \bar{x}\rvert = \lvert x_i - 10\rvert$
11	1	1
13	3	3
4	−6	6
7	−3	3
8	−2	2
6	−4	4
15	5	5
14	4	4
3	−7	7
19	9	9
Total		44

$$\text{Mean Deviation} = \frac{\sum_{i=1}^{10}\lvert x_i - \bar{x}\rvert}{n} = \frac{44}{10} = 4.4$$

3. Since, number of observations is odd.

$\therefore$ $\text{Median} = \left(\dfrac{n+1}{2}\right)\text{th term} = 6\text{th term} = 12$

x_i	$(x_i - M)$	$\lvert x_i - M\rvert$
3	−9	9
3	−9	9
5	−7	7
9	−3	3
10	−2	2
12	0	0
12	0	0
12	0	0
18	6	6
22	10	10
22	10	10
Total		56

$$\text{Mean Deviation }(M) = \frac{\sum_{i=1}\lvert x_i - M\rvert}{n} = \frac{56}{11} = 5.09$$

4. Variance $= h^2\left[\frac{1}{N}\sum f_i y_i^2 - \left(\frac{1}{N}\sum f_i y_i\right)^2\right]$

$= (10)^2\left[\frac{1}{40}\times 84 - \left(\frac{12}{40}\right)^2\right]$

$= 100\left[\frac{84}{40} - \frac{144}{1600}\right]$

$= 100\times\left[\frac{3360-144}{1600}\right]$

$= \frac{100\times 3216}{1600} = 201$

5. $\sigma^2 = \frac{\sum_{i=1}^{10} x_i^2}{N} - \left(\frac{\sum_{i=1}^{10} x_i}{N}\right)^2$

On substituting the values, we get

$\sigma^2 = \frac{1000}{10} - \left(\frac{80}{10}\right)^2$

$= 100 - 64 = 36$

$\therefore \quad \sigma = \sqrt{36} = 6$

6. Let us make the following table from the given data.

x_i	f_i	$f_i x_i$	$(x_i - \bar{x})^2 = \left(x_i - \frac{22}{7}A\right)^2$	$(x_i - \bar{x})^2 f_i$
A	2	$2A$	$(225/49)A^2$	$(450/49)A^2$
$2A$	1	$2A$	$(64/49)A^2$	$(64/49)A^2$
$3A$	1	$3A$	$(1/49)A^2$	$(1/49)A^2$
$4A$	1	$4A$	$(36/49)A^2$	$(36/49)A^2$
$5A$	1	$5A$	$(169/49)A^2$	$(169/49)A^2$
$6A$	1	$6A$	$(400/49)A^2$	$(400/49)A^2$
Total	$N=7$	$22A$		$\frac{1120}{49}A^2$

Hence, $N = \Sigma f_i = 7$ and $\Sigma f_i x_i = 22A$

$\therefore \quad \bar{x} = \frac{\sum_{i=1}^{6} f_i x_i}{N} = \frac{22A}{7}$

Now, variance $= \frac{1}{N}\sum_{i=1}^{6} f_i (x_i - \bar{x})^2$

$\Rightarrow \quad 160 = \frac{1}{7}\left(\frac{1120}{49}\right)A^2 \quad$ [$\because$ variance = 60, given]

$\Rightarrow \quad A = 7$

7. **Hint** Here, $h = 10$ and $a = 55$

Let us make the following table from the given data.

Class interval	f_i	Mid-value (x_i)	$u_i = \frac{x_i - 55}{10}$	$f_i u_i$	u_i^2	$f_i u_i^2$
0-10	3	5	−5	−15	25	75
10-20	2	15	−4	−8	16	32
20-30	4	25	−3	−12	9	36
30-40	6	35	−2	−12	4	24
40-50	5	45	−1	−5	1	5
50-60	5	55	0	0	0	0
60-70	5	65	1	5	1	5
70-80	2	75	2	4	4	8
80-90	8	85	3	24	9	72
90-100	5	95	4	20	16	80
Total	$N = 45$			1		337

Here, $N = \Sigma f_i = 45, \Sigma f_i u_i = 1, \Sigma f_i u_i^2 = 337$

$\therefore$ Mean, $\bar{x} = a + \left(\frac{1}{N}\Sigma f_i u_i\right)\times h$

and standard deviation,

$\sigma = \sqrt{h^2}\sqrt{\left\{\frac{1}{N}\Sigma f_i u_i^2 - \left(\frac{1}{N}\Sigma f_i u_i\right)^2\right\}}$

Ans. 55.22, 27.36

8. Let us make the following table from the given data.

Marks (Class -interval)	Number of students (f_i)	cf	x_i	$\lvert x_i - M\rvert = \lvert x_i - 28\rvert$	$f_i\lvert x_i - 28\rvert$
0-10	5	5	5	23	115
10-20	8	13	15	13	104
20-30	15	28	25	3	45
30-40	16	44	35	7	112
40-50	6	50	45	17	102
Total	$N = 50$				478

Here, $N = 50$, so $\frac{N}{2} = 25, cf = 13, l = 20, f = 15$ and $h = 10$

Now, Median $(M) = \frac{l + \left(\frac{N}{2} - cf\right)}{f}\times h = 20 + \frac{25-13}{15}\times 10$

$= 28$

$\therefore$ Mean deviation about median $= \frac{1}{N}(\Sigma f_i \lvert x_i - M\rvert)$

$= \frac{478}{50} = 9.56$

9. Let us make the following table from the given data.

x_i	f_i	$f_i x_i$	$(x_i - \bar{x}) = (x_i - 19)$	$(x_i - \bar{x})^2$	$f_i(x_i - \bar{x})^2$
6	2	12	−13	169	338
10	4	40	−9	81	324
14	7	98	−5	25	175
18	12	216	−1	1	12
24	8	192	5	25	200
28	4	112	9	81	324
30	3	90	11	121	363
Total	40	760			1736

$\therefore \quad \bar{x} = \frac{\Sigma f_i x_i}{N} = \frac{760}{40} = 19$

and standard deviation $= \sqrt{\frac{1}{N}[\Sigma f_i (x_i - \bar{x})^2]}$

$= \sqrt{\frac{1736}{40}} = \sqrt{43.4} = 6.59$

10. Let the assumed mean be $a = 42.5$

Let us make the table from the given data.

Class	Mid value (x_i)	f_i	$u_i = \frac{x_i - a}{5}$	u_i^2	$f_i u_i$	$f_i u_i^2$
25-30	27.5	30	−3	9	−90	270
30-35	32.5	23	−2	4	−46	92
35-40	37.5	20	−1	1	−20	20
40-45	42.5	14	0	0	0	0
45-50	47.5	10	1	1	10	10
50-55	52.5	3	2	4	6	12
Total		100			−140	404

$\therefore \quad$ Mean, $\bar{x} = a + \frac{1}{N}\Sigma f_i u_i \times h$

$= 42.5 + \frac{1}{100}(-140) \times 5$

$= 35.5$ yr

Standard deviation $= \left[\sqrt{\frac{1}{N}\Sigma f_i u_i^2 - \left(\frac{1}{N}\Sigma f_i u_i\right)^2}\right] \times h$

$= \left[\sqrt{\frac{404}{100} - \left(\frac{-140}{100}\right)^2}\right] \times 5 = 7.21$ yr

11. Let us make the table from the given data.

x_i	f_i	$d_i = x_i - 34.5$	$u_i = \frac{x_i - 34.5}{10}$	u_i^2	$f_i u_i$	$f_i u_i^2$
4.5	1	−30	−3	9	−3	9
14.5	5	−20	−2	4	−10	20
24.5	12	−10	−1	1	−12	12
34.5	22	0	0	0	0	0
44.5	17	10	1	1	17	17
54.5	9	20	2	4	18	36
64.5	4	30	3	9	12	36
Total	70				22	130

Here, assumed mean, $a = 34.5$

Variance, $\sigma^2 = \left[\frac{1}{N}\Sigma f_i u_i^2 - \left(\frac{1}{N}\Sigma f_i u_i\right)^2\right] \times h^2$

$= \left[\frac{130}{70} - \left(\frac{22}{70}\right)^2\right] \times 100$

$= 175.8$

and standard deviation $= \sqrt{\text{Variance}}$

$= \sqrt{175.8} = 13.259$

12. Let us make the table from the given data.

u	f	u^2	fu	fu^2
−2	1	4	−2	4
−1	6	1	−6	6
0	7	0	0	0
1	4	1	4	4
2	2	4	4	8
Total	20		0	22

We know that $\bar{x} = a + h\frac{\Sigma fu}{\Sigma f}$

$\Rightarrow \quad 11 = a + h \times \frac{0}{20}$

$\Rightarrow \quad a = 11$

Also, variance, $\sigma^2 = h^2\left[\frac{\Sigma fu^2}{\Sigma f} - \left(\frac{\Sigma fu}{\Sigma f}\right)^2\right]$

$\Rightarrow \quad (9.9)^2 = h^2\left[\frac{22}{20}\right] \Rightarrow h = 3$

Mid value, $u = \dfrac{x-a}{h} \Rightarrow x = a + uh$

$\therefore$ Different values of x for different values of u, we have

$11 - 2\times 3, 11 - 1\times 3, 11 - 0\times 3, 11 + 1\times 3, 11 + 2\times 3$

i.e. 5, 8, 11, 14, 17.

Thus, the original frequency table is as follows.

Class	3.5-6.5	6.5-9.5	9.5-12.5	12.5-15.5	15.5-18.5
Frequency	1	6	7	4	2

13. Let us make the table from the given data

Player A			Player B		
x_i	$x_i - 61$	$(x_i - 61)^2$	y_i	$y_i - 61$	$(y_i - 61)^2$
58	−3	9	94	33	1089
59	−2	4	26	−35	1225
60	−1	1	92	31	961
54	−7	49	65	4	16
65	4	16	96	35	1225
66	5	25	78	17	289
52	−9	81	14	−47	2209
75	14	196	34	−27	729
69	8	64	98	37	1369
52	−9	81	13	−48	2304
Σx_i = 610		$\Sigma (x_i - 61)^2$ = 526	Σy_i = 610		$\Sigma (y_i - 61)^2$ = 11416

For Player A, Mean $= \dfrac{\Sigma x_i}{n} = \dfrac{610}{10} = 61$

$$SD = \sqrt{\frac{\Sigma (x_i - 61)^2}{N}} = \sqrt{\frac{526}{10}} = 7.25$$

For Player B, Mean $= \dfrac{\Sigma y_i}{n} = \dfrac{610}{10} = 61$

$$SD = \sqrt{\frac{\Sigma (y_i - 61)^2}{N}} = \sqrt{\frac{11416}{10}} = 33.79$$

Since, SD for player A is 7.25 < SD for player B is 33.79.

Hence, player A is more consistent player.

14. Let the remaining two observations be a and b.

Given, $\bar{x} = 9$ and $\sigma^2 = 9.25$ $[\because \bar{x} = 9]$

$$\therefore \quad \frac{\text{Sum of all observations}}{\text{Number of observations}} = 9$$

$\Rightarrow$ Sum of all observations $= 9 \times 8$

$\Rightarrow$ $6 + 7 + 10 + 12 + 12 + 13 + a + b = 72$

$\Rightarrow$ $a + b = 12$ ...(i)

and $\sigma^2 = \dfrac{\Sigma x_i^2}{n} - \left(\dfrac{\Sigma x_i}{n}\right)^2 \Rightarrow \sigma^2 = \dfrac{\Sigma x_i^2}{n} - (\bar{x})^2$

$$\Rightarrow 9.25 = \frac{36 + 49 + 100 + 144 + 144 + 169 + a^2 + b^2}{8} - (9)^2$$

$\Rightarrow$ $a^2 + b^2 = 80$ ...(ii)

Now, from Eq. (i), put $b = (12 - a)$ in Eq. (ii), we get

$a^2 + (12 - a)^2 = 80 \Rightarrow a = 4$ or $a = 8$

On putting $a = 4$ or $a = 8$ in Eq. (i), we get $b = 8$ or $b = 4$

Hence, the two observations are 4 and 8.

15. Calculation of the correct arithmetic mean

Here, $n = 18, \ \bar{x} = 7$

$$\therefore \quad \sum_{i=1}^{18} x_i = 18 \times 7 = 126 \qquad \left[\because \sum_{i=1}^{n} x_i = n\bar{x}\right]$$

Since, 12 misreads as 21 in calculation.

$$\therefore \quad \text{Correct} \sum_{i=1}^{18} x_i = 126 - 21 + 12 = 117$$

$\therefore$ Correct arithmetic mean $= \dfrac{117}{18} = 6.5$

Calculation of correct standard deviation.

We know that $\sigma = \sqrt{\dfrac{\sum_{i=1}^{n} x_i^2}{n} - \left(\dfrac{\sum_{i=1}^{n} x_i}{n}\right)^2}$

or $4 = \sqrt{\dfrac{\sum_{i=1}^{18} x_i^2}{18} - (7)^2}$

$$\sum_{i=1}^{18} x_i^2 = (16 + 49)\,18 = 1170$$

$$\therefore \quad \text{Correct} \sum_{i=1}^{18} x_i^2 = 1170 - (21)^2 + (12)^2$$

$$= 1170 - 441 + 144 = 873$$

$\therefore$ Correct standard deviation

$$= \sqrt{\frac{\text{Correct} \sum_{i=1}^{n} x_i^2}{n} - (\text{Correct mean})^2}$$

$$= \sqrt{\frac{873}{18} - (6.5)^2}$$

$$= \sqrt{48.5 - 42.25}$$

$$= \sqrt{6.25} = 2.5$$

16. Let two other observations be x_1 and x_2.

Then, $1+2+6+x_1+x_2 = n \cdot \bar{x} = 5 \times 4.4$

$\Rightarrow \quad x_1 + x_2 = 22 - 9 = 13 \quad \ldots(i)$

Also, $\quad \sigma^2 = \dfrac{\sum_{i=1}^{n} x_i^2}{n} - (\bar{x})^2$

$\Rightarrow \quad 8.24 = \dfrac{\sum_{i=1}^{5} x_i^2}{5} - (4.4)^2$

$\Rightarrow \quad \sum_{i=1}^{5} x_i^2 = (8.24 + 19.36) \times 5$

$\Rightarrow \quad 1 + 4 + 36 + x_1^2 + x_2^2 = 138$

$\Rightarrow \quad x_1^2 + x_2^2 = 97$

$\Rightarrow \quad x_1^2 + (13 - x_1)^2 = 97 \quad$ [from Eq. (i)]

$\Rightarrow \quad x_1^2 + 169 - 26x_1 + x_1^2 = 97$

$\Rightarrow \quad 2x_1^2 - 26x_1 + 72 = 0$

$\Rightarrow \quad x_1^2 - 13x_1 + 36 = 0$

$\Rightarrow \quad (x_1 - 9)(x_1 - 4) = 0$

$\Rightarrow \quad x_1 - 9 = 0$ or $x_1 - 4 = 0$

$x_1 = 9$ or 4

From Eq. (i),

when $\quad x_1 = 9$, then $x_2 = 4$

when $\quad x_1 = 4$, then $x_2 = 9$

Hence, other two observations are 9 and 4.

QR Code for Quick Test 1

QR Code for Quick Test 2

Mind Map

Statistics

Measure of Dispersion
The dispersion or scatter in a data is measured on the basis of the observations and the types of measure of central tendency used.
There are following measures of dispersion
1. Range
2. Quartile deviation
3. Mean deviation
4. Standard deviation

Range
Range is the difference of maximum and minimum values of data.
Range = Maximum value – Minimum value

Mean Deviation for Ungrouped Data
Let $x_1, x_2, \ldots, x_n$ be n observations. Then, mean deviation about mean or median can be determined by following steps

I. Find the mean or median of given observations using the suitable formula.

II. Find the deviation of each observation x_i from $\bar{x}$ (mean) or M (median) and then take their absolute values i.e. $|x_i - \bar{x}|$ or $|x_i - M|$.

III. Find the sum of absolute values of deviations obtained in Step II i.e. $\sum_{i=1}^{n} |x_i - \bar{x}|$ or $\sum_{i=1}^{n} |x_i - M|$

IV. Now, find mean deviation about mean or median using the formula

$$\frac{\sum_{i=1}^{n} |x_i - \bar{x}|}{n} \text{ or } \frac{\sum_{i=1}^{n} |x_i - M|}{n}$$

where n is the number of observations.

Mean Deviation
The arithmetic mean of the absolute deviations of all the values taken from a central value or a fixed number a. The mean deviation from a is denoted as MD (a).

$$\text{MD}(a) = \frac{[\text{Sum of absolute values of deviations from } a]}{\text{Number of observations}}$$

Mean Deviation for Grouped Data

For Discrete Frequency Distribution
Let the given data have n distinct values $x_1, x_2, \ldots, x_n$ and their corresponding frequencies are $f_1, f_2, \ldots, f_n$, respectively. Then, this data can be represented in the tabular form as

x_i	x_1	x_2	x_3	...	x_n
f_i	f_1	f_2	f_3	...	f_n

and is called discrete frequency distribution.
Here, mean deviation about mean or median is given by $\frac{\sum_{i=1}^{n} f_i|x_i - A|}{N}$

Where, $N = \sum_{i=1}^{n} f_i$ = total frequency and A = mean or median.

Then, mean deviation about mean is given by

$$\text{MD} = \frac{\Sigma f_i|x_i - \bar{x}|}{\Sigma f_i}$$

and mean deviation about median is given by

$$\text{MD} = \frac{\Sigma f_i|x_i - M|}{\Sigma f_i}$$

Step Deviation or Shortcut Method for Calculating the Mean Deviation about Mean
For step deviation method, we denote the new variable by u_i and it is defined as $u_i = \frac{x_i - a}{h}$

Where, a is the assumed mean and h is the common factor or length of the class interval.

The mean $\bar{x}$ by step deviation method is given by

$$\bar{x} = a + \frac{\sum_{i=1}^{n} f_i u_i}{N} \times h$$

$\therefore$ Maen deviation $= \frac{1}{N}\sum_{i=1}^{n} f_i|x_i - \bar{x}|$

Variance
Mean of the squares of the deviations from the mean is called the variance, to be denoted by σ^2.

Standard Deviation
The positive square root of variance is called standard deviation and it is denoted by σ.

Variance and Standard Deviation of Ungrouped Data
Variance of n observations $x_1, x_2, x_3, \ldots x_n$ is given by

$$\sigma^2 = \frac{\sum_{i=1}^{n}(x_i - \bar{x})^2}{n} \text{ or } \frac{\sum_{i=1}^{n} x_i^2}{n} - \left(\frac{\sum_{i=1}^{n} x_i}{n}\right)^2$$

We know that standard deviation $= \sqrt{\text{Variance}}$

$$\therefore \quad \sigma = \sqrt{\frac{\sum_{i=1}^{n}(x_i - \bar{x})^2}{n}} \text{ or } \sqrt{\frac{\sum_{i=1}^{n} x_i^2}{n} - \left(\frac{\sum_{i=1}^{n} x_i}{n}\right)^2}$$

$$\text{or SD} = \frac{1}{n}\sqrt{n\sum_{i=1}^{n} x_i^2 - \left(\sum_{i=1}^{n} x_i\right)^2}$$

Variance and Standard Deviation of a Discrete Frequency Distribution
Let the discrete frequency distribution be $x : x_1, x_2, x_3, \ldots, x_n$ and $f : f_1, f_2, f_3, \ldots, f_n$.

Shortcut Method
Variance, $\sigma^2 = \frac{1}{N}\sum_{i=1}^{n} f_i d_i^2 - \left(\frac{\Sigma f_i d_i}{N}\right)^2$

and standard deviation,

$$\therefore \sigma = \sqrt{\frac{1}{N}\sum_{i=1}^{n} f_i d_i^2 - \left(\frac{\Sigma f_i d_i}{N}\right)^2}$$

Where, $d_i = x_i - a$ is deviation from assumed mean and a = assumed mean

Direct Method
Variance, $\sigma^2 = \frac{1}{N}\sum_{i=1}^{n} f_i(x_i - \bar{x})^2$

or $\sigma^2 = \frac{1}{N}\Sigma f_i x_i^2 - \left(\frac{\Sigma f_i x_i}{N}\right)^2$

and standard deviation,

$$\sigma = \sqrt{\frac{1}{N}\sum_{i=1}^{n} f_i(x_i - \bar{x})^2}$$

or $\sigma = \frac{1}{N}\sqrt{N\Sigma f_i x_i^2 - (\Sigma f_i x_i)^2}$

Where, $N = \sum_{i=1}^{n} f_i$

For Continuous Frequency Distribution
A continuous frequency distribution is a series in which the data is classified into different class intervals without gaps alongwith their respective frequencies.

Mean Deviation about Mean

For calculating mean deviation from mean of a continuous frequency distribution, the procedure is same as for a discrete frequency distribution. The only difference is that here, we have to obtain the mid-points of the various classes and take the deviations of these mid-points from the mean.

Mean Deviation about Median

For calculating mean deviation from median of a continuous frequency distribution, the procedure is same as about mean. The only difference is that here, we replace mean by median and median is calculated by the following formula,

$$M = l + \frac{\frac{N}{2} - cf}{f} \times h$$

l, f, h and cf are respectively the lower limit, the frequency, the width of median class and cumulative frequency of class just preceding the median class.

14

CHAPTER

Probability

As far as the law of mathematics refer to reality, they are not certain; and as far as they are certain, they do not refer to reality.

—Albert Einstein

YOUR STUDY GOALS

YOUR PRACTICE GOALS

- Objective Type Questions (37 Qs; 25 MCQs, 10 A-R, 2 Case Based)
- Subjective Type Questions (16 Qs; 5 VSA, 6 SA, 5LA)

14.1 Outcomes and Sample Space

A possible result of a random experiment is called its outcome and the set of all possible outcomes in a random experiment is called a sample space. It is generally denoted by S.

i.e. **Sample space (S) = {All possible outcomes}**

Each element of a sample space is called a **sample point** or an **event point.**

e.g. (i) On tossing a fair coin, there are two possible outcomes, namely head (H) and tail (T).

$\therefore$ The sample space S is $\{H, T\}$ i.e. $S = \{H, T\}$.

(ii) When we throw a die, then possible outcomes of this experiment are 1, 2, 3, 4, 5 or 6.

$\therefore$ The sample space S is $\{1, 2, 3, 4, 5, 6\}$

i.e. $S = \{1, 2, 3, 4, 5, 6\}$

Common Terms Related to Probability Problems

(i) **Coin** A coin has two sides, head and tail. If an experiment consists of more than one coin, then coins are considered as distinct.

- Sample space of one coin $= \{H, T\}$
- Sample space of two coins
  $= \{(H, T), (T, H), (H, H), (T, T)\}$
- Sample space of three coins
  $= \{(H,H,H), (H,H,T), (H,T,H), (T,H,H),$
  $(H,T,T), (T,H,T), (T,T,H), (T,T,T)\}$

(ii) **Die** A die has six faces marked with numbers 1, 2, 3, 4, 5 and 6 (or dots 1, 2, 3, 4, 5 and 6). If an experiment consists of more than one die, then all dice are considered as distinct.

- Sample space of a die $= \{1, 2, 3, 4, 5, 6\}$
- Sample space of two dice

$$= \begin{Bmatrix} (1,1), & (1,2), & (1,3), & (1,4), & (1,5), & (1,6), \\ (2,1), & (2,2), & (2,3), & (2,4), & (2,5), & (2,6), \\ (3,1), & (3,2), & (3,3), & (3,4), & (3,5), & (3,6), \\ (4,1), & (4,2), & (4,3), & (4,4), & (4,5), & (4,6), \\ (5,1), & (5,2), & (5,3), & (5,4), & (5,5), & (5,6), \\ (6,1), & (6,2), & (6,3), & (6,4), & (6,5), & (6,6) \end{Bmatrix}$$

(iii) **Playing Cards** A pack of playing cards has 52 cards. There are 4 suits (spade, heart, diamond and club), each having 13 cards in pack. These are of two colours, **red** (heart and diamond) and **black** (spade and club), each having 26 cards. In 13 cards of each suit, there are 3 face cards namely **king, queen and jack**, so there are total 12 face cards. Also, there are 16 honour cards, 4 of each suit namely **ace, king, queen and jack.**

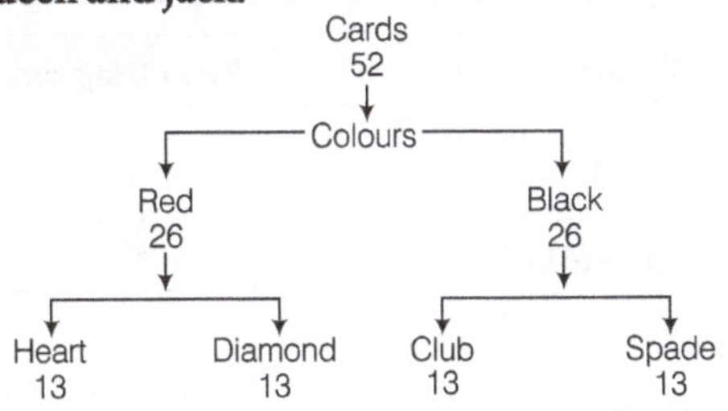

14.2 Event

Any subset E of a sample space S is called an event.
e.g. On throwing a die, we have the sample space.

$$S = \{1, 2, 3, 4, 5, 6\}$$

Then, $E = \{2, 3, 4\}$ is an event. $[\because E \subseteq S]$

We can define as many events as there are subsets of a sample space. Thus, number of events of a sample space $S = n[P(S)]$, where $P(S)$ is the power set of S.

Occurrence of an Event

An event associated with a random experiment is said to occur, if any one of the elementary events associated to it is an outcome of the experiment.

e.g. Suppose a die is thrown and let A be an event of getting an odd number.

Then, $A = \{1, 3, 5\}$

Now, if the outcome of experiment is 3, then we can say that event A has occurred.

Suppose, in another trial, outcome is 4, then we can say that event A has not occurred.

Types of Events

Events can be classified into various types on the basis of their elements.

Sure Event

The whole sample space S is also a subset of S, so it represents an event. Since, every outcome of an experiment carried out a member of S, therefore the event represented by S is called a sure or a certain event.

e.g. (i) Sun rises in the East, is a sure event.

(ii) On tossing a coin, either head or tail will occur, is a sure event.

(iii) On throwing a die, we have sample space

$$S = \{1, 2, 3, 4, 5, 6\}.$$

Let $E =$ Event of getting a number less than 7.

Then, $E = \{1, 2, 3, 4, 5, 6\}$

So, E is a sure event.

Impossible Event

The empty set ϕ is also a subset of the sample space (S), so it represents an event. Since, there is no element in the empty set, therefore it cannot occur.

That's why event ϕ is called an impossible event.
e.g. On throwing a die, we have the sample space

$$S = \{1, 2, 3, 4, 5, 6\}.$$

Let $E =$ Event of getting a number less than 1
Then, $E = \phi$

So, E is called an impossible event.

Simple Event

If an event has only one sample point, then it is called a simple or elementary event.

e.g. Let a die is thrown, then sample space

$$S = \{1, 2, 3, 4, 5, 6\}.$$

Again, let A = Event of getting $4 = \{4\}$

B = Event of getting $1 = \{1\}$

Here, A and B are simple events.

Compound Event

If an event has more than one sample point, then it is called a compound event.

e.g. On rolling a die, we have sample space

$$S = \{1, 2, 3, 4, 5, 6\}.$$

Then, the events

E: Getting an even number

F: Getting an odd number

G: Getting a multiple of 3

are all compound events. The subsets of S associated with these events are

$E = \{2, 4, 6\}, F = \{1, 3, 5\}$

and $G = \{3, 6\}$.

Each of the above subsets contain more than one sample point, therefore they are all compound events.

Algebra of Events

We know that how to combine two or more sets by using the operations on sets union ($\cup$), intersection ($\cap$) and difference ($-$). Like-wise, we can combine two or more events by using the analogous set notations.

Let A and B are two events associated with a sample space S, then

1. Complementary Event (Event 'not E') For every event E, there corresponds another event E' called the complementary event of E, which consist of those outcomes that do not correspond to the occurrence of E. E' is also called the event 'not E'.

2. The Event 'A or B' The event 'A or B' is same as the event $A \cup B$ and it contains all those elements which are either in event A or in event B or in both.

Thus, A or $B = A \cup B = \{\omega : \omega \in A \text{ or } \omega \in B\}$.

3. The Event 'A and B' The event 'A and B' is same as the event $A \cap B$ and it contains all those elements which are common to both A and B.

Thus, A and $B = A \cap B = \{\omega : \omega \in A \text{ and } \omega \in B\}$.

4. The Event 'A but not B' The events 'A but not B' is same as the event $A - B = (A \cap B')$ and it contains all those elements which are in A but not in B.

Thus, A but not $B = A - B = \{\omega : \omega \in A \text{ and } \omega \notin B\}$

In the following table, we give verbal description of some events and their equivalent set alongwith the Venn diagram.

Events	Verbal Description	Equivalent Set	Venn Diagram
(i) Neither A nor B	It is the intersection of sets $\overline{A}$ and $\overline{B}$, which contains the common elements of $\overline{A}$ and $\overline{B}$.	$\overline{A} \cap \overline{B}$ or $U - (A \cup B)$	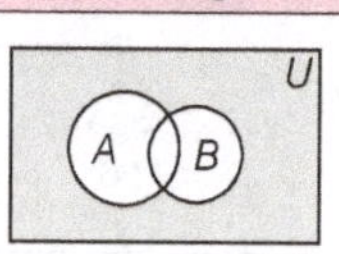
(ii) Exactly one of A and B	It is the union of one event and complement of other events. or It is the difference of union and intersection of both events.	$(A \cap \overline{B}) \cup (\overline{A} \cap B)$ or $(A \cup B) - (A \cap B)$	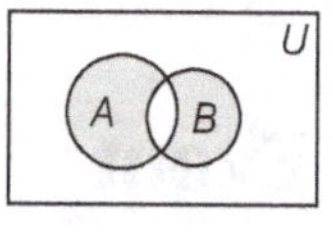

(iii)	Atleast one of A, B or C	It is a union of all three sets, which contains either A or B or C etc.	$A \cup B \cup C$	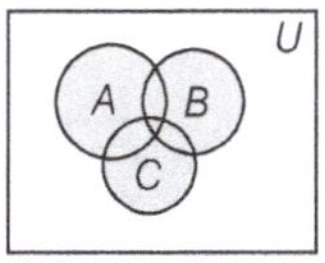
(iv)	All three of A, B and C	It is an intersection of all three sets, which contains common elements of A, B and C.	$A \cap B \cap C$	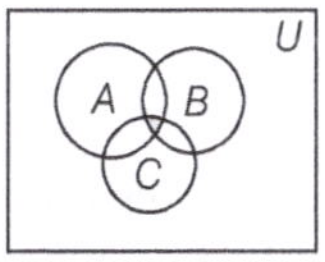
(v)	Exactly two of A, B and C	It is the union of three conditions, where in each condition we take intersection of any two events and complement of remaining third event.	$(A \cap B \cap \bar{C}) \cup (A \cap \bar{B} \cap C) \cup (\bar{A} \cap B \cap C)$	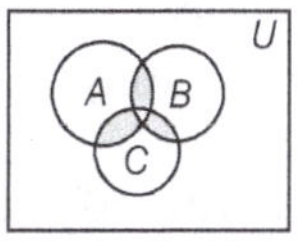

Mutually Exclusive Events

Two events are said to be mutually exclusive, if the occurrence of any one of them excludes the occurrence of other event i.e. if they cannot occur simultaneously.

Thus, two events E_1 and E_2 are said to be mutually exclusive, if $\quad E_1 \cap E_2 = \phi$

e.g. In throwing a die, we have sample space

$$S = \{1, 2, 3, 4, 5, 6\}$$

Let E_1 = Event of getting even numbers $= \{2, 4, 6\}$

E_2 = Event of getting odd numbers $= \{1, 3, 5\}$

Here, $\quad E_1 \cap E_2 = \phi$

So, E_1 and E_2 are mutually exclusive events.

In general, events $E_1, E_2, \ldots, E_n$ are said to be mutually exclusive, if they are pairwise disjoint i.e. if $E_i \cap E_j = \phi$ for all $i \neq j$.

Exhaustive Events

For a random experiment, a set of events is said to be exhaustive, if one of them necessarily occurs whenever the experiment is performed.

Let $E_1, E_2, E_3, \ldots, E_n$ be n subsets of a sample space S.

Then, events $E_1, E_2, E_3, \ldots, E_n$ are called exhaustive events, if $E_1 \cup E_2 \cup E_3 \cup \ldots \cup E_n = S$

e.g. Consider the experiment of throwing a die. We have, $S = \{1, 2, 3, 4, 5, 6\}$. Let us define the following events.

E_1 : a number less than 4 appears.

E_2 : a number greater than 2 but less than 5 appears.

E_3 : a number greater than 4 appears.

Then, $E_1 = \{1, 2, 3\}$, $E_2 = \{3, 4\}$ and $E_3 = \{5, 6\}$.

We observe that

$$E_1 \cup E_2 \cup E_3 = \{1, 2, 3\} \cup \{3, 4\} \cup \{5, 6\}$$
$$= \{1, 2, 3, 4, 5, 6\} = S$$

Thus, the events E_1, E_2 and E_3 are exhaustive events.

If $E_i \cap E_j = \phi$ for $i \neq j$ i.e. events E_i and E_j are pairwise disjoint and $\bigcup_{i=1}^{n} E_i = S$, then events $E_1, E_2, \ldots, E_n$ are called mutually exclusive and exhaustive events.

Example 1. **List all events associated with the random experiment of tossing of two coins.**

Sol. On tossing of two coins, we have the sample space

$$S = \{HH, HT, TH, TT\}$$

Here, $n(P(S)) = 2^4 = 16$

So, there are 16 events, which are defined as follows.

[if a set has n elements, then its power set has 2^n elements]

$\{\phi, \{HH\}, \{HT\}, \{TH\}, \{TT\}, \{HH, HT\}, \{HH, TH\}, \{HH, TT\}, \{HT, TH\}, \{HT, TT\}, \{TH, TT\}, \{HH, HT, TH\}, \{HH, HT, TT\}, \{HH, TH, TT\}, \{HT, TH, TT\}, \{HH, HT, TH, TT\}\}$

Example 2. **An ordinary deck of cards contains 52 cards divided into four suits. The red suits are diamonds and hearts and black suits are clubs and spades. The cards Jack (J), Queen (Q), King (K) are called face cards.**

Suppose we pick one card from the deck at random. Describe the event that the chosen card is a black face card. *NCERT Exemplar*

Sol. We have the following flowchart.

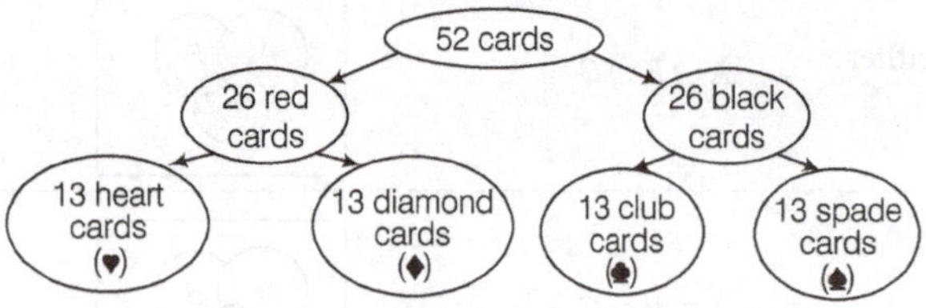

Let E be the event that the chosen card is a black face card. Then, the outcomes in E are Jack, Queen, King of spade and club cards.

Symbolically, $E = \{J♣, Q♣, K♣, J♠, Q♠, K♠\}$

Example 3. **Identify the following events as sure/impossible events.**

(i) Event of getting a number multiple of 7 in a single toss of a die.

(ii) Event of getting five kings in a pack of cards.

(iii) Event of drawing a club queen in a pack of cards.

Sol. (i) We know that in a single toss of a die only 1 to 6 numbers exist. So, multiple of 7 does not exist. Hence, it is an impossible event.

(ii) We know that in a pack of cards only four kings exist. So, five kings does not exist. Hence, it is an impossible event.

(iii) We know that in a pack of cards club queen exist. Hence, it is a sure event.

Example 4. **Two dice are rolled. Let A, B and C be the events of getting a sum of 2, a sum of 3 and a sum of 4, respectively.**

(i) Which events are elementary events?

(ii) Which events are compound events?

Sol. On throwing of two dice, we get the sample space

$$S = \begin{Bmatrix} (1,1), (1,2), (1,3), (1,4), (1,5), (1,6), \\ (2,1), (2,2), (2,3), (2,4), (2,5), (2,6), \\ (3,1), (3,2), (3,3), (3,4), (3,5), (3,6), \\ (4,1), (4,2), (4,3), (4,4), (4,5), (4,6), \\ (5,1), (5,2), (5,3), (5,4), (5,5), (5,6), \\ (6,1), (6,2), (6,3), (6,4), (6,5), (6,6) \end{Bmatrix}$$

Now, as A is the event of getting a sum of 2.

$\therefore$ $A = \{(1,1)\}$

Similarly, $B = \{(1,2), (2,1)\}$

and $C = \{(1,3), (3,1), (2,2)\}$

(i) Since, A consists a single sample point, therefore it is an elementary or simple event.

(ii) Since, both B and C contain more than one sample points, therefore each of them is compound event.

Example 5. **Two dice are thrown once. The events A, B, E are as follow.**

A: Getting an even number on first die.

B: Getting an odd number on first die.

E: Getting the sum of numbers on the dice ≥ 10.

Describe the events.

(i) A' (ii) B' (iii) E' NCERT

Sol. On throwing of two dice, we have sample space

$$S = \begin{Bmatrix} (1,1), (1,2), (1,3), (1,4), (1,5), (1,6), \\ (2,1), (2,2), (2,3), (2,4), (2,5), (2,6), \\ (3,1), (3,2), (3,3), (3,4), (3,5), (3,6), \\ (4,1), (4,2), (4,3), (4,4), (4,5), (4,6), \\ (5,1), (5,2), (5,3), (5,4), (5,5), (5,6), \\ (6,1), (6,2), (6,3), (6,4), (6,5), (6,6) \end{Bmatrix}$$

(i) A: Getting an even number on first die.

$\Rightarrow A'$: Getting an odd number on first die.

$$= \begin{Bmatrix} (1,1), (1,2), (1,3), (1,4), (1,5), (1,6), \\ (3,1), (3,2), (3,3), (3,4), (3,5), (3,6), \\ (5,1), (5,2), (5,3), (5,4), (5,5), (5,6) \end{Bmatrix}$$

(ii) B: Getting an odd number on first die.

$\Rightarrow B'$: Getting an even number on first die.

$$= \begin{Bmatrix} (2,1), (2,2), (2,3), (2,4), (2,5), (2,6), \\ (4,1), (4,2), (4,3), (4,4), (4,5), (4,6), \\ (6,1), (6,2), (6,3), (6,4), (6,5), (6,6) \end{Bmatrix}$$

(iii) E: Getting the sum of the numbers on the dice ≥ 10.

$\Rightarrow E'$: Getting the sum of the numbers on the dice < 10.

$$= \begin{Bmatrix} (1,1), (1,2), (1,3), (1,4), (1,5), (1,6), \\ (2,1), (2,2), (2,3), (2,4), (2,5), (2,6), \\ (3,1), (3,2), (3,3), (3,4), (3,5), (3,6), \\ (4,1), (4,2), (4,3), (4,4), (4,5), \\ (5,1), (5,2), (5,3), (5,4), \\ (6,1), (6,2), (6,3) \end{Bmatrix}$$

Example 6. **A die is thrown. Describe the following events.**

(i) A: a number less than 7.

(ii) B: a number greater than 7.

(iii) C: a multiple of 3.

(iv) D: a number less than 4.

(v) E: an even number greater than 4.

(vi) F: a number not less than 3.

Also, find $A \cup B, A \cap B, B \cup C, E \cap F, D \cap E, A - C, D - E, F'$ and $E \cap F'$. NCERT

Sol. When a die is thrown, then sample space

$S = \{1, 2, 3, 4, 5, 6\}$

(i) A : a number less than $7 = \{1, 2, 3, 4, 5, 6\}$

(ii) B : a number greater than $7 = \{\} = \phi$

(iii) C : a multiple of $3 = \{3, 6\}$

(iv) D : a number less than $4 = \{1, 2, 3\}$

(v) E : an even number greater than $4 = \{6\}$

(vi) F : a number not less than $3 = \{3, 4, 5, 6\}$

Now, $A \cup B$ = The elements which are in A or B or both $= \{1, 2, 3, 4, 5, 6\} \cup \phi = \{1, 2, 3, 4, 5, 6\}$

$A \cap B$ = The elements which are common in both A and $B = \{1, 2, 3, 4, 5, 6\} \cap \phi = \phi$

$B \cup C$ = The elements which are in B or C or both $= \{\} \cup \{3, 6\} = \{3, 6\}$

$E \cap F$ = The elements which are common in both E and $F = \{6\} \cap \{3, 4, 5, 6\} = \{6\}$

$D \cap E$ = The elements which are common in both D and $E = \{1, 2, 3\} \cap \{6\} = \phi$

$A - C$ = The elements which are in A but not in C $= \{1, 2, 3, 4, 5, 6\} - \{3, 6\} = \{1, 2, 4, 5\}$

$D - E$ = The elements which are in D but not in E $= \{1, 2, 3\} - \{6\} = \{1, 2, 3\}$

F' = The elements which are not in F $= (S - F) = \{1, 2, 3, 4, 5, 6\} - \{3, 4, 5, 6\} = \{1, 2\}$

and $E \cap F' = E \cap (S - F)$

$= E \cap (\{1, 2, 3, 4, 5, 6\} - \{3, 4, 5, 6\})$

$= \{6\} \cap \{1, 2\} = \phi$

Example 7. An experiment involves rolling a pair of dice and recording the numbers that comes up. Describe the following events.

A : the sum is greater than 8.

B : 2 occurs on either die.

C : the sum is atleast 7 and multiple of 3.

Which pairs of these events are mutually exclusive? NCERT

Sol. There are 36 elements in the sample space

$S = \{(x, y): x, y = 1, 2, 3, 4, 5, 6\}$

Given, A = Events of getting the sum greater than 8 i.e. events of getting the sum 9, 10, 11 and 12

$\therefore \quad A = \{(6, 3), (3, 6), (4, 5), (5, 4), (6, 4), (4, 6)$
$(5, 5), (6, 5), (5, 6), (6, 6)\}$

and B = Events of getting 2 on either die

$\therefore \quad B = \{(2, 1), (2, 2), (2, 3), (2, 4), (2, 5), (2, 6), (1, 2),$
$(3, 2), (4, 2), (5, 2), (6, 2)\}$

and C = Events of getting the sum atleast 7 and a multiple of 3, i.e. event of getting the sum 9 and $12 = \{(4, 5), (5, 4), (6, 3), (3, 6), (6, 6)\}$

Here, $A \cap B = \phi$, $B \cap C = \phi$ and $A \cap C \neq \phi$

Thus, we can say that the pairs A and B; B and C are mutually exclusive events.

Example 8. Three coins are tossed once. Let A denotes the event 'three heads show', B denotes the event 'two heads and one tail show', C denotes the event 'three tails show' and D denotes the event 'a head shows on the first coin'. Which pair of events are NCERT

(i) mutually exclusive events? (ii) simple events?

(iii) compound events?

Sol. When three coins are tossed, then the sample space is

$S = \{HHH, HHT, HTH, THH, HTT, THT, TTH, TTT\}$

Given, A = Events of getting three heads

$\therefore \quad A = \{HHH\}$

B = Events of getting two heads and one tail

$\therefore \quad B = \{HHT, HTH, THH\}$

C = Events of getting three tails

$\therefore \quad C = \{TTT\}$

and $\quad D$ = Events of getting a head on first coin.

$\therefore \quad D = \{HHH, HTT, HHT, HTH\}$

(i) Clearly, $A \cap B = \phi$, $A \cap C = \phi$,

$A \cap D = \{HHH\}$, $B \cap C = \phi$,

$B \cap D = \{HHT, HTH\}$ and $C \cap D = \phi$

Thus, we can say A and B; A and C; B and C; C and D are mutually exclusive events.

(ii) A and C are simple events, since both have only one sample point.

(iii) B and D are compound events, since they have more than one sample point.

Example 9. A coin is tossed three times, consider the following events.

A : No head appears.

B : Exactly one head appears.

and C : Atleast two heads appear.

Do they form a set of mutually exclusive and exhaustive events? NCERT

Sol. When a coin is tossed three times, then the sample space is

$S = \{HHH, HHT, HTH, THH, THT, TTH, HTT, TTT\}$

Given, A = Events of getting no head

$\therefore \quad A = \{TTT\}$

B = Events of getting exactly one head

$\therefore \quad B = \{TTH, HTT, THT\}$

and $\quad C$ = Events of getting atleast two heads

$\therefore \quad C = \{HHT, HTH, THH, HHH\}$

Clearly, $A \cap B = \phi$, $A \cap C = \phi$ and $B \cap C = \phi$

So, we can say that A, B and C are mutually exclusive events. [$\because$ A, B and C are pairwise disjoint]

Also, $A \cup B \cup C = \{TTT, TTH, HTT, THT, HHT,$
$HTH, THH, HHH\} = S$

$\therefore$ A, B and C are exhaustive events.

Hence, A, B and C form a set of mutually exclusive and exhaustive events.

Quick Test 1

1. A coin is tossed. Find the total number of elementary events and also the total number of events associated with the random experiment.
2. Two dice are rolled. Let E_1, E_2 and E_3 be the events of getting a sum of 4, 5 and 6, respectively.
   (i) Which events are elementary events?
   (ii) Which events are compound events?
3. Consider the experiment of rolling a die. Let A be the event 'getting a prime number' and B be the event 'getting an odd number'. Write the sets representing events **NCERT, Competency Based Que**
   (i) A and B. (ii) A or B.
   (iii) A but not B. (iv) not A.
4. An experiment involves tossing of two coins and recording them in the following events.
   A: No tail B: Exactly one tail
   C: Atleast one tail.
   Write the sets representing events
   (i) A and C. (ii) A not B.
5. A die is tossed. Let E denotes the event of getting a number multiple of 2 and F denotes the event of getting a number less than 7. Write the sets representing events.
   (i) $E \cup F$ (ii) $F - E$ (iii) E' (iv) $E \cap F$
6. Two dice are thrown. The events A, B and C are as follows.
   A : Getting an even number on the first die.
   B : Getting an odd number on the first die.
   C : Getting the sum of the numbers on the dice ≤ 5. **Competency Based Que**
   Describe the events.
   (i) A' (ii) not B
   (iii) A or B (iv) A and B
   (v) A but not C (vi) B or C
   (vii) B and C (viii) $A \cap B' \cap C'$
7. A coin is tossed two times, consider the following events.
   E_1 : No head appears.
   E_2 : Exactly one head appears.
   and E_3 : Atleast one head appears.
   Do they form a set of mutually exclusive and exhaustive events?
8. Three coins are tossed once. Let A denotes the event 'three tails show', B denotes the event 'one tail and two heads show', C denotes the event 'tail on the first coin'.
   Which pair of events are
   (i) mutually exclusive? (ii) simple events?
   (iii) compound events?
9. Two dice are thrown and the sum of the numbers which come up on the dice is noted. Let us consider the following events associated with this experiment. **Competency Based Que**
   A : The sum is even.
   B : The sum is a multiple of 3.
   C : The sum is less than 4.
   D : The sum is greater than 11.
   Which pairs of these events are mutually exclusive?

Answers

1. 2, 4
2. (i) No event (ii) E_1, E_2 and E_3
3. (i) {3, 5} (ii) {1,2,3,5} (iii) {2} (iv) {1,4,6}
4. (i) A and $C = A \cap C = \phi$
   (ii) A not $B = A - B = \{(H, H)\}$
5. (i) {1, 2, 3, 4, 5, 6} (ii) {1, 3, 5}
   (iii) {1, 3, 5} (iv) {2, 4, 6}
6. (i) $A' = \begin{Bmatrix} (1,1),(1,2),\ldots,(1,6), \\ (3,1),(3,2),\ldots,(3,6), \\ (5,1),(5,2),\ldots,(5,6) \end{Bmatrix} = B$
   (ii) $B' = \{(2,1),(2,2),(2,3),(2,4),(2,5),(2,6),(4,1),(4,2),(4,3),(4,4),(4,5),(4,6),(6,1),(6,2),(6,3),(6,4),(6,5),(6,6)\} = A$
   (iii) $A \cup B = \{(1,1),(1,2),(1,3),(1,4),(1,5),(1,6),(3,1),(3,2),(3,3),(3,4),(3,5),(3,6),(5,1),(5,2),(5,3),(5,4),(5,5),(5,6),(2,1),(2,2),(2,3),(2,4),(2,5),(2,6),(4,1),(4,2),(4,3),(4,4),(4,5),(4,6),(6,1),(6,2),(6,3),(6,4),(6,5),(6,6)\} = S$
   (iv) $A \cap B = \phi$
   (v) $A - C = \{(2,4),(2,5),(2,6),(4,2),(4,3),(4,4),(4,5),(4,6),(6,1),(6,2),(6,3),(6,4),(6,5),(6,6)\}$
   (vi) $B \cup C = \{(1,1),(1,2),(1,3),(1,4),(1,5),(1,6),(2,1),(2,2),(2,3),(3,1),(3,2),(3,3),(3,4),(3,5),(3,6),(4,1),(5,1),(5,2),(5,3),(5,4),(5,5),(5,6)\}$
   (vii) $B \cap C = \{(1,1),(1,2),(1,3),(1,4),(3,1),(3,2)\}$
   (viii) $A \cap B' \cap C' = \{(2,4),(2,5),(2,6),(4,2),(4,3),(4,4),(4,5),(4,6),(6,1),(6,2),(6,3),(6,4),(6,5),(6,6)\}$
7. E_1, E_2 and E_3 are not mutually exclusive but they are exhaustive events.
8. (i) A and B (ii) A (iii) B and C.
9. C and D

For Solutions, See QR Code on Pg. 412

14.3 Probability of Occurrence of an Event

A numerical value that conveys the chance of occurrence of an event, when we perform an experiment, is called the probability of that event.

Axiomatic Approach to Probability

Let S be the sample space of a random experiment. The probability P is a real valued function whose domain is the power set of S and range is the interval [0, 1] satisfying the following **axioms.**

(i) For any event $E, P(E) \geq 0$

(ii) $P(S) = 1$

(iii) If E and F are mutually exclusive events, then $P(E \cup F) = P(E) + P(F)$.

Now, let S be a sample space containing outcomes $E_1, E_2, \ldots, E_n$ i.e. $S = \{E_1, E_2, \ldots, E_n\}$.

Then, from the axiomatic definition of probability, we have

- $0 \leq P(E_i) \leq 1, \forall E_i \in S$
- For any event $A, P(A) = \sum_{i=1}^{n} P(E_i) = 1, E_i \in A$

Classical Approach to Probability

To obtain the probability of an event, we find the ratio of the number of favourable outcomes to the event, to the total number of equally likely outcomes. This theory is known as classical theory of probability.

i.e. $$P(A) = \frac{\text{Number of favourable outcomes}}{\text{Total number of possible outcomes}}$$

This is also known as theoretical probability.

Probability of Equally Likely Outcomes

The outcomes of a random experiment are said to be **equally likely,** if the chance of occurrence of each outcome is same, i.e. chances of each outcome is same. Let the sample space of an experiment is

$$S = \{s_1, s_2, \ldots, s_n\}$$

Also, let all the outcomes are occurring equally likely. i.e. $P(s_i) = p$, for all $s_i \in S$, where $0 \leq p \leq 1$

Since, by axiomatic approach to probability,

$$\sum_{i=1}^{n} P(s_i) = 1$$

$\therefore$ $$\underbrace{p + p + \cdots + p}_{(n \text{ times})} = 1$$

$\Rightarrow$ $$np = 1 \Rightarrow p = \frac{1}{n}$$

Thus, $$P(s_i) = \frac{1}{n}, \forall\, i = 1, 2, \ldots, n$$

Let E be an event associated with a sample space S, such that $n(S) = n$ and $n(E) = m$, then by axiomatic approach

$$P(E) = \underbrace{\frac{1}{n} + \frac{1}{n} + \frac{1}{n} + \ldots + \frac{1}{n}}_{m \text{ times}} = m \times \frac{1}{n} = \frac{m}{n}$$

Example 1. **Which of the following cannot be valid assignments of probabilities for outcomes of sample space. $S = \{W_1, W_2, W_3, W_4, W_5, W_6, W_7\}$?**

Assignments	W_1	W_2	W_3	W_4	W_5	W_6	W_7
(i)	0.1	0.01	0.05	0.03	0.01	0.2	0.6
(ii)	$\frac{1}{7}$	$\frac{1}{7}$	$\frac{1}{7}$	$\frac{1}{7}$	$\frac{1}{7}$	$\frac{1}{7}$	$\frac{1}{7}$
(iii)	0.1	0.2	0.3	0.4	0.5	0.6	0.7
(iv)	−0.1	0.2	0.3	0.4	−0.2	0.1	0.3
(v)	$\frac{1}{14}$	$\frac{2}{14}$	$\frac{3}{14}$	$\frac{4}{14}$	$\frac{5}{14}$	$\frac{6}{14}$	$\frac{15}{14}$

Sol. (i) Here, each of $P(W_i)$, where $i = 1$ to 7 is positive and less than 1.

Also, sum of probabilities

$$= 0.1 + 0.01 + 0.05 + 0.03 + 0.01 + 0.2 + 0.6 = 1$$

$\therefore$ This assignment is valid.

(ii) Here, each of $P(W_i)$, where $i = 1$ to 7 is positive and less than 1.

Also, sum of probabilities $= 7 \times \frac{1}{7} = 1$

$\therefore$ This assignment is also valid.

(iii) Here, each of $P(W_i)$, where $i = 1$ to 7 is positive and less than 1.

But sum of probabilities

$$= 0.1 + 0.2 + 0.3 + 0.4 + 0.5 + 0.6 + 0.7$$
$$= 2.8 \neq 1$$

$\therefore$ This assignment is not valid.

(iv) Here, $P(W_1)$ and $P(W_5)$ are negative.

$\therefore$ This assignment is not valid.

$[\because 0 \leq P(W_i) \leq 1 \,\forall\, W_i \in S]$

(v) Here, $P(W_7) > 1$.

$\therefore$ This assignment is not valid.

$[\because 0 \le P(W_i) \le 1 \, \forall \, W_i \in S]$

Hence, assignments (iii), (iv) and (v) are not valid.

EXAMPLE 2. Suppose we throw a die once. Find the probability of getting a number greater than 4.

Sol. Let A be the event of getting a number greater than 4.

Here, total number of possible outcomes = 6

Number of outcomes which are favourable to event A

$= \{5, 6\} \Rightarrow n(A) = 2$

$\therefore$ Probability of getting a number greater than 4,

$$P(A) = \frac{2}{6} = \frac{1}{3}$$

Example 3. Suppose that each child born is equally likely to be a boy or a girl. Consider a family with exactly three children. *Competency Based Que*

(i) List the eight elements in the sample space whose outcomes are all possible genders of the three children.

(ii) Write each of the following events as a set and find its probability.

(a) The event that exactly one child is a girl.

(b) The event that atleast two children are girls.

(c) The event that no child is a girl. *NCERT Exemplar*

Sol. (i) If B denotes a boy and G denotes a girl. Then, all possible genders are expressed as

$S = \{BBB, BBG, BGB, GBB, BGG, GBG, GGB, GGG\}$

(ii) (a) If E_1 denotes the events that exactly one child is a girl.

Then, $E_1 = \{BBG, BGB, GBB\} \Rightarrow P(E_1) = \frac{3}{8}$

(b) If E_2 denotes the events that atleast two children are girls.

Then, $E_2 = \{BGG, GBG, GGB, GGG\}$

$\Rightarrow P(E_2) = \frac{4}{8} = \frac{1}{2}$

(c) If E_3 denotes the event that no child is a girl.

Then, $E_3 = \{BBB\} \Rightarrow P(E_3) = \frac{1}{8}$

Example 4. Three coins are tossed once. Find the probability of getting

(i) 3 heads **(ii) atleast 2 heads**

(iii) atmost 2 tails **(iv) exactly two tails**

(v) no tail. *NCERT, Competency Based Que*

Sol. In random experiment of tossing three coins, the sample space is

$S = \{HHH, HHT, HTH, THH, HTT, THT, TTH, TTT\}$

$\Rightarrow \quad n(S) = 8$

(i) Let E_1 be the event of getting 3 heads.

Then, favourable outcomes to E_1 is $\{HHH\}$.

Thus, $n(E_1) = 1$,

$\therefore \quad P(E_1) = \frac{n(E_1)}{n(S)} = \frac{1}{8}$

(ii) Let E_2 be the event of getting atleast 2 heads.

Then, favourable outcomes to E_2 are $\{HHT, HTH, THH$ and $HHH\}$.

Thus, $n(E_2) = 4$

$\therefore \quad P(E_2) = \frac{n(E_2)}{n(S)} = \frac{4}{8} = \frac{1}{2}$

(iii) Let E_3 be the event of getting atmost 2 tails.

Then, favourable outcomes to E_3 are $\{HHH, HHT, HTH, THH, TTH, THT, HTT\}$.

Thus, $n(E_3) = 7$

$\therefore \quad P(E_3) = \frac{n(E_3)}{n(S)} = \frac{7}{8}$

(iv) Let E_4 be the event of getting exactly two tails.

Then, favourable outcomes to E_4 are $\{TTH, HTT, THT\}$.

Thus, $n(E_4) = 3$

$\therefore \quad P(E_4) = \frac{n(E_4)}{n(S)} = \frac{3}{8}$

(v) Let E_5 be the event of getting no tail.

Then, favourable outcomes to E_5 is $\{HHH\}$.

Thus, $n(E_5) = 1$

$\therefore \quad P(E_5) = \frac{n(E_5)}{n(S)} = \frac{1}{8}$

Example 5. A die is thrown. Find the probability of getting *NCERT*

(i) P (a prime number)

(ii) P(a number ≥ 3)

(iii) P(a number ≤ 1)

(iv) P(a number more than 6)

(v) P(a number less than 6)

Sol. In throwing a die, the sample space will be

$S = \{1, 2, 3, 4, 5, 6\}$

$\therefore \quad n(S) = 6$

(i) Let E_1 be the event of getting a prime number.

Then, $E_1 = \{2, 3, 5\}$

$\therefore \quad n(E_1) = 3$

Now, probability of getting a prime number,

$$P(E_1) = \frac{n(E_1)}{n(S)} = \frac{3}{6} = \frac{1}{2}$$

(ii) Let E_2 be the event of getting a number ≥ 3.

Then, $E_2 = \{3, 4, 5, 6\}$

$\therefore$ $n(E_2) = 4$

Now, probability of getting a number ≥ 3,

$$P(E_2) = \frac{n(E_2)}{n(S)} = \frac{4}{6} = \frac{2}{3}$$

(iii) Let E_3 be the event of getting a number ≤ 1.

Then, $E_3 = \{1\}$

$\therefore$ $n(E_3) = 1$

Now, probability of getting a number ≤ 1,

$$P(E_3) = \frac{n(E_3)}{n(S)} = \frac{1}{6}$$

(iv) Let E_4 be the event of getting a number more than 6.

Then, $E_4 = \{\} = \phi$

$\Rightarrow E_4$ is an impossible event.

$\therefore$ $n(E_4) = 0$

Now, probability of getting a number more than 6.

$$P(E_4) = \frac{n(E_4)}{n(S)} = \frac{0}{6} = 0$$

(v) Let E_5 be the event of getting a number less than 6.

Then, $E_5 = \{1, 2, 3, 4, 5\}$

$\therefore$ $n(E_5) = 5$

Now, probability of getting a number less than 6.

$$P(E_5) = \frac{n(E_5)}{n(S)} = \frac{5}{6}$$

Example 6. A card is selected from a pack of 52 cards.

(i) How many points are there in the sample space?

(ii) Calculate the probability that the card is an ace of spades.

(iii) Calculate the probability that the card is

(a) an ace

(b) black card *NCERT*

Sol. (i) A card is drawn at random from a pack of 52 cards, therefore there are 52 sample points in the sample space.

(ii) There is only one ace of spades, therefore number of favourable outcomes is 1.

Hence, the required probability is $\frac{1}{52}$.

(iii) (a) Since, there are four aces in a pack of 52 cards, therefore number of favourable outcomes are 4.

Hence, the required probability is $\frac{4}{52}$ or $\frac{1}{13}$.

(b) Since, there are 26 black cards in a pack of 52 cards, therefore number of favourable outcomes are 26.

Hence, the required probability is $\frac{26}{52}$ or $\frac{1}{2}$.

Example 7. Four cards are drawn at random from pack of 52 playing cards. Find the probability of getting

(i) all face cards.

(ii) two red cards and two black cards.

(iii) one card from each suit.

Sol. We have to choose four cards from pack of 52 playing cards.

Then, total number of possible outcomes = ${}^{52}C_4$

(i) We know that there are 12 face cards.

$\therefore$ Number of favourable outcomes = ${}^{12}C_4$

Hence, P (getting all face cards) = $\frac{{}^{12}C_4}{{}^{52}C_4}$

(ii) We know that there are 26 red cards and 26 black cards.

$\therefore$ Number of favourable outcomes = ${}^{26}C_2 \times {}^{26}C_2$

Hence, P (getting 2 red and 2 black cards)

$$= \frac{({}^{26}C_2)^2}{{}^{52}C_4}$$

(iii) We know that there are 4 suits each having 13 cards.

$\therefore$ Number of favourable outcomes

$$= {}^{13}C_1 \times {}^{13}C_1 \times {}^{13}C_1 \times {}^{13}C_1 = (13)^4$$

Hence, P (getting one card from each suit) = $\frac{(13)^4}{{}^{52}C_4}$

Example 8. Three letters are dictated to three persons and an envelope is addressed to each of them. The letters are inserted into the envelopes at random so that each envelope contains exactly one letter. Find the probability that atleast one letter is in its proper envelope. *NCERT, Competency Based Que*

Sol. Number of ways in which three letters put into three envelopes $= 3! = 6$

Number of ways in which one letter put in correct envelope $= {}^3C_1 \times 1 = 3$

Number of ways in which two letters put in correct envelope $= 1$

Number of ways in which atleast one letter put in correct envelope $= 3 + 1 = 4$

Thus, probability that atleast one letter put in correct envelope $= \frac{4}{6} = \frac{2}{3}$

Example 9. The number lock of a suitcase has four wheels, each labelled with 10 digits i.e. from 0 to 9. The lock opens with a sequence of four digits with no repeats. What is the probability of a person getting the right sequence to open the suitcase?

NCERT Competency Based Que

Sol. There are total 10 digits from 0 to 9.

Since, the digits cannot be repeated, so the first place may be filled in 10 ways, second place in 9 ways, third place in 8 ways and fourth place in 7 ways.

$\because$ Number of possible outcomes $=10\times9\times8\times7=5040$

Now, lock can be opened only in 1 way.

$\therefore \quad n(E)=1$

Hence, the probability of opening the lock

$$=\frac{n(E)}{n(S)}=\frac{1}{5040}$$

Example 10. On her vacations, Veena visits four cities (*A*, *B*, *C* and *D*) in a random order. What is the probability that she visits

Competency Based Que

(i) *A* before *B*?

(ii) *A* before *B* and *B* before *C*?

(iii) *A* first and *B* last?

(iv) *A* either first or second?

(v) *A* just before *B*?

NCERT

Sol. Number of ways, in which Veena can visit four cities *A*, *B*, *C* and *D*, is 4! i.e. 24.

$\therefore \quad n(S)=24$

Clearly, sample space for this experiment is

$$S=\begin{Bmatrix} ABCD, ABDC, ACBD, ACDB, ADBC, ADCB, \\ BACD, BADC, BCAD, BCDA, BDAC, BDCA, \\ CABD, CADB, CBAD, CBDA, CDAB, CDBA, \\ DABC, DACB, DCAB, DCBA, DBAC, DBCA \end{Bmatrix}$$

(i) Let E_1 be the event that Veena visits *A* before *B*. Then,

$$E_1=\begin{Bmatrix} ABCD, ABDC, ACBD, ACDB, ADBC, ADCB, \\ CABD, CADB, CDAB, DABC, DACB, DCAB \end{Bmatrix}$$

$\Rightarrow \quad n(E_1)=12$

$\therefore P(\text{she visits } A \text{ before } B)=P(E_1)=\frac{n(E_1)}{n(S)}$

$$=\frac{12}{24}=\frac{1}{2}$$

(ii) Let E_2 be the event that she visits *A* before *B* and *B* before *C*.

Then, $\quad E_2=\{ABCD, ABDC, ADBC, DABC\}$

$\Rightarrow \quad n(E_2)=4$

$\therefore P(\text{she visits } A \text{ before } B \text{ and } B \text{ before } C)=P(E_2)$

$$=\frac{n(E_2)}{n(S)}=\frac{4}{24}=\frac{1}{6}$$

(iii) Let E_3 be the event that she visits *A* first and *B* last.

Then, $E_3=\{ACDB, ADCB\} \Rightarrow n(E_3)=2$

$\therefore P(\text{she visits } A \text{ first and } B \text{ last})=P(E_3)$

$$=\frac{n(E_3)}{n(S)}=\frac{2}{24}=\frac{1}{12}$$

(iv) Let E_4 be the event that she visits *A* either first or second. Then,

$$E_4=\begin{Bmatrix} ABCD, ABDC, ACBD, ACDB, ADBC, ADCB, \\ BACD, BADC, CABD, CADB, DABC, DACB \end{Bmatrix}$$

$\Rightarrow \quad n(E_4)=12$

Hence, *P* (she visits *A* either first or second)

$$=P(E_4)=\frac{n(E_4)}{n(S)}=\frac{12}{24}=\frac{1}{2}$$

(v) Let E_5 be the event that she visits *A* just before *B*. Then,

$E_5=\{ABCD, ABDC, CABD, CDAB, DABC, DCAB\}$

$\Rightarrow \quad n(E_5)=6$

Hence, *P* (she visits *A* just before *B*)

$$=P(E_5)=\frac{n(E_5)}{n(S)}=\frac{6}{24}=\frac{1}{4}$$

Example 11. There are four men and six women on the city council. If one council member is selected for a committee at random, then how likely is it that it is a woman?

Sol. Total number of persons = 4 (men) + 6 (women) = 10

Since, a person is selected at random, therefore total number of possible outcomes = 10

Now, let *A* be the event of selecting a woman.

Then, number of outcomes favourable to *A* are 6.

Hence, the probability of selecting a woman,

$$P(A)=\frac{6}{10}=\frac{3}{5}$$

Example 12. A bag contains 9 red, 7 white and 4 black balls. A ball is drawn at random. Find the probability that the ball drawn is

(i) red (ii) white (iii) not black

Sol. Total number of balls = 9 red + 7 white + 4 black = 20

$\Rightarrow$ Total number of possible outcomes, $n(S)=20$

(i) Let E_1 be the event of getting a red ball.

Then, $n(E_1)=9$

Now, *P* (getting a red ball) = $P(E_1)$

$$=\frac{n(E_1)}{n(S)}=\frac{9}{20}$$

(ii) Let E_2 be the event of getting a white ball.
Then, $n(E_2)=7$
Now, P (getting a white ball) = $P(E_2)$
$$=\frac{n(E_2)}{n(S)}=\frac{7}{20}$$

(iii) Let E_3 be the event of getting a non-black ball.
Then, $n(E_3)=9+7=16$
Now, P (getting a non-black ball) = $P(E_3)$
$$=\frac{n(E_3)}{n(S)}=\frac{16}{20}=\frac{4}{5}$$

Quick Test 2

1. Let a sample space be $S=\{\omega_1, \omega_2, \ldots, \omega_6\}$. Which of the following assignments of probabilities to each outcome are valid?

Outcomes	ω_1	ω_2	ω_3	ω_4	ω_5	ω_6
(i)	$\frac{1}{6}$	$\frac{1}{6}$	$\frac{1}{6}$	$\frac{1}{6}$	$\frac{1}{6}$	$\frac{1}{6}$
(ii)	1	0	0	0	0	0
(iii)	$\frac{1}{8}$	$\frac{2}{3}$	$\frac{1}{3}$	$\frac{1}{3}$	$-\frac{1}{4}$	$-\frac{1}{3}$
(iv)	$\frac{1}{12}$	$\frac{1}{12}$	$\frac{1}{6}$	$\frac{1}{6}$	$\frac{1}{6}$	$\frac{3}{2}$
(v)	0.1	0.2	0.3	0.4	0.5	0.6

2. Three coins are tossed once. Find the probability of getting *Competency Based Que*
(i) 2 heads (ii) no head
(iii) 3 tails (iv) atmost two tails

3. Two dice are thrown simultaneously. Find the probability of getting
(i) an even number as the sum.
(ii) a doublet.
(iii) a doublet of even number.
(iv) a multiple of 3 as the sum.

4. A letter is chosen at random from the word 'ASSASSINATION'. Find the probability that letter is *Competency Based Que*
(i) a vowel. (ii) a consonant.

5. 4 cards are drawn from a well-shuffled deck of 52 cards. What is the probability of obtaining 3 diamonds and one spade? *NCERT*

6. A committee of two persons is selected from two men and two women. What is the probability that the committee will have *Competency Based Que*
(i) no man?
(ii) one man?
(iii) two men?

7. A bag contains 8 red, 3 white and 9 blue balls. If three balls are drawn at random, determine the probability that
(i) all the three balls are of blue colour.
(ii) all the balls are of different colours.
(iii) one is red and two are white.

8. A group consists of 3 men, 2 women and 4 children. If four persons are selected at random, then find the probability of selecting
(i) 1 man, 1 woman and 2 children.
(ii) exactly 2 children.
(iii) 2 women.

9. In a relay race, there are five teams A, B, C, D and E.
(i) What is the probability that A, B and C finish first, second and third, respectively?
(ii) What is the probability that A, B and C are first three to finish (in any order), assume that all finishing orders are equally likely? *NCERT, Competency Based Que*

10. What is the probability that in a leap year chosen at random will contain 53 Tuesday or 53 Wednesday? *Competency Based Que*

Answers

1. (i) and (ii)

2. (i) $\frac{3}{8}$ (ii) $\frac{1}{8}$ (iii) $\frac{1}{8}$ (iv) $\frac{7}{8}$

3. (i) $\frac{1}{2}$ (ii) $\frac{1}{6}$ (iii) $\frac{1}{12}$ (iv) $\frac{1}{3}$

4. (i) $\frac{6}{13}$ (ii) $\frac{7}{13}$

5. $\frac{^{13}C_3 \times {^{13}C_1}}{^{52}C_4}$

6. (i) $\frac{1}{6}$ (ii) $\frac{2}{3}$ (iii) $\frac{1}{6}$

7. (i) $\frac{7}{95}$ (ii) $\frac{18}{95}$ (iii) $\frac{2}{95}$

8. (i) $\frac{2}{7}$ (ii) $\frac{10}{21}$ (iii) $\frac{1}{6}$

9. (i) $\frac{1}{60}$ (ii) $\frac{1}{10}$

10. $\frac{3}{7}$

For Solutions, See QR Code on Pg. 412

14.4 Addition Theorems on Probability

If A and B are two events associated with a random experiment, then

$$P(A \cup B) = P(A) + P(B) - P(A \cap B)$$

i.e. $P(A \text{ or } B) = P(A) + P(B) - P(A \text{ and } B)$

It is known as **addition law of probability** for two events.

Also,

- when events A and B are mutually exclusive, then

$$P(A \cup B) = P(A) + P(B)$$

- when events A and B are mutually exclusive and exhaustive, then

$$P(A \cup B) = P(A) + P(B) = 1$$

Probability of the Event A or B or C If A, B and C are three events associated with a random experiment, then

$$P(A \cup B \cup C) = P(A) + P(B) + P(C) - P(A \cap B) - P(B \cap C) - P(C \cap A) + P(A \cap B \cap C)$$

If A, B and C are mutually exclusive events

i.e. $A \cap B = \phi, B \cap C = \phi, A \cap C = \phi, A \cap B \cap C = \phi$.

Then, $P(A \cup B \cup C) = P(A) + P(B) + P(C)$

Probability of Complementary Event Let E be any event and $\bar{E}$ (or E') be its complementary event.

Then, $P(\bar{E}) = 1 - P(E)$

The odds in favour of occurrence of the event E are defined by $P(E) : P(\bar{E})$ and the odds against the occurrence of event E are defined by $P(\bar{E}) : P(E)$.

TEACHER TIPS

- For any two events A and B, $A \subseteq B$ $\Rightarrow P(A) \leq P(B)$
- For any two events A and B,
  (i) $P(A - B) = P(A) - P(A \cap B)$ or $P(A \cap \bar{B}) = P(A) - P(A \cap B)$
  (ii) $P(\bar{A} \cap \bar{B}) = P(\overline{A \cup B})$
- $P(A \cap B) \leq P(A)$ and $P(A \cap B) \leq P(B)$ [if these conditions hold true, then events are said to be consistently defined]

Example 1. **If A and B are two events such that $P(A) = \frac{1}{4}$, $P(B) = \frac{1}{2}$ and $P(A \cap B) = \frac{1}{8}$. Then, find P (not A and not B).**

Sol. $P(\text{not } A \text{ and not } B) = P(\bar{A} \cap \bar{B})$

$$= P(\overline{A \cup B}) = 1 - P(A \cup B)$$

$$= 1 - [P(A) + P(B) - P(A \cap B)]$$

$$= 1 - \left[\frac{1}{4} + \frac{1}{2} - \frac{1}{8}\right] = 1 - \left[\frac{2 + 4 - 1}{8}\right]$$

$$= 1 - \frac{5}{8} = \frac{3}{8}$$

Example 2. **The accompanying Venn diagram shows three events A, B and C and also the probabilities of the various intersections [for instance $P(A \cap B) = 0.07$]. Determine.**

(i) $P(A)$ **(ii) $P(B \cap \bar{C})$**

(iii) $P(A \cup B)$ **(iv) $P(A \cap \bar{B})$** **(v) $P(B \cap C)$**

(vi) Probability of exactly one of the three events occurs. NCERT Exemplar

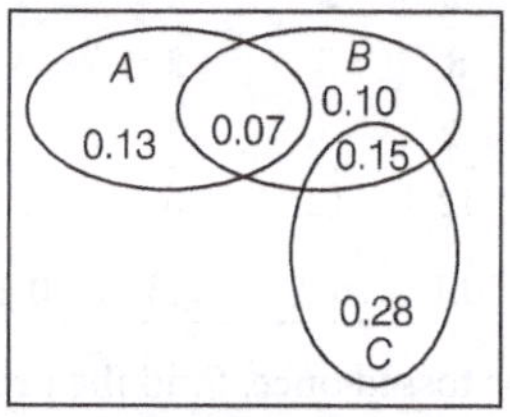

Sol. (i) $P(A) = 0.13 + 0.07 = 0.20$

(ii) $P(B \cap \bar{C}) = P(B) - P(B \cap C)$
$= (0.07 + 0.10 + 0.15) - (0.15)$
$= 0.17$

(iii) $P(A \cup B) = P(A) + P(B) - P(A \cap B)$
$= 0.20 + (0.07 + 0.10 + 0.15) - 0.07$
$= 0.20 + 0.25 = 0.45$

(iv) $P(A \cap \bar{B}) = P(A) - P(A \cap B) = 0.20 - 0.07 = 0.13$

(v) $P(B \cap C) = 0.15$

(vi) P (exactly one of the three events occurs)
$= P(A \text{ only}) + P(B \text{ only}) + P(C \text{ only})$
$= 0.13 + 0.10 + 0.28$
$= 0.51$

Example 3. **A and B are two events such that $P(A) = 0.54$, $P(B) = 0.69$ and $P(A \cap B) = 0.35$. Find**

(i) $P(A \cup B)$ **(ii) $P(A' \cap B')$**

(iii) $P(A \cap B')$ **(iv) $P(B \cap A')$** NCERT

Sol. We have, $P(A) = 0.54$, $P(B) = 0.69$ and $P(A \cap B) = 0.35$

(i) $P(A \cup B) = P(A) + P(B) - P(A \cap B)$
$= 0.54 + 0.69 - 0.35 = 0.88$

(ii) $P(A' \cap B') = P[(A \cup B)'] = 1 - P(A \cup B)$
$[\because P(E') = 1 - P(E)]$
$= 1 - 0.88 = 0.12$

(iii) $P(A \cap B') = P(A) - P(A \cap B) = 0.54 - 0.35 = 0.19$

(iv) $P(B \cap A') = P(B) - P(A \cap B) = 0.69 - 0.35 = 0.34$

Example 4. **Events E and F are such that $P(\text{not } E \text{ or not } F) = 0.25$. State whether E and F are mutually exclusive.** NCERT

Sol. We have, $P(\bar{E} \cup \bar{F}) = 0.25$

$\Rightarrow \quad P(\overline{E \cap F}) = 0.25 \quad [\because \overline{E \cap F} = \bar{E} \cup \bar{F}]$

$\Rightarrow \quad 1 - P(E \cap F) = 0.25$

$\Rightarrow \quad P(E \cap F) = 1 - 0.25 = 0.75 \neq 0$

$\therefore$ E and F are not mutually exclusive events.

Example 5. **Check whether the following probabilities $P(A)$ and $P(B)$ are consistently defined.**

(i) $P(A) = 0.5$, $P(B) = 0.7$, $P(A \cap B) = 0.6$

(ii) $P(A) = 0.5$, $P(B) = 0.4$, $P(A \cup B) = 0.8$ NCERT

Sol. (i) We have, $P(A) = 0.5$, $P(B) = 0.7$ and $P(A \cap B) = 0.6$

$\therefore P(A \cap B) \not\leq P(A)$ and $P(A \cap B) \leq P(B)$

But for the given probabilities $P(A \cap B) \not\leq P(A)$.

So, given probabilities are not consistently defined.

(ii) We have, $P(A) = 0.5$, $P(B) = 0.4$ and $P(A \cup B) = 0.8$

Clearly, $P(A \cap B) = P(A) + P(B) - P(A \cup B)$

$= 0.5 + 0.4 - 0.8 = 0.1$

Here, $P(A \cap B) \leq P(A)$ and $P(A \cap B) \leq P(B)$.

So, given probabilities are consistently defined.

Example 6. **Find the probability of getting atmost two tails or atleast two heads in a toss of three coins.**

Sol. Clearly, the sample space

$S = \{HHH, HHT, HTH, THH, TTH, THT, HTT, TTT\}$

Let A be the event of getting atmost two tails and B be the event of getting atleast two heads.

Then, $A = \{HHH, HHT, HTH, THH, TTH, THT, HTT\}$

and $B = \{HHT, HTH, THH, HHH\}$

$P(A) = \frac{7}{8}$, $P(B) = \frac{4}{8}$ and $P(A \cap B) = \frac{4}{8}$

Now, required probability $= P(A \cup B)$

$= P(A) + P(B) - P(A \cap B)$

$= \frac{7}{8} + \frac{4}{8} - \frac{4}{8} = \frac{7}{8}$

Example 7. **A coin is tossed and a die is thrown. Find the probability that the outcome will be a head or a number greater than 4.** Competency Based Que

Sol. When a coin and a die are thrown, then sample space is

$S = \{H1, H2, H3, H4, H5, H6, T1, T2, T3, T4, T5, T6\}$

$\therefore \quad n(S) = 12$

Let A: Event of getting a head.

$\Rightarrow \quad A = \{H1, H2, H3, H4, H5, H6\}$

$\therefore \quad n(A) = 6$

Then, $P(A) = \frac{n(A)}{n(S)} = \frac{6}{12}$

B: Event of getting a number greater than 4.

$\Rightarrow \quad B = \{H5, H6, T5, T6\}$

$\therefore \quad n(B) = 4$

Then, $P(B) = \frac{n(B)}{n(S)} = \frac{4}{12}$

Also, $A \cap B$ = Event of getting a head and a number greater than 4 = $\{H5, H6\}$

$\therefore \quad n(A \cap B) = 2$

Then, $P(A \cap B) = \frac{2}{12}$

Now, required probability,

$P(A \cup B) = P(A) + P(B) - P(A \cap B)$

$= \frac{6}{12} + \frac{4}{12} - \frac{2}{12} = \frac{8}{12} = \frac{2}{3}$

Example 8. **Find the probability that, when a pack of 7 cards is drawn from a well-shuffled deck of 52 cards, it contains**

(i) all kings. (ii) 3 kings. (iii) atleast 3 kings. NCERT

Sol. Let S be the sample space associated with the given random experiment. Then, $n(S) = {}^{52}C_7$

(i) Let A be the event that it contains all kings.

Then, $n(A) = {}^4C_4 \times {}^{48}C_3$

$$\Rightarrow \quad P(A) = \frac{{}^4C_4 \times {}^{48}C_3}{{}^{52}C_7}$$

$$= \frac{1 \times \frac{48 \times 47 \times 46}{3 \times 2 \times 1}}{\frac{52 \times 51 \times 50 \times 49 \times 48 \times 47 \times 46}{7 \times 6 \times 5 \times 4 \times 3 \times 2 \times 1}}$$

$$= \frac{1}{7735} \quad [\because {}^4C_4 = 1]$$

(ii) Let B be the event that it contains 3 kings.

Then, $n(B) = {}^4C_3 \times {}^{48}C_4$

$$\Rightarrow P(B) = \frac{4 \times {}^{48}C_4}{{}^{52}C_7}$$

$$= \frac{4 \times \frac{48 \times 47 \times 46 \times 45}{4 \times 3 \times 2 \times 1}}{\frac{52 \times 51 \times 50 \times 49 \times 48 \times 47 \times 46}{7 \times 6 \times 5 \times 4 \times 3 \times 2 \times 1}}$$

$$= \frac{9}{1547} \quad [\because {}^4C_3 = {}^4C_1 = 4]$$

(iii) Let C be the event that it contains atleast 3 kings.

Then, $P(C) = P(A \cup B)$

$= P(A) + P(B)$

[$\because$ A and B are mutually exclusive]

$= \frac{1}{7735} + \frac{9}{1547} = \frac{46}{7735}$

Example 9. A card is drawn from an ordinary pack and a gambler bets that it is spade or an ace. What are the odds against the winning his bet? Competency Based Que

Sol. Let S be the sample space associated with the given random experiment. Then, $n(S)=52$.

Now, let A be the event of getting a spade and B be the event of getting an ace. $A\cap B$ denotes the event of getting an ace of spade.

Then, $n(A)=13$ and $n(B)=4$

$n(A\cap B)=1$

Thus, we have $P(A)=\frac{13}{52}$, $P(B)=\frac{4}{52}$ and

$P(A\cap B)=\frac{1}{52}$

Now, P (winning the bet) = P (getting a spade or an ace)

$$=P(A\cup B)=P(A)+P(B)-P(A\cap B)$$

$$=\frac{13}{52}+\frac{4}{52}-\frac{1}{52}=\frac{16}{52}=\frac{4}{13}$$

$\Rightarrow$ P (not winning the bet) $=1-\frac{4}{13}=\frac{9}{13}$

Hence, odds against the winning his bet

$$=\frac{P(\text{not winning the bet})}{P(\text{winning the bet})}=\frac{\frac{9}{13}}{\frac{4}{13}}=\frac{9}{4}=9:4$$

Example 10. Three numbers are chosen from 1 to 20. Find the probability that they are not consecutive. NCERT Exemplar, Competency Based Que

Sol. Let S be the sample space associated with the given experiment. Then,

$n(S)={}^{20}C_3=1140$

Now, let E be the event that the chosen numbers are consecutive.

Then, $\bar{E}$ be the event that the chosen numbers are not consecutive.

Thus, $E=\{(1,2,3),(2,3,4),(3,4,5),\ldots,(18,19,20)\}$

$\Rightarrow$ $n(E)=18 \Rightarrow P(E)=\frac{18}{1140}=\frac{3}{190}$

Now, $P(\bar{E})=1-P(E)=1-\frac{3}{190}=\frac{187}{190}$

Hence, the required probability is $\frac{187}{190}$.

Example 11. In class XI of a school, 40% of the students study Mathematics and 30% study Biology. 10% of the class study both Mathematics and Biology. If a student is selected at random from the class, then find the probability that he will be studying Mathematics or Biology. NCERT

Sol. Let M be the event that selected student studies Mathematics and B be the event that selected student studies Biology.

Then, $P(M)=40\%=0.4$, $P(B)=30\%=0.3$

and $P(M\cap B)=10\%=0.1$

Now, P (he will study Mathematics or Biology)

$$=P(M\cup B)=P(M)+P(B)-P(M\cap B)$$

$$=0.4+0.3-0.1=0.6$$

Hence, the required probability is 0.6.

Example 12. The probability that a student will pass the final examination in both English and Hindi is 0.5 and the probability of passing neither is 0.1. If the probability of passing the English examination is 0.75, then what is the probability of passing the Hindi examination? NCERT, Competency Based Que

Sol. Let H and E be the events of students passing Hindi and English examination, respectively.

Given, probability of passing both examinations

$=P(H\cap E)=0.5$

and probability of passing neither examinations

$=P(H'\cap E')=P(H\cup E)'=0.1$

Then, $P(H\cup E)=1-P(H\cup E)'=1-0.1=0.9$

Also, it is given, that probability of passing English examination, $P(E)=0.75$

Now, consider $P(H\cup E)=P(H)+P(E)-P(H\cap E)$

$\Rightarrow$ $0.9=P(H)+0.75-0.5$

$\Rightarrow$ $0.9=P(H)+0.25$

$\Rightarrow$ $P(H)=0.9-0.25=0.65$

Hence, the probability of passing Hindi examination is 0.65.

Example 13. A card is drawn from a deck of 52 cards. Find the probability of getting a king or a heart or a red card. NCERT Exemplar

Sol. Let S be the sample space. Then, $n(S)=52$

Let A, B and C be the events of getting a king, a heart and a red card, respectively.

Then, $n(A)=4$, $n(B)=13$ and $n(C)=26$

$\therefore$ $P(A)=\frac{n(A)}{n(S)}=\frac{4}{52}=\frac{1}{13}$

$P(B)=\frac{n(B)}{n(S)}=\frac{13}{52}=\frac{1}{4}$, $P(C)=\frac{n(C)}{n(S)}=\frac{26}{52}=\frac{1}{2}$

Clearly, $(A\cap B)$ is the event of getting a king among hearts.

$(B\cap C)$ is the event of getting a heart among red cards.

$(A\cap C)$ is the event of getting a king among red cards.

$(A\cap B\cap C)$ is the event of getting a king of heart among the red cards.

$\Rightarrow \quad n(A \cap B) = 1,\ n(B \cap C) = 13,\ n(A \cap C) = 2$

and $\quad n(A \cap B \cap C) = 1$

$\therefore \quad P(A \cap B) = \dfrac{n(A \cap B)}{n(S)} = \dfrac{1}{52},$

$P(B \cap C) = \dfrac{n(B \cap C)}{n(S)} = \dfrac{13}{52} = \dfrac{1}{4},$

$P(A \cap C) = \dfrac{n(A \cap C)}{n(S)} = \dfrac{2}{52} = \dfrac{1}{26}$

and $\quad P(A \cap B \cap C) = \dfrac{n(A \cap B \cap C)}{n(S)} = \dfrac{1}{52}$

Now, P (getting a king or heart or red card)

$= P(A \text{ or } B \text{ or } C) = P(A \cup B \cup C)$

$= P(A) + P(B) + P(C) - P(A \cap B) - P(A \cap C)$

$\qquad - P(B \cap C) + P(A \cap B \cap C)$

$= \dfrac{1}{13} + \dfrac{1}{4} + \dfrac{1}{2} - \dfrac{1}{52} - \dfrac{1}{26} - \dfrac{1}{4} + \dfrac{1}{52} = \dfrac{28}{52} = \dfrac{7}{13}$

Quick Test 3

1. A and B are two events, such that $P(A) = 0.42, P(B) = 0.48$ and $P(A \text{ and } B) = 0.16$. Determine.

(i) $P(\text{not } A)$ (ii) $P(\text{not } B)$ (iii) P (A or B) *NCERT*

2. If E and F are events such that $P(E) = \dfrac{1}{4}$, $P(F) = \dfrac{1}{2}$ and $P(E \text{ and } F) = \dfrac{1}{8}$, then find $P(E \text{ or } F)$.

3. Check whether the following probabilities $P(A) = 0.6, P(B) = 0.5$ and $P(A \cup B) = 0.9$ are consistently defined.

4. Events E_1 and E_2 are such that $P(\text{not } E_1 \text{ or not } E_2) = 0.35$. State whether E_1 and E_2 are mutually exclusive.

5. The probabilities of three events shown in the Venn diagram is shown below.

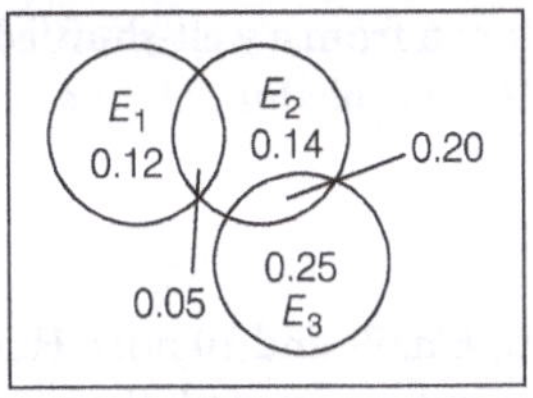

Determine.

(i) $P(E_2)$ (ii) $P(E_2 \cap \bar{E}_3)$

(iii) $P(E_1 \cup E_2)$ (iv) $P(E_1 \cap \bar{E}_2)$

6. If A and B are mutually exclusive events, such that $P(A) = 0.35$ and $P(B) = 0.45$, then find

(i) $P(A')$ (ii) $P(B')$

(iii) $P(A \cup B)$ (iv) $P(A \cap B)$

(v) $P(A \cap B')$ (vi) $P(A' \cap B')$ *NCERT Exemplar*

7. In an entrance test which is graded on the basis of two examinations, the probability of a randomly chosen student passing the first examination is 0.8 and the probability of passing the second examination is 0.7. The probability of passing atleast one of them is 0.95. What is the probability of passing both? *NCERT, Competency Based Que*

8. In any school examination, probability that Mohan passed in Chemistry is $\dfrac{1}{3}$ and the probability that he passed in Physics is $\dfrac{2}{5}$. If the probability of passing in both subjects is $\dfrac{1}{6}$, then what is the probability that Mohan will pass in atleast one of these subjects? *Competency Based Que*

9. The probability that a student will receive A, B, C and D grades are 0.50, 0.40, 0.20 and 0.15, respectively. Find the probability that a student will receive

(i) B or C grade.

(ii) atleast C grade. *Competency Based Que*

10. In a class of 60 students, 30 opted for NCC, 32 opted for NSS and 24 opted for both NCC and NSS. If one of these students is selected at random, then find the probability that *Competency Based Que*

(i) the student opted for NCC or NSS.

(ii) the student has opted for neither NCC nor NSS.

(iii) the student has opted NSS but not NCC.

Answers

1. (i) 0.58 (ii) 0.52 (iii) 0.74

2. $\dfrac{5}{8}$ **3.** Yes **4.** No

5. (i) 0.39 (ii) 0.19 (iii) 0.51 (iv) 0.12

6. (i) 0.65 (ii) 0.55 (iii) 0.8 (iv) 0 (v) 0.35 (vi) 0.2

7. 0.55 **8.** $\dfrac{17}{30}$ **9.** (i) 0.60 (ii) 0.35

10. (i) $\dfrac{19}{30}$ (ii) $\dfrac{11}{30}$ (iii) $\dfrac{2}{15}$

For Solutions, See QR Code on Pg. 412

Exam Pattern Questions

Objective Type Questions

Multiple Choice Questions

1 **A coin is tossed twice. If the second throw results in a tail, a die is thrown. The number of possible events of this experiment are**

(a) 12 (b) 14 (c) 16 (d) 18

2. **Which of the following is true?** **Competency Based Que**

I. The empty set ϕ is called an impossible event.

II. The whole sample space S is called the sure event.

(a) Only I is true. (b) Only II is true.
(c) Both I and II are true. (d) Both I and II are false.

3. **Consider the experiment of rolling a die. Let A be the event 'getting a prime number' and B be the event 'getting an odd number'. Then, which of the following is true?** **Competency Based Que**

I. A or $B = A \cup B = \{1, 2, 3\}$

II. A and $B = A \cap B = \{3, 5\}$

III. A but not $B = A - B = \{2\}$

IV. Not $A = A' = \{1, 5, 6\}$

(a) Only I is true. (b) Only II is true.
(c) II and III are true. (d) Only IV is true.

4. **A die is rolled. Let E be the event 'die shows 4' and F be the event 'die shows even number'. Then, E and F are**

(a) Mutually exclusive
(b) Exhaustive
(c) Mutually exclusive and exhaustive
(d) None of the above

5. **When an event A contains all the elements of sample space S, it represents an event called**

(a) Exhaustive event (b) Impossible event
(c) Mutually exclusive event (d) Sure event

6. **If A and B are two given events of an experiment where sample space is S such that $A \cup B = S$, then event A and B are**

(a) Exhaustive events (b) Mutually exclusive events
(c) Impossible events (d) Sure events

7. **From a deck of 52 cards, two cards are drawn. The probability that both are of same suit, is**

(a) $\frac{1}{2}$ (b) $\frac{1}{13}$

(c) $\frac{4}{17}$ (d) None of these

8. **There are 12 railways stations in a railway line. A train stops at 4 stations randomly. Then, probability that exactly three of them are consecutive is** **Competency Based Que**

(a) $\frac{17}{55}$ (b) $\frac{9}{25}$ (c) $\frac{8}{55}$ (d) $\frac{1}{55}$

9. **Three identical dice are thrown. The probability that same number will appear on each of the die is**

(a) $\frac{1}{18}$ (b) $\frac{1}{36}$ (c) $\frac{1}{6}$ (d) $\frac{5}{36}$

10. **A die is thrown, find the probability of the following events.** **Diksha Question**

A prime number will appear.

(a) $\frac{1}{2}$ (b) $\frac{1}{6}$ (c) $\frac{2}{3}$ (d) $\frac{1}{5}$

11. **A coin is tossed twice, what is the probability that atleast one tail occurs?**

(a) $\frac{1}{2}$ (b) $\frac{1}{4}$ (c) $\frac{2}{3}$ (d) $\frac{3}{4}$

12. **One card is drawn from a well-shuffled pack of 52 cards. What is the probability that a card will be a diamond?**

(a) $\frac{1}{4}$ (b) $\frac{4}{13}$ (c) $\frac{1}{52}$ (d) $\frac{2}{13}$

13. **A box contains 6 nails and 10 nuts. Half of the nails and half of the nuts are rusted. If one item is chosen at random, then the probability that it is rusted or is a nail, is ...K... . Here, K refers to**

(a) $\frac{3}{16}$ (b) $\frac{5}{16}$ (c) $\frac{11}{16}$ (d) $\frac{14}{16}$

14. **If $P(A \cup B) = P(A) + P(B)$, then the events A and B are**

(a) Exhaustive events (b) Impossible events
(c) Sure event (d) Mutually exclusive events

15. **A couple has two children. Find the probability that both children are male.** **Competency Based Que**

(a) $\frac{2}{3}$ (b) $\frac{3}{4}$ (c) $\frac{4}{5}$ (d) $\frac{1}{4}$

16. **An urn contains 6 balls of which two are red and four are black. Two balls are drawn at random. The probability that they are of different colours is** **Diksha Question**

(a) $\frac{2}{5}$ (b) $\frac{1}{15}$ (c) $\frac{8}{15}$ (d) $\frac{4}{15}$

17. Find the probability that when a hand of 7 cards are drawn from the well-shuffled deck of 52 cards, it contains all kings. **Competency Based Que**

(a) $\frac{2}{7735}$ (b) $\frac{1}{7735}$ (c) $\frac{3}{7753}$ (d) $\frac{1}{7753}$

18. Seven persons are to be seated in a row. The probability that two particular persons sit next to each other, is **Competency Based Que**

(a) $\frac{1}{3}$ (b) $\frac{1}{6}$ (c) $\frac{2}{7}$ (d) $\frac{1}{2}$

19. If A and B are any two events having $P(A \cup B) = \frac{1}{2}$ and $P(\overline{A}) = \frac{2}{3}$, then the probability of $\overline{A} \cap B$ is **Competency Based Que**

(a) $\frac{1}{2}$ (b) $\frac{2}{3}$ (c) $\frac{1}{6}$ (d) $\frac{1}{3}$

20. If M and N are any two events, the probability that at least one of them occurs is **NCERT Exemplar, Competency Based Que**

(a) $P(M) + P(N) - 2P(M \cap N)$ (b) $P(M) + P(N) - P(M \cap N)$

(c) $P(M) + P(N) + P(M \cap N)$ (d) $P(M) + P(N) + 2P(M \cap N)$

21. A fair coin with 1 marked one face and 6 on the other and a fair die both are tossed. Find the probability that the sum of numbers that turn up is 3.

(a) $\frac{1}{6}$ (b) $\frac{1}{12}$ (c) $\frac{2}{3}$ (d) $\frac{1}{4}$

22. The probability that the home team will win an upcoming football game is 0.77, the probability that it will tie the game is 0.08 and the probability that it will lose the game is **Competency Based Que**

(a) 0.15 (b) 0.23 (c) 0.92 (d) 1.15

23. If $P(A)$ is $\frac{3}{5}$. Find $P(\text{not } A)$.

(a) $\frac{2}{5}$ (b) $\frac{1}{5}$ (c) $\frac{3}{5}$ (d) $\frac{4}{5}$

24. 4 cards are drawn from a well-shuffled deck of 52 cards. Thus, the probability of obtaining 3 diamonds and one spade is

(a) $\frac{^{13}C_1 \times {}^{12}C_1}{^{52}C_1}$ (b) $\frac{^{13}C_3 \times {}^{13}C_1}{^{52}C_4}$

(c) $\frac{^{13}C_1 \times {}^{13}C_2}{^{52}C_4}$ (d) $\frac{^{13}C_3 \times {}^{13}C_2}{^{52}C_4}$

25. P and Q are two candidates seeking admission in NEET. The probability that P is selected is 0.5 and the probability that both P and Q are selected is atmost 0.3. Then, the probability of Q being selected is **Competency Based Que**

(a) ≤ 0.8 (b) ≥ 0.8

(c) $= 0.8$ (d) None of these

Assertion-Reason Based Questions

Directions (Q. Nos. 26-35) *In the questions given below are two statements labelled as Assertion (A) and Reason (R). In the context of the two statements, which one of the following is correct?*

(a) Both A and R are correct; R is the correct explanation of A.
(b) Both A and R are correct; R is not the correct explanation of A.
(c) A is correct; R is incorrect.
(d) R is correct; A is incorrect.

26. **Assertion** (A) In a throw of a dice getting a number less than 7 is a sure event.

Reason (R) If favourable outcomes of an event is equal to total outcomes of an experiment, then event is a sure event.

27. **Assertion** (A) In throw of 2 coins, the probability of getting one tail is $\frac{1}{2}$.

Reason (R) Probability of an event $= \frac{\text{Favourable outcomes for an event}}{\text{Total outcomes in experiment}}$ **Competency Based Que**

28. **Assertion** (A) If A and B are mutually exclusive events then, $A \cap B = \phi$.

Reason (R) If events A and B of an experiment, whose sample space is S are mutually exclusive and exhaustive events, then $A \cup B = S$ and $A \cap B = \phi$.

29. **Assertion** (A) If probability of winning a prize is $\frac{3}{4}$, then probability of not winning the prize is $\frac{1}{4}$.

Reason (R) For any event A, $P(A) + P(\overline{A}) = 1$.

30. **Assertion** (A) If A, B and C are three mutually exclusive and exhaustive events such that $P(B) = \frac{3}{2}P(A)$ and $P(C) = \frac{1}{2}P(B)$, then $P(A) = \frac{4}{13}$.

Reason (R) If A, B and C are mutually exclusive and exhaustive events, then $A \cup B \cup C = S$. **Competency Based Que**

31. **Assertion** (A) The probability of an event can be $\frac{3}{2}$.

Reason (R) For any event A, $0 \leq P(A) \leq 1$.

32. **Assertion** (A) From a group of 2 boys and 3 girls, two children are selected at random, then the event A of both selected children are girls in $A = \{(G_1\,G_2), (G_1\,G_3), (G_2\,G_3)\}$.

Reason (R) A subset of the sample space associated with a random experiment is called an event. **Competency Based Que**

33. Assertion (A) Consider an experiment of two dice are thrown and the sum of the numbers which come up on the dice is noted. A and B are two events such that $A =$ sum is even and $B =$ the sum is a multiple of 3. Then, A and B are mutually exclusive events.

Reason (R) Two events E and F associated with a random experiment are mutually exclusive if $E \cap F = \phi$.

34. Assertion (A) One card is drawn from a pack of 52 cards, then the probability that the card drawn is either black or king is $\frac{7}{13}$.

Reason (R) Four kings are in a pack of 52 cards.

Competency Based Que

35. Assertion (A) An urn contains 9 red, 7 white and 4 black balls. If two balls are drawn at random, then probability that one ball is white is $\frac{91}{190}$.

Reason (R) For any event E, $P(\overline{E}) = 1 - P(E)$

Competency Based Que

Case Study Based Questions

36. A group of tourist randomly selects a country for a sightseeing trip. The countries are Australia, China, Netherlands, Algeria, Romania and Argentina.

Based on the above information, answer the following questions.

(i) What is the number of total possible outcomes?

(ii) What is the probability that the name of selected country begins with R?

(iii) What is the probability that the name of selected country begins with letter A?

(iv) If $P(A) = \frac{1}{2}$, $P(B) = \frac{9}{10}$, $P(A \cup B) = 1$, then find $P(A \cap B)$.

37. Four friends Kanika, Kashvi, Priyanka and Preeti tossed three coins and reported their results as P, Q, R and W respectively.

Kanika (P) = got exactly two tail

Kashvi (Q) = got atleast two tail

Priyanka (R) = got atmost two tail

Preeti (W) = got exactly three tail

Based on the above information, answer the following questions. **Competency Based Que**

(i) Find the events which are mutually exclusive not exhaustive.

(ii) Find the events which are exhaustive.

(iii) Find the probability of event P.

(iv) Find the probability of W.

Subjective Questions

Very Short Answer Type Questions

1. An experiment has four possible outcomes A, B, C and D that are mutually exclusive. Explain why the following assignments of probabilities are not permissible?

(i) $P(A) = 0.12$, $P(B) = 0.63$, $P(C) = 0.45$, $P(D) = -0.20$

(ii) $P(A) = \frac{9}{120}$, $P(B) = \frac{45}{120}$, $P(C) = \frac{27}{120}$, $P(D) = \frac{46}{120}$

2. A coin is tossed twice. Describe the event

(i) getting no tail

(ii) getting exactly one tail

(iii) getting 2 tail

3. A bag contains 9 red and 12 white balls. One ball is drawn at random. Find the probability that the ball drawn is red.

4. In a lottery, there are 10 prizes and 25 blanks. Find the probability of getting a prize. **Competency Based Que**

5. The probability that the home team will win an upcoming game is 0.66, the probability that it will tie the game is 0.06, what is the probability that it will lose the game? **Competency Based Que**

Short Answer Type Questions

6. Five marbles are drawn from a bag which contains 7 blue marbles and 4 black marbles. What is the probability that

(i) all will be blue?

(ii) 3 will be blue and 2 black?

7. If A and B are two events associated with a random experiment such that $P(A) = 0.3$, $P(B) = 0.2$ and $P(A \cap B) = 0.1$, then find the value of $P(A \cap \overline{B})$.

Competency Based Que

8. A bag contains 9 discs of which 4 are red, 3 are blue and 2 are yellow. The discs are similar in shape and size. A disc is drawn at random from the bag. Calculate the probability that it will be
   (i) red (ii) yellow (iii) blue (iv) not blue
   (v) either red or blue **NCERT**

9. One card is drawn from a well-shuffled deck of 52 cards. Calculate the probability that the card will be
   (i) a diamond
   (ii) an ace
   (iii) a black card
   (iv) not a diamond
   (v) not an ace
   (vi) not a black card
   (vii) a red card. **NCERT Exemplar**

10. Two students Anil and Ashima appeared in an examination. The probability that Anil will qualify the examination is 0.05 and that Ashima will qualify the examination is 0.10. The probability that both will qualify the examination is 0.02. Find the probability that
   (i) both Anil and Ashima will not qualify the examination.
   (ii) atleast one of them will not qualify the examination.
   (iii) only one of them will qualify the examination. **NCERT**

11. In a single throw of three dice, find the probability of getting a total of 17 or 18.

Long Answer Type Questions

12. A team of medical students doing their internship have to assist during surgeries at a city hospital. The probabilities of surgeries rated as very complex, complex, simple or very simple respectively 0.15, 0.20, 0.31, 0.26 and 0.08. Find the probabilities that a particular surgery will be rated
   (i) complex or very complex.
   (ii) neither very complex nor very simple.
   (iii) routine or complex.
   (iv) routine or simple. **NCERT Exemplar**

13. From the employees of a company, 5 persons are elected to represent them in the managing committee of the company. Particulars of the five persons are as follows: **NCERT**

S.No.	Person	Age (in years)
1.	Male	30
2.	Male	33
3.	Female	46
4.	Female	28
5.	Male	41

A person is selected at random from this group to act as a spokespersons. What is the probability that a spokesperson will be either male or over 35 years? **Competency Based Que**

14. One of the four persons John, Rita, Aslam or Gurpreet will be promoted next month. Consequently, the sample space consists of four elementary outcomes S = {John promoted, Rita promoted, Aslam promoted, Gurpreet promoted. You are told that the chances of John's promotion is same as that of Gurpreet, Rita's chances of promotion are twice as likely as Johns. Aslam's chances are four times that of John.
   (i) Determine P (John promoted)
   P (Rita promoted);
   P (Aslam promoted)
   P (Gurpreet promoted)
   (ii) If A = {John promoted or Gurpreet promoted}, then find $P(A)$. **NCERT Exemplar**

15. Four cards are drawn at a time from a pack of 52 playing cards. Find the probability of getting all the four cards of the same unit.

16. A box contains 6 red, 4 white and 5 black balls. A person draws 4 balls from the box at random. Find the probability that among the balls drawn, there is atleast one ball of each colour. **Competency Based Que**

Hints & Solutions

Objective Type Questions

1. (b) When a coin is tossed twice, the outcomes are HH, HT, TH, TT. A coin is tossed twice and if the second throw results in a tail, a die is thrown. Out of the above 4 possibilities, on second throw tail occurs in two cases only i.e. HT, TT.

$\therefore \quad S = \{HH, TH, HT1, HT2, HT3, HT4, HT5, HT6, TT1, TT2, TT3, TT4, TT5, TT6\}$

$\therefore \quad n(S) = 14$

2. (c) The empty set ϕ is called an impossible event. The whole sample space S is called the sure event.

3. (c) If a die is rolled, then the possible outcomes are {1, 2, 3, 4, 5, 6}.

$\Rightarrow \quad S = \{1, 2, 3, 4, 5, 6\}$

$\because$ A is the event of 'getting a prime number' and B is the event of 'getting an odd number'.

So, $A = \{2, 3, 5\}$ and $B = \{1, 3, 5\}$

(i) A or $B = A \cup B = \{1, 2, 3, 5\}$

(ii) A and $B = A \cap B = \{3, 5\}$

(iii) A but not $B = A - B = \{2\}$

(iv) Not $A = A' = S - A = \{1, 4, 6\}$

4. (d) 1, 2, 3, 4, 5 and 6 are the possible outcomes when the die is thrown.

Therefore, $S = \{1, 2, 3, 4, 5, 6\}$

According to the given condition,

$E = \{4\}$

and $\quad F = \{2, 4, 6\}$

then $E \cap F = 4 \neq \phi$

$\therefore$ E and F are not mutually exclusive event.

5. (d) An even which contains all the elements of sample space S is called a sure event.

6. (a) If A and B are two given events of an experiment where sample space is S such that $A \cup B = S$, then event A and B are Exhaustive events.

7. (c) Number of suits = 4(with 13 cards each)

For any one suit, the number of ways to choose 2 cards,

$^{13}C_2 = 78$

Since, these are 4 suits, the total number of ways to choose 2 cards from the same suit is $= 4 \times 78 = 312$

Total number of cases $= {}^{52}C_2 = 1326$

$\therefore \quad$ Required probability $= \dfrac{312}{1326} = \dfrac{4}{17}$

8. (c) Let A be the event that 3 out of 4 stations are consecutive.

$\therefore \quad n(A) = 2 \times 8 + 8 \times 7 = 16 + 56 = 72$ (4 stations)

[out of 4 stations there are 2 ways in which three consecutive stations occurs]

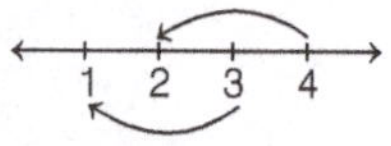

$\because \quad n(S) =$ Ways of selected 4 stations out of $12 = {}^{12}C_4$

$\therefore \quad P(A) = \dfrac{n(A)}{n(S)} = \dfrac{72}{495} = \dfrac{8}{55}$

9. (b) Total number of triplets $= 6^3 = 216$

Let $\quad A = \{(1, 1, 1), (2, 2, 2), ..., (6, 6, 6)\}$.

$\therefore \quad n(A) = 6$

$\therefore \quad P(A) = \dfrac{n(A)}{n(S)} = \dfrac{6}{216} = \dfrac{1}{36}$

10. (a) When a die is thrown, then possible outcome are 1, 2, 3, 4, 5, 6.

Sample space $(S) = \{1, 2, 3, 4, 5, 6\}$

Let E be the event getting prime numbers

$E = \{1, 3, 5\}, n(E) = 3$

$\therefore$ Required probability $P(E) = \dfrac{n(E)}{n(S)} = \dfrac{3}{6} = \dfrac{1}{2}$

11. (d) **Hint** Sample space of tossing a coin twice,

$S = \{HH, HT, TH, TT\}$

Let A be the event getting atleast one tail

$A = \{HT, TH, TT\}$

Ans. $\dfrac{3}{4}$

12. (a) There are 52 sample points in the sample space

$\therefore \quad n(S) = 52$

Let A be the event drawn card is a diamond card.

$n(A) = 13$

$\therefore$ Required probability, $P(A) = \dfrac{n(A)}{n(S)} = \dfrac{13}{52} = \dfrac{1}{4}$

13. (c) Total articles (nuts + nails) = 10 + 6 = 16

Number of rusted nails = 3

Number of rusted nuts = 5

Number of rusted articles = 8

Let N denote the event that the article drawn is a nail and R be the event denoting drawn article is rusted.

$P(N) = \dfrac{6}{16}$ and $P(R) = \dfrac{8}{16}$

Also, we have possibility that drawn nail is rusted.

$\therefore \quad P(N \cap R) = \frac{3}{16}$

$P(\text{nail or Rusted}) = P(N \cup R) = P(N) + P(R) - P(N \cap R)$

$= \frac{6}{16} + \frac{8}{16} - \frac{3}{16} = \frac{11}{16}$

14. (d) We know, $P(A \cup B) = P(A) + P(B) - P(A \cap B)$

If $\quad P(A \cup B) = P(A) + P(B)$

$\Rightarrow \quad P(A \cap B) = 0$

$\therefore$ Events are mutually exclusive.

15. (d) **Hint** Let S be the sample space,

$S = \{B_1B_2, B_1G_2, G_2B_2, G_1G_2\}$

and A be the event both children are males.

$A = \{B_1B_2\}$ **Ans.** $\frac{1}{4}$

16. Given, total number of balls = 6

$n(S) = 6$

Number of red balls = 2

and number of black balls = 4

$\therefore$ The probability that drawn balls are of different

colours $= \frac{{}^2C_1 \times {}^4C_1}{{}^6C_2} = \frac{2 \times 4}{\frac{6 \times 5}{2}} = \frac{8}{15}$ $\quad [\because {}^nC_1 = n]$

17. (b) **Hint** Let S be the sample space associated with the given random experiment. Then, $n(S) = {}^{52}C_7$

and 4 kings to be chosen out of 4 and 3 other cards are to be chosen out of remaining 48 cards.

$\therefore$ Required probability $= \frac{{}^4C_4 \times {}^{48}C_3}{{}^{52}C_7}$ **Ans.** $\frac{1}{7735}$

18. (c) Total number of ways to arrange 7 persons in a row $= 7!$

If two particular persons to be seated next to each other, then they form one group.

Now, the permutation of 6 persons = $6! \times 2!$

$\therefore$ Required probability $= \frac{6! \times 2!}{7!} = \frac{6! \times 2!}{7 \times 6!} = \frac{2}{7}$

19. (c) As, $P(A \cup B) = \frac{1}{2}$ and $P(\overline{A}) = \frac{2}{3}$

$P(A \cup B) = P(A) + P(B) - P(A \cap B)$

$P(A \cup B) = 1 - P(\overline{A}) + P(B) - P(A \cap B)$

$\frac{1}{2} = 1 - \frac{2}{3} + P(B) - P(A \cap B)$

$\frac{1}{2} = \frac{1}{3} + P(B) - P(A \cap B)$

$P(B) - P(A \cap B) = \frac{1}{6} \Rightarrow P(\overline{A} \cap B) = \frac{1}{6}$

20. (b) M and N are two events, probability that atleast one of them occurs is

$P(M \cup N) = P(M) + P(N) - P(M \cap N)$

21. (b) When a fair coin and a fair die both are tossed, then

Sample space = {(1, 1), (1, 2), (1, 3), (1, 4), (1, 5), (1, 6), (6, 1), (6, 2), (6, 3), (6, 4), (6, 5), (6, 6)}.

$n(S) = 12$

Let E be the event sum of numbers is 3.

$E = \{(1, 2)\}$

$n(E) = 1$

$\therefore$ Required probability $= \frac{1}{12}$

22. (a) Probability that the home team will win = 0.77

Probability that the home team will tie = 0.08

Probability that the home team will lose the game

$= 1 - (0.77 + 0.08) = 0.15$

23. (a) Given, $P(A) = \frac{3}{5}$

We know that

$P(A) + P(\text{not } A) = 1$

$P(\text{not } A) = 1 - P(A) = 1 - \frac{3}{5} = \frac{2}{5}$

24. (b) Let A and B be the events getting 3 diamonds and one spade.

Then, $n(A) = {}^{13}C_3$ and $n(B) = {}^{13}C_1$

Let S be the sample space associated with the given random experiment. Then, $n(S) = {}^{52}C_4$

Now, required probability $= \frac{{}^{13}C_3 \times {}^{13}C_1}{{}^{52}C_4}$

25. (a) Given, let the probability that P is selected, $P(A) = 0.5$ and the probability that P and Q both are selected,

$P(A \cap B) \le 0.3$

Now,

$P(A \cup B) \le 1$

$\Rightarrow P(A) + P(B) - n(A \cap B) \le 1$

$\Rightarrow \quad 0.5 + P(B) - 0.3 \le 1$

$\Rightarrow \quad P(B) \le 1 - 0.5 + 0.3$

$\Rightarrow \quad P(B) \le 0.8$

Hence, the probability of a being selected is less than equal to 0.8.

26. (a) Sample space of throwing a die is {1, 2, 3, 4, 5, 6}.

Since, we always obtain a number less than 7, therefore it is a sure event.

27. (a) Sample space of throwing 2 coins is $\{HH, HT, TH, TT\}$.

Favourable outcomes (getting one tail) = $\{HT, TH\}$

Number of favourable outcomes = 2

$\therefore$ Required probability

$$= \frac{\text{Number of favourable outcomes}}{\text{Total outcomes}} = \frac{2}{4} = \frac{1}{2}$$

28. (b) Two events A and B are mutually exclusive, if $A \cap B = \phi$

Two events A and B are mutually exclusive and exhaustive, if $A \cup B = S$ and $A \cap B = \phi$

29. (a) Probability of winning a prize, $P(A) = \frac{3}{4}$

We know that $P(A) + P(\overline{A}) = 1$

$$\therefore \quad P(\overline{A}) = 1 - P(A) = 1 - \frac{3}{4} = \frac{1}{4}$$

Hence, probability of not winning the prize is $\frac{1}{4}$.

30. (a) Given, $P(B) = \frac{3}{2}P(A)$ and $P(C) = \frac{1}{2}P(B)$

Let $3P(A) = 2P(B) = 4P(C) = p$

Since, A, B and C are mutually exclusive and exhaustive events.

$$\therefore \quad A \cup B \cup C = S \Rightarrow P(A \cup B \cup C) = 1$$

$$\Rightarrow \quad \frac{p}{3} + \frac{p}{2} + \frac{p}{4} = 1 \Rightarrow \frac{(4+6+3)p}{12} = 1$$

$$\Rightarrow \quad \frac{13p}{12} = 1 \Rightarrow p = \frac{12}{13}$$

$$\because \quad 3P(A) = p$$

$$\therefore \quad P(A) = \frac{12}{3 \times 13} = \frac{4}{13}$$

31. (d) We know that for any event A,

$$0 \le P(A) \le 1$$

Therefore, the probability of an event cannot be $\frac{3}{2}$.

32. (a) Sample space of the experiment is

$$S = \{(B_1\ B_2), (B_1\ G_1), (B_1 G_2), (B_1\ G_3), (B_2\ G_1), (B_2\ G_2), (B_2\ G_3), (G_1\ G_2), (G_1\ G_3), (G_2\ G_3)\}$$

A = Both the children are girls

$$= \{(G_1\ G_2), (G_1\ G_3), (G_2\ G_3)\}$$

33. (d) $\because A = \{(1, 1), (1, 3), (1, 5), (2, 2), (2, 4), (2, 6), (3, 1)$
$(3, 3), (3, 5), (4, 2), (4, 4), (4, 6), (5, 1), (5, 3), (5, 5)$
$(6, 2), (6, 4), (6, 6)\}$

$B = \{(1, 2), (2, 1), (1, 5), (5, 1), (2, 4), (4, 2), (3, 3), (3, 6)$
$(6, 3), (4, 5), (5, 4), (6, 6)\}$

$\because A \cap B = \{(1, 5), (5, 1), (2, 4), (4, 2), (3, 3), (6, 6)\}$

$A \cap B \ne \phi \Rightarrow A$ and B are not mutually exclusive events.

Hence, Assertion is incorrect but Reason is correct.

34. (b) There are 26 black cards including 2 black kings and there are 2 more kings (red colour).

$\therefore$ Total number of favourable cards = 28

$$\therefore \text{ Required probability} = \frac{{}^{28}C_1}{{}^{52}C_1} = \frac{28}{52} = \frac{7}{13}$$

35. (b) Here, 7 are white balls and 13 are non-white balls.

$$\therefore \text{ Required probability} = \frac{{}^{7}C_1 \times {}^{13}C_1}{{}^{20}C_2} = \frac{7 \times 13}{\frac{20 \times 19}{2}} = \frac{91}{190}$$

36. (i) Total possible outcomes [Australia, China, Netherlands, Algeria, Romania, Argentina]

$\therefore$ Total number of possible outcomes = 6

(ii) Total number of possible outcomes = 6

and favourable outcome = Romania

$\Rightarrow$ Number of favourable outcomes = 1

$$\therefore \quad \text{Required probability} = \frac{1}{6}$$

(iii) Total number of possible outcomes, $n(S) = 6$

Favourable outcomes = {Australia, Algeria, Argentina}

$$\Rightarrow \quad n(E) = 3$$

$$\therefore \text{ Required probability} = \frac{n(E)}{n(S)} = \frac{3}{6} = \frac{1}{2}$$

(iv) We know that

$$P(A \cup B) = P(A) + P(B) - P(A \cap B)$$

$$\Rightarrow \quad 1 = \frac{1}{2} + \frac{9}{10} - P(A \cap B)$$

$$\Rightarrow \quad P(A \cap B) = \frac{1}{2} + \frac{9}{10} - 1 = \frac{5 + 9 - 10}{10} = \frac{4}{10} = \frac{2}{5}$$

37. The sample space of tossing three coins,

$$S = \{HHH, HHT, HTH, THH, HTT, TTH, THT, TTT\}$$

According to the given conditions,

$$P = \{TTH, THT, HTT\}$$

$$Q = \{TTT, TTH, THT, HTT\}$$

$$R = \{TTH, THT, HTT, THH, HHT, HTH, HHH\}$$

$$W = \{TTT\}$$

(i) We find that

$$P \cap W = \phi \text{ and } R \cap W = \phi$$

$\therefore$ P and W are mutually exclusive

Also, R and W are mutually exclusive.

But $P \cup W \ne S$ and $R \cup W = S$

$\therefore$ Events P and W are mutually exclusive but not exhaustive.

(ii) $\because R \cup W = S$

$\therefore$ Event R and W are exhaustive.

(iii) Total number of outcomes, $n(S) = 8$

Number of favourable outcomes, $n(P) = 3$

$\therefore$ Required probability $= \frac{3}{8}$

(iv) Number of favourable outcomes, $n(W) = 1$

$\therefore$ Required probability $= \frac{1}{8}$

Subjective Questions

1. (i) Since, $P(D) = -0.20$

This is not possible as $0 \le P(A) \le 1$ for any event A.

(ii) $P(S) = P(A \cup B \cup C \cup D) = P(A) + P(B) + P(C) + P(D)$

$$= \frac{9}{120} + \frac{45}{120} + \frac{27}{120} + \frac{46}{120} = \frac{127}{120} \ne 1$$

Hence, both (i) and (ii) are not permissible.

2. A coin is tossed twice

$\therefore$ $S = \{HH, HT, TH, TT\}$

(i) getting no tail $= \{HH\}$, simple event

(ii) getting exactly one tail $= \{HT, TH\}$, compound event

(iii) getting 2 tail $= \{TT\}$, simple event

3. Number of red balls $= 9$, Number of white balls $= 12$

Total number of balls $= 9 + 12 = 21$

$$P(\text{Red ball is drawn}) = \frac{\text{Number of favourable outcomes}}{\text{Total number of outcomes}} = \frac{9}{21} = \frac{3}{7}$$

4. Number of prizes $= 10$, Number of blanks $= 25$

Total number of possible outcomes $= 10 + 25 = 35$

$\therefore$ Probability of getting a prize

$$= \frac{\text{Number of favourable outcomes}}{\text{Total number of outcomes}} = \frac{10}{35} = \frac{2}{7}$$

5. Probability that the home team will win $= 0.66$

Probability that the home team will tie $= 0.06$

Probability that the home team will lose the game

$$= 1 - (0.66 + 0.06) = 1 - 0.72 = 0.28$$

6. Given, number of blue marbles $= 7$

Number of black marbles $= 4$

Total number of marbles $= 7 + 4 = 11$

(i) Favourable outcomes $= {}^7C_5$

Total outcomes $= {}^{11}C_5$

$$\text{Probability (all will be blue)} = \frac{{}^7C_5}{{}^{11}C_5} = \frac{\frac{7!}{5!\,2!}}{\frac{11!}{5!\,6!}}$$

$$= \frac{\frac{7 \times 6}{2}}{\frac{11 \times 10 \times 9 \times 8 \times 7}{5 \times 4 \times 3 \times 2 \times 1}} = \frac{21}{11 \times 3 \times 2 \times 7} = \frac{1}{22}$$

(ii) Favourable outcomes $= {}^7C_3 \times {}^4C_2$

Probability (3 will be blue and 2 black)

$$= \frac{{}^7C_3 \times {}^4C_2}{{}^{11}C_5} = \frac{\frac{7!}{3!\,4!} \times \frac{4!}{2!\,2!}}{\frac{11!}{5!\,6!}}$$

$$= \frac{\frac{7 \times 6 \times 5}{3 \times 2} \times \frac{4 \times 3}{2}}{11 \times 3 \times 2 \times 7} = \frac{7 \times 5 \times 2 \times 3}{11 \times 3 \times 2 \times 7} = \frac{5}{11}$$

7. Given that $P(A) = 0.3, P(B) = 0.2$

and $P(A \cap B) = 0.1$

$\therefore$ $P(A \cap \overline{B}) = P(A) - P(A \cap B) = 0.3 - 0.1 = 0.2$

8. Given, number of red discs $= 4$

Number of blue discs $= 3$

Number of yellow discs $= 2$

Total number of possible outcomes, $n(S) = 9$

Let A be the event that the disc is red,

B be the event that the disc is blue

and C be the event that the disc is yellow.

So, $n(A) = 4$, $n(B) = 3$, $n(C) = 2$

(i) Probability of drawing red discs $= \frac{n(A)}{n(S)} = \frac{4}{9}$

(ii) Probability of drawing yellow discs $= \frac{n(C)}{n(S)} = \frac{2}{9}$

(iii) Probability of drawing blue discs $= \frac{n(B)}{n(S)} = \frac{3}{9} = \frac{1}{3}$

(iv) $P(\text{not blue}) = 1 - P(\text{blue}) = 1 - P(B) = 1 - \frac{1}{3} = \frac{2}{3}$

(v) Either red or blue describes the set (A or B).

Since, we can take either red or blue disc there is nothing common in A on B.

$\therefore$ A and B are mutually exclusive.

Hence, $A \cap B = \phi$

So, $P(A \cap B) = 0$

We know that

$$P(A \cup B) = P(A) + P(B) - P(A \cap B)$$

$\therefore$ $P(A \cup B) = \frac{4}{9} + \frac{1}{3} - 0 = \frac{4 + 3}{9} = \frac{7}{9}$

Hence, probability of either red or blue disc is $\frac{7}{9}$.

9. (i) Total number of cards, $n(S) = 52$

Let A be the event that card of diamond suit.

Number of diamond cards, $n(A) = 13$

$\therefore$ $P(\text{diamond}) = \frac{n(A)}{n(S)} = \frac{13}{52} = \frac{1}{4}$

(ii) Let B be the event that card is an ace.

Number of ace cards, $n(B) = 4$

$\therefore P(\text{ace}) = \frac{n(B)}{n(S)} = \frac{4}{52} = \frac{1}{13}$

(iii) Let C be the event that card is black.

Number of black cards (13 spade and 13 club),

$n(C) = 26$

$\therefore P(\text{black card}) = \frac{n(C)}{n(S)} = \frac{26}{52} = \frac{1}{2}$

(iv) From part (i),

A is the event that card is diamond.

So, A' is the event that card is not diamond.

$\therefore P(\text{not a diamond}) = P(A') = 1 - P(A) = 1 - \frac{1}{4} = \frac{3}{4}$

(v) From part (ii),

B is the event that card is an ace.

So, B' is the event that card is not an ace.

$\therefore P(\text{not an ace}) = P(B') = 1 - P(B) = 1 - \frac{1}{13} = \frac{12}{13}$

(vi) From part (iii),

C is the event that card is black.

So, C' is the event that card is not black.

$\therefore P(\text{not a black card}) = P(C') = 1 - P(C) = 1 - \frac{1}{2} = \frac{1}{2}$

(vii) Let D be the event of drawing a red card.

Number of red cards (13 heart and 13 diamond),

$n(D) = 26$

$\therefore \quad P(\text{red card}) = \frac{n(D)}{n(S)} = \frac{26}{52} = \frac{1}{2}$

10. Let A and B denote the events that Anil and Ashima will qualify the examination.

Then, $P(A) = 0.05$, $P(B) = 0.10$ and $P(A \cap B) = 0.02$

(i) P(Anil and Ashima will not qualify the exam)

$= P(A \cup B)' = 1 - P(A \cup B)$

$= 1 - [P(A) + P(B) - P(A \cap B)]$

$= 1 - [0.05 + 0.10 - 0.02]$

$= 1 - 0.13 = 0.87$

(ii) P(atleast one of them will not qualify the exam)

$= 1 - P$(both of them will qualify the exam)

$= 1 - P(A \cap B) = 1 - 0.02 = 0.98$

(iii) P(only one of them will qualify the exam)

$= P(A \cap B') + P(A' \cap B)$

$= P(A) - P(A \cap B) + P(B) - P(A \cap B)$

$[\because P(A \cap B') = P(A) - P(A \cap B)$ and $P(A' \cap B) = P(B) - P(A \cap B)]$

$= P(A) + P(B) - 2P(A \cap B)$

$= 0.05 + 0.10 - 2(0.02) = 0.15 - 0.04 = 0.11$

11. Possible outcomes in single throw of three dice,

$n(S) = 6^3 = 216$

Let E be the event of getting total of 17 or 18.

Favourable outcomes are given as

$E = \{(5, 6, 6), (6, 5, 6), (6, 6, 5), (6, 6, 6)\}$

Number of favourable outcomes, $n(E) = 4$

$\therefore \quad P(E) = \frac{n(E)}{n(S)} = \frac{4}{216} = \frac{1}{54}$

12. Let E_1, E_2, E_3, E_4 and E_5 be the event that surgeries are rated as very complex, complex, routine, simple and very simple, respectively.

$\therefore \quad P(E_1) = 0.15, \quad P(E_2) = 0.20$

$P(E_3) = 0.31, \quad P(E_4) = 0.26$

$P(E_5) = 0.08,$

(i) P(complex or very complex)

$= P(E_1 \text{ or } E_2) = P(E_1 \cup E_2)$

$= P(E_1) + P(E_2) - P(E_1 \cap E_2)$

$= 0.15 + 0.20 - 0 = 0.35$

(ii) P(neither very complex nor very simple)

$P(E_1' \cap E_5') = P(E_1 \cup E_5)' = 1 - P(E_1 \cup E_5)$

$= 1 - [P(E_1) + P(E_5)]$ [$\because$ mutually exclusive]

$= 1 - (0.15 + 0.08]$

$= 1 - 0.23 = 0.77$

(iii) $P(\text{routine or complex}) = P(E_3 \cup E_2) = P(E_3) + P(E_2)$

$= 0.31 + 0.20 = 0.51$

(iv) $P(\text{routine or simple}) = P(E_3 \cup E_4) = P(E_3) + P(E_4)$

$= 0.31 + 0.26 = 0.57$

13. Let A be the event in which spokesperson will be male and B be the event in which the spokesperson will be over 35 years of age.

Then, $P(A) = \frac{3}{5}$ and $P(B) = \frac{2}{5}$

Since, there is only one male who is over 35 years of age,

$\therefore \quad P(A \cap B) = \frac{1}{5}$

Using the formula,

$P(A \cup B) = P(A) + P(B) - P(A \cap B) = \frac{3}{5} + \frac{2}{5} - \frac{1}{5} = \frac{4}{5}$

Hence, the probability that the spokesperson will either be a male or over 35 years of age is $\frac{4}{5}$.

14. Let E_1 = John promoted

E_2 = Rita promoted

E_3 = Aslam promoted

E_4 = Gurpreet promoted

Given, sample space, $S = \{E_1, E_2, E_3, E_4\}$

It is given that chances of John's promotion is same as that of Gurpreet.

$$P(E_1) = P(E_4)$$

Rita's chances of promotion are twice as likely as John.

$$P(E_2) = 2P(E_1)$$

And Aslam's chances of promotion are four times that of John.

$$P(E_3) = 4P(E_1)$$

Now, $P(E_1) + P(E_2) + P(E_3) + P(E_4) = 1$

$\Rightarrow \quad P(E_1) + 2P(E_1) + 4P(E_1) + P(E_1) = 1 \Rightarrow 8P(E_1) = 1$

$$P(E_1) = \frac{1}{8}$$

(i) $P(\text{John promoted}) = P(E_1) = \frac{1}{8}$

$$P(\text{Rita promoted}) = P(E_2) = 2P(E_1) = 2 \times \frac{1}{8} = \frac{1}{4}$$

$$P(\text{Aslam promoted}) = P(E_3) = 4P(E_1) = 4 \times \frac{1}{8} = \frac{1}{2}$$

$$P(\text{Gurpreet promoted}) = P(E_4) = P(E_1) = \frac{1}{8}$$

(ii) Given, $A = \{\text{John promoted or Gurpreet promoted}\}$

$\Rightarrow \quad A = E_1 \cup E_4$

$\therefore \quad P(A) = P(E_1 \cup E_4) = P(E_1) + P(E_4) - P(E_1 \cap E_4)$

$= \frac{1}{8} + \frac{1}{8} - 0 \qquad [\because P(E_1 \cap E_4) = 0]$

$\therefore \quad = \frac{2}{8} = \frac{1}{4}$

15. Let S be the sample space associated with the given random experiment.

Then, $n(S) = {}^{52}C_4$

Now, let A be the event of getting all spade cards,

B be the event getting all club cards,

C be the event getting all heart cards and

D be the event of getting all diamond cards.

Then, $\quad n(A) = {}^{13}C_4, \quad n(B) = {}^{13}C_4, \quad n(C) = {}^{13}C_4$

and $\quad n(D) = {}^{13}C_4$

$\Rightarrow \quad P(A) = \frac{{}^{13}C_4}{{}^{52}C_4}, \; P(B) = \frac{{}^{13}C_4}{{}^{52}C_4}, \; P(C) = \frac{{}^{13}C_4}{{}^{52}C_4}$

and $\quad P(D) = \frac{{}^{13}C_4}{{}^{52}C_4}$

Now, required probability $= P(A \cup B \cup C \cup D)$

[$\because$ A, B, C and D are mutually exclusive events]

$$= 4 \times \frac{{}^{13}C_4}{{}^{52}C_4} = \frac{44}{4165}$$

16. Given, total number of red balls = 6,

total number of white balls = 4

and total number of black balls = 5

$\therefore \quad$ Total number of balls = 15

Let S be the sample space associated with the random experiment.

Then, $\quad n(S) = {}^{15}C_4 = 1365$

Here, atleast one balls of each colour can be drawn in one of the following ways:

(i) 1 red, 1 white and 2 black balls

(ii) 2 red, 1 white and 1 black balls

(iii) 1 red, 2 white and 1 black balls

Now, let A be the event of getting 1 red, 1 white and 2 black balls; B be the event of getting 2 red, 1 white and 1 black ball and C be the event of getting 1 red, 2 white and 1 black ball.

Then, $\quad n(A) = {}^6C_1 \times {}^4C_1 \times {}^5C_2 = 6 \times 4 \times 10 = 240$

$n(B) = {}^6C_2 \times {}^4C_1 \times {}^5C_1 = 15 \times 4 \times 5 = 300$

and $\quad n(C) = {}^6C_1 \times {}^4C_2 \times {}^5C_1 = 6 \times 6 \times 5 = 180$

$\therefore \quad P(A) = \frac{240}{1365}, P(B) = \frac{300}{1365}$ and $P(C) = \frac{180}{1365}$

Now, required probability

$= P(A \cup B \cup C) = P(A) + P(B) + P(C)$

[$\because$ A, B and C are mutually exclusive events]

$$= \frac{240}{1365} + \frac{300}{1365} + \frac{180}{1365} = \frac{720}{1365} = \frac{48}{91}$$

QR Code for Quick Test 1

QR Code for Quick Test 2

QR Code for Quick Test 3

Mind Map

Probability

Sure Event
The whole sample space S is also a subset of S, so it represents an event. Since, every outcome of an experiment carried out a member of S, therefore the event represented by S is called a sure or a certain event.

Impossible Event
The empty set ϕ is also a subset of the sample space (S), so it represents an event. Since, there is no element in the empty set, therefore it cannot occur.
That's why event ϕ is called an impossible event

Simple Event
If an event has only one sample point, then it is called a simple or elementary event.

Compound Event
If an event has more than one sample point, then it is called a compound event.

Mutually Exclusive Events
Two events are said to be mutually exclusive, if the occurrence of any one of them excludes the occurrence of other event i.e. if they cannot occur simultaneously.
Thus, two events E_1 and E_2 are said to be mutually exclusive, if $E_1 \cap E_2 = \phi$

Exhaustive Events
For a random experiment, a set of events is said to be exhaustive, if one of them necessarily occurs whenever the experiment is performed.
Let $E_1, E_2, E_3, \ldots, E_n$ be n subsets of a sample space S. Then, events $E_1, E_2, E_3, \ldots E_n$, are called exhaustive events, if
$E_1 \cup E_2 \cup E_3 \cup \ldots \cup E_n = S$

Probability of Occurrence of an Event
A numerical value that conveys the chance of occurrence of an event, when we perform an experiment, is called the probability of that event.

Event
Any subset E of a sample space S is called an event.
e.g. On throwing a die, we have the sample space.
$S = \{1, 2, 3, 4, 5, 6\}$
Then, $E = \{2, 3, 4\}$ is an event. $[\therefore E \subseteq S]$

Outcomes and Sample Space
A possible result of a random experiment is called its outcome and the set of all possible outcomes in a random experiment is called a sample space. It is generally

Probability of Complementary Event
Let E be any event and $\overline{E}$ (or E') be its complementary event.
Then, $P(\overline{E}) = 1 - P(E)$

Axiomatic Approach to Probability
Let S be the sample space of a random experiment. The probability P is a real valued function whose domain is the power set of S and range is the interval [0, 1] satisfying the following **axioms**.
(i) For any event E, $P(E) \geq 0$
(ii) $P(S) = 1$
(iii) If E and F are mutually exclusive events, then $P(E \cup F) = P(E) + P(F)$.
Now, let S be a sample space containing outcomes $E_1, E_2, \ldots, E_n$ i.e. $S = \{E_1, E_2, \ldots, En\}$.
Then,
- $0 \leq P(E_i) \leq 1, \forall E_i \in S$
- For any event A, $P(A) = \sum_{i=1}^{n} P(E_i) = 1$, $E_i \in A$

Classical Approach to Probability
To obtain the probability of an event, we find the ratio of the number of favourable outcomes to the event, to the total number of equally likely outcomes. This theory is known as classical theory of probability.
i.e. $P(A) = \dfrac{\text{Number of favourable outcomes}}{\text{Total number of possible outcomes}}$
This is also known as theoretical probability.

Probability of Equally Likely Outcomes
The outcomes of a random experiment are said to be equally likely, if the chance of occurrence of each outcome is same, i.e. chances of each outcome is same.

Addition Theorems on Probability
If A and B are two events associated with a random experiment, then
$P(A \cup B) = P(A) + P(B) - P(A \cap B)$
i.e. $P(A \text{ or } B) = P(A) + P(B) - P(A \text{ and } B)$
It is known as **addition law of probability** for two events.

Probability of the Event A or B or C
If A, B and C are three events associated with a random experiment, then
$P(A \cup B \cup C) = P(A) + P(B) + P(C) - P(A \cap B) - P(B \cap C) - P(C \cap A) + P(A \cap B \cap C)$

Topics to be Assessed Formatively

Experiment

An operation which can produce some well-defined outcomes, is known as experiment. There are two types of experiments, *viz.*

(i) **Deterministic Experiments** An experiment which gives the same result every time, when it is repeated under identical conditions, is called deterministic experiments.

(ii) **Random Experiments** In our daily life, we perform many experimental activities, where the result may not be same, when they are repeated under identical conditions. e.g. when a fair coin is tossed it may turn up a head or a tail, but we are not sure which one of these results will be actually obtained. Such experiments are called random experiments. Thus, rolling an unbiased die, drawing a card from a well-shuffled pack of cards are all examples of random experiments.

An experiment is called random experiment, if it satisfies the following two conditions.

(a) It has more than one possible outcomes.

(b) It is not possible to predict the outcomes in advance.

Outcomes and Sample Space

A possible result of a random experiment is called its outcome. The set of all possible outcomes in a random experiment is called a sample space and it is generally denoted by S, i.e.

Sample space (S) = {All possible outcomes}

Each element of a sample space is called a **sample point** or an **event point.**

e.g.

(i) On tossing a fair coin, there are two possible outcomes, namely head (H) and tail (T).

$\therefore$ The sample space S is $\{H, T\}$ i.e. $S = \{H, T\}$.

(ii) When we throw a die, then possible outcomes of this experiment are 1, 2, 3, 4, 5 or 6.

$\therefore$ The sample space S is $\{1, 2, 3, 4, 5, 6\}$

i.e. $S = \{1, 2, 3, 4, 5, 6\}$

TEACHER TIPS

Coin

A coin has two sides, head and tail. If an experiment consists of more than one coin, then coins are considered as distinct.

(i) Sample space of one coin $= \{H, T\}$

(ii) Sample space of two coins $= \{(H, T), (T, H), (H, H), (T, T)\}$

(iii) Sample space of three coins $= \{(H, H, H), (H, H, T), (H, T, H), (T, H, H), (H, T, T), (T, H, T), (T, T, H), (T, T, T)\}$

Die

A die has six faces marked with numbers 1, 2, 3, 4, 5 and 6. (or dots 1, 2, 3, 4, 5 and 6). If an experiment consists of more than one die, then all dice are considered as distinct.

(i) Sample space of a die $= \{1, 2, 3, 4, 5, 6\}$

(ii) Sample space of two dice

$$= \begin{Bmatrix} (1,1), (1,2), (1,3), (1,4), (1,5), (1,6), \\ (2,1), (2,2), (2,3), (2,4), (2,5), (2,6), \\ (3,1), (3,2), (3,3), (3,4), (3,5), (3,6), \\ (4,1), (4,2), (4,3), (4,4), (4,5), (4,6), \\ (5,1), (5,2), (5,3), (5,4), (5,5), (5,6), \\ (6,1), (6,2), (6,3), (6,4), (6,5), (6,6) \end{Bmatrix}$$

Playing Cards

A pack of playing cards has 52 cards. There are 4 suits (spade, heart, diamond and club), each having 13 cards in pack. These are of two colours, **red** (heart and diamond) and **black** (spade and club), each having 26 cards. In 13 cards of each suit, there are 3 face cards namely **king, queen and jack**, so there are total 12 face cards. Also, there are 16 honour cards, 4 of each suit namely **ace, king, queen and jack.**

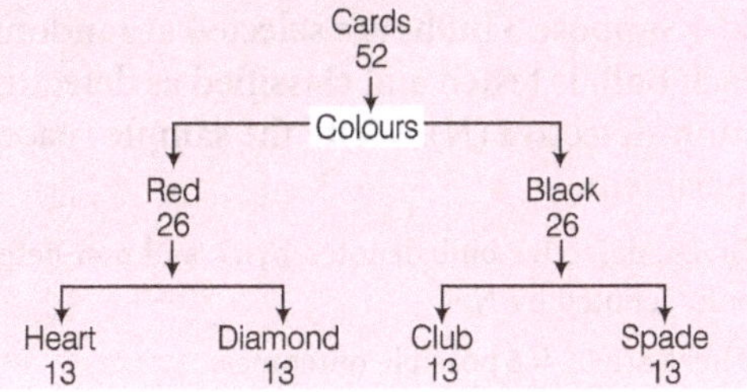

Example 1 **A bag contains 4 identical red balls and 3 identical black balls. The experiment consists of drawing one ball, then putting it into the bag and again drawing a ball. What are the possible outcomes of the experiment?**

Sol. Let R denote a red ball and B denote a black ball. Then, possible outcomes of the experiment are RR, RB, BR and BB.

Example 2 An experiment consists of recording boy-girl composition of families with 2 children.

(i) What is the sample space, if we are interested in knowing whether it is a boy or girl in the order of their births?

(ii) What is the sample space, if we are interested in the number of girls in the family?

Sol. Let B denote a boy and G denote a girl.

Then,

(i) Required sample space = $\{BB, BG, GB, GG\}$

(ii) There can be no girl, 1 girl or 2 girls.

$\therefore$ Required sample space = $\{0, 1, 2\}$

Example 3 2 boys and 2 girls are in room X and 1 boy and 3 girls are in room Y. Specify the sample space for the experiment in which a room is selected and then a person.

Sol. Let the boys and girls in room X be denoted by B_1, B_2, G_1, G_2 and those in room Y be denoted by b_1, g_1, g_2, g_3. A room out of the two can be selected in two ways; either the room X is selected or the room Y is selected. Corresponding to either case, there are four ways of selecting a person.

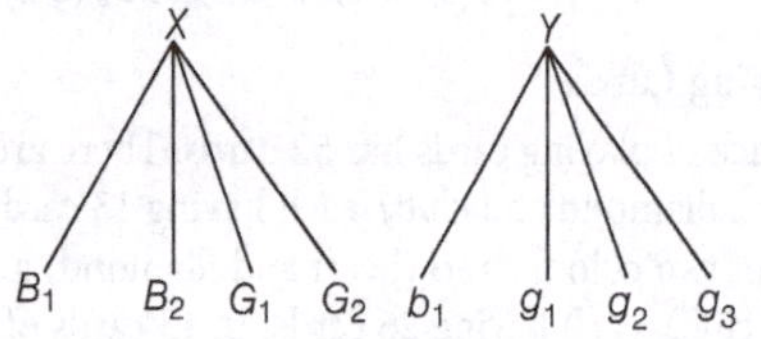

$$\therefore \quad S = \{(X, B_1), (X, B_2), (X, G_1), (X, G_2), (Y, b_1), (Y, g_1), (Y, g_2), (Y, g_3)\}$$

Example 4 Suppose 3 bulbs are selected at random from a lot. Each bulb is tested and classified as defective (D) or non-defective (N). Write the sample space for this experiment.

Sol. Given, defective bulb denoted by D and non-defective bulb denoted by N.

There are $2^3 = 8$ possible outcomes.

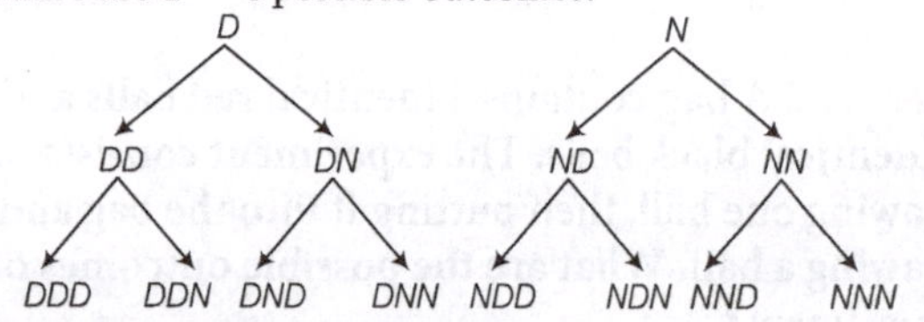

Hence, the required sample space

$= \{DDD, DDN, DND, DNN, NDD, NDN, NND, NNN\}$

EXAMPLE 5 An experiment consists of rolling a die and then tossing a coin once, if the number on the die is even. If the number on the die is odd, then coin is tossed twice. Write the sample space for this experiment.

Sol. It is a two stage experiment. When a die is rolled, then it may show up any one of the six numbers 1, 2, 3, 4, 5, 6.

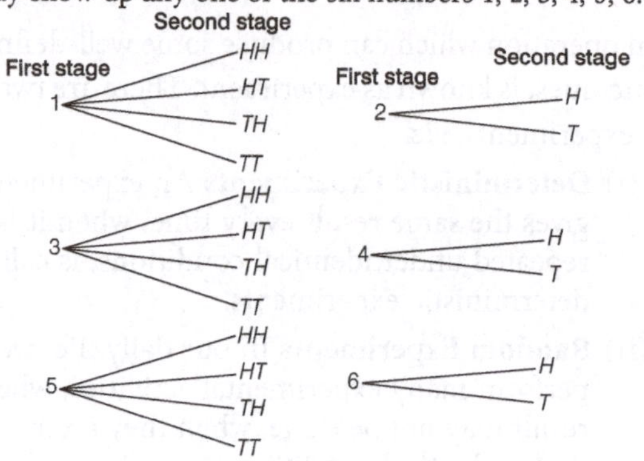

When the die shows up an odd number, a coin is tossed twice and hence, corresponding to each of 1, 3 and 5, there are four possibilities. When the die shows up an even number, a coin is tossed once and hence, corresponding to each of 2, 4 and 6, there are only two possibilities.

Hence, the sample space is

$$S = \left\{\begin{array}{l} 2H, 2T, 4H, 4T, 6H, 6T, \\ 1HH, 1HT, 1TH, 1TT, 3HH, 3HT, \\ 3TH, 3TT, 5HH, 5HT, 5TH, 5TT \end{array}\right\}$$

EXAMPLE 6 A coin is tossed. If the outcome is head, a die is thrown. If the die shows up an even number, the die is thrown again, what is the sample space for this experiment?

Sol. In tossing of a coin, if we get a tail, experiment is completed and if we get a head, the stage experiment has three stages.

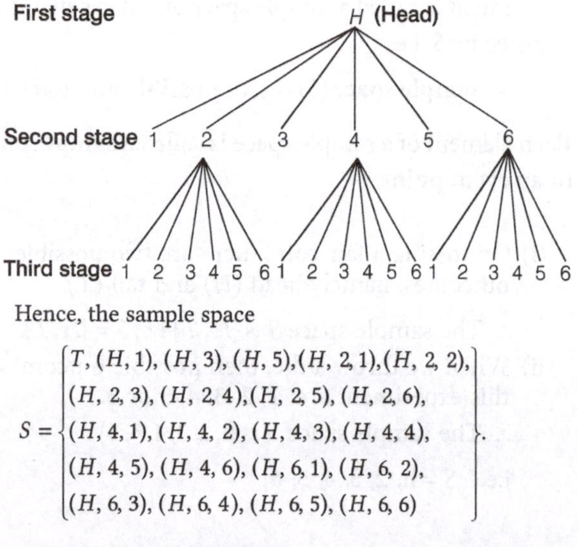

Hence, the sample space

$$S = \left\{\begin{array}{l} T, (H, 1), (H, 3), (H, 5), (H, 2, 1), (H, 2, 2), \\ (H, 2, 3), (H, 2, 4), (H, 2, 5), (H, 2, 6), \\ (H, 4, 1), (H, 4, 2), (H, 4, 3), (H, 4, 4), \\ (H, 4, 5), (H, 4, 6), (H, 6, 1), (H, 6, 2), \\ (H, 6, 3), (H, 6, 4), (H, 6, 5), (H, 6, 6) \end{array}\right\}$$

EXAMPLE 7 From a group of 2 men and 3 women, two persons are selected. Describe the sample space of this experiment.

Sol. Let 2 men be represented by M_1 and M_2 and 3 women be represented by W_1, W_2 and W_3.

Thus, sample space

$$S = \{M_1M_2, M_1W_1, M_1W_2, M_1W_3, M_2W_1, M_2W_2, M_2W_3, W_1W_2, W_1W_3, W_2W_3\}.$$

EXAMPLE 8 A die is thrown repeatedly until a six comes up. What is the sample space for this experiment?

Sol. In this experiment, 6 may comes up in 1st throw or in the 2nd throw or in the 3rd throw and so on, till 6 is obtained.

Hence, the desired sample space is

$$S = \left\{\begin{matrix} 6, (1,6), (2,6), (3,6), (4,6), (5,6), (1,1,6), \\ (1,2,6), (1,3,6), \ldots, (1,5,6), (2,1,6), \ldots, \\ (2,5,6), \ldots, (5,1,6), (5,2,6), \ldots, (5,5,6), \ldots \end{matrix}\right\}.$$

Exercise

Objective Type Questions

1 An experiment is called random experiment, if it ...X... . Here, X refers to

(a) has more than one possible outcome

(b) is not possible to predict the outcome in advance

(c) Both (a) and (b)

(d) None of the above

2 When a pair of dice (one is blue and the other is red) is rolled once, then

(a) sample space is {(1, 1), (1, 2), (1, 3), (1, 4), (1, 5), (1, 6), (2, 1), (2, 2), (2, 3), (2, 4), (2, 5), (2, 6), (3, 1), (3, 2), (3, 3), (3, 4), (3, 5), (3, 6), (4, 1), (4, 2), (4, 3) (4, 4), (4, 5), (4, 6), (5, 1), (5, 2), (5, 3), (5, 4), (5, 5), (5, 6), (6, 1), (6, 2), (6, 3), (6, 4), (6, 5), (6, 6)}

(b) total number of elements in sample space is 36

(c) Both (a) and (b)

(d) None of the above

3 One die of red colour(R), one of white colour(W) and one of blue colour(B) are placed in a bag. One die is selected at random and rolled, its colour and the number on its uppermost face is noted. Then, the sample space is

(a) $S = \left\{\begin{matrix} R1. & R2, & R3, & R4, & R5, & R6 \\ W1, & W2. & W3, & W4, & W5, & W6 \\ B1. & B2, & B3, & B4, & B5, & B6 \end{matrix}\right\}.$

(b) $S = \left\{\begin{matrix} R1, & R2 & R3, & R4, & R5 & R6 \\ W1, & W2, & W4, & W5 & & \\ B1. & B3, & B4, & B5 & B6 & \end{matrix}\right\}$

(c) Both (a) and (b)

(d) None of the above

4 A bag contains 4 identical red balls and 3 identical black balls. The experiment consists of drawing one ball, then putting it into the bag and again drawing a ball. Then, the possible outcome of this experiment is

(a) {RR, BB} (b) {RR, B,B, RR}

(c) {BB, R} (d) {RR, RB, BR, BB}

5 A bag contains one white and one red ball. A ball is drawn from the bag. If the ball drawn is white it is replaced in the bag and again a ball is drawn, otherwise, a die is tossed. The number of sample point in the sample space of above experiment is

(a) 6 (b) 7 (c) 8 (d) 9

Very Short Answer Type Questions

6 A coin is tossed and then a die is rolled only in case a head is shown on the coin Describe the sample space.

7 A coin is tossed. If it shows head, we throw a die. If it shows a tail, we toss another coin. Describe the sample space.

8 Suppose 3 bulbs are selected at random from a lot. In those of them two are defective (D) and one is non-defective (N). Write the sample space for this experiment.

Short Answer Type Questions

9 A coin is tossed. If it shows tail, we draw a ball from a box which contains 2 red and 3 black balls. If it shows head, we throw a die. Find the sample space for this experiment.

10 A coin is tossed twice. If the second throw results in a tail, a die is thrown. Describe the sample space for this experiment.

11 From a group of 2 boys and 2 girls, we select two persons to form a committee. What will be the sample space for this experiment?

12 A box contains 1 red and 3 black balls. Two balls are drawn at random in succession without replacement. Write the sample space for this experiment.

13 A coin is tossed repeatedly until a head comes up. Describe the sample space.

14 The numbers 4, 5, 6 and 7 are written separately on the form slips of the paper. The slips are put in a box and mixed thoroughly. A person draw two slips from the box, one after the other without replacement. Describe the sample space for the experiment.

15 1 boy and 2 girls are in a room X and 3 boys and 1 girl are in room Y. Write the sample space for the experiment in which a room is selected and then a person.

Answers

1. (c) **2.** (c) **3.** (a) **4.** (d)

5. (c)

6. $\{T, (H, 1), (H, 2), (H, 3), (H, 4), (H, 5), (H, 6)\}$

7. $\{(H, 1), (H, 2), (H, 3), (H, 4), (H, 5), (H, 6), (T, T), (T, H)\}$

8. $\{(D, D, N), (D, N, D), (N, D, D)\}$

9. $\{(H, 1), (H, 2), (H, 3), (H, 4), (H, 5), (H, 6), (T, R_1), (T, R_2), (T, B_1), (T, B_2), (T, B_3)\}$

10. $\{HH, TH, (HT\,1), (HT\,2), (HT\,3), (HT\,4), (HT\,5), (HT\,6), (TT\,1), (TT\,2), (TT\,3), (TT\,4), (TT\,5), (TT\,6)\}$

11. $\{(B_1, B_2), (B_1, G_1), (B_1, G_2), (B_2, G_1), (B_2, G_2), (G_1, G_2)\}$

12. $\{(R, B_1), (R, B_2), (R, B_3), (B_1, R), (B_1, B_2), (B_1, B_3), (B_2, R), (B_2, B_1), (B_2, B_3), (B_3, R), (B_3, B_1), (B_3, B_2)\}$

13. $\{H, TH, TTH, TTTH, TTTTH, \ldots\}$

14. $\{(4, 5), (4, 6), (4, 7), (5, 4), (5, 6), (5, 7), (6, 4), (6, 5), (6, 7), (7, 4), (7, 5), (7, 6)\}$

15. $\{(x, B_1), (x, G_1), (x, G_2), (y, b_1), (y, b_2), (y, b_3), (y, g_1)\}$

CHAPTERS TO BE ASSESSED FORMATIVELY

15

CHAPTER

To be Assessed Formatively

God may not play dice with the universe , but something strange is going on with the prime numbers. — *Paul Frdos*

Principle of Mathematical Induction

▼ **YOUR STUDY GOALS**

15.1 Statement
15.2 Principle of Mathematical Induction
15.3 Inductive Hypothesis
15.4 Algorithm for Principle of Mathematical Induction

▼ **YOUR PRACTICE GOALS**

- Objective Type Questions (8 Qs ; 8 MCQs)
- Subjective Type Questions (33 Qs; 12 SA(I), 19 SA(II), 2 LA)

Induction is a key aspect of scientific reasoning, where collecting and analysing data is the norm.Thus, induction means the generalisation from particular cases or facts.

In algebra or in other discipline of Mathematics, there are certain results or statements that are formulated in terms of n (where, n is a positive integer).

To prove such statements the well suited principle that is used, based on the specific technique, is the **Principle of Mathematical Induction**.

15.1 Statement

A sentence or description which can be judged either true or false is called a statement.

e.g. (i) 3 divides 9.

(ii) Jaipur is the capital of Rajasthan.

A statement involving mathematical relations is known as the **mathematical statement.** e.g. 3 divides 9.

15.2 Principle of Mathematical Induction

Suppose there is a statement $P(n)$ involving the natural number n, such that

(i) the statement is true for $n=1$ i.e. $P(1)$ is true and

(ii) if the statement is true for $n=k$ (where, k is some positive integer), then statement is also true for $n=k+1$ i.e. truth of $P(k)$ implies the truth of $P(k+1)$.

Then, $P(n)$ is true for all natural numbers n.

TEACHER TIPS

1. Property (i) is simply a statement of fact. There may be a situation when a statement is true for all $n \geq 4$. In this case, we will start from $n = 4$ and verify the result for $n = 4$, i.e. $P(4)$.
2. Property (ii) is a conditional property. It does not assert that the given statement is true for $n = k$, but only that if it is true for $n = k$, then it is also true for $n = k + 1$.
   So, to prove that the property holds, only prove the conditional proposition.
   'If the statement is true for $n = k$, then it is also true for $n = k + 1$'.

15.3 Inductive Hypothesis

The step 'if the statement is true for $n = k$, then it is also true for $n = k + 1$' is sometimes referred as the **inductive step** and the assumption in this step that statement is true for $n = k$ is called the **inductive hypothesis.**

Example 1 If $P(n)$: '$49^n + 16^n + k$ is divisible by 64 for $n \in N$' is true, then find the least negative integral value of k.

Sol. Here, the given statement is true for all $n \in N$, therefore it is true for $n = 1$ also.

So, we have $P(1): 49 + 16 + k$ is divisible by 64,

i.e. $65 + k$ is divisible be 64.

Clearly, k should be -1. [$\because 65 - 1 = 64$ is divisible by 64]

Example 2 If $P(n)$: '$3 \cdot 5^{2n+1} + 2^{3n+1}$ is divisible by λ for all $n \in N$' is true, then find the value of λ.

Sol. Here, the given statement is true for all $n \in N$.

$\therefore$ It is true for $n = 1$ and $n = 2$.

For $n = 1$, $P(1): 3 \cdot 5^3 + 2^4 = 3 \times 125 + 16 = 375 + 16 = 391$

and for $n = 2$, $P(2): 3 \cdot 5^5 + 2^7$

$= 3 \times 3125 + 128 = 9375 + 128 = 9503$

Now, the HCF of 391 and 9503 is 17. So, $3 \cdot 5^{2n+1} + 2^{3n+1}$ is divisible by 17. Hence, λ is 17.

15.4 Algorithm for Principle of Mathematical Induction

The steps used to prove a statement is true for all natural numbers by using principle of mathematical induction are given below

***Step* I** Consider the given statement as $P(n)$.

***Step* II** Put $n = 1$ and then show that it is a true statement.

***Step* III** Suppose that the statement is true for $n = k$.

***Step* IV** Show that the statement is true for $n = k+1$.

Combining the results of Steps II and IV, we can conclude by principle of mathematical induction that, $P(n)$ is true for all $n \in N$.

This algorithm is used to solve the various problems, which can be categorised into the following types

| TYPE I |

IDENTITY TYPE PROBLEMS

In this type of problems, we use principle of mathematical induction to show LHS is equal to RHS. *It can be understand with the help of following examples.*

Example 3 Prove by principle of mathematical induction that, the sum of first n natural numbers is $\dfrac{n(n+1)}{2}$.

Sol. *Step* I Let $P(n)$ be the given statement

i.e. $P(n): 1 + 2 + 3 + \ldots + n = \dfrac{n(n+1)}{2}$.

***Step* II** For $n = 1$, we have; LHS $= 1$ and RHS

$$= \frac{1 \cdot (1+1)}{2} = \frac{1 \times 2}{2} = 1$$

$\because$ LHS = RHS $\therefore P(1)$ is true.

***Step* III** Let us assume that $P(n)$ is true for $n = k$.

Then, we have $P(k)$:

$$1 + 2 + 3 + \ldots + k = \frac{k(k+1)}{2} \quad \ldots \text{(i)}$$

***Step* IV** Now, we shall prove the statement for $n = k + 1$.

For this we have to show that

$$1 + 2 + \ldots + (k+1) = \frac{(k+1)(k+1+1)}{2}$$

Consider, LHS $= 1 + 2 + \ldots + k + (k+1)$

$$= \frac{k(k+1)}{2} + (k+1) \quad \text{[using Eq. (i)]}$$

$$= (k+1)\left(\frac{k}{2}+1\right) \quad \text{[taking common } (k+1)]$$

$$= \frac{(k+1)(k+2)}{2} = \frac{(k+1)(k+1+1)}{2} = \text{RHS}$$

Thus, $P(k+1)$ is true, whenever $P(k)$ is true.

Hence, by principle of mathematical induction, $P(n)$ is true for all natural numbers n.

Example 4 **Prove by the principle of mathematical induction that**

$$1+3+3^2+\ldots+3^{n-1} = \frac{3^n-1}{2}, \text{ for all } n \in N.$$

Sol. ***Step* I** Let $P(n)$ be the given statement.

i.e. $P(n): 1 + 3 + 3^2 + \ldots + 3^{n-1} = \dfrac{3^n - 1}{2}$

***Step* II** For $n = 1$, we have

LHS $= 3^{1-1} = 3^0 = 1$ and RHS $= \dfrac{3^1 - 1}{2} = 1$

$\because$ LHS = RHS

$\therefore$ $P(1)$ is true.

***Step* III** Let us assume that $P(n)$ is true for $n = k$.

Then, we have

$$P(k): 1 + 3 + 3^2 + \ldots + 3^{k-1} = \frac{3^k - 1}{2} \quad \ldots\text{(i)}$$

***Step* IV** Now, we shall prove the statement for $n = k+1$.

For this, we have to show

$$1 + 3 + 3^2 + \ldots + 3^{k-1} + 3^k = \frac{3^{k+1} - 1}{2}$$

Then, LHS $= 1 + 3 + 3^2 + \ldots + 3^{k-1} + 3^k$

$$= \frac{3^k - 1}{2} + 3^k \quad \text{[from Eq. (i)]}$$

$$= \frac{3^k - 1 + 2\cdot 3^k}{2} = \frac{3^k(1+2) - 1}{2}$$

$$= \frac{3^k \cdot 3 - 1}{2} = \frac{3^{k+1} - 1}{2} = \text{RHS}$$

Thus, $P(k+1)$ is true, whenever $P(k)$ is true.

Hence, by principle of mathematical induction, $P(n)$ is true for all $n \in N$.

Example 5 **Prove the rule of exponents $(ab)^n = a^n b^n$ by using principle of mathematical induction for every natural number.**

Sol. ***Step* I** Let $P(n)$ be the given statement,

i.e. $P(n): (ab)^n = a^n b^n$

***Step* II** For $n = 1$, $(ab)^1 = ab = a^1 b^1$.

So, $P(1)$ is true.

***Step* III** Let $P(k)$ be true.

Then, we have $(ab)^k = a^k b^k$ ...(i)

***Step* IV** Now, we shall prove the statement for $n = k+1$.

For this, we have to show that

$$(ab)^{k+1} = a^{k+1} \cdot b^{k+1}$$

Then, LHS $= (ab)^{k+1} = (ab)^k (ab)$

$$= (a^k b^k)(ab) \quad \text{[using Eq. (i)]}$$

$$= (a^k \cdot a^1)(b^k \cdot b^1) = a^{k+1} \cdot b^{k+1}$$

Thus, $P(k+1)$ is also true, whenever $P(k)$ is true.

Hence, by principle of mathematical induction, $P(n)$ is true for all $n \in N$.

Example 6 **Using principle of mathematical induction, prove that**

$$1\cdot3 + 3\cdot5 + 5\cdot7 + \ldots + (2n-1)(2n+1) = \frac{n(4n^2 + 6n - 1)}{3}.$$

Sol. ***Step* I** Let $P(n)$ be the given statement.

Then, $P(n): 1\cdot3 + 3\cdot5 + 5\cdot7 + \ldots + (2n-1)(2n+1)$

$$= \frac{n(4n^2 + 6n - 1}{3}$$

***Step* II** For $n = 1$, we have

LHS $= 1\cdot3 = 3$ and RHS $= \dfrac{1\cdot[4(1)^2 + 6(1) - 1)]}{3} = \dfrac{9}{3} = 3$

$\because$ LHS = RHS

$\therefore$ $P(1)$ is true.

***Step* III** Let us assume that $P(n)$ is true for $n = k$.

Then, we have

$P(k): 1\cdot3 + 3\cdot5 + 5\cdot7 + \ldots + (2k-1)(2k+1)$

$$= \frac{k(4k^2 + 6k - 1)}{3} \quad \ldots\text{(i)}$$

***Step* IV** Now, we shall prove the statement for $n = k+1$.

For this, we have to show

$1\cdot3 + 3\cdot5 + 5\cdot7 + \ldots + (2k-1)(2k+1)$

$+ [(2(k+1) - 1][2(k+1) + 1)]$

$$= \frac{(k+1)[4(k+1)^2 + 6(k+1) - 1]}{3}$$

Then, LHS $= 1\cdot3 + 3\cdot5 + \ldots + (2k-1)(2k+1)$

$+ (2k + 2 - 1)(2k + 2 + 1)$

$$= \frac{k(4k^2 + 6k - 1)}{3} + (2k+1)(2k+3)$$

$$= \frac{k(4k^2 + 6k - 1) + 3(4k^2 + 6k + 2k + 3)}{3}$$

$$= \frac{k(4k^2 + 6k - 1) + 3(4k^2 + 8k + 3)}{3}$$

$$= \frac{4k^3 + 6k^2 - k + 12k^2 + 24k + 9}{3}$$

$$= \frac{4k^3 + 18k^2 + 23k + 9}{3}$$

$$= \frac{(k+1)(4k^2+14k+9)}{3}$$

$$\left[\begin{array}{l} \because k=-1 \text{ satisfy } 4k^3+18k^2+23k+9=0 \\ \therefore (k+1) \text{ is a factor of } 4k^3+18k^2+23k+9 \\ \text{Now, by long division, we can write} \\ 4k^3+18k^2+23k+9 \\ =(k+1)(4k^2+14k+9) \end{array}\right]$$

$$= \frac{(k+1)(4k^2+8k+4+6k+5)}{3}$$

$$= \frac{(k+1)[4(k^2+2k+1)+6k+6-1]}{3}$$

$$= \frac{(k+1)[4(k+1)^2+6(k+1)-1]}{3}$$

= RHS

Thus, $P(k+1)$ is true whenever $P(k)$ is true.

Hence, by principle of mathematical induction, $P(n)$ is true for all $n \in N$.

Example 7 Prove by the principle of mathematical induction that

$$1\times1!+2\times2!+3\times3!+\ldots+n\times n!=(n+1)!-1$$

for all natural numbers n.

Sol. ***Step* I** Let $P(n)$ be the given statement.

i.e. $P(n): 1\times1!+2\times2!+3\times3!+\ldots+n\times n!=(n+1)!-1.$

***Step* II** For $n=1$, we have

LHS $=1\times1!=1$

and RHS $=(1+1)!-1=2!-1=2-1=$ LHS

$\because$ LHS = RHS

$\therefore P(1)$ is true.

***Step* III** Let us assume that $P(n)$ is true for $n=k$.

Then, we have

$$P(k): 1\times1!+2\times2!+3\times3!+\ldots+k\times k! = (k+1)!-1 \quad \ldots(i)$$

***Step* IV** Now, we shall prove the statement for $n=k+1$.

For this, we have to show that

$$1\times1!+2\times2!+3\times3!+\ldots+k\times k! + (k+1)\times(k+1)! = (k+1+1)!-1$$

Then, LHS $=1\times1!+2\times2!+3\times3!+\ldots+k\times k! + (k+1)\times(k+1)!$

$=(k+1)!-1+(k+1)!\times(k+1)$ [from Eq. (i)]

$=(k+1+1)(k+1)!-1$

$=(k+2)(k+1)!-1$

$=(k+2)!-1$ $\quad [\because n(n-1)!=n!]$

Thus, $P(k+1)$ is true, whenever $P(k)$ is true.

Hence, by the principle of mathematical induction, $P(n)$ is true for all natural numbers n.

Example 8 Prove that

$$1+\frac{1}{1+2}+\frac{1}{1+2+3}+\ldots+\frac{1}{1+2+3+\ldots+n}=\frac{2n}{n+1}$$

for all $n \in N$.

Sol. ***Step* I** Let $P(n)$ be the given statement.

i.e. $P(n): 1+\frac{1}{1+2}+\frac{1}{1+2+3}+\ldots+\frac{1}{1+2+3+\ldots+n}=\frac{2n}{n+1}$

***Step* II** For $n=1$, we have

LHS $=1$

and RHS $=\frac{2\times1}{1+1}=\frac{2}{2}=1$

$\because$ LHS = RHS

$\therefore P(1)$ is true.

***Step* III** Let us assume that $P(n)$ is true for $n=k$.

Then, we have $P(k): 1+\frac{1}{1+2}+\frac{1}{1+2+3}+\ldots+\frac{1}{1+2+3+\ldots+k}=\frac{2k}{k+1} \quad \ldots(i)$

***Step* IV** Now, we shall prove the statement for $n=k+1$.

For this, we have to show that

$$1+\frac{1}{1+2}+\frac{1}{1+2+3}+\ldots+\frac{1}{1+2+3+\ldots+k}+\frac{1}{1+2+\ldots+(k+1)}=\frac{2(k+1)}{(k+1)+1}$$

Then, LHS $=\left(1+\frac{1}{1+2}+\frac{1}{1+2+3}+\ldots+\frac{1}{1+2+3+\ldots+k}\right)+\frac{1}{1+2+\ldots+(k+1)}$

$=\frac{2k}{k+1}+\frac{1}{1+2+\ldots+(k+1)}$ [from Eq. (i)]

$=\frac{2k}{k+1}+\frac{1}{\frac{(k+1)(k+2)}{2}}$

$\left[\because \text{sum of } n \text{ natural numbers} = \frac{n(n+1)}{2}\right]$

$=\frac{2k}{k+1}+\frac{2}{(k+1)(k+2)}$

$=\frac{2k(k+2)+2}{(k+1)(k+2)}=\frac{2[k^2+2k+1]}{(k+1)(k+2)}$

[taking 2 common in numerator]

$=\frac{2(k+1)^2}{(k+1)(k+2)}=\frac{2(k+1)}{k+2}=\frac{2(k+1)}{(k+1+1)}$

Thus, $P(k+1)$ is true, whenever $P(k)$ is true.

Hence, from the principle of mathematical induction, the statement is true for all natural numbers n.

Example 9 **Show that $\frac{n^5}{5}+\frac{n^3}{3}+\frac{7n}{15}$ is a natural number, for all $n \in N$.**

Sol. ***Step* I** Let $P(n)$ be the given statement.

i.e. $P(n): \frac{n^5}{5}+\frac{n^3}{3}+\frac{7n}{15}$ is a natural number for all $n \in N$.

***Step* II** For $n=1$, we have, $P\ (1): \frac{1}{5}+\frac{1}{3}+\frac{7}{15}=1 \in N$

Thus, $P(1)$ is true.

***Step* III** Let us assume that $P(k)$ is true.

i.e. $\frac{k^5}{5}+\frac{k^3}{3}+\frac{7k}{15}=\lambda$, for some $\lambda \in N$. ...(i)

***Step* IV** Now, we shall prove the statement for $n=k+1$.

For this, we have to show that

$\frac{(k+1)^5}{5}+\frac{(k+1)^3}{3}+\frac{7(k+1)}{15}$ is a natural number.

Then, $\frac{(k+1)^5}{5}+\frac{(k+1)^3}{3}+\frac{7(k+1)}{15}$

$$=\frac{1}{5}(k^5+5k^4+10k^3+10k^2+5k+1)$$
$$+\frac{1}{3}(k^3+3k^2+3k+1)+\frac{7}{15}k+\frac{7}{15}$$
$$=\left(\frac{k^5}{5}+\frac{k^3}{3}+\frac{7}{15}k\right)+(k^4+2k^3+3k^2+2k)$$
$$+\frac{1}{5}+\frac{1}{3}+\frac{7}{15}$$
$$=\lambda+k^4+2k^3+3k^2+2k+1 \quad \text{[from Eq. (i)]}$$

which is a natural number.

Thus, $P(k+1)$ is true, whenever $P(k)$ is true.

Hence, from the principle of mathematical induction, the statement is true for all natural numbers n.

Example 10 **Prove that**

$$\cos\theta\cos 2\theta\cos 2^2\theta\ldots\cos 2^{n-1}\theta=\frac{\sin 2^n\theta}{2^n\sin\theta} \text{ for all } n \in N.$$

Sol. ***Step* I** Let $P(n)$ be the given statement.

i.e. $P(n)$: $\cos\theta\cos 2\theta\cos 2^2\theta\ldots\cos 2^{n-1}\theta=\frac{\sin 2^n\theta}{2^n\sin\theta}$

***Step* II** For $n=1$, we have, LHS $=\cos\theta$

and RHS $=\frac{\sin 2\theta}{2\sin\theta}=\frac{2\sin\theta\cos\theta}{2\sin\theta}=\cos\theta$

$\because$ LHS = RHS

$\therefore$ $P(1)$ is true.

***Step* III** Let us assume that $P(k)$ is true.

i.e. $P(k)$: $\cos\theta\cos 2\theta\cos 2^2\theta\ldots\cos 2^{k-1}\theta=\frac{\sin 2^k\theta}{2^k\sin\theta}$ ...(i)

***Step* IV** Now, we shall prove the statement for $n=k+1$.

For this, we have to show that

$$\cos\theta\cos 2\theta\cos 2^2\theta\ldots\cos 2^{(k+1)-1}\theta=\frac{\sin 2^{k+1}\theta}{2^{k+1}\sin\theta}$$

Then, LHS $=\cos\theta\cos 2\theta\cos 2^2\theta\ldots\cos 2^k\theta$

$$=\cos\theta\cos 2\theta\cos 2^2\theta\ldots\cos 2^{k-1}\theta\ \cos 2^k\theta$$
$$=\frac{\sin 2^k\theta}{2^k\sin\theta}\cdot\cos 2^k\theta \quad \text{[from Eq. (i)]}$$
$$=\frac{2\cdot\sin 2^k\theta\cdot\cos 2^k\theta}{2\cdot 2^k\sin\theta}$$

[multiplying numerator and denominator by 2]

$$=\frac{\sin 2\cdot(2^k\theta)}{2^{k+1}\sin\theta} \quad [\because 2\sin\theta\cos\theta=\sin 2\theta]$$
$$=\frac{\sin 2^{k+1}\theta}{2^{k+1}\sin\theta}=\text{RHS}$$

Thus, $P(k+1)$ is true, whenever $P(k)$ is true.

Hence, by principle of mathematical induction, $P(n)$ is true for all $n \in N$.

Example 11 **Using principle of mathematical induction, prove that $\sum_{t=1}^{n}\frac{1}{t(t+1)(t+2)}=\frac{n(n+3)}{4(n+1)(n+2)}$, for all $n \in N$.**

Sol. ***Step* I** Let $P(n)$ be the given statement.

i.e. $P(n): \sum_{t=1}^{n}\frac{1}{t(t+1)(t+2)}=\frac{n(n+3)}{4(n+1)(n+2)}$

or $P(n): \frac{1}{1\cdot2\cdot3}+\ldots+\frac{1}{n(n+1)(n+2)}=\frac{n(n+3)}{4(n+1)(n+2)}$

***Step* II** For $n=1$, we have

LHS $=\frac{1}{1\cdot2\cdot3}=\frac{1}{6}$

and RHS $=\frac{1(1+3)}{4(1+1)(1+2)}=\frac{4}{4\cdot2\cdot3}=\frac{1}{6}$

$\because$ LHS = RHS

$\therefore$ $P(1)$ is true.

***Step* III** Let us assume that $P(n)$ is true for $n=k$, then we have

$$P(k): \sum_{t=1}^{k}\frac{1}{t(t+1)(t+2)}=\frac{k(k+3)}{4(k+1)(k+2)} \quad \text{...(i)}$$

***Step* IV** Now, we shall prove the statement for $n=k+1$.

For this we have to show that

$$\sum_{t=1}^{k+1}\frac{1}{t(t+1)(t+2)}=\frac{(k+1)(k+1+3)}{4(k+1+1)(k+1+2)}$$

Then, LHS $=\sum_{t=1}^{k+1}\frac{1}{t(t+1)(t+2)}$

$$=\sum_{t=1}^{k}\frac{1}{t(t+1)(t+2)}+\frac{1}{(k+1)(k+1+1)(k+1+2)}$$

$$=\frac{k(k+3)}{4(k+1)(k+2)}+\frac{1}{(k+1)(k+2)(k+3)}$$

[from Eq. (i)]

$$=\frac{1}{(k+1)(k+2)}\left[\frac{k(k+3)}{4}+\frac{1}{(k+3)}\right]$$

$$=\frac{1}{(k+1)(k+2)}\left[\frac{k(k+3)^2+4}{4(k+3)}\right]$$

$$=\frac{1}{(k+1)(k+2)}\left[\frac{k(k^2+6k+9)+4}{4(k+3)}\right]$$

$$=\frac{k^3+6k^2+9k+4}{4(k+1)(k+2)(k+3)}=\frac{(k+1)^2(k+4)}{4(k+1)(k+2)(k+3)}$$

$\because k=-1$ satisfy $k^3+6k^2+9k+4=0$

$\therefore (k+1)$ is a factor of k^3+6k^2+9k+4.

Now, by long-division method, we can write

$$k^3+6k^2+9k+4=(k+1)(k^2+5k+4)$$
$$=(k+1)(k+1)(k+4)$$

$$=\frac{(k+1)(k+4)}{4(k+2)(k+3)}=\frac{(k+1)(k+1+3)}{4(k+1+1)(k+1+2)}=\text{RHS}$$

Thus, $P(k+1)$ is true whenever $P(k)$ is true.

Hence, by principle of mathematical induction, $P(n)$ is true for all $n\in N$.

Example 12 Prove by the principle of mathematical induction that $7+77+777+\ldots+\underbrace{777\ldots7}_{n\text{ digit}}$

$$=\frac{7}{81}(10^{n+1}-9n-10),\text{ for all } n\in N.$$

***Sol.* Step I** Let $P(n)$ be the given statement,

i.e. $P(n): 7+77+777+\ldots+\underbrace{777\ldots7}_{n\text{ digit}}$

$$=\frac{7}{81}(10^{n+1}-9n-10)$$

***Step* II** For $n=1$, we have

$$\text{LHS}=7 \text{ and RHS}=\frac{7}{81}(10^{1+1}-9\cdot1-10)$$

$$=\frac{7}{81}(100-9-10)=\frac{7}{81}\times81=7$$

$\because$ LHS = RHS

$\therefore$ $P(1)$ is true.

***Step* III** Let us assume that $P(k)$ is true.

i.e. $P(k): 7+77+777+\ldots+\underbrace{777\ldots7}_{k\text{ digit}}$

$$=\frac{7}{81}[10^{k+1}-9k-10] \quad \ldots\text{(i)}$$

***Step* IV** Now, we shall prove the statement for $n=k+1$. For this, we have to show that

$$7+77+777+\ldots+\underbrace{777\ldots7}_{(k+1)\text{ digit}}$$

$$=\frac{7}{81}[10^{(k+1)+1}-9(k+1)-10]$$

Then, LHS

$$=7+77+777+\ldots+\underbrace{777\ldots7}_{k\text{ digit}}+\underbrace{777\ldots7}_{(k+1)\text{ digit}}$$

$$=\frac{7}{81}[10^{k+1}-9k-10]+7\,\underbrace{(111\ldots1)}_{(k+1)\text{ digit}} \quad \text{[from Eq. (i)]}$$

$$=\frac{7}{81}[10^{k+1}-9k-10]+\frac{7}{9}\underbrace{\{999\ldots9\}}_{(k+1)\text{ digit}}$$

$$=\frac{7}{81}[10^{k+1}-9k-10]+\frac{7}{9}(10^{k+1}-1)$$

$$\left[\because \underbrace{\{999\ldots9\}}_{(k+1)\text{ digit}}=9+90+900+\ldots+(k+1)\right.$$

$$\left.=9[1+10+100+\ldots+(k+1)]=9\left(\frac{(10)^{k+1}-1}{10-1}\right)\right]$$

$$=\frac{7}{81}[(10^{k+1}-9k-10)+9(10^{k+1}-1)]$$

$$=\frac{7}{81}[10^{k+1}-9k-10+9\cdot10^{k+1}-9]$$

$$=\frac{7}{81}[(1+9)10^{k+1}-9k-19]$$

$$=\frac{7}{81}[10\cdot10^{k+1}-9k-19]$$

$$=\frac{7}{81}[10^{(k+1)+1}-9(k+1)-10]=\text{RHS}$$

Thus, $P(k+1)$ is true, whenever $P(k)$ is true.

Hence, by principle of mathematical induction, $P(n)$ is true for all $n\in N$.

| TYPE II |

DIVISIBILITY TYPE PROBLEMS

In this type of problems, we use principle of mathematical induction to show that the given statement $P(n)$, is divisible by given number or $P(n)$ is a multiple of given number.

It can be understand with the help of following examples :

Example 13 Prove that $n(n+1)(n+5)$ is a multiple of 3, for all $n \in N$.

Sol. ***Step* I** Let $P(n)$ be the given statement

i.e. $P(n): n(n+1)(n+5)$ is a multiple of 3.

***Step* II** For $n=1$, we have

$P(1): 1(1+1)(1+5) = 1 \times 2 \times 6 = 12 = 3 \times 4$

which is a multiple of 3. So, $P(1)$ is true.

***Step* III** Let $P(k)$ be true for $n=k$.

i.e. $P(k): k(k+1)(k+5)$ is a multiple of 3.

Then, $k(k+1)(k+5) = 3\lambda$

$\Rightarrow k(k^2+5k+k+5) = 3\lambda$

$\Rightarrow \quad k^3+6k^2+5k = 3\lambda$ ...(i)

***Step* IV** Now, we shall prove the statement for $n=k+1$.

For this, we have to show that

$(k+1)(k+1+1)(k+1+5)$ is a multiple of 3.

Then, $(k+1)(k+1+1)(k+1+5)$

$= (k+1)(k+2)(k+6)$

$= (k^2+2k+k+2)(k+6)$

$= (k^2+3k+2)(k+6)$

$= k^3+6k^2+3k^2+18k+2k+12$

$= k^3+9k^2+20k+12$

$= (3\lambda-6k^2-5k)+9k^2+20k+12$

[from Eq. (i), $k^3 = 3\lambda-6k^2-5k$]

$= 3\lambda+3k^2+15k+12$

$= 3(\lambda+k^2+5k+4)$

which is a multiple of 3.

Thus, $P(k+1)$ is true, whenever $P(k)$ is true.

Hence, by principle of mathematical induction, the statement is true for all natural numbers n.

Example 14 Using the principle of mathematical induction, prove that $10^{2n-1}+1$ is divisible by 11, for all $n \in N$.

Sol. ***Step* I** Let $P(n)$ be the given statement.

i.e. $P(n): 10^{2n-1}+1$ is divisible by 11.

***Step* II** For $n=1$, we have

$P(1): 10^{2-1}+1 = 10+1 = 11$, which is divisible by 11.

Thus, $P(1)$ is true.

***Step* III** Let us assume that $P(k)$ is true.

i.e. $P(k): 10^{2k-1}+1$ is divisible by 11.

Then, $10^{2k-1}+1 = 11m$, for some $m \in N$ ...(i)

***Step* IV** Now, we shall prove the statement for $n=k+1$.

For this, we have to show that

$10^{2(k+1)-1}+1$ is divisible by 11.

Then, $10^{2(k+1)-1}+1 = 10^{2k+2-1}+1$

$= 10^{(2k-1)+2}+1 = 10^2 \cdot 10^{2k-1}+1$

$= 100\cdot(11m-1)+1$ [from Eq. (i)]

$= 100(11m)-100+1$

$= 100(11m)-99$

$= 11(100m-9)$,

which is divisible by 11.

Thus, $P(k+1)$ is divisible by 11 whenever $P(k)$ is divisible by 11.

Hence, by principle of mathematical induction, $P(n)$ is true for all $n \in N$.

Example 15 Prove by the principle of mathematical induction that for all $n \geq 2$, $6^n - 5n$ when divided by 25 the remainder is always 1.

Sol. ***Step* I** Let $P(n)$ be the given statement.

i.e. $P(n): 6^n - 5n$ when divided by 25, leaves the remainder 1.

or $P(n): 6^n - 5n = 25\lambda+1$ for some $\lambda \in N$.

***Step* II** For $n=2$, we have

LHS $= 6^2 - 5 \times 2 = 36-10 = 26$

$= 25(1)+1$

$=$ RHS, for $\lambda = 1 \in N$

Thus, $P(2)$ is true.

***Step* III** Let us assume that, $P(k)$ is true.

i.e. $P(k): 6^k - 5k = 25\lambda+1$ for some $\lambda \in N$ ...(i)

***Step* IV** Now, we shall prove the statement for $n=k+1$.

For this, we have to show that

$6^{k+1}-5(k+1) = 25\mu+1$ for some $\mu \in N$

Then, $6^{k+1}-5(k+1) = 6^k \cdot 6 - 5k - 5$

$= 6(5k+25\lambda+1)-5k-5$ [from Eq. (i)]

$= 30k+150\lambda+6-5k-5$

$= 25k+150\lambda+1 = 25(k+6\lambda)+1$

$= 25\mu+1$, where $\mu = k+6\lambda \in N$.

Thus, $P(k+1)$ is true, whenever $P(k)$ is true.

Hence, by principle of mathematical induction, $P(n)$ is true for all $n \geq 2$.

|TYPE III|

INEQUALITY TYPE PROBLEMS

In this type of problems, we have inequality symbols like $>, <, \geq$ or $\leq$. For such problems, we use the principle of mathematical induction to show that LHS and RHS satisfies the given inequality.

It can be understand with the help of following examples:

Example 16 Prove by the principle of mathematical induction that $3^n > 2^n$, for all $n \in N$.

Sol. ***Step* I** Let $P(n)$ be the given statement.

i.e. $P(n): 3^n > 2^n$

***Step* II** For $n = 1$, we have $3^1 > 2^1$

$\Rightarrow 3 > 2$, which is true.

Thus, $P(1)$ is true.

***Step* III** Let us assume that $P(k)$ is true.

i.e. $P(k): 3^k > 2^k$ ...(i)

***Step* IV** Now, we shall prove the statement for $n = k + 1$.

For this, we have to show $3^{k+1} > 2^{k+1}$

From Eq. (i), we have $3^k > 2^k$

$\therefore \quad 3^k \cdot 3 > 2^k \cdot 3$ [multiplying both sides by 3]

$\Rightarrow \quad 3^{k+1} > 2^k \cdot 3$

But $\quad 2^k \cdot 3 > 2^k \cdot 2 \Rightarrow 3^{k+1} > 2^k \cdot 2 = 2^{k+1}$

Thus, $P(k+1)$ is true, whenever $P(k)$ is true.

Hence, by principle of mathematical induction, $P(n)$ is true for all $n \in N$.

Example 17 Prove that $2n + 1 < 2^n$, for all natural numbers n ($n \geq 3$) by using principle of mathematical induction.

Sol. ***Step* I** Let $P(n)$ be the given statement.

i.e. $P(n): (2n + 1) < 2^n$

***Step* II** For $n = 3$, we have

$(2 \times 3 + 1) < 2^3 \Rightarrow 7 < 8$, which is true.

Thus, $P(1)$ is true.

***Step* III** Let us assume that $P(k)$ is true.

i.e. $P(k): (2k + 1) < 2^k$ ...(i)

***Step* IV** Now, we shall prove the statement for $n = k + 1$.

For this, we have to show that $2(k+1) + 1 < 2^{k+1}$

From Eq. (i), $(2k + 1) < 2^k$

So, $(2k + 1) + 2 < 2^k + 2$ [adding 2 on both sides]

$\Rightarrow \quad 2k + 3 < 2^k \cdot 2 \quad [\because 2^k + 2 < 2^k \cdot 2]$

$\Rightarrow \quad 2k + 3 < 2^{k+1} \Rightarrow 2(k+1) + 1 < 2^{k+1}$

Thus, $P(k+1)$ is true, whenever $P(k)$ is true.

Hence, by principle of mathematical induction $P(n)$ is true for all natural numbers, $n \geq 3$.

Example 18 Prove that $1 + 2 + 3 + \ldots + n < \frac{1}{8}(2n+1)^2$, for all natural numbers n.

Sol. ***Step* I** Let $P(n)$ be the given statement.

i.e. $P(n): 1 + 2 + \ldots + n < \frac{1}{8}(2n+1)^2$.

***Step* II** For $n = 1$, we have

$1 < \frac{1}{8}[2(1) + 1]^2 \Rightarrow 1 < \frac{9}{8}$, which is true.

Thus, $P(1)$ is true.

***Step* III** Let us assume that $P(k)$ is true.

i.e. $P(k): 1 + 2 + \ldots + k < \frac{1}{8}(2k+1)^2$ ... (i)

***Step* IV** Now, we shall prove the statement for $n = k + 1$.

For this, we have to show that

$1 + 2 + \ldots + (k+1) < \frac{1}{8}[2(k+1) + 1]^2$

From Eq. (i), we have

$$1 + 2 + \ldots + k < \frac{1}{8}(2k+1)^2$$

$$\Rightarrow 1 + 2 + \ldots + k + (k+1) < \frac{1}{8}(2k+1)^2 + (k+1)$$

[adding $(k+1)$ on both sides]

$$= \frac{1}{8}[(2k+1)^2 + 8(k+1)]$$

$$= \frac{1}{8}[4k^2 + 1 + 4k + 8k + 8] = \frac{1}{8}[4k^2 + 12k + 9]$$

$$= \frac{1}{8}(2k+3)^2 = \frac{1}{8}[2(k+1) + 1]^2$$

$$\Rightarrow 1 + 2 + \ldots + (k+1) < \frac{1}{8}[2(k+1) + 1]^2$$

Thus, $P(k+1)$ is true whenever $P(k)$ is true.

Hence, by principle of mathematical induction, $P(n)$ is true for all $n \in N$.

EXAM PATTERN QUESTIONS

Objective Type Questions

Multiple Choice Questions

1. The word induction means
(a) normalisation from particular cases or facts
(b) generalisation from particular cases or facts
(c) Both (a) and (b)
(d) None of the above

2. Application of a general case to a particular case is called
(a) formation (b) deduction
(c) induction (d) None of these

3. If $P(n)$ is a statement $(n \in N)$ such that $P(k)$ is true $\Rightarrow P(k+1)$ is true for $k \in N$, then $P(n)$ is true,
(a) for all n (b) for all $n > 1$
(c) for all $n > 2$ (d) Nothing can be said

4. Let $P(n): n^2 + n + 1 (n \in N)$ is an even integer. Therefore, $P(n)$ is true,
(a) for $n > 1$ (b) for all n
(c) for $n > 2$ (d) None of these

5. If $10^n + 3 \cdot 4^{n+2} + k$ is divisible by 9, for all $n \in N$, then the positive integral value of k is
(a) 5 (b) 3 (c) 7 (d) 1

6. For all $n \in N, 3 \cdot 5^{2n+1} + 2^{3n+1}$ is divisible by
(a) 19 (b) 17
(c) 23 (d) 25

7. If $P(n)$ is a statement $(n \in N)$ such that $P(3)$ is true. If $P(k)$ is true $\Rightarrow P(k+1)$ is true for all $k \geq 3$, then $P(n)$ is true,
(a) for all n (b) for $n \geq 3$
(c) for $n > 4$ (d) None of these

8. A student was asked to prove a statement $P(n)$ by induction. He proved that $P(k+1)$ is true, whenever $P(k)$ is true for all $k > 5 \in N$ and also proved that $P(5)$ is true. On the basis of this, he could conclude that $P(n)$ is true
(a) for all $n \in N$ (b) for all $n > 5$
(c) for all $n \geq 5$ (d) for all $n < 5$

Subjective Questions

Short Answer Type I Questions

1. Prove that the sum of first n even numbers is $n(n+1)$.

Or

Prove that $2 + 4 + 6 + 8 + \ldots + 2n = n(n+1)$.

2. Prove that $2 + 6 + 18 + \ldots + 2 \cdot 3^{n-1} = (3^n - 1)$ for all $n \in N$.

3. Prove that $1 + 2 + 2^2 + \ldots + 2^n = 2^{n+1} - 1$ for all natural numbers n.

4. Prove by the principle of mathematical induction that
$$a + ar + ar^2 + \ldots + ar^{n-1} = \frac{a(r^n - 1)}{r - 1}.$$

5. Prove by the principle of mathematical induction that
$$\frac{1}{2} + \frac{1}{4} + \frac{1}{8} + \ldots + \frac{1}{2^n} = 1 - \frac{1}{2^n}, \text{ for all } n \in N.$$

6. Using mathematical induction, prove that
$$\left(1+\frac{1}{1}\right)\left(1+\frac{1}{2}\right)\left(1+\frac{1}{3}\right)\ldots\left(1+\frac{1}{n}\right) = n+1.$$

7. Prove by mathematical induction that
$$3 \times 6 + 6 \times 9 + 9 \times 12 + \ldots + (3n) \times (3n+3) = 3n(n+1)(n+2)$$
for all $n \in N$.

8. Prove that $2^{2n} - 1$ is divisible by 3, for all natural numbers n.

9. Prove that $\sum_{t=1}^{n-1} t(t+1) = \frac{n(n-1)(n+1)}{3}$, for all natural numbers $n \geq 2$.

10. Prove that $2^n > n$ for all positive integers n.

Short Answer Type II Questions

11. Prove by mathematical induction that the sum of first n odd natural numbers is n^2.

12. For all $n \geq 1$, prove that

$$1^2 + 2^2 + 3^2 + \ldots + n^2 = \frac{n(n+1)(2n+1)}{6}.$$

13. Prove that $1\cdot 4\cdot 7 + 2\cdot 5\cdot 8 + 3\cdot 6\cdot 9 + \ldots + n(n+3)(n+6)$

$$= \frac{n}{4}(n+1)(n+6)(n+7)$$

for all natural numbers n.

14. Prove by the principle of mathematical induction that

$$\frac{1}{1\cdot 4} + \frac{1}{4\cdot 7} + \frac{1}{7\cdot 10} + \ldots + \frac{1}{(3n-2)(3n+1)} = \frac{n}{3n+1}$$

for all $n \in N$.

15. Use the principle of mathematical induction to prove that $n^3 - 7n + 3$ is divisible by 3, for all natural numbers n.

16. Using principle of mathematical induction, prove that $4^n + 15n - 1$ is divisible by 9, for all natural numbers.

17. If x and y are any two distinct integers, then prove by mathematical induction that $(x^n - y^n)$ is divisible by $(x - y)$, for all $n \in N$.

18. Prove that $n^3 - n$ is divisible by 6, for each natural number $n \geq 2$.

19. Prove that $3^{2n} - 1$ is divisible by 8, for all natural numbers n.

20. Prove by the principle of mathematical induction that for all $n \in N$, $n^2 + n$ is even natural number.

21. For all positive integer n, prove that

$$\frac{n^7}{7} + \frac{n^5}{5} + \frac{2}{3}n^3 - \frac{n}{105} \text{ is an integer.}$$

22. Prove that $(2n+7) < (n+3)^2$, for all natural numbers n.

23. Prove that $2n < (n+2)!$ for all natural numbers n.

24. Prove that $1^2 + 2^2 + \ldots + n^2 > \frac{n^3}{3}, n \in N$.

25. Using mathematical induction, prove that

$1\cdot 2 + 2\cdot 2^2 + 3\cdot 2^3 + \ldots + n\cdot 2^n = (n-1)2^{n+1} + 2$,

for all $n \in N$.

Long Answer Type Questions

26. If x is not an integral multiple of 2π, then use principle of mathematical induction to prove that

$$\cos x + \cos 2x + \cos 3x + \ldots + \cos nx = \cos\left(\frac{n+1}{2}\right)x \sin\left(\frac{nx}{2}\right) \operatorname{cosec}\left(\frac{x}{2}\right),$$

for all $n \in N$.

27. Prove by induction that for all natural numbers,

$$\sin\alpha + \sin(\alpha+\beta) + \sin(\alpha+2\beta) + \ldots + \sin[\alpha+(n-1)\beta] = \frac{\sin\left[\alpha + \left(\frac{n-1}{2}\right)\beta\right]\sin\left(\frac{n\beta}{2}\right)}{\sin\left(\frac{\beta}{2}\right)}.$$

Hints & Solutions

Objective Type Questions

1. (b) The word iduction means generalisations from particular cases or facts.

2. (b) Deduction is the application of a general case to a particular case.

3. (d) Unless, we prove $P(1)$ is true, nothing can be said.

4. (d) Given, $P(n): n^2 + n + 1$

At $n = 1, P(1): 3$, which is not an even integer.

$\therefore$ $P(1)$ is not true.

Also, $n(n+1) + 1$ is always an odd integer.

5. (a) Consider $P(n): 10^n + 3\cdot 4^{n+2} + k$ is divisible by 9, for all $n \in N$

For $n = 1$, the given statement is also true, i.e.

$10^1 + 3\cdot 4^{1+2} + k$ is divisible by 9 i.e. $202 + k$

If $202 + k$ is divisible by 9, then the least value of k must be 5.

6. (b) Consider, $P(n): 3\cdot 5^{2n+1} + 2^{3n+1}$

Putting $n = 1, 2$ in $P(n)$, we get $P(n)$ is divisible by 17 only.

7. (b) Since, $P(3)$ is true.

Assume $P(k)$ is true $\Rightarrow P(k+1)$ is true means, if $P(3)$ is true $\Rightarrow P(4)$ is true $\Rightarrow P(5)$ is true and so on.

So, statement $P(n)$ is true for all $n \geq 3$.

8. (c) Since, $P(5)$ is true and $P(k+1)$ is true whenever $P(k)$ is true.

$\therefore$ $P(x)$ is true for all $x \geq 5$.

Subjective Questions

1. Consider $P(k): 2 + 4 + 6 + 8 + \ldots + 2k = k(k+1)$

Now, $P(k+1): 2 + 4 + 6 + \ldots + 2(k) + 2(k+1)$

$$= k(k+1) + 2(k+1)$$
$$= k^2 + 3k + 2 = (k+1)(k+1)$$

2. Consider $P(k): 2 + 2\cdot 3 + 2\cdot 3^2 + \ldots + 2\cdot 3^{k-1} = (3^k - 1)$

Now, $P(k+1): 2 + 2\cdot 3 + 2\cdot 3^2 + \ldots + 2\cdot 3^{k-1} + 2\cdot 3^k$

$$= (3^k - 1) + 2\cdot 3^k$$
$$= 3\cdot 3^k - 1 = 3^{k+1} - 1$$

3. Consider $P(k)$: $1+2+2^2+\ldots+2^k=2^{k+1}-1$

Now, $P(k+1): 1+2+2^2+\ldots+2^k+2^{k+1}$
$$=2^{k+1}-1+2^{k+1}$$
$$=2^{(k+1)+1}-1$$

4. Solve as Q. 11.

5. Consider, $P(k): \frac{1}{2}+\frac{1}{2^2}+\frac{1}{2^3}+\ldots+\frac{1}{2^k}=1-\frac{1}{2^k}$

Now, solve as Q. 3.

6. Consider
$$P(k):\left(1+\frac{1}{1}\right)\left(1+\frac{1}{2}\right)\left(1+\frac{1}{3}\right)+\ldots+\left(1+\frac{1}{k}\right)=k+1$$

Now, $P(k+1):\left[\left(1+\frac{1}{1}\right)\left(1+\frac{1}{2}\right)\left(1+\frac{1}{3}\right)\ldots\left(1+\frac{1}{k}\right)\right]\left(1+\frac{1}{k+1}\right)$
$$=(k+1)\left(\frac{k+2}{k+1}\right)$$
$$=(k+1)+1$$

7. Solve as Q. 3.

8. Consider $P(k): 2^{2k}-1=3\lambda$ (say)

Now, $P(k+1): 2^{2(k+1)}-1=2^{2k}\cdot 2^2-1$
$$=(3\lambda+1)\,4-1=12\lambda+3$$
$=3(4\lambda+1)$, which is divisible by 3.

9. Consider $P(k): \sum_{t=1}^{k-1} t(t+1)=\frac{k(k-1)(k+1)}{3}, k\ge 2$

$\Rightarrow P(k): 1\cdot 2+2\cdot 3+3\cdot 4+\ldots+(k-1)k=\frac{k(k-1)(k+1)}{3}$

Now, $P(k+1): 1\cdot 2+2\cdot 3+3\cdot 4+\ldots+(k-1)k+k(k+1)$
$$=\frac{k(k-1)(k+1)}{3}+k(k+1)=\frac{k(k+1)(k+2)}{3}, k\ge 2$$

10. $2^k>k \Rightarrow 2\cdot 2^k>2k \Rightarrow 2^{k+1}>k+k>k+1$

11. Let $P(n)$ be the given statement. Then,
$$P(n): 1+3+5+\ldots+(2n-1)=n^2$$

Now, solve as Q. 1.

12. Solve as Q. 3.

13. Consider $P(k): 1\cdot 4\cdot 7+2\cdot 5\cdot 8+3\cdot 6\cdot 9+\ldots+k(k+3)(k+6)$
$$=\frac{k}{4}(k+1)(k+6)(k+7)$$

Now, $P(k+1): 1\cdot 4\cdot 7+2\cdot 5\cdot 8+3\cdot 6\cdot 9$
$$+\ldots+k(k+3)(k+6)+(k+1)(k+4)(k+7)$$
$$=\frac{k}{4}(k+1)(k+6)(k+7)+(k+1)(k+4)(k+7)$$
$$=\frac{(k+1)(k+7)}{4}(k^2+10k+16)$$
$$=(k+1)(k+7)(k+8)(k+2)$$

14. Solve as Q. 9.

15. Consider $P(k): k^3-7k+3=3\lambda$ (say)

Now, $P(k+1): (k+1)^3-7(k+1)+3$
$$=k^3+3k^2+3k+1-7k-4$$
$$=(3\lambda-3)+3k^2+3k-3=3(k^2+k-2)$$

16. Solve as Q. 18.

17. Consider $P(k): x^k-y^k=(x-y)\lambda$

Now, $P(k+1): x^{k+1}-y^{k+1}$
$$=x^k\cdot x-y^k\cdot y=[(x-y)\lambda+y^k]x-y^k\cdot y$$
$$=(x-y)\lambda x+y^k(k-y)=(x-y)(\lambda x+y^k)$$

18. Solve as Q. 10.

19. Solve as Q. 10.

20. Consider $P(k): k^2+k=2\lambda$ (say) and solve as Q. 18.

21. Consider $P(k): \frac{k^7}{7}+\frac{k^5}{5}+\frac{2}{3}k^3-\frac{k}{105}=\lambda\in I$

Now, $P(k+1): \frac{(k+1)^7}{7}+\frac{(k+1)^5}{5}+\frac{2}{3}(k+1)^3-\frac{k+1}{105}$
$$=\frac{1}{7}(k^7+7k^6+21k^5+35k^4+35k^3+21k^2+7k+1)$$
$$+\frac{1}{5}(k^5+5k^4+10k^3+10k^2+5k+1)$$
$$+\frac{2}{3}(k^3+3k^2+3k+1)-\frac{k+1}{105}$$
$$=\lambda+(k^6+3k^5+6k^4+7k^3+7k^2+4k)=\text{Integer}$$

22. Consider $P(k): (2k+7)<(k+3)^2$

$\Rightarrow \quad (2k+7)+2<(k+3)^2+2$

$\Rightarrow \quad 2(k+1)+7<(k+4)^2 \quad [\because (k+3)^2+2<(k+4)^2]$

23. $2k<(k+2)! \Rightarrow 2k+2<(k+2)!+2]$

$\Rightarrow \quad 2(k+1)<(k+2)!+2$

But, $(k+2)!+2<(k+3)!$

$\therefore \quad 2(k+1)<(k+3)!$

24. $1^2+2^2+\ldots+k^2>\frac{k^3}{3}$

$\Rightarrow 1^2+2^2+\ldots+k^2+(k+1)^2>\frac{k^3}{3}+(k+1)^2$
$$=\frac{1}{3}[k^3+3(k+1)^2]$$
$$=\frac{1}{3}[k^3+3k^2+6k+3]$$
$$=\frac{1}{3}[(k+1)^3+3k+2]\ 1^2+2^2+\ldots+k^2>\frac{1}{3}(k+1)^3$$

25. Solve as Q. 2.

26. Solve as Example 9.

27. Solve as Example 9.

ACTIVITES

ACTIVITY 1

To verify that for two sets A and B, $n(A \times B) = pq$ and the total number of relations from A to B is 2^{pq}, where $n(A) = p$ and $n(B) = q$.

Material Required

Paper, different coloured pencils.

Method of Construction

1. Take a set A_1 which has one element (say) a_1, and take another set B_1, which has one element (say) b_1.
2. Again, take a set A_2 which has two elements (say) a_1 and a_2 and take another set B_3, which has three elements (say) b_1, b_2 and b_3.
3. Similarly, take a set A_3 which has three elements (say) a_1, a_2 and a_3, and take another set B_4, which has four elements (say) b_1, b_2, b_3 and b_4.

Demonstration

1. After construction, represent all the possible correspondences of the elements of set A_1 to the elements of set B_1 visually as shown in Fig. (1.1)

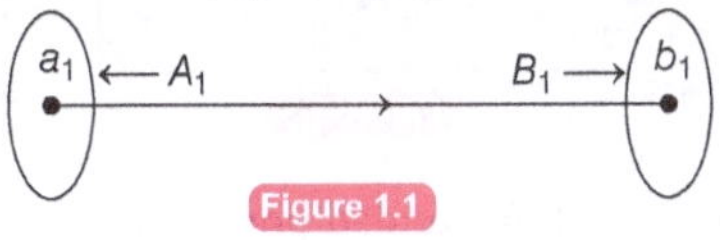

Figure 1.1

2. Again, represent all the possible correspondences of the elements of set A_2 to the elements of set B_3 visually as shown in Fig. (1.2).

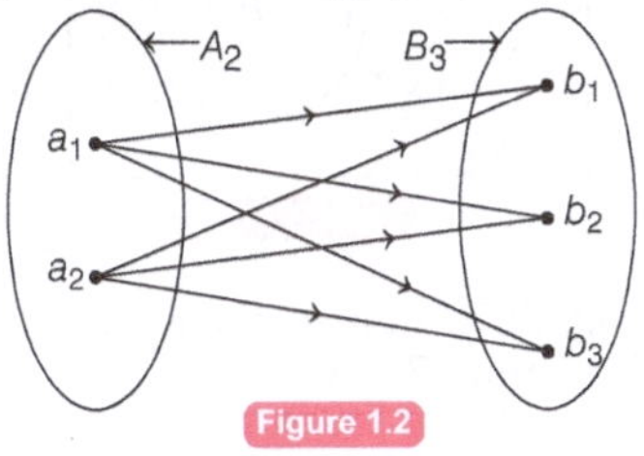

Figure 1.2

3. Represent all the possible correspondence of the elements of set A_3 to the elements of set B_4 visually as shown in Fig. (1.3)

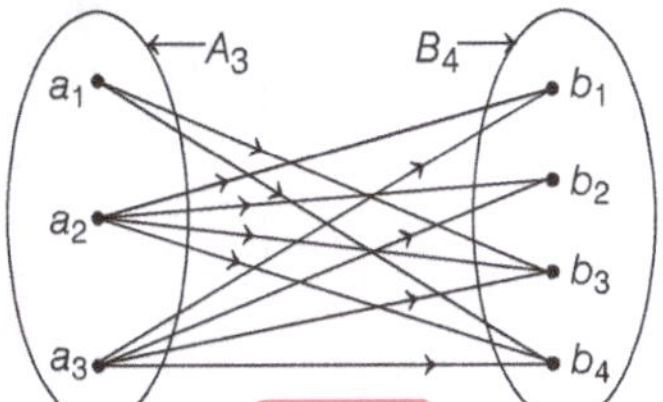

Figure 1.3

4. In the end of activity, similar visual representations can be shown between the elements of any two given sets A and B.

Observation

1. The number of arrows, i.e. the number of elements in cartesian product $(A_1 \times B_1)$ of the sets A_1 and B_1 is ...×... and the number of relations is $2^{\square}$.
2. The number of arrows, i.e. the number of elements in cartesian product $(A_2 \times B_3)$ of the sets A_2 and B_3 is ...×... and the number of relations is $2^{\square}$.
3. The number of arrows, i.e., the number of elements in cartesian product $(A_3 \times B_4)$ of the sets A_2 and B_4 is ...×... and the number of relations are $2^{\square}$.

Note We can verify this result by taking other sets $A_4, A_5, \ldots\ldots A_p$, which have elements 4, 5, ..., p, respectively, and the sets $B_5, B_6 \ldots, B_q$ which have elements 5, 6, ..., q, respectively. More precisely we arrive at the conclusion that in case of given set A containing p elements and the set B containing q elements, the total number of relations from A to B is 2^{pq}, where $n(A \times B) = n(A) \times n(B) = pq$.

VIVA-VOCE

1 Explain about the cartesian product of two sets.

Ans For two non-empty sets A and B, the cartesian product $A \times B$ is the of all ordered pairs of elements from sets A and B.

2 What do you mean by "Relation"?

Ans A relation R from set A to set B is a subset of cartesian product $A \times B$.

3 If A be a finite set having n elements, then number of relations on set A?

Ans 2^{n^2}

4 "Number of relations on A = Number of subsets of $A \times A$". Is it true?

Ans Yes, it is true.

5 "The void relation ϕ and the universal relation $A \times B$ are trivial relation from A to B." Is it true?

Ans Yes, it is true.

6 If $A = \{1, 2, 3, 4\}$, then how many relation are on A?

Ans $2^{4^2} = 1^{16}$

7 If $A = \{1, 2\}$ and $B = \{3, 4\}$, then total number of relations from A into B is ....

Ans 16

8 "If R_1 and R_2 are two relations from A to B, then $R_1 \cup R_2$, $R_1 \cap R_2$ and $R_1 - R_2$ are also relations from A to B." Is it true?

Ans Yes, it is true.

ACTIVITY 2

To represent set theoretic operations using Venn Diagrams.

Material Required

Hardboard, white thick sheet of paper, pencil, colours, scissors adhesive.

Method of Construction

1. Firstly, cut rectangular strips from a sheet of paper and paste them on a hardboard. Write the symbol U in the left/right top corner of each rectangle.
2. After that, draw two circles A and B each inside of the rectangular strips and shade/colour different portions as shown in Fig. (2.1) to Fig. (2.10).

Demonstration

1. U denotes the universal set represented by the rectangle.
2. Circles A and B represent the subsets of the universal set U as shown in the figures (2.1) to (2.10)
3. A' denote the complement of the set A and B' denote the complement of the set B as shown in the Fig. (2.3) and Fig. (2.4).
4. In the Fig. (3.1), shaded portion represents $A \cup B$.

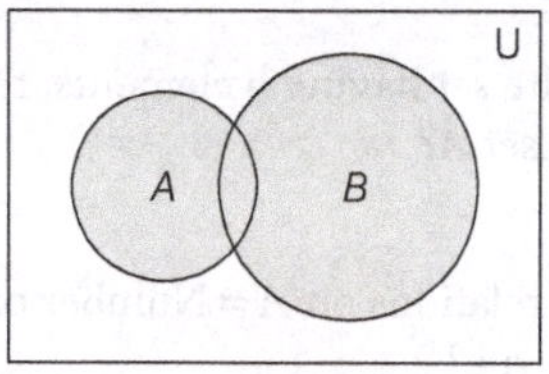

Figure 2.1

5. In the Fig. (2.2), shaded portion represents $A \cap B$.

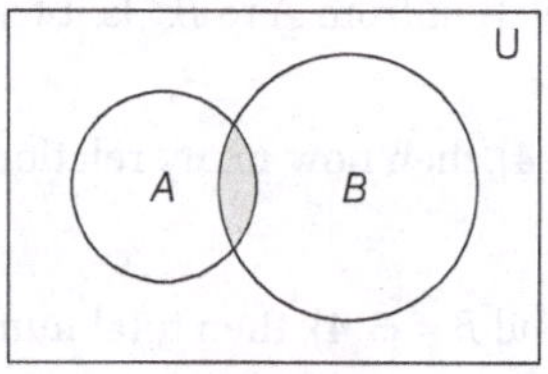

Figure 2.2

6. In the Fig. (2.3), shaded portion represents A'.

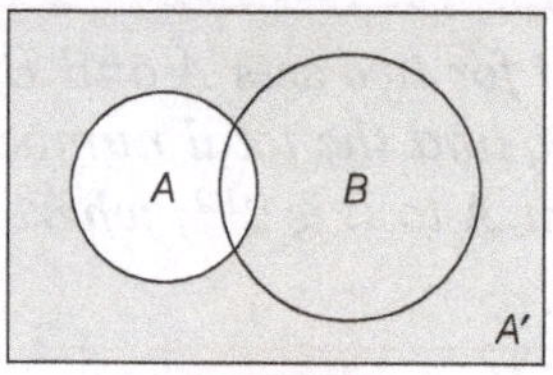

Figure 2.3

7. In the Fig. (2.4), shaded portion represents B'.

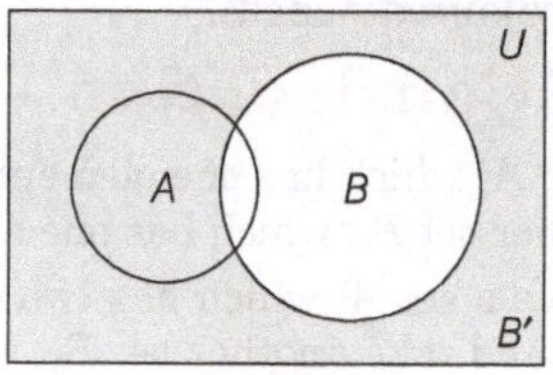

Figure 2.4

8. In the Fig. (2.5), shaded portion represents $(A \cap B)'$.

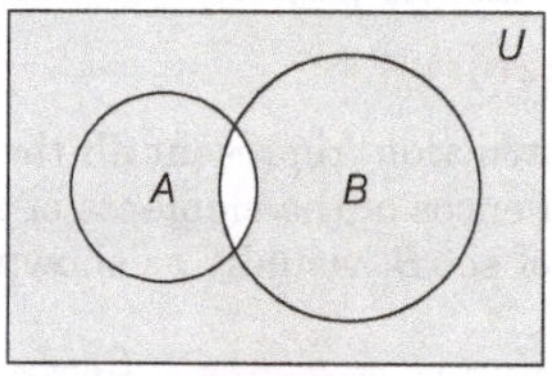

Figure 2.5

9. In the Fig. (2.6), shaded portion represents $(A \cup B)'$

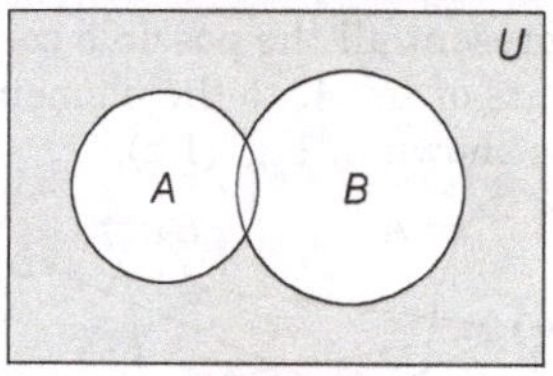

Figure 2.6

10. In the Fig. (2.7), shaded portion represents $A' \cap B$ which is similar $B - A$.

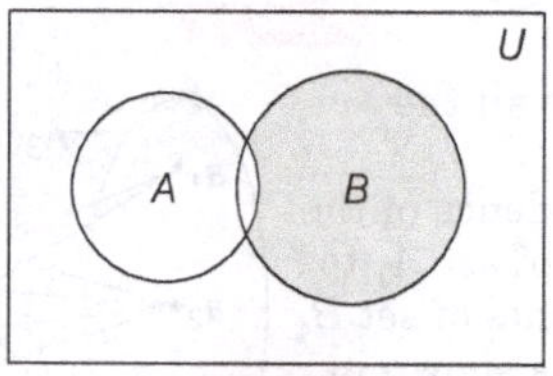

Figure 2.7

11. In the Fig. (2.8), shaded portion represents $A' \cup B$.

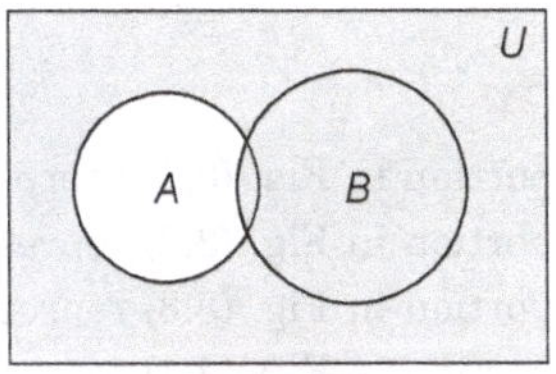

Figure 2.8

12. Fig. (2.9) shows $A \cap B = \phi$.

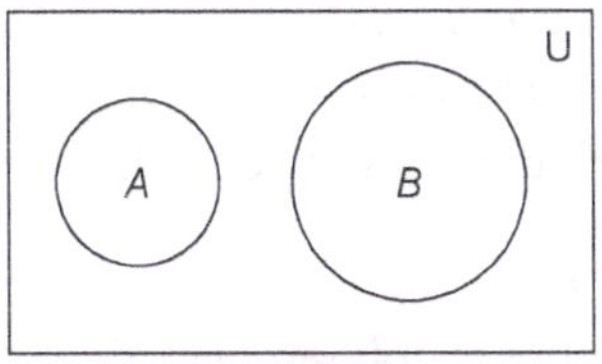

Figure 2.9

13. Fig. (2.10) shows $A \subset B$.

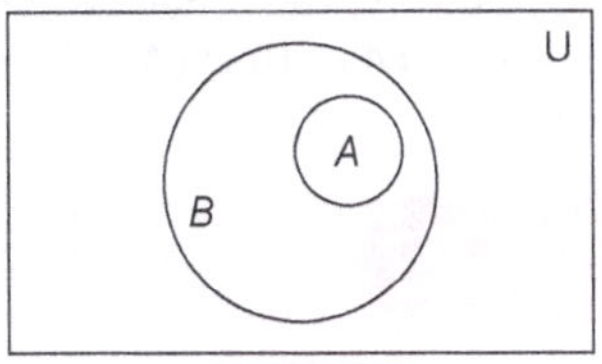

Figure 2.10

Observation

1. Coloured portion in Fig. (2.1) represents ....
2. Coloured portion in Fig. (2.2) represents ....
3. Coloured portion in Fig. (2.3) represents ....
4. Coloured portion in Fig. (2.4) represents ....
5. Coloured portion in Fig. (2.5) represents ....
6. Coloured portion in Fig. (2.6) represents ....
7. Coloured portion in Fig. (2.7) represents ....
8. Coloured portion in Fig. (2.8) represents ....
9. Fig. (2.9) shows that $(A \cap B) =$ ....
10. Fig. (2.10) represent A .... B.

Application

Set theoretic representation of Venn diagrams are used in Logic and Mathematics.

VIVA-VOCE

1 Explain about the "Union of sets".

Ans The union of A and B is the set of all those elements which belong either A or to B or to both A and B.

2 If A and B are two disjoint sets, then $n(A \cup B)$ is equals to.

Ans $n(A) + n(B)$

3 If $A = \{1, 2, 3, 4, 5\}$, then how many number of proper subsets of A is

Ans 31

4 If A and B are two sets having 4 and 7 elements respectively, then write the maximum number of elements that $A \cup B$ can have.

Ans 11

5 If $A = \{1, 2, 3\}$ and $B = \{3, 4, 5\}$, then symmetric difference of A and B is

Ans $\{1, 2, 4, 5\}$

6 Explain about the "disjoint sets".

Ans Two sets A and B are said to be disjoint, if $A \cap B = \phi$.

7 "$(A \cup B)' = \cup - (A \cup B)$". Is it true or false?

Ans True

8 If A and B are two sets, then $(A \cup B) \cap (A \cup B)$ is equal to

Ans A

ACTIVITY 3

To verify distributive law for three given non-empty sets A, B and C, that is,
$A \cup (B \cap C) = (A \cup B) \cap (A \cup C)$

Material Required

Hardboard, white thick sheets of paper, pencil, colours, scissors, adhesive.

Method of Construction

1. Firstly, cut five rectangular strips from a white sheet of paper and paste them on the hardboard in such a way that three of the rectangles are in horizontal line and two of the remaining rectangles are also placed horizontally in a line just below the above three rectangles. Write the symbol U in the left/right top corner of each rectangle as shown in Fig. (3.1) to Fig. (3.5).
2. Drawn three circles and mark them as A, B and C in each of the five rectangles as shown in the below figures.
3. Colour/shade the portions as shown in the figures.

Demonstration

1. Here, U denotes the universal set represented by the rectangle in each figure.
2. Subsets of the universal set U are represented by circles A, B and C.

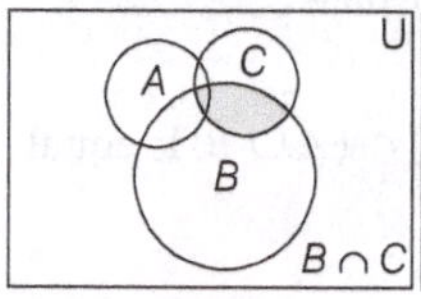

Figure 3.1

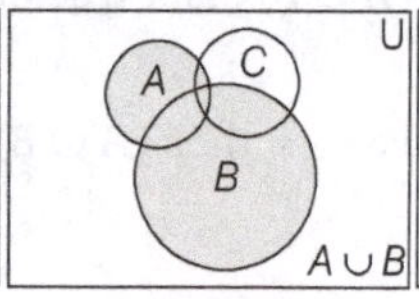

Figure 3.2

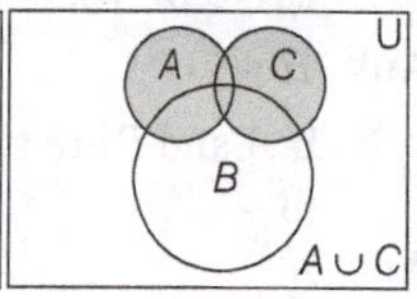

Figure 3.3

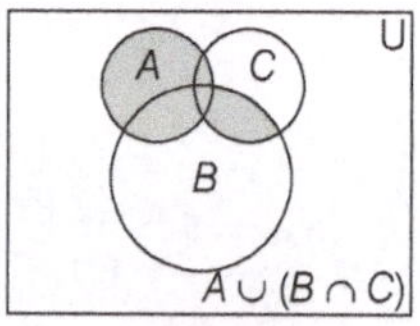

Figure 3.4

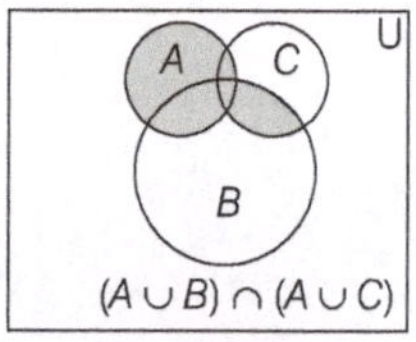

Figure 3.5

3. In Fig. 3.1 shaded portion represents $B \cap C$, shaded portions in Fig. 3.2 represents $A \cup B$. Fig. 3.3 represents $A \cup C$, Fig. 3.4 represents $A \cup (B \cap C)$ and shaded portion in Fig. 3.5 represents $(A \cup B) \cap (A \cup C)$.

Observation

1. Coloured portion in Fig. (3.1) represents ....
2. Coloured portion in Fig. (3.2) represents ....
3. Coloured portion in Fig. (3.3) represents ....
4. Coloured portion in Fig. (3.4) represents ....
5. Coloured portion in Fig. (3.5) represents ....
6. The common shaded portions in Fig. (3.4) and Fig. (3.5) are ....
7. $A \cup (B \cap C) =$ ....

Hence, the distributive law is verified.

Application

Distributivity property of set operations is used in the simplification of problems involving set operations.

Note In the same way, the other distributive law $A \cap (B \cup C) = (A \cap B) \cup (A \cap C)$ can also be verified.

VIVA-VOCE

1 Describe about the De-Morgan's law.

Ans $(A \cup B)' = A' \cap B'$ and $(A \cap B)' = A' \cup B'$

2 If A and B are two sets such that $n(A) = 115$, $n(B) = 326$, $n(A - B) = 47$, then write $n(A \cup B)$.

Ans 373

3 In which law of sets the property $A \cup A = A$ and $A \cap A = A$ satisfies?

Ans Idempotent laws

4 For any two sets A and B describe as "$A \cup (A \cap B) = A$", is it true?

Ans Yes, it is true.

5 In which law of sets the property $(A \cap B) \cap C = A \cap (B \cap C)$ satisifies?

Ans Associative law.

6 "If A, B and C are any three sets, then $A \cup (B \cap C) = (A \cup B) \cap (A \cup C)$ also known as union distributes over intersection". Is it true or false?

Ans true

7 The name of law which is apply in $A \cap \phi = \phi$.

Ans Identity law

8 Apply distributive law on $A \cap (A \cup B)$, does we get simplified form?

Ans A

ACTIVITY 4

To identify a relation and a function.

Material Required

Hardboard, battery, electric bulbs of two different colours, testing screws, tester, electrical wires and switches.

Methods of Construction

1. Firstly, take a piece of hardboard of suitable size and paste a white paper on it.
2. Drill eight holes on the left side of board in a column and mark them as A, B, C, D, E, F, G and H as shown in the Fig. (4.1).
3. Drill seven holes on the right side of the board in a column and mark them as P, Q, R, S, T, U and V as shown in the Fig (4.1).

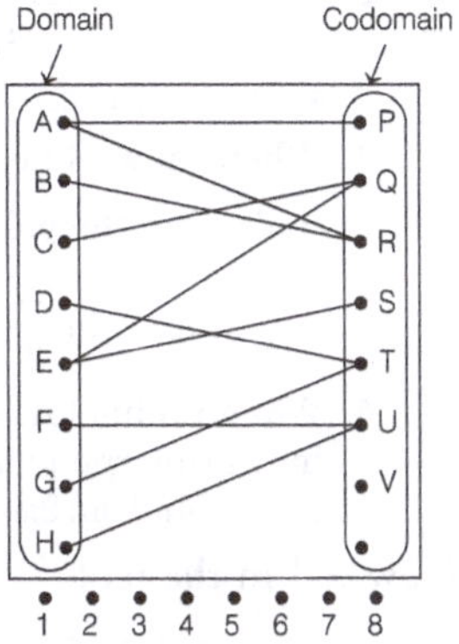

Figure 4.1

4. After that, fix bulbs of same colour in the holes A, B, C, D, E, F, G and H.
5. Similarly, fix bulbs of the other colour in the holes P, Q, R, S, T, U and V.
6. In the bottom of the board fix testing screws marked as 1, 2, 3, ..., 8.
7. Complete the electrical circuits in such a manner that a pair of corresponding bulbs, one from each column glow simultaneously.
8. These pairs of bulbs will give ordered pairs, which will constitute a relation which in turn may/may not be a function [See Fig. (4.1)].
9. A relation f from a non-empty set A to a non-empty set B is said to be a function, if every element of set A has one and only one image in set B.

Demonstration

1. Bulbs at $A, B, \ldots, H$, along the left column represent domain and bulbs along the right column at $P, Q, R, \ldots, V$ represent codomain.
2. We can obtain different order pairs out of given eight screws by using two or more testing screws. In Fig. (4.1), all the eight screws have been used to give different ordered pairs such as (A, P), (B, R), (C, Q), (A, R), (E, Q), etc.
3. By choosing different ordered pairs make different sets of ordered pairs.

Observation

1. In Fig. (4.1), ordered pairs are ....
2. These ordered pairs constitute a ....
3. The ordered pairs (A, P), (B, R), (C, Q), (E, Q), (D, T), (G, T), (F, U), (H, U) constitute a relation which is also a ....
4. The ordered pairs (B, R), (C, Q), (D, T), (E, S), (E, Q) constitute a .... which is not a ....

Application

The activity can be used to explain the concept of a relation or a function. It can also be used to explain the concept of one-one, onto functions.

VIVA-VOCE

1 Describe about the ordered pair.

Ans An ordered pair consists of two objects or elements in a given fixed order.

2 If $(3a-2, b+3) = (2a-1, 3)$, then values of a and b are.

Ans Since, $(3a-2, b+3) = (2a-1, 3)$

then by equality of ordered pairs,

$$3a - 2 = 2a - 1$$

$$\Rightarrow \quad a = 1 \text{ and } b + 3 = 3 \Rightarrow b = 0$$

3 In how many forms relation can be represented?

Ans Generally, relation is defined in four forms as following

(i) Roster form
(ii) Set builder form
(iii) By Arrow diagram
(iv) By Lattice

4 Explain about the "Inverse Relation".

Ans Let A, B be two sets and R be a relation from a set A to a set B. Then, the inverse of R, denoted by R^{-1}, is a relation from B to A and is denoted by $R^{-1} = \{(b, a) : (a, b) \in R\}$

5 Write the relation $R = \{(x, x^3) : x$ is a prime number less than 10$\}$ in Roster form.

Ans $R = \{(2, 8), (3, 27), (5, 125), (7, 343)\}$

6 If R is a relation from set $A = \{11, 12, 13\}$ to set $B = \{8, 10, 12\}$ defined by $y = x - 3$, then write R^{-1}.

Ans $\{(8, 11), (10, 13)\}$

7 If $R = \{(x, y) : x, y \in Z, x^2 + y^2 \le 4\}$ is a relation on Z, then domain of R is

Ans $\{-2, -1, 0, 1, 2\}$

8 Identify the relation $\{(2, 0), (4, 8), (2, 1), (3, 6)\}$ is function or not.

Ans It is not a function, because first element of (2, 0) and (2, 1) are same.

ACTIVITY 5

To prepare a model to illustrate the values of sine function and cosine function for different angles which are multiples of $\frac{\pi}{2}$ and π.

Material Required

A stand fitted with 0°-360° protractor and a circular plastic sheet fixed with handle which can be rotated at the centre of the protractor.

Method of Construction

1. First, take a stand fitted with 0°-360° protractor.
2. Take the radius of protractor = 1 unit.

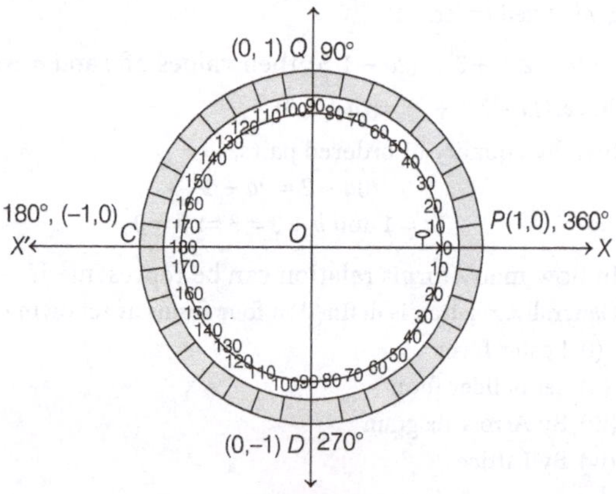

Figure 5.1

3. Draw two perpendicular lines, one joining 0°-180° line and another 90°-270° line.
4. Mark the ends of 0°-180° line as (1, 0) at 0°, (–1, 0) at 180° and that of 90°-270° line as (0, 1) at 90° (0, –1) at 270°.
5. Take a plastic circular plate and mark a line to indicate its radius and fix a handle at the outer end of the radius.
6. Fix the plastic circular plate at the centre of the protractor.

Demonstration

1. Move the circular plate in anti-clockwise direction to make different angles like $0, \frac{\pi}{2}, \pi, \frac{3\pi}{2}, 2\pi$ etc.
2. Read the values of sine and cosine function for these angles and their multiples from the perpendicular lines.

Observation

1. When radius line of circular plate is at 0° indicating the point $P(1, 0)$, $\cos 0 =$ .... and $\sin 0 =$ ....
2. When radius line of circular plate is at 90° indicating the point $Q(0, 1)$, $\cos\frac{\pi}{2} =$ .... and $\sin\frac{\pi}{2} =$ ....
3. When radius line of circular plate is at 180° indicating the point $R(-1, 0)$, $\cos \pi =$ .... and $\sin \pi =$ ....
4. When radius line of circular plate is at 270° indicating the point $S(0, -1)$, which means $\cos\frac{3\pi}{2} =$ .... and $\sin\frac{3\pi}{2} =$ ....
5. When radius line of circular plate is at 360° indicating the point again at $P(1, 0)$, $\cos 2\pi =$ .... and $\sin 2\pi =$ ....

Now, fill in the table

Trigonometric function	0	$\frac{\pi}{2}$	π	$\frac{3\pi}{2}$	2π	$\frac{5\pi}{2}$	3π	$\frac{7\pi}{2}$	4π
$\sin\theta$									
$\cos\theta$									

Application

This activity can be used to determine the values of other trigonometric functions for angles being multiple of $\frac{\pi}{2}$ and π.

VIVA-VOCE

1 Describe about the "odd function."

Ans A function $f(x)$ is an odd function, if $f(-x) = -f(x)\ \forall\ x$ in its domain.

2 Explain about the "allied angles".

Ans Two angles are said to be allied when their sum or difference is either zero or a multiple of 90°.

3 Write the trigonometric functions have positive signs in second quadrant.

Ans $\sin\theta, \cos\theta$

4 Write the maximum value of $\sin\theta$.

Ans 1

5 Write the value of $\sin(-1125°)$.

Ans We have, $\sin(-1125°) = -\sin 1125° = -\sin(90° \times 12 + 45°)$

Clearly, 1125° lies in the first quadrant. The multiple of 90° in this expression is even.

$\therefore \sin(-1125°) = -\sin(90° \times 12° + 45°) = -\sin 45° = -\dfrac{1}{\sqrt{2}}$

6 If $\cot(\alpha + \beta) = 0$, then write the value of $\sin(\alpha + 2\beta)$.

Ans $\sin\alpha$ or $\cos\beta$

7 Write the least value of $\cos^2\theta + \sec^2\theta$.

Ans 2

8 If $\tan\theta = -\dfrac{1}{\sqrt{5}}$ and θ lies in the IV quadrant, then the value of $\cos\theta$ is

Ans $\dfrac{\sqrt{5}}{\sqrt{6}}$

ACTIVITY 6

To plot the graphs of $\sin x$, $\sin 2x$, $2\sin x$ *and* $\sin\dfrac{x}{2}$, *using same coordinate axes.*

Material Required

Plyboard, squared paper, adhesive, ruler, coloured pens, eraser.

Method of Construction

1. First, take a plyboard of size 40 cm × 40 cm.
2. On the plyboard, paste a thick graph paper of size 30 cm × 30 cm.
3. Draw two mutually perpendicular lines on the squared paper, and take them as coordinate axes.
4. Graduate the two axes as shown in the Fig. (6.1).
5. Prepare the table of ordered pairs for $\sin x$, $\sin 2x$, $2\sin x$ and $\sin\dfrac{x}{2}$ for different value of x as shown in the table below:

T. ratios	0°	$\frac{\pi}{12}$	$\frac{\pi}{6}$	$\frac{\pi}{4}$	$\frac{\pi}{3}$	$\frac{5\pi}{12}$	$\frac{\pi}{2}$	$\frac{7\pi}{12}$	$\frac{2\pi}{3}$	$\frac{9\pi}{12}$	$\frac{5\pi}{6}$	$\frac{11\pi}{12}$	π
$\sin x$	0	0.26	0.50	0.71	0.86	0.97	1.00	0.97	0.86	0.71	0.50	0.26	0
$\sin 2x$	0	0.50	0.86	1.00	0.86	0.50	0	−0.5	−0.86	−1.0	−0.86	−0.50	0
$2\sin x$	0	0.52	1.00	1.42	1.72	1.94	2.00	1.94	1.72	1.42	1.00	0.52	0
$\sin\frac{x}{2}$	0	0.13	0.26	0.38	0.50	0.61	0.71	0.79	0.86	0.92	0.97	0.99	1.00

Demonstration

1. Plot the ordered pair $(x, \sin x)$, $(x, \sin 2x)$, $\left(x, \sin\dfrac{x}{2}\right)$ and $(x, 2\sin x)$ on the same coordinates axes and join the plotted ordered pairs by free hand curves in different colours as shown in the Fig. (6.1).

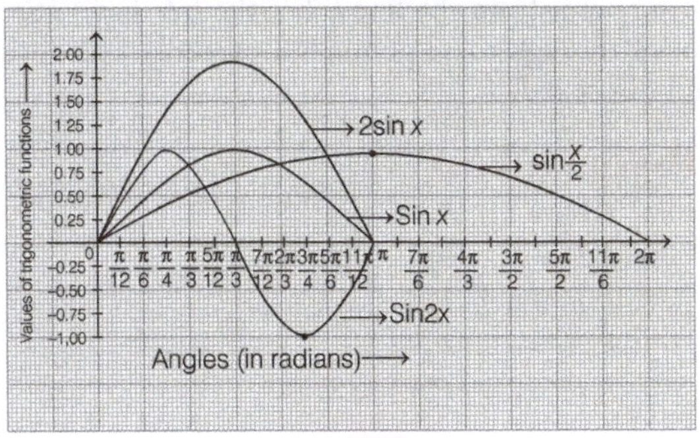

Figure 6.1

Observation

1. Graphs of $\sin x$ and $2\sin x$ are of same shape but the maximum height of the graph of $\sin x$ is .... the maximum height of the graph of ....
2. The maximum height of the graph of $\sin 2x$ is .... It is at $x =$ ....
3. The maximum height of the graph of $2\sin x$ is .... It is at $x =$ ....
4. The maximum height of the graph of $\sin\dfrac{x}{2}$ is .... It is at $\dfrac{x}{2} =$ ....
5. At $x =$ .... , $\sin x = 0$, at $x =$ .... , $\sin 2x = 0$ and $x =$ ...., $\sin\dfrac{x}{2} = 0$.
6. In the interval $[0, \pi]$, graphs of $\sin x$, $2\sin x$ and $\sin\dfrac{x}{2}$ are .... X-axis and some portion of the graph of $\sin 2x$ lies .... X-axis.

7. Graphs of $\sin x$ and $\sin 2x$ intersect at $x=$ .... in the interval $(0, \pi)$.
8. Graphs of $\sin x$ and $\sin \frac{x}{2}$ intersect at $x=$ ...... in the interval $(0, \pi)$.

Application

This activity may be used in comparing graphs of a trigonometric function of multiples and submultiples of angles.

VIVA-VOCE

1 Write any two trigonometric identities.

Ans (i) $\sin^2\theta + \cos^2\theta = 1$ (ii) $1 + \tan^2\theta = \sec^2\theta$

2 "Trigonometric ratios are same for the same angle", is it true?

Ans No, trigonometric ratios are not same for the same angle.

3 If $\sec\theta = \frac{1}{2}$, then it is true or false.

Ans false, $[\because \sec\theta \notin (-1, 1)]$

4 If $\tan\theta + \sec\theta = \sqrt{3}, 0 < \theta < \pi$, then θ is equal to

Ans $\pi/6$

5 In which quadrant tan and cot functions have positive signs?

Ans III quadrant

6 $\sin 1° > \sin 1$, is it true?

Ans No, it is false. Since, $\sin 1° < \sin 1$

7 Write the maximum and minimum values of $\sin(\sin x)$.

Ans $\sin 1, -\sin 1$

8 Write the value of $\tan\left(\frac{19\pi}{3}\right)$.

Ans We have, $\frac{19\pi}{3} = \left(\frac{19}{3} \times 180\right)^\circ = 1140° = 90° \times 12 + 60°$

Clearly, the angle lies in first quadrant.

$\therefore \tan\left(\frac{19\pi}{3}\right) = \tan(90° \times 12 + 60°) = \tan 60° = \sqrt{3}$

ACTIVITY 7

To construct a parabola.

Material Required

Cardboard, white paper, sketch pen, pencil, compass, ruler etc.

Method of Construction

1. First, take a cardboard of a convenient size and paste a white paper on it.
2. Mark a point S on the white paper on the board (see Fig. (7.1)).
3. Through the point S draw a line. Draw another line l perpendicular to the line through S at some distance λ units to the left of S.

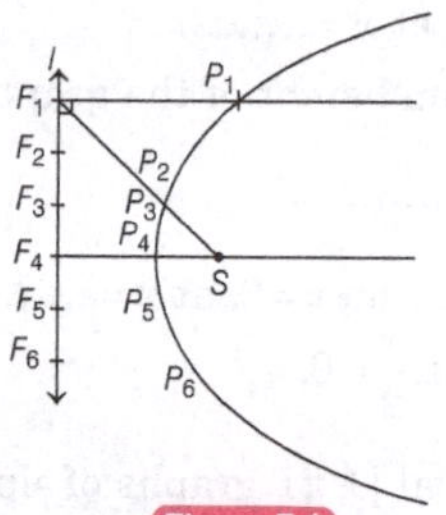

Figure 7.1

4. Take any point F_1 on the line l. Draw the perpendicular to l at this point.
5. Join F_1S and draw perpendicular bisector of F_1S meeting the perpendicular through F_1 at the point P_1.
6. Take another point F_2 on the line l and repeat the process as explained in Step 5 above to obtain the point P_2.
7. Take some more points $F_3, F_4, F_5, \ldots$ on the line l and repeat the above process to obtain points $P_3, P_4, P_5, \ldots$, respectively.
8. Draw a free hand curve through the points $P_1, P_2, P_3, P_4, \ldots$ (see Fig. (7.1))

Demonstration

The points $P_1, P_2, P_3, \ldots\ldots$ are such that the distance of each point from the fixed point S is same as the distance of the point from the line l. So, the free hand curve drawn through these points is a parabola with focus S and directrix l.

Observation

1. $P_1F_1 =$ ...... $P_1S =$ ......
2. $P_2F_2 =$ ...... $P_2S =$ ......
3. $P_3F_3 =$ ...... $P_3S =$ ......
4. $P_4F_4 =$ ...... $P_4S =$ ......
5. $P_5F_5 =$ ...... $P_5S =$ ......

6. The distance of the point P_1 from F_1 = The distance of P_1 from ....
7. The distance between the points P_2 and F_2 = The distance of P_2 from ....

   The distance of the point ...... from F_3 = The distance of the point P_3 from ....
8. Distances of the points $P_1, P_2, P_3, \ldots$ from the line l are ...... to the distances of these points from the point S.
9. Therefore, the free hand curve obtained by joining $P_1, P_2, P_3, \ldots$ is a ...... with directrix ...... and focus ....
10. Distance of the vertex P_4 and S = ...
11. Distance of the vertex of parabola from the directrix = ...

Applications

1. This activity is useful in understanding the terms related to parabola, like directrix, focus, property of the point on the parabola.
2. Parabolas have some applications in Science and Engineering.

VIVA-VOCE

1 Describe about the "conic section".

Ans A conic section is the locus of a point P which moves in such a way that its distances fram a fixed point S always bears a constant ratio to its distance from a fixed line, all being in the same plane.

2 Write the axis of the symmetry of the parabola $y^2 = x$.

Ans X-axis.

3 Write the vertex of the parabola $(y+a)^2 = 8a(x-a)$.

Ans $(a, -a)$

4 "If $e > 1$, then the conic is a parabola", is it true?

Ans No, if $e > 1$, then the conic is a hyperbola.

5 Explain about the latus rectum of a conic.

Ans The latus rectum of a conic is the chord passing through the focus and perpendicular to the axis.

6 Explain focal distance.

Ans The distance of a point $P(x, y)$ on parabola from the focus is called focal distance.

7 Write the formula for length of latus rectum of parabola.

Ans $4a$

ACTIVITY 8

To construct an ellipse using a rectangle.

Material Required

A hardboard, white paper, coloured paper, nails, nylon wire/thread, ruler, adhesive.

Method of Construction

1. Firstly, take a rectangular hardboard of a convenient size and paste a white paper on it.
2. After that, cut a rectangle $PQRS$ of suitable dimensions from a coloured paper and paste it on the hardboard.
3. Divide this rectangle into four congruent rectangles as shown in the Fig. (8.1)
4. Divide each of the sides RC and DC of the rectangle $ADCR$, into some equal parts.
5. Mark the point of subdivisions of RC as $A_1, A_2, \ldots$ and that of DC as D_1, D_2 (see Fig. (8.1))
6. Join the point A to points, $A_1, A_2, \ldots\ldots$ and draw the lines joining the point X to $D_1, D_2, \ldots\ldots$ (see fig. (8.1))
7. Mark the point of intersection of AA_1 and XD_1 as R_1, AA_2 and XD_2 as R_2 and so on.

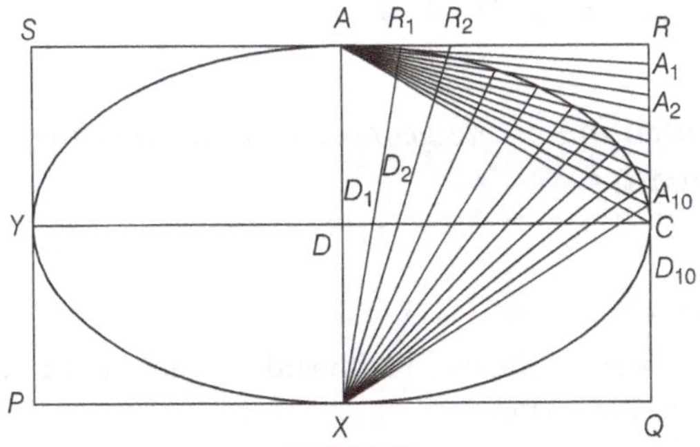

Figure 8.1

8. Fix nails at the points $R_1, R_2, \ldots\ldots R_{10}$.
9. Join the feet of nails with a nylon wire/thread, as shown in the figure.
10. Repeat the same activity for remaining three congruent rectangles and obtain a curve as shown in Fig. (8.1).

Demonstration

The curve obtained looks like an ellipse. The major axis of this ellipse is the length of the rectangle $PQRS$ and the minor axis of the ellipse is the breadth of the rectangle.

Observation

1. Length of the rectangle $PQRS$ = ...
2. Breadth of the rectangle $PQRS$ = ...
3. Major axis of the ellipse is ....
4. Minor axis of the ellipse is ....

Application

This activity may be helpful in understanding the concept such as major and minor axis of an ellipse. It is also useful in drawing elliptical designs such as in swimming pools, tables, etc.

VIVA-VOCE

1 Describe about the "ellipse".

Ans An ellipse is the locus of a point in plane which moves in the plane in such a way that the ratio of its distance from a fixed point in the same plane to its distance from a fixed straight line is always constant which is always less than unity.

2 Write the formula of eccentricity to find the eccentricity of the ellipse.

Ans $e^2 = 1 - \frac{b^2}{a^2}$ are $e = \sqrt{1 - \frac{b^2}{a^2}}$

3 If $e < 1$, then the conic is a ellipse, is it true?

Ans Yes, it is true.

4 Write the length of latusrectum for the ellipse $\frac{x^2}{a^2} + \frac{y^2}{b^2}, (a < b)$.

Ans $\frac{2a^2}{b}$.

5 Write the second focus and second directrix of the ellipse.

Ans Second focus $S'(-ae, 0)$ and second directrix is $x = -\frac{a}{e}$.

6 Write the formula of focal distance for ellipse.

Ans $2c$

7 Write the relation a, b and c.

Ans $c = \sqrt{a^2 - b^2}$

8 If $e = \frac{c}{a}$, then it is true or false.

Ans True.

ACTIVITY 9

Verification of the geometrical significance of derivative.

Material Required

Graph sheets, adhesive, hardboard, trigonometric tables, geometry box, wires.

Method of Construction

1. Firstly, paste three graph sheets on a hardboard and draw two mutually perpendicular lines representing X-axis and Y-axis on each of them.
2. After that, sketch the graph of the curve (circle) $x^2 + y^2 = 25$ on one sheet.
3. Similarly sketch the graphs of $(x-3)^2 + y^2 = 25$ and the curve $xy = 4$ (rectangular hyperbola). On the other two graph sheets.

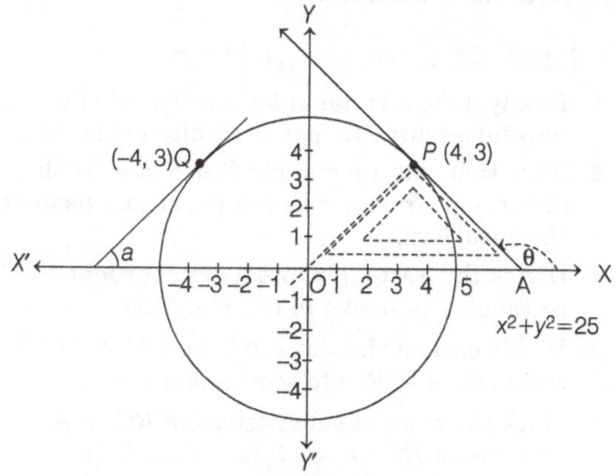

Figure 9.1

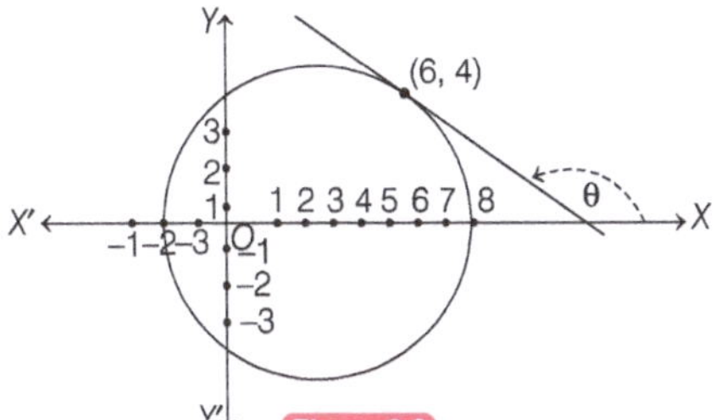

Figure 9.2

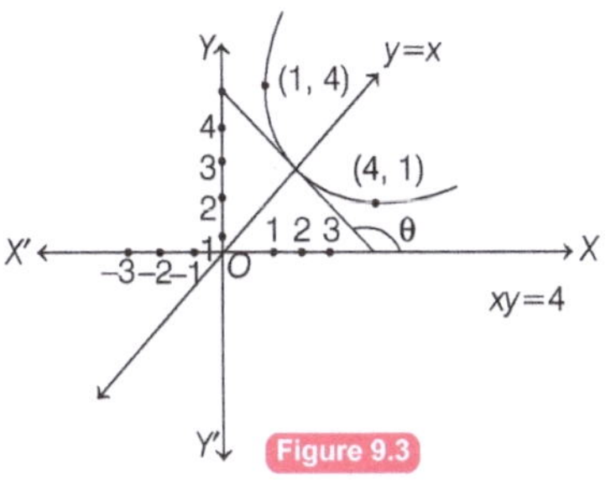

Figure 9.3

Demonstration

1. Firstly, take first sheet on which, the graph of the circle $x^2 + y^2 = 25$ has been drawn (see Fig. (9.1))
2. Take a point $P(4, 3)$ on the circle.
3. With the help of a set square, place a wire in the direction OP and other perpendicular to OP at the point P to meet X-axis at a point (say A).
4. Measure the angle between the wire and the positive direction of X-axis at A (sat θ).
5. Then find $\tan\theta$ (with the help of trigonometric tables)
   Now, $x^2 + y^2 = 25 \Rightarrow y = \sqrt{25 - x^2} \Rightarrow \dfrac{dy}{dx} = \dfrac{-x}{\sqrt{25 - x^2}}$.
   Find $\dfrac{dy}{dx}$ at the point (4, 3) and verify that $\left(\dfrac{dy}{dx}\right)$ at $(4, 3) = \tan\theta$.
6. Similarly, take another point $(-4, 3)$ on the circle. Verify that $\dfrac{dy}{dx}$ at $(-4, 3) = \tan\alpha$, where α is the angle made by the tangent to the circle at the point $(-4, 3)$ with the positive direction of X-axis. (see Fig. (9.1)).
7. Take other sheet with the graph of $(x-3)^2 + y^2 = 25$ and take the point (6, 4) on it and repeat the above process using set square and wires as shown in Fig. (9.2) i.e. verify that $\dfrac{dy}{dx}$ at $(6, 4) = \tan\theta$.
8. In the end, take the third sheet, showing the graph of the curve $xy = 4$. Take the point (2, 2) on it. Place one perpendicular side of set square along the line $y = x$ and a wire along the other side touching the curve at the point (2, 2) and find the angle made by the wire with the positive direction of X-axis as shown in Fig. (9.3). Let it be θ then, verify that $\dfrac{dy}{dx}$ at $(2, 2) = \tan\theta$.

Observation

1. For the curve $x^2 + y^2 = 25$, $\dfrac{dy}{dx}$ at the point $(3, 4) =$ .... Value of $\theta =$ ...... $\tan\theta =$ ...... $\dfrac{dy}{dx}$ at $(3, 4) =$ ....
2. For the curve $x^2 + y^2 = 25$, $\dfrac{dy}{dx}$ at $(-4, 3) =$ ...., $\tan\alpha =$ ...., $\dfrac{dy}{dx}$ at $(-4, 3) =$ ....
3. For the curve $(x-3)^2 + y^2 = 25$, $\dfrac{dy}{dx}$ at $(6, 4) =$ .... , value of $\theta =$ .... $\tan\theta =$ ...... , $\dfrac{dy}{dx}$ at $(6, 4) =$ ....
4. For the curve $xy = 4$, $\left(\dfrac{dy}{dx}\right)$ at $(2, 2) =$ ...., $\theta =$ ...., $\tan\theta =$ ....

Application

This activity can be used to verify the result that the slope of the tangent at a point is equal to the value of the derivative at that point for other curves.

Note The activity may be repeated by taking point (4, 3) on first sheet, (0, 4) on second sheet and (1, 4) on the third sheet.

VIVA-VOCE

1 Explain sandwich theorem.

Ans Let f, g and h be real functions such that $f(x) \le g(x) \le h(x)$ for all x is common domain, if $\lim\limits_{x \to a} f(x) = l = \lim\limits_{x \to a} h(x)$, then $\lim\limits_{x \to a} g(x) = l$

2 If $\lim\limits_{x \to 0} \dfrac{\sin x}{x} = 2$, then is it true or false.

Ans False

3 Write the formula for first principle of derivative.

Ans $f'(x) = \lim\limits_{x \to 0} \dfrac{f(x+h) - f(x)}{h}$

4 $\dfrac{d}{dx}(x^n) = nx^{n-1}$, which is true or false.

Ans True.

5 Write the derivative of $\sqrt{x}$.

Ans $\dfrac{d}{dx}(\sqrt{x}) = \dfrac{1}{2\sqrt{x}}$

6 Write the derivative of $\sin x$.

Ans $\cos x$

7 $\dfrac{d}{dx}(\cos x) = \sin x$, which is true or false.

Ans False.

8 Write the derivative of $\sin ax$.

Ans $a \cos ax$

Periodic Tests

Pre-Mid Term Tests

[Chapters Covered — 1-5] **10 Marks**

TEST 1

1. If $f(x) = x^2 - x^{-2}$, then $f\left(\frac{1}{x}\right)$ is equal to (1)

(a) $f(x)$ (b) $-f(x)$
(c) $2f(x)$ (d) None of these

2. $\sin(-330^\circ)$ is equal to (1)

(a) $\frac{\sqrt{3}}{2}$ (b) $\frac{1}{2}$
(c) $\frac{-1}{2}$ (d) $\frac{-\sqrt{3}}{2}$

3. Evaluate $\sqrt{-25} + 3\sqrt{-4} + 2\sqrt{-9}$ (1)

(a) $15\,i$ (b) $16\,i$
(c) $17\,i$ (d) $18\,i$

4. Find the values of x for which the functions $f(x) = 3x^2 - 1$ and $g(x) = 3 + x$ are equal. (1)

5. Find the modulus of the complex number $4 + 3i^7$. (1)

6. If $y = f(x) = \frac{1-x}{1+x}$, then show that $x = f(y)$. (1)

7. Prove that

$$\sin 2x + 2\sin 4x + \sin 6x = 4\cos^2 x \sin 4x$$ (2)

8. Write the following intervals in set builder form

(i) (5, 15) (ii) (– 3, 2]
(iii) [– 5, 12] (iv) [– 5, 16) (2)

TEST 2

1. If set A has 4 elements and the set $B = \{2, 3, 5\}$, then find the number of elements in $A \times B$. (1)

(a) 10 (b) 6 (c) 7 (d) 12

2. If $|z| = 1$, then value of $\frac{1+z}{1+\bar{z}}$ is (1)

(a) 1 (b) z (c) z^2 (d) 0

3. The set $\{x : x \in R, -2 < x \le 3\}$ in roster form is (1)

(a) [–2, 3] (b) [–2, 3) (c) (–2, 3] (d) (–2, 3)

4. If $\tan A = \frac{1}{4}$ and $\tan B = \frac{1}{5}$, then find the value of $\tan(A + B)$. (1)

5. Express $\frac{1}{1 - \cos\theta + 2i\sin\theta}$ in the form of $a + ib$. (1)

6. Let A and B be two sets such that $n(A) = 5$ and $n(B) = 2$. If a, b, c, d and e are distinct and $(a, 2), (b, 3), (c, 2), (d, 3)$ and $(e, 2)$ are elements of $A \times B$, find A and B. (1)

7. Evaluate $(1 + i)^6 + (1 - i)^3$ (2)

8. Let $f = \left\{\left(x, \frac{x^2}{1+x^2}\right) : x \in R\right\}$ be a function R into R determine the range of R. (2)

TEST 3

1. Length of an arc of a circle of radius 3 cm, if the angle subtended at the centre is 30°, is [$\pi = 3.14$] (1)

(a) 3.14 cm (b) 1.57 cm (c) 4.42 cm (d) 2.57 cm

2. Evaluate : $1 + i^2 + i^4 + i^6 + \ldots + i^{20}$. (1)

(a) 0 (b) 1 (c) 2 (d) 3

3. If $a \in N$ such that $aN = \{aN : n \in N\}$, then set $3N \cap 7N$ is (1)

(a) 21 N (b) 24 N (c) 27 N (d) 30 N

4. Let $f : R \to R$ be given by $f(x) = x^2 + 3$, then find $\{x : f(x) = 28\}$ (1)

5. Show that $\sin(40^\circ + \theta)\cos(10^\circ + \theta) - \cos(40^\circ + \theta)\sin(10^\circ + \theta) = \frac{1}{2}$ (1)

6. Find the real numbers x and y, if $(x - iy)(3 + 5i)$ is the conjugate of $(-1 - 3i)$. (1)

7. Find $\left|\frac{(1+2i)}{(2+3i)(1-i)}\right|$. (2)

8. If x and y are acute angles such that $\cos x = \frac{13}{14}$ and $\cos y = \frac{1}{7}$, then prove that $x - y = -\frac{\pi}{3}$. (2)

Answers

Test 1

1. (b) 2. (b) 3. (c) 4. $x = -1, \frac{4}{3}$ 5. 5

8. (i) $\{x : x \in R, 5 < x < 15\}$ (ii) $\{x : x \in R, -3 < x \le 2\}$ (iii) $\{x : x \in R, -5 \le x < 12\}$ (iv) $\{x : x \in R, -5 \le x < 16\}$

Test 2

1. (d) 2. (b) 3. (c) 4. $\frac{9}{19}$ 5. $\left(\frac{1-\cos\theta}{2-2\cos\theta+3\sin^2\theta}\right)+i\left(\frac{-2\sin\theta}{2-2\cos\theta+3\sin^2\theta}\right)$

6. $A = \{a, b, c, d, e\}$, $B = \{2, 3\}$ 7. $-2-10i$ 8. Range $(f) = [0, 1]$

Test 3

1. (b) 2. (b) 3. (a) 4. $x \in \{-5, 5\}$ 6. $x = \frac{6}{17}$ and $y = \frac{-7}{17}$ 7. $\sqrt{\frac{5}{26}}$

Mid Term Tests

[Chapters Covered — 1-10] **10 Marks**

TEST 1

1. Evaluate $\frac{n!}{(n-r)!}$, when $r = 3$ (1)
   (a) $n(n-1)$ (b) $n(n-1)(n-2)$
   (c) $n(n+1)(n+2)$ (d) $n(n+2)(n+3)$
2. The number of terms in the GP $2, 2\sqrt{2}, 4, \dots, 256$ is (1)
   (a) 10 (b) 15
   (c) 5 (d) 3
3. If $A = \{1, 3, 5, 7, 9\}$ and $B = \{2, 3, 5, 7, 11\}$, then value of $A - B$ is (1)
   (a) $\{2, 9\}$ (b) $\{1, 8\}$
   (c) $\{1, 9\}$ (d) None of these
4. If $\alpha = {}^mC_2$, then find the value of ${}^{\alpha}C_2$. (1)
5. Find the imaginary part of $(4-3i)(4+3i)(1+i)$. (1)
6. Let $f : R \to R$ and $g : R \to R$ defined by $f(x) = x + 1$, $g(x) = 2x - 3$, find $f + g$, $f - g$ and $\frac{f}{g}$. (1)
7. Solve $|3x - 2| \le \frac{1}{2}$. (2)
8. A student has to answer 10 questions choosing atleast 4 from each of parts A and B, if there are 6 questions in part A and 7 in part B, in how many ways can the student choose 10 questions? (2)

TEST 2

1. Value of x and y, if $x + 4iy = ix + y + 3$ (1)
   (a) $x = 4, y = 1$ (b) $x = 2, y = 2$
   (c) $x = 5, y = 2$ (d) $x = 6, y = 3$
2. If ${}^nC_{10} = {}^nC_{12}$, then value of ${}^{23}C_n$ is (1)
   (a) 21 (b) 22
   (c) 23 (d) 24
3. Slope of a line joining two points $A(1, 2)$ and $B(3, -4)$ is (1)
   (a) -3 (b) -4
   (c) -5 (d) -6
4. Find the complex conjugates of
   (i) $2 + 3i$ (ii) $3 - 4i$ (iii) $-i$ (1)
5. Solve the linear inequality $-5x + 25 > 0$. (1)
6. How many three digits are divisible by 7 ? (1)
7. If $Z_1 = 3 + i$ and $Z_2 = 1 + 4i$, then verify that $|Z_1 + Z_2| < |Z_1| + |Z_2|$. (2)
8. Find the equation of the line which passes through the point $(-4, 3)$ and the portion of the line intercepted between the axes is divided internally in the ratio 5 : 3 by this point. (2)

TEST 3

1. If ${}^nP_r = 720$ and ${}^nC_r = 120$, then the value of r is (1)
(a) 2 (b) 3 (c) 4 (d) 5

2. If $\cos A = 1/3$, then value of $\cos 2A$ is (1)
(a) $-\frac{5}{9}$ (b) $-\frac{7}{9}$ (c) -1 (d) 5

3. If $(1 + i)Z = (1 - i)\overline{Z}$, then the value of Z is (1)
(a) $-i\overline{Z}$ (b) $i\,\overline{Z}$ (c) $-\overline{Z}$ (d) $\overline{Z}$

4. If $A = \{1, 2, 3, 4\}$, $B = \{3, 4, 5, 6\}$ and $C = \{5, 6, 7, 8\}$, then find $A \cup B \cup C$. (1)

5. Solve the inequality $\frac{1}{x-4} < 0$. (1)

6. Prove that $\sum_{r=0}^{n} 5^r \cdot {}^nC_r = 6^n$ (1)

7. Find the sum of the series
$2 + 6 + 18 + 54 + \ldots + 4374$ (2)

8. Find the ratio in which the line $y = 0$, divides the line joining the points $(3, -5)$ and $(-4, 7)$. (2)

Answers

Test 1

1. (b) 2. (b) 3. (c) 4. $\frac{(m+1)\,m(m-1)\,(m-2)}{8}$ 5. 25

6. $f + g = 3x - 2;\ f - g = -x + 4$ and $\frac{f}{g} : R - \left\{\frac{3}{2}\right\} \to R,\ \frac{f}{g} = \frac{x+1}{2x-3},\ g = 2x - 3 \neq 0$ 7. $\left[\frac{1}{2}, \frac{5}{6}\right]$

8. Number of ways = ${}^6C_4 \times {}^7C_6 \times {}^6C_5 \times {}^7C_5 \times {}^6C_6 \times {}^7C_4$

Test 2

1. (a) 2. (c) 3. (a) 4. (i) $2 - 3i$ (ii) $3 + 4i$ (iii) i
5. $(-\infty, 5)$ 6. 128 8. $9x - 20y + 96 = 0$

Test 3

1. (b) 2. (b) 3. (a) 4. $\{1, 2, 3, 4, 5, 6, 7, 8\}$ 5. $(-\infty, 4)$ 7. 6560 8. 5 : 7

Post-Mid Term Tests

[Chapters Covered — 1-14] **10 Marks**

TEST 1

1. Distance between the directrices of the ellipse $\frac{x^2}{36} + \frac{y^2}{20} = 1$. (1)
(a) 9 (b) 15 (c) 18 (d) 21

2. Modulus of the complex number $1 - i$ equals to (1)
(a) $-\sqrt{2}$ (b) $\sqrt{2}$
(c) 3 (d) 2

3. If ${}^{56}P_{r+6} : {}^{54}P_{r+3} = 30800 : 1$, then the value of r is (1)
(a) 41 (b) 43
(c) 42 (d) 44

4. Find the equation of the hyperbola, whose vertices are $(0, \pm 5)$ and foci $(0, \pm 8)$. (1)

5. Find the value of x, if the distance between two points $(x, -8, 4)$ and $(3, -5, 4)$ is 5. (1)

6. Evaluate: $\lim_{x \to \frac{1}{2}} \frac{4x^2 - 1}{2x - 1}$ (1)

7. How many numbers greater than 1000000 can be formed using the digits 2, 3, 0, 3, 4, 2, 3. (2)

8. Find the mean deviation about the mean for the following data
48, 42, 55, 38, 70 (2)

TEST 2

1. If $A = \{x, y, z\}$ and $B = \{1, 2\}$, then the number of relation from A to B is (1)
(a) 32 (b) 48
(c) 64 (d) 72

2. Evaluate $\lim_{x \to 2} (x^2 - x + 1)$. (1)
(a) 2 (b) 3
(c) 4 (d) 5

3. If A and B are two events such that $P(A) = 0.54$, $P(B) = 0.69$ and $P(A \cap B) = 0.35$, then value of $P(A \cup B)$ is (1)
(a) 0.88 (b) 0.72
(c) 0.36 (d) 0.48

4. If $a\cos\theta + b\sin\theta = x$ and $a\sin\theta - b\cos\theta = y$, then prove that $a^2 + b^2 = x^2 + y^2$. (1)

5. Evaluate ${}^nC_r + 2 \cdot {}^nC_{r-1} + {}^nC_{r-2}$ (1)

6. If a, b, c, d are in GP, then prove that $a + b, b + c, c + d$ are also in GP. (1)

7. Solve $\frac{x+8}{x-2} \geq 0$ (2)

8. Find the coordinates of the focus and the vertex, equation of directrix and length of latusrectum of the following
(i) $y^2 = 10x$
(ii) $5y^2 = -16x$ (2)

TEST 3

1. Solve $2x - 5 \leq \frac{(4x-7)}{3}$. (1)
(a) $[-4, \infty)$ (b) $(-4, \infty)$
(c) $(-\infty, 4)$ (d) $(-\infty, 4]$

2. Evaluate $\lim_{x \to 1} \frac{(x-1)^2 + 3x^2}{(x^4+1)^2}$ (1)
(a) $\frac{3}{4}$ (b) $\frac{7}{4}$ (c) $\frac{9}{4}$ (d) $\frac{11}{4}$

3. A fair coin is tossed four times. What is the probability that almost three tail occur? (1)
(a) $\frac{14}{15}$ (b) $\frac{15}{18}$ (c) $\frac{15}{16}$ (d) $\frac{9}{10}$

4. Write the set $A = \{3^n - 2^n : n \in N \text{ and } 1 \leq n \leq 5\}$ in roster form. (1)

5. The following relates to sample of size 60, $\Sigma x^2 = 18000$ and $\Sigma x = 960$, find the variance. (1)

6. Find the domain of the function
$$f(x) = \frac{x}{x^2 + 3x + 2}$$ (1)

7. Find the conjugate of
$$Z = (3 + 2i)^2$$ (2)

8. Find the probability that, when a hand of 7 cards is drawn from a well-shuffled deck of 52 cards, it contains
(i) all kings
(ii) 3 kings
(iii) at least 3 kings (2)

Answers

Test 1

1. (c) **2.** (b) **3.** (a) **4.** $\frac{y^2}{25} - \frac{x^2}{39} = 1$ **5.** $x = 7$ as -1 **6.** 2
7. 360 **8.** 9.52

Test 2

1. (c) **2.** (b) **3.** (a) **5.** ${}^{n+2}C_r$ **7.** $(-\infty, -8] \cup (2, \infty)$
8. (i) $F\left(\frac{5}{2}, 0\right)$; $O(0, 0)$; $2x + 5 = 0$; 10 units (ii) $F\left(-\frac{4}{5}, 0\right)$; $O(0, 0)$; $5x - 4 = 0$; $\frac{16}{5}$ units

Test 3

1. (d) **2.** (a) **3.** (c) **4.** $\{1, 5, 19, 65, 211\}$ **5.** 44
6. $R - \{-1, -2\}$ **7.** $5 - 12i$ **8.** (i) $\frac{1}{7735}$ (ii) $\frac{9}{1547}$ (iii) $\frac{46}{7735}$

3 SAMPLE QUESTIONS PAPERS

SAMPLE QUESTION PAPER 1

A Highly Simulated Sample Question Paper for CBSE Class XI Examination

MATHEMATICS (Unsolved)

Time : 3 Hours **Max. Marks : 80**

General Instructions

(i) This question paper contains - five sections **A, B, C, D** and **E**. Each section is compulsory. However, there are internal choices in some questions.
(ii) **Section A** has 18 **MCQ's** and 02 **Assertion-Reason** based questions of 1 mark each.
(iii) **Section B** has 5 **Very Short Answer (VSA)-type** questions of 2 marks each.
(iv) **Section C** has 6 **Short Answer (SA)-type** questions of 3 marks each.
(v) **Section D** has 4 **Long Answer (LA)-type** questions of 5 marks each.
(vi) **Section E** has 3 **source based/case based/passage based/integrated units of assessment** of 4 marks each with sub parts.

SECTION A

Multiple Choice Questions

Each question carries 1 mark

1. Let $A = \{1, 2, 3, 4, 5, 6, 7, 8, 9, 10\}$ and $B = \{2, 3, 5, 7\}$. Then, which of the following is true?
(a) $A \cap B = A$ (b) $A \cap B = B$
(c) $A \cap B \not\subset B$ (d) None of these

2. The number of combination of 4 different objects A, B, C, D taken 2 at a time is
(a) 4 (b) 6 (c) 7 (d) 8

3. The marks obtained by 7 students are 8, 9, 11, 13, 14, 15, 21. The variance of these marks is
(a) 18.20 (b) 16.29 (c) 16.80 (d) 16.10

4. A coin is tossed twice. Then, the probability that atleast one tail occurs is
(a) $\frac{1}{4}$ (b) 1 (c) $\frac{1}{3}$ (d) $\frac{3}{4}$

5. Find a point on the X-axis which is equidistant from the points (7, 6) and (3, 4).
(a) $\left(0, \frac{15}{2}\right)$ (b) $\left(\frac{15}{2}, 0\right)$ (c) $\left(\frac{15}{2}, \frac{15}{2}\right)$ (d) $\left(\frac{-15}{2}, 0\right)$

6. Find the values of x and y, if $x + i(3x - y) = 3 - 6i$.
(a) $x = 3, y = 15$ (b) $x = 15, y = 3$
(c) $x = 0, y = 15$ (d) $x = 3, y = 7$

7. Evaluate $\lim_{x \to 2} \frac{x-2}{\sqrt[3]{x} - \sqrt[3]{2}}$.
(a) $2(3^{2/3})$ (b) 2
(c) $3(2^{2/3})$ (d) None of these

8. If the odds against the occurrence of an event are 3 : 5. Find the probability of occurrence of the event.
(a) $\frac{5}{8}$ (b) $\frac{8}{5}$ (c) 0 (d) 1

9. Set builder form of set $A = \{3, 6, 9, 12, 15\}$ is
(a) $A = \{x : x$ is a natural number multiple of 3 and $x < 18\}$
(b) $A = \{x : x$ is a natural number$\}$
(c) $A = \{x = x = n^2, n \in N\}$
(d) $A = \{x : x$ is a prime number$\}$

10. Let $f : R \to R$ be given by $f(x) = x^2 + 3$. The value of $\{x : f(x) = 28\}$ is
(a) $x \in \{-5, 5\}$ (b) $x \in \{-6, 6\}$
(c) $x \in \{-4, 4\}$ (d) None of these

11. If $\tan x = \frac{3}{4}$ and x lies in III quadrant, then value of $\sec x$ is
(a) $-\frac{4}{5}$ (b) $\frac{4}{5}$ (c) $\frac{5}{4}$ (d) $-\frac{5}{4}$

12. Which of the following are correct?
(a) $\sin 1^\circ > \sin 1$
(b) $\sin 1^\circ < \sin 1$
(c) $\sin 1^\circ = \sin 1$
(d) None of the above

13. The value of a and b such that $2a + 4bi$ and $2i$ represent the same complex number.
(a) $a = 0, b = \frac{1}{2}$ (b) $a = 0, b = -\frac{1}{2}$
(c) $a = \frac{1}{2}, b = 0$ (d) $a = \frac{1}{2}, b = \frac{1}{2}$

14. Solve the inequalities $2 \le 3x - 4 \le 5$
(a) [2,3] (b) [3, 4]
(c) (3, 4) (d) None of these

15. The value of $\frac{20!}{18!(20-18)!}$ is
(a) 180 (b) 200
(c) 210 (d) 190

16. If ${}^nP_4 = 360$, then the value of n is
(a) 5 (b) 4
(c) 6 (d) 0

17. The distance between the points $A(2, -3)$ and $B(6, -3)$ is
(a) 1 unit (b) 4 units
(c) 0 unit (d) – 4 units

18. Mean deviation about the median for the data. 3, 9, 5, 3, 12, 10, 18, 4, 7, 19, 21 is
(a) 4.27 (b) 5.24
(c) 5.27 (d) 4.24

Assertion- Reason Based Questions

Directions (Q. Nos. 19-20) *In the following questions, a statement of Assertion (A) is followed by a statement of Reason (R). Choose the correct answer out of the following choices.*

(a) Both (A) and (R) are true; (R) is the correct explanation of (A).
(b) Both (A) and (R) are true; (R) is not the correct explanation of (A).
(c) (A) is true; (R) is false.
(d) (A) is false; (R) is true.

19. **Assertion (A)** The value of $\lim_{x \to 2^-} (x^2 - x + 1)$ is 3.
Reason (R) $\lim_{x \to a} [f(x) \cdot g(x)] = \lim_{x \to a} f(x) \cdot \lim_{x \to a} g(x)$.

20. **Assertion (A)** The 12th term of a GP, whose 8th term is 192 and common ratio is 2, is 3072.
Reason (R) If a is the first term and r is the common ratio of a GP such that $|r| < 1$, then sum of an infinite GP is given by $S = \frac{a}{1+r}$.

SECTION B

[This section comprises of very short answer type questions (VSA) of 2 marks each]

21. Prove that $\frac{\sin 2\theta}{1 + \cos 2\theta} = \tan\theta$.

Or

Find the value of $2 \sin 52\frac{1}{2}^\circ \sin 7\frac{1}{2}^\circ$.

22. Find the mean deviation from the mean for the following data.

6.5, 5, 5.25, 5.5, 4.75, 4.5, 6.25, 7.75, 9

23. If $f(x) = \begin{cases} x + 2, & x \le -1 \\ cx^2, & x > -1 \end{cases}$, then find c when $\lim_{x \to -1} f(x)$ exists.

Or

Let $f(x) = \begin{cases} 2x + 3, \text{ if } x \le 0 \\ 3(x+1), \text{ if } x > 0 \end{cases}$, then evaluate $\lim_{x \to 1} f(x)$.

24. Find the conjugate of $(6 + 5i)^2$.

25. Insert three numbers between $\frac{1}{3}$ and 432 so that the resulting sequence is a GP.

SECTION C

[This section comprises of short answer type questions (SA) of 3 marks each]

26. Let A and B be two sets. Prove that $(A - B) \cup B = A$ if and only if $B \subset A$.

27. Find the equation of the straight lines which passes through the origin and trisect the intercept of line $3x + 4y = 12$ between the axes.

28. Prove that $\cot 7\frac{1}{2}^\circ = \tan 82\frac{1}{2}^\circ = (\sqrt{3} + \sqrt{2})(\sqrt{2} + 1)$.

Or

Find the value of $\tan(\alpha + \beta)$, if

$\cot \alpha = \frac{1}{2}, \alpha \in \left(\pi, \frac{3\pi}{2}\right)$ and $\sec \beta = \frac{-5}{3}, \beta \in \left(\frac{\pi}{2}, \pi\right)$.

29. Find the mean and variance for the following frequency distribution

Class	0-30	30-60	60-90	90-120	120-150	150 -180	180 -210
Frequency	2	3	5	10	3	5	2

Or

The mean and standard deviation of six observations are 8 and 4, respectively. If each observation is multiplied by 3, find the mean and new standard deviation of the resulting observations.

30. The probabilities that a student will receive A, B, C or D grade are 0.40, 0.35, 0.15 and 0.10, respectively. Find the probability that a student will receive

(i) B or C grade (ii) atleast C grade.

Or

A bag contains 9 red, 7 white and 4 black balls. A ball is drawn at random. Find the probability that the ball drawn is

(i) red (ii) white (iii) not black

31. Prove that $\cos 2\theta \cos\frac{\theta}{2} - \cos 3\theta \cos\frac{9\theta}{2} = \sin 5\theta \sin\frac{5\theta}{2}$.

SECTION D

[This section comprises of long answer type questions (LA) of 5 marks each]

32. For any two sets A and B, prove that

$$A \cup B = A \cap B \Leftrightarrow A = B.$$

33. Evaluate ${}^{20}C_5 + \sum_{r=2}^{5} {}^{25-r}C_4$.

Or

Using binomial theorem, expand the expression $\left(\frac{2x}{3} - \frac{3}{2x}\right)^4$.

34. Prove that

$$(\cos A - \cos B)^2 + (\sin A - \sin B)^2 = 4\sin^2\frac{A-B}{2}.$$

35. If $A = \{1, 2, 3\}$, $B = \{3, 4\}$ and $C = \{4, 5, 6\}$, then find

(i) $A \times (B \cap C)$ (ii) $(A \times B) \cap (A \times C)$

Or

If $A = \{1, 2, 3\}$, $B = \{2, 3, 4\}$, $C = \{1, 3, 4\}$ and $D = \{2, 4, 5\}$, then verify that

$(A \times B) \cap (C \times D) = (A \cap C) \times (B \cap D)$.

SECTION E

[This section comprises 3 case-study/passage based questions of 4 marks each with sub parts. The first two case study questions have three sub parts (i), (ii), (iii) of marks 1, 1, 2 respectively. The third case study question has two sub parts (i), (ii) of 2 marks each]

36. Using trigonometric ratio of 30° and 45°, answer the following questions.

(i) Find the value of sin 15°.

(ii) Find the value of cos 15°.

(iii) Find the value of tan 15°.

Or

Find the value of sin 75°.

37. Consider $z = a + ib$ and $\bar{z} = a - ib$, where a and b are real numbers, are conjugates of each other.

Based on the above information, answer the following questions.

(i) Find $|z| + |\bar{z}|$.

(ii) If the complex numbers $-3 + i(x^2y)$ and $x^2 + y + 4i$, where x and y are real, are conjugate to each other, then find the number of ordered pairs (x, y).

(iii) Let $z = x^2 - 7x - 9yi$ such that $\bar{z} = y^2 i + 20i - 12$, then find the number of ordered pairs (x, y).

Or

Find the number of real values of x such that $\sin x + i\cos 2x$ and $\cos x - i\sin 2x$ are conjugate to each other.

38. On her vacation, Sania visits four cities. New Delhi, Lucknow, Agra, Meerut in a random order.

On the basis of above information, answer the following questions.

(i) What is the probability that she visits New Delhi before Lucknow?

(ii) What is the probability she visit New Delhi before Lucknow and Lucknow before Agra?

SAMPLE QUESTION PAPER 2

A Highly Simulated Sample Question Paper for CBSE Class XI Examination

MATHEMATICS (Unsolved)

Time : 3 Hours **Max. Marks : 80**

General Instructions

(i) This question paper contains - five sections **A, B, C, D** and **E.** Each section is compulsory. However, there are internal choices in some questions.
(ii) **Section A** has 18 **MCQ's** and 02 **Assertion-Reason** based questions of 1 mark each.
(iii) **Section B** has 5 **Very Short Answer (VSA)-type** questions of 2 marks each.
(iv) **Section C** has 6 **Short Answer (SA)-type** questions of 3 marks each.
(v) **Section D** has 4 **Long Answer (LA)-type** questions of 5 marks each.
(vi) **Section E** has 3 **source based/case based/passage based/integrated units of assessment** of 4 marks each with sub parts.

SECTION A

Multiple Choice Questions

Each question carries 1 mark

1. Which of the following represent the union of two sets A and B?
(a) $\{x : x \in A\}$ (b) $\{x : x \in A \text{ and } x \in B\}$
(c) $\{x : x \in B\}$ (d) $\{x : x \in A \text{ or } x \in B\}$

2. If the variance of a distribution is 29.16, then its standard deviation is
(a) 5.2 (b) 5.3
(c) 5.4 (d) None of these

3. If $x, y \in R$, then $x + iy$ is a non-real complex number, if
(a) $x = 0$ (b) $y = 0$ (c) $x \neq 0$ (d) $y \neq 0$

4. If $\frac{x^2}{49} - \frac{y^2}{9} = 1$, then the foci of this hyperbola, is
(a) (58, 0) (b) $(\pm\sqrt{58}, 1)$
(c) $(\pm\sqrt{58}, 0)$ (d) (58, 1)

5. The derivative of $f(x) = 3$ at $x = 0$ and at $x = 3$ are
(a) negative (b) zero
(c) different (d) not defined

6. If $\cos A = \frac{2}{3}$, then the value of $\cos 2A$ is
(a) $-\frac{1}{3}$ (b) $-\frac{1}{9}$ (c) $\frac{2}{3}$ (d) $\frac{3}{2}$

7. Solve $x^2 + 4 = 0$
(a) $\pm 2i$ (b) $2i$ (c) $-i$ (d) i

8. The variance of the data 6, 5, 9, 13, 12, 8 and 10 is
(a) $\frac{51}{7}$ (b) $\frac{13}{7}$ (c) $\frac{52}{7}$ (d) $\frac{30}{7}$

9. Evaluate $\sin 930°$.
(a) $-\frac{\sqrt{3}}{2}$ (b) $-\frac{1}{2}$ (c) -1 (d) $-\sqrt{3}$

10. If $\tan x = \frac{3}{4}$ and x lies in 3rd quadrant, then the value of $\sec x$ is
(a) $-\frac{1}{2}$ (b) $\frac{1}{2}$ (c) $-\frac{5}{4}$ (d) $\frac{1}{\sqrt{2}}$

11. Which of the following option define 'imaginary number'?
(a) Square root of any number
(b) Square root of positive number
(c) Square root of negative number
(d) Cube root of number

12. Evaluate $\frac{i^2 + i^4 + i^6 + i^7}{1 + i^2 + i^3}$.
(a) $1 - i$ (b) $i - 1$ (c) $1 + i$ (d) i^2

13. Solve the inequality $3(x - 1) \leq 2(x - 3)$ for real x.
(a) $(-\infty, -3)$ (b) $(-\infty, 3]$
(c) $(-\infty, -3]$ (d) $[3, \infty)$

14. The number of terms in the expansion $(1-3x+3x^2-x^3)^9$ is equal to

(a) 27 (b) 9 (c) 10 (d) 28

15. For what value of x are the numbers $(x+9),(x-6)$ and 4 are in GP?

(a) 0, or 24 (b) 0 or 16 (c) 1 or 24 (d) 1 or 16

16. Find the centre and radius of the circle of the equation $(x+5)^2+(y-3)^2=36$.

(a) $(5,-3)$; 6 (b) $(-5,-3)$; 6
(c) $(-5, 3)$; 6 (d) None of these

17. The three numbers representing the perpendicular distances of the point from three mutually perpendicular planes are called the

(a) coordinates of the point with respect to the two coordinate planes
(b) coordinates of the origin with respect to the three coordinate planes
(c) coordinates of the point with reference to the three coordinate planes
(d) None of the above

18. Which of the following is/are true about the range of the data?

I. It helps to find the variability in the observations on the basis of maximum and minimum value of observations.
II. Range of series = Minimum value − Maximum value
III. It tells us about the dispersion of the data from a measure of central tendency.

(a) Only I is true (b) II and III are true
(c) I and II are true (d) All are true

Assertion- Reason Based Questions

Directions (Q. Nos. 19-20) *In the following questions, a statement of Assertion (A) is followed by a statement of Reason (R). Choose the correct answer out of the following choices.*

(a) Both (A) and (R) are true; (R) is the correct explanation of (A)
(b) Both (A) and (R) are true; (R) is not the correct explanation of (A).
(c) (A) is true; (R) is false.
(d) (A) is false; (R) is true.

19. **Assertion (A)** If A and B are two events such that $P(A)=\frac{1}{4}, P(B)=\frac{1}{2}$ and $P(A \text{ and } B)=\frac{1}{8}$, then $P(A \text{ or } B)$ equal to $\frac{5}{8}$.

Reason (R) Events E and F are such that $P(\text{not } E \text{ or not } F)=0.25$, then E and F are not mutually exclusive.

20. **Assertion (A)** The number of possible orientations of parabola is 1.

Reason (R) The equation of the circle with centre $(-3,2)$ and radius 4 units is $(x+3)^2+(y-2)^2=16$.

SECTION B

[This section comprises of very short answer type questions (VSA) of 2 marks each]

21. Solve the inequality $-3 \le 4-\frac{7x}{2} \le 18$.

Or

Find the solution set of the linear inequalities $-15<\frac{3(x-2)}{5} \le 0$.

22. Find the domain and range of the function $f(x)=1-|x-2|$.

23. Find the centre and radius of the circle whose equation is $x^2+y^2-4x-8y-45=0$.

Or

Find the eccentricity of the hyperbola, the length of whose conjugate axis is $\frac{3}{4}$ of the length of transverse axis.

24. Insert two numbers between 9 and 243, so that the resulting sequence is in GP.

25. Prove that $\frac{\cos 2\theta}{1+\sin 2\theta}=\tan\left(\frac{\pi}{4}-\theta\right)$.

SECTION C

[This section comprises of short answer type questions (SA) of 3 marks each]

26. Find the mean deviation about the median of the following frequency distribution.

Class	0-6	6-12	12-18	18-24	24-30
Frequency	8	10	12	9	5

27. The cable of a uniformly loaded bridge hangs in the form of a parabola. The roadway which is horizontal and 200 m long is supported by vertical wires attached to the cable, the longest wire being 30 m and the shortest being 5 m.

Find the length of the supporting wire attached to the roadway 20 m from the middle.

28. Evaluate $\lim_{x\to 0}\frac{\sin ax + bx}{ax + \sin bx}, a, b, a + b \neq 0.$

Or

Evaluate $\lim_{x\to 0}\frac{(1+x)^6 - 1}{(1+x)^2 - 1}.$

29. Prove that $\frac{\cos x}{1-\sin x} = \tan\left(\frac{\pi}{4}+\frac{x}{2}\right).$

Or

If $A + B + C = \pi$, then prove that

$$\cot\frac{A}{2} + \cot\frac{B}{2} + \cot\frac{C}{2} = \cot\frac{A}{2}\cot\frac{B}{2}\cot\frac{C}{2}.$$

30. A bag contains tickets numbered from 1 to 20. Two tickets are drawn. Find the probability that both the tickets have prime numbers on them.

Or

One urn contains two black balls (labelled B_1 and B_2) and one white ball. A second urn contains one black ball and two white balls (labelled W_1 and W_2). Suppose, the following experiment is performed. One of the two urns is chosen at random. Next a ball is randomly chosen from the urn. Then, a second ball is chosen at random from the same urn without replacing the first ball.

(i) Write the sample space showing all possible outcomes.

(ii) What is the probability that two black balls are chosen?

(iii) What is the probability that two balls of opposite colour are chosen?

31. From 12 books, find the difference between number of ways a selection of 5 books, when one specified book is always excluded and one specified book is always included.

SECTION D

[This section comprises of long answer type questions (LA) of 5 marks each]

32. If $f(x) = x^3 - \frac{1}{x^3}$, show that $f(x) + f\left(\frac{1}{x}\right) = 0.$

33. Reduce $\left(\frac{1}{1-4i} - \frac{2}{1+i}\right)\left(\frac{3-4i}{5+i}\right)$ to the standard form.

Or

If $|z+1| = z + 2(1+i)$, then find the value of z.

34. Differentiate $e^{\sqrt{\tan x}}$ by first principle.

35. Let $A = \{-3, -2, -1, 4\}$, and $f : A \to Z$ given by $f(x) = x^2 + x + 2$. Find the range of f and pre-images of 6 and 4.

Or

If $A = \{4^n - 3n - 1, n \in N\}$ and $B = \{9(n-1) : n \in N\}$, then show that $A \subset B$.

SECTION E

[This section comprises 3 case-study/passage based questions of 4 marks each with sub parts. The first two case study questions have three sub parts (i), (ii), (iii) of marks 1, 1, 2 respectively. The third case study question has two sub parts (i), (ii) of 2 marks each]

36. If $a < b < c$ are numbers lying between 2 and 18 such that

(a) $a + b + c = 25$

(b) $2, a, b$ are three consecutive terms of an AP.

(c) $b, c, 18$ are three consecutive terms of a GP.

Based on the above information, answer the following questions.

(i) Find the relation between b and c.

(ii) Find the value of a.

(iii) Find the possible value of b.

Or

Find the possible value of c

37. Let $f : R - \{0, 1\} \to R$ be a function satisfying the relation $f(x) + f\left(\frac{x-1}{x}\right) = x$ for all $x \in R - \{0, 1\}$.

Based on the above information, answer the following questions.

(i) Find $f\left(\frac{x-1}{x}\right) + f\left(\frac{1}{1-x}\right)$.

(ii) Find $f\left(\frac{1}{1-x}\right) + f(x)$.

(iii) Find the value of $f(x)$.

Or

Find the value of $f(-1)$.

38. In a class test of class XI, a teacher asked to students to consider $A + B = \frac{\pi}{4}$, where A and B are acute angles.

Based on the above information, answer the following questions.

(i) Find the value of $(1 + \tan A)(1 + \tan B)$.

(ii) Find the value of $(\cot A - 1)(\cot B - 1)$.

SAMPLE QUESTION PAPER 3

A Highly Simulated Sample Question Paper for CBSE Class XI Examination

MATHEMATICS (Unsolved)

Time : 3 Hours **Max. Marks : 80**

General Instructions

(i) This question paper contains - five sections **A, B, C, D** and **E**. Each section is compulsory. However, there are internal choices in some questions.
(ii) **Section A** has 18 **MCQ's** and 02 **Assertion-Reason** based questions of 1 mark each.
(iii) **Section B** has 5 **Very Short Answer (VSA)-type** questions of 2 marks each.
(iv) **Section C** has 6 **Short Answer (SA)-type** questions of 3 marks each.
(v) **Section D** has 4 **Long Answer (LA)-type** questions of 5 marks each.
(vi) **Section E** has 3 **source based/case based/passage based/integrated units of assessment** of 4 marks each with sub parts.

SECTION A

Multiple Choice Questions

Each question carries 1 mark

1. The equation of line which passes through $(-1, 2)$ and having slope $\frac{2}{5}$, is

(a) $2x + 5y + 12 = 0$ (b) $2x - 5y + 12 = 0$
(c) $3x + 5y + 9 = 0$ (d) $3x - 5y + 10 = 0$

2. The value of $\lim_{x \to 1}(3x^3 + 5x^2 - 4x + 3)$ is

(a) 5 (b) 6 (c) 7 (d) 8

3. If $A = \{1, 2, 3\}$, $B = \{2, 3, 8\}$, $C = \{3, 8\}$, then the value of $(A - B) \cap (B - C)$ is

(a) $\{1, 2\}$ (b) ϕ
(c) $\{1, 2, 3\}$ (d) $\{2, 3, 5\}$

4. If the equation $ax^2 + 2bxy + y^2 + 3ax + 2by + 1 = 0$ shows a circle, then the value of a and b is

(a) $a = 1, b = 1$ (b) $a = 1, b = -1$
(c) $a = 1, b = 0$ (d) $a = 0, b = 1$

5. If out of 400 ships sailing in sea, 40 ships sink, then the probability of not sinking of a ship is

(a) $\frac{1}{10}$ (b) $\frac{9}{10}$
(c) 1 (d) None of these

6. Express $\frac{1}{3 - 4i}$ in $x + iy$ form.

(a) $\frac{4}{25} + i\frac{3}{25}$ (b) $-\frac{3}{25} + i\frac{4}{25}$
(c) $\frac{3}{25} - i\frac{4}{25}$ (d) $\frac{3}{25} + i\frac{4}{25}$

7. Represent hyperbola $3y^2 - x^2 = 108$ in standard form.

(a) $\frac{y^2}{36} - \frac{x^2}{108} = 1$ (b) $\frac{y^2}{108} - \frac{x^2}{36} = 1$
(c) $3y^2 - x^2 = 108$ (d) Both (a) and (c)

8. The empty set is represented by

(I) ϕ (II) $\{\phi\}$ (III) $\{\}$ (IV) $\{\}\{\}$

(a) I and II (b) I and III
(c) II and III (d) I and IV

9. If $f(x) = \frac{1}{2 - \sin 3x}$, then range (f) is equal to

(a) $[-1, 1]$ (b) $\left[-\frac{1}{3}, \frac{1}{3}\right]$ (c) $\left[\frac{1}{3}, 1\right]$ (d) $\left[-1, -\frac{1}{3}\right]$

10. If $f(x) = \log\left(\frac{1+x}{1-x}\right)$ and $g(x) = \frac{3x + x^3}{1 + 3x^2}$, then $f[g(x)]$ equal to

(a) $-f(x)$ (b) $3[f(x)]$
(c) $[f(x)]^3$ (d) $-3[f(x)]$

11. If $\theta + \phi = d$ and $\tan\theta = p\tan\phi$, then $\sin(\theta - \phi)$ is equal to

(a) $\frac{p-1}{p+1}\sin\alpha$ (b) $\frac{p-1}{p+1}\cos\alpha$

(c) $\frac{p-1}{p+1}$ (d) $\frac{p+1}{p-1}$

12. Ravi goes to market with ₹ 200 to buy rice, which is available in packets of 1 kg. The price of one packet of rice is ₹ 30. If x denotes the number of packets or rice which he buys, then the total amount spent by him is ₹ $30x$. Then, mathematical formulation of the given problem is

(a) $30x > 200$ (b) $30x < 200$

(c) $30x \le 200$ (d) $30x \ge 200$

13. The sum of all the numbers that can be formed with the digits 2, 3, 4, 5 taken all at a time, is

(a) 93321 (b) 93322 (c) 93323 (d) 93324

14. ${}^{15}C_8 + {}^{15}C_9 - {}^{15}C_6 - {}^{15}C_7$ is equal to

(a) 0 (b) 2 (c) 1 (d) 3

15. The value of $\sum_{K=1}^{10}(1+2^K)$ is

(a) 2085 (b) 2805 (c) 2056 (d) 2508

16. If $y = 1 + \frac{x}{1!} + \frac{x^2}{2!} + \frac{x^3}{3!} + \ldots + \frac{x^n}{n!}$, then the value of $\frac{dy}{dx} - y + \frac{x^n}{n!}$ is

(a) 0 (b) n

(c) $2n$ (d) None of these

17. The variance of the following data

x_i	6	10	14	18	24	28	30
f_i	2	4	7	12	8	4	3

(a) 41.5 (b) 40.6 (c) 43.4 (d) 45.2

18. If A and B are mutually exclusive events, then

(a) $P(A) \le P(\overline{B})$ (b) $P(A) \ge P(\overline{B})$

(c) $P(A) < P(\overline{B})$ (d) None of these

Assertion- Reason Based Questions

Directions (Q. Nos. 19-20) *In the following questions, a statement of Assertion (A) is followed by a statement of Reason (R). Choose the correct answer out of the following choices.*

(a) Both (A) and (R) are true; (R) is the correct explanation of (A)

(b) Both (A) and (R) are true; (R) is not the correct explanation of (A).

(c) (A) is true; (R) is false.

(d) (A) is false; (R) is true.

19. **Assertion (A)** The total number of terms in the expansion of $(a+x)^{51} - (a-x)^{51}$ after simplification is 51.

Reason (R) If n is odd, then there are two middle terms.

20. **Assertion (A)** The length of the foot of perpendicular drawn from the point $P(3, 4, 5)$ on Y-axis is $\sqrt{34}$ units.

Reason (R) Three points A, B and C are said to be collinear, if $AB + BC = AC$ (as shown below).

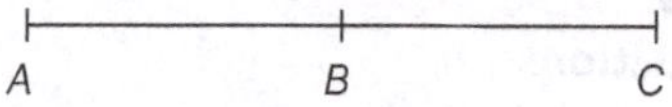

SECTION B

[This section comprises of very short answer type questions (VSA) of 2 marks each]

21. If $\cos\theta = \frac{-3}{5}$ and $\pi < \theta < \frac{3\pi}{2}$, then find the value of $\frac{\operatorname{cosec}\theta + \cot\theta}{\sec\theta - \tan\theta}$.

Or

If $\sin\theta = \frac{3}{5}$ and $\cos\phi = \frac{-12}{13}$, where θ and ϕ both lie in second quadrant, then find the value of $\cos(\theta + \phi)$.

22. Find the mean deviation about median of the following data.

Class interval	0-10	10-20	20-30	30-40	40-50	50-60
Frequency	6	7	15	16	4	2

23. How many different words can be formed with the letters of the word 'INDEPENDENT'?

Or

How many numbers can be formed with the digits 1, 2, 3, 4, 3, 2 and 1 so that odd digits always occupy the odd places?

24. If $f(x) = x + \frac{1}{x}$, then prove that $[f(x)]^3 = f(x^3) + 3f\left(\frac{1}{x}\right)$.

25. Prove that $i^{107} + i^{112} + i^{117} + i^{122} = 0$.

SECTION C

[This section comprises of short answer type questions (SA) of 3 marks each]

26. In a survey of 100 students, the number of students studying the various languages were found to be English only 18, English but not Hindi 23, English and Sanskrit 8, English 26, Sanskrit 48, Sanskrit and Hindi 8, no language 24. Then,

(i) how many students were studying Hindi?

(ii) how many students were studying both English and Hindi?

(iii) which language is study by most of the students?

27. Prove that the centres of the three circles

$$x^2 + y^2 - 4x - 6y - 12 = 0,$$
$$x^2 + y^2 + 2x + 4y - 5 = 0$$

and $x^2 + y^2 - 10x - 16y + 7 = 0$ are collinear.

28. Evaluate $\lim_{x \to \sqrt{10}} \frac{\sqrt{7-2x} - (\sqrt{5} - \sqrt{2})}{x^2 - 10}$.

Or

Find the derivative $\sqrt{\sin x}$ from first principle.

29. Find the mean deviation from the median for the following data

x_i	74	89	42	54	91	94	35
f_i	20	12	2	4	5	3	4

Or

The mean and standard deviation of a group of 100 observations were found to be 20 and 3, respectively. Later on, it was found that three observations were incorrect, which are recorded by 21, 21 and 18. Find the mean and standard deviation, if the incorrect observations are omitted.

30. The probabilities of happening of two events A and B are 0.25 and 0.50, respectively. If the probability of happening of A and B together is 0.14, then find the probability that neither A nor B occurs.

Or

A box contains 100 bolts and 50 nuts. It is given that 50% bolts and 50% nuts are rusted. Two objects are selected from the box at random. Find the probability that either both are bolts or both are rusted.

31. (i) Solve : $x + 5 > 2(x + 1)$ and $2 - x < 3(x + 2)$

(ii) The water acidity in a pool is considered normal when the average pH reading of three daily measurements in between 6.2 and 6.8, if the first two pH readings are 6.46 and 6.84, find the range of pH values for the third reading that will result in the acidity level.

SECTION D

[This section comprises of long answer type questions (LA) of 5 marks each]

32. Find the derivative of $\frac{2}{x+1} - \frac{x^2}{3x-1}$ with respect to x.

33. Determine the domain and range of the following relations.

(i) $R_1 = \left\{\left(x, \frac{1}{x}\right) : 0 < x < 6, x \in N\right\}$

(ii) $R_2 = \{(x, x^2) : x \text{ is prime number less than } 10\}$

Or

Using operations on set of three sets A, B and C, prove the following.

(i) $(A - B) \cup A = A$

(ii) $(A - B) \cap (B - A) = \phi$

(iii) $A - (A - B) = A \cap B$

34. Prove that $\sum_{r=0}^{n} {}^nC_r \cdot 3^r = 4^n$.

35. If $f(x) = 3\sin x$ and $g(x) = \cos^2 x$, then find the value of $(f + g)\left(\frac{\pi}{6}\right)$.

Or

If $A + B = 225°$, then prove that

$$\frac{\cot A}{1 + \cot A} \cdot \frac{\cot B}{1 + \cot B} = \frac{1}{2}.$$

SECTION E

[This section comprises 3 case-study/passage based questions of 4 marks each with sub parts. The first two case study questions have three sub parts (i), (ii), (iii) of marks 1, 1, 2 respectively. The third case study question has two sub parts (i) (ii) of 2 marks each]

36. The side of a given square is 10 cm. The mid-points of its sides are joined to form a new square. Again, the mid-points of the sides of this new square are joined to form another square. This process is continued indefinitely.

Based on the above information, answer the following questions.

(i) Find the common ratio of the areas of formed squares.

(ii) What will be the common ratio of the perimeters of the formed squares?

(iii) Find the sum of the perimeters of the squares formed.

Or

Find the sum of the areas of the squares formed.

37. In a group of 25 students aged between 16 yr and 18 yr, it was found that 15 students play cricket, 12 students play tennis, 11 students play football, 5 students play both cricket and football, 9 students play both cricket and tennis, 4 students play tennis and football and 3 students play all the three games.

Based on the above information, answer the following questions.

(i) Find the number of students in the group who play only football.

(ii) Find the number of students in the group who play only cricket.

(iii) Find the number of students in the group who play only tennis.

Or

Find the number of students in the group who play all three games.

38. Rajiv construct two right angled triangles in the fourth quadrant in such a way that the measure of triangle gives $\cos A = \frac{4}{5}$ and $\cos B = \frac{12}{13}$, where $\frac{3\pi}{2} < A$ and $B > 2\pi$.

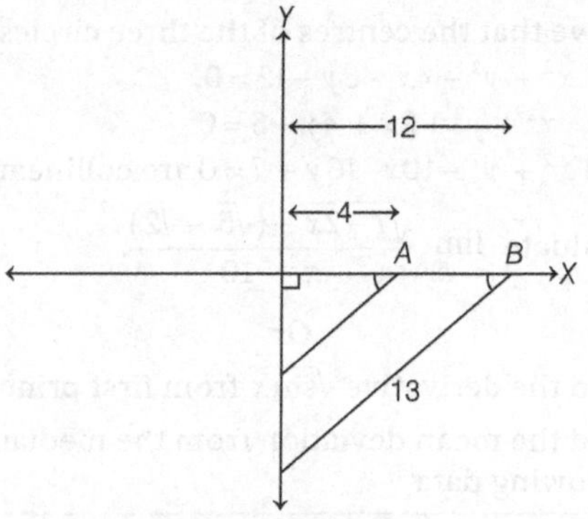

Based on the above information, answer the following questions.

(i) Find the value of $\cos(A + B)$.

(ii) Find the value of $\sin(A - B)$.

Answers

Sample Paper 1

1. (b) 2. (b) 3. (b) 4. (d)
5. (b) 6. (a) 7. (c) 8. (a)
9. (a) 10. (a) 11. (d) 12. (b)
13. (a) 14. (a) 15. (d) 16. (c)
17. (b) 18. (c) 19. (b) 20. (c)

21. *Or* $\frac{\sqrt{2}-1}{2}$

22. 1.1

23. $c=1$ *Or* 6

24. $11-60i$

25. 2, 12, 72

27. $3x-8y=0$ and $3x-2y=0$

28. *Or* $\frac{2}{11}$

29. Mean = 107 and Variance = 2276
Or Mean = 24 and Standard deviation = 12

30. (i) 0.5 (ii) 0.25 *Or* (i) $\frac{9}{20}$ (ii) $\frac{7}{20}$ (iii) $\frac{4}{5}$

33. 42504 *Or* $\frac{16}{81}x^4-\frac{16}{9}x^2+6-\frac{9}{x^2}+\frac{81}{16x^4}$

35. (i) {(1, 4), (2, 4), (3, 4)} (ii) {(1, 4), (2, 4), (3, 4)}

36. (i) $\frac{\sqrt{3}-1}{2\sqrt{2}}$ (ii) $\frac{\sqrt{3}+1}{2\sqrt{2}}$ (iii) $\frac{\sqrt{3}-1}{\sqrt{3}+1}$ *Or* $\frac{\sqrt{3}+1}{2\sqrt{2}}$

37. (i) $2\sqrt{a^2-b^2}$ (ii) 2 (iii) 4 *Or* 0

38. (i) $\frac{1}{2}$ (ii) $\frac{1}{6}$

Sample Paper 2

1. (d) 2. (c) 3. (d) 4. (c)
5. (b) 6. (b) 7. (a) 8. (c)
9. (b) 10. (c) 11. (c) 12. (a)
13. (c) 14. (d) 15. (b) 16. (c)
17. (c) 18. (a) 19. (b) 20. (d)

21. [−4, 2] *Or* (−23, 2]

22. Domain = $(-\infty, \infty)$ and Range = $(-\infty, 1]$

23. (2, 4), $\sqrt{65}$ *Or* $\frac{5}{4}$

24. 27, 81

26. 6.318

27. 6 m

28. 1 *Or* 3

30. $\frac{14}{95}$ *Or* (i) $\{I\,B_1B_2, I\,B_1W, I\,B_2\,B_1, I\,B_2\,W, I\,WB_1, I\,WB_2, II\,BW_1, II\,BW_2, II\,W_1\,B, II\,W_1W_2, II\,W_2B, II\,W_2W_1\}$
(ii) $\frac{1}{6}$ (iii) $\frac{2}{3}$

31. 132

33. $\frac{307}{442}+\frac{599}{442}i$ *Or* $\frac{1}{2}-2i$

34. $\frac{e^{\sqrt{\tan x}}}{2\sqrt{\tan x}}.\sec^2 x$

35. Range = {8, 4, 2, 22}
and no pre-image of 6, pre-image of 4 is −2

36. (i) $c^2=18b$ (ii) 5 (iii) 8 *Or* 12

37. (i) $1-\frac{1}{x}$ (ii) $\frac{1}{1-x}$ (iii) $\frac{1}{2}\left(x+\frac{1}{1-x}-\frac{x-1}{x}\right)$ *Or* $-\frac{5}{4}$

38. (i) 2 (ii) 2

Sample Paper 3

1. (b) 2. (c) 3. (b) 4. (c)
5. (b) 6. (d) 7. (d) 8. (b)
9. (c) 10. (b) 11. (a) 12. (b)
13. (d) 14. (a) 15. (c) 16. (a)
17. (c) 18. (a) 19. (d) 20. (b)

21. $\frac{1}{6}$ *Or* $\frac{33}{65}$ 22. 10.16 23. $\frac{11!}{3!2!3!}$ *Or* 18

26. (i) 18 (ii) 3 (iii) Sanskrit

28. $-\left(\frac{\sqrt{5}+\sqrt{2}}{6\sqrt{10}}\right)$ *Or* $\frac{\cos x}{2\sqrt{\sin x}}$

29. 12.5 *Or* Mean = 20, Standard deviation = 3.03

30. 0.39 *Or* 0.58

31. (i) (−1, 3) (ii) Between 5.3 and 7.1

32. $-\frac{2}{(x+1)^2}-\frac{x(3x-2)}{(3x-1)^2}$

33. (i) Domain = {1, 2, 3, 4, 5} and Range = $\left\{1, \frac{1}{2}, \frac{1}{3}, \frac{1}{4}, \frac{1}{5}\right\}$
(ii) Domain = {2, 3, 5, 7} and Range = {4, 9, 25, 49}

35. $\frac{9}{4}$

36. (i) $\frac{1}{2}$ (ii) $\frac{1}{\sqrt{2}}$ (iii) $80+40\sqrt{2}$ cm or 200 cm^2

37. (i) 5 (ii) 4 (iii) 2 *Or* 3

38. (i) $\frac{33}{65}$ (ii) $\frac{-16}{65}$